Structure and Function of Antibodies

Structure and Function of Antibodies

Editors

Roy Jefferis
Koichi Kato
William R. (Bill) Strohl

MDPI • Basel • Beijing • Wuhan • Barcelona • Belgrade • Manchester • Tokyo • Cluj • Tianjin

MDPI

Editors

Roy Jefferis
University of Birmingham
UK

Koichi Kato
National Institutes of Natural
Sciences
Japano

William R. (Bill) Strohl
BiStro Biotech Consulting
USA

Editorial Office
MDPI
St. Alban-Anlage 66
4052 Basel, Switzerland

This is a reprint of articles from the Special Issue published online in the open access journal *Antibodies* (ISSN 2073-4468) (available at: https://www.mdpi.com/journal/antibodies/special_issues/Structure).

For citation purposes, cite each article independently as indicated on the article page online and as indicated below:

LastName, A.A.; LastName, B.B.; LastName, C.C. Article Title. *Journal Name* **Year**, *Volume Number*, Page Range.

ISBN 978-3-03943-897-6 (Hbk)
ISBN 978-3-03943-898-3 (PDF)

Cover image courtesy of Lila M. Strohl.

Contents

About the Editors

Roy Jefferis Following a BS.c. and Ph.D. in Chemistry (1964) at University of Birmingham, Royston Jefferis joined the laboratory of Dennis Stanworth in the Medical School (Birmingham) to research the structure and function of antibody molecules. Following a sabbatical (1968–1969) at the University of California San Diego, spent in the laboratory of Russell Doolittle, he returned to the Medical School to establish his own research group. Access to monoclonal IgG proteins, isolated from the sera of patients with multiple myeloma, allowed the elucidation of relationships between antibody structure and function. A principal feature was the profound impact that glycosylation has on the mechanisms of action (MoA) of IgG antibodies, in vitro and in vivo—leading to numerous consultancies with pharmaceutical companies. He was appointed Professor of Molecular Immunology in 1992 and Emeritus Professor, on retirement, in 2006. In consideration of the published research he was awarded the DS.c. (University of Birmingham, 1987); elected Fellow of the Royal College of Pathologists (FRCPath; 1987) and Member of the Royal College of Physicians (MRCP, 2007).

Koichi Kato received his Ph.D. in 1991 at Graduate School of Pharmaceutical Sciences, the University of Tokyo under the supervision of Prof. Yoji Arata, and continued his research as Assistant Professor and Senior Lecturer in the same institution. In 2000, he moved to Nagoya City University as a full Professor. In 2008, he moved to Okazaki Institute for Integrative Bioscience, National Institutes of Natural Sciences, holding a Professor position at Nagoya City University. Since 2018, he became the first director of Exploratory Research Center on Life and Living Systems, National Institutes of Natural Sciences. His research interests include structural analyses of antibodies by NMR spectroscopy and other biophysical and biochemical methods. Based on his accomplishments, he was awarded the Pharmaceutical Society of Japan Award for Young Scientists in 2000, the Pharmaceutical Society of Japan Award for Divisional Scientific Promotions as well as the 48th Baelz Prize in 2011.

William R. (Bill) Strohl owns BiStro Biotech Consulting LLC, started in 2016 to help biotechnology companies grow and expand their capabilities. Prior to retiring from Johnson & Johnson in 2016, Dr. Strohl was VP and head of Janssen BioTherapeutics, Janssen R&D, J&J, where he lead the discovery and early development of Antibody Therapeutics and other Biologics. Previously, Dr. Strohl was the head of Biologics Research, the discovery arm of Janssen BioTherapeutics. At J&J, Dr. Strohl and his team placed more than two dozen innovative therapeutic proteins into development, many of which are still in clinical trials. Before joining J&J, Dr. Strohl was at Merck and Co. from 1997 to 2008, leading Natural Products Biology, founding a Microbial Vaccines department, and then later, leading the Biologics discovery efforts. Dr. Strohl started his career rising from assistant to full professor, at Department of Microbiology and Program of Biochemistry, Ohio State University, from 1980 to 1997. Dr. Strohl has over 140 publications and 17 issued patents. Additionally, Dr. Strohl has edited three books, acted as the guest editor for four journal issues dedicated to topics in areas of his expertise, and wrote a book entitled "Therapeutic Antibody Engineering: Current and Future Advances Driving the Strongest Growth Area in the Pharma Industry" (published 2012).

Preface to "Structure and Function of Antibodies"

Biologics are established as potent additions to the drug armamentaria, and their number is set to multiply (explode!) over the next decade and beyond. They are challenging and expensive to produce and represent the largest cost and cost growth area for healthcare budgets. A majority of biologics are large proteins that undergo post-translational modifications (PTMs) to the peptide backbone structure as they transit from the ribosome to the endoplasmic reticulum, during passage through the Golgi apparatus and at the site of activity. Principal amongst these modifications is the addition and processing of complex oligosaccharide moieties to yield a glycoprotein (GP). Since the addition of defined oligosaccharide chains is essential to the function of GPs, they are necessarily produced in mammalian cell lines (Chinese Hamster Ovary: CHO; murine cell lines: NSO and Sp2/0; and human cell lines: HEK293 and Per.C6).

The first GP approved by the European Medicines Agency (EMA) and Federal Drug Administration (FDA) was the murine monoclonal antibody muromonab (1986), produced in mouse hybridoma cells, for treatment of liver transplant patients undergoing an episode of acute organ rejection; however, it could only be successfully employed as an "emergency" "one-off" treatment because patients developed a vigorous anti-mouse IgG immune response that precluded repeat exposure.

The therapeutic era for monoclonal antibodies (mAbs) may be identified with the development of chimeric mouse/human mAbs comprising the variable regions of a mouse antibody combined with the constant regions of human IgG1 to yield a molecule that is ~30% mouse and ~70% human in structure (e.g., Remicade, Rituxan). Chimeric antibodies exhibited a significant reduction in immunogenicity, and a majority of patients could be repeatedly dosed. The next advance was the generation of "humanized" antibodies by transplantation of structural elements of a chimeric mouse/human antibody that formed the antigen-binding site (paratope) into a human V region framework (e.g., Avastin), generating mAbs that are >90% human in structure. This technology has, in turn, been replaced by the development of protocols for the generation of "fully human" antibodies (e.g., adalimumab/Humira) encoded by human genes. The Humira antibody was generated following random recombination of a diverse library of human heavy and light chains and selection for high-affinity binding to the target antigen, tumor necrosis factor-alpha (TNFα); the technique was pioneered by one of the 2018 Nobel Laureates for Chemistry: Prof. Sir Gregory Winter (Cambridge University). Although fully human mAbs are encoded by human Ig genes, the drug product is produced in a nonhuman cell line that may introduce nonhuman structural characteristics (PTMs), particularly for the IgG-Fc oligosaccharide. In addition, the very uniqueness of an individual mAb structure, reflecting its unique paratope specificity, may be "seen" as nonself by the immune system of some patients within an outbred human population and trigger an antidrug/antitherapeutic antibody (ADA/ATA) response.

Data Mining

The hallmark of a humoral immune response is an ability to produce antibodies expressing exquisite specificity for a seemingly infinite number of unique antigenic determinants (epitopes); it is currently estimated that the naive antibody repertoire has the potential to encode at least 10^{12} unique sequences generated by gene recombination events. The secondary immune response results in a rapid expansion of antibody-specific cells, and this is coupled with somatic hypermutation which, together with random heavy–light chain pairing, extends the estimated repertoire to 10^{16}–10^{18} unique sequences. These numbers far exceed the number of B lymphocytes present in the human body at any given time ($\sim 5 \times 10^9$); however, the potential repertoire is continuously sampled since the lifespan of a B cell is 1–8 weeks. Currently, the extent of the repertoire is being sampled with deep sequencing of peripheral B-cell DNA and/or RNA, employing next-generation sequencing (NGS) protocols. These studies are providing masses of new data, the analysis of which will allow quantitation of germ-line gene usage and the diversity of expressed antibody repertoires within and between population groups. Additionally, germ-line variable (v) gene usage may allow patient stratification, e.g., in B-cell leukemias, depending on clinical response(s) achieved with defined treatment protocols. Meaningful analysis of these data required the development of new protocols for the storage, access, and analysis of data and a new discipline at the interface between immunogenetics and bioinformatics, i.e., immunoinformatics. Such tools were pioneered, and have been continuously developed, by Marie-Paule and Gerard Lefranc with the establishment of the ImMunoGeneTics (IMGT) information system, reviewed in the first chapter. The concepts of IMGT provide insights for antibody V and C domain structure and function that may be exploited in immune repertoire analysis, antibody humanization, and antibody engineering to modulate effector functions. Evolution has exploited multiple pathways to generate antibody repertoires, and variants to those established for humans and the mouse, e.g., camelid VH gene repertoires and nurse shark IgN antibodies, are providing additional insights that are being exploited for the generation of novel therapeutic constructs. The above summarizes developments within the IgG isotypes; however, the IgM, IgA, and IgE isotypes are being evaluated to exploit their unique effector properties, as addressed within this book.

Structural Fidelity of mAbs

Structural studies of proteins isolated from human serum, and other bodily fluids, reveal macro- and microstructural heterogeneity as a consequence of PTMs. Within the individual, these structures would be "seen" as "self" by the immune system and result in immune tolerance. However, production of an externally sourced mAb in nonhuman (CHO) cell lines could result in loss of structural fidelity and the generation of variants having altered biologic activities and/or immunogenic potential. Consequently, regulatory authorities require mAbs to be comprehensively characterized by the use of "state of the art" orthogonal physicochemical techniques that allow "critical quality attributes" (CQA) to be defined. The contribution of Beck and Liu identifies and quantitates common and uncommon PTMs both for nascent mAb samples and following exposure to accelerated storage conditions. If approved for clinical use, the structural characteristics of the drug-substance/product submitted define the therapeutic and must be maintained throughout its life cycle; improvements in manufacturing protocols may be adopted if evidence of maintained efficacy is presented and approved by the appropriate legislating authority.

IgG Charge: Practical and Biological Implications

The overall charge of a protein influences multiple biological activities, from the rate of translation to folding, stability, and interactions with cognate effector molecules. Proteins exhibit charge heterogeneity that is determined by its immediate milieu, e.g., pH. The contribution of Yang et al. reports a study in which the pI and charge heterogeneity of 13 IgG anti-IL13 antibodies, embracing the four IgG subclasses, and their Fab/Fc fragments were determined. The results illustrate that current methods of calculating (predicting) charge are inadequate as they vary considerably from measured values. The charge heterogeneity profile of an antibody can be a selective property that informs of conditions favorable for downstream processing, formulation, and appropriate function in vivo, both at the site of administration and sites of biological activity.

Human IgG Subclasses and Their Mechanisms of Action (MoAs)

An antibody may be protective and deliver therapeutic benefit due to its binding specificity for its target, e.g., neutralizing an exogenous bacterial toxin or endogenous TNFα; however, when the target is a bacterium, virus, or cancer cell, MoAs that result in neutralization, removal, and destruction of the immune complex (IC) formed are essential. Multiple MoAs are mediated by leucocytes that bear cell surface receptors (FcγR) specific to the IgG heavy chain Fc region. There are three families of FcγR (FcγRI, FcγRII, and FcγRIII) that are differentially expressed by human leucocytes that may also bind the IgG subclasses differentially. The cross-linking of multiple FcγRs results in leucocyte activation with the release of toxic agents and/or ingestion (phagocytosis); ICs may also activate the C1 component of the complement system to trigger a cascade of enzymatic reactions resulting in the formation of a membrane attack complex (MAC) that inserts into the cellular membrane with the formation of pores that allow the ingress of water and egress of cellular constituents. Molecules released from the complement cascade also adhere to the IC and engage complement receptors expressed on leucocytes to further enhance cellular activation. Additional IgG-Fc binding ligands are discussed in the contribution of de Taeye et al., and their differential activation by the four IgG subclasses is assessed. These data are of critical importance when selecting the IgG subclass for a potential mAb therapeutic since it determines the MoA(s) activated and consequently may influence clinical outcomes.

David vs. Goliath: The Structure, Function, and Clinical Prospects of Antibody Fragments

In health, IgG antibodies equilibrate between the blood, extravascular space, and lymph, affording access for humoral immune protection of cells and surface tissue. However, diffusion into solid tissue is compromised by the relatively large size of the IgG molecule. Fortunately, the organization of the immunoglobulin genes, in which each domain is encoded by a distinct exon, allows for the generation of multiple "novel" constructs and the selection of customized therapeutics. The Fab fragment retains full antigen (epitope) binding activity whilst not activating downstream inflammatory functions. This has been exploited for the generation of the TNFα inhibitory Fab therapeutic Cimzia (certolizumab), the biological half-life of which is extended by the addition of a polyethylene glycol "tail". The library of antigen-binding antibody fragments available is limited only by the collective insight and may be exploited to deliver selected downstream effector functions, as addressed in the contribution of Bates and Power.

Antibody Structure and Function: The Basis for Engineering Therapeutics

Antibody modeling assessment studies have been undertaken to gain insight into the quality of the results of antibody structure prediction software. These blinded studies involved providing multiple antibody structure prediction software groups with the primary sequence of Fv regions for which structures had been determined but were not publicly available. Once the predictions were completed by the participants, the results were submitted to the organizers and the models were assessed and compared with the unpublished structures. In a second study, after the prediction of the structures of the entire Fv were completed, the participants were provided with the Fv structures without their CDR-H3s. The structures of the CDR-H3s were then predicted and submitted. Each of the methods applied in these studies had different strengths and weaknesses. Overall, the second antibody assessment revealed an improved quality of the models with incremental improvement in the accuracy of the predictions from the first assessment. Further development to improve these methods is clearly warranted.

The Fc domain has been engineered to optimize effector function, clustering, and Fc receptor engagement. In general, antibody engineering of both the Fab and Fc domains is an indispensable part of the drug development process and as such will continue to advance as more and more antibodies are considered for therapeutic use. Despite great progress in the methods of antibody engineering, new approaches are in high demand. One of the remaining goals is to improve the accuracy of computational methods, which will allow for the prediction of point mutations that improve affinity and other properties of interest. New approaches are continually being developed to create antibody-based molecules that are superior in their potency, specificity, localization, and safety.

Dynamic Views of the Fc Region of Immunoglobulin G

The predominant "read-out" of antibody activity is the effector functions that can inactivate, remove, and destroy a target through the induction of multiple downstream inflammatory mechanisms; ultimately, these activities have to be downregulated to re-establish hemostasis. Whilst there is evidence that binding antigen may induce structural changes within the paratope that may be transmitted, by allosteric mechanisms, to the hinge and Fc regions, a majority of dynamic studies have focused on the IgG-Fc in attempts to rationalize the binding and activation of >12 disparate ligands and multiple peptides (de Taeye et al., this volume). Additionally, binding affinities, and consequent amplitude of activation, can vary between polymorphic variants and their glycoform profiles. The article of Yanaka et al. interprets data obtained from long-timescale molecular dynamic simulations coupled with experiments in solution and demonstrates that conformational space of IgG-Fc is optimally restricted by dynamic intramolecular-interaction networks involving the glycan to be compatible with its binding of multiple ligands. These data suggest that antibody engineering may be targeted to control the allosteric network, thereby selectively modulating interactions with specific ligands and hence biological outcomes.

Design and Production of Bispecific Antibodies

Conventional mAbs bind their target to form immune complexes that bind cell surface Fc receptors expressed on effector cells. Given the relatively low affinity of IgG-Fc/Ffc receptor interactions, it was anticipated that recruitment of effector cells may be enhanced if a bispecific antibody could be created in which one paratope has specificity for target and the other for an effector cell surface molecule, e.g., anti-CD20 + anti-CD3. Two common formats of bispecific antibodies have entered the market, full-length IgG antibodies and single-chain (scFv); the latter allows for better penetration into solid tissue tumors. Comprehensive coverage of possible pitfalls and solutions is offered in the contribution of Wang et al.

Bispecific T-Cell Redirection versus Chimeric Antigen Receptor (CAR)-T Cells as Approaches to Kill Cancer Cells

Two old, yet remarkably modern, T-cell based therapeutic antibody strategies are now taking a front seat in the war against cancer cells. The fundamental approaches and the base technologies underlying both T-cell redirecting bispecific antibodies (TRBAs) and chimeric antigen receptor (CAR)-T cells date from the early 1990s, but technical difficulties limited their deployment to a few laboratories and small, investigator-initiated clinical trials for over a decade. Both TRBAs and CAR-T cell therapeutic strategies, however, experienced a dynamic rebirth over the past dozen or so years, resulting in two regulatory approvals for each approach and hundreds of clinical candidates that are attempting to address significant unmet medical needs. So far, both TRBAs and CAR-Ts have shown remarkable activity in hematological cancers such as B-cell chronic lymphocytic leukemia (B-CLL), the highly aggressive diffuse large- B-cell lymphoma (DLBCL), and the more indolent follicular lymphoma (FL), but they have not yet demonstrated that level of clinical activity for solid tumors. Newer approaches, including concomitant use of TRBAs or CAR-T cells with checkpoint inhibitors and/or T-cell stimulating cytokines, provide hope for expanded use and clinical efficacy. Clinical-stage TRBAs come in many sizes, shapes, and flavors, including the dual single-chain Fv (scFv)-styled "bispecific T-cell engager" (BiTE); half-life extended versions of BiTEs; several types of asymmetric heterodimeric bivalent, bispecific IgG-like antibodies; and 2:1 T-cell bispecifics (TCBs), to name a few platforms. Current CAR-T cells are predominantly autologous, meaning that T-cells are retrieved from the patient, manipulated ex vivo to incorporate the tumor-targeting CAR, and then re-administered to that same patient. This is a laborious, expensive, and time-consuming process that, while successful, is not optimal. A few allogeneic, or "off-the-shelf", CAR-T cells are beginning to find their way into the clinic. These cells need to be engineered so that they do not cause graft-versus-host or host-versus-graft (rejection) responses. Additionally, because they will ultimately be invisible to the immune system, there need to be highly efficient methods to eliminate or turn off allogeneic CAR-T cells. The advantages to this approach, however, are centralized manufacturing, cost of goods, time to treatment, and treatment consistency, so it is likely that allogeneic CAR-T cells are the ultimate future for CAR-Ts. The biggest hurdle facing both TRBAs and CAR-Ts is the safety/efficacy window, which in most cases is defined by the treatment-related overproduction of proinflammatory cytokines, i.e., cytokine release syndrome (CRS). CRS accompanies both TRBA and CAR-T treatment and, at least to date, significantly limits dosing and complicates treatment paradigms. Efforts to find TRBAs and/or CAR-Ts that do not induce CRS are ongoing, and the first hopes for success have been seen in a few very recent preclinical studies. The contribution in this volume by Strohl and Naso describes the present and future states of TRBAs and CAR-T and points out how the fate for the therapeutic use of both TRBAs and CAR-Ts will rest on the ability to manage or eliminate CRS and how success in this endeavor may drive a new era in the treatment of cancer.

IgE Antibodies: From Structure to Function and Clinical Translation

When I (RJ) transferred from the Chemistry Department, at the University of Birmingham, to the Medical School, only IgM, IgG, and IgA antibodies had been definitively identified, and allergic reactions were mediated by so-called "reagenic" antibodies that, following isolation and purification, were most closely associated with IgA. At this time my mentor Dennis Stanworth published a review under the title "Structure of reagins" that opened with the statement: "It is perhaps somewhat premature to talk in terms of structure as far as reagins are concerned, considering that virtually nothing is yet known about their chemistry". However, at that time a Swedish patient newly diagnosed with multiple myeloma was shown to have a serum protein demonstrated to have a four-chain antibody structure that, serologically, was neither IgM, IgG, nor IgA. A collaborative study between Dennis, Hans Bennich, and Gunner Johansson (Uppsala) resulted in the demonstration that this protein could inhibit immediate hypersensitivity reactions (IHRs) in monkey skin; i.e., it had the properties of a reagin. Subsequently, the eminent immunologist John Humphrey offered Dennis Stanworth his back to demonstrate that the myeloma protein (ND) inhibited IHRs in humans also.

Similar studies in the laboratory of Kimishige Ishizaka resulted in the establishment of the IgE antibody class. Soon the high-affinity FcεRI (type I Fcε receptor) expressed on multiple leucocyte types was described; however, it took more years to uncover the complex series of events that followed the binding of IgE to leucocytes expressing FcεRI and the downstream events that followed in vivo. Current knowledge and understanding of the structure and function of IgE are elegantly presented in the contribution from Brian Sutton and colleagues. Our current understanding not only contributes to the development of treatments for the alleviation of allergic reactions but also offers the potential to develop IgE-class antibodies as therapeutics.

IgM Antibodies: Structure, Function, and Developability

Following the demonstration of IgM (macroglobulin) antibodies in humans, evolutionary studies demonstrated IgM to be the primordial antibody, at least in jawed vertebrates. It is expressed as the antigen receptor on the surface of B lymphocytes which when cross-linked by antigen stimulates differentiation and division to generate plasma cells secreting antigen-specific pentameric IgM antibodies. Whilst germline-encoded IgM antibodies are of low affinity, the multiple binding sites, available on B cells and the IgM pentamer, provide for high-avidity binding. Early experiences of monoclonal IgM antibodies, such as human paraproteins or mouse mAbs, showed them to exhibit relatively low solubility, with a marked propensity for aggregation (precipitation); however, in vivo, their immune complexes activate effector activities, e.g., complement activation. Given their unique functions, they are being developed as an addition to the immunotherapeutic armory. This is challenging for the biopharmaceutical industry since, in addition to their large size, their heavy chains express five glycosylation sites bearing simple or bi, tri-, and tetra-antennary glycans. This contribution provides a comprehensive overview of the history, challenges, and current developments for this antibody isotype and proffers a bright future for the use of engineered IgM antibodies in the treatment of a variety of human diseases, exemplified by the engineered CD20 x CD3 bispecific antibody IGM-2323.

IgA Antibodies: Structure, Function, and Developability

The human body produces more IgA antibody/day ($\sim$60 mg) than all other isotypes combined. Serum IgA, at a concentration of 2–3 mg/mL, is present mostly as a four-chain monomer; however, as the predominant immunoglobulin secreted at mucous surfaces, it is present as a complex dimeric form that includes the two additional proteins, namely a J (joining) chain and a secretory component (SC). It is estimated that the mucosal surface occupies an area in the order of 400 m^2 (the area of a doubles tennis court is 260 m^2). There are two subclasses of IgA, namely IgA1 and IgA2, that differ significantly in protein and oligosaccharide structure. The IgA1 isotype predominates in serum and upper regions of the GI tract whilst IgA2 predominates in the lower regions of the GI tract, possibly due to its more compact structure and consequent resistance to enzyme cleavage. Plasma cells of the GI tract synthesize IgA having a peptide extension at the C-terminus of the heavy chain that binds the J chain to allow generation of (IgA)2JSC dimers. Both IgA1 and IgA2 have N-linked oligosaccharide moieties attached within the Fcα region and a further one attached within the tail piece; the IgA2 heavy chain has two additional N-linked oligosaccharides in the Cα1 and Cα2 regions whilst IgA1 has multiple O-linked oligosaccharides attached within the hinge region. This potential for structural heterogeneity may impact the developability of IgA therapeutics; however, the IgA isotype is attracting interest within the biopharmaceutical industry, particularly where protection at mucosal surfaces is required.

Immunogenicity of Innovative and Biosimilar Monoclonal Antibodies

Endogenous human proteins and glycoproteins (Ps/GPs) are structurally heterogeneous but recognized as self by an individual's immune system to induce tolerance; however, recombinant P/GP molecules are necessarily produced in heterologous systems and undergo rigorous purification processes giving rise to structural variants that may be nonself to individual patients and potentially immunogenic. The addition of human type oligosaccharides may be critical to function, whilst the addition of nonhuman sugar residues can render biologics immunogenic. A particular concern is the structure of oligosaccharides attached by the hamster and murine cell lines that provide the dominant production platform. Early experiences of mouse-derived mAbs being immunogenic in humans provided a driving force for the sequential development of chimeric, humanized, and "fully" human antibodies; however, the first "fully" human mAb adalimumab has proved to exhibit significant immunogenicity, at least in a proportion of patients. The structural heterogeneity of adalimumab was considered to be a consequence of the unique structure of its paratope (idiotype) and the production platform adopted.

Concluding Remarks

Initially, it was thought that due to the intrinsic complexity of biologics it would not be possible to develop "generic" biologics, e.g., mAbs, and that the innovator company might enjoy patent protection throughout the lifetime of the drug. However, a "quantum leap" in the resolution and sensitivity of physicochemical techniques available for the characterization of approved biologics allows for comparisons between an innovator product and an intended biosimilar biologic. This, together with guidelines developed by regulatory authorities, has provided for the successful development and approval of biosimilar biologics. The next challenge was, and is, to improve on nature: a step that must be approached with more than caution, including humility. We are already "awash" with "novel" and/or innovative constructs that have not stood the test of time on an evolutionary timescale. However, the impact of an individual's unique physiology on any "novel" in vivo intervention is increasingly being appreciated and documented. Unfortunately, our current tutor is the strip of nucleotide bases that constitute the SARS-CoV-2 RNA virus responsible for COVID-19 (Marshall, 2020).

Marshall, M. The lasting misery of coronavirus long-haulers. *Nature* **2020**, *585*, 339–341. doi: https://doi.org/10.1038/d41586-020-02598-6.

Roy Jefferis, Koichi Kato, William R. (Bill) Strohl
Editors

antibodies

MDPI

Review

IMGT® and 30 Years of Immunoinformatics Insight in Antibody V and C Domain Structure and Function

Marie-Paule Lefranc * and Gérard Lefranc

IMGT®, the international ImMunoGeneTics information system®, University of Montpellier, CNRS, Laboratoire d'ImmunoGénétique Moléculaire LIGM, Institut de Génétique Humaine IGH, UMR 9002 CNRS-UM, 141 rue de la Cardonille, 34396 Montpellier CEDEX 5, France; glefranc@univ-montp2.fr
* Correspondence: Marie-Paule.Lefranc@igh.cnrs.fr

Received: 13 March 2019; Accepted: 9 April 2019; Published: 11 April 2019

Abstract: At the 10th Human Genome Mapping (HGM10) Workshop, in New Haven, for the first time, immunoglobulin (IG) or antibody and T cell receptor (TR) variable (V), diversity (D), joining (J), and constant (C) genes were officially recognized as 'genes', as were the conventional genes. Under these HGM auspices, IMGT®, the international ImMunoGeneTics information system®, was created in June 1989 at Montpellier (University of Montpellier and CNRS). The creation of IMGT® marked the birth of immunoinformatics, a new science, at the interface between immunogenetics and bioinformatics. The accuracy and the consistency between genes and alleles, sequences, and three-dimensional (3D) structures are based on the IMGT Scientific chart rules generated from the IMGT-ONTOLOGY axioms and concepts: IMGT standardized keywords (IDENTIFICATION), IMGT gene and allele nomenclature (CLASSIFICATION), IMGT standardized labels (DESCRIPTION), IMGT unique numbering and IMGT Collier de Perles (NUMEROTATION). These concepts provide IMGT® immunoinformatics insights for antibody V and C domain structure and function, used for the standardized description in IMGT® web resources, databases and tools, immune repertoires analysis, single cell and/or high-throughput sequencing (HTS, NGS), antibody humanization, and antibody engineering in relation with effector properties.

Keywords: IMGT; immunoinformatics; immunogenetics; IMGT-ONTOLOGY; IMGT Collier de Perles; IMGT unique numbering; immunoglobulin; antibody; paratope; complementarity determining region

1. Introduction

IMGT®, the international ImMunoGeneTics information system® (http://www.imgt.org), was created in June 1989 at Montpellier, by Marie-Paule Lefranc (University of Montpellier and CNRS) to characterize the genes and alleles of the antigen receptors, immunoglobulins (IG) or antibodies [1] and T cell receptors (TR) [2] and to manage the huge and complex diversity of the adaptive immune responses of the jawed vertebrates (or *gnathostomata*) from fishes to humans [3]. The creation of IMGT® marked the birth of immunoinformatics, a new science at the interface between immunogenetics and bioinformatics [3]. The variable (V), diversity (D), joining (J), and constant (C) genes of the antigen receptors were officially recognized as 'genes', as were the conventional genes, at the 10th Human Genome Mapping (HGM10) Workshop, in New Haven, allowing IG and TR gene and allele classification. The IMGT® databases and tools, built on the IMGT-ONTOLOGY axioms and concepts, bridge the gap between genes, sequences and three-dimensional (3D) structures [3]. The data accuracy and consistency are based on the IMGT Scientific chart rules generated from the axioms and concepts: IMGT® standardized keywords (IDENTIFICATION axiom, concepts of identification), IMGT® gene and allele nomenclature (CLASSIFICATION axiom, concepts of classification), IMGT® standardized labels (DESCRIPTION axiom, concepts of description), IMGT unique numbering and IMGT Collier de Perles (NUMEROTATION axiom, concepts of numerotation) [3].

The antigen receptor IG and TR variable domains form a huge repertoire of 2.10^{12} different specificities per individual. Owing to the particularities of their synthesis that involve DNA rearrangements, there was a need for a systematic and coherent numbering of the amino acids and codons, whatever the molecule, configuration or chain type. The IMGT unique numbering was therefore a breakthrough in immunogenetics and immunoinformatics when it was defined for the first time in 1997 for the variable (V) domain [4–6]. The IMGT unique numbering bridges the gap between amino acid and codon sequences of any V and C and their two-dimensional (2D) and three-dimensional (3D) structures and has been fundamental in the creation of the IMGT Collier de Perles graphical representation [4–6]. Both concepts have allowed the standardization of the description of mutations, amino acid changes, polymorphisms, and contact analysis in the IMGT® databases, tools and Web resources (http://www.imgt.org) [3].

The IMGT unique numbering was created by taking into account the high conservation of the structure of the V domain and by integrating the knowledge acquired by the analysis of multiple sources: alignment of more than 5000 sequences, literature data on the framework (FR) and complementarity determining regions (CDR), structural data from X-ray diffraction studies and characterization of the CDR hypervariable loops [4–6]. The standardized delimitation of the FR-IMGT and CDR-IMGT was defined based on the longest CDR1-IMGT and CDR2-IMGT found in the IMGT® multiple alignments of the germline IG and TR genes and, for the rearranged CDR3-IMGT, on statistical analysis of the IG and TR rearrangements [4–6]. The IMGT unique numbering, originally defined for the numerotation of the IG and TR V-DOMAIN [4], was rapidly extended to the V-LIKE-DOMAIN of the immunoglobulin superfamily (IgSF) other than IG and TR [5,6], then to the constant (C) domain (C-DOMAIN of IG and TR and C-LIKE-DOMAIN of IgSF other than IG and TR) [7]. Based on the same concepts, and despite a very different structure of the groove (G) domain, the IMGT unique numbering for G domain was successfully set up for the G-DOMAIN of major histocompatibility (MH) proteins and G-LIKE-DOMAIN of MhSF other than MH [8].

2. IMGT Unique Numbering for V Domain

2.1. V Domain Strands and Loops

The V domain strands and loops and their IMGT® positions and lengths, based on the IMGT unique numbering for V domain (V-DOMAIN of IG and TR and V-LIKE-DOMAIN) [6], are shown in Table 1.

The V domain (V-DOMAIN and V-LIKE-DOMAIN) is composed of the A-STRAND of fifteen (or fourteen if gap at position 10) amino acids (positions 1 to 15), the B-STRAND of eleven amino acids (positions 16 to 26) with the first conserved cysteine (1st-CYS) at position 23, the BC-LOOP (positions 27 to 38; the longest BC loops have 12 amino acids), the C-STRAND of eight amino acids (positions 39 to 46) with the tryptophan (CONSERVED-TRP) at position 41, the C'-STRAND of nine amino acids (positions 47 to 55), the C'C''-LOOP (positions 56 to 65; the longest C'C'' loops have 10 amino acids), the C''-STRAND of nine (or eight if gap at position 73) amino acids (positions 66 to 74), the D-STRAND of ten (or eight if gaps at positions 81 and 82) amino acids (positions 75 to 84), the E-STRAND of twelve amino acids (positions 85 to 96) with a conserved hydrophobic amino acid at position 89, the F-STRAND of eight amino acids (positions 97 to 104) with the second conserved cystein (2nd-CYS) at position 104, the FG-LOOP (positions 105 to 117; these positions correspond to a FG loop of 13 amino acids) and the G-STRAND of eleven (or ten) amino acids (positions 118 to 128) (Table 1, Figure 1). In the IG and TR V-DOMAIN, the G-STRAND is the C-terminal part of the J-REGION, with J-PHE or J-TRP 118 and the canonical motif F/W-G-X-G (J-MOTIF) at positions 118–121 [6] (Table 1).

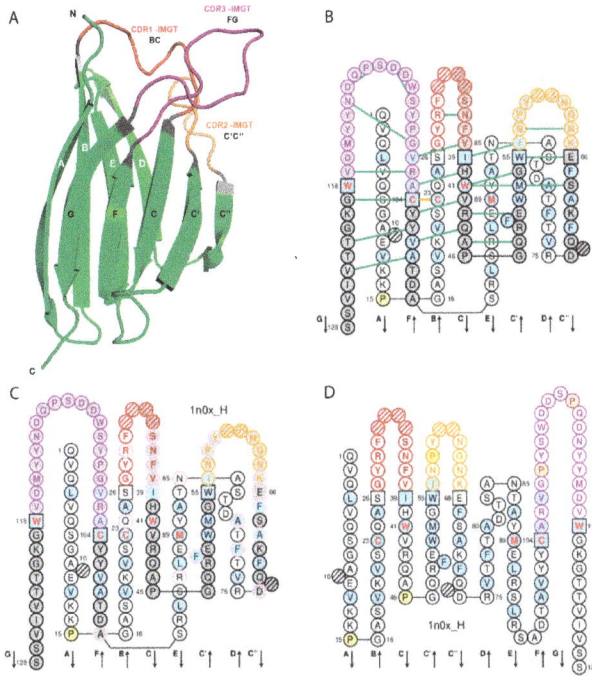

Figure 1. Variable (V) domain. An immunoglobulin (IG) variable heavy VH (V-DOMAIN) is shown as example. Reproduced with permission from IMGT®, the international ImMunoGeneTics information system®, http://www.imgt.org. (**A**) 3D structure ribbon representation with the IMGT strand and loop delimitations [6]. (**B**) IMGT Collier de Perles on two layers with hydrogen bonds. The IMGT Collier de Perles on two layers show, in the forefront, the GFCC'C'' strands (forming the sheet located at the interface VH/VL of the IG) and, in the back, the ABED strands. The IMGT Collier de Perles with hydrogen bonds (green lines online, only shown here for the GFCC'C'' sheet) is generated by the IMGT/Collier-de-Perles tool integrated in IMGT/3Dstructure-DB, from experimental 3D structure data [9–11]. (**C**) IMGT Collier de Perles on two layers generated from IMGT/DomainGapAlign [10,12,13]. Pink circles (online) indicate amino acid changes compared to the closest genes and alleles from the IMGT reference directory. (**D**) IMGT Collier de Perles on one layer. Amino acids are shown in the one-letter abbreviation. All proline (P) are shown online in yellow. IMGT anchors are in square. Hatched circles are IMGT gaps according to the IMGT unique numbering for V domain [6,14]. Positions with bold (online red) letters indicate the four conserved positions that are common to a V domain and to a C domain: 23 (1st-CYS), 41 (CONSERVED-TRP), 89 (hydrophobic), 104 (2nd-CYS) [4–7,14], and the fifth conserved position, 118 (J-TRP or J-PHE) which is specific to a V-DOMAIN and belongs to the motif F/W-G-X-G that characterizes the J-REGION [6,14] (Table 2). The hydrophobic amino acids (hydropathy index with positive value: I, V, L, F, C, M, A) and tryptophan (W) [15] found at a given position in more than 50% of sequences are shown (online with a blue background color). Arrows indicate the direction of the beta strands and their designations in 3D structures. IMGT color menu for the CDR-IMGT of a V-DOMAIN indicates the type of rearrangement, V-D-J (for a VH here, red, orange and purple) or V-J (for V-KAPPA or V-LAMBDA (not shown), blue, green and greenblue) [1]. The identifier of the chain to which the VH domain belongs is 1n0x_H (from the *Homo sapiens* b12 Fab) in IMGT/3Dstructure-DB (http://www.imgt.org). The CDR-IMGT lengths of this VH are [8.8.20] and the FR-IMGT are [25.17.38.11]. The 3D ribbon representation was obtained using PyMOL (http://www.pymol.org) and 'IMGT numbering comparison' of 1n0x_H (VH) from IMGT/3Dstructure-DB (http://www.imgt.org).

Table 1. V domain strands and loops, IMGT (IMGT®, the international ImMunoGeneTics information system®) positions and lengths, based on the IMGT unique numbering for V domain (V-DOMAIN and V-LIKE-DOMAIN) [6]. FR: framework; CDR: complementarity determining regions.

V Domain Strands and Loops [a]	IMGT Positions	Lengths [b]	Characteristic Residue@Position [c]	V-DOMAIN FR-IMGT and CDR-IMGT
A-STRAND	1–15	15 (14 if gap at 10)		FR1-IMGT
B-STRAND	16–26	11	1st-CYS 23	
BC-LOOP	27–38	12 (or less)		CDR1-IMGT
C-STRAND	39–46	8	CONSERVED-TRP 41	FR2-IMGT
C'-STRAND	47–55	9		
C'C"-LOOP	56–65	10 (or less)		CDR2-IMGT
C"-STRAND	66–74	9 (or 8 if gap at 73)		FR3-IMGT
D-STRAND	75–84	10 (or 8 if gaps at 81, 82)		
E-STRAND	85–96	12	hydrophobic 89	
F-STRAND	97–104	8	2nd-CYS 104	
FG-LOOP	105–117	13 (or less, or more)		CDR3-IMGT
G-STRAND	118–128	11 (or 10)	V-DOMAIN J-PHE 118 or J-TRP 118 [d]	FR4-IMGT

[a] IMGT® labels (concepts of description) are written in capital letters. [b] in number of amino acids (or codons). [c] Residue@Position is a IMGT® concept of numerotation that numbers the position of a given residue (or that of a conserved property amino acid class), based on the IMGT unique numbering. [d] In the IG and TR V-DOMAIN, the G-STRAND (or FR4-IMGT) is the C-terminal part of the J-REGION, with J-PHE or J-TRP 118 and the canonical motif F/W-G-X-G (J-MOTIF) at positions 118–121.

Table 2. C domain strands, turns and loops, IMGT positions and lengths, based on the IMGT unique numbering for C domain (C-DOMAIN and C-LIKE-DOMAIN) [7].

C Domain Strands, Turns and Loops [a]	IMGT Positions	Lengths [b]	Characteristic Residue@Position [c]
A-STRAND	1–15	15 (14 if gap at 10)	
AB-TURN	15.1–15.3	0–3	
B-STRAND	16–26	11	1st-CYS 23
BC-LOOP	27–31 34–38	10 (or less)	
C-STRAND	39–45	7	CONSERVED-TRP 41
CD-STRAND	45.1–45.9	0–9	
D-STRAND	77–84	8 (or 7 if gap at 82)	
DE-TURN	84.1–84.7 85.1–85.7	0–14	
E-STRAND	85–96	12	hydrophobic 89
EF-TURN	96.1–96.2	0–2	
F-STRAND	97–104	8	2nd-CYS 104
FG-LOOP	105–117	13 (or less, or more)	
G-STRAND	118–128	11 (or less)	

[a] IMGT® labels (concepts of description) are written in capital letters. [b] in number of amino acids (or codons). [c] Residue@Position is a IMGT® concept of numerotation that numbers the position of a given residue (or that of a conserved property amino acid class), based on the IMGT unique numbering.

In the IG and TR V-DOMAIN, the structurally conserved antiparallel beta strands are also designated as framework regions (FR-IMGT) whereas the loops are designated as complementarity determining regions (CDR-IMGT) [6]. Strands A and B correspond to the FR1-IMGT (positions 1 to 26), strands C and C' to the FR2-IMGT (positions 39 to 55), strands C", D, E, and F to the FR3-IMGT (positions 66 to 104) and strand G to the FR4-IMGT (positions 118 to 128). The BC, C'C", and FG loops correspond to the CDR1-IMGT, CDR2-IMGT and CDR3-IMGT, respectively [6] (Table 1, Figure 1).

IMGT anchors belong to the strands (or FR-IMGT) and represent 'anchors' supporting the three BC, C'C" and FG loops (or CDR-IMGT). V domain anchor positions are positions 26 and 39, 55 and 66, and 104 and 118, shown in square in IMGT Colliers de Perles. In a V-DOMAIN, the 2nd-CYS at position 104 (F strand) and J-PHE or J-TRP at position 118 (G strand) are anchors of the FG loop (or CDR3-IMGT) [5,6].

The loop length (number of amino acids (or codons), that is number of occupied positions) is a crucial and original concept of IMGT-ONTOLOGY [3]. The lengths of the CDR1-IMGT (BC), CDR2-IMGT (C'C"), and CDR3-IMGT (FG) characterize the V-DOMAIN. Thus, the length of the three CDR-IMGT (loops) is shown, in number of amino acids (or codons), into brackets and separated by dots. For example [8.8.20] means that the CDR1-IMGT (BC), CDR2-IMGT (C'C"), and CDR3-IMGT (FG) have lengths of 8, 8, and 20 amino acids (or codons), respectively. The JUNCTION of an IG or TR V-DOMAIN includes the anchors 104 and 118 and is therefore two amino acids longer than the corresponding CDR3-IMGT (positions 105–117) [5,6].

2.2. IMGT Gaps and Additional Positions

IMGT gaps are shown by dots in IMGT Protein displays [1,2,9,10] and by hatched circles or squares in IMGT Colliers de Perles for V domain and correspond to unoccupied positions according to the IMGT unique numbering for V domain [6] (Figure 1). For BC, C'C" or FG loops shorter than 12, 10, and 13 amino acids, respectively, gaps are created at the apex (positions hatched in IMGT Collier de Perles, or not shown in structural data representations). The gaps are placed at the apex of the loop with an equal number of amino acids (or codons) on both sides if the loop length is an even number, or with one more amino acid (or codon) in the left part if it is an odd number. For example, for FG loops shorter than 13 amino acids, gaps are created from the apex of the loop, in the following order: 111, 112, 110, 113, 109, etc. (IMGT® http://www.imgt.org, IMGT Scientific chart > Numbering > IMGT unique numbering for V-DOMAIN and V-LIKE-DOMAIN).

For CDR3-IMGT (FG) loop longer than 13 amino acids, additional positions are created, between positions 111 and 112 at the top of the loop, in the following order: 112.1, 111.1, 112.2, 111.2, 112.3, etc. [6] (IMGT® http://www.imgt.org, IMGT Scientific chart > Numbering > IMGT unique numbering for V-DOMAIN and V-LIKE-DOMAIN).

3. IMGT Unique Numbering for C Domain

3.1. C Domain Strands, Loops, and Turns

The C domain strands, turns and loops and their IMGT positions and lengths, based on the IMGT unique numbering for C domain (C-DOMAIN of IG and TR and C-LIKE-DOMAIN) [7], are shown in Table 2.

The C domain (C-DOMAIN and C-LIKE-DOMAIN) is composed by the A-STRAND of fifteen (or fourteen if gap at 10) amino acids (positions 1 to 15), the AB-TURN (additional positions 15.1, 15.2, and 15.3; the longest AB-TURN have 3 amino acids), the B-STRAND of eleven amino acids (positions 16 to 26) with the 1st-CYS at position 23, the BC-LOOP (positions 27 to 31, 34 to 38), the C-STRAND of seven amino acids (positions 39 to 45) with the CONSERVED-TRP at position 41, the CD-STRAND of one to nine amino acids (additional positions 45.1 to 45.9), the D-STRAND of eight (or seven if gap at 82) amino acids (positions 77 to 84), the DE-TURN (additional positions 84.1 to 84.7 and 85.7 to 85.1, corresponding to 14 amino acids), the E-STRAND of twelve amino acids (positions 85 to 96) with a conserved hydrophobic amino acid at position 89, the EF-TURN (additional positions 96.1 and 96.2, corresponding to 2 amino acids), the F-STRAND of eight amino acids (positions 97 to 104) with the 2nd-CYS at position 104, the FG-LOOP (positions 105 to 117, these positions corresponding to a FG loop of 13 amino acids), and the G-STRAND of eleven (or less) amino acids (positions 118 to 128) [7] (Table 2, Figure 2).

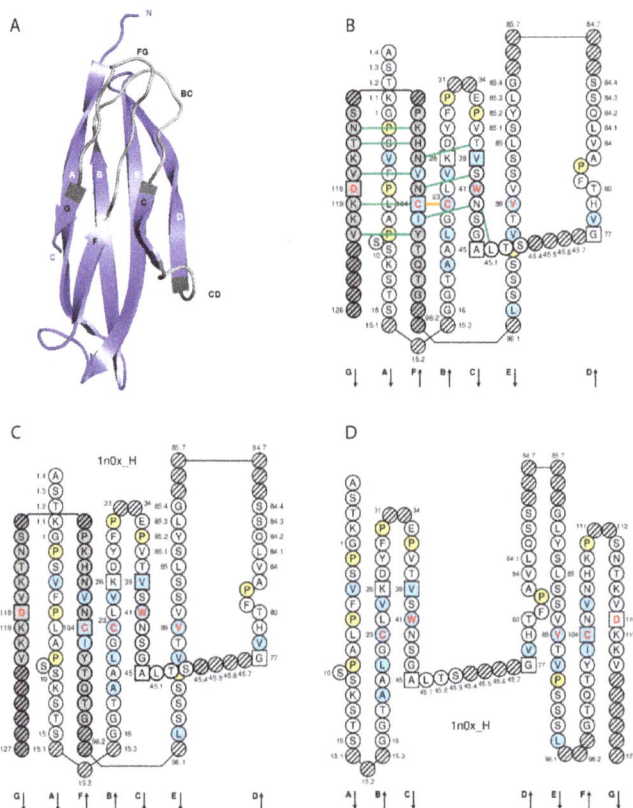

Figure 2. Constant (C) domain. An IG IGHG1 CH1 (C-DOMAIN) is shown as example. Reproduced with permission from IMGT®, the international ImMunoGeneTics information system®, http://www.imgt.org. (**A**) 3D structure ribbon representation with the IMGT strand and loop delimitations [7]. (**B**) IMGT Collier de Perles on two layers with hydrogen bonds. The IMGT Colliers de Perles on two layers show, in the forefront, the GFC strands and, in the back, the ABED strands (located at the interface CH1/CL of the IG), linked by the CD transversal strand. The IMGT Collier de Perles with hydrogen bonds (green lines online, only shown here for the GFC sheet) is generated by the IMGT/Collier-de-Perles tool integrated in IMGT/3Dstructure-DB, from experimental 3D structure data [9–11]. (**C**) IMGT Collier de Perles on two layers from IMGT/DomainGapAlign [10,12,13]. (**D**) IMGT Colliers de Perles on one layer. Amino acids are shown in the one-letter abbreviation. All proline (P) are shown online in yellow. IMGT anchors are in square. Hatched circles are IMGT gaps according to the IMGT unique numbering for C domain [7,14]. Positions with bold (online red) letters indicate the four conserved positions that are common to a V domain and to a C domain: 23 (1st-CYS), 41 (CONSERVED-TRP), 89 (hydrophobic), 104 (2nd-CYS) [4–7,14], and position 118 which is only conserved in V-DOMAIN. The identifier of the chain to which the CH1 domain belongs is 1n0x_H (from the *Homo sapiens* b12 Fab, in IMGT/3Dstructure-DB, http://www.imgt.org). The 3D ribbon representation was obtained using PyMOL and 'IMGT numbering comparison' of 1n0x_H (CH1) from IMGT/3Dstructure-DB (http://www.imgt.org).

IMGT anchors belong to the strands and represent, for the C domains, anchors for the BC and FG loops and by extension to the CD strand (as C domains do not have the C'-C" loop) [7]. Anchor positions are shown in square in IMGT Colliers de Perles. C domain anchor positions are positions 26 and 39, 45 and 77 (anchors of the CD strand), and 104 and 118 [7].

3.2. C Domain and V Domain Comparison

The A-STRAND and B-STRAND of the C domain are similar to those of the V domain [7]. The longest BC-LOOP of the C domain have 10 amino acids (missing positions 32 and 33), instead of 12 amino acids in the V domain. The C-STRAND and the D-STRAND of the C domain are shorter of one position (46) and two positions (75, 76), respectively, compared to those of the V domain. The transversal CD-STRAND is a characteristic of the C domain (a V domain has instead two antiparallel beta strands C'-STRAND and C"-STRAND linked by the C'C"-LOOP). The E-STRAND, F-STRAND and G-STRAND of the C domain are similar to those of the V domain [7] (IMGT® http://www.imgt.org, IMGT Scientific chart > Numbering > IMGT unique numbering for C-DOMAIN and C-LIKE-DOMAIN).

3.3. IMGT Gaps and Additional Positions

IMGT gaps are shown by dots in IMGT Protein displays and by hatched circles or squares in IMGT Colliers de Perles for C domain and correspond to unoccupied positions according to the IMGT unique numbering for C domain [7].

The longest BC-LOOP of the C domain have 10 amino acids (missing positions 32 and 33, that are a feature of the C domain are not shown in the IMGT Colliers de Perles and IMGT Protein displays for C domain). For BC loops shorter than 10 amino acids, gaps are created from the apex in the following order 34, 31, 35, 30, 36, etc. The FG-LOOP of the C domain is similar to that of the V domain. Gaps for FG loops shorter than 13 amino acids and additional positions for FG loops longer than 13 amino acids, are created following the same rules as those of the V domain.

Additional positions in the C domain define the AB-TURN, DE-TURN and EF-TURN (Table 2). For AB-TURN shorter than 3 amino acids, gaps are created (hatched in IMGT Colliers de Perles, or not shown in structural data representations) in a decreasing ordinal manner. For DE-TURN shorter than 14 amino acids, gaps are created in the following order: 85.7, 84.7, 85.6, 84.6, 85.5, etc. For EF-TURN shorter than 2 amino acids, gaps are created in the following order: 96.2, 96.1 [7].

4. IMGT® V and C Domain Insight for Antibody Humanization and Engineering

4.1. Antibody Humanization

4.1.1. CDR-IMGT Delimitation for Grafting

The objective of antibody humanization is to graft at the DNA level the CDR of an antibody V domain, from mouse (or other species) and of a given specificity, onto a human V domain framework, thus preserving the specificity of the original (murine or other species) antibody while decreasing its immunogenicity [16]. IMGT/DomainGapAlign [10,12,13] is the reference tool for antibody humanization design based on CDR grafting: (i) it precisely defines the CDR1-IMGT, CDR2-IMGT and CDR3-IMGT to be grafted, and (ii) it helps selecting the most appropriate human FR-IMGT by providing the alignment of the mouse (or other species) V-DOMAIN amino acid sequence with the closest germline *Homo sapiens* V-REGION and J-REGION.

Analyses performed on humanized therapeutic antibodies underline the importance of a correct delimitation of the CDR and FR. As an example, two amino acid changes were required in the first version of the humanized VH of alemtuzumab, in order to restore the specificity and affinity of the original rat antibody. The positions of these amino acid changes (S28>F and S35>F) are now known to be located in the CDR1-IMGT and should have been directly grafted, but at the time of this mAb humanization they were considered as belonging to the FR according to the Kabat numbering [17]. In contrast, positions 66–74 were, at the same time, considered as belonging to the CDR according to the Kabat numbering, whereas they clearly belong to the FR2-IMGT and the corresponding sequence should have been 'human' instead of being grafted from the 'rat' sequence.

4.1.2. Amino Acid Interactions between FR-IMGT and CDR-IMGT

IMGT Colliers de Perles from crystallized 3D structures in IMGT/3Dstructure-DB [9–11] highlight two conserved hydrogen bonds between FR-IMGT and CDR-IMGT positions: the first one between FR2-IMGT 39 and CDR2-IMGT 56 (or 57), and the second one between FR2-IMGT 40 and CDR3-IMGT 105 (Figure 1B). Antibody engineering and humanization should therefore preserve these bondings which stabilize the loops. It is also worthwhile to note that, in VH CDR3, the stem of the CDR3 loop is stabilized by a conserved salt bridge between R106 (arginine contributed by the 3'V-REGION) and D116 (aspartate contributed by the 5'J-REGION of the *Homo sapiens* IGHJ2, IGHJ3, IGHJ4, IGHJ5 or IGHJ6 (IMGT® http://www.imgt.org, IMGT Repertoire > Alignments of alleles > IGHJ > human (*Homo sapiens*) Overview).

4.1.3. V-DOMAIN Contact Analysis and Paratope

The amino acids of the V-DOMAIN CDR-IMGT involved in the contacts with the antigen can be visualized in IMGT/3Dstructure-DB Contact analysis [9–11] which provides extensive information on the atom pair contacts. Domain pair contacts ('DomPair') provide information on the contacts between a pair of partners (for examples, between the VH domain of motavizumab (3ixt_H chain) and the ligand (3ixt_P chain), or between the V-KAPPA domain of motavizumab (3ixt_L chain) and the ligand (3ixt_P chain) (Figure 3) [9–11]. Clicking on R@P gives access to the IMGT Residue@Position cards [9–11].

IMGT/3Dstructure-DB Domain pair contacts

Contacts of | Domain | Chain |
VH [D1] 3ixt_H *with* | Domain | Chain |
(Ligand) 3ixt_P

Summary:

Residue pair contacts	Number of residues			Atom pair contact types			
	Total	From 1	From 2	Total	Polar	Hydrogen	Nonpolar
32	24	11	13	289	29	3	260

List of the Residue@Position pair contacts:
Click 'R@P' for IMGT Residue@Position cards

	Order					Order					Atom pair contact types			
	IMGT Num	Residue		Domain	Chain		IMGT Num	Residue		Domain	Chain	Total	Polar	Hydrogen	Nonpolar
R@P	35	ALA	A	VH [D1]	3ixt_H	R@P	2	ASN	N	(Ligand)	3ixt_P	2	0	0	2
R@P	35	ALA	A	VH [D1]	3ixt_H	R@P	3	SER	S	(Ligand)	3ixt_P	7	0	0	7
R@P	35	ALA	A	VH [D1]	3ixt_H	R@P	6	LEU	L	(Ligand)	3ixt_P	6	0	0	6
R@P	36	GLY	G	VH [D1]	3ixt_H	R@P	6	LEU	L	(Ligand)	3ixt_P	3	0	0	3
R@P	57	TRP	W	VH [D1]	3ixt_H	R@P	10	ASN	N	(Ligand)	3ixt_P	4	0	0	4
R@P	58	TRP	W	VH [D1]	3ixt_H	R@P	3	SER	S	(Ligand)	3ixt_P	3	1	0	2
R@P	58	TRP	W	VH [D1]	3ixt_H	R@P	6	LEU	L	(Ligand)	3ixt_P	7	0	0	7
R@P	58	TRP	W	VH [D1]	3ixt_H	R@P	7	SER	S	(Ligand)	3ixt_P	12	1	0	11
R@P	58	TRP	W	VH [D1]	3ixt_H	R@P	10	ASN	N	(Ligand)	3ixt_P	11	0	0	11
R@P	59	ASP	D	VH [D1]	3ixt_H	R@P	7	SER	S	(Ligand)	3ixt_P	2	1	0	1
R@P	59	ASP	D	VH [D1]	3ixt_H	R@P	10	ASN	N	(Ligand)	3ixt_P	10	3	1	7
R@P	59	ASP	D	VH [D1]	3ixt_H	R@P	11	ASP	D	(Ligand)	3ixt_P	3	1	0	2
R@P	64	LYS	K	VH [D1]	3ixt_H	R@P	10	ASN	N	(Ligand)	3ixt_P	19	3	1	16
R@P	64	LYS	K	VH [D1]	3ixt_H	R@P	11	ASP	D	(Ligand)	3ixt_P	8	3	0	5
R@P	64	LYS	K	VH [D1]	3ixt_H	R@P	12	MET	M	(Ligand)	3ixt_P	1	1	0	0
R@P	64	LYS	K	VH [D1]	3ixt_H	R@P	19	LYS	K	(Ligand)	3ixt_P	2	1	0	1
R@P	109	ILE	I	VH [D1]	3ixt_H	R@P	6	LEU	L	(Ligand)	3ixt_P	4	0	0	4
R@P	109	ILE	I	VH [D1]	3ixt_H	R@P	9	ILE	I	(Ligand)	3ixt_P	2	0	0	2
R@P	109	ILE	I	VH [D1]	3ixt_H	R@P	10	ASN	N	(Ligand)	3ixt_P	3	0	0	3
R@P	109	ILE	I	VH [D1]	3ixt_H	R@P	19	LYS	K	(Ligand)	3ixt_P	9	1	0	8
R@P	109	ILE	I	VH [D1]	3ixt_H	R@P	20	LYS	K	(Ligand)	3ixt_P	12	2	0	10
R@P	109	ILE	I	VH [D1]	3ixt_H	R@P	23	SER	S	(Ligand)	3ixt_P	12	1	0	11
R@P	110	PHE	F	VH [D1]	3ixt_H	R@P	6	LEU	L	(Ligand)	3ixt_P	8	0	0	8
R@P	110	PHE	F	VH [D1]	3ixt_H	R@P	20	LYS	K	(Ligand)	3ixt_P	17	2	0	15
R@P	110	PHE	F	VH [D1]	3ixt_H	R@P	23	SER	S	(Ligand)	3ixt_P	22	2	0	20
R@P	110	PHE	F	VH [D1]	3ixt_H	R@P	24	ASN	N	(Ligand)	3ixt_P	21	3	1	18
R@P	112	ASN	N	VH [D1]	3ixt_H	R@P	20	LYS	K	(Ligand)	3ixt_P	13	0	0	13
R@P	112	ASN	N	VH [D1]	3ixt_H	R@P	24	ASN	N	(Ligand)	3ixt_P	2	2	0	0
R@P	113	PHE	F	VH [D1]	3ixt_H	R@P	16	ASN	N	(Ligand)	3ixt_P	35	0	0	35
R@P	113	PHE	F	VH [D1]	3ixt_H	R@P	19	LYS	K	(Ligand)	3ixt_P	6	0	0	6
R@P	113	PHE	F	VH [D1]	3ixt_H	R@P	20	LYS	K	(Ligand)	3ixt_P	20	1	0	19
R@P	114	TYR	Y	VH [D1]	3ixt_H	R@P	20	LYS	K	(Ligand)	3ixt_P	3	0	0	3

A

Figure 3. *Cont.*

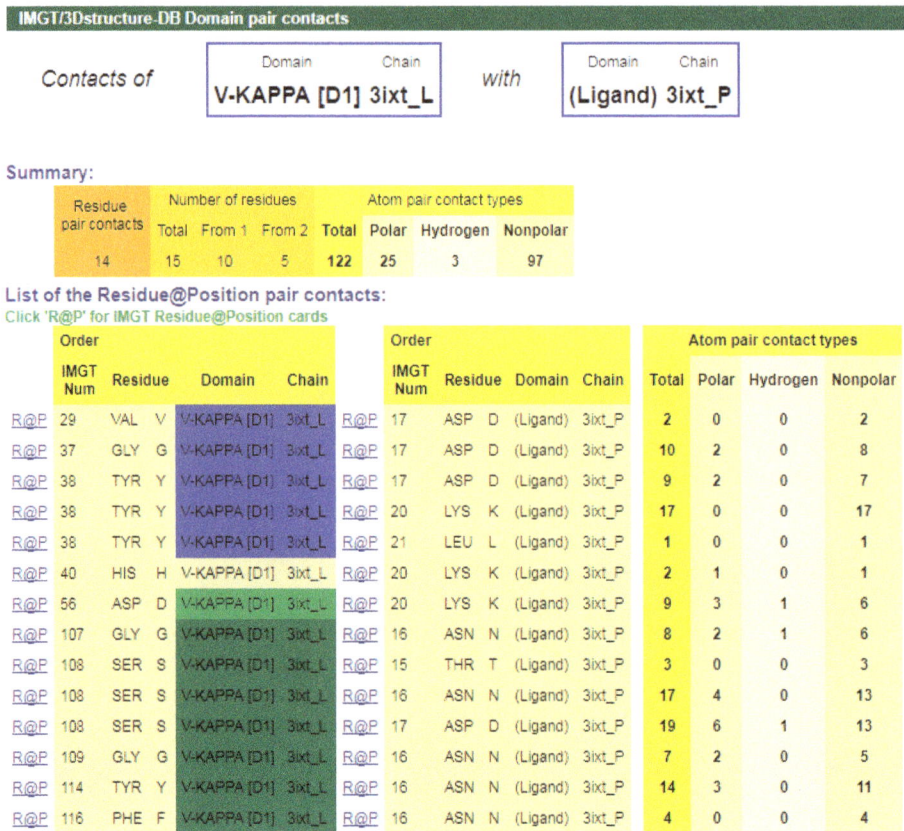

IMGT/3Dstructure-DB Domain pair contacts

Contacts of **V-KAPPA [D1] 3ixt_L** with **(Ligand) 3ixt_P**

Summary:

Residue pair contacts	Number of residues			Atom pair contact types			
	Total	From 1	From 2	Total	Polar	Hydrogen	Nonpolar
14	15	10	5	122	25	3	97

List of the Residue@Position pair contacts:
Click 'R@P' for IMGT Residue@Position cards

	Order					Order				Atom pair contact types			
	IMGT Num	Residue	Domain	Chain		IMGT Num	Residue	Domain	Chain	Total	Polar	Hydrogen	Nonpolar
R@P	29	VAL V	V-KAPPA [D1]	3ixt_L	R@P	17	ASP D	(Ligand)	3ixt_P	2	0	0	2
R@P	37	GLY G	V-KAPPA [D1]	3ixt_L	R@P	17	ASP D	(Ligand)	3ixt_P	10	2	0	8
R@P	38	TYR Y	V-KAPPA [D1]	3ixt_L	R@P	17	ASP D	(Ligand)	3ixt_P	9	2	0	7
R@P	38	TYR Y	V-KAPPA [D1]	3ixt_L	R@P	20	LYS K	(Ligand)	3ixt_P	17	0	0	17
R@P	38	TYR Y	V-KAPPA [D1]	3ixt_L	R@P	21	LEU L	(Ligand)	3ixt_P	1	0	0	1
R@P	40	HIS H	V-KAPPA [D1]	3ixt_L	R@P	20	LYS K	(Ligand)	3ixt_P	2	1	0	1
R@P	56	ASP D	V-KAPPA [D1]	3ixt_L	R@P	20	LYS K	(Ligand)	3ixt_P	9	3	1	6
R@P	107	GLY G	V-KAPPA [D1]	3ixt_L	R@P	16	ASN N	(Ligand)	3ixt_P	8	2	1	6
R@P	108	SER S	V-KAPPA [D1]	3ixt_L	R@P	15	THR T	(Ligand)	3ixt_P	3	0	0	3
R@P	108	SER S	V-KAPPA [D1]	3ixt_L	R@P	16	ASN N	(Ligand)	3ixt_P	17	4	0	13
R@P	108	SER S	V-KAPPA [D1]	3ixt_L	R@P	17	ASP D	(Ligand)	3ixt_P	19	6	1	13
R@P	109	GLY G	V-KAPPA [D1]	3ixt_L	R@P	16	ASN N	(Ligand)	3ixt_P	7	2	0	5
R@P	114	TYR Y	V-KAPPA [D1]	3ixt_L	R@P	16	ASN N	(Ligand)	3ixt_P	14	3	0	11
R@P	116	PHE F	V-KAPPA [D1]	3ixt_L	R@P	16	ASN N	(Ligand)	3ixt_P	4	0	0	4

B

Figure 3. V-DOMAIN Contact analysis results. Reproduced with permission from IMGT®, the international ImMunoGeneTics information system®, http://www.imgt.org. (**A**) IMGT/3Dstructure-DB domain pair contacts between the VH domain of motavizumab (3ixt_H) and the Fusion glycoprotein F1 (ligand) (3ixt_P). (**B**) IMGT/3Dstructure-DB Domain pair contacts between the V-KAPPA domain of motavizumab (3ixt_L) and the Fusion glycoprotein F1 (ligand) (3ixt_P). 'Polar', 'Hydrogen bonds', and 'Nonpolar' were selected prior to display, in 'Atom contact types'. Amino acids belonging to the CDR1-IMGT, CDR2-IMGT and CDR3-IMGT are colored according to the IMGT color menu (red, orange, and purple, respectively, for VH; blue, light green and green, respectively, for V-KAPPA). In this 3D structure, all but one of the amino acids contacting the antigen belong to the CDR-IMGT. Clicking on R@P gives access to the IMGT Residue@Position cards [9–11].

The IG paratope of 3ixt (motavizumab Fab) comprises AA of the VH (3ixt_H chain) and of the V-KAPPA (3ixt_L chain). Fifteen AA of the IG, eight from VH and seven from V-KAPPA, form the paratope (IMGT® http://www.imgt.org, IMGT/3Dstructure-DB > Query 3ixt > Paratope and epitope). The IMGT Colliers de Perles show that eight (out of the eight VH positions of the paratope) belong to the VH CDR-IMGT: A35 (CDR1-IMGT); W58, D59, and K64 (CDR2-IMGT); I109, F110, N112 and F113 (CDR3-IMGT), and that similarly seven (out of the seven V-KAPPA positions of the paratope) belong to the V-KAPPA CDR-IMGT: G37 and Y38 (CDR1-IMGT); D56 (CDR2-IMGT); G107, S108, G109, and Y114 (CDR3-IMGT) [9–11].

4.1.4. Potential Immunogenicity and Physicochemical Properties

The number of amino acid differences in the FR-IMGT and CDR-IMGT is one of the criteria to evaluate the potential immunogenicity. The framework of a VH domain comprises 91 positions (25, 17, 38, and 11 positions for FR1-, FR2-, FR3-, and FR4-IMGT, respectively), whereas the framework of a VL domain comprises 89 positions (26, 17, 36, 10 positions for FR1-, FR2-, FR3-, and FR4-IMGT, respectively) [9–11]. The amino acid (AA) changes are described for the hydropathy (three classes), volume (five classes) and physicochemical properties (11 classes) [15] (IMGT® http://www.imgt.org, IMGT Aide-mémoire > IMGT classes of the 20 common amino acids). S40 > G (+ + -) means that the two AA involved in the change (S > G) at codon 40 belong to the same hydropathy (+) and volume (+) classes but to different physicochemical properties (-) classes [15]. This qualification of AA replacement has led to the identification of four types of AA changes: very similar (+ + +), similar (+ + -, + - +), dissimilar (- - +, - + -, + - -), and very dissimilar (- - -).

4.1.5. V-DOMAIN CDR-IMGT Lengths and Canonical Structures

For V-DOMAIN comparison including sequences and structures, the CDR1-IMGT and CDR2-IMGT lengths are more informative than the "canonical structures" (IMGT® http://www.imgt.org, IMGT Repertoire (IG and TR) > 2D and 3D structures > CDR1-IMGT (summary) and correspondence with "canonical structures": human *(Homo sapiens)* and mouse *(Mus musculus)* Immunoglobulins; ibid CDR2-IMGT). Indeed, (1) most identified (15 out of 19) canonical structures correspond to a given CDR-IMGT length, (2) only two CDR-IMGT lengths have two canonical structures (CDR1-IMGT of nine AA of IGLV, and CDR2-IMGT of eight AA of IGHV), (3) canonical structures have not been identified for every CDR-IMGT length, (4) many 'variants' are described in the literature, based only on sequences and without experimental evidence, (5) canonical structures cannot be identified for CDR3 owing to their diversity in lengths and sequences and to their flexibility, and (6) canonical structure identification is reliable only if 3D structures are known [14]. Thus, the CDR-IMGT length is the most accurate way to define the three CDR, while working on sequences, that information being completed with characteristics Residue@Position, if necessary [9–11].

4.2. IGHG1 Alleles and G1m Allotypes

Allotypes are polymorphic markers of an IG subclass that correspond to amino acid changes and are detected serologically by antibody reagents [18]. In therapeutic antibodies (human, humanized, or chimeric), allotypes may represent potential immunogenic residues [19], as demonstrated by the presence of antibodies in individuals immunized against these allotypes [18]. The allotypes of the human heavy gamma chains of the IgG are designated as Gm (for gamma marker). The allotypes G1m, G2m, and G3m are carried by the constant region of the gamma1, gamma2 and gamma3 chains, encoded by the IGHG1, IGHG2 and IGHG3 genes, respectively [18]. The gamma1 chains may express four G1m alleles (combinations of G1m allotypes): G1m3, G1m3,1, G1m17,1, and G1m17,1,2 (and in Negroid populations three additional G1m alleles, G1m17,1,27, G1m17,1,28, and G1m17,1,27,28) [18] (Table 3). The C region of the G1m3,1, G1m17,1, and G1m17,1,2 chains differ from that of the G1m3 chains by two, three and four amino acids, respectively [18]. The correspondence between the G1m alleles and IGHG1 alleles is shown in Table 3. Thus, IGHG1*01, IGHG1*02 and IGHG1*05 are G1m17,1, IGHG1*03 is G1m3, IGHG1*04 is G1m17,1,27 and IGHG1*08p is G1m3,1. In the IGHG1 CH1, the lysine at position 120 (K120) in strand G corresponds to the G1m17 allotype [18] (Figure 2D). The isoleucine I103 (strand F) is specific of the gamma1 chain isotype. If an arginine is expressed at position 120 (R120), the simultaneous presence of R120 and I103 corresponds to the expression of the G1m3 allotype [18]. For the gamma3 and gamma4 isotypes (which also have R120 but T in 103), R120 only corresponds to the expression of the nG1m17 isoallotype (an isoallotype or nGm is detected by antibody reagents that identify this marker as an allotype in one IgG subclass and as an isotype for other subclasses) [18]. In the IGHG1 CH3, the aspartate D12 and leucine L14 (strand A) correspond to G1m1, whereas glutamate

E12 and methionine M14 correspond to the nG1m1 isoallotype [18] (Table 3). A glycine at position 110 corresponds to G1m2, whereas an alanine does not correspond to any allotype (G1m2-negative chain) (Table 3). Therapeutic antibodies are most frequently of the IgG1 isotype, and to avoid a potential immunogenicity, the constant region of the gamma1 chains are often engineered to replace the G1m3 allotype by the less immunogenic G1m17 (CH1 R120 > K) (G1m17 is more extensively found in different populations) [18].

Table 3. Correspondence between the IGHG1 alleles and G1m alleles.

IGHG1 alleles	G1m alleles[a]		IMGT amino acid positions[b]								Populations [18]
	allotypes	isoallotype s[c]	CH1				CH3				
			103	120	12	14	101	110	115	116	
				G1m17/ nG1m17	G1m1/nG 1m1		/G1m 27	/G1m2	/G1m28		
				G1m3[d]						-	
IGHG1*01[e], IGHG1*02[e], IGHG1*05[e]	G1m17,1		I	K	D	L	V	A	H	Y	Caucasoid Negroid Mongoloid
IGHG1*03	G1m3	nG1m1, nG1m17	I	R	E	M	V	A	H	Y	Caucasoid
IGHG1*04	G1m17,1,27		I	K	D	L	I	A	H	Y	Negroid
IGHG1*05p[f]	G1m17,1,28		I	K	D	L	V	A	R	Y	Negroid
IGHG1*06p[f]	G1m,17,1,27, 28		I	K	D	L	I	A	R	Y	Negroid
IGHG1*07p[f]	G1m17,1,2		I	K	D	L	V	G	H	Y	Caucasoid Mongoloid
IGHG1*08p[f]	G1m3,1	nG1m17	I	R	D	L	V	A	H	Y	Mongoloid

[a] In Negroid populations, the G1m17,1 allele frequently includes G1m27 and/or G1m28, leading to three additional G1m alleles, G1m17,1,27, G1m17,1,28 and G1m17,1,27,28 [18]. [b] Amino acids corresponding to G1m allotypes are shown in bold. [c] The nG1m1 and nG1m17 isoallotypes present on the Gm1-negative and Gm-17 negative gamma-1 chains (and on other gamma chains) are shown in italics. [d] The presence of R120 is detected by anti-nG1m17 antibodies whereas the simultaneous presence of I103 and R120 in the gamma1 chains is detected by anti-Gm3 antibodies [18]. [e] The IGHG1*01, IGHG1*02 and IGHG1*05 alleles only differ at the nucleotide level (codon 85.1 in CH2 of *02 and *05 differs from *01, codon 19 in CH1 and codon 117 in CH3 of *05 differ from *01 and *02). [f] IGHG1*05p, IGHG1*06p, IGHG1*07p and IGHG1*08p amino acids are expected [18] but not yet sequenced at the nucleotide level and therefore these alleles are not shown in IMGT Repertoire, Alignments of alleles: *Homo sapiens* IGHG1 (http://www.imgt.org).

4.3. Only-Heavy-Chain Antibodies

4.3.1. Dromedary IgG2 and IgG3 Only-Heavy-Chain Antibodies

Two IgG antibody formats are expressed in the dromedary or Arabian camel (*Camelus dromedarius*) and in Camelidae in general: the conventional IG (with two identical heavy gamma chains associated to two identical light chains) and the 'only-heavy-chain' IG (no light chain, and only two identical heavy gamma chains lacking CH1) [20]. The Camdro (for *Camelus dromedarius* in the 6-letter species abbreviation) IGHV3 genes belong to two sets based on four amino acid changes which are characteristic of each set [21]. The first set of IGHV3 genes is expressed in conventional tetrameric IgG1 that constitute 25% of circulating antibodies. The second set is expressed in the only-heavy-chain antibodies, IgG2 and IgG3 that constitute 75% of the circulating antibodies [20]. The four amino acid changes are located in the FR2-IMGT at positions 42, 49, 50 and 52, the first position 42 is in the C strand and the three others (49, 50 and 52) in the C' strand (Figure 1). They belong to the (GFCC'C") sheet at the hydrophobic VH-VL interface in conventional antibodies of Camelidae as well as of any vertebrate species whereas, in camelid only-heavy-chain antibodies (no light chains, and therefore no VL), these positions are exposed to the environment with, through evolution, a selection of hydrophilic amino acids.

The respective heavy gamma2 and gamma3 chains are both characterized by the absence of the CH1 domain owing to a splicing defect [22]. It is the absence of CH1 which is responsible for the lack of association of the light chains. Only-heavy-chain antibodies is a feature of the Camelidae IG as they have also been found in the Bactrian camel (*Camelus bactrianus*) of Central Asia and in the llama (*Lama glama*) and alpaca (*Vicugna pacos*) of South America. The genetic event (splicing defect) responsible for the lack of CH1 occurred in their common ancestor before the radiation between the 'camelini' and 'lamini', dating approximately 11 million years (Ma) ago.

The V domains of Camelidae only-heavy-chain antibodies have characteristics for potential pharmaceutical applications (e.g., easy production and selection of single-domain format, extended CDR3 with novel specificities and binding to protein clefts). They are designated as VH_H when they have to be distinguished from conventional VH (the sequence criteria is based on the four amino acids at positions 42, 49, 50 and 52). The term 'nanobody' initially used for describing a single-domain format antibody is not equivalent to VH_H, as it has been used for V domains other than VH_H and for constructs containing more than one V domain (VH and/or VH_H) (e.g., caplacizumab, ozoralizumab) (IMGT® http://www.imgt.org, IMGT Repertoire > Locus and Genes > Gene tables; ibid., The IMGT Biotechnology page > Characteristics of the camelidae (camel, llama) antibody synthesis; ibid. IMGT/mAb-DB > caplacizumab; ibid. IMGT/mAb-DB > ozoralizumab).

4.3.2. Human Heavy Chain Diseases (HCD)

The camelidae only-heavy-chain antibodies synthesis is remarkably reminiscent of what is observed in human heavy chain diseases (HCD). These proliferative disorders of B lymphoid cells produce truncated monoclonal immunoglobulin heavy chains which lack associated light chains. In most HCD, the absence of the heavy chain CH1 domain by deletion or splicing defect may be responsible for the lack of assembly of the light chain [23]. Similar observations have also been reported in mouse variants [23]. (IMGT® http://www.imgt.org, IMGT Education > Tutorials > Molecular defects in Immunoglobulin Heavy Chain Diseases (HCDs))

4.3.3. Nurse Shark IgN

A convergence mechanism in evolution is observed in nurse shark (*Ginglymostoma cirratum*, 'Gincir' in the 6-letter species abbreviation) IgN antibodies (previously IgNAR, 'immunoglobulin new antigen receptor') [24] which are only-heavy-chain antibodies (homodimeric heavy nu chains without CH1, and no associated light chains). The IGHV genes expressed in the Gincir heavy nu chains belong to the IGHV2 subgroup and are characterized by the absence of the CDR2-IMGT owing to a deletion that encompasses position 54 to 67. The Gincir IGH genes are organized in duplicated cassettes, and those that express IgN comprise Gincir IGHV2 subgroup genes and an IGHN constant gene. (IMGT® http://www.imgt.org, IMGT Repertoire (IG and TR) > Protein displays: nurse shark (*Ginglymostoma cirratum*) IGHV).

5. IGHG CH Properties and Antibody Engineering

5.1. N-Linked Glycosylation Site CH2 N84.4

A N-linked glycosylation site is present in the CH2 domain of the constant region of the human IG heavy chains of the four IgG isotypes. The N-linked glycosylation site belongs to the classical N-glycosylation motif N-X-S/T (where N is asparagine, X any amino acid except proline, S serine, T threonine) and is defined as CH2 N84.4. As shown in the IMGT Collier de Perles, this asparagine is localized at the DE turn. The IMGT unique numbering has the advantage of identifying the C domain (here, CH2) and, in the domain, the amino acid and its localization (here, N84.4) which can be visualized in the IMGT Collier de Perles and correlated with the 3D structure [25–27] (IMGT® http://www.imgt.org, The IMGT Biotechnology page > Glycosylation (IMGT Lexique)).

5.2. Interface Ball-and-Socket-Like Joints

The 3D structure comparison, between *Homo sapiens* IGHG1 Fc and IGHG2 Fc, of the CH2 and CH3 domain interface revealed that in all IGHG Fc the movement of the CH2 results from a pivoting around a highly conserved ball-and-socket-like joint [28]. Using the IMGT numbering, the CH2 L15 side chain (last position of the A strand, next to the AB turn) (the ball) interacts with a pocket (the socket) formed by CH3 M107, H108, E109, and H115 (FG loop) [25]. These amino acids are well conserved between the gamma isotypes and the IGHG genes and alleles except for IGHG3 H115 that shows a polymorphism associated to different G3m allotypes [18]. This ball-and-socket-like joint is a structural feature similar but reversed to that previously described at the VH and CH1 domain interface [29], in which the VH L12, T125 and S127 form the socket whereas the CH1 F29 and P30 (BC loop) form the ball.

5.3. Knobs-Into-Holes CH3 for the Obtaining of Bispecific Antibodies

The knobs-into-holes methodology has been proposed for obtaining bispecific antibodies [30]. The aim is to increase interactions between the CH3 domain of two gamma1 chains that belong to antibodies with a different specificity. Two amino acids, CH3 T22 (B strand) and Y86 (E strand), which belong to the [ABED] sheet, at the interface of the two *Homo sapiens* IGHG1 CH3 domains [25], were selected for amino acid changes. Interactions of these two amino acids are described in 'Contact analysis' in IMGT/3Dstructure-DB [9–11]. The knobs-into-holes methodology consists of an amino acid change on one CH3 domain (e.g., T22>Y) that creates a knob, and another amino acid change on the other CH3 domain (e.g., Y86>T) that creates a hole, thus favoring increased interactions between the CH3 of the two gamma1 chains at both positions 22 and 86 [30] (IMGT® http://www.imgt.org, The IMGT Biotechnology page > Knobs-into-holes).

5.4. IGHG Engineered Variants and Effector Properties

Amino acids in the IGHG constant regions of the IG heavy chains are frequently engineered to modify the effector properties of the therapeutic monoclonal antibodies. Amino acids changes are engineered at positions involved in antibody-dependent cellular (ADCC), antibody-dependent cellular phagocytosis (ADCP), complement-dependent cytotoxicity (CDC), half-life increase, half-IG exchange, and B cell inhibition by coengagement of antigen and FcγR on the same cell (IMGT® http://www.imgt.org, The IMGT Biotechnology page > Amino acid positions involved in ADCC, ADCP, CDC, half-life and half-IG exchange).

The IMGT engineered variant nomenclature (Table 4) has been set up for an easier comparison between engineered antibodies. The IMGT engineered variant name comprises the species, the gene name, the letter 'v' with a number (e.g., *Homo sapiens* IGHG1v1), and then the domain(s) with AA change(s) defined by the letter of the novel AA and position in the domain (e.g., CH2, P1.4). The IMGT engineered variants are classified by comparison with the allele *01 of the gene and, if the effects are independent on the alleles, as a reference for the description of the amino acid (AA) changes for the other alleles. In those cases, the same variant (v) number is used for any allele of the same gene in the same species.

Table 4. IMGT engineered variant nomenclature. Examples of IGHG1 variants involved in antibody-dependent cellular (ADCC), antibody-dependent cellular phagocytosis (ADCP), complement-dependent cytotoxicity (CDC), half-life increase, half-IG exchange, B cell inhibition, and knobs-into-holes are shown. Amino acid positions, correspondence with the EU numbering and bibliographical references are quoted at http://www.imgt.org/IMGTeducation/Tutorials/IGandBcells/_UK/IGproperties/Tableau3.html.

		IGHG gene variant description				Property modifications		
IMGT engineered variant nomenclature	CH2	AA and IMGT position in CH2 of IGHG gene variant	CH3	AA and IMGT position in CH3 of IGHG gene variant	ADCC enhancement or reduction, ADCP enhancement, B cell inhibition	CDC enhancement or reduction	Half-IG exchange reduction, Half-life increase, Knobs-into-holes
Homsap IGHG1v1	CH2	P1.4			ADCC reduction		
Homsap IGHG1v2	CH2	V1.3			ADCC reduction		
Homsap IGHG1v3	CH2	A1.2			ADCC reduction		
Homsap IGHG1v4	CH2	A114			ADCC reduction	CDC reduction	
Homsap IGHG1v5	CH2	W109			ADCC reduction	CDC enhancement	
Homsap IGHG1v6	CH2	A85.4, A118, A119			ADCC enhancement		
Homsap IGHG1v7	CH2	D3, E117			ADCC enhancement		
Homsap IGHG1v8	CH2	D3, L115, E117			ADCC enhancement	CDC reduction	
Homsap IGHG1v9	CH2	L7, P83, L85.2, I88	CH3	L83	ADCC enhancement		
Homsap IGHG1v10	CH2	Y1.3, Q1.2, W1.1, M3, D30, E34, A85.4			ADCC enhancement		
Homsap IGHG1v11	CH2	E34, D109, M115, E119			ADCC enhancement		
Homsap IGHG1v12	CH2	A1.1, D3, L115, E117			ADCC enhancement		
Homsap IGHG1v13	CH2	A1.1, D3, E117			ADCP enhancement		
Homsap IGHG1v14	CH2	A1.3, A1.2			ADCC reduction	CDC reduction	

Table 4. *Cont.*

IMGT engineered variant nomenclature	IGHG gene variant description				Property modifications		
	CH2	AA and IMGT position in CH2 of IGHG gene variant	CH3	AA and IMGT position in CH3 of IGHG gene variant	ADCC enhancement or reduction, ADCP enhancement, B cell inhibition	CDC enhancement or reduction	Half-IG exchange reduction, Half-life increase, Knobs-into-holes
Homsap IGHG1v15	CH2	S118				CDC enhancement	
Homsap IGHG1v16	CH2	W109, S118				CDC enhancement	
Homsap IGHG1v17	CH2	E29, F30, T107				CDC enhancement	
Homsap IGHG1v18			CH3	R1, G109, Y120		CDC enhancement	
Homsap IGHG1v19	CH2	A34				CDC reduction	
Homsap IGHG1v20	CH2	A105				CDC reduction	
Homsap IGHG1v21	CH2	Y15.1, T16, E18					Half-life increase
Homsap IGHG1v22	CH2	Y15.1, T16, E18	CH3	K113, F114, H116			Half-life increase
Homsap IGHG1v23	CH2	E1.2			ADCC reduction	CDC reduction	
Homsap IGHG1v24			CH3	L107, S114			Half-life increase
Homsap IGHG1v25	CH2	E29, F113			B cell inhibition		
Homsap IGHG1v26			CH3	Y22, T86			Knobs-into-holes

6. Conclusions

IMGT®, created in 1989 with the official recognition of IG and TR genes, is at the origin of immunoinformatics [3]. The concepts of classification (nomenclature and IG and TR gene and allele names, CLASSIFICATION axiom) were soon followed by the concepts of identification (standardized IMGT keywords, IDENTIFICATION axiom) and the concepts of description (standardized IMGT labels, DESCRIPTION axiom) which led to the implementation of IMGT/LIGM-DB, the first IMGT sequence database demonstrated online at the 9th International Congress of Immunology (ICI), San Francisco (USA), in July 1995. It took two more years to conceive the concepts of numerotation, IMGT unique numbering and IMGT Collier de Perles (NUMEROTATION axiom) which bridge sequences and structures of V and C domain (at the amino acid and codon levels) [3]. Interestingly, the first IMGT Collier de Perles, created manually in December 2007, not only identified conflicts between the SEQRES and ATOM lines of the PDB file but also the absence of a serine at position 93, demonstrating that indeed sequence and structure were bridged using the IMGT unique numbering (http://www.imgt.org/IMGTrepertoire/2D-3Dstruct/2D-representations/mouse/IG/E5.2Fv/ighV-D-J_E5_2Fv.html).

The IMGT® databases, tools and web resources have been built to manage immunogenetics knowledge and immunoinformatics, based on the IMGT Scientific chart rules generated from the IMGT-ONTOLOGY axioms and concepts [3]. Nowadays, IMGT® provides standardized and integrated databases, tools and web resources for IG and TR, from gene to structure and function [31–43]. The same concepts and insights for the V and C domain, are used for all vertebrate species with jaws (*gnathostomata*), from fishes to humans, providing a unique resource whatever the antigen receptor, the chain type and the taxon, for study of the adaptive immune response [3]. IG repertoire analysis and therapeutic antibody development represent two major current fields of immunoinformatics, involving V and C domains, in fundamental, pharmaceutical and medical research. High throughput (HTS) data obtained by NGS has made IMGT® standardization, developed originally to handle the huge diversity of the immune repertoires, more needed as ever. Since October 2010, the IMGT/HighV-QUEST web portal has been a paradigm for the characterization of the V domain diversity and expression and the identification of the IMGT clonotypes (AA) [44–46]. Statistical comparison of the V domain and IMGT clonotype (AA) diversity and expression between two sets can be performed using the IMGT/StatClonotype package [47,48]. NGS analysis of V domain provides immunoprofiling in normal (infectious diseases, vaccination, aging) or pathological (leukemias, lymphomas, myelomas, immunodeficiencies) conditions. An IMGT/HighV-QUEST novel functionality includes, with the same high-quality criteria, the analysis of the two V domains of single chain Fragment variable (scFv) from phage display combinatorial libraries) [49–51].

The therapeutic monoclonal antibody engineering field represents the most promising potential in medicine. Standardized genomic and expressed sequence, structure and interaction analysis of IG is crucial for a better molecular understanding and comparison of the mAb specificity, affinity, half-life, Fc effector properties, and potential immunogenicity. IMGT/3Dstructure-DB provides a standardized description and antibody structure/contact analysis characterization, at the V and C domain level, at the chain level (with the 'chimeric' and 'humanized' added as 'taxon'), and at the receptor level. Amino acids (or codons) changes (either polymorphic or resulting from engineering are identified. The structural unit is the V or C domain, with for regions (hinge, linker, CHS). This modular characterization per domain (and/or region) provides a great flexibility and is applicable to any novel format of antibody engineering [52–59]. IMGT concepts have been integrated in the Encyclopedia of Systems Biology [60–63]. The CDR-IMGT lengths are now required for mAb INN applications and are included in the World Health Organization International Nonproprietary Name WHO INN definitions [64], bringing a new level of standardized information in the comparative analysis of therapeutic antibodies.

Availability and Citation: Authors who use IMGT® databases and tools are encouraged to cite this article and to quote the IMGT® Home page, http://www.imgt.org. Online access to IMGT® databases and tools are freely available for academics and under licences and contracts for companies.

Funding: IMGT® was funded in part by the BIOMED1 (BIOCT930038), Biotechnology BIOTECH2 (BIO4CT960037), fifth PCRDT Quality of Life and Management of Living Resources (QLG2-2000-01287), and sixth PCRDT Information Science and Technology (ImmunoGrid, FP6 IST-028069) programmes of the European Union (EU). IMGT® received financial support from the GIS IBiSA, BioCampus Montpellier, the Région Occitanie (Grand Plateau Technique pour la Recherche (GPTR)), the Agence Nationale de la recherche (ANR) and the Labex MabImprove (ANR-10-LABX-53-01). IMGT® is currently supported by the Centre National de la Recherche Scientifique (CNRS), the Ministère de l'Enseignement Supérieur, de la Recherche et de l'Innovation (MESRI) and the University of Montpellier.

Acknowledgments: We are grateful to the IMGT® team for its constant motivation. IMGT® is a registered trademark of CNRS. IMGT® is a member of the International Medical Informatics Association (IMIA) and of the Global Alliance for Genomics and Health (GA4GH).

Conflicts of Interest: The authors declare no conflict of interest. The funders had no role in the design of the study; in the collection, analyses, or interpretation of data; in the writing of the manuscript, or in the decision to publish the results.

References

1. Lefranc, M.-P.; Lefranc, G. *The Immunoglobulin FactsBook*; Academic Press: London, UK, 2001.
2. Lefranc, M.-P.; Lefranc, G. *The T Cell Receptor FactsBook*; Academic Press: London, UK, 2001.
3. Lefranc, M.-P. Immunoglobulin and T Cell Receptor Genes: IMGT® and the Birth and Rise of Immunoinformatics. *Front. Immunol.* **2014**, *5*, 22. [CrossRef] [PubMed]
4. Lefranc, M.-P. Unique database numberings system for immunogenetic analysis. *Immunol. Today* **1997**, *18*, 509. [CrossRef]
5. Lefranc, M.-P. The IMGT unique numbering for Immunoglobulins, T cell receptors and Ig-like domains. *Immunologist* **1999**, *7*, 132–136.
6. Lefranc, M.-P.; Pommie, C.; Ruiz, M.; Giudicelli, V.; Foulquier, E.; Truong, L.; Thouvenin-Contet, V.; Lefranc, G. IMGT unique numbering for immunoglobulin and T cell receptor variable domains and Ig superfamily V-like domains. *Dev. Comp. Immunol.* **2003**, *27*, 55–77. [CrossRef]
7. Lefranc, M.-P.; Pommie, C.; Kaas, Q.; Duprat, E.; Bosc, N.; Guiraudou, D.; Jean, C.; Ruiz, M.; Da Piedade, I.; Rouard, M.; et al. IMGT unique numbering for immunoglobulin and T cell receptor constant domains and Ig superfamily C-like domains. *Dev. Comp. Immunol.* **2005**, *29*, 185–203. [CrossRef]
8. Lefranc, M.-P.; Duprat, E.; Kaas, Q.; Tranne, M.; Thiriot, A.; Lefranc, G. IMGT unique numbering for MHC groove G-DOMAIN and MHC superfamily (MhcSF) G-LIKE-DOMAIN. *Dev. Comp. Immunol.* **2005**, *29*, 917–938.
9. Kaas, Q.; Ruiz, M.; Lefranc, M.-P. IMGT/3Dstructure-DB and IMGT/StructuralQuery, a database and a tool for immunoglobulin, T cell receptor and MHC structural data. *Acids Res.* **2004**, *32*, D208–D210. [CrossRef] [PubMed]
10. Ehrenmann, F.; Kaas, Q.; Lefranc, M.-P. IMGT/3Dstructure-DB and IMGT/DomainGapAlign: A database and a tool for immunoglobulins or antibodies, T cell receptors, MHC, IgSF and MhcSF. *Nucl. Acids Res.* **2010**, *38*, D301–D307. [CrossRef] [PubMed]
11. Ehrenmann, F.; Lefranc, M.-P. IMGT/3Dstructure-DB: Querying the IMGT Database for 3D Structures in Immunology and Immunoinformatics (IG or Antibodies, TR, MH, RPI, and FPIA). *Cold Spring Harb. Protoc.* **2011**, *2011*, 750–761. [CrossRef] [PubMed]
12. Ehrenmann, F.; Lefranc, M.-P. IMGT/DomainGapAlign: IMGT Standardized Analysis of Amino Acid Sequences of Variable, Constant, and Groove Domains (IG, TR, MH, IgSF, MhSF). *Cold Spring Harb. Protoc.* **2011**, *2011*, 737–749. [CrossRef]
13. Ehrenmann, F.; Lefranc, M.-P. IMGT/DomainGapAlign: The IMGT® tool for the analysis of IG, TR, MHC, IgSF and MhSF domain amino acid polymorphism. In *Immunogenetics*; Christiansen, F., Tait, B., Eds.; Humana Press: Totowa, NJ, USA, 2012; pp. 605–633.
14. Lefranc, M.-P. IMGT unique numbering for the Variable (V), Constant (C), and Groove (G) domains of IG, TR, MH, IgSF, and MhSF. *Cold Spring Harb. Protoc.* **2011**, *6*, 633–642. [CrossRef]
15. Pommie, C.; Levadoux, S.; Sabatier, R.; Lefranc, G.; Lefranc, M.-P. IMGT standardized criteria for statistical analysis of immunoglobulin V-REGION amino acid properties. *J. Mol. Recognit.* **2004**, *17*, 17–32. [CrossRef]
16. Riechmann, L.; Clark, M.; Waldmann, H.; Winter, G. Reshaping human antibodies for therapy. *Nature* **1988**, *332*, 323–327. [CrossRef]
17. Kabat, E.A.; Wu, T.T.; Perry, H.M.; Gottesman, K.S.; Foeller, C. *Sequences of Proteins of Immunological Interest*; U.S. Department of Health and Human Services (USDHHS), National Institute of Health NIH Publication: Washington, DC, USA, 1991.
18. Lefranc, M.-P.; Lefranc, G. Human Gm, Km, and Am Allotypes and Their Molecular Characterization: A Remarkable Demonstration of Polymorphism. *Methods Mol. Biol.* **2012**, *882*, 635–680.
19. Jefferis, R.; Lefranc, M.-P. Human immunoglobulin allotypes: Possible implications for immunogenicity. *MAbs* **2009**, *1*, 332–338. [CrossRef]
20. Hamers-Casterman, C.; Atarhouch, T.; Muyldermans, S.; Robinson, G.; Hammers, C.; Songa, E.B.; Bendahman, N.; Hammers, R. Naturally occurring antibodies devoid of light chains. *Nature* **1993**, *363*, 446–448. [CrossRef]
21. Nguyen, V.K.; Hamers, R.; Wyns, L.; Muyldermans, S.; Niu, J.; Miao, R.; Huang, S.; Chang, J.; Davis-Dusenbery, B.N.; Kashima, R.; et al. Camel heavy-chain antibodies: Diverse germline VHH and specific mechanisms enlarge the antigen-binding repertoire. *EMBO J.* **2000**, *19*, 921–930. [CrossRef]

22. Nguyen, V.K.; Hamers, R.; Wyns, L.; Muyldermans, S. Loss of splice consensus signal is responsible for the removal of the entire CH1 domain of the functional camel IgG2a heavy chain antibodies. *Mol. Immunol.* **1999**, *36*, 515–524. [CrossRef]

23. Lefranc, M.-P.; Lefranc, G. The constant region genes of the immunoglobulin heavy chains. *Mol. Genet.* **1988**, *7*, 39–45.

24. Greenberg, A.S.; Avila, D.; Hughes, M.; Hughes, A.; McKinney, E.C.; Flajnik, M.F. A new antigen receptor gene family that undergoes rearrangement and extensive somatic diversification in sharks. *Nature* **1995**, *374*, 168–173. [CrossRef]

25. Lefranc, M.-P. How to Use IMGT® for Therapeutic Antibody Engineering. *Handb. Ther. Antib.* **2014**, *1*, 229–264.

26. Lefranc, M.-P. IMGT® immunoglobulin repertoire analysis and antibody humanization. In *Molecular Biology of B Cells*, 2nd ed.; Alt, F.W., Honjo, T., Radbruch, A., Reth, M., Eds.; Academic Press: London, UK, 2014; Chapter 26; pp. 481–514.

27. Shirai, H.; Prades, C.; Vita, R.; Marcatili, P.; Popovic, B.; Xu, J.; Overington, J.P.; Hirayama, K.; Soga, S.; Tsunoyama, K.; et al. Antibody informatics for drug discovery. *Biochim. Biophys. BBA-Proteins Proteom.* **2014**, *1844*, 2002–2015. [CrossRef]

28. Teplyakov, A.; Zhao, Y.; Malia, T.J.; Obmolova, G.; Gilliland, G.L. IgG2 Fc structure and the dynamic features of the IgG CH2–CH3 interface. *Mol. Immunol.* **2013**, *56*, 131–139. [CrossRef]

29. Lesk, A.M.; Chothia, C. Elbow motion in the immunoglobulins involves a molecular ball-and-socket joint. *Nature* **1988**, *335*, 188–190. [CrossRef]

30. Ridgway, J.B.; Presta, L.G.; Carter, P. 'Knobs-into-holes' engineering of antibody CH3 domains for heavy chain heterodimerization. *Eng. Des. Sel.* **1996**, *9*, 617–621. [CrossRef]

31. Lefranc, M.-P. IMGT, the international ImMunoGeneTics information system. In *Immunoinformatics: Bioinformatic Strategies for Better Understanding of Immune Function*; Bock, G., Goode, J., Eds.; Novartis Foundation Symposium; John Wiley and Sons: Chichester, UK, 2003; Volume 254, p. 126, discussion 136–142, 216–222, 250–252.

32. Lefranc, M.-P.; Giudicelli, V.; Ginestoux, C.; Chaume, D. IMGT, the international ImMunoGeneTics information system: The reference in immunoinformatics. *Stud. Health Technol. Inform.* **2003**, *95*, 74–79.

33. Lefranc, M.-P. IMGT databases, web resources and tools for immunoglobulin and T cell receptor sequence analysis. *Leukemia* **2003**, *17*, 260–266. [CrossRef]

34. Lefranc, M.-P. IMGT, the international ImMunoGenetics information system®. In *Antibody Engineering Methods and Protocols*, 2nd ed.; Humana Press: Totowa, NJ, USA, 2004; pp. 27–49.

35. Lefranc, M.-P. IMGT-ONTOLOGY and IMGT databases, tools and Web resources for immunogenetics and immunoinformatics. *Mol. Immunol.* **2004**, *40*, 647–659. [CrossRef]

36. Lefranc, M.-P. IMGT, the international ImMunoGeneTics information system®: A standardized approach for immunogenetics and immunoinformatics. *Immunome Res.* **2005**, *1*, 3. [CrossRef]

37. Lefranc, M.-P. IMGT®, the International ImMunoGeneTics Information System® for Immunoinformatics. In *Immunoinformatics: Predicting Immunogenicity In Silico*; Flowers, D.R., Ed.; Humana Press: Totowa, NJ, USA, 2007; Volume 409, pp. 19–42.

38. Lefranc, M.-P. IMGT-ONTOLOGY, IMGT® databases, tools and Web resources for Immunoinformatics. In *Immunoinformatics*; Schoenbach, C., Ranganathan, S., Brusic, V., Eds.; Immunomics Reviews, Series of Springer Science and Business Media LLC; Springer: New York, NY, USA, 2008; Volume 1, Chapter 1; pp. 1–18.

39. Regnier, L.; Lefranc, M.-P.; Giudicelli, V.; Duroux, P. IMGT, a system and an ontology that bridge biological and computational spheres in bioinformatics. *Brief. Bioinform.* **2008**, *9*, 263–275.

40. Lefranc, M.-P. IMGT®, the International ImMunoGeneTics Information System® for Immunoinformatics. *Mol. Biotechnol.* **2008**, *40*, 101–111. [CrossRef]

41. Lefranc, M.-P. IMGT, the International ImMunoGeneTics Information System. *Cold Spring Harb. Protoc.* **2011**, *2011*, 595–603. [CrossRef]

42. Lefranc, M.-P.; Giudicelli, V.; Duroux, P.; Jabado-Michaloud, J.; Folch, G.; Aouinti, S.; Carillon, E.; Duvergey, H.; Houles, A.; Paysan-Lafosse, T.; et al. IMGT®, the international ImMunoGeneTics information system® 25 years on. *Nucleic Acids Res.* **2015**, *43*, D413–D422. [CrossRef]

43. Lefranc, M.-P. Antibody Informatics: IMGT, the International ImMunoGeneTics Information System. In *Antibodies for Infectious Diseases*; Crowe, J., Boraschi, D., Rappuoli, R., Eds.; ASM Press: Washington, DC, USA, 2015; pp. 363–379.

44. Alamyar, E.; Giudicelli, V.; Shuo, L.; Duroux, P.; Lefranc, M.-P. IMGT/HighV-QUEST: The IMGT® web portal for immunoglobulin (IG) or antibody and T cell receptor (TR) analysis from NGS high throughput and deep sequencing. *Immunome Res.* **2012**, *8*, 26.

45. Li, S.; Lefranc, M.-P.; Miles, J.J.; Alamyar, E.; Giudicelli, V.; Duroux, P.; Freeman, J.D.; Corbin, V.D.A.; Scheerlinck, J.-P.; Frohman, M.A.; et al. IMGT/HighV QUEST paradigm for T cell receptor IMGT clonotype diversity and next generation repertoire immunoprofiling. *Nat. Commun.* **2013**, *4*, 2333. [CrossRef]

46. Giudicelli, V.; Duroux, P.; Lavoie, A.; Aouinti, S.; Lefranc, M.-P.; Kossida, S. From IMGT-ONTOLOGY to IMGT/HighV-QUEST for NGS immunoglobulin (IG) and T cell receptor (TR) repertoires in autoimmune and infectious diseases. *Autoimmune Infec. Dis.* **2015**, *1*, 1–15. [CrossRef]

47. Aouinti, S.; Malouche, D.; Giudicelli, V.; Kossida, S.; Lefranc, M.-P. IMGT/HighV-QUEST statistical significance of IMGT clonotype (AA) diversity per gene for standardized comparisons of next generation sequencing immunoprofiles of immunoglobulins and T cell receptors. *PLoS ONE* **2015**, *10*, e0142353. [CrossRef]

48. Aouinti, S.; Giudicelli, V.; Duroux, P.; Malouche, D.; Kossida, S.; Lefranc, M.-P. IMGT/StatClonotype for Pairwise Evaluation and Visualization of NGS IG and TR IMGT Clonotype (AA) Diversity or Expression from IMGT/HighV-QUEST. *Front. Immunol.* **2016**, *7*, 339. [CrossRef]

49. Giudicelli, V.; Duroux, P.; Kossida, S.; Lefranc, M.-P. IG and TR single chain fragment variable (scFv) sequence analysis: A new advanced functionality of IMGT/V-QUEST and IMGT/HighV-QUEST. *BMC Immunol.* **2017**, *18*, 35. [CrossRef]

50. Hemadou, A.; Giudicelli, V.; Smith, M.L.; Lefranc, M.-P.; Duroux, P.; Kossida, S.; Heiner, C.; Hepler, N.L.; Kuijpers, J.; Groppi, A.; et al. Pacific Biosciences Sequencing and IMGT/HighV-QUEST Analysis of Full-Length Single Chain Fragment Variable from an In Vivo Selected Phage-Display Combinatorial Library. *Front. Immunol.* **2017**, *8*, 1796. [CrossRef]

51. Han, S.Y.; Antoine, A.; Howard, D.; Chang, B.; Chang, W.S.; Slein, M.; Deikus, G.; Kossida, S.; Duroux, P.; Lefranc, M.-P.; et al. Coupling of Single Molecule, Long Read Sequencing with IMGT/HighV-QUEST Analysis Expedites Identification of SIV gp140-Specific Antibodies from scFv Phage Display Libraries. *Front. Immunol.* **2018**, *9*, 329. [CrossRef]

52. Lefranc, M.-P. *Antibody Databases and Tools: The IMGT® Experience*; Wiley: Hoboken, NJ, USA, 2009; Volume 4, pp. 91–114.

53. Lefranc, M.-P. Antibody databases: IMGT®, a French platform of world-wide interest. [in French] Bases de données anticorps: IMGT®, une plate-forme française d'intérêt mondial. *Médecine* **2009**, *25*, 1020–1023. [CrossRef]

54. Ehrenmann, F.; Duroux, P.; Giudicelli, V.; Lefranc, M.-P. *Standardized Sequence and Structure Analysis of Antibody Using IMGT®*; Springer Nature: Basingstoke, UK, 2010; Volume 2, Chapter 2; pp. 11–31.

55. Lefranc, M.-P.; Ehrenmann, F.; Ginestoux, C.; Giudicelli, V.; Duroux, P. Use of IMGT® Databases and Tools for Antibody Engineering and Humanization. In *Antibody Engineering*; Humana Press: Totowa, NJ, USA, 2012; pp. 3–37.

56. Alamyar, E.; Giudicelli, V.; Duroux, P.; Lefranc, M.-P. Antibody V and C Domain Sequence, Structure, and Interaction Analysis with Special Reference to IMGT®. *Methods Mol. Biol.* **2014**, *1131*, 337–381.

57. Lefranc, M.-P. Immunoinformatics of the V, C, and G Domains: IMGT® Definitive System for IG, TR and IgSF, MH, and MhSF. *Methods Mol. Biol.* **2014**, *1184*, 59–107.

58. Marillet, S.; Lefranc, M.-P.; Boudinot, P.; Cazals, F. Novel Structural Parameters of Ig–Ag Complexes Yield a Quantitative Description of Interaction Specificity and Binding Affinity. *Front. Immunol.* **2017**, *8*, 34. [CrossRef]

59. Lefranc, M.-P.; Ehrenmann, F.; Kossida, S.; Giudicelli, V.; Duroux, P. Use of IMGT® Databases and Tools for Antibody Engineering and Humanization. *Microinjection* **2018**, *1827*, 35–69. [CrossRef]

60. Lefranc, M.-P. IMGT® Information System. In *Encyclopedia of Systems Biology*; Dubitzky, W., Wolkenhauer, O., Cho, K.-H., Yokota, H., Eds.; Springer Science+Business Media, LLC: New York, NY, USA, 2013; pp. 959–964. [CrossRef]

61. Giudicelli, V.; Lefranc, M.-P. IMGT-ONTOLOGY. In *Encyclopedia of Systems Biology*; Dubitzky, W., Wolkenhauer, O., Cho, K.-H., Yokota, H., Eds.; Springer Science+Business Media, LLC: New York, NY, USA, 2013; pp. 964–972.

62. Lefranc, M.-P. IMGT unique numbering. In *Encyclopedia of Systems Biology*; Dubitzky, W., Wolkenhauer, O., Cho, K.-H., Yokota, H., Eds.; Springer Science+Business Media, LLC: New York, NY, USA, 2013; pp. 952–959.

63. Lefranc, M.-P. IMGT Collier de Perles. In *Encyclopedia of Systems Biology*; Dubitzky, W., Wolkenhauer, O., Cho, K.-H., Yokota, H., Eds.; Springer Science+Business Media, LLC: New York, NY, USA, 2013; pp. 944–952.

64. Lefranc, M.-P. Antibody nomenclature: From IMGT-ONTOLOGY to INN definition. *MAbs* **2011**, *3*, 1–2. [CrossRef]

antibodies

MDPI

Review

Macro- and Micro-Heterogeneity of Natural and Recombinant IgG Antibodies

Alain Beck [1],* and Hongcheng Liu [2],*

[1] Biologics CMC and developability, IRPF, Center d'immunologie Pierre Fabre, St Julien-en-Genevois CEDEX, 74160 Saint-Julien en Genevois, France
[2] Anokion, 50 Hampshire Street, Suite 402, Cambridge, MA 02139, USA
* Correspondence: alain.beck@pierre-fabre.com (A.B.); hongcheng.liu@anokion.com (H.L.)

Received: 22 December 2018; Accepted: 13 February 2019; Published: 19 February 2019

Abstract: Recombinant monoclonal antibodies (mAbs) intended for therapeutic usage are required to be thoroughly characterized, which has promoted an extensive effort towards the understanding of the structures and heterogeneity of this major class of molecules. Batch consistency and comparability are highly relevant to the successful pharmaceutical development of mAbs and related products. Small structural modifications that contribute to molecule variants (or proteoforms) differing in size, charge or hydrophobicity have been identified. These modifications may impact (or not) the stability, pharmacokinetics, and efficacy of mAbs. The presence of the same type of modifications as found in endogenous immunoglobulin G (IgG) can substantially lower the safety risks of mAbs. The knowledge of modifications is also critical to the ranking of critical quality attributes (CQAs) of the drug and define the Quality Target Product Profile (QTPP). This review provides a summary of the current understanding of post-translational and physico-chemical modifications identified in recombinant mAbs and endogenous IgGs at physiological conditions.

Keywords: critical quality attributes; comparability; developability; glycosylation; quality target product profile; mass spectrometry; post-translational modifications; proteoforms; safety

1. Introduction

Recombinant monoclonal antibodies are heterogeneous due to post-translational modifications (PTMs) and physico-chemical transformations that could occur during their entire life-span. Understanding of the mechanisms and the ways to control the heterogeneity are essential to the successful clinical development of monoclonal antibody (mAb) therapeutics. Based on International Conference on Harmonization (ICH) Q6B, mAb variants can be classified as either "Product-related substances" or "Product-related impurities". Product-related substances are defined as "Molecular variants of the desired product formed during manufacturer and/or storage which are active and have no deleterious effect on the safety and efficacy of the drug product. These variants possess properties comparable to the desired product and are not considered impurities." Product-related impurities are defined as "Molecular variants of the desired products (e.g., precursors, certain degradation products arising during manufacture and/or storage) which do not have properties comparable to those of the desired product with respect to activity, efficacy, and safety." Therefore, mAb variants are required to be thoroughly characterized to determine their chemical nature and impact on stability, activity, efficacy, and safety.

Because process changes are inevitable during process development, optimization and scale-up, a thorough understanding of mAb variants is also critical to demonstrating comparability between batches. The acceptance criteria to establish comparability for product-related impurities are more stringent than that of product-related substances (ICH Q5E). Failure to demonstrate the presence of the same type of modifications at comparable levels in post-change materials may require additional preclinical or clinical studies, due to safety concerns. Furthermore, mAb variants with different modifications might impact long-term stability and, thus, shelf-life, efficacy, and safety.

Therapeutic mAbs have evolved from a murine origin, to chimeric, and humanized or fully human to reduce immunogenicity, based on amino acid sequence homology. Generally, human-like modifications, identified as such by their presence in natural Immunoglobulin Gs (IgGs), pose a lower risk of immunogenicity.

This review focuses on the current understanding of the various types of modifications of mAbs, that can occur during manufacturing, storage, and post-administration in vivo or during clinical trials. Known modifications of human endogenous IgGs are also discussed. An overall comparison between the different modifications found in mAbs versus natural IgGs is presented in Table 1.

Table 1. Micro-heterogeneity natural IgGs and recombinant mAbs.

Modifications	Natural	Recombinant	Resulting Heterogeneity
N-terminal modifications			
PyroGlu	100% pyroGlu	Varied levels	Mass, charge for Gln to pyroGlu
Truncation	Not expected	Rare and low	Mass
Signal peptides	Not expected	Low	Mass and charge
Asn deamidation	Substantial level	Common, varied levels	Mass and charge
Asp isomerization	Not expected	Common, varied levels	Charge and hydrophobicity
Succinimide	Not expected	Common, varied levels	Mass, charge, and hydrophobicity
Oxidation	Low	Met, Trp, Cys, His	Mass and hydrophobicity
Cysteine related modifications			
Free cysteine	Low	Low	Mass, charge and hydrophobicity
Alternative disulfide bond linkage	Common	Common	Charge
Trisulfide bond	Extremely low	Low	Mass and charge
Thioether	Low	Low	Mass
Glycosylation	Common	Common	Mass and charge
Glycation	Common	Common	Mass and charge
C-terminal modifications			
C-terminal Lys	Complete removal	Common, varied levels	Mass, charge and hydrophobicity
C-terminal modifications	Not detected	Low varied levels	Mass and charge

2. N-Terminal Modifications

N-terminal pyroglutamate (pyroGlu) is a common mAb modification resulting mainly from a non-enzymatic cyclization of N-terminal glutamine (Gln) [1–5]. At a much lower rate, N-terminal glutamate (Glu) can also be converted to pyroGlu [6–8]. Various environmental factors, such as buffer composition, pH, and temperature during cell culture and purification, can impact the conversion rates, which accounts for the varied levels of N-terminal pyroGlu found in mAbs [1–5]. Conversion of Glu to pyroGlu does not contribute as extensively to N-terminal heterogeneity as does the more commonly observed Gln to pyroGlu conversion because of the dramatic difference in the conversion rates. Cyclization of N-terminal Gln or Glu to pyroGlu reduces the molecular weight of a mAb by 17 Da or 18 Da, respectively. MAbs with the original Gln are more basic than those with pyroGlu [1,2,9], though, the presence of N-terminal pyroGlu has no impact on mAb structure and function [2,8]. The same conversion from Gln to pyroGlu is expected for mAbs in circulation because of the non-enzymatic nature of this reaction. N-terminal Glu has also been shown to be converted to pyroGlu in circulation [6].

Another common N-terminal modification is the incomplete removal of light chain or heavy chain signal peptides, which results in mAbs with truncated signal peptides of varying sizes [2,9–14]. The presence of signal peptides with different number and type of amino acids adds mass heterogeneity to mAbs. Interestingly, mAbs with signal peptides have been detected mainly as basic species [2,9,12,13,15], but rarely as acidic species [16]. The presence of low levels of signal peptides has no impact on potency [2,9,13] and pharmacokinetics (PK) in rats [13].

Although it is less common, N-terminal truncation has been reported. A combination of a murine signal peptide and antibody lambda light chain causes an alternative cleavage of the signal peptide resulting in a mAb with the loss of three amino acids from the light chain [17]. With the use of a specific tag to label N-terminal primary amine in combination with liquid chromatography mass spectrometry (LC-MS), an mAb variant with the loss of one amino acid from the light chain was observed [11].

Natural IgG contains approximately 1.8-mole pyroGlu/mole IgG [6]. Based on the different reaction rates, it is expected that most of the pyroGlu originates from N-terminal Gln, rather than Glu. Assuming that the stress condition for massive production of mAbs from ex vivo expression in host cell lines is the cause of N-terminal signal peptides and truncation, they are not expected to occur and have not been reported for natural IgGs.

3. Asn Deamidation

Asparagine (Asn) deamidation is almost a ubiquitous modification of mAbs and has been well studied because of its contribution to heterogeneity, and its potential impact on potency and immunogenicity. Asn residues in the complementarity determining regions (CDRs) are inherently susceptible to deamidation because of their relatively higher flexibility and exposure to solvents than at other locations [2,18–22]. Deamidation in CDRs can cause a substantial loss of potency [20,22–24]. In addition to deamidation in the CDRs, deamidation also occurs in susceptible Asn residues in the constant region. The most widely observed deamidation site is located in the fragment crystallizable (Fc) region within the amino acid sequence of SNGQPENNY [2,25–29]. Deamidation in the constant regions other than within the commonly observed sequence has also been reported [25,30]. When measured by differential scanning calorimetry, the fragment antigen binding (Fab) fragment with the deamidation product, isoaspartate (isoAsp), is less stable compared to Fab with the original Asn residue [22]. Deamidation increases the molecular weight of mAbs by 1 Da and generates acidic species [2,12,13,20–22,31,32]. Variants containing deamidation products are less hydrophobic than those with the original Asn residues [20,32]. Deamidation does not impact in vivo clearance [21,26]. Deamidation of Asn residues continues to occur in vivo in the CDRs [19,21,33,34] and Fc region [26] of mAbs. The in vivo deamidation kinetics can be fully predicted via in vitro stress studies under physiological conditions [19,26], which indicates the same non-enzymatic mechanism. It should be noted that it is important to optimize digestion procedures and distinguish procedure-induced artifact versus real for the accurate determination of Asn deamidation levels.

Natural human IgG has 23% deamidation at the conserved site in the Fc region, which is consistent with the molecules' in vivo half-life [26]. The presence of high levels of deamidation in natural human IgG suggests that deamidation, at least at the conserved site, is not foreign to the immune system and, therefore, would not present an increased risk of immunogenicity.

4. Asp Isomerization

Aspartate (Asp) isomerization has been commonly observed in mAbs in CDRs due to higher levels of flexibility and exposure [15,18,20,35–41]. Isomerization of Asp in CDRs has been shown to cause a decrease in antigen binding affinity [20,35,36,39,42]. Since there is no charge difference between Asp and isoAsp, the observed decrease in potency is probably caused by conformational changes due to the introduction of a methyl group into the peptide backbone. Isomerization does not change mAb molecular weight; however, depending on the specific location, isomerization can either generate acidic [12] or basic [20,43] species. Similarly, isomerization could result in mAbs or their Fab fragments becoming either more [15,35,37,44] or less [20,45] hydrophobic. Isomerization is a non-enzymatic reaction with an optimal pH of around 5 [23,41]. Under physiological conditions, isomerization was not found to increase for 34 days [26]. Therefore, the level of isomerization is expected to be low in natural IgGs.

5. Succinimide

Succinimide is the reaction intermediate of both Asn deamidation and Asp isomerization and is commonly detected in CDRs [15,20,23,34,35,37,41,42,46]. The presence of succinimide in the CDR has been demonstrated to cause a decrease in potency [23,34,35,42]. Succinimide as a deamidation intermediate has also been detected in the conserved susceptible Asn deamidation site [25,28]. MAb variants containing succinimide from Asp isomerization have been shown to become more acidic [43] or basic [15,20,23,41], due to direct charge difference and conformational changes. Similarly, mAb or Fab with succinimide as isomerization intermediate could be more [35,37,44] or less hydrophobic [15]. It is worth mentioning that mAb variants containing succinimide, which alters the molecular weight difference by only 18 Da, have been reported to appear as a back shoulder of the main peak by size exclusion chromatography (SEC) [34,47], suggesting a substantial conformational difference. It has been shown that the succinimide residue contained in mAb was converted to Asp and isoAsp after administration to monkeys [34]. Due to its instability under physiological pH, succinimide is not expected to be detected in natural human IgG.

6. Oxidation

Methionine (Met) residue is the most commonly observed amino acid that is susceptible to oxidation in mAbs. Studies have shown oxidation of Met in the heavy chain CDR2 [39] or the frame work region [48]. Met oxidation did not show a negative impact on antigen binding in either case. Two conserved Met residues close to the heavy chain constant domain 2 (CH2)-CH3 domain interface have been shown to be susceptible to oxidation [48–51]. The addition of one oxygen atom increases mAb molecular weight by 16 Da. As expected, mAb variants with oxidized Met are less hydrophobic compared to the non-oxidized molecules [44,45,52,53]. Interestingly, one mAb with the Fc conserved Met oxidized appeared to be more basic [51], while, another mAb with an oxidized Met in the Fab region appeared to be more acidic [31]. Oxidation of the two conserved Met residues in the Fc region caused conformational changes mainly in the CH2 domain [49,54] along with a host of negative impacts, including decreased thermal stability, [48–50,55] increased aggregation [49,55], decreased complement-dependent cytotoxicity (CDC) [48], decreased binding affinity to neonatal Fc receptor (FcRn) [48,56,57] and shorter in vivo half-life [58].

Oxidation has also been observed at tryptophan (Trp) residues in mAbs [59–62]. Trp residues in CDRs are more susceptible to oxidation due to a higher level of solvent exposure [63]. Oxidation of Trp generates a number of species, the major ones having molecular weight increases of 16 Da, and 32 Da [62,64]. MAb variants with oxidized Trp are less hydrophobic [52]. Oxidation of Trp residues in the CDRs can lead to reduced potency, decreased thermal stability, and increased aggregation propensity [59–61]. Trp oxidation has also been demonstrated to cause yellow coloration of the mAb solution, [64] due to kynurenine formation.

Oxidation of Met and several other amino acids has been detected in natural human IgG [65,66]. Oxidative stress under various pathological conditions and the resulting reactive oxygen species are expected to cause oxidation of susceptible residues in natural IgGs, as one of the most abundant proteins in circulation.

7. Cysteine and Disulfide Bond

Theoretically, all cysteine residues of mAbs should be involved in the formation of either intra- or inter-chain disulfide bonds in a well-defined linkage pattern. However, several variants that deviate from the well-established IgG disulfide bond structure have been discovered. These variants include the presence of free cysteine (Cys) residues, alternative disulfide bond linkage (scrambling), trisulfide bonding, the formation of thioether, and cysteine racemization.

The presence of free cysteine can be classified into three scenarios. The first scenario is the widely-reported occurrence of free cysteine residues [67–71]. These free cysteines have been shown to lower thermal stability [67] and increase the formation of reducible covalent aggregates [72–74]. The second scenario is the detection of relatively high levels of free Cys often due to the incomplete formation of a particular disulfide bond, mainly in the heavy chain variable domain [4,37,44,75,76] or the disulfide bond between the light chain and heavy chain [13,77]. The incomplete variable domain disulfide bond reduces the potency of one mAb [44], but has no impact on a different mAb [75]. MAb variants with the incomplete heavy chain variable domain disulfide bond were separated as acidic species in one case [75], but basic species in the other case [4], indicating that a structural change was likely the cause of different chromatographic behaviors. MAb variants with the incomplete variable domain disulfide bond are more hydrophobic [15,37,44]. MAb variants without the disulfide bond between the light chain and heavy chain are enriched in the acidic species [13,77]. The incomplete heavy chain variable domain disulfide bond can be reformed in vivo [4]. In the third scenario, the mAbs contained an extra non-canonical cysteine residue, mostly in the CDRs. The extra Cys in mAbs can be modified by small thiol containing compounds, such as free cysteines [78–81], and glutathione, [80,81] or oxidized to form cysteine sulfinic or sulfonic acid [81]. Modification of Cys introduces molecular weight heterogeneity. In addition, mAbs with modified Cys are less hydrophobic compared to unmodified molecules [81]. Cysteinylation increases mAb molecular weight, causes the formation of acidic species, and decreases antigen binding [78]. Modification of the extra Cys residue also causes lower expression titer, decreased thermal stability, and higher propensity towards aggregation [78,80].

The alternative disulfide bond linkage was first discovered in IgG4 molecules, where the formation of two inter-heavy chain disulfide bonds is in equilibrium with the formation of two intra-chain disulfide bonds [82–85]. The direct outcome of this equilibrium is the formation of bispecific antibodies, which has been reported for both recombinant monoclonal antibodies and natural antibodies [82,84]. Mutation of the IgG4 hinge region amino acid sequence, CPSC, to the IgG1 amino acid sequence, CPPC, can eliminate the Fab-exchange phenomena [85–87], which has been employed as a strategy to create stable mAb therapeutics based on the IgG4 framework. Later, the alternative disulfide bond linkage in the hinge region of IgG2 antibodies was discovered, both in recombinant and in natural human IgG2 [88]. Different IgG2 isoforms showed a subtle difference in structure and thermal stability [89]. While having no difference in molecular weights, the three disulfide isoforms, A, B, and A/B, can be differentiated using several analytical methods. By ion-exchange chromatography, the B isoform appeared to be more acidic than A/B, followed by A [43,88]. By reversed-phase chromatography, the B isoform eluted from columns earlier than A/B, followed by A [89–91]. By capillary electrophoresis sodium dodecyl sulfate (CE-SDS), the A isoform migrated faster than A/B, whereas the B isoform migrated the slowest [88]. Depending on the specific molecule, different isoforms may or may not have an impact on potency [89]. The conversion from A to B through the A/B isoform continues in mAbs in circulation [91].

The thioether linkage was first discovered in an IgG1 antibody as a non-reducible species using reducing sodium dodecyl sulfate polyacrylamide gel electrophoresis (SDS-PAGE) and CE-SDS [92] Later, it was found that the thioether between the light and heavy chains can also be formed in mAbs in vivo and in human natural IgGs [93]. The rate of thioether formation in IgG1 containing the lambda chain is faster than the conversion rate in IgG1 containing the kappa chain [93]. The formation of thioether reduces mAb molecular weight by 32 Da due to the loss of a sulfur atom.

Trisulfide bond was first discovered in an IgG2 mAb [94]. It was later found that trisulfide bond occurs in all classes of mAbs and natural human IgGs, mainly between the light chain and heavy chain [93,95,96]. Trisulfide bond formation can be controlled by changing the feeding strategy [97] or removed by a cysteine wash step during the protein A chromatography step [98]. Trisulfide bond increases mAb molecular weight by 32 Da. A mAb variant with a trisulfide bond appeared to be more acidic compared to mAb with the typical disulfide bond pattern [94]. The presence of a trisulfide bond has no impact on antigen binding [95,98] or thermal stability [94].

The cysteine residues located in the heavy chains that are involved in the formation of the light and heavy chain disulfide bonds were also found to exist in the D form [99]. A detailed study showed that racemization occurred in both the heavy and light chain cysteine residues in IgG1 lambda, but only the heavy chain in IgG1 kappa in both mAbs and human natural IgG1 [100]. The level of cysteine racemization is much lower in IgG2 [100]. As both thioether and racemization are catalyzed by basic condition and involve the same disulfide bonds, a general base-catalyzed mechanism was proposed, where beta elimination of the disulfide bond results in the formation of a dehydroalanine residue and the dehydroalanine residue can either form a thioether bond or revert to the disulfide bond, where chirality is regained to result in a mixture of D- and L- cysteine residues [100].

Most of the modifications identified in mAbs, including free cysteine [67,101,102], alternative disulfide bond linkage for IgG4 [82,84] and IgG2 [88], thioether [93], trisulfide bond [93,95,96], and D-Cys from racemization [100] have also been reported in natural human IgGs. However, cysteinylation and the presence of incomplete disulfide bonds have not been reported in natural human IgGs. Given all the negative impacts of cysteinylation, this modification may have been eliminated from natural IgG during evolution. The same could be true for the presence of a single pair of incomplete disulfide bonds.

8. Glycosylation

Similar to natural IgG molecules, mAbs are N-glycosylated at the conserved Asn residues in the CH2 domain. In addition, mAbs may have N-linked oligosaccharides in the Fab region [103–105]. Heterogeneity related to these oligosaccharides arises mainly from galactosylation, fucosylation, and sialylation of the biantennary complex oligosaccharides. The presence of low abundance oligosaccharides, such as high mannose (Man), hybrid and bisecting oligosaccharides adds further heterogeneity to mAbs.

The three major glycoforms are the core-fucosylated structures with either zero (G0F), one (G1F) or two (G2F) galactose [104,106–111]. Galactose adds an additional mass of 162 Da. However, galactosylation has not been reported to cause mAb heterogeneity in charge or hydrophobicity. The slight separation of mAb variants with different levels of galactosylation is probably caused by conformational differences since galactose should not change the charge properties [77]. Galactosylation can cause subtle conformational changes around the glycosylation site [54,112–115]. Conflicting results have been reported regarding biological functions, but it is generally agreed that galactose might slightly impact CDC, but not antibody-dependent cellular cytotoxicity (ADCC) [106,113–122]. Galactosylation has no impact on mAb stability [113,114,123,124], nor half-life [103,117,125–128].

Because the absence of the core-fucose can result in enhanced ADCC [122,129], the level of core-fucose has attracted great attention for the development of mAb therapeutics, especially to establish comparability or biosimilarity. The attachment of fucose adds a mass of 146 Da. Besides mass heterogeneity, fucose has not been reported to have an impact on charge and hydrophobicity. The addition of a fucose only has a subtle impact on mAb structure [54,130,131]. For mAbs without the core-fucose, animal studies on half-life have shown conflicting results [132,133]. However, the half-life was found to be as expected in human studies [125,127]. The level of core-fucose needs to be evaluated based on the target and therapeutic goal to balance the risk versus benefit [134].

Besides the complex oligosaccharides, high mannose oligosaccharides have been commonly observed in mAbs [108,109,135]. In addition to mass heterogeneity, mAbs with high mannose oligosaccharides demonstrate a slightly different chromatographic separation when using Protein A or Protein G columns [136]. High mannose oligosaccharides cause a subtle conformational change and increase the flexibility of the CH2 domain [137,138]. Although high mannose decreased the thermal stability of mAbs, it had no impact on long-term stability [131], or aggregation propensity under accelerated conditions [124]. High mannose shows increased Fc gamma receptor binding and ADCC, due to the absence of core-fucose [139]. IgG with high mannose showed reduced activities mediated by the first subcomponent of the C1 complex (C1q) binding [140,141]. High mannose oligosaccharides with greater than five mannose residues are rapidly converted into a structure with only 5 mannose residues (Man5) in human circulation [125]. MAbs with high mannose are cleared at a faster rate compared to those with complex oligosaccharides in animals and humans [126,127,139,140,142].

Sialic acid and alpha 1,3-galactose are two low abundant oligosaccharides that require special attention primarily because of safety concerns. In general, the level of sialic acid of mAbs that are associated with the conserved Fc glycosylation site is low [103,109,135]. However, substantial amounts of sialic acid have been found in mAbs containing a Fab glycosylation site [103,135]. Sialic acid adds mass heterogeneity and generates acidic species [2,13,27], but does not impact antigen binding [2,13,117,119,143,144] and clearance [13,125,128]. Sialic acid has been shown to cause subtle conformational changes that are local to the glycosylation sites [137,138,145–147]. Studies have demonstrated that sialic acid exerts no or a negative impact on ADCC and CDC [117,119,143,144]. Among the two types of sialic acids, N-Acetylneuraminic acid (NANA) and N-Glycolylneuraminic acid (NGNA), the latter, which is commonly found in mAbs from murine cell lines [2,103,104,108], has been linked to immunogenicity [148]. Similar to sialic acid, mAbs expressed in murine cell lines may contain low levels of alpha 1,3 galactose when associated with the Fc [3,103,104,108,109,149] and at relatively higher levels for mAbs containing Fab glycosylation [135]. Alpha1,3 galactose is also considered immunogenic [150] when associated with Fab [151].

Several other types of oligosaccharides including hybrid, bisecting, and smaller structures, such as those lacking outer arm N-acetylglucosamine (GlcNAc) residues, are present in mAbs at extremely low levels. Hybrid, bisecting, and smaller oligosaccharides could cause a subtle conformational change [138] and have a minimal impact on ADCC [118,122,137,152] and clearance [125,126]. Because of their extremely low levels, these types of oligosaccharides are not expected to have a substantial impact on mAb therapeutic development from the safety and efficacy point of view.

The absence of oligosaccharides also contributes to mAb heterogeneity, though, at low levels [153–156]. MAbs lacking oligosaccharides showed significant conformational changes [157,158], decreased thermal stability [112,113,131,159,160], and increased aggregation propensity [123,161]. The absence of oligosaccharides has a substantial impact on ADCC and CDC [113,119,157]. Initially, animal studies showed that the absence of oligosaccharides either caused a faster clearance [157,162,163] or had no impact on half-life [127,157,160,164,165]. However, later human trials demonstrated that aglycosylated mAbs had a normal half-life [164].

Human IgG contains similar major oligosaccharide structures but higher structural diversity [108,109,166]. The levels of bisecting and sialic acid are higher in natural human IgGs compared to mAbs [108,109,166], while, high mannose oligosaccharides in natural human IgG are extremely low at approximately 0.1% [109,167]. Human IgGs have also been shown to have less than 0.2% aglycosylation [167]. NGNA and alpha 1,3 galactose, are absent from natural human IgGs [108,109,166,168].

9. Glycation

Glycation is a non-enzymatic reaction between reducing sugars and the primary amine of the lysine (Lys) side chain or the N-terminus of the light chain or heavy chain [169–171]. Glycation mainly occurs during cell culture as sugars are used as nutrients [169], and, to a lesser degree, during storage or accelerated conditions due to decomposition of the non-reducing sugars used in formulation [172,173]. A slightly increased level of glycation has been reported during the course of administration when a diluent containing sugars is used [174]. Glycation increases mAb molecular weight by 162 Da with each site of glycation and generates acidic species due to loss of positive charges of Lys side chains or N-termini [13,27,43,169,171]. Glycation also increases the aggregation propensity under accelerated condition [173]. Glycation in the CDRs has not been shown to decrease antigen binding [13,169,171], and even a substantial level of glycation does not affect Fc gamma, FcRn, and protein A binding [175]. Advanced glycation end products (AGEs) contribute to product coloration [176]. The presence of glycation does not impact PK in rats [13]. Glycation of mAbs continues to occur in circulation in humans at a rate that can be predicted via in vitro incubation under physiological conditions [175].

As expected, glycation has been detected in endogenous human IgG [175], further supporting the simple reaction mechanism between circulating IgGs and sugars in vivo.

10. C-Terminal Modifications

Mostly, mAbs are synthesized with the heavy chain C-terminal Lys, which can be removed during cell culture due to carboxypeptidase activity [177]. Incomplete removal results in mAbs with either zero, one or two C-terminal lysine at various levels [2,3,12,178,179]. When analyzed by mass spectrometry, heterogeneity caused by C-terminal lysine is reflected by peaks that differ in molecular weight by 128 Da. C-terminal Lys is a common cause of the generation of basic species [1,9,12,43,178–180]. The presence of C-terminal Lys results in the formation of less hydrophobic mAb variants [35,45]. C-terminal Lys does not impact mAb structure, stability, or biological functions including PK [2,9,13,180–182], though, one study demonstrated that the removal of C-terminal Lys is required for optimal CDC [183]. Interestingly, inclusion of the C-terminal Lys codon may impact mAb titer of cell culture [184]. C-terminal Lys can be rapidly removed from mAbs during circulation with a half-life of 62 minutes [185].

C-terminal amidation was first discovered in a recombinant monoclonal IgG1 antibody [186]. Later, it was found that C-terminal amidation is as common as C-terminal Lys removal [187]. C-terminal amidation is catalyzed by peptidylglycine alpha-amidating monooxygenase (PAM) [187]. The level of C-terminal amidation can be modulated by changing the copper concentration in the cell culture media [188] or via genetic engineering to reduce PAM activity [189]. Compared to mAbs without C-terminal Lys, a loss of glycine and conversion of the newly exposed amino acid carboxyl group to an amide group results in a net molecular weight decrease of 58 Da. MAbs with C-terminal amidation are separated as basic species [43,186,188]. MAb variants without C-terminal Lys or without both Lys and Gly showed no difference in structure, stability, function, and PK [181].

The overall level of C-terminal amidation in natural human IgG is extremely low, approximately 0.02% or lower [185,187].

11. Uncommon Modifications

Several of the reported modifications only contribute to mAb heterogeneity at very low levels or in only a limited number of cases.

Low level of sequence variation has been observed for several mAbs [190–198], which is expected to be the norm rather than the exception because of the inherent errors in protein transcription and translation. An mAb variant with the heavy chains containing amino acids that were coded by part of the intron sequence was also found [3]. Recombination between light chain and heavy chain sequences has been reported to result in a minor mAb species where the heavy chain containing a portion of the light chain sequence [199].

Aside from the few cases of amino acid variation, several rare chemical modifications can occur at various stages. Methylglyoxal generated during cell culture has been shown to modify an mAb at arginine (Arg) residues, resulting in molecular weight increases by 54 Da or 72 Da and generation of acidic species [200]. Metals can catalyze oxidative carbonylation of several surface-exposed residues including Arg, Proline (Pro), Lys, and Thr [201]. When exposed to light, histidine (His) can be oxidized [202], which can further lead to His–His cross-linking [203]. Cysteinylation, which frequently occurs at non-canonical cysteine residues, has also been reported at canonical cysteine residues in IgG2 [12,96] and is probably due to the relative instability of the IgG2 disulfide bond linkage around the hinge region. The presence of tyrosine sulfation resulted in the formation of a distinct acidic peak for a mAb expressed in Chinese hamster ovary (CHO) cells [204]. Modification of light chain and heavy chain N-termini by maleuric acid has been detected in a mAb expressed in transgenic goats [205]. During storage, the N-terminal primary amine or lysine side chain of mAbs can be modified by citric acid or its degradation products [206,207]. In addition to glycosylation of the conserved Asn residues in the Fc region or glycosylation of Asn in the consensus sequence in the variable domains, O-fucosylation of a serine residue in the light chain CDR1 [208] and N-glycosylation of Asn in non-consensus sequence and Gln [209,210] have also been reported.

MAbs expressed in mammalian cell lines have been extensively characterized. However, novel modifications are expected whenever new cell culture media or formulations are used. The use of alternative expression systems is also expected to lead to novel modifications that are specific to the selected organism. Novel and non-clinically qualified modifications naturally bear higher safety risks, and, thus, requires thorough evaluation.

12. Heterogeneity in the Broader Scheme

12.1. Stability

ICH Q6B states that "degradation of drug substance and drug product, which may occur during storage, should be considered when establishing specifications." ICH Q6B also discusses the concept of "Release limits vs shelf-life limits", where tighter release limits will ensure that product at the end of shelf life can meet the acceptance criteria to maintain safety and efficacy.

Regarding stability, the aforementioned PTMs can be classified into two categories. The first category includes modifications that are catalyzed by enzymatic reactions. Those modifications include signal peptides, various glycoforms, C-terminal Lys removal, and C-terminal amidation. These types of modifications are not expected to continue during storage because of the lack of their respective enzymes in the drug substance and drug product. However, the levels of these modifications can potentially impact other degradation pathways, and, thus, stability. For example, the subtle conformational difference in mAbs with various oligosaccharides and the substantial conformational difference caused by the lack of oligosaccharides are expected, at least in theory, to impact other modifications by modulating surface exposure and inter-molecule interactions. The second category includes modifications that are dependent only on environmental factors, such as pH, temperature, and light exposure. Modifications in this category include N-terminal Gln and Glu cyclization, deamidation, isomerization, succinimide intermediate formation, oxidation, cysteine and disulfide bond related modifications, and glycation. Modifications in this category are expected to continue to occur during storage.

Overall, PTMs are, either indirectly or directly, linked to mAb stability. Detailed characterization of drug substances at the time of lot release and understanding of the degradation pathways derived from forced degradation, and stability studies can ensure mAb stability during shelf-life for consistent safety and efficacy.

12.2. Comparability and Biosimilarity

Comparability is required when process changes are introduced, which is inevitable during development. Q5E states that "The demonstration of comparability does not necessarily mean that the quality attributes of the pre-change and post-change product are identical, but that they are highly similar and that the existing knowledge is sufficiently predictive to ensure that any differences in quality attributes have no adverse impact upon safety or efficacy of the drug product". Scientific understanding of the chemical nature of PTMs and their impact on safety and efficacy is critical to establishing comparability, especially when a quality attribute is outside of the historical range.

MAb heterogeneity is also central to the development of biosimilar products. Given the requirement that the primary sequence of the originator and a biosimilar product should be identical, it becomes clear that similarity is mainly dependent on various PTMs.

In-depth characterization of mAb heterogeneity plays an essential role in establishing comparability and biosimilarity. The National Institute of Standards and Technology mAb (NISTmAb) tryptic peptide spectral library can be used as a good reference for those detailed comparisons [211], as it contains an extensive list of modifications, including the commonly observed analytical artifacts, which should be differentiated from true modifications.

12.3. Antibody-Drug Conjugate

Antibody-drug conjugates (ADCs) take advantage of the specificities of mAbs to deliver functional molecules to targets, and commonly, high toxicity compounds, to cancer cells [212]. MAb heterogeneity, thus, becomes an integral characteristic of ADCs, and exerts similar impact on structure, and stability. The microenvironment of the conjugation sites including solvent accessibility and charges has been demonstrated to have a substantial impact on the in vivo stability and activity of ADCs [213]. The presence of trisulfide bonds, for example, has also been shown to affect conjugation and the resulting drug-to-antibody ratio (DAR) [16,214]. Higher levels of heterogeneity have been reported for ADCs based on IgG2 mAbs, which are known for their various disulfide bond isoforms and difference in disulfide bond accessibility [215].

13. Conclusions

Heterogeneity is recognized as a common feature of mAbs due to modifications that cause IgG variants or proteoforms that differ in molecular weight, charge or hydrophobicity. MAb variants are required to be evaluated to establish their structure–function and safety relationships. In addition, different variants may have (or not) different impacts on stability, which is ultimately linked to safety and efficacy.

A wealth of information has been accumulated over the past decades. Such knowledge can be generally used to define the quality target product profile and applied to the assessment of developability of clinical candidates during the early phase of pharmaceutical development. Later in development, molecule-specific modifications are observed and managed throughout the lifecycle of the selected mAb.

Funding: This research received no external funding.

Conflicts of Interest: The authors declare no conflict of interest. Alain Beck is an employee of Institut de Recherche Pierre Fabre. Hongcheng Liu is an employee of Anokion.

Abbreviations

ADC	Antibody-drug conjugate
ADCC	Antibody-dependent cellular cytotoxicity
AGE	Advanced glycation end product
Asn	Asparagine
Asp	Aspartate
Arg	Arginine
C1q	First subcomponent of the C1 complex
CDC	Complement-dependent cytotoxicity
CDR	Complementarity determining region
CE-SDS	Capillary electrophoresis sodium dodecyl sulfate
CH2	Heavy chain constant domain 2
CHO	Chinese hamster ovary
CQA	Critical quality attribute
Cys	Cysteine
Da	Dalton
Fab	Fragment antigen binding
Fc	Fragment crystallizable
FcRn	Neonatal Fc receptor
GlcNAc	N-acetylglucosamine
Gln	Glutamine
Glu	Glutamate
Gly	Glycine
His	Histidine
ICH	International Conference on Harmonization

IgG	Immunoglobulin G
IsoAsp	Isoaspartate
LC-MS	Liquid chromatography mass spectrometry
Lys	Lysine
mAb	Monoclonal antibody
Man	Mannose
Met	Methionine
NANA	N-Acetylneuraminic acid
NGNA	N-Glycolylneuraminic acid
PAM	Peptidylglycine alpha-amidating monooxygenase
PK	Pharmacokinetics
Pro	Proline
PTM	Posttranslational modification
PyroGlu	Pyroglutamate
QTPP	Quality target product profile
SDS-PAGE	Sodium dodecyl sulfate polyacrylamide gel electrophoresis
Thr	Threonine
Trp	Tryptophan

References

1. Moorhouse, K.G.; Nashabeh, W.; Deveney, J.; Bjork, N.S.; Mulkerrin, M.G.; Ryskamp, T. Validation of an HPLC method for the analysis of the charge heterogeneity of the recombinant monoclonal antibody IDEC-C2B8 after papain digestion. *J. Pharm. Biomed. Anal.* **1997**, *16*, 593–603. [CrossRef]

2. Lyubarskaya, Y.; Houde, D.; Woodard, J.; Murphy, D.; Mhatre, R. Analysis of recombinant monoclonal antibody isoforms by electrospray ionization mass spectrometry as a strategy for streamlining characterization of recombinant monoclonal antibody charge heterogeneity. *Anal. Biochem.* **2006**, *348*, 24–39. [CrossRef] [PubMed]

3. Beck, A.; Bussat, M.C.; Zorn, N.; Robillard, V.; Klinguer-Hamour, C.; Chenu, S.; Goetsch, L.; Corvaïa, N.; Van Dorsselaer, A.; Haeuw, J.F. Characterization by liquid chromatography combined with mass spectrometry of monoclonal anti-IGF-1 receptor antibodies produced in CHO and NS0 cells. *J. Chromatogr. B Anal. Technol. Biomed. Life Sci.* **2005**, *819*, 203–218. [CrossRef] [PubMed]

4. Ouellette, D.; Alessandri, L.; Chin, A.; Grinnell, C.; Tarcsa, E.; Radziejewski, C.; Correia, I. Studies in serum support rapid formation of disulfide bond between unpaired cysteine residues in the VH domain of an immunoglobulin G1 molecule. *Anal. Biochem.* **2010**, *397*, 37–47. [CrossRef] [PubMed]

5. Dick, L.W., Jr.; Kim, C.; Qiu, D.; Cheng, K.C. Determination of the origin of the N-terminal pyro-glutamate variation in monoclonal antibodies using model peptides. *Biotechnol. Bioeng.* **2007**, *97*, 544–553. [CrossRef] [PubMed]

6. Liu, Y.D.; Goetze, A.M.; Bass, R.B.; Flynn, G.C. N-terminal glutamate to pyroglutamate conversion in vivo for human IgG2 antibodies. *J. Biol. Chem.* **2011**, *286*, 11211–11217. [CrossRef] [PubMed]

7. Chelius, D.; Jing, K.; Lueras, A.; Rehder, D.S.; Dillon, T.M.; Vizel, A.; Rajan, R.S.; Li, T.; Treuheit, M.J.; Bondarenko, P.V. Formation of pyroglutamic acid from N-terminal glutamic acid in immunoglobulin gamma antibodies. *Anal. Chem.* **2006**, *78*, 2370–2376. [CrossRef] [PubMed]

8. Yu, L.; Vizel, A.; Huff, M.B.; Young, M.; Remmele, R.L., Jr.; He, B. Investigation of N-terminal glutamate cyclization of recombinant monoclonal antibody in formulation development. *J. Pharm. Biomed. Anal.* **2006**, *42*, 455–463. [CrossRef] [PubMed]

9. Meert, C.D.; Brady, L.J.; Guo, A.; Balland, A. Characterization of antibody charge heterogeneity resolved by preparative immobilized pH gradients. *Anal. Chem.* **2010**, *82*, 3510–3518. [CrossRef]

10. Ying, H.; Liu, H. Identification of an alternative signal peptide cleavage site of mouse monoclonal antibodies by mass spectrometry. *Immunol. Lett.* **2007**, *111*, 66–68. [CrossRef]

11. Ayoub, D.; Bertaccini, D.; Diemer, H.; Wagner-Rousset, E.; Colas, O.; Cianferani, S.; Van Dorsselaer, A.; Beck, A.; Schaeffer-Reiss, C. Characterization of the N-terminal heterogeneities of monoclonal antibodies using in-gel charge derivatization of alpha-amines and LC-MS/MS. *Anal. Chem.* **2015**, *87*, 3784–3790. [CrossRef] [PubMed]

12. Neill, A.; Nowak, C.; Patel, R.; Ponniah, G.; Gonzalez, N.; Miano, D.; Liu, H. Characterization of Recombinant Monoclonal Antibody Charge Variants Using OFFGEL Fractionation, Weak Anion Exchange Chromatography, and Mass Spectrometry. *Anal. Chem.* **2015**, *87*, 6204–6211. [CrossRef] [PubMed]

13. Khawli, L.A.; Goswami, S.; Hutchinson, R.; Kwong, Z.W.; Yang, J.; Wang, X.; Yao, Z.; Sreedhara, A.; Cano, T.; Tesar, D.; et al. Charge variants in IgG1: Isolation, characterization, in vitro binding properties and pharmacokinetics in rats. *MAbs* **2010**, *2*, 613–624. [CrossRef]

14. Kotia, R.B.; Raghani, A.R. Analysis of monoclonal antibody product heterogeneity resulting from alternate cleavage sites of signal peptide. *Anal. Biochem.* **2010**, *399*, 190–195. [CrossRef] [PubMed]

15. Sreedhara, A.; Cordoba, A.; Zhu, Q.; Kwong, J.; Liu, J. Characterization of the isomerization products of aspartate residues at two different sites in a monoclonal antibody. *Pharm. Res.* **2012**, *29*, 187–197. [CrossRef] [PubMed]

16. Liu, H.; Ren, W.; Zong, L.; Zhang, J.; Wang, Y. Characterization of recombinant monoclonal antibody charge variants using WCX chromatography, icIEF and LC-MS/MS. *Anal. Biochem.* **2018**, *564–565*, 1–12. [CrossRef] [PubMed]

17. Gibson, S.J.; Bond, N.J.; Milne, S.; Lewis, A.; Sheriff, A.; Pettman, G.; Pradhan, R.; Higazi, D.R.; Hatton, D. N-terminal or signal peptide sequence engineering prevents truncation of human monoclonal antibody light chains. *Biotechnol. Bioeng.* **2017**, *114*, 1970–1977. [CrossRef] [PubMed]

18. Sydow, J.F.; Lipsmeier, F.; Larraillet, V.; Hilger, M.; Mautz, B.; Molhoj, M.; Kuentzer, J.; Klostermann, S.; Schoch, J.; Voelger, H.R.; et al. Structure-based prediction of asparagine and aspartate degradation sites in antibody variable regions. *PLoS ONE* **2014**, *9*, e100736. [CrossRef]

19. Tran, J.C.; Tran, D.; Hilderbrand, A.; Andersen, N.; Huang, T.; Reif, K.; Hotzel, I.; Stefanich, E.G.; Liu, Y.; Wang, J. Automated Affinity Capture and On-Tip Digestion to Accurately Quantitate in Vivo Deamidation of Therapeutic Antibodies. *Anal. Chem.* **2016**, *88*, 11521–11526. [CrossRef]

20. Harris, R.J.; Kabakoff, B.; Macchi, F.D.; Shen, F.J.; Kwong, M.; Andya, J.D.; Shire, S.J.; Bjork, N.; Totpal, K.; Chen, A.B. Identification of multiple sources of charge heterogeneity in a recombinant antibody. *J. Chromatogr. B Biomed. Sci. Appl.* **2001**, *752*, 233–245. [CrossRef]

21. Huang, L.; Lu, J.; Wroblewski, V.J.; Beals, J.M.; Riggin, R.M. In vivo deamidation characterization of monoclonal antibody by LC/MS/MS. *Anal. Chem.* **2005**, *77*, 1432–1439. [CrossRef] [PubMed]

22. Vlasak, J.; Bussat, M.C.; Wang, S.; Wagner-Rousset, E.; Schaefer, M.; Klinguer-Hamour, C.; Kirchmeier, M.; Corvaïa, N.; Ionescu, R.; Beck, A. Identification and characterization of asparagine deamidation in the light chain CDR1 of a humanized IgG1 antibody. *Anal. Biochem.* **2009**, *392*, 145–154. [CrossRef] [PubMed]

23. Yan, B.; Steen, S.; Hambly, D.; Valliere-Douglass, J.; Vanden Bos, T.; Smallwood, S.; Yates, Z.; Arroll, T.; Han, Y.; Gadgil, H.; et al. Succinimide formation at Asn 55 in the complementarity determining region of a recombinant monoclonal antibody IgG1 heavy chain. *J. Pharm. Sci.* **2009**, *98*, 3509–3521. [CrossRef] [PubMed]

24. Yang, X.; Xu, W.; Dukleska, S.; Benchaar, S.; Mengisen, S.; Antochshuk, V.; Cheung, J.; Mann, L.; Babadjanova, Z.; Rowand, J.; et al. Developability studies before initiation of process development: Improving manufacturability of monoclonal antibodies. *MAbs* **2013**, *5*, 787–794. [CrossRef] [PubMed]

25. Chelius, D.; Rehder, D.S.; Bondarenko, P.V. Identification and characterization of deamidation sites in the conserved regions of human immunoglobulin gamma antibodies. *Anal. Chem.* **2005**, *77*, 6004–6011. [CrossRef] [PubMed]

26. Liu, Y.D.; van Enk, J.Z.; Flynn, G.C. Human antibody Fc deamidation in vivo. *Biologicals* **2009**, *37*, 313–322. [CrossRef] [PubMed]

27. Xiao, Z.; Yin, X.; Han, L.; Sun, B.; Shen, Z.; Liu, W.; Yu, F. A comprehensive approach for evaluating charge heterogeneity in biosimilars. *Eur. J. Pharm. Sci.* **2018**, *115*, 19–24. [CrossRef]

28. Sinha, S.; Zhang, L.; Duan, S.; Williams, T.D.; Vlasak, J.; Ionescu, R.; Elizabeth, M.T. Effect of protein structure on deamidation rate in the Fc fragment of an IgG1 monoclonal antibody. *Protein Sci.* **2009**, *18*, 1573–1584. [CrossRef]

29. Gaza-Bulseco, G.; Li, B.; Bulseco, A.; Liu, H.C. Method to differentiate asn deamidation that occurred prior to and during sample preparation of a monoclonal antibody. *Anal. Chem.* **2008**, *80*, 9491–9498. [CrossRef]

30. Zhang, Y.T.; Hu, J.; Pace, A.L.; Wong, R.; Wang, Y.J.; Kao, Y.H. Characterization of asparagine 330 deamidation in an Fc-fragment of IgG1 using cation exchange chromatography and peptide mapping. *J. Chromatogr. B Anal. Technol. Biomed. Life Sci.* **2014**, *965*, 65–71. [CrossRef]

31. Ponniah, G.; Kita, A.; Nowak, C.; Neill, A.; Kori, Y.; Rajendran, S.; Liu, H. Characterization of the acidic species of a monoclonal antibody using weak cation exchange chromatography and LC-MS. *Anal. Chem.* **2015**, *87*, 9084–9092. [CrossRef] [PubMed]

32. King, C.; Patel, R.; Ponniah, G.; Nowak, C.; Neill, A.; Gu, Z.; Liu, H. Characterization of recombinant monoclonal antibody variants detected by hydrophobic interaction chromatography and imaged capillary isoelectric focusing electrophoresis. *J. Chromatogr. B Anal. Technol. Biomed. Life Sci.* **2018**, *1085*, 96–103. [CrossRef] [PubMed]

33. Bults, P.; Bischoff, R.; Bakker, H.; Gietema, J.A.; van de Merbel, N.C. LC-MS/MS-Based Monitoring of In Vivo Protein Biotransformation: Quantitative Determination of Trastuzumab and Its Deamidation Products in Human Plasma. *Anal. Chem.* **2016**, *88*, 1871–1877. [CrossRef] [PubMed]

34. Ouellette, D.; Chumsae, C.; Clabbers, A.; Radziejewski, C.; Correia, I. Comparison of the in vitro and in vivo stability of a succinimide intermediate observed on a therapeutic IgG1 molecule. *MAbs* **2013**, *5*, 432–444. [CrossRef] [PubMed]

35. Cacia, J.; Keck, R.; Presta, L.G.; Frenz, J. Isomerization of an aspartic acid residue in the complementarity-determining regions of a recombinant antibody to human IgE: Identification and effect on binding affinity. *Biochemistry* **1996**, *35*, 1897–1903. [CrossRef] [PubMed]

36. Rehder, D.S.; Chelius, D.; McAuley, A.; Dillon, T.M.; Xiao, G.; Crouse-Zeineddini, J.; Vardanyan, L.; Perico, N.; Mukku, V.; Brems, D.N.; et al. Isomerization of a single aspartyl residue of anti-epidermal growth factor receptor immunoglobulin gamma2 antibody highlights the role avidity plays in antibody activity. *Biochemistry* **2008**, *47*, 2518–2530. [CrossRef] [PubMed]

37. Wakankar, A.A.; Borchardt, R.T.; Eigenbrot, C.; Shia, S.; Wang, Y.J.; Shire, S.J.; Liu, J.L. Aspartate isomerization in the complementarity-determining regions of two closely related monoclonal antibodies. *Biochemistry* **2007**, *46*, 1534–1544. [CrossRef] [PubMed]

38. Wakankar, A.A.; Liu, J.; Vandervelde, D.; Wang, Y.J.; Shire, S.J.; Borchardt, R.T. The effect of cosolutes on the isomerization of aspartic acid residues and conformational stability in a monoclonal antibody. *J. Pharm. Sci.* **2007**, *96*, 1708–1718. [CrossRef] [PubMed]

39. Yan, Y.; Wei, H.; Fu, Y.; Jusuf, S.; Zeng, M.; Ludwig, R.; Krystek, S.R., Jr.; Chen, G.; Tao, L.; Das, T.K. Isomerization and Oxidation in the Complementarity-Determining Regions of a Monoclonal Antibody: A Study of the Modification-Structure-Function Correlations by Hydrogen-Deuterium Exchange Mass Spectrometry. *Anal. Chem.* **2016**, *88*, 2041–2050. [CrossRef] [PubMed]

40. Xiao, G.; Bondarenko, P.V. Identification and quantification of degradations in the Asp-Asp motifs of a recombinant monoclonal antibody. *J. Pharm. Biomed. Anal.* **2008**, *47*, 23–30. [CrossRef] [PubMed]

41. Chu, G.C.; Chelius, D.; Xiao, G.; Khor, H.K.; Coulibaly, S.; Bondarenko, P.V. Accumulation of succinimide in a recombinant monoclonal antibody in mildly acidic buffers under elevated temperatures. *Pharm. Res.* **2007**, *24*, 1145–1156. [CrossRef] [PubMed]

42. Valliere-Douglass, J.; Jones, L.; Shpektor, D.; Kodama, P.; Wallace, A.; Balland, A.; Bailey, R.; Zhang, Y. Separation and characterization of an IgG2 antibody containing a cyclic imide in CDR1 of light chain by hydrophobic interaction chromatography and mass spectrometry. *Anal. Chem.* **2008**, *80*, 3168–3174. [CrossRef] [PubMed]

43. Ponniah, G.; Nowak, C.; Neill, A.; Liu, H. Characterization of charge variants of a monoclonal antibody using weak anion exchange chromatography at subunit levels. *Anal. Biochem.* **2017**, *520*, 49–57. [CrossRef]

44. Harris, R.J. Heterogeneity of recombinant antibodies: Linking structure to function. *Dev. Biol. (Basel)* **2005**, *122*, 117–127.

45. Valliere-Douglass, J.; Wallace, A.; Balland, A. Separation of populations of antibody variants by fine tuning of hydrophobic-interaction chromatography operating conditions. *J. Chromatogr. A* **2008**, *1214*, 81–89. [CrossRef]

46. Huang, H.Z.; Nichols, A.; Liu, D. Direct identification and quantification of aspartyl succinimide in an IgG2 mAb by RapiGest assisted digestion. *Anal. Chem.* **2009**, *81*, 1686–1692. [CrossRef] [PubMed]

47. Nowak, C.; Ponniah, G.; Neill, A.; Liu, H. Characterization of succinimide stability during trypsin digestion for LC-MS analysis. *Anal. Biochem.* **2017**, *526*, 1–8. [CrossRef] [PubMed]

48. Mo, J.; Yan, Q.; So, C.K.; Soden, T.; Lewis, M.J.; Hu, P. Understanding the Impact of Methionine Oxidation on the Biological Functions of IgG1 Antibodies Using Hydrogen/Deuterium Exchange Mass Spectrometry. *Anal. Chem.* **2016**, *88*, 9495–9502. [CrossRef]

49. Liu, D.; Ren, D.; Huang, H.; Dankberg, J.; Rosenfeld, R.; Cocco, M.J.; Li, L.; Brems, D.N.; Remmele, R.L. Structure and stability changes of human IgG1 Fc as a consequence of methionine oxidation. *Biochemistry* **2008**, *47*, 5088–5100. [CrossRef]

50. Liu, H.; Gaza-Bulseco, G.; Xiang, T.; Chumsae, C. Structural effect of deglycosylation and methionine oxidation on a recombinant monoclonal antibody. *Mol. Immunol.* **2008**, *45*, 701–708. [CrossRef]

51. Chumsae, C.; Gaza-Bulseco, G.; Sun, J.; Liu, H. Comparison of methionine oxidation in thermal stability and chemically stressed samples of a fully human monoclonal antibody. *J. Chromatogr. B Anal. Technol. Biomed. Life Sci.* **2007**, *850*, 285–294. [CrossRef] [PubMed]

52. Boyd, D.; Kaschak, T.; Yan, B. HIC resolution of an IgG1 with an oxidized Trp in a complementarity determining region. *J. Chromatogr. B Anal. Technol. Biomed. Life Sci.* **2011**, *879*, 955–960. [CrossRef] [PubMed]

53. Lam, X.M.; Yang, J.Y.; Cleland, J.L. Antioxidants for prevention of methionine oxidation in recombinant monoclonal antibody HER2. *J. Pharm. Sci.* **1997**, *86*, 1250–1255. [CrossRef] [PubMed]

54. Houde, D.; Peng, Y.; Berkowitz, S.A.; Engen, J.R. Post-translational modifications differentially affect IgG1 conformation and receptor binding. *Mol. Cell Proteom.* **2010**, *9*, 1716–1728. [CrossRef] [PubMed]

55. Zhang, A.; Hu, P.; MacGregor, P.; Xue, Y.; Fan, H.; Suchecki, P.; Olszewski, L.; Liu, A. Understanding the conformational impact of chemical modifications on monoclonal antibodies with diverse sequence variation using hydrogen/deuterium exchange mass spectrometry and structural modeling. *Anal. Chem.* **2014**, *86*, 3468–3475. [CrossRef] [PubMed]

56. Bertolotti-Ciarlet, A.; Wang, W.; Lownes, R.; Pristatsky, P.; Fang, Y.; McKelvey, T.; Li, Y.; Li, Y.; Drummond, J.; Prueksaritanont, T.; et al. Impact of methionine oxidation on the binding of human IgG1 to Fc Rn and Fc gamma receptors. *Mol. Immunol.* **2009**, *46*, 1878–1882. [CrossRef] [PubMed]

57. Pan, H.; Chen, K.; Chu, L.; Kinderman, F.; Apostol, I.; Huang, G. Methionine oxidation in human IgG2 Fc decreases binding affinities to protein A. and FcRn. *Protein Sci.* **2009**, *18*, 424–433. [CrossRef]

58. Wang, W.; Vlasak, J.; Li, Y.; Pristatsky, P.; Fang, Y.; Pittman, T.; Roman, J.; Wang, Y.; Prueksaritanont, T.; Ionescu, R. Impact of methionine oxidation in human IgG1 Fc on serum half-life of monoclonal antibodies. *Mol. Immunol.* **2011**, *48*, 860–866. [CrossRef]

59. Dashivets, T.; Stracke, J.; Dengl, S.; Knaupp, A.; Pollmann, J.; Buchner, J.; Schlothauer, T. Oxidation in the complementarity-determining regions differentially influences the properties of therapeutic antibodies. *MAbs* **2016**, *8*, 1525–1535. [CrossRef]

60. Qi, P.; Volkin, D.B.; Zhao, H.; Nedved, M.L.; Hughes, R.; Bass, R.; Yi, S.C.; Panek, M.E.; Wang, D.; Dalmonte, P. Characterization of the photodegradation of a human IgG1 monoclonal antibody formulated as a high-concentration liquid dosage form. *J. Pharm. Sci.* **2009**, *98*, 3117–3130. [CrossRef]

61. Wei, Z.; Feng, J.; Lin, H.Y.; Mullapudi, S.; Bishop, E.; Tous, G.I.; Casas-Finet, J.; Hakki, F.; Strouse, R.; Schenerman, M.A. Identification of a single tryptophan residue as critical for binding activity in a humanized monoclonal antibody against respiratory syncytial virus. *Anal. Chem.* **2007**, *79*, 2797–2805. [CrossRef] [PubMed]

62. Nowak, C.; Ponniah, G.; Cheng, G.; Kita, A.; Neill, A.; Kori, Y.; Liu, H. Liquid chromatography-fluorescence and liquid chromatography-mass spectrometry detection of tryptophan degradation products of a recombinant monoclonal antibody. *Anal. Biochem.* **2016**, *496*, 4–8. [CrossRef] [PubMed]

63. Sharma, V.K.; Patapoff, T.W.; Kabakoff, B.; Pai, S.; Hilario, E.; Zhang, B.; Charlene, L.; Oleg, B.; Robert, F.K.; Ilya, C.; et al. In silico selection of therapeutic antibodies for development: Viscosity, clearance, and chemical stability. *Proc. Natl. Acad. Sci. USA* **2014**, *111*, 18601–18606. [CrossRef] [PubMed]

64. Li, Y.; Polozova, A.; Gruia, F.; Feng, J. Characterization of the degradation products of a color-changed monoclonal antibody: Tryptophan-derived chromophores. *Anal. Chem.* **2014**, *86*, 6850–6857. [CrossRef] [PubMed]

65. Jasin, H.E. Oxidative modification of inflammatory synovial fluid immunoglobulin G. *Inflammation* **1993**, *17*, 167–181. [CrossRef] [PubMed]

66. Lunec, J.; Blake, D.R.; McCleary, S.J.; Brailsford, S.; Bacon, P.A. Self-perpetuating mechanisms of immunoglobulin G aggregation in rheumatoid inflammation. *J. Clin. Investig.* **1985**, *76*, 2084–2090. [CrossRef]

67. Lacy, E.R.; Baker, M.; Brigham-Burke, M. Free sulfhydryl measurement as an indicator of antibody stability. *Anal. Biochem.* **2008**, *382*, 66–68. [CrossRef]

68. Chumsae, C.; Gaza-Bulseco, G.; Liu, H. Identification and localization of unpaired cysteine residues in monoclonal antibodies by fluorescence labeling and mass spectrometry. *Anal. Chem.* **2009**, *81*, 6449–6457. [CrossRef]

69. Xiang, T.; Chumsae, C.; Liu, H. Localization and quantitation of free sulfhydryl in recombinant monoclonal antibodies by differential labeling with 12C and 13C iodoacetic acid and LC-MS analysis. *Anal. Chem.* **2009**, *81*, 8101–8108. [CrossRef]

70. Cheng, Y.; Chen, M.T.; Patterson, L.C.; Yu, X.C.; Zhang, Y.T.; Burgess, B.L.; Chen, Y. Domain-specific free thiol variant characterization of an IgG1 by reversed-phase high-performance liquid chromatography mass spectrometry. *Anal. Biochem.* **2017**, *519*, 8–14. [CrossRef]

71. Zhang, W.; Czupryn, M.J. Free sulfhydryl in recombinant monoclonal antibodies. *Biotechnol. Prog.* **2002**, *18*, 509–513. [CrossRef] [PubMed]

72. Brych, S.R.; Gokarn, Y.R.; Hultgen, H.; Stevenson, R.J.; Rajan, R.; Matsumura, M. Characterization of antibody aggregation: Role of buried, unpaired cysteines in particle formation. *J. Pharm. Sci.* **2010**, *99*, 764–781. [CrossRef] [PubMed]

73. Huh, J.H.; White, A.J.; Brych, S.R.; Franey, H.; Matsumura, M. The identification of free cysteine residues within antibodies and a potential role for free cysteine residues in covalent aggregation because of agitation stress. *J. Pharm. Sci.* **2013**, *102*, 1701–1711. [CrossRef] [PubMed]

74. Van Buren, N.; Rehder, D.; Gadgil, H.; Matsumura, M.; Jacob, J. Elucidation of two major aggregation pathways in an IgG2 antibody. *J. Pharm. Sci.* **2009**, *98*, 3013–3030. [CrossRef] [PubMed]

75. Zhang, T.; Zhang, J.; Hewitt, D.; Tran, B.; Gao, X.; Qiu, Z.J.; Tejada, M.; Gazzano-Santoro, H.; Kao, Y.H. Identification and characterization of buried unpaired cysteines in a recombinant monoclonal IgG1 antibody. *Anal. Chem.* **2012**, *84*, 7112–7123. [CrossRef]

76. Chaderjian, W.B.; Chin, E.T.; Harris, R.J.; Etcheverry, T.M. Effect of copper sulfate on performance of a serum-free CHO cell culture process and the level of free thiol in the recombinant antibody expressed. *Biotechnol. Prog.* **2005**, *21*, 550–553. [CrossRef]

77. Miao, S.; Xie, P.; Zou, M.; Fan, L.; Liu, X.; Zhou, Y.; Zhao, L.; Ding, D.; Wang, H.; Tan, W.S. Identification of multiple sources of the acidic charge variants in an IgG1 monoclonal antibody. *Appl. Microbiol. Biotechnol.* **2017**, *101*, 5627–5638. [CrossRef]

78. Banks, D.D.; Gadgil, H.S.; Pipes, G.D.; Bondarenko, P.V.; Hobbs, V.; Scavezze, J.L.; Kim, J.; Jiang, X.R.; Mukku, V.; Dillon, T.M. Removal of cysteinylation from an unpaired sulfhydryl in the variable region of a recombinant monoclonal IgG1 antibody improves homogeneity, stability, and biological activity. *J. Pharm. Sci.* **2008**, *97*, 775–790. [CrossRef]

79. Gadgil, H.S.; Bondarenko, P.V.; Pipes, G.D.; Dillon, T.M.; Banks, D.; Abel, J.; Kleemann, G.R.; Treuheit, M.J. Identification of cysteinylation of a free cysteine in the Fab region of a recombinant monoclonal IgG1 antibody using Lys-C limited proteolysis coupled with LC/MS analysis. *Anal. Biochem.* **2006**, *355*, 165–174. [CrossRef]

80. Buchanan, A.; Clementel, V.; Woods, R.; Harn, N.; Bowen, M.A.; Mo, W.; Popovic, B.; Bishop, S.M.; Dall'Acqua, W.; Minter, R.; et al. Engineering a therapeutic IgG molecule to address cysteinylation, aggregation and enhance thermal stability and expression. *MAbs* **2013**, *5*, 255–262. [CrossRef]

81. McSherry, T.; McSherry, J.; Ozaeta, P.; Longenecker, K.; Ramsay, C.; Fishpaugh, J.; Allen, S. Cysteinylation of a monoclonal antibody leads to its inactivation. *MAbs* **2016**, *8*, 718–725. [CrossRef] [PubMed]

82. Schuurman, J.; Van Ree, R.; Perdok, G.J.; Van Doorn, H.R.; Tan, K.Y.; Aalberse, R.C. Normal human immunoglobulin G4 is bispecific: It has two different antigen-combining sites. *Immunology* **1999**, *97*, 693–698. [CrossRef] [PubMed]

83. Aalberse, R.C.; Schuurman, J. IgG4 breaking the rules. *Immunology* **2002**, *105*, 9–19. [CrossRef] [PubMed]

84. van der Neut Kolfschoten, M.; Schuurman, J.; Losen, M.; Bleeker, W.K.; Martinez-Martinez, P.; Vermeulen, E.; den Bleker, T.H.; Wiegman, L.; Vink, T.; Aarden, L.A.; et al. Anti-inflammatory activity of human IgG4 antibodies by dynamic Fab arm exchange. *Science* **2007**, *317*, 1554–1557. [CrossRef] [PubMed]

85. Schuurman, J.; Perdok, G.J.; Gorter, A.D.; Aalberse, R.C. The inter-heavy chain disulfide bonds of IgG4 are in equilibrium with intra-chain disulfide bonds. *Mol. Immunol.* **2001**, *38*, 1–8. [CrossRef]

86. Bloom, J.W.; Madanat, M.S.; Marriott, D.; Wong, T.; Chan, S.Y. Intrachain disulfide bond in the core hinge region of human IgG4. *Protein Sci.* **1997**, *6*, 407–415. [CrossRef]

87. Angal, S.; King, D.J.; Bodmer, M.W.; Turner, A.; Lawson, A.D.; Roberts, G.; Pedley, B.; Adair, J.R. A single amino acid substitution abolishes the heterogeneity of chimeric mouse/human (IgG4) antibody. *Mol. Immunol.* **1993**, *30*, 105–108. [CrossRef]

88. Wypych, J.; Li, M.; Guo, A.; Zhang, Z.; Martinez, T.; Allen, M.J.; Fodor, S.; Kelner, D.N.; Flynn, G.C.; Liu, Y.D.; et al. Human IgG2 antibodies display disulfide-mediated structural isoforms. *J. Biol. Chem.* **2008**, *283*, 16194–16205. [CrossRef]

89. Dillon, T.M.; Ricci, M.S.; Vezina, C.; Flynn, G.C.; Liu, Y.D.; Rehder, D.S.; Plant, M.; Henkle, B.; Li, Y.; Deechongkit, S.; et al. Structural and functional characterization of disulfide isoforms of the human IgG2 subclass. *J. Biol. Chem.* **2008**, *283*, 16206–16215. [CrossRef]

90. Dillon, T.M.; Bondarenko, P.V.; Rehder, D.S.; Pipes, G.D.; Kleemann, G.R.; Ricci, M.S. Optimization of a reversed-phase high-performance liquid chromatography/mass spectrometry method for characterizing recombinant antibody heterogeneity and stability. *J. Chromatogr. A* **2006**, *1120*, 112–120. [CrossRef]

91. Liu, Y.D.; Chen, X.; Enk, J.Z.; Plant, M.; Dillon, T.M.; Flynn, G.C. Human IgG2 antibody disulfide rearrangement in vivo. *J. Biol. Chem.* **2008**, *283*, 29266–29272. [CrossRef] [PubMed]

92. Tous, G.I.; Wei, Z.; Feng, J.; Bilbulian, S.; Bowen, S.; Smith, J.; Strouse, R.; McGeehan, P.; Casas-Finet, J.; Schenerman, M.A. Characterization of a novel modification to monoclonal antibodies: Thioether cross-link of heavy and light chains. *Anal. Chem.* **2005**, *77*, 2675–2682. [CrossRef] [PubMed]

93. Zhang, Q.; Schenauer, M.R.; McCarter, J.D.; Flynn, G.C. IgG1 thioether bond formation in vivo. *J. Biol. Chem.* **2013**, *288*, 16371–16382. [CrossRef] [PubMed]

94. Pristatsky, P.; Cohen, S.L.; Krantz, D.; Acevedo, J.; Ionescu, R.; Vlasak, J. Evidence for trisulfide bonds in a recombinant variant of a human IgG2 monoclonal antibody. *Anal. Chem.* **2009**, *81*, 6148–6155. [CrossRef] [PubMed]

95. Gu, S.; Wen, D.; Weinreb, P.H.; Sun, Y.; Zhang, L.; Foley, S.F.; Kshirsagar, R.; Evans, D.; Mi, S.; Meier, W.; et al. Characterization of trisulfide modification in antibodies. *Anal. Biochem.* **2010**, *400*, 89–98. [CrossRef] [PubMed]

96. Kita, A.; Ponniah, G.; Nowak, C.; Liu, H. Characterization of Cysteinylation and Trisulfide Bonds in a Recombinant Monoclonal Antibody. *Anal. Chem.* **2016**, *88*, 5430–5437. [CrossRef] [PubMed]

97. Kshirsagar, R.; McElearney, K.; Gilbert, A.; Sinacore, M.; Ryll, T. Controlling trisulfide modification in recombinant monoclonal antibody produced in fed-batch cell culture. *Biotechnol. Bioeng.* **2012**, *109*, 2523–2532. [CrossRef]

98. Aono, H.; Wen, D.; Zang, L.; Houde, D.; Pepinsky, R.B.; Evans, D.R. Efficient on-column conversion of IgG1 trisulfide linkages to native disulfides in tandem with Protein A affinity chromatography. *J. Chromatogr. A* **2010**, *1217*, 5225–5232. [CrossRef]

99. Amano, M.; Hasegawa, J.; Kobayashi, N.; Kishi, N.; Nakazawa, T.; Uchiyama, S.; Fukui, K. Specific racemization of heavy-chain cysteine-220 in the hinge region of immunoglobulin gamma 1 as a possible cause of degradation during storage. *Anal. Chem.* **2011**, *83*, 3857–3864. [CrossRef]

100. Zhang, Q.; Flynn, G.C. Cysteine racemization on IgG heavy and light chains. *J. Biol. Chem.* **2013**, *288*, 34325–34335. [CrossRef]

101. Gevondyan, N.M.; Volynskaia, A.M.; Gevondyan, V.S. Four free cysteine residues found in human IgG1 of healthy donors. *Biochemistry (Mosc.)* **2006**, *71*, 279–284. [CrossRef] [PubMed]

102. Schauenstein, E.; Dachs, F.; Reiter, M.; Gombotz, H.; List, W. Labile disulfide bonds and free thiol groups in human IgG. I. Assignment to IgG1 and IgG2 subclasses. *Int. Arch. Allergy Appl. Immunol.* **1986**, *80*, 174–179. [CrossRef] [PubMed]

103. Huang, L.; Biolsi, S.; Bales, K.R.; Kuchibhotla, U. Impact of variable domain glycosylation on antibody clearance: An LC/MS characterization. *Anal. Biochem.* **2006**, *349*, 197–207. [CrossRef] [PubMed]

104. Qian, J.; Liu, T.; Yang, L.; Daus, A.; Crowley, R.; Zhou, Q. Structural characterization of N-linked oligosaccharides on monoclonal antibody cetuximab by the combination of orthogonal matrix-assisted laser desorption/ionization hybrid quadrupole-quadrupole time-of-flight tandem mass spectrometry and sequential enzymatic digestion. *Anal. Biochem.* **2007**, *364*, 8–18. [PubMed]

105. Lim, A.; Reed-Bogan, A.; Harmon, B.J. Glycosylation profiling of a therapeutic recombinant monoclonal antibody with two N-linked glycosylation sites using liquid chromatography coupled to a hybrid quadrupole time-of-flight mass spectrometer. *Anal. Biochem.* **2008**, *375*, 163–172. [CrossRef] [PubMed]

106. Raju, T.S.; Jordan, R.E. Galactosylation variations in marketed therapeutic antibodies. *MAbs* **2012**, *4*, 385–391. [CrossRef]

107. Schiestl, M.; Stangler, T.; Torella, C.; Cepeljnik, T.; Toll, H.; Grau, R. Acceptable changes in quality attributes of glycosylated biopharmaceuticals. *Nat. Biotechnol.* **2011**, *29*, 310–312. [CrossRef]

108. Maeda, E.; Kita, S.; Kinoshita, M.; Urakami, K.; Hayakawa, T.; Kakehi, K. Analysis of nonhuman N-glycans as the minor constituents in recombinant monoclonal antibody pharmaceuticals. *Anal. Chem.* **2012**, *84*, 2373–2379. [CrossRef]

109. Stadlmann, J.; Pabst, M.; Kolarich, D.; Kunert, R.; Altmann, F. Analysis of immunoglobulin glycosylation by LC-ESI-MS of glycopeptides and oligosaccharides. *Proteomics* **2008**, *8*, 2858–2871. [CrossRef]

110. Kamoda, S.; Ishikawa, R.; Kakehi, K. Capillary electrophoresis with laser-induced fluorescence detection for detailed studies on N-linked oligosaccharide profile of therapeutic recombinant monoclonal antibodies. *J. Chromatogr. A* **2006**, *1133*, 332–339. [CrossRef]

111. Giorgetti, J.; D'Atri, V.; Canonge, J.; Lechner, A.; Guillarme, D.; Colas, O.; Wagner-Rousset, E.; Beck, A.; Leize-Wagner, E.; François, Y.-N. Monoclonal antibody N-glycosylation profiling using capillary electrophoresis—Mass spectrometry: Assessment and method validation. *Talanta* **2018**, *178*, 530–537. [CrossRef] [PubMed]

112. Ghirlando, R.; Lund, J.; Goodall, M.; Jefferis, R. Glycosylation of human IgG-Fc: Influences on structure revealed by differential scanning micro-calorimetry. *Immunol. Lett.* **1999**, *68*, 47–52. [CrossRef]

113. Mimura, Y.; Church, S.; Ghirlando, R.; Ashton, P.R.; Dong, S.; Goodall, M.; Lund, J.; Jefferis, R. The influence of glycosylation on the thermal stability and effector function expression of human IgG1-Fc: Properties of a series of truncated glycoforms. *Mol. Immunol.* **2000**, *37*, 697–706. [CrossRef]

114. Mimura, Y.; Sondermann, P.; Ghirlando, R.; Lund, J.; Young, S.P.; Goodall, M.; Jefferis, R. Role of oligosaccharide residues of IgG1-Fc in Fc gamma RIIb binding. *J. Biol. Chem.* **2001**, *276*, 45539–45547. [CrossRef] [PubMed]

115. Yamaguchi, Y.; Nishimura, M.; Nagano, M.; Yagi, H.; Sasakawa, H.; Uchida, K.; Shitara, K.; Kato, K. Glycoform-dependent conformational alteration of the Fc region of human immunoglobulin G1 as revealed by NMR spectroscopy. *Biochim. Biophys. Acta* **2006**, *1760*, 693–700. [CrossRef] [PubMed]

116. Raju, T.S. Terminal sugars of Fc glycans influence antibody effector functions of IgGs. *Curr. Opin. Immunol.* **2008**, *20*, 471–478. [CrossRef] [PubMed]

117. Wright, A.; Morrison, S.L. Effect of C2-associated carbohydrate structure on Ig effector function: Studies with chimeric mouse-human IgG1 antibodies in glycosylation mutants of Chinese hamster ovary cells. *J. Immunol.* **1998**, *160*, 3393–3402.

118. Hodoniczky, J.; Zheng, Y.Z.; James, D.C. Control of recombinant monoclonal antibody effector functions by Fc N-glycan remodeling in vitro. *Biotechnol. Prog.* **2005**, *21*, 1644–1652. [CrossRef]

119. Boyd, P.N.; Lines, A.C.; Patel, A.K. The effect of the removal of sialic acid, galactose and total carbohydrate on the functional activity of Campath-1H. *Mol. Immunol.* **1995**, *32*, 1311–1318. [CrossRef]

120. Nimmerjahn, F.; Anthony, R.M.; Ravetch, J.V. Agalactosylated IgG antibodies depend on cellular Fc receptors for in vivo activity. *Proc. Natl. Acad. Sci. USA* **2007**, *104*, 8433–8437. [CrossRef]

121. Groenink, J.; Spijker, J.; van den Herik-Oudijk, I.E.; Boeije, L.; Rook, G.; Aarden, L.; Smeenk, R.; van de Winkel, J.G.; van den Broek, M.F. On the interaction between agalactosyl IgG and Fc gamma receptors. *Eur. J. Immunol.* **1996**, *26*, 1404–1407. [CrossRef] [PubMed]

122. Shinkawa, T.; Nakamura, K.; Yamane, N.; Shoji-Hosaka, E.; Kanda, Y.; Sakurada, M.; Uchida, K.; Anazawa, H.; Satoh, M.; Yamasaki, M.; et al. The absence of fucose but not the presence of galactose or bisecting N-acetylglucosamine of human IgG1 complex-type oligosaccharides shows the critical role of enhancing antibody-dependent cellular cytotoxicity. *J. Biol. Chem.* **2003**, *278*, 3466–3473. [CrossRef] [PubMed]

123. Onitsuka, M.; Kawaguchi, A.; Asano, R.; Kumagai, I.; Honda, K.; Ohtake, H.; Omasa, T. Glycosylation analysis of an aggregated antibody produced by Chinese hamster ovary cells in bioreactor culture. *J. Biosci. Bioeng.* **2014**, *117*, 639–644. [CrossRef] [PubMed]

124. Lu, Y.; Westland, K.; Ma, Y.H.; Gadgil, H. Evaluation of effects of Fc domain high-mannose glycan on antibody stability. *J. Pharm. Sci.* **2012**, *101*, 4107–4117. [CrossRef] [PubMed]

125. Chen, X.; Liu, Y.D.; Flynn, G.C. The effect of Fc glycan forms on human IgG2 antibody clearance in humans. *Glycobiology* **2009**, *19*, 240–249. [CrossRef] [PubMed]

126. Alessandri, L.; Ouellette, D.; Acquah, A.; Rieser, M.; Leblond, D.; Saltarelli, M.; Radziejewski, C.; Fujimori, T.; Correia, I. Increased serum clearance of oligomannose species present on a human IgG1 molecule. *MAbs* **2012**, *4*, 509–520. [CrossRef] [PubMed]

127. Goetze, A.M.; Liu, Y.D.; Zhang, Z.; Shah, B.; Lee, E.; Bondarenko, P.V.; Flynn, G.C. High-mannose glycans on the Fc region of therapeutic IgG antibodies increase serum clearance in humans. *Glycobiology* **2011**, *21*, 949–959. [CrossRef] [PubMed]

128. Millward, T.A.; Heitzmann, M.; Bill, K.; Langle, U.; Schumacher, P.; Forrer, K. Effect of constant and variable domain glycosylation on pharmacokinetics of therapeutic antibodies in mice. *Biologicals* **2008**, *36*, 41–47. [CrossRef]

129. Shields, R.L.; Lai, J.; Keck, R.; O'Connell, L.Y.; Hong, K.; Meng, Y.G.; Weikert, S.H.; Presta, L.G. Lack of fucose on human IgG1 N-linked oligosaccharide improves binding to human Fcgamma RIII and antibody-dependent cellular toxicity. *J. Biol. Chem.* **2002**, *277*, 26733–26740. [CrossRef]

130. Matsumiya, S.; Yamaguchi, Y.; Saito, J.; Nagano, M.; Sasakawa, H.; Otaki, S.; Satoh, M.; Shitara, K.; Kato, K. Structural comparison of fucosylated and nonfucosylated Fc fragments of human immunoglobulin G1. *J. Mol. Biol.* **2007**, *368*, 767–779. [CrossRef]

131. Zheng, K.; Yarmarkovich, M.; Bantog, C.; Bayer, R.; Patapoff, T.W. Influence of glycosylation pattern on the molecular properties of monoclonal antibodies. *MAbs* **2014**, *6*, 649–658. [CrossRef] [PubMed]

132. Junttila, T.T.; Parsons, K.; Olsson, C.; Lu, Y.; Xin, Y.; Theriault, J.; Crocker, L.; Pabonan, O.; Baginski, T.; Meng, G.; et al. Superior in vivo efficacy of afucosylated trastuzumab in the treatment of HER2-amplified breast cancer. *Cancer Res.* **2010**, *70*, 4481–4489. [CrossRef]

133. Kanda, Y.; Yamada, T.; Mori, K.; Okazaki, A.; Inoue, M.; Kitajima-Miyama, K.; Kuni-Kamochi, R.; Nakano, R.; Yano, K.; Kakita, S.; et al. Comparison of biological activity among nonfucosylated therapeutic IgG1 antibodies with three different N-linked Fc oligosaccharides: The high-mannose, hybrid, and complex types. *Glycobiology* **2007**, *17*, 104–118. [CrossRef] [PubMed]

134. Jiang, X.R.; Song, A.; Bergelson, S.; Arroll, T.; Parekh, B.; May, K.; Chung, S.; Strouse, R.; Mire-Sluis, A.; Schenerman, M. Advances in the assessment and control of the effector functions of therapeutic antibodies. *Nat. Rev. Drug Discov.* **2011**, *10*, 101–111. [CrossRef] [PubMed]

135. Largy, E.; Cantais, F.; Van Vyncht, G.; Beck, A.; Delobel, A. Orthogonal liquid chromatography-mass spectrometry methods for the comprehensive characterization of therapeutic glycoproteins, from released glycans to intact protein level. *J. Chromatogr. A* **2017**, *1498*, 128–146. [CrossRef] [PubMed]

136. Gaza-Bulseco, G.; Hickman, K.; Sinicropi-Yao, S.; Hurkmans, K.; Chumsae, C.; Liu, H. Effect of the conserved oligosaccharides of recombinant monoclonal antibodies on the separation by protein A and protein G chromatography. *J. Chromatogr. A* **2009**, *1216*, 2382–2387. [CrossRef] [PubMed]

137. Falck, D.; Jansen, B.C.; Plomp, R.; Reusch, D.; Haberger, M.; Wuhrer, M. Glycoforms of Immunoglobulin G Based Biopharmaceuticals Are Differentially Cleaved by Trypsin Due to the Glycoform Influence on Higher-Order Structure. *J. Proteome Res.* **2015**, *14*, 4019–4028. [CrossRef] [PubMed]

138. Fang, J.; Richardson, J.; Du, Z.; Zhang, Z. Effect of Fc-Glycan Structure on the Conformational Stability of IgG Revealed by Hydrogen/Deuterium Exchange and Limited Proteolysis. *Biochemistry* **2016**, *55*, 860–868. [CrossRef]

139. Yu, M.; Brown, D.; Reed, C.; Chung, S.; Lutman, J.; Stefanich, E.; Wong, A.; Stephan, J.P.; Bayer, R. Production, characterization, and pharmacokinetic properties of antibodies with N-linked mannose-5 glycans. *MAbs* **2012**, *4*, 475–487. [CrossRef]

140. Wright, A.; Morrison, S.L. Effect of altered CH2-associated carbohydrate structure on the functional properties and in vivo fate of chimeric mouse-human immunoglobulin G1. *J. Exp. Med.* **1994**, *180*, 1087–1096. [CrossRef]

141. Zhou, Q.; Shankara, S.; Roy, A.; Qiu, H.; Estes, S.; McVie-Wylie, A.; Culm-Merdek, K.; Park, A.; Pan, C.; Edmunds, T. Development of a simple and rapid method for producing non-fucosylated oligomannose containing antibodies with increased effector function. *Biotechnol. Bioeng.* **2008**, *99*, 652–665. [CrossRef] [PubMed]

142. Liu, Y.D.; Flynn, G.C. Effect of high mannose glycan pairing on IgG antibody clearance. *Biologicals* **2016**, *44*, 163–169. [CrossRef] [PubMed]

143. Naso, M.F.; Tam, S.H.; Scallon, B.J.; Raju, T.S. Engineering host cell lines to reduce terminal sialylation of secreted antibodies. *MAbs* **2010**, *2*, 519–527. [CrossRef] [PubMed]

144. Scallon, B.J.; Tam, S.H.; McCarthy, S.G.; Cai, A.N.; Raju, T.S. Higher levels of sialylated Fc glycans in immunoglobulin G molecules can adversely impact functionality. *Mol. Immunol.* **2007**, *44*, 1524–1534. [CrossRef] [PubMed]

145. Ahmed, A.A.; Giddens, J.; Pincetic, A.; Lomino, J.V.; Ravetch, J.V.; Wang, L.X.; Bjorkman, P.J. Structural characterization of anti-inflammatory immunoglobulin G Fc proteins. *J. Mol. Biol.* **2014**, *426*, 3166–3179. [CrossRef] [PubMed]

146. Barb, A.W.; Meng, L.; Gao, Z.; Johnson, R.W.; Moremen, K.W.; Prestegard, J.H. NMR characterization of immunoglobulin G Fc glycan motion on enzymatic sialylation. *Biochemistry* **2012**, *51*, 4618–4626. [CrossRef] [PubMed]

147. Crispin, M.; Yu, X.; Bowden, T.A. Crystal structure of sialylated IgG Fc: Implications for the mechanism of intravenous immunoglobulin therapy. *Proc. Natl. Acad. Sci. USA* **2013**, *110*, E3544–E3546. [CrossRef] [PubMed]

148. Ghaderi, D.; Taylor, R.E.; Padler-Karavani, V.; Diaz, S.; Varki, A. Implications of the presence of N-glycolylneuraminic acid in recombinant therapeutic glycoproteins. *Nat. Biotechnol.* **2010**, *28*, 863–867. [CrossRef] [PubMed]

149. Sheeley, D.M.; Merrill, B.M.; Taylor, L.C. Characterization of monoclonal antibody glycosylation: Comparison of expression systems and identification of terminal alpha-linked galactose. *Anal. Biochem.* **1997**, *247*, 102–110. [CrossRef]

150. Chung, C.H.; Mirakhur, B.; Chan, E.; Le, Q.T.; Berlin, J.; Morse, M.; Murphy, B.A.; Satinover, S.M.; Hosen, J.; Mauro, D.; et al. Cetuximab-induced anaphylaxis and IgE specific for galactose-alpha-1,3-galactose. *N. Engl. J. Med.* **2008**, *358*, 1109–1117. [CrossRef]

151. Lammerts van Bueren, J.J.; Rispens, T.; Verploegen, S.; van der Palen-Merkus, T.; Stapel, S.; Workman, L.J.; James, H.; van Berkel, P.H.; van de Winkel, J.G.; Platts-Mills, T.A.; et al. Anti-galactose-alpha-1,3-galactose IgE from allergic patients does not bind alpha-galactosylated glycans on intact therapeutic antibody Fc domains. *Nat. Biotechnol.* **2011**, *29*, 574–576. [CrossRef] [PubMed]

152. Umana, P.; Jean-Mairet, J.; Moudry, R.; Amstutz, H.; Bailey, J.E. Engineered glycoforms of an antineuroblastoma IgG1 with optimized antibody-dependent cellular cytotoxic activity. *Nat. Biotechnol.* **1999**, *17*, 176–180. [CrossRef] [PubMed]

153. Chen, X.; Tang, K.; Lee, M.; Flynn, G.C. Microchip assays for screening monoclonal antibody product quality. *Electrophoresis* **2008**, *29*, 4993–5002. [CrossRef] [PubMed]

154. Rustandi, R.R.; Washabaugh, M.W.; Wang, Y. Applications of CE SDS gel in development of biopharmaceutical antibody-based products. *Electrophoresis* **2008**, *29*, 3612–3620. [CrossRef] [PubMed]

155. Salas-Solano, O.; Tomlinson, B.; Du, S.; Parker, M.; Strahan, A.; Ma, S. Optimization and validation of a quantitative capillary electrophoresis sodium dodecyl sulfate method for quality control and stability monitoring of monoclonal antibodies. *Anal. Chem.* **2006**, *78*, 6583–6594. [CrossRef] [PubMed]

156. Smith, M.T.; Zhang, S.; Adams, T.; DiPaolo, B.; Dally, J. Establishment and validation of a microfluidic capillary gel electrophoresis platform method for purity analysis of therapeutic monoclonal antibodies. *Electrophoresis* **2017**, *38*, 1353–1365. [CrossRef] [PubMed]

157. Tao, M.H.; Morrison, S.L. Studies of aglycosylated chimeric mouse-human IgG. Role of carbohydrate in the structure and effector functions mediated by the human IgG constant region. *J. Immunol.* **1989**, *143*, 2595–2601.

158. Raju, T.S.; Scallon, B.J. Glycosylation in the Fc domain of IgG increases resistance to proteolytic cleavage by papain. *Biochem. Biophys. Res. Commun.* **2006**, *341*, 797–803. [CrossRef]

159. More, A.S.; Toprani, V.M.; Okbazghi, S.Z.; Kim, J.H.; Joshi, S.B.; Middaugh, C.R.; Tolbert, T.J.; Volkin, D.B. Correlating the Impact of Well-Defined Oligosaccharide Structures on Physical Stability Profiles of IgG1-Fc Glycoforms. *J. Pharm. Sci.* **2016**, *105*, 588–601. [CrossRef]

160. Hristodorov, D.; Fischer, R.; Joerissen, H.; Muller-Tiemann, B.; Apeler, H.; Linden, L. Generation and comparative characterization of glycosylated and aglycosylated human IgG1 antibodies. *Mol. Biotechnol.* **2013**, *53*, 326–335. [CrossRef]

161. Kayser, V.; Chennamsetty, N.; Voynov, V.; Forrer, K.; Helk, B.; Trout, B.L. Glycosylation influences on the aggregation propensity of therapeutic monoclonal antibodies. *Biotechnol. J.* **2011**, *6*, 38–44. [CrossRef] [PubMed]

162. Wawrzynczak, E.J.; Cumber, A.J.; Parnell, G.D.; Jones, P.T.; Winter, G. Blood clearance in the rat of a recombinant mouse monoclonal antibody lacking the N-linked oligosaccharide side chains of the CH2 domains. *Mol. Immunol.* **1992**, *29*, 213–220. [CrossRef]

163. Wawrzynczak, E.J.; Parnell, G.D.; Cumber, A.J.; Jones, P.T.; Winter, G. Blood clearance in the mouse of an aglycosyl recombinant monoclonal antibody. *Biochem. Soc. Trans.* **1989**, *17*, 1061–1062. [CrossRef] [PubMed]

164. Hristodorov, D.; Fischer, R.; Linden, L. With or without sugar? (A)glycosylation of therapeutic antibodies. *Mol. Biotechnol.* **2013**, *54*, 1056–1068. [CrossRef] [PubMed]

165. Ng, C.M.; Stefanich, E.; Anand, B.S.; Fielder, P.J.; Vaickus, L. Pharmacokinetics/pharmacodynamics of nondepleting anti-CD4 monoclonal antibody (TRX1) in healthy human volunteers. *Pharm. Res.* **2006**, *23*, 95–103. [CrossRef] [PubMed]

166. Routier, F.H.; Hounsell, E.F.; Rudd, P.M.; Takahashi, N.; Bond, A.; Hay, F.C.; Alavi, A.; Axford, J.S.; Jefferis, R. Quantitation of the oligosaccharides of human serum IgG from patients with rheumatoid arthritis: A critical evaluation of different methods. *J. Immunol. Methods* **1998**, *213*, 113–130. [CrossRef]

167. Flynn, G.C.; Chen, X.; Liu, Y.D.; Shah, B.; Zhang, Z. Naturally occurring glycan forms of human immunoglobulins G1 and G2. *Mol. Immunol.* **2010**, *47*, 2074–2082. [CrossRef]

168. Raju, T.S.; Briggs, J.B.; Borge, S.M.; Jones, A.J. Species-specific variation in glycosylation of IgG: Evidence for the species-specific sialylation and branch-specific galactosylation and importance for engineering recombinant glycoprotein therapeutics. *Glycobiology* **2000**, *10*, 477–486. [CrossRef]

169. Quan, C.; Alcala, E.; Petkovska, I.; Matthews, D.; Canova-Davis, E.; Taticek, R.; Ma, S. A study in glycation of a therapeutic recombinant humanized monoclonal antibody: Where it is, how it got there, and how it affects charge-based behavior. *Anal. Biochem.* **2008**, *373*, 179–191. [CrossRef]

170. Zhang, B.; Yang, Y.; Yuk, I.; Pai, R.; McKay, P.; Eigenbrot, C.; Dennis, M.; Katta, V.; Francissen, K.C. Unveiling a glycation hot spot in a recombinant humanized monoclonal antibody. *Anal. Chem.* **2008**, *80*, 2379–2390. [CrossRef]

171. Miller, A.K.; Hambly, D.M.; Kerwin, B.A.; Treuheit, M.J.; Gadgil, H.S. Characterization of site-specific glycation during process development of a human therapeutic monoclonal antibody. *J. Pharm. Sci.* **2011**, *100*, 2543–2550. [CrossRef]

172. Gadgil, H.S.; Bondarenko, P.V.; Pipes, G.; Rehder, D.; McAuley, A.; Perico, N.; Dillon, T.; Ricci, M.; Treuheit, M. The LC/MS analysis of glycation of IgG molecules in sucrose containing formulations. *J. Pharm. Sci.* **2007**, *96*, 2607–2621. [CrossRef] [PubMed]

173. Banks, D.D.; Hambly, D.M.; Scavezze, J.L.; Siska, C.C.; Stackhouse, N.L.; Gadgil, H.S. The effect of sucrose hydrolysis on the stability of protein therapeutics during accelerated formulation studies. *J. Pharm. Sci.* **2009**, *98*, 4501–4510. [CrossRef] [PubMed]

174. Fischer, S.; Hoernschemeyer, J.; Mahler, H.C. Glycation during storage and administration of monoclonal antibody formulations. *Eur. J. Pharm. Biopharm.* **2008**, *70*, 42–50. [CrossRef]

175. Goetze, A.M.; Liu, Y.D.; Arroll, T.; Chu, L.; Flynn, G.C. Rates and impact of human antibody glycation in vivo. *Glycobiology* **2012**, *22*, 221–234. [CrossRef] [PubMed]

176. Butko, M.; Pallat, H.; Cordoba, A.; Yu, X.C. Recombinant antibody color resulting from advanced glycation end product modifications. *Anal. Chem.* **2014**, *86*, 9816–9823. [CrossRef] [PubMed]

177. Harris, R.J. Processing of C-terminal lysine and arginine residues of proteins isolated from mammalian cell culture. *J. Chromatogr. A* **1995**, *705*, 129–134. [CrossRef]

178. Dick, L.W., Jr.; Qiu, D.; Mahon, D.; Adamo, M.; Cheng, K.C. C-terminal lysine variants in fully human monoclonal antibodies: Investigation of test methods and possible causes. *Biotechnol. Bioeng.* **2008**, *100*, 1132–1143. [CrossRef]

179. Santora, L.C.; Krull, I.S.; Grant, K. Characterization of recombinant human monoclonal tissue necrosis factor-alpha antibody using cation-exchange HPLC and capillary isoelectric focusing. *Anal. Biochem.* **1999**, *275*, 98–108. [CrossRef]

180. Antes, B.; Amon, S.; Rizzi, A.; Wiederkum, S.; Kainer, M.; Szolar, O.; Fido, M.; Kircheis, R.; Nechansky, A. Analysis of lysine clipping of a humanized Lewis-Y specific IgG antibody and its relation to Fc-mediated effector function. *J. Chromatogr. B Anal. Technol. Biomed. Life Sci.* **2007**, *852*, 250–256. [CrossRef]

181. Jiang, G.; Yu, C.; Yadav, D.B.; Hu, Z.; Amurao, A.; Duenas, E.; Wong, M.; Iverson, M.; Zheng, K.; Lam, X.; et al. Evaluation of Heavy-Chain C-Terminal Deletion on Product Quality and Pharmacokinetics of Monoclonal Antibodies. *J. Pharm. Sci.* **2016**, *105*, 2066–2072. [CrossRef] [PubMed]

182. Liu, H.; Bulseco, G.G.; Sun, J. Effect of posttranslational modifications on the thermal stability of a recombinant monoclonal antibody. *Immunol. Lett.* **2006**, *106*, 144–153. [CrossRef] [PubMed]

183. Van den Bremer, E.T.; Beurskens, F.J.; Voorhorst, M.; Engelberts, P.J.; de Jong, R.N.; van der Boom, B.G.; Cook, E.M.; Lindorfer, M.A.; Taylor, R.P.; van Berkel, P.H.; et al. Human IgG is produced in a pro-form that requires clipping of C-terminal lysines for maximal complement activation. *MAbs* **2015**, *7*, 672–680. [CrossRef]

184. Hu, Z.; Tang, D.; Misaghi, S.; Jiang, G.; Yu, C.; Yim, M.; Shaw, D.; Snedecor, B.; Laird, M.W.; Shen, A. Evaluation of heavy chain C-terminal deletions on productivity and product quality of monoclonal antibodies in Chinese hamster ovary (CHO) cells. *Biotechnol. Prog.* **2017**, *33*, 786–794. [CrossRef] [PubMed]

185. Cai, B.; Pan, H.; Flynn, G.C. C-terminal lysine processing of human immunoglobulin G2 heavy chain in vivo. *Biotechnol. Bioeng.* **2011**, *108*, 404–412. [CrossRef]

186. Johnson, K.A.; Paisley-Flango, K.; Tangarone, B.S.; Porter, T.J.; Rouse, J.C. Cation exchange-HPLC and mass spectrometry reveal C-terminal amidation of an IgG1 heavy chain. *Anal. Biochem.* **2007**, *360*, 75–83. [CrossRef]

187. Tsubaki, M.; Terashima, I.; Kamata, K.; Koga, A. C-terminal modification of monoclonal antibody drugs: Amidated species as a general product-related substance. *Int. J. Biol. Macromol.* **2013**, *52*, 139–147. [CrossRef]

188. Kaschak, T.; Boyd, D.; Lu, F.; Derfus, G.; Kluck, B.; Nogal, B.; Emery, C.; Summers, C.; Zheng, K.; Bayer, R.; et al. Characterization of the basic charge variants of a human IgG1: Effect of copper concentration in cell culture media. *MAbs* **2011**, *3*, 577–583. [CrossRef]

189. Skulj, M.; Pezdirec, D.; Gaser, D.; Kreft, M.; Zorec, R. Reduction in C-terminal amidated species of recombinant monoclonal antibodies by genetic modification of CHO cells. *BMC Biotechnol.* **2014**, *14*, 76. [CrossRef]

190. Harris, R.J.; Murnane, A.A.; Utter, S.L.; Wagner, K.L.; Cox, E.T.; Polastri, G.D.; Helder, J.C.; Sliwkowski, M.B. Assessing genetic heterogeneity in production cell lines: Detection by peptide mapping of a low level Tyr to Gln sequence variant in a recombinant antibody. *Biotechnology (N. Y.)* **1993**, *11*, 1293–1297. [CrossRef]

191. Yu, X.C.; Borisov, O.V.; Alvarez, M.; Michels, D.A.; Wang, Y.J.; Ling, V. Identification of codon-specific serine to asparagine mistranslation in recombinant monoclonal antibodies by high-resolution mass spectrometry. *Anal. Chem.* **2009**, *81*, 9282–9290. [CrossRef] [PubMed]

192. Wen, D.; Vecchi, M.M.; Gu, S.; Su, L.; Dolnikova, J.; Huang, Y.M.; Foley, S.F.; Garber, E.; Pederson, N.; Meier, W. Discovery and investigation of misincorporation of serine at asparagine positions in recombinant proteins expressed in Chinese hamster ovary cells. *J. Biol. Chem.* **2009**, *284*, 32686–32694. [CrossRef] [PubMed]

193. Khetan, A.; Huang, Y.M.; Dolnikova, J.; Pederson, N.E.; Wen, D.; Yusuf-Makagiansar, H.; Chen, P.; Ryll, T. Control of misincorporation of serine for asparagine during antibody production using CHO cells. *Biotechnol. Bioeng.* **2010**, *107*, 116–123. [CrossRef] [PubMed]

194. Yang, Y.; Strahan, A.; Li, C.; Shen, A.; Liu, H.; Ouyang, J.; Katta, V.; Francissen, K.; Zhang, B. Detecting low level sequence variants in recombinant monoclonal antibodies. *MAbs* **2010**, *2*, 285–298. [CrossRef] [PubMed]

195. Ren, D.; Zhang, J.; Pritchett, R.; Liu, H.; Kyauk, J.; Luo, J.; Amanullah, A. Detection and identification of a serine to arginine sequence variant in a therapeutic monoclonal antibody. *J. Chromatogr. B Anal. Technol. Biomed. Life Sci.* **2011**, *879*, 2877–2884. [CrossRef] [PubMed]

196. Fu, J.; Bongers, J.; Tao, L.; Huang, D.; Ludwig, R.; Huang, Y.; Qian, Y.; Basch, J.; Goldstein, J.; Krishnan, R.; et al. Characterization and identification of alanine to serine sequence variants in an IgG4 monoclonal antibody produced in mammalian cell lines. *J. Chromatogr. B Anal. Technol. Biomed. Life Sci.* **2012**, *908*, 1–8. [CrossRef] [PubMed]

197. Feeney, L.; Carvalhal, V.; Yu, X.C.; Chan, B.; Michels, D.A.; Wang, Y.J.; Shen, A.; Ressl, J.; Dusel, B.; Laird, M.W. Eliminating tyrosine sequence variants in CHO cell lines producing recombinant monoclonal antibodies. *Biotechnol. Bioeng.* **2013**, *110*, 1087–1097. [CrossRef]

198. Guo, D.; Gao, A.; Michels, D.A.; Feeney, L.; Eng, M.; Chan, B.; Laird, M.W.; Zhang, B.; Yu, X.C.; Joly., J. Mechanisms of unintended amino acid sequence changes in recombinant monoclonal antibodies expressed in Chinese Hamster Ovary (CHO) cells. *Biotechnol. Bioeng.* **2010**, *107*, 163–171. [CrossRef]

199. Wan, M.; Shiau, F.Y.; Gordon, W.; Wang, G.Y. Variant antibody identification by peptide mapping. *Biotechnol. Bioeng.* **1999**, *62*, 485–488. [CrossRef]

200. Chumsae, C.; Gifford, K.; Lian, W.; Liu, H.; Radziejewski, C.H.; Zhou, Z.S. Arginine modifications by methylglyoxal: Discovery in a recombinant monoclonal antibody and contribution to acidic species. *Anal. Chem.* **2013**, *85*, 11401–11409. [CrossRef]

201. Yang, Y.; Stella, C.; Wang, W.; Schoneich, C.; Gennaro, L. Characterization of oxidative carbonylation on recombinant monoclonal antibodies. *Anal. Chem.* **2014**, *86*, 4799–4806. [CrossRef] [PubMed]

202. Amano, M.; Kobayashi, N.; Yabuta, M.; Uchiyama, S.; Fukui, K. Detection of histidine oxidation in a monoclonal immunoglobulin gamma (IgG) 1 antibody. *Anal. Chem.* **2014**, *86*, 7536–7543. [CrossRef] [PubMed]

203. Liu, M.; Zhang, Z.; Cheetham, J.; Ren, D.; Zhou, Z.S. Discovery and characterization of a photo-oxidative histidine-histidine cross-link in IgG1 antibody utilizing (1)(8)O-labeling and mass spectrometry. *Anal. Chem.* **2014**, *86*, 4940–4948. [CrossRef] [PubMed]

204. Zhao, J.; Saunders, J.; Schussler, S.D.; Rios, S.; Insaidoo, F.K.; Fridman, A.L.; Li, H.; Liu, Y.H. Characterization of a novel modification of a CHO-produced mAb: Evidence for the presence of tyrosine sulfation. *MAbs* **2017**, *9*, 985–995. [CrossRef] [PubMed]

205. Santora, L.C.; Stanley, K.; Krull, I.S.; Grant, K. Characterization of maleuric acid derivatives on transgenic human monoclonal antibody due to post-secretional modifications in goat milk. *Biomed. Chromatogr.* **2006**, *20*, 843–856. [CrossRef]

206. Chumsae, C.; Zhou, L.L.; Shen, Y.; Wohlgemuth, J.; Fung, E.; Burton, R.; Radziejewski, C.; Zhou, Z.S. Discovery of a chemical modification by citric acid in a recombinant monoclonal antibody. *Anal. Chem.* **2014**, *86*, 8932–8936. [CrossRef] [PubMed]

207. Valliere-Douglass, J.F.; Connell-Crowley, L.; Jensen, R.; Schnier, P.D.; Trilisky, E.; Leith, M.; Follstad, B.D.; Kerr, J.; Lewis, N.; Vunnum, S. Photochemical degradation of citrate buffers leads to covalent acetonation of recombinant protein therapeutics. *Protein Sci.* **2010**, *19*, 2152–2163. [CrossRef]

208. Valliere-Douglass, J.F.; Brady, L.J.; Farnsworth, C.; Pace, D.; Balland, A.; Wallace, A.; Wang, W.; Treuheit, M.J.; Yan, B. O-fucosylation of an antibody light chain: Characterization of a modification occurring on an IgG1 molecule. *Glycobiology* **2009**, *19*, 144–152. [CrossRef]

209. Valliere-Douglass, J.F.; Eakin, C.M.; Wallace, A.; Ketchem, R.R.; Wang, W.; Treuheit, M.J.; Balland, A. Glutamine-linked and non-consensus asparagine-linked oligosaccharides present in human recombinant antibodies define novel protein glycosylation motifs. *J. Biol. Chem.* **2010**, *285*, 16012–16022. [CrossRef]

210. Valliere-Douglass, J.F.; Kodama, P.; Mujacic, M.; Brady, L.J.; Wang, W.; Wallace, A.; Yan, B.; Reddy, P.; Treuheit, M.J.; Balland, A. Asparagine-linked oligosaccharides present on a non-consensus amino acid sequence in the CH1 domain of human antibodies. *J. Biol. Chem.* **2009**, *284*, 32493–32506. [CrossRef]

211. Dong, Q.; Liang, Y.; Yan, X.; Markey, S.P.; Mirokhin, Y.A.; Tchekhovskoi, D.V.; Bukhari, T.H.; Stein, S.E. The NISTmAb tryptic peptide spectral library for monoclonal antibody characterization. *MAbs* **2018**, *10*, 354–369. [CrossRef] [PubMed]

212. Beck, A.; Goetsch, L.; Dumontet, C.; Corvaia, N. Strategies and challenges for the next generation of antibody-drug conjugates. *Nat. Rev. Drug Discov.* **2017**, *16*, 315–337. [CrossRef] [PubMed]

213. Shen, B.Q.; Xu, K.; Liu, L.; Raab, H.; Bhakta, S.; Kenrick, M.; Parsons-Reponte, K.L.; Tien, J.; Yu, S.F.; Mai, E.; et al. Conjugation site modulates the in vivo stability and therapeutic activity of antibody-drug conjugates. *Nat. Biotechnol.* **2012**, *30*, 184–189. [CrossRef] [PubMed]

214. Cumnock, K.; Tully, T.; Cornell, C.; Hutchinson, M.; Gorrell, J.; Skidmore, K.; Chen, Y.; Jacobson, F. Trisulfide modification impacts the reduction step in antibody-drug conjugation process. *Bioconjug. Chem.* **2013**, *24*, 1154–1160. [CrossRef] [PubMed]

215. Liu-Shin, L.; Fung, A.; Malhotra, A.; Ratnaswamy, G. Influence of disulfide bond isoforms on drug conjugation sites in cysteine-linked IgG2 antibody-drug conjugates. *MAbs* **2018**, *10*, 583–595. [CrossRef] [PubMed]

antibodies

MDPI

Article

IgG Charge: Practical and Biological Implications

Danlin Yang [1,2], Rachel Kroe-Barrett [1], Sanjaya Singh [1,2] and Thomas Laue [3,*]

[1] Biotherapeutics Discovery Research, Boehringer Ingelheim Pharmaceuticals, Inc., Ridgefield, CT 06877, USA; dyang55@its.jnj.com (D.Y.); rachel.kroe-barrett@boehringer-ingelheim.com (R.K.-B.); ssing207@ITS.JNJ.com (S.S.)

[2] Janssen BioTherapeutics, Janssen Research & Development, LLC, Spring House, PA 19477, USA

[3] Department of Molecular, Cellular and Biomedical Sciences, University of New Hampshire, Durham, NH 03861, USA

[*] Correspondence: tom.laue@unh.edu; Tel.: +1-603-978-5579

Received: 1 February 2019; Accepted: 6 March 2019; Published: 14 March 2019

Abstract: Practically, IgG charge can contribute significantly to thermodynamic nonideality, and hence to solubility and viscosity. Biologically, IgG charge isomers exhibit differences in clearance and potency. It has been known since the 1930s that all immunoglobulins carry a weak negative charge in physiological solvents. However, there has been no systematic exploration of this fundamental property. Accurate charge measurements have been made using membrane confined electrophoresis in two solvents (pH 5.0 and pH 7.4) on a panel of twelve mAb IgGs, as well as their F(ab')$_2$ and Fc fragments. The following observations were made at pH 5.0: (1) the measured charge differs from the calculated charge by ~40 for the intact IgGs, and by ~20 for the Fcs; (2) the intact IgG charge depends on both Fv and Fc sequences, but does not equal the sum of the F(ab')$_2$ and Fc charge; (3) the Fc charge is consistent within a class. In phosphate buffered saline, pH 7.4: (1) the intact IgG charges ranged from 0 to -13; (2) the F(ab')$_2$ fragments are nearly neutral for IgG1s and IgG2s, and about -5 for some of the IgG4s; (3) all Fc fragments are weakly anionic, with IgG1 < IgG2 < IgG4; (4) the charge on the intact IgGs does not equal the sum of the F(ab')$_2$ and Fc charge. In no case is the calculated charge, based solely on H$^+$ binding, remotely close to the measured charge. Some mAbs carried a charge in physiological salt that was outside the range observed for serum-purified human poly IgG. To best match physiological properties, a therapeutic mAb should have a measured charge that falls within the range observed for serum-derived human IgGs. A thermodynamically rigorous, concentration-dependent protein–protein interaction parameter is introduced. Based on readily measured properties, interaction curves may be generated to aid in the selection of proteins and solvent conditions. Example curves are provided.

Keywords: analytical electrophoresis; IgG subclasses; monoclonal IgG; protein charge; protein–protein interactions

---

1. Introduction

It is known that charge and charge distribution are important contributors to protein solubility and solution viscosity [1–11]. In general, increased charge correlates with higher solubility and lower viscosity because charge–charge repulsion weakens protein–protein interactions [12]. Experimentally, nonideality is quantified by the thermodynamic second virial coefficient (B$_{22}$ or A$_2$), with B$_{22}$ > 0 corresponding to net repulsion and B$_{22}$ < 0 corresponding to net attraction between molecules. Molecules possessing the same sign net charge will repel, while those having opposite charge will attract.

However, net charge alone does not fully capture the effects of charge on B$_{22}$. In particular, dipole moments resulting from asymmetric charge distributions can lead to orientation-dependent protein–protein attraction due to charge–dipole and dipole–dipole interactions, which decrease

B_{22} [5,9]. If B_{22} is < 0, highly viscous [5,7–9] or opalescent [2] solutions may result at high protein concentrations. Recent work suggests that there may be weak, promiscuous attractive interactions between IgGs [13,14]. These attractive interactions may or may not be entirely electrostatic in origin (e.g., weak hydrophobic interactions could contribute), though the salt and temperature dependence suggest electrostatic attractions are involved. Regardless of their origin, it has been suggested that the weak attraction (apparent monomer–dimer Kds of 10^{-4}–10^{-3} M [13–15]) may reflect the cooperative free energy needed for effector functions [14].

In addition to the importance of charge in the development of high concentration therapeutic formulations, mAb charge may influence in vivo processes. For example, neonatal Fc receptor (FcRn)-independent clearance rates are lower for mAbs with lower pI values than those with higher pI values [16–18], presumably due to decreased nonspecific cell surface binding [16,17,19,20]. Furthermore, basic charge variants of mAbs display stronger binding to the FcγRIIIa receptor and increased antibody-dependent cellular cytotoxicity response compared to more acidic charge variants [21,22]. Finally, there is an increasing body of evidence suggesting that IgG sialylation may impact therapeutic efficacy [23] and IgG function [24]. Together these in vivo and in vitro data show that mAb charge correlates with physical and biological consequences and highlight the need to understand what governs IgG charge.

The majority of biotherapeutic mAbs exhibit pIs ≥ 8 [25], and carry a positive charge under formulation conditions (typically pH 5–6) [2–4]. However, it has been known for over 80 years that all serum proteins, including the immunoglobulins, carry a net negative charge under physiological conditions [26]. Furthermore, IgGs from several species are anionic in the pH 5–6 range [27,28]. More recently, it was shown that freshly prepared human polyclonal IgGs have a Debye–Hückel–Henry charge, Z_{DHH} [26], between −3 and −9 [14]. The narrow range of charge is somewhat surprising since isoelectric focusing analysis of the same sample yielded pIs covering the pH range from less than 4 to greater than 10 [14]. There is no published charge data for mAbs in physiological solvents. Consequently, it is not known whether the charge on therapeutic mAbs falls into the rather narrow range observed for normal human IgGs. It is apparent that a systematic analysis of the charge on mAbs would be useful.

Presented here are charge measurements on twelve anti IL-13 IgGs. Using membrane confined electrophoresis, MCE, data have been acquired for three IgGs, mAb 1, mAb 2, and mAb 3, that bind to different IL-13 epitopes [14]. For each mAb, Z_{DHH} has been measured for four subclasses, IgG1, IgG2, IgG4, and IgG4Pro. Furthermore, the charge on the Fc and F(ab')$_2$ fragments was measured to determine whether the intact IgG charge is the sum of the Fc and F(ab')$_2$ fragment charges, and to assess how the charge is distributed over the IgG structure. Finally, the charge on the IgGs and their fragments were measured at both pH 5.0 and pH 7.4 to determine how the charge varies between formulation and physiological conditions. The results illustrate how little is known about protein charge and demonstrates the power of analytical electrophoresis in assessing this fundamental property.

Theoretical Basics of Protein Charge

Protein charge contributes significantly to a variety of biochemical, biophysical, and biological phenomena [29]. Thermodynamically, charge is a system property that depends on temperature, pressure, salt concentration, salt type, and pH [12]. At present there is no way to calculate protein charge accurately. However, charge may be measured with both precision and accuracy [26,30,31]. Of the measurement methods, membrane confined electrophoresis [32,33] is the most accurate and flexible [26,34].

There are a variety of charge descriptions (e.g., ζ potential, $Z_{effective}$, Z_{DHH}) [26]. While each description is useful, here we will use Z_{DHH}, which is the unitless valence resulting from the ratio of the protein charge (in coulombs) to the proton unit charge (e.g., Ca^{2+} has a valence of +2, Cl^- has a valence of −1). Calculation of Z_{DHH} from the free-boundary electrophoretic mobility removes the effects of electrophoresis and the Debye-Hückel solvent ion cloud [26,32,35]. Thus, Z_{DHH} reflects any changes in protein charge that accompany changes in solvent pH, salt type or salt concentration [26].

Even though proton binding to proteins has been studied extensively [36,37], it has been difficult to reconcile values calculated from amino acid side chain pKas with measurements [5,38,39]. Shifts in the pKas due to net protein charge, proximity of charged residues and protein flexibility are known to occur [36,37,40–43]. Though H$^+$ binding contributes to protein charge, Z_{DHH} reflects binding by all solvent ions (e.g., Na$^+$, PO$_4{}^2$, Cl$^-$) and not just H$^+$. It has been known for over 60 years that proteins bind monovalent ions, and bind anions to a greater extent than cations [12,44]. Two non-exclusive models have emerged for the mechanism of anion binding. One model focuses on the tendency for anions to accumulate preferentially at hydrophobic surfaces [38,45]. Based on NMR data, the other model suggests that anion binding may involve amide protons [46].

Because ion binding and dissociation occur rapidly, Z_{DHH} values are time averages. The extent of fluctuation about the mean value is proportional to the change in charge with ion chemical potential (i.e., the slope of the curve of Z versus log[X]) [36]. If the titration curve is flat (i.e., dZ/dlog[X] ~ 0), there will be very little charge variation, and the charge distribution about the average value will be narrow. A steep titration curve, however, indicates large charge variations which, particularly if they swing around neutrality, result in the inter-molecular attractions that reduce solubility and cause higher viscosities. Thus, measurement of Z_{DHH} as a function of solvent ion concentration (including pH) may be helpful in finding solvent conditions that optimize solubility and viscosity.

2. Materials and Methods

2.1. Monoclonal and Human Serum IgGs

Twelve anti-IL13 IgGs comprising three unique variable regions, each constructed as four human IgG subclasses, IgG1, IgG2, wild-type IgG4(Ser^{222}), and a hinge mutant IgG4(Pro^{222}), were made from stable NS0 cell line at Boehringer Ingelheim. Human serum derived from male AB plasma was purchased from Sigma-Aldrich, St. Louis, MO, USA (cat# H4522). The IgGs were purified by ÄKTA affinity chromatography system and MabSelect Sure resin (GE Healthcare, Chicago, IL, USA) following standard methods [47]. The quality of the purified mAb IgGs and their fragments generated by subsequent enzymatic digestion was evaluated by analytical size-exclusion ultra-performance liquid chromatography (SE-UPLC) using a BEH200 column on the Waters Acquity UPLC system (Waters Corporation, Milford, MA, USA). The mobile phase buffer consisted of 50 mM sodium phosphate (pH 6.8), 200 mM arginine, and 0.05% sodium azide. For each sample run, 10 µg of material was injected onto the column with the running flow rate at 0.5 mL/min for 5 min.

2.2. IgG Fragmentation

A FragIT kit with individual spin columns containing the active IdeS, a cycstein protease secreted by Streptococcus pyogenes covalently coupled to agarose beads was used (Genovis, cat# A2-FR2-025). After the IgG sample was buffer exchanged into the cleavage buffer (10 mM sodium phosphate, 150 mM NaCl) and the column was equilibrated with the cleavage buffer, the IgG-enzyme mixture was incubated at 37 °C for an hour on an orbital shaker. The digested fragments were separated from the immobilized enzyme, followed by the purification of F(ab')$_2$ using a supplied CaptureSelect column containing Fc affinity matrix (Thermo Fisher Scientific, Waltham, MA, USA). Upon the collection of the F(ab')$_2$ in the flow-through, the Fc was eluted using the 0.1 M glycine (pH 3.0) elution buffer and immediately neutralized by adding 10% v/v of 1 M Tris (pH 8.0).

2.3. Sample Preparation

Each sample was dialyzed into desired buffers at 4–10 °C overnight using Zeba desalting columns (Thermo Fischer), after which the concentration was determined using appropriate extinction coefficients in NanoDrop™ 8000 Spectrophotometer (Thermo Fischer). Two solvents were used: 10 mM sodium acetate, 50 mM NaCl, pH 5.0; and Dulbecco's PBS (pH 7.4) containing 8 mM sodium phosphate dibasic, 1.5 mM potassium phosphate monobasic, 2.7 mM KCl, and 138 mM NaCl. The acetate buffer was prepared by diluting chemicals purchased from Sigma into distilled deionized water from a Milli-Q Plus filtration system (Millipore, Burlington, MA, USA) and titrating to the desired pH 5.0 with 10 N NaOH solution. For all measurements, the sample solutions were used within a week of preparation and stored at 4 °C between measurements.

2.4. Liquid Chromatography Mass Spectrometry (LC-MS)

The sequences of the purified mAbs and respective F(ab')$_2$ and Fc fragments were evaluated by LC-MS using a PoroShell 300SB-C8 column (5 µm, 75 × 1.0 mm) on the Agilent HPLC system followed by analysis in the Agilent 6210 time-of-flight mass spectrometer (Agilent Technologies). The composition of the mobile phase A was 99% water, 1% acetonitrile, and 0.1% formic acid, and that of mobile phase B was 95% acetonitrile, 5% water, and 0.1% formic acid. The gradient started with 20% B at 0 min and increased to 85% B at 10 min with the constant flow rate of 50 µL/min. Each sample was subjected to a native run, a reduced run after incubation with TCEP (Sigma), and a deglycosylated run after incubation with TCEP and PNGase F (New England Biolabs). The MassHunter Qualitative Analysis program (version B.06.00, Agilent, Santa Clara, CA, USA) was used to deconvolute the raw data.

2.5. Analytical Ultracentrifugation (AUC)

The solution properties of the purified mAbs and cleaved F(ab')$_2$ and Fc were evaluated by sedimentation velocity experiments in an Optima XL-I AUC equipped with absorbance optics (Beckman Coulter, Brea, CA, USA). Each sample was prepared in three concentrations with 1:3 serial dilutions starting from 0.5 mg/mL in the corresponding buffer, and 400 µL of the prepared solution was loaded into the sample chamber, whereas buffer was loaded into the reference chamber of an AUC cell assembled with standard double-sector centerpieces and quartz windows. The experiments were conducted at 20 °C using an An60Ti 4-hole rotor spinning at 40,000 rpm. The sedimentation process was monitored by collecting absorbance data at 280 nm wavelength and 30-µm radial increments. The collected data was analyzed using the SEDANAL software by which the apparent sedimentation coefficient distribution g(s*) was derived [48]. The resulting analysis was initially plotted as g(s*) vs. s* in which the areas under the peaks provided the concentration for the boundary corresponding to each peak in the distribution. The weight average sedimentation coefficient (s_w) was computed by selecting a range over which to do the average on the plots. The plots were concentration-normalized to enable the inspection for reversible interactions. The Stokes radius, R_s, which is used for Z_{DHH} calculation is derived from the Svedberg equation:

$$R_s = \frac{M(1 - \bar{v}\rho)}{sN_A 6\pi\eta} \tag{1}$$

where M is the molar mass, $\bar{v}$ is the partial specific volume, ρ is the solvent density, s is the sedimentation coefficient, N_A is the Avogadro's number, and η is the viscosity of the solvent.

2.6. Imaged Capillary Isoelectric Focusing (icIEF)

The pI and charge heterogeneity of the IgG samples were determined on an iCE3 system (Protein Simple) [49,50]. Briefly, the pH gradient was created by an ampholyte mixture consisted of 44% (v/v) of 1% methylcellulose, 1.25% (v/v) of pharmalyte 3–10 solution, 3.75% (v/v) µL of pharmalyte 5–8 solution, 1.25% (v/v) of servalyte 9–11 solution, 0.63% (v/v) of pI marker pH 6.14, 0.63% (v/v) of pI marker pH 8.79, 6.3% (v/v) of 200 mM iminodiacetic acid, and 43% (v/v) of water. After sample preparation at 1 mg/mL in DI water, 40 µL of the diluted sample was mixed with 160 µL of ampholyte mixture and centrifuged for 5 min. The operating protocol used an initial potential of 1500 volts for 1 min, followed by a potential of 3000 volts for 20 min. For samples containing highly basic species, pI markers at pH 7.55 and pH 9.77 (0.63% v/v) and a focus period of 10 min at 3000 volts was used.

Separation was monitored at 280 nm, and the data analyzed using the iCE CFR software to calibrate the pI values and to select the markers. Subsequently, the data files were exported to Empower for analysis using the cIEF processing method.

2.7. Membrane-Confined Electrophoresis (MCE) and Z_{DHH} Determinations

Protein valence was measured in the MCE instrument (Spin Analytical, Inc., Berwick, Me, USA), which provides a direct measurement of the electrophoretic mobility (μ) to derive the Z_{eff} and the Z_{DHH} [32,33]. In each experiment, 20 µL of sample at 1 mg/mL was loaded into a $2 \times 2 \times 4$ mm quartz cuvette whose ends were sealed with semipermeable membranes (MWCO 3 kDa, Spectra/Por Biotech grade). An electric field was applied (4.3 V/cm for IgG, 8.5 V/cm for F(ab')$_2$ and Fc, and 19.8 V/cm for serum IgGs) longitudinally across the cell. The applied electric field, E, is a function of the applied current, i, the buffer conductivity (κ, 5.8 mS for 10 mM acetate, 50 mM NaCl [pH 5.0] and 16.8 mS for PBS [pH 7.4]), and the cross-sectional area of the cuvette, A, as $E = \frac{i}{\kappa A}$. Image scans of the cuvette were acquired with 25 µm resolution at 280 nm every 10–20 s. Time difference analysis provided an apparent electrophoretic mobility distribution, g(μ) versus μ, uncorrected for diffusion. Values of μ were converted to charge using the Spin Analytical software:

$$Z_{eff} = \frac{\mu}{fe} \tag{2}$$

$$Z_{DHH} = Z_{eff} \frac{1 + \kappa_D a}{H(\kappa_D a)} \tag{3}$$

where μ is the electrophoretic mobility, f is the translational frictional coefficient, e is the elementary proton charge, k_D is the inverse Debye length, a is the sum of the Stokes radius of the macromolecule and its counterion (0.18 nm for Cl^{-1} and 0.122 nm for Na$^+$), and $H(\kappa_D a)$ is Henry's function that accounts for electrophoretic effects. For reference, under the experimental conditions used here, $\kappa_D a \sim 2$ and $H(\kappa_D a) \sim 1.1$, though exact values are calculated for each experiment.

2.8. Calculated Charge, Z_{cal}, and Calculated Isoelectric pH, pI_{Cal}

Sednterp was used to calculate pI values, pI_{Cal}, as well as the H$^+$ titration curve from which Z_{cal} was determined [51]. These calculations are based on the amino acid composition and use pKa values from Edsall and Wyman [52]. It was assumed that the N-terminal amino groups were not blocked.

2.9. Dynamic Light Scattering (DLS) and k_D Determinations

A DynaPro Plate Reader (Wyatt Technology, Santa Barbara, CA, USA) running Dynamics (version 7.4.0.72) was used to determine the diffusion interaction parameter, k_D. Each sample was prepared at 5 concentrations ranging from 10 mg/mL to 0.625 mg/mL in 2-fold serial dilutions. 35 µL of each solution was added to a 384-well UV-Star Clear Microplate (Greiner Bio-One), spun in a centrifuge for 2 min to remove air bubbles and then placed into the plate reader. The experiment was started after the temperature inside the reader reached 20 °C. A total of 10 acquisitions at 20 s per acquisition were obtained for each sample. A well image was acquired after the last acquisition measurement to look for bubbles or deposited aggregates. The mutual diffusion coefficient (D_m) was plotted against the sample concentration $D_m = D_0(1 + k_D C)$, with D_0 and k_D determined by linear regression analysis using GraphPad Prism (version 7.03). The error for k_D was determined by calculating the propagation of the standard error of the coefficients from the linear regression.

2.10. Calculation of the Protein–Protein Interaction Curve

Thermodynamic nonideality reflects a balance of repulsive ($B_{22} > 0$) and attractive interactions ($B_{22} < 0$) between molecules. Only two protein characteristics contribute to positive B_{22} values, charge–charge repulsion (when the molecules have the same sign charge) and excluded volume (always repulsive). The contribution charge–charge repulsion, including the impact of the Debye-Hückel counterion cloud, may be calculated from $B_Z = \frac{1000 Z_2^2 \bar{v}_1}{4 m_3 M_2^2}\left(\frac{1+2\kappa r_s}{(1+\kappa r_s)^2}\right)$, where Z_2^2 is the square of the protein charge (i.e., Z_{DHH}), $\bar{v}_1$ the solvent partial specific volume (mL/g), M_2 the protein molecular weight (g/mol), m_3 the salt molality (mol/kg), κ the inverse Debye length (cm^{-1}) and r_s the solvated protein radius (cm) [12,44]. The excluded volume includes contributions from the shape of the molecule (in this case, using the axial ratio) and the hydration layer [12,44]. The excluded volume contribution is $B_{Ex} = \frac{8VN_A}{2M_2^2}$, where V is the solvated protein volume (mL/particle) and N_A is Avogadro's number [12,44]. The overall repulsive nonideality, B_{22}, is the sum of these two contributions. The weak attractive interactions observed for IgGs may be expressed in terms of a dimer dissociation, e.g., $IgG_2 \leftrightarrow IgG + IgG$, with the strength given by the dissociation constant given by $K_d = \frac{[IgG]^2}{[IgG_2]}$. At any concentration, the weight-average molecular weight, M_w, of a monomer–dimer mixture may be calculated knowing K_d. For systems exhibiting only repulsive interactions, the slope of a graph of $1/M_w$ versus concentration, C (in g/mL), B_{22} (in ml-mole/g^2), will be positive. Often, a graph of M_1/M_w, where M_1 is the monomer molar mass, is used to 'normalize' the data, in which case the slope of the line is $M_1 \cdot B_{22}$ and is called A_2 (in mL/g) in the literature. In either case, for purely repulsive systems over a wide concentration range, B_{22} or A_2 are positive and constant. For a system that exhibits a mass-action association, M_w increases with concentration ($1/M_w$ or M_1/M_w decrease, producing a negative curve). However, even in the face of self-association, B_{22} (or A_2) remain constant and positive, and push the curve in the opposite direction of self-association. Thus, for systems exhibiting both repulsive nonideality and weak self-association, unusual curves may result, starting with a negative slope at low C and winding up with a positive slope at higher C. The slope of the $1/M_w$ or M_1/M_w curve at each concentration, then, is an apparent B_{22}, $B_{22\text{-}app}$, or A_2, $A_{2\text{-}app}$. It is important to note that both $B_{22\text{-}app}$ and $A_{2\text{-}app}$ are thermodynamic parameters and represent useful protein–protein interaction parameters. For this work, data are presented as $A_{2\text{-}app}$. $A_{2\text{-}app}$ is > 0 for net repulsion, <0 for net attraction and =0 for a thermodynamically ideal system.

3. Results

3.1. Solution Properties of IgGs and Their Fragments

All purified IgGs were subjected to purity characterization by SE-UPLC, sequence identify and glycoform distribution analysis by LC-MS. As summarized in Table 1, all purified materials contain >99% monomer content and were confirmed by sequence to be in the expected IgG subclass with typical distribution of G0F, G1F, and G2F asparagine (N)-linked glycoforms.

The IgGs also displayed homogeneous solution properties within each mAb group in either pH 5.0 acetate and pH 7.4 PBS as illustrated in Figure 1 by the overlapping g(s*) curves. The weight-average sedimentation coefficients (s_w) are mAb1 6.37 ± 0.06, mAb2 6.37 ± 0.05, and mAb3 6.43 ± 0.09 in pH 5.0 acetate, and mAb1 6.28 ± 0.04, mAb2 6.27 ± 0.07, and mAb3 6.31 ± 0.06 in pH 7.4 PBS. These s_w values are consistent with the molecular weight of ~150 kDa IgG antibodies.

IgG cleavage sites and fragment purity are presented in Table 2. The solution homogeneity of each cleaved fragment was assessed by SV-AUC. All IgG fragments showed sedimentation distribution profiles like that in Figure 2 for mAb 1, where the superposition of the three concentrations of F(ab')$_2$ and Fc samples indicate homogeneity and the absence of self-association. The weight-average sedimentation coefficients (s_w) from the Fc evaluations are 3.45 ± 0.02, 3.46 ± 0.02, and 3.38 ± 0.18 for IgG1, IgG2, and IgG4/IgG4Pro, respectively. These values are consistent with the molecular weight of ~50 kDa, which indicates the Fc fragment remains a homodimer in solution despite cleavage below the hinge region. The s_w from the F(ab')$_2$ evaluations are 4.86 ± 0.01, 5.14 ± 0.06, 4.90 ± 0.02, and 4.95 ± 0.01 for IgG1, IgG2, IgG4, and IgG4Pro, respectively. These values are consistent with the molecular weight of ~100 kDa, which is expected for a bivalent Fab.

Table 1. Evaluation of IgG quality.

ID	Subclass	Monomer (%)	Mass (Da)	Glycoform Level (%) *		
				G0F	G1F	G2F
mAb 1	IgG1	>99	148,480	49	43	8
	IgG2	>99	147,913	52	39	9
	IgG4	>99	148,190	45	39	16
	IgG4Pro	>99	148,210	43	40	17
mAb 2	IgG1	>99	148,301	45	42	13
	IgG2	>99	147,734	50	41	9
	IgG4	>99	148,012	43	43	14
	IgG4Pro	>99	148,032	20	50	30
mAb 3	IgG1	>99	149,959	30	52	18
	IgG2	>99	149,231	45	41	14
	IgG4	>99	149,507	49	43	8
	IgG4Pro	>99	149,529	25	52	25

* N-acetylglucosamine ■; mannose ●; galactose ▶; ☆ fucose.

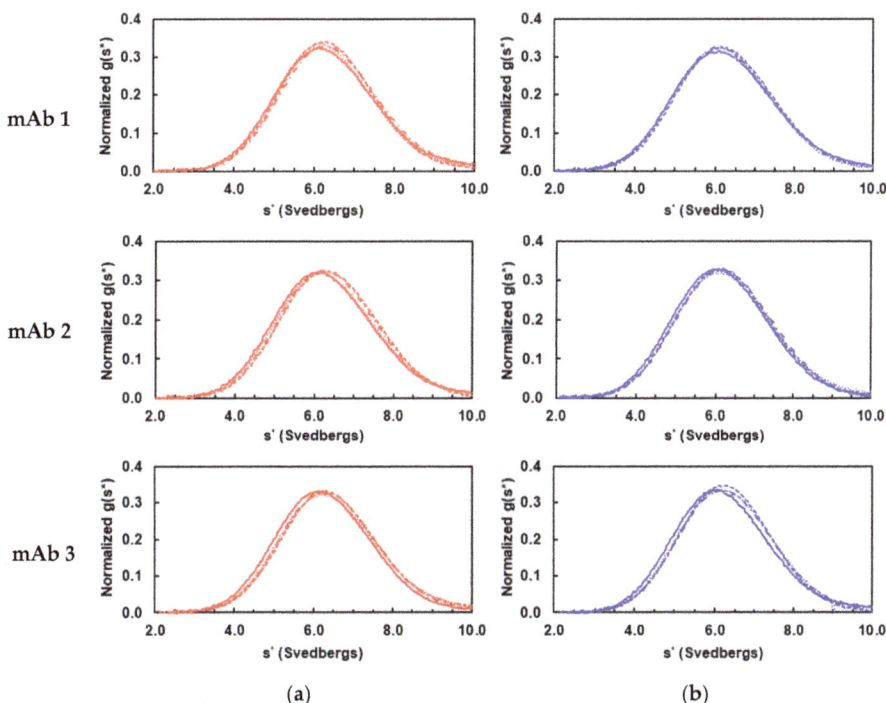

Figure 1. Sedimentation velocity analysis of IgG subclasses from mAb1, mAb2, and mAb3 in (**a**) pH 5.0 acetate (red) and (**b**) pH 7.4 PBS (blue) solutions. Normalized g(s*) sedimentation distributions are obtained from IgG1 (solid line), IgG2 (dotted line), IgG4 (dashed line), and IgG4Pro (dot-dashed line) in both buffers. The curves are superimposed on each other in both panels.

Table 2. Quality of IgG fragments from IdeS digestion.

Subclass	V Region	Cleaved Site	F(ab')$_2$ Purity (%)	Fc Purity (%)
IgG1	mAb 1		95	
	mAb 2	CPPCPAPELLG/GPSVF	100	100
	mAb 3		100	
IgG2	mAb 1		100	
	mAb 2	CPPCPAPPVA/GPSVF	100	98
	mAb 3		100	
IgG4	mAb 1		95	
	mAb 2	CPSCPAPELLG/GPSVF	95	97 *
	mAb 3		97	
IgG4Pro	mAb 1		97	
	mAb 2	CPPCPAPELLG/GPSVF	100	97 *
	mAb 3		100	

* The cleaved Fc is identical between IgG4Pro and IgG4 because the enzymatic digest occurred below the hinge region.

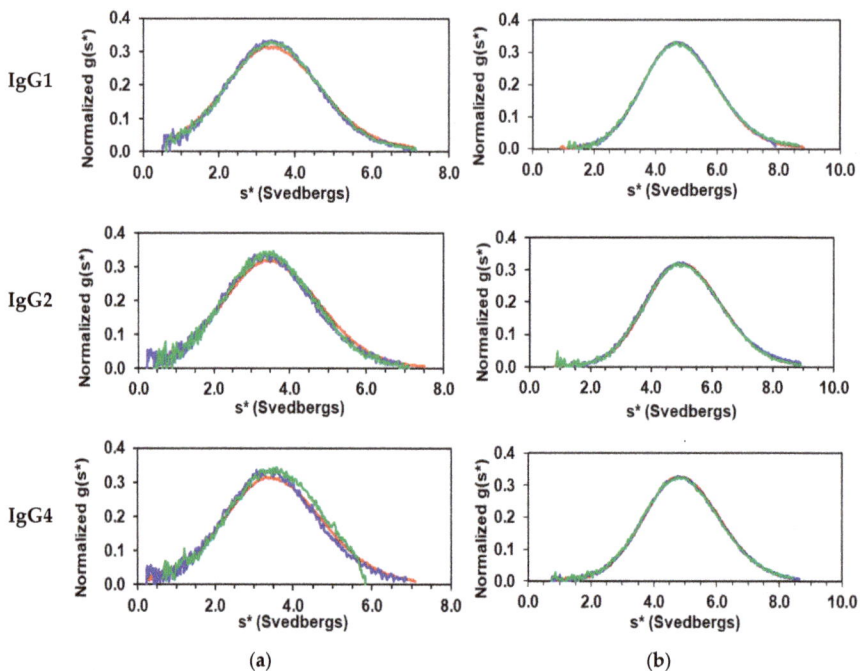

Figure 2. Sedimentation velocity analysis of IgG1, IgG2, and IgG4 cleaved (a) Fc and (b) F(ab')$_2$ from mAb 1 in pH 5.0 acetate. Normalized g(s*) sedimentation distributions obtained with the concentration of 0.5 mg/mL (red), 0.167 mg/mL (blue), and 0.056 mg/mL (green). The graph for IgG4Pro F(ab)'$_2$ is not shown because it is indistinguishable from IgG4. Refer to text for the s_w values.

3.2. Isoelectric Point and Correlation to Calculated Values

All IgGs exhibited pI profiles similar to that in Figure 3 for mAb 1 IgG1. Three-peaks are observed, acidic, main and basic. The pI values for each IgG are presented in Table 3, along with the calculated pI. For each mAb, the subclass pIs followed the trend: IgG1 > IgG2 > IgG4, with those of IgG4 and IgG4Pro being identical. The measured main species pI and the calculated pI are correlated (Figure 4), though the intercept (−1) suggests that pI$_{Cal}$ corresponds to the more acidic species.

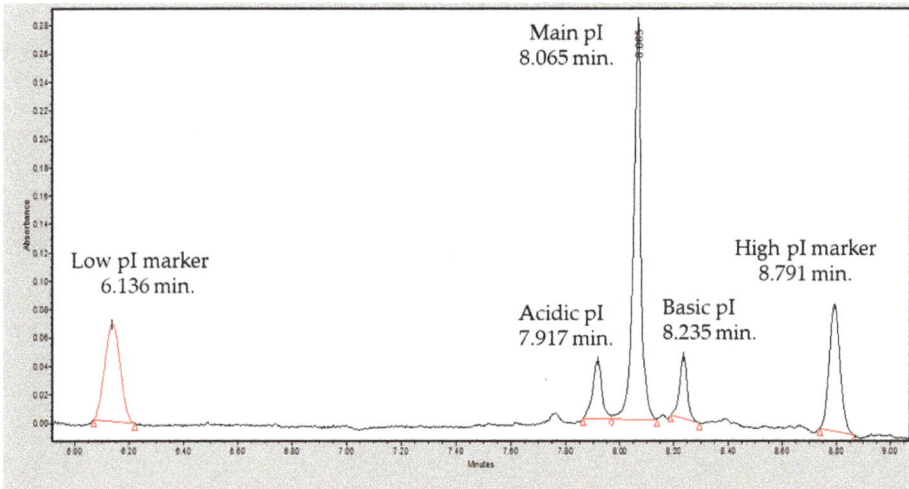

Figure 3. Electrophoretogram image of mAb1 IgG1. The peaks to the left and to the right of the main peak indicates acidic and basic charge variant, respectively.

Table 3. Measured and calculated pI values of IgG.

ID	Subclass	pI_{cal}	pI_{icIEF}		
			Acidic Peak	Main Peak	Basic Peak
mAb 1	IgG1	7.7	7.9	8.1	8.2
	IgG2	6.9	6.9	7.0	7.3
	IgG4	6.6	6.2	6.3	6.5
	IgG4Pro	6.6	6.2	6.3	6.5
mAb 2	IgG1	8.2	8.2	8.4	8.6
	IgG2	7.3	7.9	8.0	8.2
	IgG4	7.0	7.4	7.6	7.7
	IgG4Pro	7.0	7.4	7.6	7.7
mAb 3	IgG1	8.2	8.2	8.4	8.6
	IgG2	7.4	7.2	8.0	8.1
	IgG4	7.1	7.5	7.7	7.8
	IgG4Pro	7.1	7.5	7.7	7.8

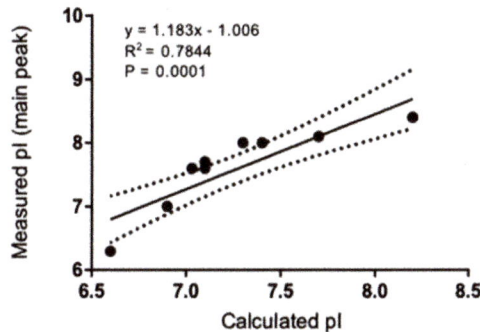

Figure 4. Linear regression analysis and correlation between experimental pI as measured by icIEF and theoretical pI calculated from the IgG sequence. Dotted lines indicate the 95% confidence interval.

3.3. Net Charge of IgGs and Fragments in Formulation and Physiological Solutions

Using MCE, the electrophoretic mobility was determined for each IgG and its cleaved F(ab')$_2$ and Fc in pH 5.0 acetate and pH 7.4 PBS as illustrated in Figure 5. By applying the Debye–Hückel approximation to correct for the solvent shielding effects, Henry's function to correct for electrophoretic effects, and using the sum of the measured protein Stokes radius and its counterion, the Z_{DHH} distribution may be calculated from the electrophoretic mobility (Figure 5, right-hand panels).

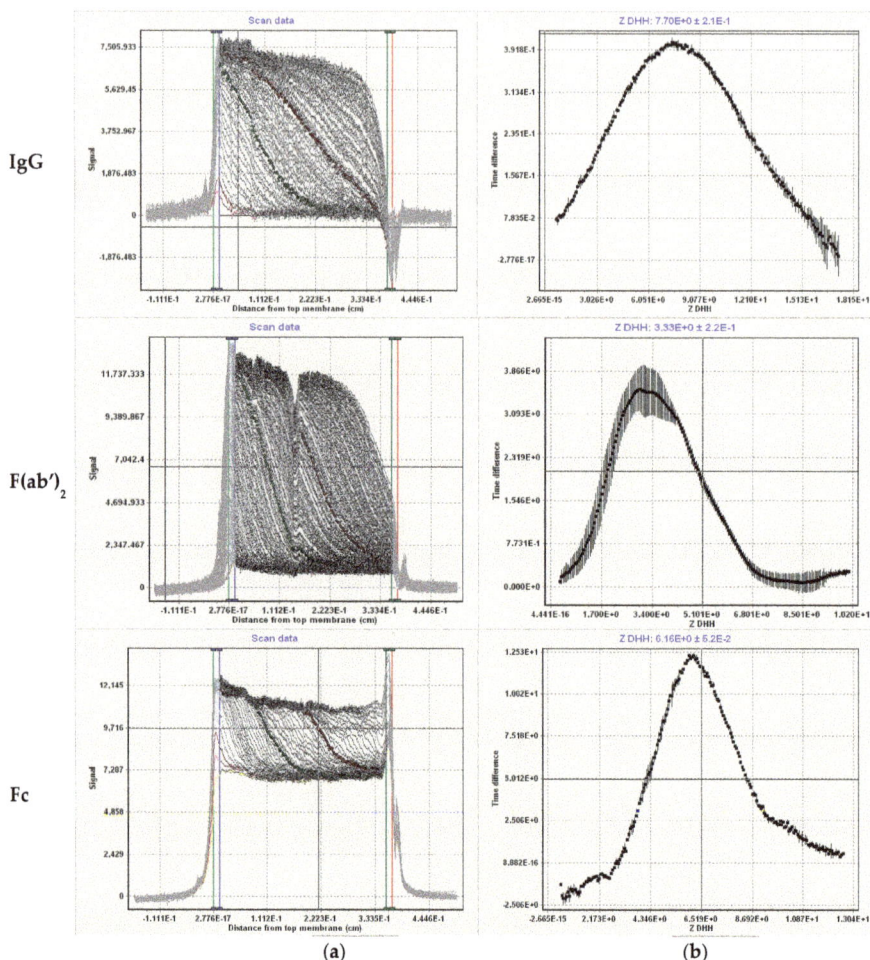

(a)   (b)

Figure 5. Z_{DHH} determination of IgG, F(ab')$_2$, and Fc by Membrane-Confined Electrophoresis (MCE) in pH 5.0 acetate. (**a**) Raw MCE scans over time during electrophoresis. The data (left panel) shows the light intensity (I, vertical axis) as a function of the distance moved from the membrane (cm, horizontal). Time difference curves ($\Delta I / \Delta t$) are calculated from data between the green and red highlighted scans. The electrophoretic mobility distribution is calculated from distance moved from the membrane, x, divided by the product of the electric field, E, and average elapsed time for the middle scan $\bar{t}$, $\mu = \frac{x}{E \cdot \bar{t}}$. (**b**) The vertical axis shows the time derivative ($\Delta I / \Delta t$) of the intensity data in panel (**a**) as a function of Z_{DHH} (horizontal axis). Z_{DHH} was calculated from the mobility using T = 20 °C; viscosity = 0.98 cp; conductance = 16.8 mS; E = −19.8 V/cm, D = 78; counterion radius, 0.18 nm; Stokes radius, 5.5 nm. The peak Z_{DHH} position is displayed above the curve.

Tables 4 and 5 summarize the Z_{DHH} measurements, as well as the calculated charge, Z_{cal}, in pH 5.0 acetate and pH 7.4 PBS, respectively. A 0 charge was assigned if no boundary formed during electrophoresis regardless of the E field direction or magnitude. In acetate pH 5.0 all IgGs and their fragments are cationic (Table 4). However, in all cases the measured Z_{DHH} is substantially lower than Z_{cal}. In PBS pH 7.4 (Table 5), all intact IgGs are neutral (mAb 2/IgG1) or anionic, despite the fact the Z_{cal} is cationic in some cases. For all mAbs, Z_{DHH} decreases with subclass in the rank order of IgG1 > IgG2 > IgG4.

Table 4. Measured and calculated Z values of IgG, F(ab')$_2$, and Fc in pH 5.0 acetate.

ID	Subclass	IgG		F(ab')$_2$		Fc *	
		Z_{DHH}	Z_{cal}	Z_{DHH}	Z_{cal}	Z_{DHH}	Z_{cal}
mAb 1	IgG1	7.7 ± 0.2	57.3	3.3 ± 0.2	31.2	6.2 ± 0.1	26.30
	IgG2	3.9 ± 0.1	50.0	0	25.9	4.9 ± 0.1	24.30
	IgG4	1.4 ± 0.2	46.7	1.3 ± 0.1	27.9	0.45 ± 0.1	18.98
	IgG4Pro	1.4 ± 0.8	46.7	1.5 ± 0.2	27.9	0.45 ± 0.1	18.98
mAb 2	IgG1	10.6 ± 0.1	61.0	8.6 ± 0.2	34.9	6.2 ± 0.1	26.30
	IgG2	10.1 ± 0.2	53.7	4.7 ± 0.1	29.6	4.9 ± 0.1	24.30
	IgG4	5.6 ± 0.2	50.4	6.2 ± 0.1	31.6	0.45 ± 0.1	18.98
	IgG4Pro	5.6 ± 0.2	50.4	6.2 ± 0.1	31.6	0.45 ± 0.1	18.98
mAb 3	IgG1	12.5 ± 0.1	65.8	9.4 ± 0.1	39.6	6.2 ± 0.1	26.30
	IgG2	10.3 ± 0.2	58.5	5.3 ± 0.2	34.3	4.9 ± 0.1	24.30
	IgG4	7.7 ± 0.2	55.1	7.1 ± 0.1	36.3	0.45 ± 0.1	18.98
	IgG4Pro	7.8 ± 0.2	55.1	7.3 ± 0.1	36.3	0.45 ± 0.1	18.98

* The value from each subclass is identical across the mAb set because it was measured on pooled Fc samples from the three mAb digestions.

Table 5. Measured and calculated Z values of IgG, F(ab')$_2$, and Fc in pH 7.4 PBS.

ID	Subclass	IgG		F(ab')$_2$		Fc *	
		Z_{DHH}	Z_{cal}	Z_{DHH}	Z_{cal}	Z_{DHH}	Z_{cal}
mAb 1	IgG1	−5.6 ± 0.1	1.8	0	−0.48	−2.8 ± 0.1	1.50
	IgG2	−7.7 ± 0.6	−4.4	0	−4.59	−6.0 ± 0.6	−0.48
	IgG4	−10.6 ± 0.5	−6.5	−4.3 ± 0.8	−2.61	−10.4 ± 0.3	−4.60
	IgG4Pro	−13 ± 0.3	−6.5	−5.05 ± 0.5	−2.61	−10.4 ± 0.3	−4.60
mAb 2	IgG1	0	5.8	0	3.5	−2.8 ± 0.1	1.50
	IgG2	−3.2 ± 0.2	−0.4	0	−0.61	−6.0 ± 0.6	−0.48
	IgG4	−7.4 ± 0.2	−2.5	0	1.38	−10.4 ± 0.3	−4.60
	IgG4Pro	−9.6 ± 0.4	−2.5	0	1.38	−10.4 ± 0.3	−4.60
mAb 3	IgG1	−5.3 ± 0.5	6.0	0	3.45	−2.8 ± 0.1	1.50
	IgG2	−6.1 ± 0.3	−0.1	0	−0.36	−6.0 ± 0.6	−0.48
	IgG4	−6.1 ± 0.2	−2.2	0	1.63	−10.4 ± 0.3	−4.60
	IgG4Pro	−10.7 ± 0.4	−2.2	0	1.63	−10.4 ± 0.3	−4.60

* The value from each subclass is identical across the mAb set because it was measured on pooled Fc samples from the three mAb digestions.

While Z_{DHH} and Z_{cal} are correlated in either solvent (Figure 6), the slope is about 1/2–3/4 of what would be expected if there were a 1:1 correspondence between the expected H^+ uptake/release and Z_{DHH}. These data are consistent with a model in which an anion is bound for every 1.3–2 H^+ bound. Similarly, Z_{DHH} for the intact IgGs correlates with the sum of Z_{DHH} from fragments (Figure 7), albeit with a slope that is about $\frac{1}{2}$ of that expected if the charge on the fragments simply summed. We have no mechanism or explanation for the data in Figure 7 and present them here in the hope that they will encourage future work.

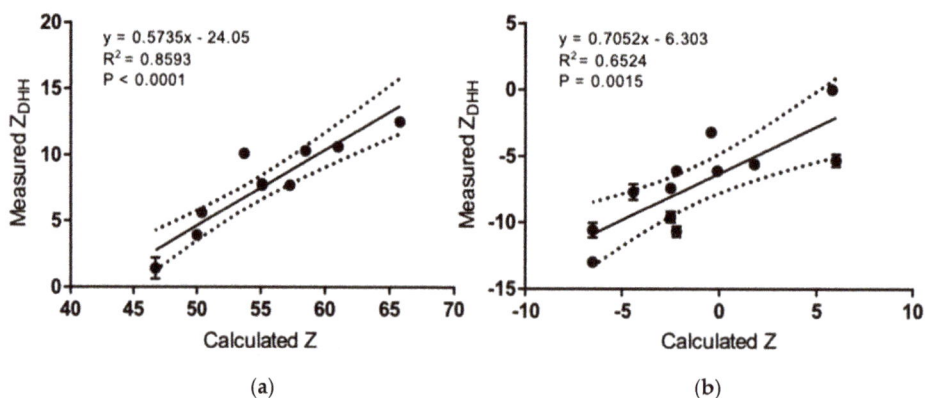

Figure 6. Linear regression analysis and correlation between experimental Z_{DHH} measured by MCE and theoretical Z calculated from the IgG sequence. (**a**) pH 5.0 acetate. (**b**) pH 7.4 PBS. Dotted lines indicate the 95% confidence interval.

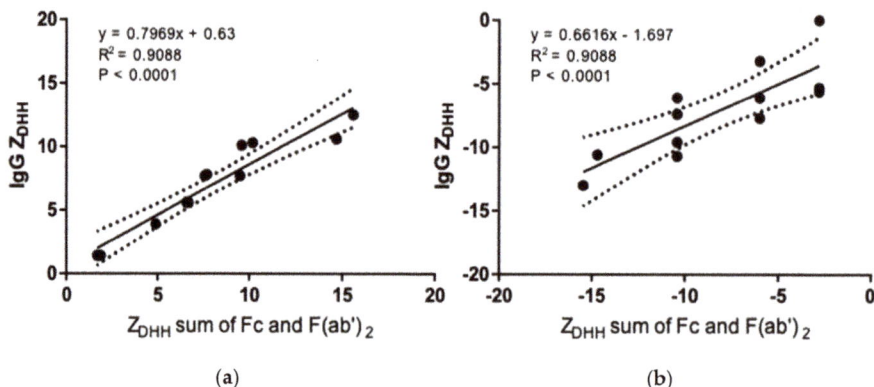

Figure 7. Linear regression analysis and correlation between Z_{DHH} measured from intact IgG and the sum of Z_{DHH} from the fragments. (**a**) pH 5.0 acetate. (**b**) pH 7.4 PBS. Dotted lines indicate the 95% confidence interval.

4. Discussion

Protein charge is a fundamental property that directly influences its structure, stability, solubility, and ability to interact with other macromolecules [53]. Charge–charge repulsion is important for overcoming the attractive forces that lead to high viscosities in high-concentration protein solutions [54]. Because protein charge can vary with solvent conditions, it is a *system* property rather than a property of the protein. The systematic analysis of twelve mAbs and their F(ab')₂ and Fc fragments provides several insights into IgG charge and raises several fundamental questions about our understanding of protein charge.

Charge–charge repulsion contributes to thermodynamic nonideality and, consequently, the colloidal stability of protein solutions [12]. It is clear from the data in Tables 4 and 5 that charge calculations based solely on H⁺ binding lead to highly inaccurate estimates of IgG charge. Thus, even though there is a correlation between the measured and calculated charge (Figure 6), charge calculations should not be considered reliable. Given its potential importance to colloidal stability, it is important to determine the impact of charge on nonideality.

At low to moderate protein concentrations (<~15 mg/mL), the net sum of all repulsive and attractive interactions is described by the second virial coefficient, B_{22} or A_2. The diffusion interaction parameter, k_D, is related to and often used as a stand-in for these quantities [55], with more positive values of k_D correlating with more positive values of B_{22}, i.e., greater repulsive interactions. If charge–charge repulsion contributes significantly to nonideality, there should be a positive correlation of charge with k_D. Figure 8 shows the correlation of Z_{DHH} with the diffusion interaction parameter, k_D. Under formulation conditions (Figure 6, panel a) increasing Z_{DHH} correlates with increased repulsive interaction (i.e., k_D becomes more positive). This suggests that charge measurements may be a useful parameter for selecting candidate mAbs for development. It should be noted that it is the effective charge, Z_{eff}, rather than Z_{DHH}, that impacts thermodynamic nonideality [26]. This distinction is important because Z_{eff} includes the contribution of the solvent ions, with Z_{eff} decreasing (i.e., repulsive interactions decreasing) as salt concentration is increased [12]. Because salt diminishes charge–charge interactions, thus reducing colloidal stability, it should be no surprise that most mAbs are manufactured and formulated in low-salt solvents.

While charge does contribute to nonideality under formulation conditions, there is no correlation between Z_{DHH} and k_D under physiological conditions (Figure 8b). This result means that it is unfavorable solvent displacement energies that keep mAbs in solution, for all other protein–protein interactions are attractive [56]. Similarly, it is likely that it is the protein solvation shell that dominates the solubility of serum IgG.

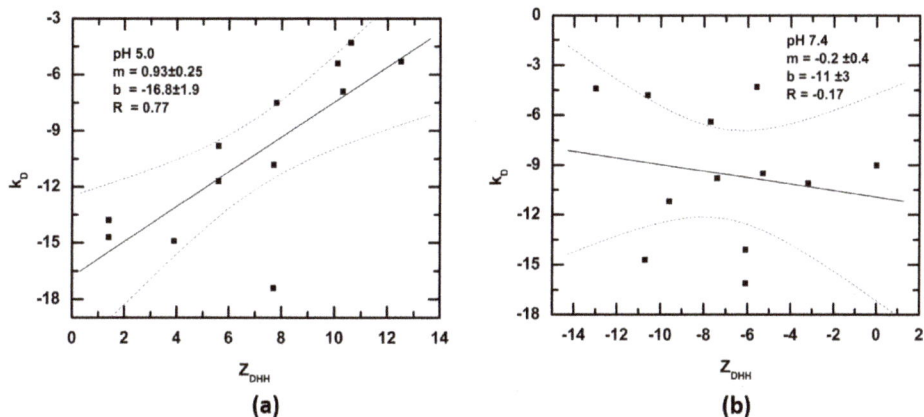

Figure 8. Linear regression analysis and correlation between Z_{DHH} measured for intact IgG and the concentration-dependence of the diffusion coefficient, k_D. (**a**) pH 5.0, acetate buffer, (**b**) pH 7.4 PBS. Dotted lines indicate the 95% confidence interval.

One surprising result of our work is that freshly prepared human IgG exhibits a rather narrow Z_{DHH} distribution in physiological solvent (from approximately -10 to -2, Figure 9), even though isoelectric focusing shows that the same sample has species ranging from pI < 4 to pI > 10 [14]. This exact same Z_{DHH} range may be calculated from electrophoretic mobility measurements published 80 years ago [27]. It would seem from these results that IgGs exhibit charge homeostasis. The mechanism for this homeostasis is not clear. None of the mAbs contained anionic carbohydrates (Table 1), so it is not possible to determine whether the addition of, say, sialic acid would result in a more anionic IgG under physiological conditions. Given the narrowness of the human IgG charge distribution under physiological conditions, it seems likely that sialylation contributes specifically to interactions rather than merely impacting the global charge.

Figure 9 shows that most, but not all, of the mAbs in this study exhibit Z_{DHH} that fall in the range for human serum poly IgG. It is not clear whether there are any physiological or medical consequences associated with a mAb Z_{DHH} that falls outside the normal physiological range. Thus, these results are presented in the hopes of stimulating further research.

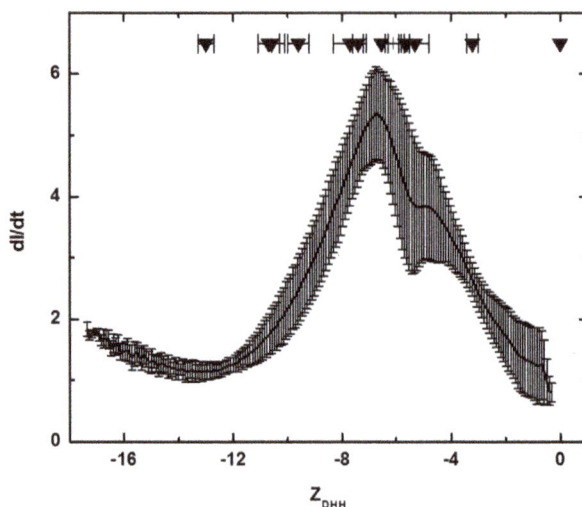

Figure 9. Z_{DHH} distribution for freshly prepared human IgG in DPBS. Z_{DHH} was calculated for T = 20 °C, viscosity = 0.98 cp, electric field = −14.88 V/cm, ionic strength = 0.167 M, conductivity = 16.6 ms, protein radius = 5.5 nm, counterion radius = 0.18 nm, D = 78. The Z_{DHH} for the twelve intact IgGs in this study are noted (inverted triangles) along with bars indicating the measurement uncertainty.

Since both aggregation and high viscosity are reflections of protein–protein interactions, it would be useful to have a rigorous means of determining whether an IgG (or solvent condition) is good, bad or indifferent. We suggest that the apparent thermodynamic nonideality ($dB_{22\text{-}app}/dC$ or $dA_{2\text{-}app}/dC$) might fulfill this need. To calculate $A_{2\text{-}app}$, several quantities are required (see Figure 10 legend), but each of these values are tabulated, easily calculated or readily obtained experimentally. A dimer dissociation constant of 1 mM was used to mimic the attractive interactions in all cases. This value of K_d is at the upper range of what has been found experimentally [13,15]. If stronger attractive interactions are used (e.g., 300 µM rather than 1 mM), the range where interactions are net attractive is more extensive. A more complete report on determining and using this interaction parameter is being developed.

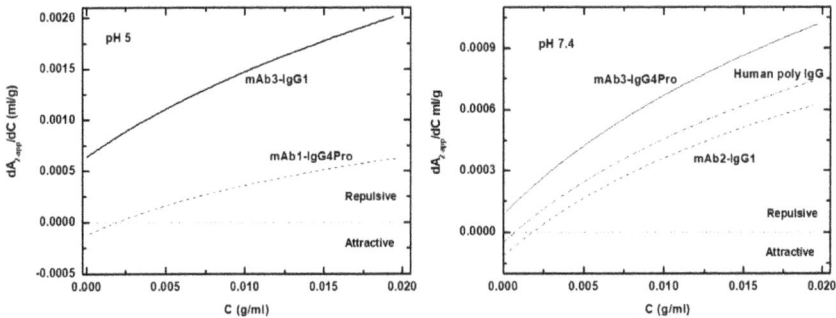

Figure 10. Protein–protein-interaction parameter dA_{2-app}/dC for pH 5 and pH 7.4 data. Note the concentrations are present in g/mL (cgs units) in order to be consistent with the derivation of the equations [12], and correspond to a concentration range of 0–20 mg/mL. The parameters used to generate these curves are: M_1 = 150,000 g/mole, hydrated radius 4.39 nm, hydration 0.3 g-H_2O/g-protein, axial ratio 5, monomer–dimer K_d 1 × 10^{-3} M, protein partial specific volume = 0.73 mL/g, solvent partial specific volume 0.993 mL/g, temperature 20 °C, and solvent density 1.0 g/mL. The salt concentration for pH 5 was set to 60 mMolal, and for pH 7.4 to 150 mMolal. For either condition, curves for the mAbs having the lowest Z_{DHH} (dashed lines) and highest Z_{DHH} (solid lines) are shown. For pH 7.4, a curve for human poly IgG (dash-dot) is shown. The horizontal dotted line at 0 corresponds to ideal conditions, with values less than zero corresponding to net attraction and greater than 0 to net repulsion.

5. Conclusions

Charge is a fundamental property of antibodies and is important in providing colloidally stable mAb solutions during their development, manufacture and formulation. At this time, protein charge cannot be calculated with any accuracy even using the most detailed structural information and the most sophisticated algorithms. Protein charge, however, is readily measured with accuracy and precision. In this first systematic and comprehensive examination of the charge on IgGs it is clear that: (1) IgGs exhibit charge homeostasis in physiological solvent, (2) they appear to bind significant quantities of anions, (3) anion binding will contribute to the desolvation energy, thus preventing IgG aggregation, (4) mAb charge measurements may be useful in selecting candidate molecules for development and (5) mAb charge measurements under physiological conditions may be useful in determining whether a candidate molecule falls within the normal range for human IgGs.

Author Contributions: Conceptualization, D.Y., T.L., S.S., and R.K.-B.; Methodology, T.L. and D.Y.; Analysis, D.Y. T.L.; Investigation, D.Y.; Resources, R.K.-B.; Writing—Original Draft Preparation, D.Y. and T.L.; Writing—Review and Editing, R.K.-B. and S.S.; Supervision, R.K.-B. and T.L.; Project Administration, R.K.-B. and T.L.; Funding Acquisition, R.K.-B.

Funding: This research was funded by Boehringer-Ingelhem.

Acknowledgments: The authors wish to thank Boehringer-Ingelhem for supporting the doctorate research of Danlin Yang, a portion of which is published here. Special thanks to her Ph.D. committee members, David Hayes and Christopher Roberts, who encouraged this work and offered helpful advice. We also are thankful for the encouragement and interest expressed by the members of the Biomolecular Interactions Technology Center (BITC). This paper is dedicated to the memory of Eric and Betty Laue.

Conflicts of Interest: The authors declare no conflict of interest. The funders had no role in the design of the study; in the collection, analyses, or interpretation of data; in the writing of the manuscript, or in the decision to publish the results. D.Y. and S.S. were employed by Boehringer Ingelheim Pharmaceuticals at the time of this research, they are now employed by Janssen Biotherapeutics. R.K.-B. is employed by Boehringer Ingelheim Pharmaceuticals. T.L. was employed by the University of New Hampshire at the time of this research and now is retired from that institution.

Antibodies **2019**, *8*, 24

References

1. Cohn, E.J. Studies in the physical chemistry of the proteins: I. The solubility of certain proteins at their isoelectric points. *J. Gen. Physiol.* **1922**, *4*, 697–722. [CrossRef] [PubMed]
2. Raut, A.S.; Kalonia, D.S. Opalescence in Monoclonal Antibody Solutions and Its Correlation with Intermolecular Interactions in Dilute and Concentrated Solutions. *J. Pharm. Sci.* **2015**, *104*, 1263–1274. [CrossRef]
3. Shire, S.J.; Shahrokh, Z.; Liu, J. Challenges in the development of high protein concentration formulations. *J. Pharm. Sci.* **2004**, *93*, 1390–1402. [CrossRef] [PubMed]
4. Li, L.; Kumar, S.; Buck, P.M.; Burns, C.; Lavoie, J.; Singh, S.K.; Warne, N.W.; Nichols, P.; Luksha, N.; Boardman, D. Concentration dependent viscosity of monoclonal antibody solutions: Explaining experimental behavior in terms of molecular properties. *Pharm. Res.* **2014**, *31*, 3161–3178. [CrossRef] [PubMed]
5. Yadav, S.; Laue, T.M.; Kalonia, D.S.; Singh, S.N.; Shire, S.J. The Influence of Charge Distribution on Self-Association and Viscosity Behavior of Monoclonal Antibody Solutions. *Mol. Pharm.* **2012**, *9*, 791–802. [CrossRef] [PubMed]
6. Chi, E.Y.; Krishnan, S.; Randolph, T.W.; Carpenter, J.F. Physical stability of proteins in aqueous solution: Mechanism and driving forces in nonnative protein aggregation. *Pharm. Res.* **2003**, *20*, 1325–1336. [CrossRef] [PubMed]
7. Neergaard, M.S.; Kalonia, D.S.; Parshad, H.; Nielsen, A.D.; Møller, E.H.; van de Weert, M. Viscosity of high concentration protein formulations of monoclonal antibodies of the IgG1 and IgG4 subclass—Prediction of viscosity through protein–protein interaction measurements. *Eur. J. Pharm. Sci.* **2013**, *49*, 400–410. [CrossRef] [PubMed]
8. Pindrus, M.A.; Shire, S.J.; Yadav, S.; Kalonia, D.S. The Effect of Low Ionic Strength on Diffusion and Viscosity of Monoclonal Antibodies. *Mol. Pharm.* **2018**, *15*, 3133–3142. [CrossRef]
9. Singh, S.N.; Yadav, S.; Shire, S.J.; Kalonia, D.S. Dipole-Dipole Interaction in Antibody Solutions: Correlation with Viscosity Behavior at High Concentration. *Pharm. Res.* **2014**, *31*, 2549–2558. [CrossRef]
10. Olsen, S.N.; Andersen, K.B.; Randolph, T.W.; Carpenter, J.F.; Westh, P. Role of electrostatic repulsion on colloidal stability of Bacillus halmapalus alpha-amylase. *Biochim. Biophys. Acta* **2009**, *1794*, 1058–1065. [CrossRef]
11. Connolly, B.D.; Petry, C.; Yadav, S.; Demeule, B.; Ciaccio, N.; Moore, J.M.R.; Shire, S.J.; Gokarn, Y.R. Weak Interactions Govern the Viscosity of Concentrated Antibody Solutions: High-Throughput Analysis Using the Diffusion Interaction Parameter. *Biophys. J.* **2012**, *103*, 69–78. [CrossRef]
12. Tanford, C. *Physical Chemistry of Macromolecules*, 1st ed.; John Wiley & Sons Inc.: New York, NY, USA, 1961; ISBN 978-0-471-84447-1.
13. Wright, R.T.; Hayes, D.B.; Stafford, W.F.; Sherwood, P.J.; Correia, J.J. Characterization of therapeutic antibodies in the presence of human serum proteins by AU-FDS analytical ultracentrifugation. *Anal. Biochem.* **2018**, *550*, 72–83. [CrossRef] [PubMed]
14. Yang, D.; Correia, J.J.; Iii, W.F.S.; Roberts, C.J.; Singh, S.; Hayes, D.; Kroe-Barrett, R.; Nixon, A.; Laue, T.M. Weak IgG self- and hetero-association characterized by fluorescence analytical ultracentrifugation. *Protein Sci.* **2018**, *27*, 1334–1348. [CrossRef] [PubMed]
15. Hopkins, M.M.; Lambert, C.L.; Bee, J.S.; Parupudi, A.; Bain, D.L. Determination of Interaction Parameters for Reversibly Self-Associating Antibodies: A Comparative Analysis. *J. Pharm. Sci.* **2018**, *107*, 1820–1830. [CrossRef] [PubMed]
16. Kelly, R.L.; Yu, Y.; Sun, T.; Caffry, I.; Lynaugh, H.; Brown, M.; Jain, T.; Xu, Y.; Wittrup, K.D. Target-independent variable region mediated effects on antibody clearance can be FcRn independent. *mAbs* **2016**, *8*, 1269–1275. [CrossRef] [PubMed]
17. Leipold, D.; Prabhu, S. Pharmacokinetic and pharmacodynamic considerations in the design of therapeutic antibodies. *Clin. Transl. Sci.* **2018**. [CrossRef]
18. Bas, M.; Terrier, A.; Jacque, E.; Dehenne, A.; Pochet-Béghin, V.; Beghin, C.; Dezetter, A.-S.; Dupont, G.; Engrand, A.; Beaufils, B.; et al. Fc Sialylation Prolongs Serum Half-Life of Therapeutic Antibodies. *J. Immunol.* **2019**, *202*, 1582–1594. [CrossRef]
19. Boswell, C.A.; Tesar, D.B.; Mukhyala, K.; Theil, F.-P.; Fielder, P.J.; Khawli, L.A. Effects of charge on antibody tissue distribution and pharmacokinetics. *Bioconjug. Chem.* **2010**, *21*, 2153–2163. [CrossRef]

20. Igawa, T.; Tsunoda, H.; Tachibana, T.; Maeda, A.; Mimoto, F.; Moriyama, C.; Nanami, M.; Sekimori, Y.; Nabuchi, Y.; Aso, Y.; et al. Reduced elimination of IgG antibodies by engineering the variable region. *Protein Eng. Des. Select.* **2010**, *23*, 385–392. [CrossRef]

21. Hintersteiner, B.; Lingg, N.; Janzek, E.; Mutschlechner, O.; Loibner, H.; Jungbauer, A. Microheterogeneity of therapeutic monoclonal antibodies is governed by changes in the surface charge of the protein. *Biotechnol. J.* **2016**, *11*, 1617–1627. [CrossRef]

22. Hintersteiner, B.; Lingg, N.; Zhang, P.; Woen, S.; Hoi, K.M.; Stranner, S.; Wiederkum, S.; Mutschlechner, O.; Schuster, M.; Loibner, H.; et al. Charge heterogeneity: Basic antibody charge variants with increased binding to Fc receptors. *mAbs* **2016**, *8*, 1548–1560. [CrossRef]

23. Zhang, G.; Massaad, C.A.; Gao, T.; Pillai, L.; Bogdanova, N.; Ghauri, S.; Sheikh, K.A. Sialylated intravenous immunoglobulin suppress anti-ganglioside antibody mediated nerve injury. *Exp. Neurol.* **2016**, *282*, 49–55. [CrossRef]

24. Lardinois, O.M.; Deterding, L.J.; Hess, J.J.; Poulton, C.J.; Henderson, C.D.; Jennette, J.C.; Nachman, P.H.; Falk, R.J. Immunoglobulins G from patients with ANCA-associated vasculitis are atypically glycosylated in both the Fc and Fab regions and the relation to disease activity. *PLoS ONE* **2019**, *14*, e0213215. [CrossRef] [PubMed]

25. Goyon, A.; Excoffier, M.; Janin-Bussat, M.-C.; Bobaly, B.; Fekete, S.; Guillarme, D.; Beck, A. Determination of isoelectric points and relative charge variants of 23 therapeutic monoclonal antibodies. *J. Chromatogr. B* **2017**, *1065–1066*, 119–128. [CrossRef]

26. Filoti, D.I.; Shire, S.J.; Yadav, S.; Laue, T.M. Comparative study of analytical techniques for determining protein charge. *J. Pharm. Sci.* **2015**, *104*, 2123–2131. [CrossRef]

27. Tiselius, A.; Kabat, E.A. An electrophoretic study of immune sera and purified antibody preparations. *J. Exp. Med.* **1939**, *69*, 119–131. [CrossRef]

28. Tiselius, A.; Kabat, E.A. Electrophoresis of Immune Serum. *Science* **1938**, *87*, 416–417. [CrossRef]

29. Mathews, C.K.; van Holde, K.E.; Appling, D.R.; Anthony-Cahill, S.J. *Biochemistry*, 4th ed.; Pearson: Toronto, ON, Canada, 2012; ISBN 978-0-13-800464-4.

30. Moody, T.P.; Kingsbury, J.S.; Durant, J.A.; Wilson, T.J.; Chase, S.F.; Laue, T.M. Valence and anion binding of bovine ribonuclease A between pH 6 and 8. *Anal. Biochem.* **2005**, *336*, 243–252. [CrossRef]

31. Her, C.; Filoti, D.I.; McLean, M.A.; Sligar, S.G.; Alexander Ross, J.B.; Steele, H.; Laue, T.M. The Charge Properties of Phospholipid Nanodiscs. *Biophys. J.* **2016**, *111*, 989–998. [CrossRef]

32. Ridgeway, T.M.; Hayes, D.B.; Moody, T.P.; Wilson, T.J.; Anderson, A.L.; Levasseur, J.H.; Demaine, P.D.; Kenty, B.E.; Laue, T.M. An apparatus for membrane-confined analytical electrophoresis. *Electrophoresis* **1998**, *19*, 1611–1619. [CrossRef]

33. Laue, T.M.; Shepard, H.K.; Ridgeway, T.M.; Moody, T.P.; Wilson, T.J. Membrane-confined analytical electrophoresis. *Methods Enzymol.* **1998**, *295*, 494–518. [PubMed]

34. Kyne, C.; Jordon, K.; Filoti, D.I.; Laue, T.M.; Crowley, P.B. Protein charge determination and implications for interactions in cell extracts. *Protein Sci. Publ. Protein Soc.* **2017**, *26*, 258–267. [CrossRef] [PubMed]

35. Moody, T.P.; Shepard, H.K. Nonequilibrium thermodynamics of membrane-confined electrophoresis. *Biophys. Chem.* **2004**, *108*, 51–76. [CrossRef]

36. Edsall, J.T.; Wyman, J. Chapter 9—Polybasic Acids, Bases, and Ampholytes, Including Proteins. In *Biophysical Chemistry*; Edsall, J.T., Wyman, J., Eds.; Academic Press: Cambridge, MA, USA, 1958; pp. 477–549, ISBN 978-1-4832-2946-1.

37. Pace, C.N.; Grimsley, G.R.; Scholtz, J.M. Protein ionizable groups: pK values and their contribution to protein stability and solubility. *J. Biol. Chem.* **2009**, *284*, 13285–13289. [CrossRef]

38. Gokarn, Y.R.; Fesinmeyer, R.M.; Saluja, A.; Razinkov, V.; Chase, S.F.; Laue, T.M.; Brems, D.N. Effective charge measurements reveal selective and preferential accumulation of anions, but not cations, at the protein surface in dilute salt solutions: Effective Charge Measurements Reveal Direct Anion-Protein Interactions. *Protein Sci.* **2011**, *20*, 580–587. [CrossRef] [PubMed]

39. Kukić, P.; Nielsen, J.E. Electrostatics in proteins and protein-ligand complexes. *Future Med. Chem.* **2010**, *2*, 647–666. [CrossRef] [PubMed]

40. Kumar, S.; Nussinov, R. Close-range electrostatic interactions in proteins. *Chembiochem* **2002**, *3*, 604–617. [CrossRef]

41. Shi, C.; Wallace, J.A.; Shen, J.K. Thermodynamic Coupling of Protonation and Conformational Equilibria in Proteins: Theory and Simulation. *Biophys. J.* **2012**, *102*, 1590–1597. [CrossRef]

42. Swails, J.M.; York, D.M.; Roitberg, A.E. Constant pH Replica Exchange Molecular Dynamics in Explicit Solvent Using Discrete Protonation States: Implementation, Testing, and Validation. *J. Chem. Theory Comput.* **2014**, *10*, 1341–1352. [CrossRef]

43. Oliveira, A.S.F.; Campos, S.R.R.; Baptista, A.M.; Soares, C.M. Coupling between protonation and conformation in cytochrome c oxidase: Insights from constant-pH MD simulations. *Biochim. Biophys. Acta BBA Bioenergy* **2016**, *1857*, 759–771. [CrossRef]

44. Harding, S.E.; Horton, J.C.; Jones, S.; Thornton, J.M.; Winzor, D.J. COVOL: An Interactive Program for Evaluating Second Virial Coefficients from the Triaxial Shape or Dimensions of Rigid Macromolecules. *Biophys. J.* **1999**, *76*, 2432–2438. [CrossRef]

45. Collins, K.D.; Neilson, G.W.; Enderby, J.E. Ions in water: Characterizing the forces that control chemical processes and biological structure. *Biophys. Chem.* **2007**, *128*, 95–104. [CrossRef] [PubMed]

46. Miao, L.; Qin, H.; Koehl, P.; Song, J. Selective and specific ion binding on proteins at physiologically-relevant concentrations. *FEBS Lett.* **2011**, *585*, 3126–3132. [CrossRef] [PubMed]

47. Harlow, E.; Lane, D.P. *Antibodies: A Laboratory Manual*, 1st ed.; Cold Spring Harbor Laboratory Press: Cold Spring Harbor, NY, USA, 1988; ISBN 978-0-87969-314-5.

48. Stafford, W.F.; Braswell, E.H. Sedimentation velocity, multi-speed method for analyzing polydisperse solutions. *Biophys. Chem.* **2004**, *108*, 273–279. [CrossRef] [PubMed]

49. Bjellqvist, B.; Ek, K.; Giorgio Righetti, P.; Gianazza, E.; Görg, A.; Westermeier, R.; Postel, W. Isoelectric focusing in immobilized pH gradients: Principle, methodology and some applications. *J. Biochem. Biophys. Methods* **1982**, *6*, 317–339. [CrossRef]

50. Mao, Q.; Pawliszyn, J. Capillary isoelectric focusing with whole column imaging detection for analysis of proteins and peptides. *J. Biochem. Biophys. Methods* **1999**, *39*, 93–110. [CrossRef]

51. Hayes, D.; Laue, T.; Philo, J. *Program Sednterp: Sedimentation Interpretation Program*, version 1.09; Alliance Protein Laboratories: Thousand Oaks, CA, USA, 1995.

52. Edsall, J.T.; Wyman, J. *Biophysical Chemistry*; Academic Press: New York, NY, USA, 1958; ISBN 978-0-12-232201-3.

53. Laue, T. Charge matters. *Biophys. Rev.* **2016**, *8*, 287–289. [CrossRef]

54. Tomar, D.S.; Kumar, S.; Singh, S.K.; Goswami, S.; Li, L. Molecular basis of high viscosity in concentrated antibody solutions: Strategies for high concentration drug product development. *mAbs* **2016**, *8*, 216–228. [CrossRef]

55. Saluja, A.; Fesinmeyer, R.M.; Hogan, S.; Brems, D.N.; Gokarn, Y.R. Diffusion and sedimentation interaction parameters for measuring the second virial coefficient and their utility as predictors of protein aggregation. *Biophys. J.* **2010**, *99*, 2657–2665. [CrossRef]

56. Laue, T. Proximity energies: A framework for understanding concentrated solutions. *J. Mol. Recognit. JMR* **2012**, *25*, 165–173. [CrossRef]

antibodies

MDPI

Review

The Ligands for Human IgG and Their Effector Functions

Steven W. de Taeye [1,2,*], Theo Rispens [1] and Gestur Vidarsson [2]

[1] Sanquin Research, Dept Immunopathology and Landsteiner Laboratory, Amsterdam UMC,
University of Amsterdam, 1066 CX Amsterdam, The Netherlands; t.rispens@sanquin.nl

[2] Sanquin Research, Dept Experimental Immunohematology and Landsteiner Laboratory, Amsterdam UMC,
University of Amsterdam, 1066 CX Amsterdam, The Netherlands; g.vidarsson@sanquin.nl

* Correspondence: s.detaeye@sanquin.nl

Received: 26 March 2019; Accepted: 18 April 2019; Published: 25 April 2019

Abstract: Activation of the humoral immune system is initiated when antibodies recognize an antigen and trigger effector functions through the interaction with Fc engaging molecules. The most abundant immunoglobulin isotype in serum is Immunoglobulin G (IgG), which is involved in many humoral immune responses, strongly interacting with effector molecules. The IgG subclass, allotype, and glycosylation pattern, among other factors, determine the interaction strength of the IgG-Fc domain with these Fc engaging molecules, and thereby the potential strength of their effector potential. The molecules responsible for the effector phase include the classical IgG-Fc receptors (FcγR), the neonatal Fc-receptor (FcRn), the Tripartite motif-containing protein 21 (TRIM21), the first component of the classical complement cascade (C1), and possibly, the Fc-receptor-like receptors (FcRL4/5). Here we provide an overview of the interactions of IgG with effector molecules and discuss how natural variation on the antibody and effector molecule side shapes the biological activities of antibodies. The increasing knowledge on the Fc-mediated effector functions of antibodies drives the development of better therapeutic antibodies for cancer immunotherapy or treatment of autoimmune diseases.

Keywords: Antibodies; IgG; Fc effector molecules; allotypes; glycosylation

1. Introduction

The human adaptive humoral immune system is dependent on antigen recognition via the B cell receptor on naïve B cells, which initiates B cell maturation and eventually production of antibodies by plasmablasts and plasma cells. IgM is the initial antibody class that is made when naïve B cells are activated and can be found as a membrane-bound B cell receptor (BCR) on naïve B cells together with IgD. Like all immunoglobulins, the basic secreted unit is a dimer of two identical heavy chains, each coupled to identical light chains. For IgM, five such units associate together with a Joining (J) chain forming a pentamer, which is a strong activator of the classical complement pathway [1]. Class switching from the initial IgM isotype allows the humoral immune system to engage with each antigen in a specific manner, with unique effector mechanisms being imprinted by each class (IgM, IgG, IgA, IgE, and IgD). Additionally, IgA and IgG are further subdivided in two and four subclasses, respectively (IgA1-2 and IgG1-4). Although the IgA subclasses seem to have similar if not identical effector functions, the abundance at different locations (serum/mucosa) is very different. The effector functions of IgG subclasses are very different and will be a major topic of this review.

During the onset of initial class switching in a given B cell any class switching event is theoretically possible from IgM to any other isotype. However, further sequential class switching events are dependent on the order of the Ig heavy chain constant genes on chromosome 14 (IgM, IgD, IgG3, IgG1, IgA1, IgG2, IgG4, IgE, and IgA2) [2]. This is because of genetic excisions of constant regions, e.g., the exons encoding for IgM, IgD, and IgG3 constant regions are deleted after a class switch event from

IgM to IgG1, preventing descendants of the proliferating B cell from generating IgG3. These class switching events of naïve B cells in the germinal center during clonal expansion are not completely random, but are regulated through signals received from T-helper cells and antigen presenting cells (APC). Cytokines produced by T-helper cells and signaling via toll-like receptors (TLR) on B cells initiate class switching of antigen specific B cells via activation-induced deaminase (AID) activity [3].

All the immunoglobulin isotypes have their own biodistribution, function and are often elicited upon specific triggers. IgD, for example, may be found in a secreted form, mostly in the tonsils, but its function remains enigmatic [4,5]. IgE is known to interact with mast cells to trigger the release of histamine mostly through the high affinity IgE-Fc Receptor I (FcεRI), but it also interacts with the atypical FcεRII (CD23), a c-type lectin. IgA has differential function depending on whether it is secretory IgA (SIgA) or serum IgA. SIgA is a dimeric form containing the J-chain (also found in IgM) that is associated with the extracellular domain of the polymeric Ig-Receptor (pIgR), which cleaves off after the transcytosis of dimeric IgA by the pIgR on epithelial cell of the mucosa [6]. Only serum IgA, which is monomeric and not associated with the J-chain, can bind and activate the myeloid IgA-receptor FcαRI efficiently and trigger a strong cellular response [7–9]. These isotypes—IgA, IgE, and IgD—generally do not activate complement, and therefore rely on other mechanisms to carry out their function [5,10,11]. Thus detailed discussion of these isotypes is beyond the scope of this chapter where we will focus on the biology of IgG subclasses.

2. Immunoglobulin G (IgG)

In the majority of humoral antibody responses, whether it is the protection against viral or cellular pathogens, IgG-mediated effector functions are involved. This includes humoral responses in allo- or autoimmune diseases. IgG1 is the most abundant antibody in human sera, followed by IgG2, IgG3, and IgG4 respectively [12]. Although the IgG subclasses are more than 90% identical on the amino acid level, each IgG subclass has a unique profile with respect to structure, antigen binding, immune complex formation, complement activation, triggering of FcγR, half-life, and placental transport [12] (Figure 1). IgG1, IgG3, and to some extent, IgG4 are generally formed against protein antigens, while IgG2 is the major subclass formed against repetitive T cell-independent polysaccharide structures found on encapsulated bacteria [13]. IgG3 is often the first subclass to form, which is followed by IgG1 responses that later dominate. The development of IgG4 responses is often the outcome of repeated or prolonged antigen exposure, although class switching from IgM expressing naïve B cells to IgG4 is possible [14]. The unusually weak CH3–CH3 interactions in the Fc domain of IgG4 and the redox sensitive disulfide bonds in the hinge of IgG4 facilitate exchange of two half-molecules (each consisting of one heavy and one light chain from a single IgG4 molecule), which enables the formation of bispecific IgG4 molecules [14–16]. For IgG2, two isoforms (IgG2 A/B) exist as a result of different disulfide bonding in the Fab and hinge domain, which determines the rigidity of the Fab domains when engaging antigen [17]. Functionally, IgG1 and IgG3 are strong inducers of Fc-mediated effector mechanisms, such as antibody-dependent cellular cytotoxicity (ADCC), complement dependent cytotoxicity (CDC), and antibody-dependent cellular phagocytosis (ADCP). IgG2 and IgG4, on the other hand, generally induce more subtle responses, although IgG2 has been shown to be quite capable of inducing good complement and Fc-receptor-mediated responses against epitopes of high-density such as polysaccharides [18,19]. This capacity of IgG2 may be related also to the peculiar rigidity of the hinge, shown to result in super agonistic antibodies and triggering strong signaling when targeting immune costimulatory receptors such as CD40 [20]. Below, the interaction of different ligands with human IgG subclasses is discussed as well as the effector functions triggered via these interactions.

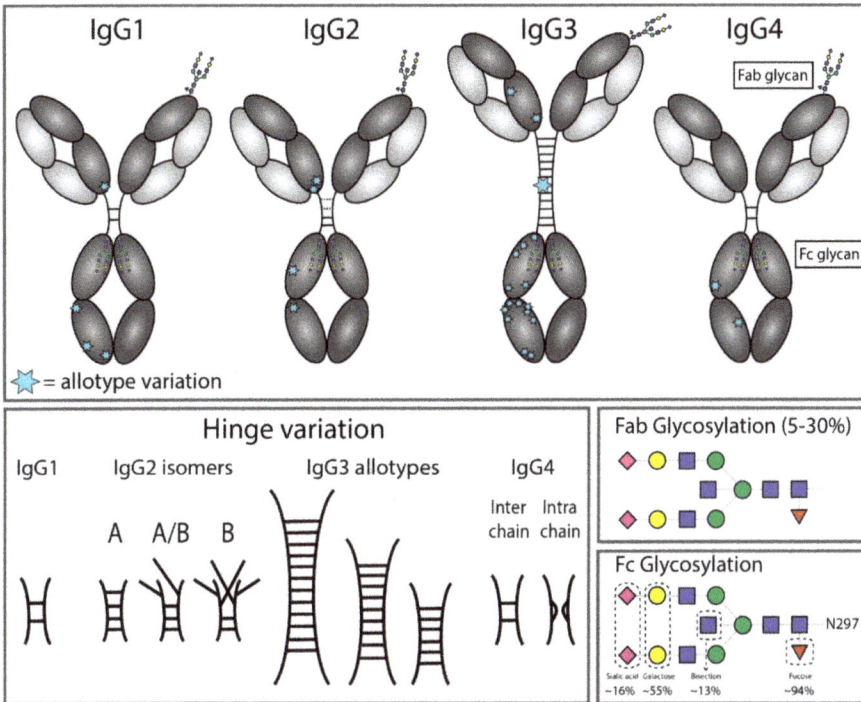

Figure 1. Structural variation of immunoglobulin G subclasses. A structural representation of the immunoglobulin G (IgG) subclasses and the variation within these subclasses, including allotype variation, hinge variation, and glycosylation. The variation originating from allotypic polymorphisms in the immunoglobulin heavy gamma (IGHG) constant domain is indicated by blue stars. Except for the star representing the variation in hinge length between IgG3 allotypes, each blue star indicates amino acid variation at one particular residue in the constant domain. Fab glycosylation is indicated and is present in 5–30% of antibodies in serum, depending on subclass and antigen specificity. The glycoform of the N297 Fc glycan is highly variable, for which the frequency of each glycan moiety on IgG antibodies in human serum is indicated.

3. IgG-Fc-Engaging Effector Molecules

The Fc domain of antibodies is the target for many proteins, including receptors on myeloid cells and thereby serves as a ligand for adaptor molecules (Figure 2). Many biological activities of antibodies are dependent on the interaction with these effector molecules, comprising Fc gamma receptors (FcγR) [21], two members of the Fc receptor-Like (FcRL) family (FcRL4 and FcRL5) [22], complement components (C1q) [23], neonatal Fc receptor (FcRn) [24], and Tripartite motif-containing protein 21 (TRIM21) [25].

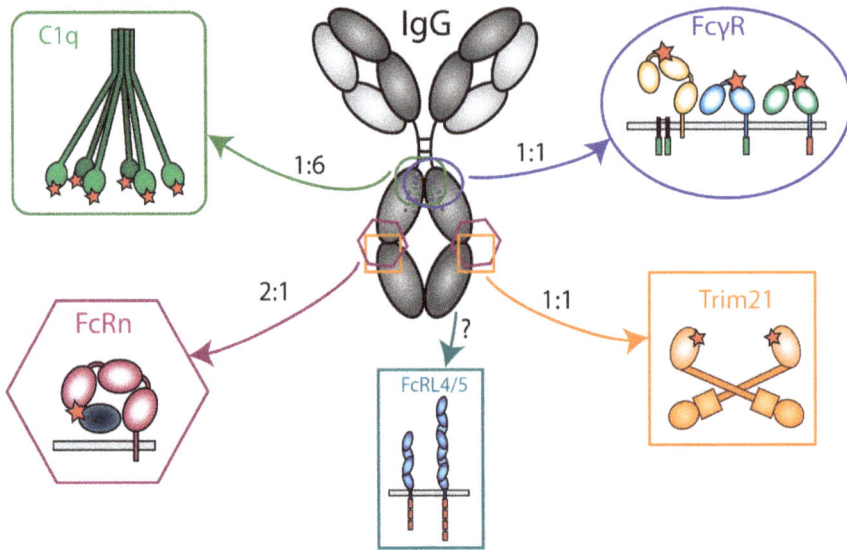

Figure 2. Interaction of IgG with Fc effector molecules. Schematic representation of IgG- and all Fc-engaging molecules (Complement component (C1q), Fc gamma receptors (FcγR), the Neonatal Fc receptor (FcRn), Tripartite motif 21 (Trim21), and Fc receptor-like (FcRL) molecules 4 and 5) through which antibodies exert their biological activity. For each ligand the binding site on IgG and the stoichiometry of the interaction with IgG is indicated. The red stars represent the binding site of IgG on the Fc effector molecules.

3.1. Fc-Receptors

Human myeloid, NK, and some lymphoid cells express Fc gamma receptors (FcγR), which sense antibody-opsonized particles and exert their specific effector mechanisms upon recognition and clustering of the Fc receptors [26]. Based on monovalent IgG:FcγR binding studies, FcγR were classified into high-affinity (FcγRIa) and low-affinity (FcγRIIa, IIb, IIc, IIIa, and IIIb) receptors [27]. This classification is somewhat of an oversimplification, as, for example, the affinity of IgG1 to FcγRIIIa can approach that of FcγRI depending of fucosylation in the IgG1-Fc (see below). These affinities also are not always indicative of their differential functionalities as it is the cross-linking of these Fc-receptors, brought about by engagement with immune complexes or opsonized pathogens with multiple IgG molecules, which enables the initiation of signaling. [28,29].

Most FcγR associate with an intracellular immunoreceptor tyrosine-based activation motif (ITAM), which is either directly found in the cytoplasmic domain (FcγRIIa and FcγRIIc-ORF) or through the associated FcRγ-chain (FcγRIa and FcγRIIIa). The exceptions are FcγRIIIb, which is GPI-linked, and FcγRIIb, which has an immunoreceptor tyrosine-based inhibition motif (ITIM). The latter is therefore the only receptor with inhibitory activity, and the only Fc receptor expressed on B cells [30]. The ratio of activating and inhibitory (A:I) FcγR expression on immune cells is thought to determine the antibody threshold necessary to activate the effector cell and induce ADCC or ADCP [31]. The FcγR expression pattern is highly variable between different immune cells. NK cells, for example, only express the low affinity FcγRIIIa receptor, while macrophages and monocytes express multiple receptors (FcγRIa, IIa, IIb, and IIIa) [26]. Depending on the Fc receptor a range of different effector functions can be triggered via the interaction with IgG, for example, binding of FcγRIIIa to IgG opsonized viruses or infected cells facilitates cross-linking of FcγRIIIa, which initiates ADCC of the target cell.

Fc gamma receptors bind the Fc domain of IgG in a 1:1 stoichiometry via interactions with the lower hinge (residues 234–238), the CH2 domain (residues 265, 297, 298, 299, and 329), and the N297 Fc glycan [32]. All FcγRs bind to IgG via their second extracellular domain, which shows great structural homology between the Fc receptors (root mean square deviation of atomic positions <1.0 Å). While all low-affinity receptors have two extracellular domains, the high-affinity receptor FcγRIa consists of three domains. The interaction of antibodies with various FcγRs is influenced by the IgG subclass. IgG1 and IgG3 bind efficiently to all Fc gamma receptors, contributing to their overall strong effector function profile [28]. The affinity of IgG4 for FcγRIa is two-fold lower compared to IgG1 and IgG3 (K_A 3.4×10^7 M^{-1}) [28]. IgG4 binds very weakly to the other FcγRs, which only leads to activation in situations where multivalency/avidity are involved [28]. IgG2 lacks a leucine at position 235 in the low hinge of Fc that is critical for binding to the high affinity receptor FcγRIa. This may therefore be an important reason why IgG2 does not bind FcγRIa [32]. Binding of IgG2 to FcγRIIa and FcγRIIIa is of low affinity (K_A of $10e^6$ and $10e^5$ M^{-1}, respectively), which is functionally relevant in the recognition of IgG2 immune complexes particularly through FcγRIIa [28,33]. Of note, IgG2, which is almost exclusively found as fucosylated species in humans, does show elevated binding to FcγRIIIa when afucosylated [33]. However, despite measurable binding to FcγRIIIa, afucosylated IgG2 only showed a slight albeit not significant increase in ADCC by NK cells (<5% killing) using IgG2-anti-Rhesus D opsonized RBC [33].

FcγR are highly polymorphic, thus their exact composition differs from person to person, and the ethnic makeup is also variable [21]. Not all FcγR-allotypic variation seems to have functional consequences, but a few polymorphisms are particularly noteworthy. Polymorphic variants of FcγRIIa (131H/R) and FcγRIIIa (158F/V) have different binding affinities to IgG. Thus, in contrast to monomeric IgG1 and IgG3, IgG2 has a particularly strong preference for FcγRIIa-131H compared to the FcγRIIa-131R variant [28,33]. The polymorphic variant FcγRIIIa 158V binds IgG1 and IgG3 with a 5-fold stronger affinity compared to FcγRIIIa 158F [33]. One allotypic variation results in the lack of expression of the FCGR2C gene, which is a pseudogene in most individuals and most ethnic groups [34]. However, in ~7–15% of individuals of European origin, FcγRIIc (FCGR2C-ORF) is expressed on some immune cells, including NK cells and perhaps B cells [35–37]. FcγRIIc expression depends on the presence of a single nucleotide polymorphism in exon 3 of FCGR2C, which normally encodes a stop codon (FCGR2C-Stop) [38]. Curiously, in some individuals with a FCGRIIC-STOP allele, FcγRIIb expression has been found on NK cells, which is normally absent for this cell type. Although the reason is unknown, this phenotype is accompanied by a genetic deletion of the FCGR2C and FCGR3B genes adjacent to FCGR2B on one chromosome, perhaps because this results in a net replacement of the FCGR2B promotor with the promotor of FCGR2C [36].

In line with the differences in interaction with the IgG subclasses, FcγR polymorphisms were found to correlate with IgG-dependent diseases, such as in allo- and autoimmunity [39–41], and with outcome of treatment in therapeutic antibody regimens that trigger FcγR for its therapeutic effect [21]. RA patients receiving rituximab treatment targeting CD20 on B cells generally respond better when bearing the higher-affinity FcγRIIIa 158V variant [42,43]. This advantage of expressing polymorphic variant FcγRIIIa 158V was less conclusive in other patient groups receiving rituximab, for example patients with non-Hodgkin's lymphoma [44,45]. In addition to single nucleotide polymorphisms in the FCGR-gene locus that alter interaction with the IgG Fc domain, copy number variation (CNV) also influences FcγR expression in a gene dosage fashion [36,46,47].

3.2. DC-SIGN and CD23

In addition to the type I Fc receptors that we discussed above, a second class of Fc receptors (type II) has been described to bind to the CH2:CH3 interface of sialylated IgG Fc [48,49]. These are the C-type lectin homologs DC-SIGN and CD23, of which the latter is the low-affinity receptor for IgE, which has been proposed to embody another group of Fc engaging molecules. The sialylated Fc fraction of IVIG was proposed to be responsible for the anti-inflammatory mechanism of IVIG, through binding of these type II Fc receptors. Structural studies of sialylated Fc revealed that the overall conformation is comparable to nonsialylated Fc, implying that the CH2:CH3 interface is not altered when the Fc-glycan is sialylated [50]. In fact, we and others found no detectable binding of human DC-SIGN and CD23 to human IgG Fc by FACS or SPR [51,52]. There is thus discrepancy between those reporting a biological effect of IgG through these receptors [48,49,53,54] and those finding no effect [50–52,55]. However, the overall consensus seems now to indicate that DC-SIGN and CD23 are not bona fide receptors for human IgG.

3.3. FcRn

The neonatal Fc receptor (FcRn) belongs to the family of MHC-related proteins and enables transcytosis of IgG over the maternal placenta. In addition, it mediates the long half-life of IgG. FcRn is rather ubiquitously expressed, including endothelial cells aligning our vessel walls, with a particularly high expression on myeloid cells [56]. Binding of IgG by FcRn occurs exclusively at low pH in endosomes (pH < 6.5) after IgG is engulfed into endosomes via pinocytosis. Thereafter, FcRn shuttles IgG back to the cell membrane, where IgG dissociates from FcRn after fusion with the surface at a pH of 7.4 in the extracellular fluid. This pH-dependent interaction results from the binding of FcRn with two histidines at the CH2:CH3 interface (H310 and H435), which become protonated at low pH (pKa around 7.4) [57,58]. This property of the IgG–FcRn interaction is critical to allow for pH-dependent FcRn-mediated shuttling of IgG. This increases the half-life of IgG1, IgG2, and IgG4 to ~21 days. Most IgG3 allotypes lack the histidine at position 435, and instead have an arginine (pKa ~12.5) at this position, which is always positively charged, independent of the physiological pH found in endosomes or extracellular medium. As a consequence, it is recycled to a much lesser extent than the other subclasses, and therefore has a much shorter half-life of ~7 days [59] and has less transport across the placenta [60,61]. However, some IgG3 allotypes express the histidine at position 435 (homologous to the other subclasses) and therefore have an increased half-life and transcytosis [59–61]. Interestingly, charge variations in the Fab domains of IgG have recently also been found to affect the half-life of IgG, suggesting that domains outside the binding site affect pinocytosis levels and/or in vivo FcRn binding kinetics [62,63]. In the proposed model, this effect may be due to the positively charged Fab domains affecting the interaction with the negatively charged cell membranes through the primary interaction site in the IgG-Fc.

3.4. TRIM21

Tripartite motif 21 (TRIM21) is an intracellular cytosolic IgG receptor which recognizes the CH2:CH3 interface, a binding site partly overlapping with that of FcRn. TRIM21 is a dimeric molecule, containing two PRYSPRY binding domains, which bind both CH2:CH3 elbow domains in a pincer-like interaction [25]. One PRYSPRY domain binds IgG with a reported affinity of 130nM [64,65]. However, the affinity of TRIM21 for IgG increases ~250-fold when dimeric TRIM21 binds both IgG heavy chain CH2:CH3 domains simultaneously (reported affinity being 0.6nM) [66]. A histidine at position 433 in the Fc domain of IgG is critical for binding of TRIM21 to the IgG Fc domain [67]. Entry of an IgG-opsonized non-enveloped virus or intracellular bacteria is recognized by TRIM21 and binding of TRIM21 triggers polyubiquitinylation of the opsonized particles and proteasomal degradation, as well as transcriptional activation of several immune regulator genes through the NF-κB, AP-1, and several IRF genes [25,67,68]. This process is described as antibody-dependent intracellular neutralization (ADIN), and extends the effector functions of antibodies to the intracellular compartment of cells. Whether strong immune signaling is initiated via activation of transcription factors, depends on the affinity of the antibody for the antigen/pathogen [67]. Viral particles opsonized with high affinity antibodies trigger transcriptional activation of immune regulators and production of cytokines [67]. So far, the evidence seem to suggest that early immune responses against cytosolic pathogens, probably non-enveloped viruses in particular, can be counteracted by the early immune response as TRIM21 also recognizes IgM and IgA [25]. Recognition of cytosolic immunoglobulins by TRIM21 may also function as a back-up mechanism in secondary infection for regular extracellular antibody mediated effector mechanisms when normal cellular compartmentalization has been compromised.

3.5. FcRL

Fc receptor-like (FcRL) molecules are part of the immunoglobulin superfamily (IgSF). In the human context, six transmembrane receptors—FcRL1–6—are described, consisting of three to nine extracellular Ig-like domains and several ITIM and/or ITAM signaling molecules intracellularly. These receptors are expressed on B cells, and the expression pattern of FcRLs varies during the different stages of B cell development [69]. In addition, FcRL3 and FcRL6 are also expressed on certain NK and T cell subsets [70–72]. The ligands of FcRL molecules and the processes in which signaling via FcRL is involved are not completely understood. However, FcRL4 (four extracellular immunoglobulin domains) and FcRL5 (eight extracellular immunoglobulins) were found to bind immunoglobulins with low affinity. FcRL4 specifically binds IgA, IgG3, and IgG4, while FcRL5 is able to bind all IgG subclasses [73,74]. The exact binding epitope of FcRL5 on IgG was not determined. However a complex binding interaction was proposed, in which both the Fc and Fab domains of IgG are involved in the interaction with FcRL5 [74]. Each receptor express several classical and atypical ITIM-motifs, as well as a potential ITAM-motif [75]. Both receptors are found on B cells suggesting a role in regulation of humoral immune responses [76]. One recent study suggested that FcRL5 can induce either inhibiting or activating signals to B cell receptor signaling pathways, depending on coligation with complement receptor 2 (CD21) [77]. This seems to suggest that complement C3 deposition may convert an inherent inhibitory signal to an activating signal and positively stimulate the humoral immune response. However, the exact mechanism and significance for human immune responses has not been elucidated.

3.6. Complement (C1q)

Binding of C1q to monomeric IgG requires clustering of IgG to establish a multivalent interaction platform. This platform resembles that of an IgM molecule, which consists mostly of a pentameric structure, but can also form a hexameric structure, in the absence of a Joining (J) chain, and binds C1q in a 1:1 stoichiometry when bound to antigen [1,78]. Recently, it has been determined that IgG bound to a membrane structure is able to assume a hexameric configuration in complex with C1 thereby initiating complement activation. The structural characterization of an IgG1 hexamer in complex with C1q unraveled some important features of this interaction [79]. Two major interaction sites determine binding of the globular head domain of C1q to IgG Fc. The BC loop (residues 266–272) and DE loop (294–300) of one CH2 domain are part of the first site and the FG loop (325–331) of the other CH2 domain is the second site [79]. Hexamer formation does not take place in solution to a measurable degree because the interactions between the individual IgG Fc tails are very weak. However, upon binding a cellular surface, multiple antibodies are confined into a limited space, favoring Fc–Fc interactions between these IgG molecules, which is expected to be even more favorable in the presence of C1q. IgG hexamer formation is impaired when C-terminal lysines on IgG enforce charge repulsion between IgG molecules. The C-terminal clipping of these lysines residues by plasma carboxypeptidase is necessary to facilitate efficient hexamer formation and complement activation by IgG [80]. IgG is produced in a proform and C1q binding is also modulated by the glycoform of the N297 Fc glycan, with increasing galactosylation stimulating better C1q binding and downstream activation (C3, C4 deposition, and CDC) [81–83]. Multivalent binding of IgG Fc condenses the C1q arms, which drives the rearrangements of the C1r2s2 proteases allowing catalytic activity of these complement components [79]. In this particular structure an IgG1 hexamer was studied, although a structurally similar principle should apply to IgG3 despite its significantly longer hinge. Binding and activation of C1q is in general believed to be more efficient with IgG3 compared to IgG1, although in certain cases CDC with IgG1 sometimes seems to outperform IgG3 [84–86]. The molecular reasons for this remain unresolved.

4. IgG Allotypes

Similar to FcγR, allotypic variations also exist for antibody heavy chains, especially for IgG. These polymorphisms form another layer of variation that may influence functional and structural features on top of what we know for the IgG subclasses [12]. Multiple studies found IgG polymorphisms to be associated with susceptibility to infectious diseases and autoimmune diseases, suggesting that IgG allotypes affect the humoral antibody response [87–92]. However, the structural or functional characteristics of IgG allotypes underlying these associations have not yet been elucidated.

The IgG allotypic background of an individual correlates with the individual IgG—subclass plasma levels [93,94]. A possible explanation may be the result of the formation of noncoding transcripts or RNA transcripts with unfavorable codon composition that impedes transcription and/or translation. In addition, IgG polymorphisms may also be associated with altered class switching efficiency, through variations within the noncoding switch regions, which would subsequently affect serum concentrations [93,95,96]. In a recent study, Shattock and colleagues showed that IgG1 allotypic variants were associated with the subclass distribution (IgG1/IgG2) of an HIV-specific antibody response, illustrating the association of IgG polymorphisms with the tendency for particular IgG subclass switching [97].

In addition to the association of IgG allotypes with antibody expression and IgG class switching in B cells, the Fc-mediated effector functions may be different between IgG allotypes. Previous studies already identified IgG3 allotypes with less stable CH3–CH3 interactions and an IgG4 allotype lacking the capacity to exchange half-molecules [16,59,98,99]. Furthermore, a particular IgG3 isoallotype (IMGT: IGHG3-17, -18, and -19) expressing a histidine at position 435 in the CH3 domain was found to improve pH-dependent binding to the neonatal Fc receptor (FcRn) and therefore showed an half-life that resembled that of IgG1 antibodies [59]. Also, infants of mothers carrying this IgG3 polymorphic variant were found to have an increased protection against malaria, since the malaria specific IgG3 antibodies crossed the placental membrane more efficiently as a result of increased binding to FcRn [60,61].

Future studies will shed light on the effector functions of the various IgG allotypes and the potential implications in susceptibility to infectious diseases or translation to antibody-based therapeutics. Interestingly, reactivity of monoclonal or polyclonal anti-IgG antibodies with all IgG allotypes indicated that some monoclonal antibodies, either subclass- or isotype-specific do not recognize all allotypic variants, a phenomenon described as 'serological blind spots' [100,101]. In addition, subclass-specific polyclonal anti-IgG were found to react with isoallotypic variants of another subclass, which could lead to misinterpretation of IgG subclass responses and can be of great importance, not only for scientific interpretations of immune responses, but also for critical diagnostic conclusion that leads to life-and-death decisions for patients [100].

5. IgG Glycosylation

Both heavy chains of the IgG express an N297-linked glycan in the CH2-domain in the Fc regions, which have a role in stability of the Fc domain and in the interaction with FcγR and possibly C1q [102]. During antibody production in the plasmablasts, the presence and activity of glycan-processing enzymes determines the composition of Fc-glycans, which results in a heterogeneous glycosylation pattern on antibodies.

Fucosylated complex glycans, with low-to-intermediate levels of galactosylation and low sialylation are most commonly found in serum IgG-Fc. Although not commonly found in plasma, afucosylated glycan-species of the N297 glycan (found in ~6% of plasma IgG) were previously found to increase the binding strength of IgG-Fc to FcγRIIIa [81,103,104]. This leads also to enhanced ADCC and has already been exploited in some therapeutic antibodies to improve their effector functions [105]. The molecular reason for the enhanced binding to the FcγRIII-family of receptors has been enigmatic and has still not fully been uncovered, although it is known to depend on a glycan found at position 162 only in human FcγRIII (a and b) (and conserved in other species, e.g., mice have FγRIV also with N162) [106]. Recent work based on structural modeling provided evidence that the number of conformations sampled by the N162-glycan is reduced by the presence of the fucose in the IgG Fc glycan [107,108]. This work suggests that the fucose moiety on the Fc glycan may be affecting the N162-glycan mobility which partially inhibits FcγRIIIa/b to engage in binding. Thus, without the fucose, the IgG-glycan can more effectively make room for effective FcγRIIIa/b binding. These notions are also supported by the fact that removing the N162 glycan of FcγRIIIa also increases the affinity to IgG which is no longer affected by the fucosylation status of the IgG [106–109].

In recent years glycoengineering has been utilized to develop next generation afucosylated therapeutic antibodies with enhanced ADCC activity. For example, an afucosylated anti-CD20 antibody has been approved to treat patients with B cell lymphomas [105]. In addition to fucosylation, galactosylation and sialylation of the N297 glycan were also described to modulate IgG effector functions [33,81–83,110]. Whereas galactosylation of IgG seems to increase complement activation, sialylation has both been reported to decrease CDC activity of rituximab-IgG anti-CD20, but in other cases to increase RBC lysis of anti-D IgG1 [81,83]. As the architecture of the immune complex formed by the anti-CD20 and anti-D may differ, this may offer an explanation for the differential outcome [111]. Both studies were carried out using normal human serum as complement source (albeit at different concentrations 5% and 10%). This may also offer a potential explanation as difference in serum composition (e.g., endogenous IgG glycosylation status and immune complex formation) may affect C1q binding to the intended target as we recently suggested [112].

Increased complement activation by IgG Fc galactosylation may also seem at odds with findings in several autoimmune diseases, where a low degree of galactosylation of total IgG-antibodies was found to be associated with disease progression [112]. Recently, we put forward a model that may explain how this is possible, taking into account the relative difference between the glycosylation of the pathogenic antibodies (in most cases not determined) and total IgG that may affect the threshold of immune activation [112]. This is because the bulk of aspecific IgG will always account for significant occupation of both FcγR but also partly C1q.

In addition to the conserved N-linked glycan in the Fc-domain [113], potential N-glycosylation sites are also present in the variable domain of antibodies. These Fab glycans have been described to modulate antibody stability, but also antigen binding directly [114,115]. As such, we recently postulated these sites to be a fundamental enhancement to the generation of antibody diversification—on top of VDJ recombination and somatic hypermutation leading to amino acid changes [114]. Fab glycosylation was found to be isotype and subclass-specific and associated with several autoimmune diseases including rheumatoid arthritis (RA) and primary Sjögren's syndrome [116–118]. Anticitrullinated protein antibodies (ACPA) are formed in the majority of RA patients and ACPA positive patients have an increased risk for rapid disease progression [119]. ACPA more frequently harbor Fab glycans compared to total serum IgG, suggesting that B cells producing auto-antibodies with Fab glycans are positively selected during affinity maturation [120]. In addition, the dominant Ig-producing cells in parotid glands of Primary Sjögren's syndrome were found to frequently express Fab glycans in the variable domain of the heavy chain [121]. In addition, it is possible that the existence of Fab-glycans may also affect regulation and affect the threshold for B cell activation through co-cross-linking of lectins in either cis or trans [122]. All in all, it is clear that both B cell biology, the humoral repertoire composition and the effector phase of antibodies is regulated through both Fab- and Fc-glycosylation.

6. Antibody Fc engineering

In recent years impressive progress has been made in the application of antibody-based therapeutics in various fields including B cell lymphomas, solid tumors and in autoimmune diseases. A popular strategy has been the generation of afucosylated antibody therapeutics to improve effector function of therapeutic antibodies for tumor immunotherapy. Both glycoengineering and protein engineering have rendered IgG Fc domains with enhanced binding to activating Fc receptors and reduced binding to inhibitory receptor FcγRIIb [81,123–126].

When the primary mode of action of a therapeutic antibody is to block the activity of a molecule, such as a proinflammatory cytokine, antibodies lacking Fc-mediated effector functions, 'Fc dead', may be desired to prevent activation and inflammation during treatment [126]. This can sometimes partly be achieved by generating antibodies with an IgG4 Fc domain, IgG2 Fc domain, combination of both or an N297-glycan deficient Fc domain. Even better is to engineer IgG-Fc variants that have no FcγR or C1q binding activity at all [127–129]. It should be cautioned that engineered antibodies may also express additional modes of action, e.g. afucosylated anti-TNF antibodies might have enhanced therapeutic potential in inflammatory bowel diseases. Although the exact mechanism is not known, it seems to require stimulation of wound healing macrophages through either TNF-anti-TNF- or membrane bound TNF-anti-TNF-complexes interacting with FcγRIIIa on CD206+ macrophages [130].

Another strategy to improve the efficacy of therapeutic antibodies is based on enhancing the half-life of antibodies by increasing the affinity for FcRn at low pH. Structure-based design of Fc fragments with improved affinity for FcRn at low pH was found to increase the half-life of therapeutic antibodies in vivo [131–133]. To stimulate the clearance of harmful auto-antibodies, FcRn blocking antibodies or Fc-fragments have been developed. Most of these antibodies bind with increased affinity to FcRn and in an pH independent fashion, thereby blocking the receptor for recycling of serum IgG [134–136]. Alternatively anti-FcRn antibodies are designed to bind to FcRn at the interaction site with IgG, blocking IgG recycling [137,138].

In addition to the structural details that determine the interaction of IgG with Fc engaging effector molecules, the context in which these interactions occur are similarly important and very relevant for the implication of antibody-based therapeutics. For example antigen density and mobility on the cell surface of target cells determine whether the bound antibodies can sufficiently trigger Fc gamma receptor cross-linking, which is a prerequisite for ADCC and ADCP. Furthermore the distance of the antigen from the cellular surface was found to be important in the initiation of effector mechanism by antibodies [139,140]. Antigen positioned close to the membrane allows for a stronger interaction between effector cell and target cell, which drives a more efficient ADCC or ADCP [139].

Beyond tweaking the interaction between IgG and Fc engaging molecules, Fc engineering has also been extended to the generation of bispecific antibodies. This has been realized by swapping one half (e.g., one pair of a heavy and a light chain) of a specific IgG antibody with another half of an IgG molecule with a different specificity. This allows the resulting molecule to bind two different antigens simultaneously and gives bispecific antibodies several advantages. This has for example enabled the application for antibody therapeutics including dual epitope targeting and recruitment of T cells to targeted malignant cells. Two examples of strategies to produce bispecific antibodies are controlled Fab arm exchange (cFAE) and the knob-into-hole (KIH) design [141–143]. The latter strategy is based on the coexpression of two antibodies, one with a knob (bulky amino acid) and one with a hole in the CH3:CH3 interface. Coexpression of the knob and hole heavy chains with a common light chain followed by protA affinity chromatography leads to 95% heterodimerization efficiency [141,142]. The other strategy (cFAE) is also known as the DuoBody platform [143]. For each application of a bispecific antibody, whether that is cancer immunotherapy or neutralizing infectious agents, the desired features are different. This is why many different bispecific antibodies are developed based on full IgG (bsIgG) or single chain/variable domain only (scFv) [144,145].

In conclusion, antibodies come in all shapes and sizes and interact with a variety of ligands to mediate effector functions. For potential protection or therapeutic applications, the appropriate format that fits the target is likely to be of utmost importance. This is further complicated by the presence of regulatory inhibitory molecules/receptor–ligand pairs found in the immunological synapse regulating myeloid and NK cell activities [146]. For prophylactic immunotherapies with antibodies, there other factors that are important to consider. This may be especially in the tumor microenvironment where checkpoint receptor–ligand receptor pairs, which can be anti-inflammatory, must be overcome before therapeutic antibodies can be of beneficial value. This can be achieved by applying more potent engineered antibodies and/or by applying a combination of antibodies targeting both tumors and checkpoint inhibitors. This will affect both myeloid and lymphoid regulatory cells and secretion profiles stimulating or inhibiting cytokines [147,148]. The variety in antibodies in terms of isotype, allotype, subclass, glycosylation profile, and specificity, together with the number of Fc engaging molecules expressed on immune cells through which effector functions are exerted, illustrate the complexity and plasticity of the antibody response. Elucidating the interactions of antibodies with Fc engaging molecules is of crucial importance in the development of antibody therapeutics.

Author Contributions: S.W.d.T. wrote the paper and designed the figures, G.V. wrote and edited the paper, T.R. edited the paper.

Funding: This research received no external funding.

Conflicts of Interest: The authors declare no conflict of interest.

References

1. Hiramoto, E.; Tsutsumi, A.; Suzuki, R.; Matsuoka, S.; Arai, S.; Kikkawa, M.; Miyazaki, T. The IgM pentamer is an asymmetric pentagon with an open groove that binds the AIM protein. *Sci. Adv.* **2018**, *4*, eaau1199. [CrossRef]

2. Berkowska, M.A.; Driessen, G.J.A.; Bikos, V.; Grosserichter-Wagener, C.; Stamatopoulos, K.; Cerutti, A.; He, B.; Biermann, K.; Lange, J.F.; van der Burg, M.; et al. Human memory B cells originate from three distinct germinal center-dependent and -independent maturation pathways. *Blood* **2011**, *118*, 2150–2158. [CrossRef]

3. Methot, S.P.; Di Noia, J.M. Molecular Mechanisms of Somatic Hypermutation and Class Switch Recombination. *Adv. Immunol.* **2017**, *133*, 37–87. [PubMed]

4. Shan, M.; Carrillo, J.; Yeste, A.; Gutzeit, C.; Segura-Garzón, D.; Walland, A.C.; Pybus, M.; Grasset, E.K.; Yeiser, J.R.; Matthews, D.B.; et al. Secreted IgD Amplifies Humoral T Helper 2 Cell Responses by Binding Basophils via Galectin-9 and CD44. *Immunity* **2018**, *49*, 709–724. [CrossRef] [PubMed]

5. Gutzeit, C.; Chen, K.; Cerutti, A. The enigmatic function of IgD: some answers at last. *Eur. J. Immunol.* **2018**, *48*, 1101–1113. [CrossRef]

6. Horton, R.E.; Vidarsson, G. Antibodies and Their Receptors: Different Potential Roles in Mucosal Defense. *Front. Immunol.* **2013**, *4*, 200. [CrossRef] [PubMed]

7. Herr, A.B.; Ballister, E.R.; Bjorkman, P.J. Insights into IgA-mediated immune responses from the crystal structures of human FcαRI and its complex with IgA1-Fc. *Nature* **2003**, *423*, 614–620. [CrossRef]

8. van Egmond, M.; van Garderen, E.; van Spriel, A.B.; Damen, C.A.; van Amersfoort, E.S.; van Zandbergen, G.; van Hattum, J.; Kuiper, J.; van de Winkel, J.G.J. FcαRI-positive liver Kupffer cells: Reappraisal of the function of immunoglobulin A in immunity. *Nat. Med.* **2000**, *6*, 680–685. [CrossRef]

9. Vidarsson, G.; van Der Pol, W.L.; van Den Elsen, J.M.; Vilé, H.; Jansen, M.; Duijs, J.; Morton, H.C.; Boel, E.; Daha, M.R.; Corthésy, B.; et al. Activity of human IgG and IgA subclasses in immune defense against Neisseria meningitidis serogroup B. *J. Immunol.* **2001**, *166*, 6250–6256. [CrossRef] [PubMed]

10. Sutton, B.J.; Davies, A.M. Structure and dynamics of IgE-receptor interactions: FcεRI and CD23/FcεRII. *Immunol. Rev.* **2015**, *268*, 222–235. [CrossRef] [PubMed]

11. Bunker, J.J.; Bendelac, A. IgA Responses to Microbiota. *Immunity* **2018**, *49*, 211–224. [CrossRef] [PubMed]

12. Vidarsson, G.; Dekkers, G.; Rispens, T. IgG subclasses and allotypes: from structure to effector functions. *Front. Immunol.* **2014**, *5*, 520. [CrossRef] [PubMed]

13. Ferrante, A.; Beard, L.J.; Feldman, R.G. IgG subclass distribution of antibodies to bacterial and viral antigens. *Pediatr. Infect. Dis. J.* **1990**, *9*, 516–524. [CrossRef] [PubMed]

14. Lighaam, L.; Rispens, T. The Immunobiology of Immunoglobulin G4. *Semin. Liver Dis.* **2016**, *36*, 200–215. [CrossRef] [PubMed]

15. van der Neut Kolfschoten, M.; Schuurman, J.; Losen, M.; Bleeker, W.K.; Martínez-Martínez, P.; Vermeulen, E.; den Bleker, T.H.; Wiegman, L.; Vink, T.; Aarden, L.A.; et al. Anti-inflammatory activity of immunoglobulin G resulting from Fc sialylation. *Science* **2007**, *313*, 670–673.

16. Rispens, T.; Davies, A.M.; Ooijevaar-de Heer, P.; Absalah, S.; Bende, O.; Sutton, B.J.; Vidarsson, G.; Aalberse, R.C. Dynamics of inter-heavy chain interactions in human immunoglobulin G (IgG) subclasses studied by kinetic Fab arm exchange. *J. Biol. Chem.* **2014**, *289*, 6098–6109. [CrossRef]

17. Wypych, J.; Li, M.; Guo, A.; Zhang, Z.; Martinez, T.; Allen, M.J.; Fodor, S.; Kelner, D.N.; Flynn, G.C.; Liu, Y.D.; et al. Human IgG2 antibodies display disulfide-mediated structural isoforms. *J. Biol. Chem.* **2008**, *283*, 16194–16205. [CrossRef]

18. Barrett, D.J.; Ayoub, E.M. IgG2 subclass restriction of antibody to pneumococcal polysaccharides. *Clin. Exp. Immunol.* **1986**, *63*, 127–134.

19. Saeland, E.; Vidarsson, G.; Leusen, J.H.W.; Van Garderen, E.; Nahm, M.H.; Vile-Weekhout, H.; Walraven, V.; Stemerding, A.M.; Verbeek, J.S.; Rijkers, G.T.; et al. Central role of complement in passive protection by human IgG1 and IgG2 anti-pneumococcal antibodies in mice. *J. Immunol.* **2003**, *170*, 6158–6164. [CrossRef]

20. White, A.L.; Chan, H.T.C.; French, R.R.; Willoughby, J.; Mockridge, C.I.; Roghanian, A.; Penfold, C.A.; Booth, S.G.; Dodhy, A.; Polak, M.E.; et al. Conformation of the Human Immunoglobulin G2 Hinge Imparts Superagonistic Properties to Immunostimulatory Anticancer Antibodies. *Cancer Cell* **2015**, *27*, 138–148. [CrossRef]

21. Hargreaves, C.E.; Rose-Zerilli, M.J.J.; Machado, L.R.; Iriyama, C.; Hollox, E.J.; Cragg, M.S.; Strefford, J.C. Fcγ receptors: genetic variation, function, and disease. *Immunol. Rev.* **2015**, *268*, 6–24. [CrossRef] [PubMed]

22. Li, F.J.; Won, W.J.; Becker, E.J.; Easlick, J.L.; Tabengwa, E.M.; Li, R.; Shakhmatov, M.; Honjo, K.; Burrows, P.D.; Davis, R.S. Emerging Roles for the FCRL Family Members in Lymphocyte Biology and Disease. *In Curr. Top. Microbiol. Immunol.* **2014**, *382*, 29–50.

23. Lu, J.; Kishore, U. C1 Complex: An Adaptable Proteolytic Module for Complement and Non-Complement Functions. *Front. Immunol.* **2017**, *8*, 592. [CrossRef] [PubMed]

24. Stapleton, N.M.; Einarsdóttir, H.K.; Stemerding, A.M.; Vidarsson, G. The multiple facets of FcRn in immunity. *Immunol. Rev.* **2015**, *268*, 253–268. [CrossRef]

25. Foss, S.; Watkinson, R.; Sandlie, I.; James, L.C.; Andersen, J.T. TRIM21: a cytosolic Fc receptor with broad antibody isotype specificity. *Immunol. Rev.* **2015**, *268*, 328–339. [CrossRef] [PubMed]

26. Gillis, C.; Gouel-Chéron, A.; Jönsson, F.; Bruhns, P. Contribution of Human FcγRs to Disease with Evidence from Human Polymorphisms and Transgenic Animal Studies. *Front. Immunol.* **2014**, *5*, 254. [CrossRef] [PubMed]

27. Daëron, M. F c RECEPTOR BIOLOGY. *Annu. Rev. Immunol.* **1997**, *15*, 203–234. [CrossRef]

28. Bruhns, P.; Iannascoli, B.; England, P.; Mancardi, D.A.; Fernandez, N.; Jorieux, S.; Daëron, M. Specificity and affinity of human Fcgamma receptors and their polymorphic variants for human IgG subclasses. *Blood* **2009**, *113*, 3716–3725. [CrossRef]

29. Vidarsson, G.; van de Winkel, J.G. Fc receptor and complement receptor-mediated phagocytosis in host defence. *Curr. Opin. Infect. Dis.* **1998**, *11*, 271–278. [CrossRef]

30. Veri, M.-C.; Gorlatov, S.; Li, H.; Burke, S.; Johnson, S.; Stavenhagen, J.; Stein, K.E.; Bonvini, E.; Koenig, S. Monoclonal antibodies capable of discriminating the human inhibitory Fc?-receptor IIB (CD32B) from the activating Fc?-receptor IIA (CD32A): biochemical, biological and functional characterization. *Immunology* **2007**, *121*, 392–404. [CrossRef]

31. Nimmerjahn, F.; Ravetch, J. V Divergent Immunoglobulin G Subclass Activity Through Selective Fc Receptor Binding. *Science* **2005**, *310*, 1510–1512. [CrossRef]

32. Caaveiro, J.M.M.; Kiyoshi, M.; Tsumoto, K. Structural analysis of Fc/FcγR complexes: a blueprint for antibody design. *Immunol. Rev.* **2015**, *268*, 201–221. [CrossRef]

33. Bruggeman, C.W.; Dekkers, G.; Bentlage, A.E.H.; Treffers, L.W.; Nagelkerke, S.Q.; Lissenberg-Thunnissen, S.; Koeleman, C.A.M.; Wuhrer, M.; van den Berg, T.K.; Rispens, T.; et al. Enhanced effector functions due to antibody defucosylation depend on the effector cell Fcγ receptor profile. *J. Immunol.* **2017**, *199*, 204–211. [CrossRef]

34. Treffers, L.W.; van Houdt, M.; Bruggeman, C.W.; Heineke, M.H.; Zhao, X.W.; van der Heijden, J.; Nagelkerke, S.Q.; Verkuijlen, P.J.J.H.; Geissler, J.; Lissenberg-Thunnissen, S.; et al. FcγRIIIb Restricts Antibody-Dependent Destruction of Cancer Cells by Human Neutrophils. *Front. Immunol.* **2019**, *9*, 3124. [CrossRef]

35. Meinderts, S.M.; Sins, J.W.R.; Fijnvandraat, K.; Nagelkerke, S.Q.; Geissler, J.; Tanck, M.W.; Bruggeman, C.; Biemond, B.J.; Rijneveld, A.W.; Kerkhoffs, J.-L.H.; et al. Nonclassical *FCGR2C* haplotype is associated with protection from red blood cell alloimmunization in sickle cell disease. *Blood* **2017**, *130*, 2121–2130. [CrossRef]

36. van der Heijden, J.; Breunis, W.B.; Geissler, J.; de Boer, M.; van den Berg, T.K.; Kuijpers, T.W. Phenotypic Variation in IgG Receptors by Nonclassical FCGR2C Alleles. *J. Immunol.* **2012**, *188*, 1318–1324. [CrossRef]

37. Li, X.; Wu, J.; Ptacek, T.; Redden, D.T.; Brown, E.E.; Alarcón, G.S.; Ramsey-Goldman, R.; Petri, M.A.; Reveille, J.D.; Kaslow, R.A.; et al. Allelic-Dependent Expression of an Activating Fc Receptor on B Cells Enhances Humoral Immune Responses. *Sci. Transl. Med.* **2013**, *5*, 216ra175. [CrossRef]

38. Metes, D.; Ernst, L.K.; Chambers, W.H.; Sulica, A.; Herberman, R.B.; Morel, P.A. Expression of functional CD32 molecules on human NK cells is determined by an allelic polymorphism of the FcgammaRIIC gene. *Blood* **1998**, *91*, 2369–2380.

39. Breunis, W.B.; van Mirre, E.; Bruin, M.; Geissler, J.; de Boer, M.; Peters, M.; Roos, D.; de Haas, M.; Koene, H.R.; Kuijpers, T.W. Copy number variation of the activating FCGR2C gene predisposes to idiopathic thrombocytopenic purpura. *Blood* **2008**, *111*, 1029–1038. [CrossRef]

40. Stegmann, T.C.; Veldhuisen, B.; Nagelkerke, S.Q.; Winkelhorst, D.; Schonewille, H.; Verduin, E.P.; Kuijpers, T.W.; de Haas, M.; Vidarsson, G.; van der Schoot, C.E. RhIg-prophylaxis is not influenced by FCGR2/3 polymorphisms involved in red blood cell clearance. *Blood* **2017**, *129*, 1045–1048. [CrossRef]

41. Miescher, S.; Spycher, M.O.; Amstutz, H.; de Haas, M.; Kleijer, M.; Kalus, U.J.; Radtke, H.; Hubsch, A.; Andresen, I.; Martin, R.M.; et al. A single recombinant anti-RhD IgG prevents RhD immunization: association of RhD-positive red blood cell clearance rate with polymorphisms in the FcγRIIA and FcγIIIA genes. *Blood* **2004**, *103*, 4028–4035. [CrossRef]

42. Ruyssen-Witrand, A.; Rouanet, S.; Combe, B.; Dougados, M.; Le Loët, X.; Sibilia, J.; Tebib, J.; Mariette, X.; Constantin, A. Fcγ receptor type IIIA polymorphism influences treatment outcomes in patients with rheumatoid arthritis treated with rituximab. *Ann. Rheum. Dis.* **2012**, *71*, 875–877. [CrossRef]

43. Lee, Y.H.; Bae, S.-C.; Song, G.G. Functional FCGR3A 158 V/F and IL-6 −174 C/G polymorphisms predict response to biologic therapy in patients with rheumatoid arthritis: a meta-analysis. *Rheumatol. Int.* **2014**, *34*, 1409–1415. [CrossRef]

44. Liu, D.; Tian, Y.; Sun, D.; Sun, H.; Jin, Y.; Dong, M. The FCGR3A polymorphism predicts the response to rituximab-based therapy in patients with non-Hodgkin lymphoma: a meta-analysis. *Ann. Hematol.* **2016**, *95*, 1483–1490. [CrossRef]

45. Carlotti, E.; Palumbo, G.A.; Oldani, E.; Tibullo, D.; Salmoiraghi, S.; Rossi, A.; Golay, J.; Pulsoni, A.; Foà, R.; Rambaldi, A. FcgammaRIIIA and FcgammaRIIA polymorphisms do not predict clinical outcome of follicular non-Hodgkin's lymphoma patients treated with sequential CHOP and rituximab. *Haematologica* **2007**, *92*, 1127–1130. [CrossRef]

46. van der Heijden, J.; Nagelkerke, S.; Zhao, X.; Geissler, J.; Rispens, T.; van den Berg, T.K.; Kuijpers, T.W. Haplotypes of Fc RIIa and Fc RIIIb Polymorphic Variants Influence IgG-Mediated Responses in Neutrophils. *J. Immunol.* **2014**, *192*, 2715–2721. [CrossRef]

47. Treffers, L.W.; Zhao, X.W.; van der Heijden, J.; Nagelkerke, S.Q.; van Rees, D.J.; Gonzalez, P.; Geissler, J.; Verkuijlen, P.; van Houdt, M.; de Boer, M.; et al. Genetic variation of human neutrophil Fcγ receptors and SIRPα in antibody-dependent cellular cytotoxicity towards cancer cells. *Eur. J. Immunol.* **2018**, *48*, 344–354. [CrossRef]

48. Sondermann, P.; Pincetic, A.; Maamary, J.; Lammens, K.; Ravetch, J.V. General mechanism for modulating immunoglobulin effector function. *Proc. Natl. Acad. Sci. USA* **2013**, *110*, 9868–9872. [CrossRef]

49. Anthony, R.M.; Wermeling, F.; Karlsson, M.C.I.; Ravetch, J.V. Identification of a receptor required for the anti-inflammatory activity of IVIG. *Proc. Natl. Acad. Sci. USA* **2008**, *105*, 19571–19578. [CrossRef]

50. Crispin, M.; Yu, X.; Bowden, T.A. Crystal structure of sialylated IgG Fc: Implications for the mechanism of intravenous immunoglobulin therapy. *Proc. Natl. Acad. Sci. USA* **2013**, *110*, E3544–E3546. [CrossRef]

51. Yu, X.; Vasiljevic, S.; Mitchell, D.A.; Crispin, M.; Scanlan, C.N. Dissecting the Molecular Mechanism of IVIg Therapy: The Interaction between Serum IgG and DC-SIGN is Independent of Antibody Glycoform or Fc Domain. *J. Mol. Biol.* **2013**, *425*, 1253–1258. [CrossRef]

52. Temming, A.R.; Dekkers, G.; van de Bovenkamp, F.S.; Plomp, H.R.; Bentlage, A.E.H.; Szittner, Z.; Derksen, N.I.L.; Wuhrer, M.; Rispens, T.; Vidarsson, G. Human DC-SIGN and CD23 do not interact with human IgG. Submitted.

53. Anthony, R.M.; Wermeling, F.; Ravetch, J.V. Novel roles for the IgG Fc glycan. *Ann. N. Y. Acad. Sci.* **2012**, *1253*, 170–180. [CrossRef]

54. Ahmed, A.A.; Giddens, J.; Pincetic, A.; Lomino, J.V.; Ravetch, J.V.; Wang, L.-X.; Bjorkman, P.J. Structural Characterization of Anti-Inflammatory Immunoglobulin G Fc Proteins. *J. Mol. Biol.* **2014**, *426*, 3166–3179. [CrossRef]

55. Guhr, T.; Bloem, J.; Derksen, N.I.L.; Wuhrer, M.; Koenderman, A.H.L.; Aalberse, R.C.; Rispens, T. Enrichment of sialylated IgG by lectin fractionation does not enhance the efficacy of immunoglobulin G in a murine model of immune thrombocytopenia. *PLoS One* **2011**, *6*, e21246. [CrossRef]

56. Vidarsson, G.; Stemerding, A.M.; Stapleton, N.M.; Spliethoff, S.E.; Janssen, H.; Rebers, F.E.; de Haas, M.; van de Winkel, J.G. FcRn: an IgG receptor on phagocytes with a novel role in phagocytosis. *Blood* **2006**, *108*, 3573–3579. [CrossRef]

57. Shimizu, A.; Honzawa, M.; Ito, S.; Miyazaki, T.; Matsumoto, H.; Nakamura, H.; Michaelsen, T.E.; Arata, Y. H NMR studies of the Fc region of human IgG1 and IgG3 immunoglobulins: assignment of histidine resonances in the CH3 domain and identification of IgG3 protein carrying G3m(st) allotypes. *Mol. Immunol.* **1983**, *20*, 141–148. [CrossRef]

58. Vaughn, D.E.; Bjorkman, P.J. Structural basis of pH-dependent antibody binding by the neonatal Fc receptor. *Structure* **1998**, *6*, 63–73. [CrossRef]

59. Stapleton, N.M.; Andersen, J.T.; Stemerding, A.M.; Bjarnarson, S.P.; Verheul, R.C.; Gerritsen, J.; Zhao, Y.; Kleijer, M.; Sandlie, I.; de Haas, M.; et al. Competition for FcRn-mediated transport gives rise to short half-life of human IgG3 and offers therapeutic potential. *Nat. Commun.* **2011**, *2*, 599. [CrossRef]

60. Einarsdottir, H.; Ji, Y.; Visser, R.; Mo, C.; Luo, G.; Scherjon, S.; van der Schoot, C.E.; Vidarsson, G. H435-containing immunoglobulin G3 allotypes are transported efficiently across the human placenta: implications for alloantibody-mediated diseases of the newborn. *Transfusion* **2014**, *54*, 665–671. [CrossRef]

61. Dechavanne, C.; Dechavanne, S.; Sadissou, I.; Lokossou, A.G.; Alvarado, F.; Dambrun, M.; Moutairou, K.; Courtin, D.; Nuel, G.; Garcia, A.; et al. Associations between an IgG3 polymorphism in the binding domain for FcRn, transplacental transfer of malaria-specific IgG3, and protection against Plasmodium falciparum malaria during infancy: A birth cohort study in Benin. *PLoS Med.* **2017**, *14*, e1002403. [CrossRef]

62. Schoch, A.; Kettenberger, H.; Mundigl, O.; Winter, G.; Engert, J.; Heinrich, J.; Emrich, T. Charge-mediated influence of the antibody variable domain on FcRn-dependent pharmacokinetics. *Proc. Natl. Acad. Sci. USA* **2015**, *112*, 5997–6002. [CrossRef]

63. Monnet, C.; Jorieux, S.; Urbain, R.; Fournier, N.; Bouayadi, K.; De Romeuf, C.; Behrens, C.K.; Fontayne, A.; Mondon, P. Selection of IgG Variants with Increased FcRn Binding Using Random and Directed Mutagenesis: Impact on Effector Functions. *Front. Immunol.* **2015**, *6*, 39. [CrossRef]

64. James, L.C.; Keeble, A.H.; Khan, Z.; Rhodes, D.A.; Trowsdale, J. Structural basis for PRYSPRY-mediated tripartite motif (TRIM) protein function. *Proc. Natl. Acad. Sci. USA* **2007**, *104*, 6200–6205. [CrossRef]

65. Keeble, A.H.; Khan, Z.; Forster, A.; James, L.C. TRIM21 is an IgG receptor that is structurally, thermodynamically, and kinetically conserved. *Proc. Natl. Acad. Sci. USA* **2008**, *105*, 6045–6050. [CrossRef]

66. Mallery, D.L.; McEwan, W.A.; Bidgood, S.R.; Towers, G.J.; Johnson, C.M.; James, L.C. Antibodies mediate intracellular immunity through tripartite motif-containing 21 (TRIM21). *Proc. Natl. Acad. Sci. USA* **2010**, *107*, 19985–19990. [CrossRef]

67. Foss, S.; Watkinson, R.E.; Grevys, A.; McAdam, M.B.; Bern, M.; Høydahl, L.S.; Dalhus, B.; Michaelsen, T.E.; Sandlie, I.; James, L.C.; et al. TRIM21 Immune Signaling Is More Sensitive to Antibody Affinity Than Its Neutralization Activity. *J. Immunol.* **2016**, *196*, 3452–3459. [CrossRef]

68. Fletcher, A.J.; Mallery, D.L.; Watkinson, R.E.; Dickson, C.F.; James, L.C. Sequential ubiquitination and deubiquitination enzymes synchronize the dual sensor and effector functions of TRIM21. *Proc. Natl. Acad. Sci. USA* **2015**, *112*, 10014–10019. [CrossRef]

69. Rostamzadeh, D.; Kazemi, T.; Amirghofran, Z.; Shabani, M. Update on Fc receptor-like (FCRL) family: new immunoregulatory players in health and diseases. *Expert Opin. Ther. Targets* **2018**, *22*, 487–502. [CrossRef]

70. Swainson, L.A.; Mold, J.E.; Bajpai, U.D.; McCune, J.M. Expression of the Autoimmune Susceptibility Gene FcRL3 on Human Regulatory T Cells Is Associated with Dysfunction and High Levels of Programmed Cell Death-1. *J. Immunol.* **2010**, *184*, 3639–3647. [CrossRef]

71. Bin Dhuban, K.; d'Hennezel, E.; Nashi, E.; Bar-Or, A.; Rieder, S.; Shevach, E.M.; Nagata, S.; Piccirillo, C.A. Coexpression of TIGIT and FCRL3 Identifies Helios + Human Memory Regulatory T Cells. *J. Immunol.* **2015**, *194*, 3687–3696. [CrossRef]

72. Kulemzin, S.V.; Zamoshnikova, A.Y.; Yurchenko, M.Y.; Vitak, N.Y.; Najakshin, A.M.; Fayngerts, S.A.; Chikaev, N.A.; Reshetnikova, E.S.; Kashirina, N.M.; Peclo, M.M.; et al. FCRL6 receptor: Expression and associated proteins. *Immunol. Lett.* **2011**, *134*, 174–182. [CrossRef]

73. Wilson, T.J.; Fuchs, A.; Colonna, M. Cutting Edge: Human FcRL4 and FcRL5 Are Receptors for IgA and IgG. *J. Immunol.* **2012**, *188*, 4741–4745. [CrossRef]

74. Franco, A.; Damdinsuren, B.; Ise, T.; Dement-Brown, J.; Li, H.; Nagata, S.; Tolnay, M. Human Fc receptor-like 5 binds intact IgG via mechanisms distinct from those of Fc receptors. *J. Immunol.* **2013**, *190*, 5739–5746. [CrossRef]

75. Miller, I.; Hatzivassiliou, G.; Cattoretti, G.; Mendelsohn, C.; Dalla-Favera, R. IRTAs: a new family of immunoglobulinlike receptors differentially expressed in B cells. *Blood* **2002**, *99*, 2662–2669. [CrossRef]

76. Haga, C.L.; Ehrhardt, G.R.A.; Boohaker, R.J.; Davis, R.S.; Cooper, M.D. Fc receptor-like 5 inhibits B cell activation via SHP-1 tyrosine phosphatase recruitment. *Proc. Natl. Acad. Sci. USA* **2007**, *104*, 9770–9775. [CrossRef]

77. Franco, A.; Kraus, Z.; Li, H.; Seibert, N.; Dement-Brown, J.; Tolnay, M. CD21 and FCRL5 form a receptor complex with robust B-cell activating capacity. *Int. Immunol.* **2018**, *30*, 569–578. [CrossRef]

78. Niles, M.J.; Matsuuchi, L.; Koshland, M.E. Polymer IgM assembly and secretion in lymphoid and nonlymphoid cell lines: evidence that J chain is required for pentamer IgM synthesis. *Proc. Natl. Acad. Sci. USA* **1995**, *92*, 2884–2888. [CrossRef]

79. Ugurlar, D.; Howes, S.C.; de Kreuk, B.-J.; Koning, R.I.; de Jong, R.N.; Beurskens, F.J.; Schuurman, J.; Koster, A.J.; Sharp, T.H.; Parren, P.W.H.I.; et al. Structures of C1-IgG1 provide insights into how danger pattern recognition activates complement. *Science* **2018**, *359*, 794–797. [CrossRef]

80. van den Bremer, E.T.; Beurskens, F.J.; Voorhorst, M.; Engelberts, P.J.; de Jong, R.N.; van der Boom, B.G.; Cook, E.M.; Lindorfer, M.A.; Taylor, R.P.; van Berkel, P.H.; et al. Human IgG is produced in a pro-form that requires clipping of C-terminal lysines for maximal complement activation. *MAbs* **2015**, *7*, 672–680. [CrossRef]

81. Dekkers, G.; Treffers, L.; Plomp, R.; Bentlage, A.E.H.; de Boer, M.; Koeleman, C.A.M.; Lissenberg-Thunnissen, S.N.; Visser, R.; Brouwer, M.; Mok, J.Y.; et al. Decoding the human immunoglobulin G-glycan repertoire reveals a spectrum of Fc-receptor- and complement-mediated-effector activities. *Front. Immunol.* **2017**, *8*. [CrossRef]

82. Quast, I.; Keller, C.W.; Maurer, M.A.; Giddens, J.P.; Tackenberg, B.; Wang, L.-X.; Münz, C.; Nimmerjahn, F.; Dalakas, M.C.; Lünemann, J.D. Sialylation of IgG Fc domain impairs complement-dependent cytotoxicity. *J. Clin. Invest.* **2015**, *125*, 4160–4170. [CrossRef]

83. Peschke, B.; Keller, C.W.; Weber, P.; Quast, I.; Lünemann, J.D. Fc-Galactosylation of Human Immunoglobulin Gamma Isotypes Improves C1q Binding and Enhances Complement-Dependent Cytotoxicity. *Front. Immunol.* **2017**, *8*, 646. [CrossRef]

84. Brüggemann, M.; Williams, G.T.; Bindon, C.I.; Clark, M.R.; Walker, M.R.; Jefferis, R.; Waldmann, H.; Neuberger, M.S. Comparison of the effector functions of human immunoglobulins using a matched set of chimeric antibodies. *J. Exp. Med.* **1987**, *166*, 1351–1361. [CrossRef]

85. Natsume, A.; In, M.; Takamura, H.; Nakagawa, T.; Shimizu, Y.; Kitajima, K.; Wakitani, M.; Ohta, S.; Satoh, M.; Shitara, K.; et al. Engineered antibodies of IgG1/IgG3 mixed isotype with enhanced cytotoxic activities. *Cancer Res.* **2008**, *68*, 3863–3872. [CrossRef]

86. Giuntini, S.; Granoff, D.M.; Beernink, P.T.; Ihle, O.; Bratlie, D.; Michaelsen, T.E. Human IgG1, IgG3, and IgG3 Hinge-Truncated Mutants Show Different Protection Capabilities against Meningococci Depending on the Target Antigen and Epitope Specificity. *Clin. Vaccine Immunol.* **2016**, *23*, 698–706. [CrossRef]

87. Oxelius, V.-A.; Pandey, J.P. Human immunoglobulin constant heavy G chain (IGHG) (Fcγ) (GM) genes, defining innate variants of IgG molecules and B cells, have impact on disease and therapy. *Clin. Immunol.* **2013**, *149*, 475–486. [CrossRef]

88. Oxelius, V.-A. Immunoglobulin constant heavy G subclass chain genes in asthma and allergy. *Immunol. Res.* **2008**, *40*, 179–191. [CrossRef]

89. Oxelius, V.A.; Carlsson, A.M.; Aurivillius, M. Alternative G1m, G2m and G3m allotypes of IGHG genes correlate with atopic and nonatopic pathways of immune regulation in children with bronchial asthma. *Int. Arch. Allergy Immunol.* **1998**, *115*, 215–219. [CrossRef]

90. O'Hanlon, T.P.; Rider, L.G.; Schiffenbauer, A.; Targoff, I.N.; Malley, K.; Pandey, J.P.; Miller, F.W. Immunoglobulin gene polymorphisms are susceptibility factors in clinical and autoantibody subgroups of the idiopathic inflammatory myopathies. *Arthritis Rheum.* **2008**, *58*, 3239–3246. [CrossRef]

91. Pandey, J.P.; Namboodiri, A.M. Genetic variants of IgG1 antibodies and FcγRIIIa receptors influence the magnitude of antibody-dependent cell-mediated cytotoxicity against prostate cancer cells. *Oncoimmunology* **2014**, *3*, e27317. [CrossRef]

92. Pandey, J.P.; Luo, Y.; Elston, R.C.; Wu, Y.; Philp, F.H.; Astemborski, J.; Thomas, D.L.; Netski, D.M. Immunoglobulin allotypes influence IgG antibody responses to hepatitis C virus envelope proteins E1 and E2. *Hum. Immunol.* **2008**, *69*, 158–164. [CrossRef]

93. Seppälä, I.J.; Sarvas, H.; Mäkelä, O. Low concentrations of Gm allotypic subsets G3 mg and G1 mf in homozygotes and heterozygotes. *J. Immunol.* **1993**, *151*, 2529–2537.

94. Pan, Q.; Petit-Frére, C.; Hammarström, L. An allotype-associated polymorphism in the γ3 promoter determines the germ-line γ3 transcriptional rate but does not influence switching and subsequent IgG3 production. *Eur. J. Immunol.* **2000**, *30*, 2388–2393. [CrossRef]

95. Jonsson, S.; Sveinbjornsson, G.; de Lapuente Portilla, A.L.; Swaminathan, B.; Plomp, R.; Dekkers, G.; Ajore, R.; Ali, M.; Bentlage, A.E.H.; Elmér, E.; et al. Identification of sequence variants influencing immunoglobulin levels. *Nat. Genet.* **2017**, *49*, 1182–1191. [CrossRef]

96. Pan, Q.; Hammarström, L. Molecular basis of IgG subclass deficiency. *Immunol. Rev.* **2000**, *178*, 99–110. [CrossRef]

97. Kratochvil, S.; McKay, P.F.; Chung, A.W.; Kent, S.J.; Gilmore, J.; Shattock, R.J. Immunoglobulin G1 allotype influences antibody subclass distribution in response to HIV gp140 vaccination. *Front. Immunol.* **2017**, *8*, 1883. [CrossRef]

98. Ternant, D.; Arnoult, C.; Pugnière, M.; Dhommée, C.; Drocourt, D.; Perouzel, E.; Passot, C.; Baroukh, N.; Mulleman, D.; Tiraby, G.; et al. IgG1 Allotypes Influence the Pharmacokinetics of Therapeutic Monoclonal Antibodies through FcRn Binding. *J. Immunol.* **2016**, *196*, 607–613. [CrossRef]

99. Brusco, A.; Saviozzi, S.; Cinque, F.; DeMarchi, M.; Boccazzi, C.; de Lange, G.; van Leeuwen, A.M.; Carbonara, A.O. Molecular characterization of immunoglobulin G4 gene isoallotypes. *Eur. J. Immunogenet.* **1998**, *25*, 349–355. [CrossRef]

100. Howie, H.L.; Delaney, M.; Wang, X.; Er, L.S.; Kapp, L.; Lebedev, J.N.; Zimring, J.C. Errors in data interpretation from genetic variation of human analytes. *JCI Insight* **2017**, *2*, 1–9. [CrossRef]

101. Howie, H.L.; Delaney, M.; Wang, X.; Er, L.S.; Vidarsson, G.; Stegmann, T.C.; Kapp, L.; Lebedev, J.N.; Wu, Y.; AuBuchon, J.P.; et al. Serological blind spots for variants of human IgG3 and IgG4 by a commonly used anti-immunoglobulin reagent. *Transfusion* **2016**, *56*, 2953–2962. [CrossRef]

102. Subedi, G.P.; Barb, A.W. The Structural Role of Antibody N-Glycosylation in Receptor Interactions. *Structure* **2015**, *23*, 1573–1583. [CrossRef] [PubMed]

103. Niwa, R.; Natsume, A.; Uehara, A.; Wakitani, M.; Iida, S.; Uchida, K.; Satoh, M.; Shitara, K. IgG subclass-independent improvement of antibody-dependent cellular cytotoxicity by fucose removal from Asn297-linked oligosaccharides. *J. Immunol. Methods* **2005**, *306*, 151–160. [CrossRef] [PubMed]

104. Shinkawa, T.; Nakamura, K.; Yamane, N.; Shoji-Hosaka, E.; Kanda, Y.; Sakurada, M.; Uchida, K.; Anazawa, H.; Satoh, M.; Yamasaki, M.; et al. The Absence of Fucose but Not the Presence of Galactose or Bisecting *N*-Acetylglucosamine of Human IgG1 Complex-type Oligosaccharides Shows the Critical Role of Enhancing Antibody-dependent Cellular Cytotoxicity. *J. Biol. Chem.* **2003**, *278*, 3466–3473. [CrossRef] [PubMed]

105. Pereira, N.A.; Chan, K.F.; Lin, P.C.; Song, Z. The "less-is-more" in therapeutic antibodies: Afucosylated anti-cancer antibodies with enhanced antibody-dependent cellular cytotoxicity. *MAbs* **2018**, *10*, 693–711. [CrossRef]

106. Dekkers, G.; Bentlage, A.E.H.; Plomp, R.; Visser, R.; Koeleman, C.A.M.; Beentjes, A.; Mok, J.Y.; van Esch, W.J.E.; Wuhrer, M.; Rispens, T.; et al. Conserved FcγR- glycan discriminates between fucosylated and afucosylated IgG in humans and mice. *Mol. Immunol.* **2018**, *94*, 54–60. [CrossRef] [PubMed]

107. Patel, K.R.; Roberts, J.T.; Subedi, G.P.; Barb, A.W. Restricted processing of CD16a/Fc receptor IIIa N-glycans from primary human NK cells impacts structure and function. *J. Biol. Chem.* **2018**, *293*, 3477–3489. [CrossRef]

108. Falconer, D.J.; Subedi, G.P.; Marcella, A.M.; Barb, A.W. Antibody fucosylation lowers the FcγRIIIa/CD16a affinity by limiting the conformations sampled by the N162-glycan. *ACS Chem. Biol.* **2018**, *13*, 2179–2189. [CrossRef]

109. Ferrara, C.; Grau, S.; Jäger, C.; Sondermann, P.; Brünker, P.; Waldhauer, I.; Hennig, M.; Ruf, A.; Rufer, A.C.; Stihle, M.; et al. Unique carbohydrate-carbohydrate interactions are required for high affinity binding between FcgammaRIII and antibodies lacking core fucose. *Proc. Natl. Acad. Sci. USA* **2011**, *108*, 12669–12674. [CrossRef]

110. Thomann, M.; Reckermann, K.; Reusch, D.; Prasser, J.; Tejada, M.L. Fc-galactosylation modulates antibody-dependent cellular cytotoxicity of therapeutic antibodies. *Mol. Immunol.* **2016**, *73*, 69–75. [CrossRef]

111. Beers, S.A.; Chan, C.H.T.; French, R.R.; Cragg, M.S.; Glennie, M.J. CD20 as a Target for Therapeutic Type I and II Monoclonal Antibodies. *Semin. Hematol.* **2010**, *47*, 107–114. [CrossRef]

112. Dekkers, G.; Rispens, T.; Vidarsson, G. Novel Concepts of Altered Immunoglobulin G Galactosylation in Autoimmune Diseases. *Front. Immunol.* **2018**, *9*, 553. [CrossRef]

113. Pincetic, A.; Bournazos, S.; DiLillo, D.J.; Maamary, J.; Wang, T.T.; Dahan, R.; Fiebiger, B.-M.; Ravetch, J. V Type I and type II Fc receptors regulate innate and adaptive immunity. *Nat. Immunol.* **2014**, *15*, 707–716. [CrossRef]

114. van de Bovenkamp, F.S.; Derksen, N.I.L.; Ooijevaar-de Heer, P.; van Schie, K.A.; Kruithof, S.; Berkowska, M.A.; van der Schoot, C.E.; IJspeert, H.; van der Burg, M.; Gils, A.; et al. Adaptive antibody diversification through *N* -linked glycosylation of the immunoglobulin variable region. *Proc. Natl. Acad. Sci. USA* **2018**, *115*, 1901–1906. [CrossRef]

115. van de Bovenkamp, F.S.; Derksen, N.I.L.; van Breemen, M.J.; de Taeye, S.W.; Ooijevaar-de Heer, P.; Sanders, R.W.; Rispens, T. Variable Domain N-Linked Glycans Acquired During Antigen-Specific Immune Responses Can Contribute to Immunoglobulin G Antibody Stability. *Front. Immunol.* **2018**, *9*, 740. [CrossRef]

116. Hamza, N.; Hershberg, U.; Kallenberg, C.G.M.; Vissink, A.; Spijkervet, F.K.L.; Bootsma, H.; Kroese, F.G.M.; Bos, N.A. Ig Gene Analysis Reveals Altered Selective Pressures on Ig-Producing Cells in Parotid Glands of Primary Sjögren's Syndrome Patients. *J. Immunol.* **2015**, *194*, 514–521. [CrossRef]

117. Youings, A.; Chang, S.C.; Dwek, R.A.; Scragg, I.G. Site-specific glycosylation of human immunoglobulin G is altered in four rheumatoid arthritis patients. *Biochem. J.* **1996**, *314*, 621–630. [CrossRef]

118. Koers, J.; Derksen, N.I.L.; Ooijevaar-de Heer, P.; Nota, B.; van de Bovenkamp, F.S.; Vidarsson, G.; Rispens, T. Biased N-Glycosylation Site Distribution and Acquisition across the Antibody V Region during B Cell Maturation. *J. Immunol.* **2019**, *202*, 2220–2228. [CrossRef]

119. Willemze, A.; Trouw, L.A.; Toes, R.E.M.; Huizinga, T.W.J. The influence of ACPA status and characteristics on the course of RA. *Nat. Rev. Rheumatol.* **2012**, *8*, 144–152. [CrossRef]

120. Rombouts, Y.; Willemze, A.; van Beers, J.J.B.C.; Shi, J.; Kerkman, P.F.; van Toorn, L.; Janssen, G.M.C.; Zaldumbide, A.; Hoeben, R.C.; Pruijn, G.J.M.; et al. Extensive glycosylation of ACPA-IgG variable domains modulates binding to citrullinated antigens in rheumatoid arthritis. *Ann. Rheum. Dis.* **2016**, *75*, 578–585. [CrossRef]

121. Visser, A.; Doorenspleet, M.E.; de Vries, N.; Spijkervet, F.K.L.; Vissink, A.; Bende, R.J.; Bootsma, H.; Kroese, F.G.M.; Bos, N.A. Acquisition of N-Glycosylation Sites in Immunoglobulin Heavy Chain Genes During Local Expansion in Parotid Salivary Glands of Primary Sjögren Patients. *Front. Immunol.* **2018**, *9*, 491. [CrossRef]

122. van de Bovenkamp, F.S.; Hafkenscheid, L.; Rispens, T.; Rombouts, Y. The Emerging Importance of IgG Fab Glycosylation in Immunity. *J. Immunol.* **2016**, *196*. [CrossRef]

123. Stavenhagen, J.B.; Gorlatov, S.; Tuaillon, N.; Rankin, C.T.; Li, H.; Burke, S.; Huang, L.; Vijh, S.; Johnson, S.; Bonvini, E.; et al. Fc optimization of therapeutic antibodies enhances their ability to kill tumor cells in vitro and controls tumor expansion in vivo via low-affinity activating Fcgamma receptors. *Cancer Res.* **2007**, *67*, 8882–8890. [CrossRef]

124. Nordstrom, J.L.; Gorlatov, S.; Zhang, W.; Yang, Y.; Huang, L.; Burke, S.; Li, H.; Ciccarone, V.; Zhang, T.; Stavenhagen, J.; et al. Anti-tumor activity and toxicokinetics analysis of MGAH22, an anti-HER2 monoclonal antibody with enhanced Fcγ receptor binding properties. *Breast Cancer Res.* **2011**, *13*, R123. [CrossRef]

125. Romain, G.; Senyukov, V.; Rey-Villamizar, N.; Merouane, A.; Kelton, W.; Liadi, I.; Mahendra, A.; Charab, W.; Georgiou, G.; Roysam, B.; et al. Antibody Fc engineering improves frequency and promotes kinetic boosting of serial killing mediated by NK cells. *Blood* **2014**, *124*, 3241–3249. [CrossRef]

126. Bruhns, P.; Jönsson, F. Mouse and human FcR effector functions. *Immunol. Rev.* **2015**, *268*, 25–51. [CrossRef]

127. Vafa, O.; Gilliland, G.L.; Brezski, R.J.; Strake, B.; Wilkinson, T.; Lacy, E.R.; Scallon, B.; Teplyakov, A.; Malia, T.J.; Strohl, W.R. An engineered Fc variant of an IgG eliminates all immune effector functions via structural perturbations. *Methods* **2014**, *65*, 114–126. [CrossRef]

128. An, Z.; Forrest, G.; Moore, R.; Cukan, M.; Haytko, P.; Huang, L.; Vitelli, S.; Zhao, J.Z.; Lu, P.; Hua, J.; et al. IgG2m4, an engineered antibody isotype with reduced Fc function. *MAbs* **2009**, *1*, 572–579. [CrossRef]

129. Stapleton, N.M.; Armstrong-Fisher, S.S.; Andersen, J.T.; van der Schoot, C.E.; Porter, C.; Page, K.R.; Falconer, D.; de Haas, M.; Williamson, L.M.; Clark, M.R.; et al. Human IgG lacking effector functions demonstrate lower FcRn-binding and reduced transplacental transport. *Mol. Immunol.* **2018**, *95*, 1–9. [CrossRef]

130. Bloemendaal, F.M.; Levin, A.D.; Wildenberg, M.E.; Koelink, P.J.; McRae, B.L.; Salfeld, J.; Lum, J.; van der Neut Kolfschoten, M.; Claassens, J.W.; Visser, R.; et al. Anti–Tumor Necrosis Factor With a Glyco-Engineered Fc-Region Has Increased Efficacy in Mice With Colitis. *Gastroenterology* **2017**, *153*, 1351–1362.e4. [CrossRef]

131. Zalevsky, J.; Chamberlain, A.K.; Horton, H.M.; Karki, S.; Leung, I.W.L.; Sproule, T.J.; Lazar, G.A.; Roopenian, D.C.; Desjarlais, J.R. Enhanced antibody half-life improves in vivo activity. *Nat. Biotechnol.* **2010**, *28*, 157–159. [CrossRef]

132. Monnet, C.; Jorieux, S.; Souyris, N.; Zaki, O.; Jacquet, A.; Fournier, N.; Crozet, F.; de Romeuf, C.; Bouayadi, K.; Urbain, R.; et al. Combined glyco- and protein-Fc engineering simultaneously enhance cytotoxicity and half-life of a therapeutic antibody. *MAbs* **2014**, *6*, 422–436. [CrossRef]

133. Dall'Acqua, W.F.; Kiener, P.A.; Wu, H. Properties of Human IgG1s Engineered for Enhanced Binding to the Neonatal Fc Receptor (FcRn). *J. Biol. Chem.* **2006**, *281*, 23514–23524. [CrossRef]

134. Patel, D.A.; Puig-Canto, A.; Challa, D.K.; Montoyo, H.P.; Ober, R.J.; Ward, E.S. Neonatal Fc Receptor Blockade by Fc Engineering Ameliorates Arthritis in a Murine Model. *J. Immunol.* **2011**, *187*, 1015–1022. [CrossRef]

135. Ling, L.E.; Hillson, J.L.; Tiessen, R.G.; Bosje, T.; Iersel, M.P.; Nix, D.J.; Markowitz, L.; Cilfone, N.A.; Duffner, J.; Streisand, J.B.; et al. M281, an Anti-FcRn Antibody: Pharmacodynamics, Pharmacokinetics, and Safety Across the Full Range of IgG Reduction in a First-in-Human Study. *Clin. Pharmacol. Ther.* **2019**, *105*, 1031–1039. [CrossRef]

136. Kiessling, P.; Lledo-Garcia, R.; Watanabe, S.; Langdon, G.; Tran, D.; Bari, M.; Christodoulou, L.; Jones, E.; Price, G.; Smith, B.; et al. The FcRn inhibitor rozanolixizumab reduces human serum IgG concentration: A randomized phase 1 study. *Sci. Transl. Med.* **2017**, *9*, eaan1208. [CrossRef]

137. Ward, E.S.; Ober, R.J. Targeting FcRn to Generate Antibody-Based Therapeutics. *Trends Pharmacol. Sci.* **2018**, *39*, 892–904. [CrossRef]

138. Ulrichts, P.; Guglietta, A.; Dreier, T.; van Bragt, T.; Hanssens, V.; Hofman, E.; Vankerckhoven, B.; Verheesen, P.; Ongenae, N.; Lykhopiy, V.; et al. Neonatal Fc receptor antagonist efgartigimod safely and sustainably reduces IgGs in humans. *J. Clin. Investig.* **2018**, *128*, 4372–4386. [CrossRef]

139. Cleary, K.L.S.; Chan, H.T.C.; James, S.; Glennie, M.J.; Cragg, M.S. Antibody distance from the cell membrane regulates antibody effector mechanisms. *J. Immunol.* **2017**, *198*, 3999–4011. [CrossRef]

140. Rösner, T.; Derer, S.; Kellner, C.; Dechant, M.; Lohse, S.; Vidarsson, G.; Peipp, M.; Valerius, T. An IgG3 switch variant of rituximab mediates enhanced complement-dependent cytotoxicity against tumour cells with low CD20 expression levels. *Br. J. Haematol.* **2013**, *161*, 282–286. [CrossRef]

141. Carter, P. Bispecific human IgG by design. *J. Immunol. Methods* **2001**, *248*, 7–15. [CrossRef]

142. Merchant, A.M.; Zhu, Z.; Yuan, J.Q.; Goddard, A.; Adams, C.W.; Presta, L.G.; Carter, P. An efficient route to human bispecific IgG. *Nat. Biotechnol.* **1998**, *16*, 677–681. [CrossRef]

143. Labrijn, A.F.; Meesters, J.I.; de Goeij, B.E.C.G.; van den Bremer, E.T.J.; Neijssen, J.; van Kampen, M.D.; Strumane, K.; Verploegen, S.; Kundu, A.; Gramer, M.J.; et al. Efficient generation of stable bispecific IgG1 by controlled Fab-arm exchange. *Proc. Natl. Acad. Sci. USA* **2013**, *110*, 5145–5150. [CrossRef]

144. Brinkmann, U.; Kontermann, R.E. The making of bispecific antibodies. *MAbs* **2017**, *9*, 182–212. [CrossRef]

145. Sedykh, S.E.; Prinz, V.V.; Buneva, V.N.; Nevinsky, G.A. Bispecific antibodies: design, therapy, perspectives. *Drug Des. Devel. Ther.* **2018**, *12*, 195. [CrossRef]

146. Bakalar, M.H.; Joffe, A.M.; Schmid, E.M.; Son, S.; Podolski, M.; Fletcher, D.A. Size-dependent segregation controls macrophage phagocytosis of antibody-opsonized targets. *Cell* **2018**, *174*, 131–142. [CrossRef]

147. Bohn, T.; Rapp, S.; Luther, N.; Klein, M.; Bruehl, T.-J.; Kojima, N.; Aranda Lopez, P.; Hahlbrock, J.; Muth, S.; Endo, S.; et al. Tumor immunoevasion via acidosis-dependent induction of regulatory tumor-associated macrophages. *Nat. Immunol.* **2018**, *19*, 1319–1329. [CrossRef]

148. Li, H.Y.; McSharry, M.; Bullock, B.; Nguyen, T.T.; Kwak, J.; Poczobutt, J.M.; Sippel, T.R.; Heasley, L.E.; Weiser-Evans, M.C.; Clambey, E.T.; et al. The Tumor Microenvironment Regulates Sensitivity of Murine Lung Tumors to PD-1/PD-L1 Antibody Blockade. *Cancer Immunol. Res.* **2017**, *5*, 767–777. [CrossRef]

antibodies

MDPI

Review

David vs. Goliath: The Structure, Function, and Clinical Prospects of Antibody Fragments

Adam Bates and Christine A. Power *

Biopharm Molecular Discovery, GlaxoSmithKline, Hertfordshire SG1 2NY, UK; adam.x.bates@gsk.com
* Correspondence: christine.a.power@gsk.com; Tel.: +44-1438-766246

Received: 1 February 2019; Accepted: 2 April 2019; Published: 9 April 2019

Abstract: Since the licensing of the first monoclonal antibody therapy in 1986, monoclonal antibodies have become the largest class of biopharmaceuticals with over 80 antibodies currently approved for a variety of disease indications. The development of smaller, antigen binding antibody fragments, derived from conventional antibodies or produced recombinantly, has been growing at a fast pace. Antibody fragments can be used on their own or linked to other molecules to generate numerous possibilities for bispecific, multi-specific, multimeric, or multifunctional molecules, and to achieve a variety of biological effects. They offer several advantages over full-length monoclonal antibodies, particularly a lower cost of goods, and because of their small size they can penetrate tissues, access challenging epitopes, and have potentially reduced immunogenicity. In this review, we will discuss the structure, production, and mechanism of action of EMA/FDA-approved fragments and of those in clinical and pre-clinical development. We will also discuss current topics of interest surrounding the potential use of antibody fragments for intracellular targeting and blood–brain barrier (BBB) penetration.

Keywords: ADC; antibody fragments; BiTE®; diabodies; domain antibodies; fab; ImmTAC®; Nanobody®; scFv; TandAb; V-NAR

---

1. Introduction

Since the licensing of the first monoclonal antibody (mAb) therapy, Orthoclone™ (OKT3), in 1986, the specificity, flexibility, and diversity of antibodies and antibody derivatives has led to their becoming the largest class of biopharmaceuticals [1]. Monoclonal antibodies have a history of safe use, a strong scientific basis, and a high degree of technical feasibility. As of October 2018, over 80 antibodies were marketed or approved by the EMA/FDA for multiple disease indications including cancer, inflammation/autoimmunity, transplantation, infectious and cardiovascular diseases, haematology, allergy, and ophthalmology. Over 500 more, including 2nd generation products and novel antibody formats, are currently in clinical trials around the world [1–3].

The modular nature of antibodies, both structurally and functionally, allows for the generation of smaller antigen binding fragments, such as fragment antigen binding (Fab), the single chain fragment variable (scFv), single-domain antibodies, and the fragment crystallizable (Fc) domain, through molecular cloning, antibody engineering, and even enzymatic methods. Antibody fragments can then be used on their own or linked to other molecules or fragments to generate bispecific, multi-specific, multimeric, or multifunctional molecules to achieve a variety of biological effects [4].

Antibody fragments can offer several advantages over the use of conventional antibodies. For example, they can be produced easily, generally using microbial expression systems, which results in faster cultivation, higher yields, and lower production costs [5]. Their small size allows access to challenging, cryptic epitopes, and tumour penetration, they have reduced immunogenicity, and the lack of Fc limits bystander activation of the immune system [6]. On the other hand, their smaller size results in faster renal excretion, which may require higher doses and/or more frequent dosing regimens in vivo unless mitigated by the addition of half-life extension moieties such as polyethylene glycol or albumin binding fragments.

In this review, we describe the history, structure, formats, mechanisms of action, and production of some of the most common antibody fragments. We also discuss some of the formats currently being tested in clinical and non-clinical settings, as well as briefly touching on future applications of this expanding class of biopharmaceuticals.

2. Antibody Fragment Formats

2.1. Fragment Variable (Fv)-Based Formats

2.1.1. Single Chain Fragment Variable (scFv)

The single chain fragment variable (scFv) was first described in 1988 by Bird et al. [7] and comprises the variable regions of the light chain (VL) and heavy chain (VH) of an antibody linked by a flexible peptide, which is most commonly glycine- and serine-rich with dispersed hydrophilic residues (Figure 1), to produce a single chain protein with an affinity for its antigen comparable to that of the parental mAb. The sequence and length of the ideal linker may differ between scFvs in order to optimise affinity for the antigen and thermostability. It is believed that the linker must span a ~3.5 nm distance between the VL and VH without disrupting the formation of the antigen binding site [8].

Figure 1. The single chain fragment variable format. The C-terminus of the light chain (VL) is linked the N-terminus of the heavy chain (VH) by a flexible glycine- and serine- rich linker.

The scFv has several advantages over a conventional mAb. Firstly, their small size of approximately 27 kDa makes them ideal for large-scale production in microbial systems [9]. Additionally, as the VL and VH coding sequences are genetically linked in a single transcript, there is no need to balance the expression of the light chain (LC) and heavy chain (HC). This allows fragments to be produced more quickly, in higher yields, and at lower costs than full-sized mAbs, which generally require mammalian expression systems [10]. Their small size also facilitates tissue penetration and access to cryptic epitopes, making them especially useful for tumour penetration in cancer immunotherapy [11,12]. The lack of an Fc region removes the risk of bystander immune cell activation and antibody effector functions such as antibody-dependent cellular cytotoxicity (ADCC), antibody-dependent cellular phagocytosis (ADCP), or complement-dependent cytotoxicity (CDC), allowing the molecule to bind its target without activation of the host's immune system [6]. This may be advantageous or disadvantageous depending on the context.

The lack of an Fc domain also brings about several disadvantages, mainly low thermostability compared to the parental mAb, a greater propensity for aggregation, therefore increasing the risk of immunogenicity, and a shorter half-life due to a lack of FcRn-mediated recycling. This can lead to the need for higher and more frequent dosing [4,13,14]. Fusion of the scFv to albumin or polyethylene glycol (PEG) can be used to improve half-life. However, such fusions can offset the advantages that an scFv holds over a mAb due to cost and the increase in size [15,16]. Examples of scFvs in clinical development include gancotamab (Merrimack Pharma), pexelizumab (Alexion), and Novartis's brolucizumab [17].

2.1.2. Tandem scFvs

Tandem scFvs such as Micromet AG's solitomab [17] are an adaptation of the scFv format. When two different scFvs are linked, the tandem scFv is arguably the simplest bispecific antibody platform [18]. A tandem scFv links two or more scFvs through helical peptide linkers in the orientation $NH_2-VL_1-VH_1-(linker-VL_2-VH_2)_n-COOH$, resulting in a single chain bivalent and bi-specific molecule encoded by a single gene [19] (Figure 2). These can be used to target one antigen with increased avidity, to target two distinct antigens simultaneously, or even to target albumin, thus increasing half-life [20].

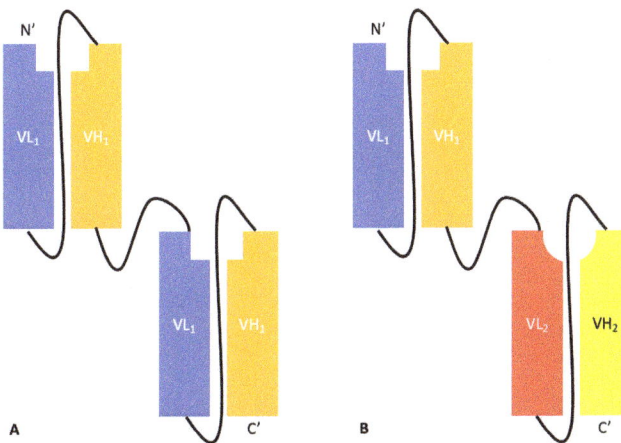

Figure 2. The tandem scFv platform. (**A**) A monospecific bivalent tandem scFv composed of two identical scFvs joined by a helical linker. (**B**) A bispecific bivalent scFv composed of two different scFvs joined by a helical linker.

2.1.3. Diabodies, DART®s, and TandAbs

Diabodies are bivalent dimers formed from two chains, each containing a VH and a VL domain. The two domains within a chain are separated by a pentameric glycine-rich linker (G$_4$S) that is too short to facilitate intrachain dimerization leading to two chains dimerising in a head-to-tail arrangement. By using two different chains with the same orientation, the first containing the VH of Antibody 1 and the VL of Antibody 2, and the second containing the VH of Antibody 2 and the VL of Antibody 1, bispecific bivalent dimers are produced [21] (Figure 3). A study comparing different formats of bispecific diabodies showed that not all possible diabody formats retain binding to both antigens, highlighting the importance of domain arrangement and orientation [22].

Over time, additional modifications have been made to the diabody format to further improve stability. Dual affinity re-targeting proteins (DART®s) (Figure 3) developed by MacroGenics, contain an interdomain disulphide bond for increased stability that results in a structure that is rigid and compact [23]. MacroGenics currently have four DART®s in phase I clinical trials for oncology, autoimmune disease, and HIV infection [24].

Tri- and tetra-valent molecules with a structure similar to that of a diabody can be produced by linking three or four variable domains together in a single chain. Affimed specialise in the development of TandAbs, tetravalent bispecific molecules composed of two diabodies fused in a linear fashion (Figure 3), through their proprietary ROCK® platform. There are currently five ongoing clinical trials involving a selection of T-cell and natural killer (NK) cell engagers for the treatment Hodgkin's lymphoma and acute lymphocytic leukaemia (ALL) [25].

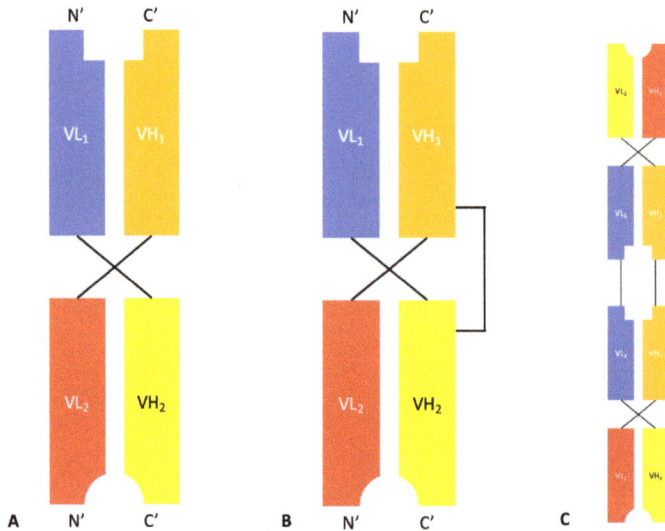

Figure 3. The structures of diabody, DART®, and TandAb fragments. (**A**) A bispecific diabody composed of two different chains, each containing a VL and VH from different antibodies, in a head-to-tail arrangement. (**B**) A bispecific dual affinity re-targeting (DART®) protein containing two distinct polypeptide chains held together by non-covalent interactions and a disulphide bond. (**C**) A TandAb composed of two diabodies linked in a linear arrangement to produce a tetravalent bispecific molecule.

A major advantage of diabodies and tandem scFvs is their bivalency and ability to bring two targets into proximity. Furthermore, diabodies and tandem scFvs are bispecific-compatible formats, which has made them promising molecular formats for cancer immunotherapy, e.g., bispecific T-cell engagers (BiTE®s) [26,27].

2.1.4. Bispecific Fv Fusion Antibodies with an Fc Domain

Although not fragments in their own right, there are many bispecific formats such as the IgG-scFv, which uses an scFv to target an additional epitope or antigen (Figure 4). As well as being a flexible format that allows for the production of multivalent multi-specific molecules with a modifiable effector function, these molecules can easily be produced recombinantly [18,28,29]. Istiratumab, an IgG-scFv developed by Merrimack Pharmaceuticals, is currently in clinical trials for hepatocellular carcinoma and pancreatic cancer [17]. Fab-scFv-Fc (Figure 4) is another bispecific format that uses an scFv to grant specificity for a second epitope. The format is being used in molecules such as Zymeworks' ZW25, an anti-Her2/Her2 bispecific (biparatopic) antibody in phase I clinical trials for the treatment of Her2 expressing cancers. Xencor have four bispecific molecules with this format in phase I clinical trials for the treatment of acute myeloid leukaemia (AML), B-cell tumours, and neuroendocrine tumours [30].

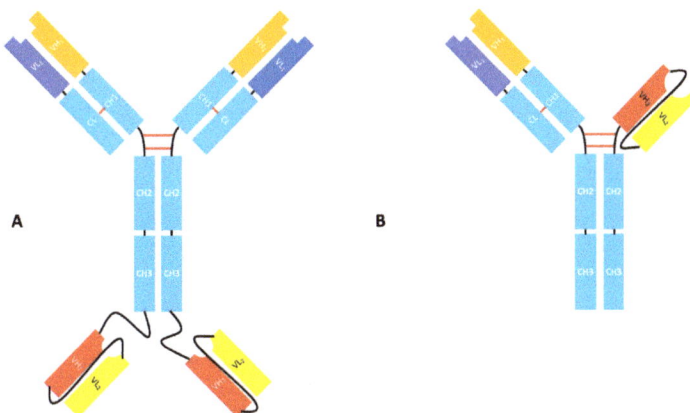

Figure 4. scFv fusion bispecific formats with an Fc domain. (**A**) IgG-scFv. Canonical IgG with a scFv fused to the C-terminus of the CH3 domain to produce a tetravalent bispecific molecule. (**B**) Fab-scFv-Fc IgG with one IgG Fab arm exchanged for a scFv.

2.2. Fab Based Formats

Fab & F(ab')$_2$ Formats

The fragment of antigen binding (Fab) (Figure 5) was the first therapeutic antibody fragment format and remains one of the most successful, laying claim to eight molecules entering clinical trials pre-1995 and comprising ~49% of all antibody fragments that have entered clinical trials [4,31]. There are currently three FDA-approved Fabs: abciximab (Reopro®), idarucizumab (Praxbind®), and ranibizumab (Leucentis®) [17].

Fab fragments are composed of an antibody light chain (VL + CL domains) linked by a disulphide bond to the antibody heavy chain VH and CH1 domains; the molecular format is monovalent and monospecific and retains the parental antibody's ability to bind its antigen with high specificity and affinity. Fabs share many of the characteristics of scFvs. Although Fabs are not as small, and therefore presumably not as good at penetrating tissue as scFvs [12], with a mass of ~50 kDa they are much smaller than mAbs (the MW of IgG is ~150 kDa). Fabs lack an Fc domain, reducing the risk of immune cell bystander activation and non-specific binding [6] and allowing for easier production at the expense of increased aggregation, lower stability, and reduced half-life [5,32]. Fab fragments are more stable than their scFv counterparts due to the mutual stabilisation that occurs between the VH/VL and CH1/CL interfaces [33]. They also have the advantage of being completely native structures and as such they avoid the time and resources required to engineer an ideal linker and are therefore less likely to be immunogenic.

F(ab')$_2$ fragments are composed of two Fab fragments held together by an Ig hinge region and have a molecular mass of ~110 kDa (Figure 5). F(ab')$_2$ fragments are bivalent, giving them increased avidity compared to Fabs [32], but their larger size may lead to reduced tissue penetration. However, their tissue penetration is still superior to that of a full-sized mAb [12]. They can be generated by the enzymatic digestion of full-length antibodies or by the expression of the recombinant F(ab')$_2$ in mammalian cells.

Like scFvs, Fab fragments suffer from a reduced half-life compared to their parental mAbs due to the lack of an Fc domain. This removes the possibility of FcRn-mediated recycling leading to rapid degradation of the Fab-antigen complex after absorption by macropinocytosis [34–36]. The half-life of Fab fragments can be extended in a similar manner to the half-life extension of scFvs, typically by conjugation to PEG or fusion to an albumin binding protein [16,37,38]. These Fab conjugate proteins have had some success with multiple molecules in clinical trials or having already received

FDA approval such as certolizumab pegol (Cimzia®), a marketed PEGylated anti-TNFα Fab for rheumatoid arthritis.

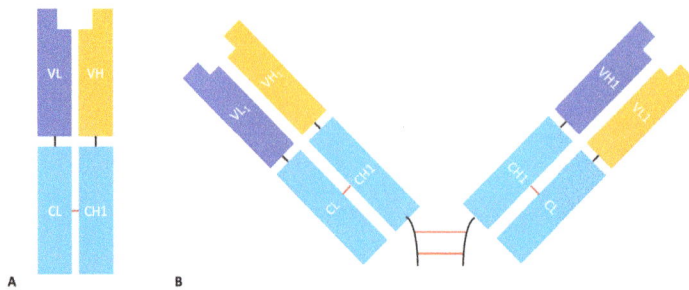

Figure 5. The structure of Fab and F(ab')$_2$ fragments. (**A**) Fab fragment composed of an LC (containing VL and CL) linked to an Fd (containing VH and CH1) by a disulphide bond between the CL and CH1 domains. (**B**) F(ab')$_2$ fragment composed of two Fab fragments joined by an IgG hinge region.

2.3. Single-Domain Antibodies

2.3.1. Nanobodies

Nanobodies® (Nbs), a class of antibody fragment developed by Ablynx, are recombinantly expressed antigen binding VHH domains from heavy-chain IgG, a type of immunoglobulin found in camelids [39]. With an MW of 12–15 kDa, Nbs are one of the smallest naturally occurring antigen binding fragments. Heavy-chain IgG (Figure 6) is devoid of light chain, lacks the CH1 domain found in mammalian immunoglobulins, and therefore binds its antigen bivalently solely through two VHH domains. Isolated VHH domains retain the ability to bind their antigen and are robust under stringent conditions. Nbs can resist a wide pH range and high temperatures and have been shown to tolerate the presence of organic solvents (although these characteristics are not present in all Nbs). They are also highly soluble and, due to their small size and extended CDRH3 loop, can rapidly penetrate tissue and access cryptic epitopes [40–42]. Such molecules are also easy to produce recombinantly without issues relating to inter-domain interactions, such as those found in scFvs and Fab fragments [43]. Although not of human origin and frequently 'humanised', Nbs are rarely immunogenic due to their small size and similarities with the human VH3 gene family [43].

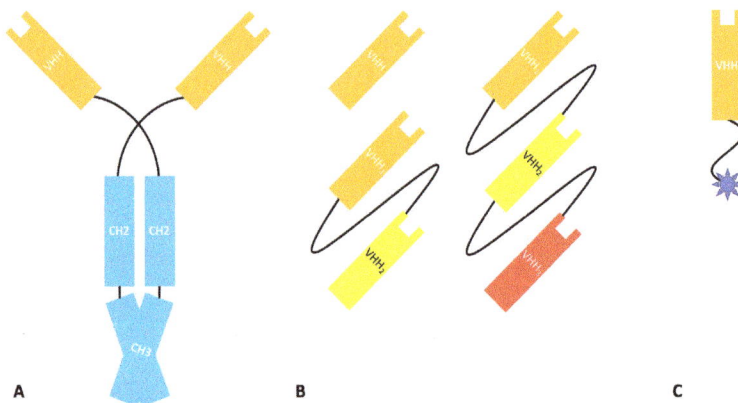

Figure 6. Camelid heavy-chain IgG and Nanobody® fragments. (**A**) The structure of heavy-chain IgG, composed of two heavy chains, each containing a VHH domain, a CH2 domain, and a CH3 domain. (**B**) Mono-, bi-, and tri-valent Nb formats with each VHH having a different antigen specificity. (**C**) A Nanobody® drug conjugate.

Fusion of Nbs that bind different epitopes allows the creation of multivalent molecules with high affinity or potency. The monomeric behaviour, good solubility, and modular nature of Nbs also allow them to be easily fused to other Nbs with different antigen specificities or covalently linked to other molecules [44].

Unfortunately, the small size of Nbs is below renal cut-off, leading to rapid renal clearance with a half-life of approximately 2 h, making them particularly unsuitable for chronic use in many therapeutic areas [43,45]. However, the fast clearance is a beneficial property for in vivo diagnostics (see Section 5). Strategies to extend their half-life, including PEGylation, conjugation to the Fc domain of conventional antibodies, and coupling to abundant serum proteins such as human serum albumin (HSA), apolipoprotein L1, and β-Lactamase, have been explored. Fusion of Nbs to an anti-albumin VHH has been validated in the clinic [43]. Fusion of a VHH to the Fc of IgG has also been investigated as a means to produce bispecific tetravalent molecules [46].

The favourable characteristics of VHH domains has led to excitement over their potential. With 5 Nbs currently in clinical trials, and the recent approval of caplacizumab (Cablivi^{TM}), a bivalent VHH targeting von Willebrand factor (vWF), by the EMA in October 2018 and the FDA in February 2019 for thrombotic thrombocytopenic purpura and thrombosis, this excitement is not surprising [17].

Analogous to the camelid nanobodies is the single variable new antigen receptor (V-NAR), domain antibody fragments obtained from cartilaginous fishes such as sharks (Figure 7) [47]. Whilst V-NARs share some similarities in terms of their structure and properties with domain antibodies, they are smaller (11 kDa) and more stable proteins. Contrary to mammalian variable domains, V-NAR domains have only two complementarity determining regions, CDR1 and CDR3. However, they also contain two mutation-prone regions named HV2 and HV4; the latter has been shown to contribute to antigen binding [48,49]. As V-NARs are evolutionarily derived from a non-antibody lineage, this arguably places them outside the complex and competitive antibody patent landscape. The V-NAR format is being developed by Elasmogen (UK) under the name of soloMERs^{TM}.

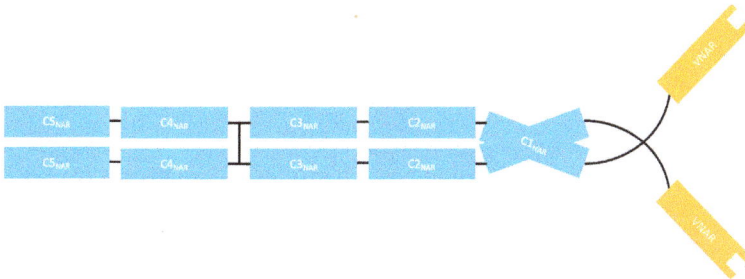

Figure 7. Shark heavy chain antibody (Ig-NAR), a dimer of heavy chains containing five constant domains and the antigen binding variable nucleotide antigen receptor (V-NAR).

2.3.2. Domain Antibodies

Domain antibodies (dAbs) are fully human unpaired variable domains (either VH or VL), which have been engineered to prevent dimerization whilst maintaining the specificity and affinity of a canonical antigen binding site (VH and VL). This engineering commonly involves 'camelisation', in which the hydrophobic residues usually found at the VH/VL interface are substituted for hydrophilic residues found in camelid VHH and in the extension of the CDRH3 [50,51]. Similar in size and structure, these molecules possess many of the advantageous properties of Nbs, including high thermostability, high solubility, and a short half-life and are amenable to conjugation/fusion and high-yield microbial expression [51]. Although naked dAbs may have some application as therapeutic molecules (see Sections 4.1 and 4.2), dAbs have been most extensively investigated as fusion proteins to other moieties, such as full-length antibodies to enable specificity for a second antigen [18,52], an

Fc domain (e.g., Placulumab), or with an anti-albumin dAb (e.g., GSK/Domantis' AlbudAb®s) [53] (Figure 8).

Albudab®s (Figure 8) contain an anti-HSA binding domain which can increase the half-life of the species to 19 days, the half-life of HSA. GSK2374697 was an AlbudAb® developed for type 2 diabetes which phase I clinical trials were reached. It was composed of exendin-4, a GLP-1 mimetic peptide isolated from Gila monster saliva, linked to an anti-HSA dAb as a single transcript. The anti-HSA dAb extended the half-life of the peptide from 30 min to 6–10 days [53].

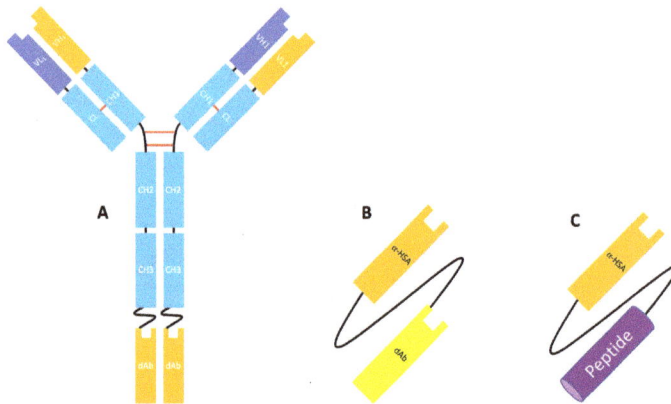

Figure 8. The uses of domain antibodies (dAbs). (**A**) IgG-dAb (also called a mAb-dAb). IgG with a dAb fused to the C terminus of each heavy chain to produce a bispecific tetravalent molecule. (**B**) Tandem dAb with an anti-HSA domain. The dAb against the target of interest is linked to an anti-HSA dAb to improve half-life. (**C**) AlbuDab®. A peptide linked to an anti-HSA dAb to improve the peptide's half-life.

3. The Production of Antibody Fragments

3.1. Expression

Conventional full-length mAbs contain an N-linked glycosylation site in the CH2 domain of the heavy chain which is important for stability, preventing aggregation, and effector function [54]. Aberrant glycosylation can cause unfavourable molecular properties and be highly immunogenic. As such, the ability to produce post-translational modifications that closely resemble those that occur naturally necessitates expression in mammalian cell lines such as Chinese hamster ovary (CHO) cells, human embryonic kidney (HEK) cells, and to a lesser extent NS0 murine myeloma cells and human PER.C6™ cells [55].

Unlike full-length mAbs, antibody fragments are not generally glycosylated. Fragments are therefore amenable to production in microbial systems, allowing for faster, cheaper production in cell lines that are easier to cultivate and manipulate [10].

3.1.1. Escherichia coli

Escherichia coli was the first microbial system used for the production of biopharmaceuticals and is still used for the production of ~40% of biopharmaceuticals available today, including certolizumab pegol and ranibizumab [10,17]. As a prokaryote, *E. coli*'s rapid growth in inexpensive media, well understood genetics, and high manipulability make it an ideal expression system when glycosylation is not required.

Traditionally, the gene(s) of interest was placed on a self-replicating, high copy number plasmid under the control of a promoter such as the bacteriophage T7, lac operon (*lac*), or tryptophan (*trp*) promoter alongside a selectable marker. However, the greater volumetric productivity that can be achieved using this method is offset by the inhibition of cell growth and cell death caused by the metabolic burden of antibody fragment production. Recently, marker-free and plasmid-free expression systems have been developed to reduce this metabolic burden and allow for greater overall yields [10,56,57].

There are two main established approaches for the production of antibody fragments in *E. coli*, the first of which is to express the protein of interest in *E. coli*'s reducing cytoplasm. This method allows for high yields of protein, but the reducing conditions are not permissive for disulphide bond formation, and inclusion bodies are regularly formed. The protein must then be re-folded after purification which can be time-consuming, inefficient, and costly [10]. This issue can be somewhat alleviated by the co-expression of chaperones to facilitate correct protein folding [58]. The second pathway involves targeting, by fusion to an N-terminal leader peptide such as Pel B, the protein of interest to the oxidising periplasm, where disulphide bonds can readily form [59,60]. This second method does result in lower yields; therefore, if efficient re-folding of the antibody fragment is possible, cytoplasmic expression may be preferable.

Finally, due to the ability of *E. coli* to grow at high cell densities, antibody fragments are commonly produced in high cell density cultures grown in a stirred tank reactor using a fed-batch method [61].

3.1.2. *Saccharomyces cerevisiae*

Saccharomyces cerevisiae was the first yeast used as an expression system for recombinant proteins. Its high genomic stability, manipulability, and ease of cultivation have made it a strong choice for the expression of antibody fragments. Additionally, *S. cerevisiae* is used in a well-established production method of llama VHH fragments, consistently giving yields of hundreds of milligrams per litre [62].

There are three commonly used vector systems with *S. cerevisiae*. Firstly, a yeast episomal plasmid containing an origin of replication allows for gene expression with a high plasmid copy number without genomic integration [10]. Secondly, yeast centromeric plasmids containing a self-replicating sequence allow for gene expression with a single or low plasmid copy number without genomic integration [10]. Finally, yeast integrative plasmids do not contain an origin of replication but rather are incorporated into the yeast's genome, leading to improved process quality and stability at the cost of expression levels [10,63]. However, methods have been developed to circumvent this issue such as targeted integration of the gene of interest at the ribosomal DNA locus, a highly-transcribed region [64]. Glyceraldehyde-3-phosphate dehydrogenase (GAPDH), alcohol dehydrogenase 1 (ADH1), and phosphoglycerate kinase 1 (PGK1) are promotors from *S. cerevisiae*'s native glycolytic pathway that are commonly used to achieve high expression levels, and methods have been developed to allow the co-expression of multiple genes on self-replicating plasmids making *S. cerevisiae* suitable for the expression of multi-gene fragments such as Fabs [65,66].

A major issue that plagues *S. cerevisiae* is endoplasmic reticulum (ER) misfolding and inefficient trafficking of the protein of interest leading to the accumulation of misfolded protein in the ER or vacuolar-like structures [10]. Although *S. cerevisiae* has proven to be an excellent expression system for VHH domains [41], this issue is particularly apparent with the more hydrophobic scFvs [67]; however, simultaneous overexpression of chaperones and foldases has been shown to facilitate scFv secretion [68].

Similar to *E. coli*, *S. cerevisiae* is usually grown in glucose-limited fed-batch culture [62]. The limited glucose helps prevent the depletion of oxygen and switch to fermentative metabolism, which leads to the undesirable production of toxic metabolites [69].

3.1.3. *Pichia pastoris*

Pichia pastoris can be used as an alternative to *S. cerevisiae* and uses an integrated vector to achieve stable expression of the protein of interest.

P. pastoris metabolises methanol as its sole source of carbon and, as such, expresses large amounts of alcohol oxidase. The alcohol oxidase promoter was commonly used to express proteins of interest, but expression was hard to control. More adjustable promotors are now being investigated [70].

The genome of several strains of *P. pastoris* have been published online along with a genome scale metabolic model allowing for straightforward engineering and strain optimisation [71].

The preference of *P. pastoris* for respiratory over fermentative growth allows it to be grown to much higher cell densities than *S. cerevisiae* using inexpensive media, usually using a fed-batch method [72]. A *P. pastoris* system has been used to produce ALX-0171, a trimeric nanobody being developed for RSV infection [73], scFv-h3D6, an anti Aβ antibody fragment which is being developed for Alzheimer's disease [9], and scFvTEG4-2c against platelet anti-αIIbβ3, for potential use as an imaging agent for atherosclerosis [74].

3.1.4. Cell-Free Expression Systems

There has recently been significant interest in cell-free production of antibody fragments. Approaches using *E. coli* cell lysates [75], CHO cell lysates [76], and insect cell lysates [77] have been successfully used to produce antibody fragments. Cell-free expression systems have been shown to be fast, reliable, flexible, and scalable [78]. Of note is the ability to directly input linear DNA encoding the protein of interest rather than constructing a complex plasmid for transformation and the selection of transformed cells. Additionally, the lack of a phospholipid bilayer barrier allows for the simple addition of resources required for the production of polypeptides, such as amino acids and ATP, and supplements, such as chaperones, the prokaryotic disulphide bond isomerase disulphide bond c (Dsbc), and oxidised/reduced glutathione rather than the co-expression or simultaneous over-expression required with cell-based systems.

Although not usually suitable for the commercial scale production of recombinant proteins, cell-free systems have been successfully used to produce functional scFvs [79] and more recently, using both prokaryotic and eukaryotic systems, functional Fabs [78,80]. Pioneered by companies like Sutro Biopharma, cell-free systems have been used to produce scFvs conjugated to moieties, such as granulocyte-monocyte colony stimulating factor (GM-CSF) and interleukin 1β-derived peptide [81] in a scalable manner, and can be used for the incorporation of non-canonical amino acids for site-directed conjugation, making cell-free expression an optimal system for protein engineering [82]. Such non-canonical amino acids are often toxic to cells or are unable to cross the cell membrane, making them difficult to incorporate whilst using cell-based systems. Although not currently a preferred method, the rise of antibody drug conjugates and improvements to cell-free expression systems may soon increase the desirability of cell-free expression systems for antibody fragment production. Sutro's cell-free antibody production system was used to make STRO-001, a novel CD74 targeting antibody drug conjugate in phase I clinical trials for B-cell malignancies including multiple myeloma and non-Hodgkin's lymphoma [83].

3.2. Enzymatic Cleavage

Although recombinant expression is the method most commonly used for the production of antibody fragments, fragments such as Fab, F(ab')₂, and Fc can be easily produced from their parent mAb by enzymatic cleavage using commercially available enzymes. Enzymatic cleavage can be preferable to recombinant expression for the production of certain fragment formats, such as F(ab')₂, which has a propensity to aggregate when expressed recombinantly due to the hinge region. This was originally done using papain, which cleaves just above of the hinge region to produce two Fab fragments and a hinge-CH2-CH3 fragment [84], or pepsin, which cleaves just below the hinge region to produce an F(ab')₂ and an Fc fragment [85]. Companies like Genovis now produce improved versions of these enzymes, allowing for a more efficient production of antibody fragments from the parental mAb [86].

3.3. Purification

The purification of antibody fragments is somewhat more complicated than the purification of full-sized mAbs, owing to the lack of an Fc domain which facilitates efficient purification by Protein A or Protein G affinity chromatography [87,88]. However, VH 3 family containing fragments can be purified using Protein A [89]. Currently, there are no 'toolbox' or generic approaches to the production of pharmaceutical antibody fragments, with the current approved fragment therapies being purified using different combinations of chromatographic and non-chromatographic techniques [90]. In theory, an antibody fragment could be captured selectively using its antigen fixed to a resin. Whilst this may be used in some cases, antigens are not always readily available and prior production, purification, and fixation of said antigen would have a significant impact on the cost of goods.

The unique nature of fragments and the lack of large conserved regions pose an additional challenge to the development of generic purification approaches. To this end, micro-fluidic approaches have been tested to quickly determine the ideal binding conditions of specific fragments to inform future purification attempts [91].

If the expression system used expresses the fragment in the cytoplasm or periplasm rather than secreting it, the fragments are first freed by lysing the cells, and proteins are then refolded if necessary. Once this has been completed, the fragment can be purified using one or more of the methods described in the following sections.

A simplified workflow for the purification of applicable fragments by affinity chromatography is shown in Figure 9.

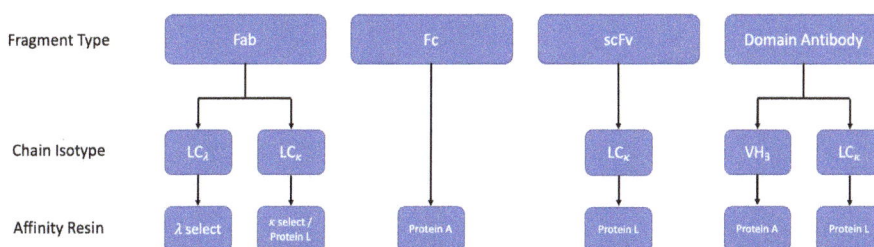

Figure 9. Simplified purification workflow for different fragment formats. A series of stages that can be used to purify common antibody fragments based on the presence of certain domains. The workflow typically includes affinity chromatography (where applicable) followed by multi-modal polishing stages, which may include size exclusion chromatography (SEC), cation (CIEX) or anion (AIEX) exchange, and hydrophobic interaction chromatography (HIC). Antibody fragments can also be purified using cation exchange chromatography (CIEX) or immobilised metal affinity chromatography (IMAC) as the initial capture step.

3.3.1. Protein L Affinity Chromatography

Protein L is a cell wall-associated protein isolated from *Peptostreptococcus magnus* [92], which binds strongly to human Ig LCs, scFv, and Fab fragments [93]. Protein L targets a site that lies within the variable region of κ 1, 3, or 4 light chains, allowing it to capture a wide range of mAbs and fragment formats. However, should the fragment of interest be derived from a λ mAb or from the κ 2 sub-family, protein L would be unable to capture the fragment [93]. In cases such as these, commercially available κ- and λ-select resins may be suitable alternatives.

3.3.2. Affinity Tags

As antibody fragments are most commonly produced recombinantly, they can easily be generated with affinity tags such as hexa-histidine (6HIS), glutathione-S transferase (GST), or mannose binding protein (MBP) using a cleavable linker to allow purification by immobilised metal affinity chromatography (IMAC) or other affinity-based methods [94]. Such techniques would allow for the selective capture of the desired fragment regardless of the Ig germline family. However, affinity tags are not a blanket solution. Removal of the affinity tag by proteolytic cleavage may leave residual amino acids, which may cause issues with aggregation, misfolding, and immunogenicity. It is also important to consider the host's cellular proteases when designing the linker to avoid premature cleavage and production of irrecoverable material [95,96]. Additionally, a study carried out by Das et al. showed that Protein L affinity chromatography, where applicable, was a more robust and versatile method for the purification of scFvs than IMAC using a (6HIS) tag [97].

3.3.3. Other Chromatographic Methods

As well as the affinity-based methods described above, other chromatographic methods such as size exclusion chromatography, ion exchange chromatography, and multi-modal chromatography are used to separate the desired fragment from contaminants based on size and isoelectric point [98,99]. Ion exchange chromatography usually takes the form of cation-exchange chromatography (CIEX) and has the added advantage of being able to separate charge variants [100]. In order to achieve the high purity required for pharmaceutical applications, these chromatographic methods are used in conjunction with each other, with affinity-based methods, and/or with non-chromatographic methods [90]. More recently, multi-modal approaches have been used in the purification of Fab fragments and have been shown to be superior to traditional CIEX resins due to increased salt tolerance and Fab binding [91,101].

4. Antibody Fragments in the Clinic

4.1. Oncology

The majority of antibody fragments currently being developed in the clinic are for oncological applications. In addition to the generic characteristics of antibody fragments that make them attractive as immunotherapies, e.g., their small size, which grants them superior tissue and tumour penetration compared to a conventional mAb [12], and the lack of an Fc domain that reduces non-specific activation of innate immune cells, there are many mechanisms of action that are unique to a specific format. The diversity of formats being investigated for their therapeutic potential in oncology is astounding, but the majority of fragments are reformatted as bispecific molecules combining an anti-CD3 binding moiety with an anti-tumour binding domain. For more detailed information, the reader is referred to two recent excellent reviews by Wu et al. and Velasquez et al. [102,103]. Here we will only cover the more established formats and describe the over-arching pathways that they exploit.

4.1.1. BiTE®s

Bispecific T-cell engagers (BiTE®s) originally developed by Micromet, are a specific class of tandem scFv used to redirect cytotoxic T-cells to tumours. The induced response is highly selective for the target tumour cells, more so than can be achieved by radio- or chemotherapy. The hope is that this selectivity will lead to reduced off-target effects (Figure 10) [26].

BiTE®s contain two antigen binding sites. The first is directed against a tumour antigen, whilst the second is directed against the T-cell receptor (TCR) signalling complex CD3. Simultaneous binding of the BiTE® to CD3 and the tumour antigen (e.g., CD19) bypasses pMHC restriction and induces T-cell activation, cytokine production, the formation of cytolytic immunological synapses leading to a tumour-directed cytotoxic response, and the activation of other host immune responses [26,104]. The use of a monovalent anti-CD3 is thought to be important in limiting off-target immune activating functions that can lead to cytokine release syndrome and cytokine storm, a problem seen with some of the early anti-CD3 mAbs in the clinic such as muromonab (OKT3), which eventually led to its withdrawal.

There is currently one licenced BiTE® which received FDA approval in 2017. Blinatumomab (Blincyto®), developed by Amgen and Astellas Pharma Inc., is an anti-CD19/CD3 BiTE®, used for the treatment of Non-Hodgkin's lymphoma and Philadelphia chromosome negative acute lymphoblastic leukaemia. It is also in clinical trials for a number of additional indications. Other tumour antigens are currently being trialled as targets for anti-CD3 containing BiTE®s including BCMA, CD33, CEA, HER2, EGFR, and EpCAM [26,102].

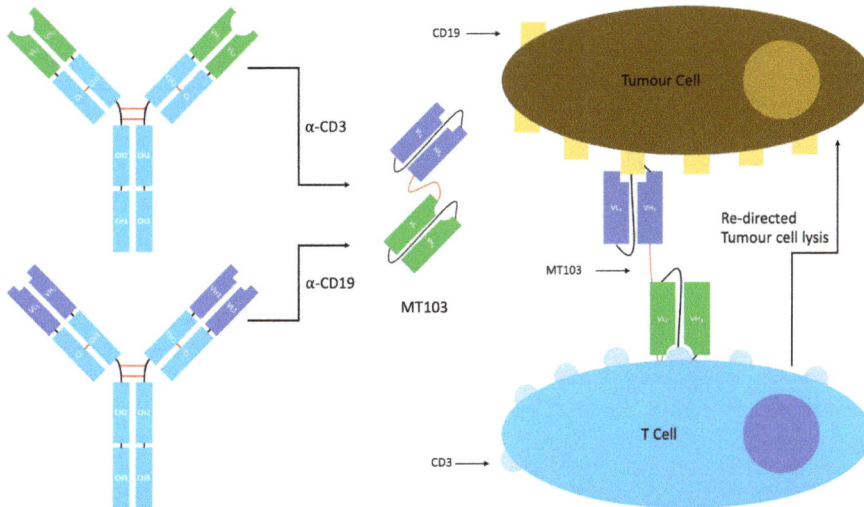

Figure 10. The structure and mechanism of action of MT103, an α-CD3/α-CD19 bispecific T-cell engager (BiTE®). The antigen binding site of each parental antibody is isolated and converted into an scFv format. The two scFvs are then joined by a flexible peptide linker to produce a bispecific moiety. The anti-CD19 scFv binds to tumour cells whilst the anti-CD3 scFv will bind passing T-cells, re-directing them to attack the tumour cell.

4.1.2. BiKEs & TriKEs

Bispecific killer cell engagers (BiKEs) and trispecific killer cell engagers (TriKEs) are bi- or tri-specific tandem scFvs used to redirect natural killer (NK) cells via an anti-CD16 scFv. They work in a similar fashion to BiTE®s. Although there are currently no BiKEs or TriKEs in clinical trials, BiKEs are currently being developed by Sanofi in collaboration with Innate Pharma using Innate Pharma's anti-CD335 (NKp46) antibody to redirect NK cells [105].

Anti-CD16/CD33 BiKEs showed promise in an in vitro study treating myelodysplastic syndrome (MDS) [106]. Here, the anti-CD16 scFv was used to activate depleted NK cells which were redirected against myeloid-derived suppressor cells (MDSCs) expressing CD33 by the anti-CD33 scFv. The treatment was shown to reduce the immunosuppression of NK cells by MDSCs, induce MDSC cell lysis, and induce optimal MDS-NK cell function regardless of disease stage.

More recently, an anti-CD16-IL-15-anti-CD33 TriKE was also shown to overcome the cancer induced immunosuppression observed in MDS and AML [107].

4.1.3. DART®s

Developed by MacroGenics, the DART® platform has been used to produce five anti-cancer molecules, four of which entered phase I clinical trials for the treatment of AML/MDS, solid tumours, or colorectal cancer [24].

The DART® platform is compatible with several modalities. Of the five anti-cancer DARTs listed, three include an anti-CD3 binding moiety to re-direct T-cells towards cells expressing the cancer antigen complimentary to the second binding site in the same fashion as a BiTE®. The other two DARTs, which target PD-1/LAG3 or PD-1/CTLA-4, block pathways involved in T-cell inhibition leading to an enhanced T-cell response against tumour cells [108].

Although BiTE®s and DART®s can exploit the same pathways, a 2011 study comparing a CD19/CD3 bispecific molecule in DART® and BiTE® formats showed increased potency of the DART® compared to the BiTE®, which was not accompanied by an increase in non-specific T-cell activation of CD19- cell lysis in vitro [109]. The study also trialled an anti- CD19/TCR DART®, which showed activity in an in vivo xenograft mouse model and was virtually identical in vitro to the anti-CD19/CD3 DART®, describing another potential mechanism of action for the DART® platform.

4.1.4. ImmTAC®s

The immune mobilising monoclonal T-cell receptors against cancer (ImmTAC®) is a novel scFv-TCR chimeric format developed by Immunocore. ImmTAC®s are peptide-HLA-specific, dimeric affinity-enhanced soluble TCRs containing an artificial disulphide bond, joined by a peptide linker to an anti-CD3 scFv (Figure 11). The TCR portion binds with picomolar affinity to the target cell expressing the peptide antigen in the context of MHC. The anti-CD3 portion then binds with nanomolar affinity to passing T-cells that are recruited to kill the target cell. The binding of multiple CD3 surface proteins by multiple ImmTAC®s on a single T-cell causes T-cell activation leading to an immune response against the target cell/tissue [110]. Importantly, ImmTAC®s hold the potential to overcome T-cell tolerance and the low affinity of native TCRs for cancer antigen/MHC complexes and mediate an enduring immune response [111]. An ImmTAC® targeting gp100 is being tested in two phase I trials for uveal melanoma. An ImmTAC® against NY ESO entered phase I clinical trials in 2018 for a variety of cancer indications, and IMC-C101C against melanoma associated antigen 4 is planned to enter phase I in 2019 [112].

As a side note, immune-mobilising monoclonal T-cell receptors against virus antigens (ImmTAV®s) are similar to ImmTAC®s in that they are composed of an affinity-enhanced soluble TCR linked to an anti-CD3 scFv. However, as the name would suggest, the TCR is designed to bind specific viral antigens as opposed to cancer antigens. These molecules are being investigated as a novel class of HIV therapy [113].

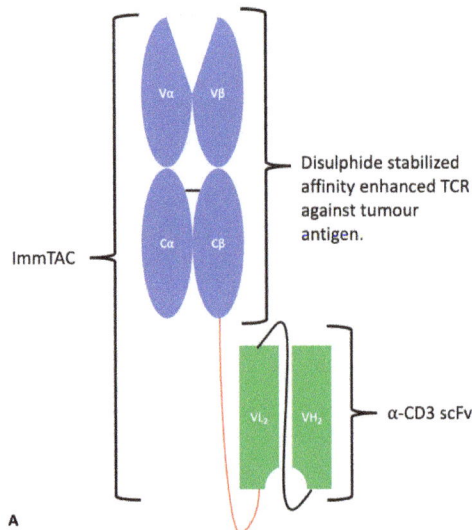

Figure 11. *Cont.*

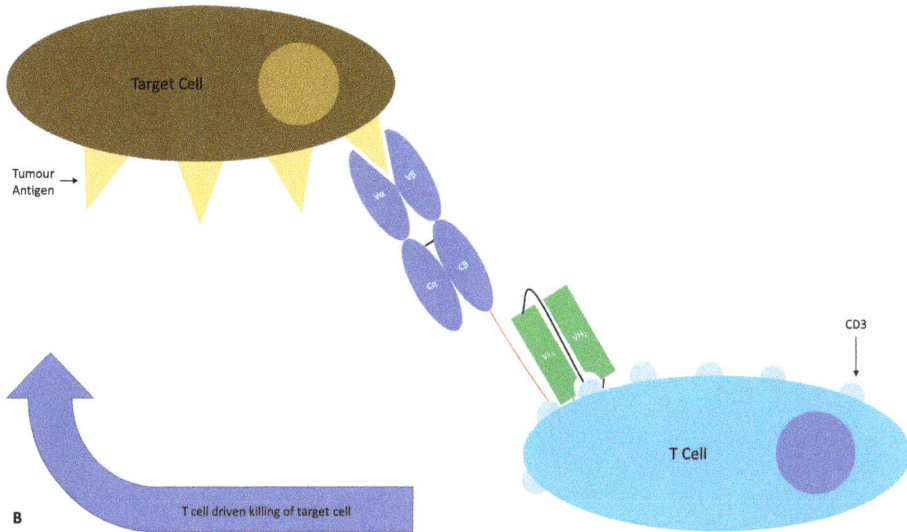

Figure 11. The structure and mechanism of action of ImmTAC®s. (**A**) The structure of an ImmTAC®, comprising an α-CD3 scFv linked to a disulphide-stabilised, affinity-enhanced soluble T-cell receptor. (**B**) The T-cell receptor binds its target with picomolar affinity causing the ImmTAC® to cluster on the target cell. The anti-CD3 scFv then recruits passing T-cells by binding CD3 with nanomolar affinity. The clustering of CD3 on the T-cell leads to activation and re-direction of the T-cell to produce an immune response against the target cell.

4.1.5. Nanobodies®

Two nanobodies reached phase I clinical trials for oncology indications: ALX-0651, which targets CXCR4 for multiple myeloma and non-Hodgkin's lymphoma, and TAS266 (Ablynx/Novartis), which targets the death receptor DR5 for solid tumours. However, their development has not been pursued [44]. On the contrary, there are a large number of nanobodies in preclinical development for a variety of other cancers. The reader is referred to the excellent review by Steeland et al. (2016) [44].

4.1.6. Antibody Fragment-Drug Conjugates

In addition to PEGylation or fusion to albumin binding antibody fragments to improve half-life and pharmacokinetics, a wide range of effector moieties, including cellular toxins, radioisotopes, cytokines, and enzymes, have been conjugated to Fab, scFv, and Nb fragments. In 2010, there were 12 scFv and 12 Fab conjugates in clinical trials worldwide [4,40,114,115]. Citatuzumab bogatox, an anti-epithelial cell adhesion molecule (EpCAM) Fab conjugated to bouganin, developed by Viventia Biotechnologies Inc., is currently in phase I clinical trials for the treatment of solid tumours. Naptumomab estafenatox, an anti-trophoblast glycoprotein 5T4 (TBGP) that Fab fused to *Staphylococcus aureus* enterotoxin E, developed by Active Biotech, is currently in phase III clinical trials for renal cell carcinoma and non-small cell lung carcinoma [17].

Conjugation to cytokines is also an effective way of enhancing anti-tumour efficacy. Philogen (Switzerland) are developing a number of immunocytokines for oncology indications including Fibromun, a scFv (L19) against the tumour antigen EDB fused to TNF, Darleukin, which contains L19 scFv fused to IL-2, Teleukin, which contains a vascular targeting antibody F16 linked to IL-2, and Dodekin which contains two subunits of the immunomodulatory payload IL-12 fused to a human vascular targeting antibody in tandem diabody format [116].

Radioimmunotherapy (RIT) is the combination of radiation therapy with Ab immunotherapy and has become an attractive strategy in cancer treatment because it allows for the selective destruction of cancer cells and is less pervasive than radiotherapy. The Ab recognises and binds the surface of the primary tumour site and disseminated disease tissue and thereby delivers high doses of radiation directly to the tumour without significant damage to healthy tissue. Recent examples are the generation of radiolabelled antibodies for the radioimmunotherapy of multiple myeloma [117] and radio-iodinated anti-HER2 Nanobody® for breast cancer [118].

Although the effector moieties add another mechanism through which the antibody fragments can mediate a therapeutic effect, antibody drug conjugates are not easy to develop and optimise. Many factors need to be considered, including what antibody/fragment is used, what to conjugate, what linker/chemistry to use, and the ratio of naked to conjugated antibodies [119,120].

4.2. Autoimmune and Inflammatory Diseases

Autoimmune diseases are chronic and potentially life-threatening, and antibody therapies are extremely expensive because they usually require intensive, life-long treatment. The lower production costs of antibody fragments and potential reduced immunogenicity due to their small size renders the use of antibody fragments with half-life extension moieties as a viable alternative to full-length antibodies. Furthermore, like for cancer immunotherapies, the development of antibody fragments for the treatment of autoimmune diseases has been growing at a fast pace and there are numerous possibilities for bispecific targeting.

One of the first antibody fragments to be marketed for an autoimmune disease indication was Certolizumab pegol (Cimzia®), a pegylated Fab targeting TNF developed by UCB (Belgium), approved by the FDA for the treatment of Crohn's disease in 2008. It has subsequently been approved for rheumatoid arthritis, psoriatic arthritis, and ankylosing spondylitis. Two other Fabs are in clinical trials: FR104 (OSE/Janssen) against CD28 in phase II for RA, and Dapirolizumab, an anti-CD40L Fab developed by UCB in phase II for SLE.

There are four nanobodies reported to be in clinical development: ALX-0061 (Vobarilizumab) is a monovalent Nanobody® against IL6R linked to a half-life extending Nb against HSA from Ablynx in phase II trials for RA and SLE [121,122]. ALX-0761 (Merck/Ablynx) is a bispecific Nanobody® targeting IL17A/IL17F linked to an HSA Nb in phase Ib for psoriasis [123], and Ozoralizumab or ATN-103 (Taisho) is a Nanobody® against TNF. Some success has been reported for ATN-103 in a phase II interventional long-term safety study in subjects with RA at week 48 [124]. ATN-192 is a pegylated version of ATN-103, which is in phase I clinical trials [125]. Several Nbs are also reported in pre-clinical development for autoimmune disease indications [44].

One scFv format currently being evaluated in a phase II clinical study for the treatment of RA is Dekavil or F8IL10 (Philogen). It is a fully human fusion protein composed of the vascular targeting scFv antibody F8 fused to the cytokine interleukin-10. A number of other immunocytokines fused to scFvs are also in preclinical development—the reader is referred to the Philogen website for further information [126].

MacroGenics are developing MGD-010, a DART® targeting CD32B and CD79B on B-cells for the treatment of autoimmune disorders. CD32B is a checkpoint molecule expressed on B lymphocytes that, when co-ligated with CD79B (a component of the B-cell antigen receptor complex), delivers a co-inhibitory signal that dampens B-cell activation. The intended mechanism of MGD010 is to modulate the function of human B-cells while avoiding their depletion. MGD010 completed phase I clinical trials in 2016 and has now been licensed to PreventionBio, who will evaluate the safety and efficacy of MGD-010, now called PRV-3279, in a phase Ib trial, expected to commence in the second half of 2019.

Another fragment in development for a wide variety of autoimmune disease indications is ARGX-113. This is an IgG1 Fc-fragment developed by ArGenX for the treatment of patients with high levels of circulating pathogenic IgG, found in acquired thrombotic thrombocytopenic purpura (aTTP), SLE, MS, or myasthenia gravis (MG). ARGX-113 binds to the neonatal Fc receptor (FcRn) and blocks IgG recycling, resulting in clearance of autoreactive antibodies through lysosomal degradation.

ARGX-113 showed statistically significant improvement in a phase II clinical trial on patients with MG [127] and is currently being evaluated for efficacy, safety, and tolerability in a randomised, double-blind, placebo-controlled, multicentre phase III trial in patients with MG having generalised muscle weakness [128].

The last category of antibody fragments tested in clinical trials for autoimmune and inflammatory diseases are dAbs. Lulizumab pegol, a pegylated Domain Antibody® targeting CD28 developed by Bristol-Myers Squibb, was evaluated in a phase II trial in subjects with active systemic lupus erythematosus (SLE). There was no significant difference between lulizumab and placebo for the primary (BICLA response rate) or secondary endpoints at week 24, although PD activity was observed [129].

GSK has developed two inhalable anti-TNFR1 VH domain antibodies for selective antagonism of TNF in the lung interstitium for acute respiratory distress syndrome (ARDS)/acute lung inflammation (ALI). Delivery of antibody fragments directly into the lung by inhalation has great potential for treatment of inflammatory lung diseases, the advantages being rapid onset of action, reduced systemic exposure, lower doses, as well as needle-less administration. GSK1995057 was tested in a phase I clinical trial in healthy volunteers. It was not developed further because of the presence of naturally occurring, pre-existing anti-drug antibodies (ADAs) which could lead to early neutralising anti-drug-antibody responses [130]. GSK1995057 was subsequently engineered by adding a C-terminal alanine residue to render it less susceptible to ADAs. The resultant GSK2862277 was tested in a phase I trial and was well tolerated when administered both as an orally inhaled aerosol and by iv route. A phase II placebo-controlled randomised trial in patients that were undergoing esophagectomy surgery and were at risk to develop ARDS has been completed [131] with results expected in 2019.

4.3. Other Clinical Applications

While oncology and autoimmune disease are two major areas in which antibody fragments have become a prominent class of therapeutic molecules, there are several other disease areas in which, although not as dominant, fragments are being evaluated.

4.3.1. Ophthalmic Indications

Antibody fragments such as Fabs and scFvs, unlike full-length antibodies, have been shown to be able to penetrate the cornea and pass into the eye and achieve clinically useful concentrations in the anterior chamber over a reasonable time-span following topical administration [132] but to date there are no reports of this route of administration being tested in the clinic. Most are administered by direct injection into the eye (intravitreal route).

The most common eye disorder treated with antibodies or antibody fragments is age-related macular degeneration (AMD), which is the leading cause of irreversible blindness in people aged 50 years or older, in the developed world. For AMD, the antibody fragments are applied directly to the eye via the intravitreal route. Extremely high local drug concentrations can be achieved in the eye with minimal risk of systemic side effects.

Ranibizumab (Lucentis®) is an anti-angiogenic monoclonal antibody fragment targeting VEGF-A, derived from the same parental mouse antibody as bevacizumab. It was approved in 2006 for wet AMD and subsequently in 2012 and 2015 for diabetic macular oedema and diabetic retinopathy, respectively.

Lampalizumab (Roche), a Fab against complement factor D, entered phase III clinical trials for geographic atrophy, an advanced form of age-related macular degeneration [133,134] in 2014 but failed to meet primary endpoints [135].

Brolucizumab (Alcon/Novartis) is a scFv targeting VEGF that is currently in phase III for wet AMD [136,137].

A number of antibody fragments are also in preclinical development for eye indications [138,139]. Elasmogen is developing V-NARs such as ELN/21, an ICOSL G-binding soloMER™, in preclinical development for posterior uveitis and corneal graft rejection, and ELN/12, an anti-VEGF soloMER^{TM} for AMD [140]. Abzyme has a bivalent nanobody targeting VEGF and TfR for wet AMD.

4.3.2. Infectious Diseases

Only three full-length mAbs have been approved for the treatment of infectious diseases: Synagis® (Palivizumab) for RSV infection, Abthrax® (Raxibacumab) against anthrax, and Zinplava^{TM} (Bezlotoxumab) against *C. difficile* (although technically the latter two mAbs neutralise bacterial toxins—see below) and there are currently over 60 mAbs in various stages of clinical trials for the treatment of infectious diseases including Ebola, hepatitis B, and respiratory syncytial virus (RSV). The development of antibody fragments in infectious diseases is under-exploited, likely due to their lack of effector function. To our knowledge, there are only three molecules currently in clinical trials and no approved therapies. Afelimomab is an F(ab′)$_2$ in phase III trials for sepsis toxic shock [17]. Rivabazumab pegol is a pegylated Fab in phase II for the treatment of chronic *Pseudomonas aeruginosa* infection [17]. The third is ALX-0171, a trivalent Nanobody® (VHH$_3$) in phase III for the treatment of RSV infection [17]. ALX-0171 binds to RSV F protein. The potency of the trivalent ALX-0171 against RSV-A and RSV-B strains was found to be several thousand-fold higher than that of the monovalent nanobody. It is also the first Nanobody® treatment developed for delivery directly into the lungs, the site of RSV infection, by nebulisation [141]. Unfortunately, Sanofi decided to stop development ALX-0171 in Feb 2019 [142].

Although ALX-0171 remains the only Nb to reach clinical trials in this therapy area, there have been many in vivo and in vitro studies investigating the use of nanobodies against a wide range of bacteria, viruses, parasites, and fungi, including rotavirus, norovirus, HIV-1, *Helicobacter pylori*, *Trypanosoma brucei*, and *Plasmodium falciparum*. The reader is directed to two recent reviews by Steeland et al. and Wilken and McPherson [44,141].

4.3.3. Anti-Toxins and Anti-Venoms

Traditionally, the treatment for envenoming has been the transfusion of serum from immunised animals. Primarily containing IgG, Fab and F(ab′)$_2$ fragments, much of which will not be specific for the venom, serotherapy can have several undesirable affects including IgE-mediated and non-IgE-mediated early adverse reactions, anaphylaxis and serum sickness [143]. The ability of antibody fragments to rapidly penetrate tissue, their lack of effector function, and their retained specificity and affinity for their antigen makes them promising candidates for anti-venoms. In addition, they can easily be produced recombinantly in a homogeneous form.

Although none have progressed to the clinic as yet, many Nbs have been generated that have shown the ability to neutralise toxins/venoms in in vitro and in vivo models. Venoms from *Androctonus australis* (Fat tailed scorpion), *Hemiscorpius lepturus* (Iranian scorpion), and *Naja kaouthia* (monocled cobra) and toxins from *Clostridium difficile*, *Vibrio cholera*, *Bacillus anthracis*, and *E. coli*, including Shiga toxins 1 and 2, have all been neutralised [44,144].

5. Non-Therapeutic Uses

Imaging & Diagnostics

The use of antibodies for molecular imaging is well established. In essence, their high affinity and specificity make them ideal for the detection of a specific surface protein in vivo or in vitro. Additionally, their large size allows for their conjugation to radioisotopes, fluorescent molecules, or even enzymes without inhibiting binding to their target [145].

However, conventional antibodies are by no means perfect for in vivo imaging. Their long half-lives and low rates of clearance necessitates a several-day waiting period to obtain an acceptable signal-to-noise ratio, exposing patients to excessive radiation from radioisotopes. In addition, potential off-target immune effects, conferred by the Fc domain, are undesirable. Antibody fragments, either produced recombinantly or by enzymatic cleavage, provide a solution to these downsides. Fragments lack an Fc domain, thus removing their immune activating potential and reducing their half-life simultaneously. This reduces the risk of disturbing the system being visualised, allows for rapid high-contrast imaging, and reduces radiation exposure [145]. In addition, their small size allows for better tissue distribution and provides more options with regard to epitope choice [12]. Nanobodies targeting CAIX and HER2 have been used for optical imaging of pre-invasive breast cancer, which requires a high tumour to background ratio [146,147].

Diabodies, tribodies, and tetrabodies also have potential uses in applications such as radioimmunotherapy and diagnostic in vivo imaging [148,149]. In addition, fluorescently labelled nanobodies have used for real-time analysis of epithelial mesenchymal transition [150].

The applications of single-domain antibodies in in vivo imaging and diagnostics are not restricted to oncology. In 2014, an anti-VCAM1 single-domain antibody fragment was shown to be an accurate and reproducible tool for the imaging of atherosclerotic lesions [151,152].

6. Future Opportunities

6.1. Neurodegenerative Diseases

Antibody therapies have traditionally been thought to be of limited relevance in the treatment of neurodegenerative disease due to the miniscule proportion of antibodies in circulation that can cross the blood–brain barrier (BBB) [153]. The four monoclonal antibodies approved for the treatment of multiple sclerosis—natalizumab (Tysabri®), alemtuzumab (Lemtrada®), daclizumab (Zinbryta®), and ocrelizumab (Ocrevus®)—are thought to mediate their effects primarily in the periphery [154]. There are also two antibodies against alpha synuclein (PRX002/RO7046015 from Roche and BIIB-054 from Biogen) entering clinical trials for Parkinson's disease. No antibody fragments have yet progressed into the clinic.

The causal mechanism of many neurodegenerative diseases, including Alzheimer's and Huntington's disease, involves aggregation of misfolded protein [155], and it has been shown that it is possible to raise antibodies that can neutralise these toxic aggregates. One possible approach to circumvent the BBB challenge is to express antibody fragments, termed intrabodies, within the cells of the brain. Although this approach may be far from reaching clinical trials, partially due to its invasive nature, it has been shown to be efficacious in some in vivo models [156]. Penetration of the BBB is discussed in Section 6.2.

6.2. Cell and Tissue Specific Antibody Delivery

Bispecific antibodies where one specificity is used to target the antibody to a specific tissue or cellular compartment and the second specificity is used to target the antigen of interest would have great advantages in limiting off-target effects due to systemic administration. Conditions in which the target of interest is located in the central nervous system (CNS) are particularly challenging, however, as most antibodies are generally unable to penetrate the BBB. Receptor-mediated transcytosis (RMT) is an example of a macromolecule transport system that is employed by cells of the BBB to supply essential proteins to the brain. This system can be utilised to deliver biologic payloads, such as antibodies, across the BBB. Increased brain penetration of therapeutic antibodies can be achieved by engineering bispecific antibodies in which one antibody binding specificity recognises a BBB receptor that undergoes RMT from the circulatory compartment into brain parenchyma, and the second binding specificity recognises a therapeutic target within the CNS. Anti-transferrin receptor (TfR)-based bispecific antibodies have previously shown promise for boosting antibody uptake in the brain [157]. Abzyme Therapeutics are now exploiting modular anti-TfR antibodies (nanobodies) and TFR-directed bispecifics capable of overcoming the blood–brain barrier for treatment of CNS disorders [158].

Some fragment formats, including F(ab')$_2$ and basic VHHs, have been demonstrated to cross the BBB with relative efficiency, although they may not necessarily accumulate to therapeutic concentration due to rapid clearance [159–161]. Engineering these formats to bind key proteins in the causal mechanism of neurodegenerative diseases will likely be challenging, but their existence shows the significant progress that has been made in recent years. In 2001, Muruganandam et al. identified two single-domain antibody fragments of camelid origin capable of crossing the BBB endothelium, FC5 and FC44, by phenotypic panning of a naive llama single-domain antibody phage display library [162]. Recently, Farrington et al. engineered FC5 as a mono- and bivalent fusion with the human Fc domain and showed up to a 30-fold enhanced apparent brain exposure (derived from serum and cerebrospinal fluid pharmacokinetic profiles) compared with control domain antibody-Fc fusions after systemic dosing in rats [163]. This study demonstrates that modular incorporation of FC5 as the BBB carrier arm in bispecific antibodies or antibody-drug conjugates may have potential use in the development of pharmacologically active biotherapeutics for CNS indications. As an alternative, Caljon et al. suggested grafting CDR loops of BBB penetrating Nbs onto an as-yet undiscovered scaffold to combine BBB penetration and high antigen specificity with the desired pharmacokinetic properties [161].

6.3. Intracellular Targeting

Antibodies have proven to be effective at modulating a wide variety of disease associated molecules belonging to different target classes, but there are still hundreds of disease-associated intracellular targets that are inaccessible to antibodies and undruggable with small molecules and that include phosphatases, E3 ubiquitin ligases, GTPases, and transcription factors. The majority of antibodies are unable to penetrate into cells. Thus, while small molecules drugs can easily penetrate cell membranes to hit intracellular targets, they often lack specificity, in particular when multiple targets have similar binding pockets. Furthermore, their small size makes them ineffective at blocking certain protein–protein interactions where large interfaces are involved. Antibodies or antibody fragments can solve the specificity problem and effectively block protein–protein interactions, but they have been largely restricted to targets in the extracellular milieu because they cannot cross the lipid bilayer. It has therefore been the 'holy grail' of many pharmaceutical companies to combine the targeting power of monoclonal antibodies with the cell-penetrating abilities of small molecules. Even if cell penetration can be achieved, a further complication is the need for tissue and cellular specificity to limit off-target effects, although one could argue that cell-specific action can be reached on the intracellular antigen level.

A number of different approaches have been described in the literature including transfection, cell penetrating peptides (CPPs), fragments of bacterial toxins, nanocarriers (lipid-based, polymer-based, and virus- and virus-like particle-based), and physical methods such as microinjection and electroporation and have recently been extensively reviewed [164]. Below we briefly describe some of the most promising approaches that could be applicable for the systemic delivery of antibody fragments.

A cell-penetrating, intracellular targeting antibody needs to bind to a receptor on the cell surface and trigger internalization, for example, by receptor-mediated endocytosis. Once internalised, the antibody needs to be able to escape from the endosome in order to bind to/neutralise its intracellular target in the cytosol, where the majority of potential therapeutic targets are concentrated, or in the nucleus.

A frequently used approach for intracellular delivery is the fusion of an antibody or antibody fragments including scFv, Fab, and nanobody to a CPP. Early studies showed that HIV tat and other CPPs such as the membrane translocating sequence from Kaposi fibroblast growth factor, the Antennapedia protein transduction domain, the Penetration of the Drosophila homeodomain, nona-arginine, and certain oligonucleotides could cross the plasma membrane to enter cells. However, their mechanism of intracellular delivery was unclear, and it is unknown if/how CPPs and CPP-conjugated antibody fragments are released from endocytotic vesicles, making efficient endosomal escape one of the limitations of this approach. Other limitations include the lack of optimisation for intracellular localization and the lack of cell or tissue-specific targeting. The latter may be solved by addition of a cell-targeting ligand to the antibody fragment. This was elegantly demonstrated by the authors in [165], who fused a cancer-cell specific, 17-amino acid peptide (BR2) to anti-mutated K-ras scFv, and demonstrated significant and cancer-cell-selective effects in vitro.

Most currently used CPPs seem to contain a disproportionately high number of positively charged lysine and/or arginine residues and have a high theoretical net charge. These positively charged residues facilitate interaction with negatively charged cell surface proteoglycans, such as sulfated proteoglycans like HSPG, ultimately enabling cellular uptake. These findings suggested that polycationic protein resurfacing could endow cell penetration properties, the downside being that extensive mutation of most proteins would lead to a loss of functional activity. To this end, Bruce et al. recently showed that nanobodies are amenable to cationic resurfacing [166]. Structural analysis of a GFP-binding nanobody revealed a number of solvent exposed residues that were not within the CDRs. Polycationic resurfacing of these solvent exposed residues resulted in a new protein that expressed well in *E. coli*, retained affinity for GFP, and penetrates mammalian cells. Analogous mutation of HER2-or β-lactamase-binding nanobodies also resulted in well-expressed nanobodies that exhibited potent cell-penetrating properties, and the majority of the internalised proteins were found to reside in the cytosol, although the mechanism of uptake remains unclear.

Probably one of the most exciting recent technology developments in the field of intracellular targeting has come from Orum Therapeutics (South Korea). Orum have developed a cell-penetrating antibody technology that uses a cell surface receptor-specific cyclic peptide fused to an antibody targeting activated K-ras that has been engineered from a naturally occurring autoimmune antibody able to penetrate into cells. The peptide, which confers cell-type specificity, is genetically linked to the light chain variable domain of the antibody. The light chain contains a sequence motif in L-CDR3, which enables the antibody to escape from acidified endosomes into the cytosol, where the heavy chain of the antibody is able to engage with and neutralise its intracellular target, activated K-ras [167]. Using this strategy, a humanised IgG1 format antibody named iMab RT11-i fused to a tumour-homing αv integrin binding RGD10 cyclic peptide was developed, and this had significant anti-tumour effects in vivo in a tumour xenograft model in mice, demonstrating the feasibility of this approach for the cytosolic delivery of antibodies. Based on this approach, it is conceivable that antibody fragments, e.g., nanobodies or domain antibodies with different targeting specificities, could be combined to achieve a similar effect.

7. Conclusions

In this review we have covered the structure of a range of antibody fragments, from the isolated domains of canonical IgG to nanobodies, their production in microbial prokaryotic and eukaryotic, cell-free systems, and their application in and outside of the clinic. We have discussed the main advantages of each fragment format with a focus on size and effector function and have highlighted the main mechanisms of action through which these fragments mediate their therapeutic effects.

Since the approval of muromonab in 1986, antibodies have rapidly expanded to become a major class of therapeutic molecule and are essential to the way we treat many diseases. Therapies have also diversified from the canonical mAb to a wide range of fragments and antibody-drug conjugate formats, which can offer context-dependent improvements to full-sized mAbs, the most common of which are discussed here. There is nothing to suggest that this trend will not continue with new formats and artificial frameworks (not discussed in this review) constantly being developed and gradually making their way into the clinic.

New formats offer exciting opportunities to expand the uses of antibodies into previously uncharted territory. For example, an orally taken antibody for the treatment of neurological diseases or antibodies against intracellular targets were once thought impossible due to an antibody's large size and susceptibility to acidic proteases. Small, highly stable, even at low pH, nanobodies that are easy to link in series may potentially offer a way to access previously undruggable targets and tissues whilst being completely un-invasive. Fragments allow us to break central tolerance and re-target a host's immune cells against cells displaying previously unrecognised cancer antigens in tissues too far removed from circulation for conventional antibodies to access. Fragments allow us to discover new antibodies at an astounding rate, create a vast array of multi-specific molecules, and rapidly search for indicators of disease almost anywhere in the body.

Although still not without their limitations and complications, the future of antibody-derived fragments undoubtedly looks bright. The diverse range of formats and modifications available combined with yet to be explored sequence space may allow us to overcome the challenges that we face in this modern era, from diseases of the wealthy and old aged, to the infectious and transmissible.

Funding: This research received no external funding.

Acknowledgments: Abthrax® is a registered trademark of Human Genome Science, Inc. BiTE® is a registered trademark of Micromet, Inc., Amgen; Blincyto® is a registered trademark of Amgen Inc.; Cablivi™ is a registered trademark of Ablynx; Cimzia® is a registered trademark of UCB Pharma, S.A.; DART® is a registered trademark of MacroGenics; ImmTAC® and ImmTAV® are registered trademarks of Immunocore Ltd.; Lemtrada® is a registered trademark of the Genzyme Corporation; Leucentis® is a registered trademark of Novartis AG; Ocrevus® is a registered trademark of Genentech, Inc.; Orthoclone™ is a registered trademark of Johnson & Johnson; PER.C6™ is a registered trademark of Janssen Vaccines & Prevention B.V.; Praxbind® is a registered trademark of Boehringer Ingelheim; ReoPro® is a registered trademark of Eli Lilly and Company; ROCK® is a registered trademark of Affimed; SoloMER™ is a registered trademark of Elasmogen Ltd.; Synagis® is a registered trademark of the AstraZeneca group of companies; The Domain Antibody® and AlbudAb® platforms are trademarks of GlaxoSmithKline plc.; The Nanobody® platform is a registered trademark of Ablynx; Tysabri® and Zinbryta® are registered trademarks of Biogen MA Inc.; Zinplava™ is a registered trademark of Merck Sharpe & Dohme corp.

Conflicts of Interest: The authors declare no conflicts of interest. A.B. is a complementary worker on assignment at GSK. C.A.P. is a full time employee of GSK.

Abbreviations

6HIS	hexa-histidine
ADA	anti-drug antibody
ADCC	antibody dependant cellular cytotoxicity
ADCP	antibody dependant cellular phagocytosis
ADH1	alcohol dehydrogenase 1
AIEX	anion exchange chromatography
ALI	acute lung inflammation
ARDS	acute respiratory distress syndrome
aTTP	acquired thrombotic thrombocytopenic purpura
ALL	acute lymphoblastic leukaemia
AMD	age-related macular degeneration
AML	acute myeloid leukaemia
BBB	blood–brain barrier
BCMA	B-cell maturation antigen
BiKE	bispecific natural killer cell engager
BiTE®	bispecific T-cell engager
CDC	complement-dependent cytotoxicity
CDR	complementarity determining region
CEA	carcinoembryonic antigen
CHO	Chinese hamster ovary
CIEX	cation exchange chromatography
CL	constant domain of immunoglobulin light chain
CNS	central nervous system
CPP	cell penetrating peptide
dAb	Domain antibody®
DART®	dual affinity re-targeting protein

EpCAM	epithelial cell adhesion molecule
ER	endoplasmic reticulum
Fab	fragment of antigen binding
Fc	fragment crystallizable
FcRn	neonatal fragment crystallizable receptor
Fv	fragment variable
GAPDH	glyceraldehyde-3-phosphate dehydrogenase
GFP	green fluorescent protein
GM-CSF	granulocyte-monocyte colony stimulating factor
GST	glutathione-S transferase
HAVH	human anti-VH
HEK	human embryonic kidney
HER2	human epidermal growth factor 2
HIC	hydrophobic interaction chromatography
HIV	human immunodeficiency virus
HLA	human leukocyte antigen
IgG	immunoglobulin gamma
IMAC	immobilised metal affinity chromatography
ImmTAC®	immune mobilising monoclonal t-cell receptors against cancer
ImmTAV®	immune mobilising monoclonal t-cell receptors against virus antigens
mAb	monoclonal antibody
MBP	mannose binding protein
MDS	myelodysplastic syndrome
MDSC	myeloid derived suppressor cell
MG	myasthenia gravis
MHC	major histocompatibility complex
MS	multiple sclerosis
Nb	Nanobody®
NK cell	Natural killer cell
NY ESO	New York esophageal squamous cell carcinoma
PD	pharmacodynamic
PEG	polyethylene glycol
PGK1	phosphoglycerate kinase 1
RMT	receptor-mediated endocytosis
RSV	respiratory syncytial virus
scFv	single chain fragment variable
SEC	size exclusion chromatography
SLE	systemic lupus erythematosus
TBGP	anti-trophoblast glycoprotein 5T4
TCR	T-cell receptor
TfR	transferrin receptor
TNF	tumour necrosis factor
IL	interleukin
TNFR1	tumour necrosis factor receptor 1
TriKE	trispecific natural killer cell engager
VEGF	vascular endothelial growth factor
VH	variable domain of immunoglobulin heavy chain
VL	variable domain of immunoglobulin light chain
V-NAR	variable new antigen receptor
vWF	von Willebrand factor

References

1. Ecker, D.M.; Jones, S.D.; Levine, H.L. The therapeutic monoclonal antibody market. *MAbs* **2015**, *7*. [CrossRef] [PubMed]

2. Müller, D.; Kontermann, R.E. Bispecific Antibodies. In *Handbook of Therapeutic Antibodies*, 2nd ed.; Wiley-Blackwell: Hoboken, NJ, USA, 2014; ISBN 9783527682423.

3. Drake, P.M.; Rabuka, D. An emerging playbook for antibody-drug conjugates: Lessons from the laboratory and clinic suggest a strategy for improving efficacy and safety. *Curr. Opin. Chem. Biol.* **2015**, *28*, 174–180. [CrossRef] [PubMed]

4. Nelson, A.L. Antibody fragments: Hope and hype. *MAbs* **2010**, *2*, 77–83. [CrossRef]

5. Fernandes, J.C. Therapeutic application of antibody fragments in autoimmune diseases: Current state and prospects. *Drug Discov. Today* **2018**, *23*, 1996–2002. [CrossRef] [PubMed]

6. Kholodenko, R.V.; Kalinovsky, D.V.; Doronin, I.I.; Ponomarev, E.D.; Kholodenko, I. V Antibody Fragments as Potential Biopharmaceuticals for Cancer Therapy: Success and Limitations. *Curr. Med. Chem.* **2017**. [CrossRef] [PubMed]

7. Bird, R.E.; Hardman, K.D.; Jacobson, J.W.; Johnson, S.; Kaufman, B.M.; Lee, S.M.; Lee, T.; Pope, S.H.; Riordan, G.S.; Whitlow, M. Single-chain antigen-binding proteins. *Science* **1988**. [CrossRef]

8. Huston, J.S.; Levinson, D.; Mudgett-Hunter, M.; Tai, M.S.; Novotný, J.; Margolies, M.N.; Ridge, R.J.; Bruccoleri, R.E.; Haber, E.; Crea, R. Protein engineering of antibody binding sites: Recovery of specific activity in an anti-digoxin single-chain Fv analogue produced in *Escherichia coli*. *Proc. Natl. Acad. Sci. USA* **1988**, *85*, 5879–5883. [CrossRef] [PubMed]

9. Montoliu-Gaya, L.; Esquerda-Canals, G.; Bronsoms, S.; Villegas, S. Production of an anti-Aβ antibody fragment in Pichia pastoris and in vitro and in vivo validation of its therapeutic effect. *PLoS ONE* **2017**. [CrossRef] [PubMed]

10. Spadiut, O.; Capone, S.; Krainer, F.; Glieder, A.; Herwig, C. Microbials for the production of monoclonal antibodies and antibody fragments. *Trends Biotechnol.* **2014**, *32*, 54–60. [CrossRef]

11. Yokota, T.; Milenic, D.E.; Whitlow, M.; Schlom, J. Rapid Tumor Penetration of a Single-Chain Fv and Comparison with Other Immunoglobulin Forms. *Cancer Res.* **1992**. [CrossRef]

12. Li, Z.; Krippendorff, B.F.; Sharma, S.; Walz, A.C.; Lavé, T.; Shah, D.K. Influence of molecular size on tissue distribution of antibody fragments. *MAbs* **2016**. [CrossRef]

13. Cumber, A.J.; Ward, E.S.; Winter, G.; Parnell, G.D.; Wawrzynczak, E.J. Comparative stabilities in vitro and in vivo of a recombinant mouse antibody FvCys fragment and a bisFvCys conjugate. *J. Immunol.* **1992**, *149*, 120–126.

14. Sanz, L.; Cuesta, Á.M.; Compte, M.; Álvarez-Vallina, L. Antibody engineering: Facing new challenges in cancer therapy. *Acta Pharmacol. Sin.* **2005**, *26*, 641–648. [CrossRef]

15. Jain, A.; Jain, S. PEGylation: An approach for drug delivery. A review. *Crit. Rev. Drug Carr. Syst.* **2008**. [CrossRef]

16. Müller, D.; Karle, A.; Meißburger, B.; Höfig, I.; Stork, R.; Kontermann, R.E. Improved pharmacokinetics of recombinant bispecific antibody molecules by fusion to human serum albumin. *J. Biol. Chem.* **2007**. [CrossRef]

17. Poiron, C.; Wu, Y.; Ginestoux, C.; Ehrenmann, F.; Duroux, P.; Lefranc, M. IMGT®, the international ImMunoGeneTics information system®. *Nucleic Acids Res.* **2008**, *37* (Suppl. S1), D1006–D1012.

18. Brinkmann, U.; Kontermann, R.E. The making of bispecific antibodies. *MAbs* **2017**, *9*, 182–212. [CrossRef] [PubMed]

19. Hayden, M.S.; Linsley, P.S.; Gayle, M.A.; Bajorath, J.; Brady, W.A.; Norris, N.A.; Fell, H.P.; Ledbetter, J.A.; Gilliland, L.K. Single-chain mono- and bispecific antibody derivatives with novel biological properties and antitumour activity from a COS cell transient expression system. *Ther. Immunol.* **1994**, *1*, 3–15. [PubMed]

20. Holt, L.J.; Basran, A.; Jones, K.; Chorlton, J.; Jespers, L.S.; Brewis, N.D.; Tomlinson, I.M. Anti-serum albumin domain antibodies for extending the half-lives of short lived drugs. *Protein Eng. Des. Sel.* **2008**. [CrossRef]

21. Holliger, P. "Diabodies": Small Bivalent and Bispecific Antibody Fragments. *Proc. Natl. Acad. Sci. USA* **1993**. [CrossRef]

22. Lu, D.; Jimenez, X.; Witte, L.; Zhu, Z. The effect of variable domain orientation and arrangement on the antigen-binding activity of a recombinant human bispecific diabody. *Biochem. Biophys. Res. Commun.* **2004**. [CrossRef]

23. Kipriyanov, S.M.; Moldenhauer, G.; Braunagel, M.; Reusch, U.; Cochlovius, B.; Le Gall, F.; Kouprianova, O.A.; Von Der Lieth, C.W.; Little, M. Effect of domain order on the activity of bacterially produced bispecific single-chain Fv antibodies. *J. Mol. Biol.* **2003**. [CrossRef]

24. MacroGenics Pipeline. Available online: https://www.macrogenics.com/pipeline/ (accessed on 16 January 2019).

25. Affimed Pipeline. Available online: https://www.affimed.com/pipeline/ (accessed on 17 January 2019).

26. Huehls, A.M.; Coupet, T.A.; Sentman, C.L. Bispecific T-cell engagers for cancer immunotherapy. *Immunol. Cell Biol.* **2015**, *93*, 290–296. [CrossRef]

27. Kipriyanov, S.M.; Moldenhauer, G.; Schuhmacher, J.; Cochlovius, B.; Von Der Lieth, C.W.; Matys, E.R.; Little, M. Bispecific tandem diabody for tumor therapy with improved antigen binding and pharmacokinetics. *J. Mol. Biol.* **1999**. [CrossRef]

28. Wu, C.; Ying, H.; Grinnell, C.; Bryant, S.; Miller, R.; Clabbers, A.; Bose, S.; McCarthy, D.; Zhu, R.R.; Santora, L.; et al. Simultaneous targeting of multiple disease mediators by a dual-variable-domain immunoglobulin. *Nat. Biotechnol.* **2007**. [CrossRef]

29. Metz, S.; Haas, A.K.; Daub, K.; Croasdale, R.; Stracke, J.; Lau, W.; Georges, G.; Josel, H.-P.; Dziadek, S.; Hopfner, K.-P.; et al. Bispecific digoxigenin-binding antibodies for targeted payload delivery. *Proc. Natl. Acad. Sci. USA* **2011**. [CrossRef]

30. Xencor Pipeline. Available online: https://www.xencor.com/pipeline/ (accessed on 17 January 2019).

31. Nelson, A.L.; Dhimolea, E.; Reichert, J.M. Development trends for human monoclonal antibody therapeutics. *Nat. Rev. Drug Discov.* **2010**, *9*, 767–774. [CrossRef]

32. Bazin-Redureau, M.I.; Renard, C.B.; Scherrmann, J.M.G. Pharmacokinetics of heterologous and homologous immunoglobulin G, F(ab′)2 and Fab after intravenous administration in the rat. *J. Pharm. Pharmacol.* **1997**. [CrossRef]

33. Röthlisberger, D.; Honegger, A.; Plückthun, A. Domain interactions in the Fab fragment: A comparative evaluation of the single-chain Fv and Fab format engineered with variable domains of different stability. *J. Mol. Biol.* **2005**. [CrossRef]

34. Simister, N.E.; Mostov, K.E. An Fc receptor structurally related to MHC class I antigens. *Nature* **1989**. [CrossRef]

35. Ober, R.J.; Martinez, C.; Lai, X.; Zhou, J.; Ward, E.S. Exocytosis of IgG as mediated by the receptor, FcRn: An analysis at the single-molecule level. *Proc. Natl. Acad. Sci. USA* **2004**. [CrossRef]

36. Rodewald, R.; Kraehenbuhl, J.P. Receptor-mediated transport of IgG. *J. Cell Biol.* **1984**, *99*, 159s–164s. [CrossRef]

37. Chapman, A.P.; Antoniw, P.; Spitali, M.; West, S.; Stephens, S.; King, D.J. Therapeutic antibody fragments with prolonged in vivo half-lives. *Nat. Biotechnol.* **1999**. [CrossRef]

38. Schreiber, S. Certolizumab pegol for the treatment of Crohn's disease. *Ther. Adv. Gastroenterol.* **2011**, *357*, 228–238. [CrossRef]

39. Hamers-Casterman, C.; Atarhouch, T.; Muyldermans, S.; Robinson, G.; Hamers, C.; Songa, E.B.; Bendahman, N.; Hammers, R. Naturally occurring antibodies devoid of light chains. *Nature* **1993**. [CrossRef]

40. Cortez-Retamozo, V.; Backmann, N.; Senter, P.D.; Wernery, U.; De Baetselier, P.; Muyldermans, S.; Revets, H. Efficient Cancer Therapy with a Nanobody-Based Conjugate. *Cancer Res.* **2004**. [CrossRef]

41. Van Der Linden, R.H.J.; Frenken, L.G.J.; De Geus, B.; Harmsen, M.M.; Ruuls, R.C.; Stok, W.; De Ron, L.; Wilson, S.; Davis, P.; Verrips, C.T. Comparison of physical chemical properties of llama V(HH) antibody fragments and mouse monoclonal antibodies. *Biochim. Biophys. Acta Protein Struct. Mol. Enzymol.* **1999**. [CrossRef]

42. De Genst, E.; Silence, K.; Decanniere, K.; Conrath, K.; Loris, R.; Kinne, J.; Muyldermans, S.; Wyns, L. Molecular basis for the preferential cleft recognition by dromedary heavy-chain antibodies. *Proc. Natl. Acad. Sci. USA* **2006**. [CrossRef]

43. Harmsen, M.M.; De Haard, H.J. Properties, production, and applications of camelid single-domain antibody fragments. *Appl. Microbiol. Biotechnol.* **2007**, *77*, 13–22. [CrossRef]

44. Steeland, S.; Vandenbroucke, R.E.; Libert, C. Nanobodies as therapeutics: Big opportunities for small antibodies. *Drug Discov. Today* **2016**, *21*, 1076–1113. [CrossRef]

45. Cortez-Retamozo, V.; Lauwereys, M.; Hassanzadeh Gh., G.; Gobert, M.; Conrath, K.; Muyldermans, S.; De Baetselier, P.; Revets, H. Efficient tumor targeting by single-domain antibody fragments of camels. *Int. J. Cancer* **2002**. [CrossRef]

46. Shen, J.; Vil, M.D.; Jimenez, X.; Iacolina, M.; Zhang, H.; Zhu, Z. Single variable domain-IgG fusion: A novel recombinant approach to Fc domain-containing bispecific antibodies. *J. Biol. Chem.* **2006**. [CrossRef]

47. Greenberg, A.S.; Avila, D.; Hughes, M.; Hughes, A.; McKinney, E.C.; Flajnik, M.F. A new antigen receptor gene family that undergoes rearrangement and extensive somatic diversification in sharks. *Nature* **1995**. [CrossRef]

48. Stanfield, R.L.; Dooley, H.; Verdino, P.; Flajnik, M.F.; Wilson, I.A. Maturation of Shark Single-domain (IgNAR) Antibodies: Evidence for Induced-fit Binding. *J. Mol. Biol.* **2007**. [CrossRef]

49. Zielonka, S.; Empting, M.; Grzeschik, J.; Könning, D.; Barelle, C.J.; Kolmar, H. Structural insights and biomedical potential of IgNAR scaffolds from sharks. *MAbs* **2015**, *7*, 15–25. [CrossRef]

50. Davies, J.; Riechmann, L. "Camelising" human antibody fragments: NMR studies on VH domains. *FEBS Lett.* **1994**. [CrossRef]

51. Holt, L.J.; Herring, C.; Jespers, L.S.; Woolven, B.P.; Tomlinson, I.M. Domain antibodies: Proteins for therapy. *Trends Biotechnol.* **2003**, *21*, 484–490. [CrossRef]

52. Scott, M.J.; Lee, J.A.; Wake, M.S.; Batt, K.V.; Wattam, T.A.; Hiles, I.D.; Batuwangala, T.D.; Ashman, C.I.; Steward, M. 'In-Format' screening of a novel bispecific antibody format reveals significant potency improvements relative to unformatted molecules. *MAbs* **2017**. [CrossRef]

53. O'Connor-Semmes, R.L.; Lin, J.; Hodge, R.J.; Andrews, S.; Chism, J.; Choudhury, A.; Nunez, D.J. GSK2374697, a Novel Albumin-Binding Domain Antibody (AlbudAb), Extends Systemic Exposure of Exendin-4: First Study in Humans—PK/PD and Safety. *Clin. Pharmacol. Ther.* **2014**. [CrossRef]

54. Jefferis, R. Recombinant antibody therapeutics: The impact of glycosylation on mechanisms of action. *Trends Pharmacol. Sci.* **2009**, *30*, 356–362. [CrossRef]

55. Li, F.; Vijayasankaran, N.; Shen, A.; Kiss, R.; Amanullah, A. Cell culture processes for monoclonal antibody production. *MAbs* **2010**, *2*, 466–479. [CrossRef]

56. Striedner, G.; Pfaffenzeller, I.; Markus, L.; Nemecek, S.; Grabherr, R.; Bayer, K. Plasmid-free T7-based Escherichia coli expression systems. *Biotechnol. Bioeng.* **2010**. [CrossRef]

57. Mairhofer, J.; Cserjan-Puschmann, M.; Striedner, G.; Nöbauer, K.; Razzazi-Fazeli, E.; Grabherr, R. Marker-free plasmids for gene therapeutic applications—Lack of antibiotic resistance gene substantially improves the manufacturing process. *J. Biotechnol.* **2010**. [CrossRef]

58. Sonoda, H.; Kumada, Y.; Katsuda, T.; Yamaji, H. Effects of cytoplasmic and periplasmic chaperones on secretory production of single-chain Fv antibody in Escherichia coli. *J. Biosci. Bioeng.* **2011**. [CrossRef]

59. Yuan, J.; Zweers, J.C.; Van Dijl, J.M.; Dalbey, R.E. Protein transport across and into cell membranes in bacteria and archaea. *Cell. Mol. Life Sci.* **2010**, *67*, 179–199. [CrossRef]

60. Levy, R.; Ahluwalia, K.; Bohmann, D.J.; Giang, H.M.; Schwimmer, L.J.; Issafras, H.; Reddy, N.B.; Chan, C.; Horwitz, A.H.; Takeuchi, T. Enhancement of antibody fragment secretion into the Escherichia coli periplasm by co-expression with the peptidyl prolyl isomerase, FkpA, in the cytoplasm. *J. Immunol. Methods* **2013**. [CrossRef]

61. Jalalirad, R. Production of antibody fragment (Fab) throughout *Escherichia coli* fed-batch fermentation process: Changes in titre, location and form of product. *Electron. J. Biotechnol.* **2013**. [CrossRef]

62. Gorlani, A.; De Haard, H.; Verrips, T. Expression of VHHs in saccharomyces cerevisiae. *Methods Mol. Biol.* **2012**. [CrossRef]

63. Chee, M.K.; Haase, S.B. New and Redesigned pRS Plasmid Shuttle Vectors for Genetic Manipulation of *Saccharomyces cerevisiae. G3 Genes Genomes Genet.* **2012**. [CrossRef]

64. Leite, F.C.B.; dos Anjos, R.S.G.; Basilio, A.C.M.; Leal, G.F.C.; Simões, D.A.; de Morais, M.A. Construction of integrative plasmids suitable for genetic modification of industrial strains of Saccharomyces cerevisiae. *Plasmid* **2013**. [CrossRef]

65. Partow, S.; Siewers, V.; Bjørn, S.; Nielsen, J.; Maury, J. Characterization of different promoters for designing a new expression vector in Saccharomyces cerevisiae. *Yeast* **2010**. [CrossRef]

66. Maury, J.; Asadollahi, M.A.; Møller, K.; Schalk, M.; Clark, A.; Formenti, L.R.; Nielsen, J. Reconstruction of a bacterial isoprenoid biosynthetic pathway in Saccharomyces cerevisiae. *FEBS Lett.* **2008**. [CrossRef]

67. Joosten, V.; Lokman, C.; van den Hondel, C.A.M.J.J.; Punt, P.J. The production of antibody fragments and antibody fusion proteins by yeasts and filamentous fungi. *Microb. Cell Fact.* **2003**, *2*. [CrossRef]

68. Xu, P.; Raden, D.; Doyle, F.J.; Robinson, A.S. Analysis of unfolded protein response during single-chain antibody expression in Saccharomyces cerevisiae reveals different roles for BiP and PDI in folding. *Metab. Eng.* **2005**. [CrossRef]

69. Ferndahl, C.; Bonander, N.; Logez, C.; Wagner, R.; Gustafsson, L.; Larsson, C.; Hedfalk, K.; Darby, R.A.J.; Bill, R.M. Increasing cell biomass in Saccharomyces cerevisiae increases recombinant protein yield: The use of a respiratory strain as a microbial cell factory. *Microb. Cell Fact.* **2010**. [CrossRef]

70. Delic, M.; Mattanovich, D.; Gasser, B. Repressible promoters—A novel tool to generate conditional mutants in Pichia pastoris. *Microb. Cell Fact.* **2013**. [CrossRef]

71. Sohn, S.B.; Graf, A.B.; Kim, T.Y.; Gasser, B.; Maurer, M.; Ferrer, P.; Mattanovich, D.; Lee, S.Y. Genome-scale metabolic model of methylotrophic yeast Pichia pastoris and its use for in silico analysis of heterologous protein production. *Biotechnol. J.* **2010**. [CrossRef]

72. Jahic, M.; Rotticci-Mulder, J.; Martinelle, M.; Hult, K.; Enfors, S.O. Modeling of growth and energy metabolism of Pichia pastoris producing a fusion protein. *Bioprocess Biosyst. Eng.* **2001**. [CrossRef]

73. Detalle, L.; Stohr, T.; Palomo, C.; Piedra, P.A.; Gilbert, B.E.; Mas, V.; Millar, A.; Power, U.F.; Stortelers, C.; Allosery, K.; et al. Generation and characterization of ALX-0171, a potent novel therapeutic nanobody for the treatment of respiratory syncytial virus infection. *Antimicrob. Agents Chemother.* **2016**. [CrossRef]

74. Vallet-Courbin, A.; Larivière, M.; Hocquellet, A.; Hemadou, A.; Parimala, S.N.; Laroche-Traineau, J.; Santarelli, X.; Clofent-Sanchez, G.; Jacobin-Valat, M.J.; Noubhani, A. A recombinant human anti-platelet SCFV antibody produced in pichia pastoris for atheroma targeting. *PLoS ONE* **2017**. [CrossRef]

75. Oh, I.S.; Lee, J.C.; Lee, M.S.; Chung, J.H.; Kim, D.M. Cell-free production of functional antibody fragments. *Bioprocess Biosyst. Eng.* **2010**. [CrossRef]

76. Stech, M.; Nikolaeva, O.; Thoring, L.; Stöcklein, W.F.M.; Wüstenhagen, D.A.; Hust, M.; Dübel, S.; Kubick, S. Cell-free synthesis of functional antibodies using a coupled in vitro transcription-Translation system based on CHO cell lysates. *Sci. Rep.* **2017**. [CrossRef]

77. Stech, M.; Hust, M.; Schulze, C.; Dübel, S.; Kubick, S. Cell-free eukaryotic systems for the production, engineering, and modification of scFv antibody fragments. *Eng. Life Sci.* **2014**. [CrossRef]

78. Stech, M.; Kubick, S. Cell-Free Synthesis Meets Antibody Production: A Review. *Antibodies* **2015**. [CrossRef]

79. Ryabova, L.A.; Desplancq, D.; Spirin, A.S.; Plückthun, A. Functional antibody production using cell-free translation: Effects of protein disulfide isomerase and chaperones. *Nat. Biotechnol.* **1997**. [CrossRef]

80. Jiang, X.; Ookubo, Y.; Fujii, I.; Nakano, H.; Yamane, T. Expression of Fab fragment of catalytic antibody 6D9 in an *Escherichia coli* in vitro coupled transcription/translation system. *FEBS Lett.* **2002**. [CrossRef]

81. Kanter, G.; Yang, J.; Voloshin, A.; Levy, S.; Swartz, J.R.; Levy, R. Cell-free production of scFv fusion proteins: An efficient approach for personalized lymphoma vaccines. *Blood* **2007**. [CrossRef]

82. Shimizu, Y.; Kuruma, Y.; Ying, B.W.; Umekage, S.; Ueda, T. Cell-free translation systems for protein engineering. *FEBS J.* **2006**, *273*, 4133–4140. [CrossRef]

83. STR001—Clinical Trial: NCT03424603. Available online: https://clinicaltrials.gov/ct2/show/NCT03424603 (accessed on 15 January 2019).

84. Wang, A.C.; Wang, I.Y. Cleavage sites of human IgGl immunoglobulin by papain. *Immunochemistry* **1977**. [CrossRef]

85. Jones, R.G.A.; Landon, J. Enhanced pepsin digestion: A novel process for purifying antibody F(ab')2 fragments in high yield from serum. *J. Immunol. Methods* **2002**. [CrossRef]

86. Genovis Website. Available online: https://www.genovis.com (accessed on 17 January 2019).

87. Hober, S.; Nord, K.; Linhult, M. Protein A chromatography for antibody purification. *J. Chromatogr. B Anal. Technol. Biomed. Life Sci.* **2007**, *848*, 40–47. [CrossRef]

88. Grodzki, A.C.; Berenstein, E. Antibody Purification: Affinity Chromatography—Protein A and Protein G Sepharose. In *Immunocytochemical Methods and Protocols*; Humana Press: New York, NY, USA, 2009.

89. Roben, P.W.; Salem, A.N.; Silverman, G.J. VH3 family antibodies bind domain D of staphylococcal protein A. *J. Immunol.* **1995**, *154*, 6437–6445.

90. Rodrigo, G.; Gruvegård, M.; Van Alstine, J. Antibody Fragments and Their Purification by Protein L Affinity Chromatography. *Antibodies* **2015**. [CrossRef]

91. Nascimento, A.; Pinto, I.F.; Chu, V.; Aires-Barros, M.R.; Conde, J.P.; Azevedo, A.M. Studies on the purification of antibody fragments. *Sep. Purif. Technol.* **2018**. [CrossRef]

92. Björck, L. A Novel Bacterial Cell Wall Protein with Affinity for Ig L Chains. *J. Immunol.* **1988**, *140*, 1194–1197.

93. De Château, M.; Nilson, B.H.K.; Erntell, M.; Myhre, E.; Magnusson, C.G.M.; Åkerström, B.; Björck, L. On the Interaction between Protein L and Immunoglobulins of Various Mammalian Species. *Scand. J. Immunol.* **1993**. [CrossRef]

94. Lichty, J.J.; Malecki, J.L.; Agnew, H.D.; Michelson-Horowitz, D.J.; Tan, S. Comparison of affinity tags for protein purification. *Protein Expr. Purif.* **2005**. [CrossRef]

95. Goel, A.; Colcher, D.; Koo, J.S.; Booth, B.J.M.; Pavlinkova, G.; Batra, S.K. Relative position of the hexahistidine tag effects binding properties of a tumor-associated single-chain Fv construct. *Biochim. Biophys. Acta Gen. Subj.* **2000**. [CrossRef]

96. Schmeisser, H.; Kontsek, P.; Esposito, D.; Gillette, W.; Schreiber, G.; Zoon, K.C. Binding Characteristics of IFN-alpha Subvariants to IFNAR2-EC and Influence of the 6-Histidine Tag. *J. Interferon Cytokine Res.* **2006**. [CrossRef]

97. Das, D.; Allen, T.M.; Suresh, M.R. Comparative evaluation of two purification methods of anti-CD19-c-myc-His6-Cys scFv. *Protein Expr. Purif.* **2005**. [CrossRef]

98. Liu, H.; Gaza-Bulseco, G.; Chumsae, C. Analysis of Reduced Monoclonal Antibodies Using Size Exclusion Chromatography Coupled with Mass Spectrometry. *J. Am. Soc. Mass Spectrom.* **2009**. [CrossRef]

99. Ljunglöf, A.; Lacki, K.M.; Mueller, J.; Harinarayan, C.; van Reis, R.; Fahrner, R.; Van Alstine, J.M. Ion exchange chromatography of antibody fragments. *Biotechnol. Bioeng.* **2007**. [CrossRef]

100. Lee, H.J.; Lee, C.M.; Kim, K.; Yoo, J.M.; Kang, S.M.; Ha, G.S.; Park, M.K.; Choi, M.A.; Lee, D.E.; Seong, B.L. Purification of antibody fragments for the reduction of charge variants using cation exchange chromatography. *J. Chromatogr. B Anal. Technol. Biomed. Life Sci.* **2018**. [CrossRef]

101. Karkov, H.S.; Krogh, B.O.; Woo, J.; Parimal, S.; Ahmadian, H.; Cramer, S.M. Investigation of protein selectivity in multimodal chromatography using in silico designed Fab fragment variants. *Biotechnol. Bioeng.* **2015**. [CrossRef]

102. Wu, Z.; Cheung, N.V. T-cell engaging bispecific antibody (T-BsAb): From technology to therapeutics. *Pharmacol. Ther.* **2018**, *182*, 161–175. [CrossRef]

103. Velasquez, M.P.; Bonifant, C.L.; Gottschalk, S. Redirecting T-cells to hematological malignancies with bispecific antibodies. *Blood* **2018**, *131*, 30–38. [CrossRef]

104. Offner, S.; Hofmeister, R.; Romaniuk, A.; Kufer, P.; Baeuerle, P.A. Induction of regular cytolytic T-cell synapses by bispecific single-chain antibody constructs on MHC class I-negative tumor cells. *Mol. Immunol.* **2006**. [CrossRef]

105. Bispecific Antibodies Technology: NK Cells Engagers—Innate Pharma. Available online: https://www.innate-pharma.com/en/pipeline/bispecific-antibodies-technology-nk-cells-engagers (accessed on 18 January 2019).

106. Gleason, M.K.; Ross, J.A.; Warlick, E.D.; Lund, T.C.; Verneris, M.R.; Wiernik, A.; Spellman, S.; Haagenson, M.D.; Lenvik, A.J.; Litzow, M.R.; et al. CD16xCD33 bispecific killer cell engager (BiKE) activates NK cells against primary MDS and MDSC CD33+ targets. *Blood* **2014**. [CrossRef]

107. Felices, M.; Sarhan, D.; Brandt, L.; Guldevall, K.; McElmurry, R.; Lenvik, A.; Chu, S.; Tolar, J.; Taras, E.; Spellman, S.R.; et al. CD16-IL15-CD33 Trispecific Killer Engager (TriKE) Overcomes Cancer-Induced Immune Suppression and Induces Natural Killer Cell-Mediated Control of MDS and AML Via Enhanced Killing Kinetics. *Blood* **2016**, *128*, 4291.

108. Lichtenegger, F.S.; Rothe, M.; Schnorfeil, F.M.; Deiser, K.; Krupka, C.; Augsberger, C.; Schlüter, M.; Neitz, J.; Subklewe, M. Targeting LAG-3 and PD-1 to enhance T-cell activation by antigen-presenting cells. *Front. Immunol.* **2018**. [CrossRef]

109. Moore, P.A.; Zhang, W.; Rainey, G.J.; Burke, S.; Li, H.; Huang, L.; Gorlatov, S.; Veri, M.C.; Aggarwal, S.; Yang, Y.; et al. Application of dual affinity retargeting molecules to achieve optimal redirected T-cell killing of B-cell lymphoma. *Blood* **2011**. [CrossRef]

110. Oates, J.; Jakobsen, B.K. ImmTACs: Novel bi-specific agents for targeted cancer therapy. *Oncoimmunology* **2013**, *2*, e22891. [CrossRef]

111. Oates, J.; Hassan, N.J.; Jakobsen, B.K. ImmTACs for targeted cancer therapy: Why, what, how, and which. *Mol. Immunol.* **2015**, *67*, 67–74. [CrossRef]

112. Immunocore Pipeline. Available online: https://www.immunocore.com/pipeline (accessed on 15 January 2019).

113. Zaia, J.A. A new agent in the strategy to cure AIDS. *Mol. Ther.* **2016**, *24*, 1894–1896. [CrossRef]

114. Safdari, Y.; Ahmadzadeh, V. Use of Single-Chain Antibody Derivatives for Targeted Drug Delivery. *Mol. Med.* **2016**. [CrossRef]

115. Lu, Z.R.; Kopekov, P.; Kopeek, J. Polymerizable Fab' antibody fragments for targeting of anticancer drugs. *Nat. Biotechnol.* **1999**. [CrossRef]

116. Philogen Pipeline. Available online: http://www.philogen.com/en/products/pipeline/pipeline_16.html (accessed on 17 January 2019).

117. Lemaire, M.; D'Huyvetter, M.; Lahoutte, T.; Van Valckenborgh, E.; Menu, E.; De Bruyne, E.; Kronenberger, P.; Wernery, U.; Muyldermans, S.; Devoogdt, N.; et al. Imaging and radioimmunotherapy of multiple myeloma with anti-idiotypic Nanobodies. *Leukemia* **2014**, *28*, 444–447. [CrossRef]

118. Pruszynski, M.; Koumarianou, E.; Vaidyanathan, G.; Revets, H.; Devoogdt, N.; Lahoutte, T.; Zalutsky, M.R. Targeting breast carcinoma with radioiodinated anti-HER2 Nanobody. *Nucl. Med. Biol.* **2013**. [CrossRef]

119. Shen, B.Q.; Xu, K.; Liu, L.; Raab, H.; Bhakta, S.; Kenrick, M.; Parsons-Reponte, K.L.; Tien, J.; Yu, S.F.; Mai, E.; et al. Conjugation site modulates the in vivo stability and therapeutic activity of antibody-drug conjugates. *Nat. Biotechnol.* **2012**. [CrossRef]

120. Beck, A.; Goetsch, L.; Dumontet, C.; Corvaïa, N. Strategies and challenges for the next generation of antibody-drug conjugates. *Nat. Rev. Drug Discov.* **2017**, *16*, 315–337. [CrossRef]

121. ALX-0061—Clinical Trial: NCT02287922. Available online: https://clinicaltrials.gov/ct2/show/NCT02287922 (accessed on 21 January 2019).

122. ALX-0061—Clinical Trial: NCT02437890. Available online: https://clinicaltrials.gov/ct2/show/NCT02437890 (accessed on 21 January 2019).

123. ALX-0761—Clinical Trial: NCT02156466. Available online: https://clinicaltrials.gov/ct2/show/NCT02156466 (accessed on 21 January 2019).

124. ATN-103—Clinical trial: NCT01063803. Available online: https://clinicaltrials.gov/ct2/show/NCT01063803 (accessed on 21 January 2019).

125. ATN-192—Clinical Trial: NCT01284036. Available online: https://clinicaltrials.gov/ct2/show/NCT01284036 (accessed on 21 January 2019).

126. Philogen Website. Available online: http://www.philogen.com/en/ (accessed on 21 January 2019).

127. Ulrichts, P.; Cousin, T.; Dreier, T.; de Haard, H.; Leupin, N. Argx-113, a novel Fc-based approach for antibody-induced pathologies such as primary immune thrombocytopenia. *Blood* **2016**, *128*, 4919.

128. ARGX-113—Clinical Trial NCT03669588. Available online: https://clinicaltrials.gov/ct2/show/NCT03669588 (accessed on 29 January 2019).

129. Merrill, J.T.; Shevell, D.E.; Duchesne, D.; Nowak, M.; Kundu, S.; Girgis, I.G.; Hu, Y.S.; Nadler, S.G.; Banerjee, S.; Throup, J. An Anti-CD28 Domain Antibody, Lulizumab, in Systemic Lupus Erythematosus: Results of a Phase II Study. In *Arthritis & Rheumatology*; Wiley: Hoboken, NJ, USA, 2018.

130. Cordy, J.C.; Morley, P.J.; Wright, T.J.; Birchler, M.A.; Lewis, A.P.; Emmins, R.; Chen, Y.Z.; Powley, W.M.; Bareille, P.J.; Wilson, R.; et al. Specificity of human anti-variable heavy (VH) chain autoantibodies and impact on the design and clinical testing of a VH domain antibody antagonist of tumour necrosis factor-α receptor 1. *Clin. Exp. Immunol.* **2015**. [CrossRef]

131. GSK286227—Clinical Trial: NCT02221037. Available online: https://clinicaltrials.gov/ct2/show/NCT02221037 (accessed on 29 January 2019).

132. Thiel, M.A.; Coster, D.J.; Standfield, S.D.; Brereton, H.M.; Mavrangelos, C.; Zola, H.; Taylor, S.; Yusim, A.; Williams, K.A. Penetration of engineered antibody fragments into the eye. *Clin. Exp. Immunol.* **2002**. [CrossRef]

133. Lampalizumab—Clinical Trial: NCT02247531. Available online: https://clinicaltrials.gov/ct2/show/NCT02247531 (accessed on 24 January 2019).

134. Lampalizumab—Clinical Trial: NCT02247479. Available online: https://clinicaltrials.gov/ct2/show/NCT02247479 (accessed on 24 January 2019).

135. Holz, F.G.; Sadda, S.R.; Busbee, B.; Chew, E.Y.; Mitchell, P.; Tufail, A.; Brittain, C.; Ferrara, D.; Gray, S.; Honigberg, L.; et al. Efficacy and safety of lampalizumab for geographic atrophy due to age-related macular degeneration: Chroma and spectri phase 3 randomized clinical trials. *JAMA Ophthalmol.* **2018**. [CrossRef]

136. RTH 258—Clinical Trial: NCT02307682. Available online: https://www.google.com/url?sa=t&rct=j&q=&esrc=s&source=web&cd=1&cad=rja&uact=8&ved=2ahUKEwiNsqmMi5PgAhUS2qQKHbi1BZoQFjAAegQICRAB&url=https%3A%2F%2Fclinicaltrials.gov%2Fct2%2Fshow%2FNCT02307682&usg=AOvVaw2oDAJKgexjOb5qZHcOVqJy (accessed on 24 January 2018).

137. RTH 258—Clinical Trial: NCT02434328. Available online: https://www.google.com/url?sa=t&rct=j&q=&esrc=s&source=web&cd=1&cad=rja&uact=8&ved=2ahUKEwiMjoKbi5PgAhVO3KQKHa2cDKoQFjAAegQIChAB&url=https%3A%2F%2Fclinicaltrials.gov%2Fct2%2Fshow%2FNCT02434328&usg=AOvVaw1Uobk9ivR8iGx_t6gl-Lec (accessed on 24 January 2019).

138. Camacho-Villegas, T.A.; Mata-González, M.T.; García-Ubbelohd, W.; Núñez-García, L.; Elosua, C.; Paniagua-Solis, J.F.; Licea-Navarro, A.F. Intraocular penetration of a vNAR: In vivo and in vitro VEGF165 neutralization. *Mar. Drugs* **2018**. [CrossRef] [PubMed]

139. Kovaleva, M.; Johnson, K.; Steven, J.; Barelle, C.J.; Porter, A. Therapeutic potential of shark anti-ICOSL VNAR domains is exemplified in a murine model of autoimmune non-infectious uveitis. *Front. Immunol.* **2017**. [CrossRef] [PubMed]

140. Elasmogen Website. Available online: www.elasmogen.com (accessed on 24 January 2019).

141. Wilken, L.; McPherson, A. Application of camelid heavy-chain variable domains (VHHs) in prevention and treatment of bacterial and viral infections. *Int. Rev. Immunol.* **2018**, *37*, 69–76. [CrossRef] [PubMed]

142. AdisInsight ALX-0171. Available online: https://adisinsight.springer.com/drugs/800035341 (accessed on 24 January 2019).

143. Laustsen, A.H.; María Gutiérrez, J.; Knudsen, C.; Johansen, K.H.; Bermúdez-Méndez, E.; Cerni, F.A.; Jürgensen, J.A.; Ledsgaard, L.; Martos-Esteban, A.; Øhlenschlæger, M.; et al. Pros and cons of different therapeutic antibody formats for recombinant antivenom development. *Toxicon* **2018**. [CrossRef] [PubMed]

144. Tremblay, J.M.; Mukherjee, J.; Leysath, C.E.; Debatis, M.; Ofori, K.; Baldwin, K.; Boucher, C.; Peters, R.; Beamer, G.; Sheoran, A.; et al. A single VHH-based toxin-neutralizing agent and an effector antibody protect mice against challenge with Shiga toxins 1 and 2. *Infect. Immun.* **2013**. [CrossRef]

145. Freise, A.C.; Wu, A.M. In vivo imaging with antibodies and engineered fragments. *Mol. Immunol.* **2015**, *67*, 142–152. [CrossRef] [PubMed]

146. van Brussel, A.S.A.; Adams, A.; Oliveira, S.; Dorresteijn, B.; El Khattabi, M.; Vermeulen, J.F.; van der Wall, E.; Mali, W.P.T.M.; Derksen, P.W.B.; van Diest, P.J.; et al. Hypoxia-Targeting Fluorescent Nanobodies for Optical Molecular Imaging of Pre-Invasive Breast Cancer. *Mol. Imaging Biol.* **2016**. [CrossRef]

147. Kijanka, M.M.; van Brussel, A.S.A.; van der Wall, E.; Mali, W.P.T.M.; van Diest, P.J.; van Bergen en Henegouwen, P.M.P.; Oliveira, S. Optical imaging of pre-invasive breast cancer with a combination of VHHs targeting CAIX and HER2 increases contrast and facilitates tumour characterization. *EJNMMI Res.* **2016**. [CrossRef]

148. Viola-Villegas, N.T.; Sevak, K.K.; Carlin, S.D.; Doran, M.G.; Evans, H.W.; Bartlett, D.W.; Wu, A.M.; Lewis, J.S. Noninvasive imaging of PSMA in prostate tumors with89Zr-Labeled huJ591 engineered antibody fragments: The faster alternatives. *Mol. Pharm.* **2014**. [CrossRef]

149. Raubitschek, A.A.; Tsai, S.-W.; Shively, J.E.; Yazaki, P.J.; Williams, L.E.; Ikle', D.N.; Wu, A.M.; Wong, J.Y.C. Tumor Targeting of Radiometal Labeled Anti-CEA Recombinant T84.66 Diabody and T84.66 Minibody: Comparison to Radioiodinated Fragments. *Bioconjug. Chem.* **2002**. [CrossRef]

150. Maier, J.; Traenkle, B.; Rothbauer, U. Real-time analysis of epithelial-mesenchymal transition using fluorescent single-domain antibodies. *Sci. Rep.* **2015**. [CrossRef]

151. Broisat, A.; Hernot, S.; Toczek, J.; De Vos, J.; Riou, L.M.; Martin, S.; Ahmadi, M.; Thielens, N.; Wernery, U.; Caveliers, V.; et al. Nanobodies targeting mouse/human VCAM1 for the nuclear imaging of atherosclerotic lesions. *Circ. Res.* **2012**. [CrossRef]

152. Broisat, A.; Toczek, J.; Dumas, L.S.; Ahmadi, M.; Bacot, S.; Perret, P.; Slimani, L.; Barone-Rochette, G.; Soubies, A.; Devoogdt, N.; et al. 99mTc-cAbVCAM1-5 Imaging Is a Sensitive and Reproducible Tool for the Detection of Inflamed Atherosclerotic Lesions in Mice. *J. Nucl. Med.* **2014**. [CrossRef]

153. Yu, Y.J.; Watts, R.J. Developing Therapeutic Antibodies for Neurodegenerative Disease. *Neurotherapeutics* **2013**, *10*, 459–472. [CrossRef]

154. Carter, P.J.; Lazar, G.A. Next generation antibody drugs: Pursuit of the "high-hanging fruit". *Nat. Rev. Drug Discov.* **2018**, *17*, 197–223. [CrossRef]

155. Ross, C.A.; Poirier, M.A. Protein aggregation and neurodegenerative disease. *Nat. Med.* **2004**. [CrossRef]

156. Messer, A.; Lynch, S.M.; Butler, D.C. Developing intrabodies for the therapeutic suppression of neurodegenerative pathology. *Expert Opin. Biol. Ther.* **2009**. [CrossRef]

157. Pardridge, W.M. Targeted delivery of protein and gene medicines through the blood-brain barrier. *Clin. Pharmacol. Ther.* **2015**, *97*, 347–361. [CrossRef]

158. Abzyme Website. Available online: www.abzymetx.com (accessed on 21 January 2019).

159. Li, T.; Bourgeois, J.P.; Celli, S.; Glacial, F.; Le Sourd, A.M.; Mecheri, S.; Weksler, B.; Romero, I.; Couraud, P.O.; Rougeon, F.; et al. Cell-penetrating anti-GFAP VHH and corresponding fluorescent fusion protein VHH-GFP spontaneously cross the blood-brain barrier and specifically recognize astrocytes: Application to brain imaging. *FASEB J.* **2012**. [CrossRef]

160. Poduslo, J.F.; Ramakrishnan, M.; Holasek, S.S.; Ramirez-Alvarado, M.; Kandimalla, K.K.; Gilles, E.J.; Curran, G.L.; Wengenack, T.M. In vivo targeting of antibody fragments to the nervous system for Alzheimer's disease immunotherapy and molecular imaging of amyloid plaques. *J. Neurochem.* **2007**. [CrossRef]

161. Caljon, G.; Caveliers, V.; Lahoutte, T.; Stijlemans, B.; Ghassabeh, G.H.; Van Den Abbeele, J.; Smolders, I.; De Baetselier, P.; Michotte, Y.; Muyldermans, S.; et al. Using microdialysis to analyse the passage of monovalent nanobodies through the blood-brain barrier. *Br. J. Pharmacol.* **2012**. [CrossRef]

162. Muruganandam, A.; Tanha, J.; Narang, S.; Stanimirovic, D. Selection of phage-displayed llama single-domain antibodies that transmigrate across human blood-brain barrier endothelium. *FASEB J.* **2002**. [CrossRef]

163. Farrington, G.K.; Caram-Salas, N.; Haqqani, A.S.; Brunette, E.; Eldredge, J.; Pepinsky, B.; Antognetti, G.; Baumann, E.; Ding, W.; Garber, E.; et al. A novel platform for engineering blood-brain barrier-crossing bispecific biologics. *FASEB J.* **2014**. [CrossRef]

164. Slastnikova, T.A.; Ulasov, A.V.; Rosenkranz, A.A.; Sobolev, A.S. Targeted intracellular delivery of antibodies: The state of the art. *Front. Pharmacol.* **2018**. [CrossRef]

165. Lim, K.J.; Sung, B.H.; Shin, J.R.; Lee, Y.W.; Kim, D.J.; Yang, K.S.; Kim, S.C. A Cancer Specific Cell-Penetrating Peptide, BR2, for the Efficient Delivery of an scFv into Cancer Cells. *PLoS ONE* **2013**. [CrossRef]

166. Bruce, V.J.; Lopez-Islas, M.; McNaughton, B.R. Resurfaced cell-penetrating nanobodies: A potentially general scaffold for intracellularly targeted protein discovery. *Protein Sci.* **2016**. [CrossRef]

167. Shin, S.M.; Choi, D.K.; Jung, K.; Bae, J.; Kim, J.S.; Park, S.W.; Song, K.H.; Kim, Y.S. Antibody targeting intracellular oncogenic Ras mutants exerts anti-tumour effects after systemic administration. *Nat. Commun.* **2017**. [CrossRef]

antibodies

MDPI

Review

Antibody Structure and Function: The Basis for Engineering Therapeutics

Mark L. Chiu [1,*], Dennis R. Goulet [2], Alexey Teplyakov [3] and Gary L. Gilliland [3]

[1] Drug Product Development Science, Janssen Research & Development, LLC, Malvern, PA 19355, USA

[2] Department of Medicinal Chemistry, University of Washington, P.O. Box 357610, Seattle, WA 98195-7610, USA; dennisrgoulet@gmail.com

[3] Biologics Research, Janssen Research & Development, LLC, Spring House, PA 19477, USA; dr.alexey.teplyakov@gmail.com (A.T.); garygilliland911@gmail.com (G.L.G.)

* Correspondence: mchiu@its.jnj.com

Received: 16 October 2019; Accepted: 28 November 2019; Published: 3 December 2019

Abstract: Antibodies and antibody-derived macromolecules have established themselves as the mainstay in protein-based therapeutic molecules (biologics). Our knowledge of the structure–function relationships of antibodies provides a platform for protein engineering that has been exploited to generate a wide range of biologics for a host of therapeutic indications. In this review, our basic understanding of the antibody structure is described along with how that knowledge has leveraged the engineering of antibody and antibody-related therapeutics having the appropriate antigen affinity, effector function, and biophysical properties. The platforms examined include the development of antibodies, antibody fragments, bispecific antibody, and antibody fusion products, whose efficacy and manufacturability can be improved via humanization, affinity modulation, and stability enhancement. We also review the design and selection of binding arms, and avidity modulation. Different strategies of preparing bispecific and multispecific molecules for an array of therapeutic applications are included.

Keywords: antibody engineering; therapeutic biologics

1. Introduction

Currently, all antibodies and antibody-derived macromolecules being developed for a wide spectrum of therapeutic indications [1,2] require protein engineering. The engineering approaches being used are based on our knowledge of protein structure and, in particular, our knowledge of how the structures are linked to their function [3]. Our knowledge of the three-dimensional structure of antibodies has emerged from crystallographic studies reported from numerous laboratories beginning in the 1970s. At present, the Protein Data Bank (PDB) [4] contains over 3500 structures of antibody fragments (Fabs, Fvs, scFvs, and Fcs), as well as a small number of intact antibody structures. The structural data includes complexes of these molecules with proteins, other macromolecules, peptides, and haptens. The overall structure of antibodies, including the folding pattern of the individual domains and basic features of the antigen-combining sites, has been the subject of several reviews [3,5–8].

Human immunoglobulins are Y-shaped proteins composed of two identical light chains (LCs) and two identical heavy chains (HCs). In natural systems, the pairing of one LC with one HC associates with another identical heterodimer to form the intact immunoglobulin. The HC and LC of the heterodimer are linked through disulfide bonds. The two HCs of the heterotetramer are also linked by disulfide bridges. Human LCs can be one of two functionally similar classes, κ or λ. Both LC classes have two domains, a constant domain (CL) and a variable domain (VL). In comparison, human antibody HCs can be one of five isotypes, IgA, IgD, IgE, IgG, and IgM, each with an independent role in the adaptive immune system. IgAs, IgDs, and IgGs have three constant (C) and one variable (V) domains. IgEs and IgMs have one variable and four constant domains. The IgA and IgM isotopes have an additional J-chain, which allows the formation of dimers and pentamers, respectively. The other isotypes are monomeric (a monomer is defined here as a pair of HC-LCs.).

The general features of antibodies described below will focus on the IgG1 framework. Our knowledge of how antibody structure relates to function is being exploited to create antibodies and antibody-related biologics with the appropriate functional and biophysical properties to address specific therapeutic needs. The engineering approaches applied to antibodies, antibody fragments, antibody, and antibody fusion products include effector function engineering, antibody humanization, affinity modulation, and stability enhancement to improve efficacy and manufacturability.

1.1. Overall Features of the Immunoglobulin

The intact antibody molecule shown in Figure 1 has three functional components, two Fragment antigen binding domains (Fabs) and the fragment crystallizable (Fc), with the two Fabs linked to the Fc by a hinge region that allows the Fabs a large degree of conformation flexibility relative to the Fc. Each of the Fabs have identical antigen-binding sites (or what is often called antigen-combining sites) for binding to a specific target antigen. The Fv region of the Fab is composed of a pair of variable domains (VH and VL) contributed by the HC and LC. In contrast, the glycosylated Fc region binds to a variety of receptor molecules providing the effector function profile that dictates how the antibody interacts with other components of the adaptive and humoral immune system.

All the domains of heavy and light chains are approximately 110 amino acid residues in length whose conformations have been termed the "immunoglobulin fold" (Figure 2) [9,10]. The fold is comprised of two tightly packed anti-parallel β-sheets. One of the two β-sheets of the C domains has four β-strands, ↓A ↑B ↓E ↑D, and the other three β-strands, ↓C ↑F ↓G. The overall fold is often referred to as a Greek key barrel. The two β-sheets are covalently linked together by an intra-domain disulfide bridge formed between two cysteine residues in the ↑B and ↑F β-strands. The C domains are in general compact, with short loops connecting the β-strands. The two β-sheets pack together using the non-covalent interactions of the side chains of amino acid residues on the complementary faces.

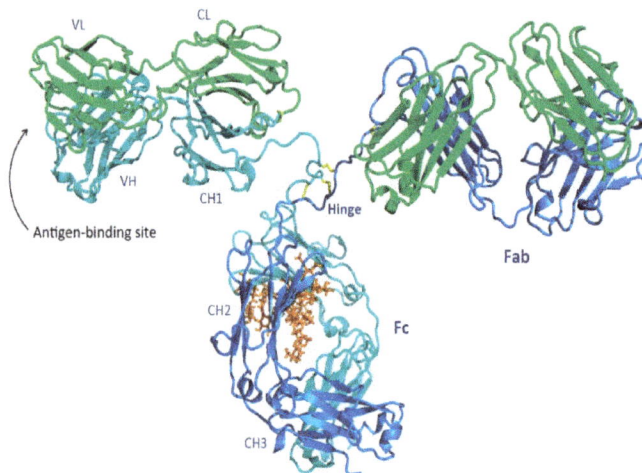

Figure 1. A ribbon representation of an intact IgG, Protein Data Bank (PDB) id: 1igt [11], which is a mouse IgG2a isotype. The light chains are green, the heavy chains are cyan and blue, the glycan is orange sticks, and the interchain disulfides are yellow sticks.

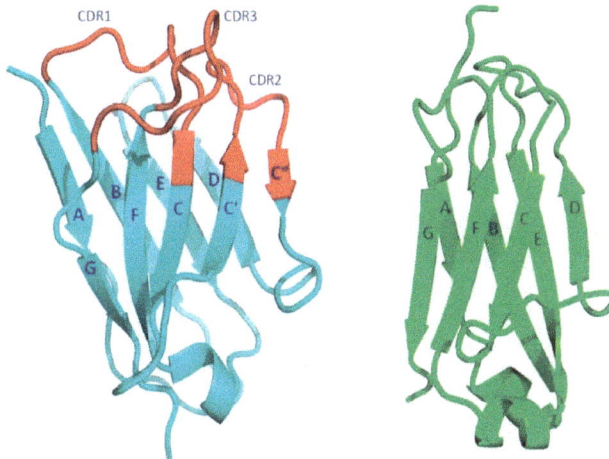

Figure 2. The immunoglobulin fold. The left ribbon image (cyan and red) of the heavy-chain variable (VH) domain illustrates the V domain immunoglobulin folding pattern (VH of Fab 388, PDBid 5i1a) [12]. The V domain complementarity-determining regions (CDRs) are shown in red. The right ribbon image (green) illustrates the similar folding pattern of a typical C domain (CL of Fab 5844, PDBid: 5i18 [12].

The V domains of the immunoglobulin structure, which interact with the target antigen, are at the N-termini of the HCs and LCs. These domain structures are like that of the C domains but with some differences. The two β-sheets have a configuration like that found in the C domain. The four-stranded β-sheet, formed from four β-strands, ↓A ↑B ↓E ↑D is like the corresponding β-sheet in the C domain. The other β-sheet has five β-strands, ↓C″ ↑C′ ↓C ↑F ↓B, instead of the three found in the C domain. An insertion of two β-strands, ↓C″ ↑C′ is present between β-strands ↓C and ↑D. Just as in the C domain, an intra-domain disulfide bridge is formed between β-strands ↑B and ↑F. The less-compact V domains in general have longer loops connecting the β-strands.

1.2. Fab Region

1.2.1. Fab Overall Features

The Fab regions of an immunoglobulin are formed by the pairing of VL and CL of the LCs with VH and CH1 of the HCs. The pairing of VL and VH, form the antigen-binding site. The two β-sheets formed with β-strands ↓C″ ↑C′ ↓C ↑F ↓B pack together, forming a barrel-like structure that aligns the connecting loops (complementarity determining regions or CDRs, see below) and forming the antigen-binding site. In contrast, the CH1 and CL domains pack tightly in an almost perpendicular mode using the complementary faces of the opposite ↓A ↑B ↓E ↑D β-sheet.

The overall arrangement of the HC and LC domains of the Fab are characterized by what is called the elbow bend or elbow angle. This is defined by the angle between the pseudo-two-fold axes relating the two pairs of domains (VH, VL and CH1, CL) [10,13]. The switch region, an extended polypeptide chain, connects the V and C domains. The orientation of the V domains with respect to the C domains is referred to as the elbow angle or elbow bend, which can vary significantly. In an early survey of Fabs with kappa (κ) light chains, the angle was shown to vary from 116° to 226° [14]. Fabs with lambda (λ) light chains have a wider range of angles, indicating higher levels of flexibility. This may result from the presence of an extra amino acid residue (usually a glycine) present in the switch region of λ LCs. An early analysis of the elbow motion in Fabs discovered a conserved feature that is referred to as a molecular ball-and-socket joint [15]. This occurs in the HC at the interface between VH and CH1. The ball consists of conserved amino acid residues Phe148 and Pro149 in VH and the socket is formed by conserved amino acid residues Leu/Val11, Thr110, and Ser112 in the CH1 domain.

This interaction could restrict the elbow angle to a maximum of 180°. However, larger angles were reported for subsequent Fab structures (e.g., [16]) in which the ball and socket move apart, allowing elbow angles >180° [14].

1.2.2. The Fab Antigen-Binding Site

The antigen-binding site is formed by the pairing of the Fab VH and VL with the N-terminal region designated as the Fv region. As shown in Figure 3, each domain contributes three complementarity-determining regions CDR-L1, CDR-L2, and CDR-L3 for VL and CDR-H1, CDR-H2, and CDR-H3 for VH. These hypervariable regions were identified by early amino acid sequence variability analyses [17,18] that pre-dated our knowledge of the structure of the antibodies. The six CDR loops are in proximity to each other, resulting from the orientation of VL and VH after the formation of the Fv. This is a result of the packing of the β-sheets composed of the ↓C″ ↑C′ ↓C ↑F ↓B from the two domains. This configuration brings the three CDRs of the VL and VH domains together to form the antigen-binding site. The strands of the two β-sheets and the non-hypervariable loops are referred as to framework regions (FRs).

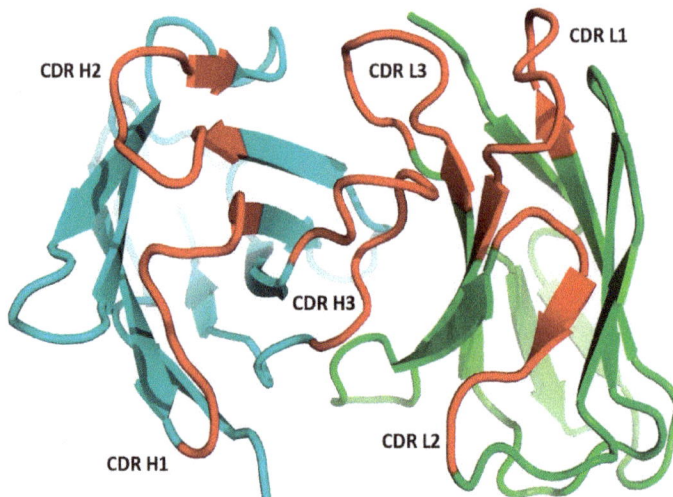

Figure 3. The Ab Fv region with the VH in cyan and the VL in green. The Martin CDRs are highlighted in red (Fv of Fab 388, PDBid: 5i1a) [12].

Both the number of amino acid residues and the sequences can vary for the CDRs. Genetic recombination of the V, D, and J gene segments for VH and V and J gene segments for VL with subsequent somatic hypermutation in mature B cells accounts for antibody CDR sequence diversity. In the two domains, the CDRs are composed of amino acid residues in the loops connecting the framework β-strands ↑B and ↓C for CDR-L1 and CDR-H1, ↑C′ and ↓C″ for CDR-L2 and CDR-H2, and ↑F and ↓G for CDR-L3 and CDR-H3.

The Fv amino acid residues in contact with the antigen have been called specificity-determining residues (SDRs) [19]. Antibodies in complex with haptens, proteins, or peptides show distinctive SDR patterns [19,20]. Anti-hapten antibodies have small and deep binding pockets at the VH–VL interface. The antigen-binding sites specific for peptides are groove-shaped depressions between VH and VL, while anti-protein antibodies tend to have extended and larger binding sites compared to those of the other two classes of antibodies. These structural features of antibody recognition sites for different classes of antigens have been employed in the development of productive synthetic antibody libraries for the specific recognition of haptens [21], peptides [22], and proteins [23].

1.2.3. Relationship between Binding and Affinity

The antigen binding of antibodies often results in conformational changes in the contact surface areas of both the antibody and the antigen. These events have been studied in detail by many laboratories in the structure determinations of both an antibody fragment (Fabs or Fvs) alone and in complex with its antigen (for reviews, see [8,24]. When discussing antigen–antibody interactions, the general modes of binding are cited: Lock and key, induced fit, and conformational selection. In the lock and key model, the two molecules interact in a manner that minimizes changes in the conformations of the two protein surfaces from that observed in the unbound and bound states. Thus, the backbone conformations of the antibody and antigen are essentially the same in both the unbound and bound states. In contrast, the conformational changes for the antibody and antigen in the induced-fit mode can be quite extensive. Both the side chain and backbone atoms in the contact region can undergo conformational changes after the binding takes place, especially in the CDR regions. Of all the CDRs, the CDR-H3 most often has changes in conformations when the unbound and bound structures are compared. In addition, differences in the orientation of VL with respect to VH are often seen. Lastly, the Fab elbow angle may differ in the two forms. It has been suggested that the induced-fit mode of binding introduces plasticity into the antigen-binding site, expanding antibody diversity beyond that resulting from amino acid residue changes [25]. In the conformational selection model, the antigen samples a population of different conformational states prior to binding [26,27]. Antibody binding can then depend on pre-activation states of the antigen, which can be affected by the microenvironment around the antigen [28]. Sorting out the kinetics of target engagement also provides a guideline of how to optimize pharmacology. Understanding this aspect of binding can drive the development of better *in situ* antibody therapeutic design [29]. This also serves as a reminder that binding affinity may not be directly linked with pharmacology [30].

1.2.4. Canonical Structures of the CDRs

An early structural analysis of antigen-binding sites of the small set of structures of immunoglobulin fragments available at the time revealed that the conformations of five out of the six hypervariable loops or CDRs had a limited set of main-chain conformations or 'canonical structures' [31,32]. The canonical structure model implied a paradigm shift in the field, replacing the notion that each antibody has unique hypervariable loop conformations. A canonical structure is defined by the loop length, the conformation of the loop, and the conserved amino acid residues within the hypervariable loop and FRs. Based on this model, studies of antibody sequences indicated that from the total number of possible combinations of canonical structures only a few occur [33–35]. This suggested that structural restrictions at the antigen-binding site may affect antigen recognition. Subsequent work [36] reported that the hypervariable loop lengths are the primary determining factor of the antigen-binding site topography, as they are the primary factor determining the canonical structures [31,37].

This early work was extended to include conformational analysis of the CDRs of 17 high-resolution antibody fragments [37]. The CDRs of the light chain CDR-L1, CDR-L2, and CDR-L3 were all found to have preferred sets of canonical structures based on the length and amino acid sequence composition. This was also found for CDRs of the heavy chain CDR-H1 and CDR-H2, but not for heavy chain CDR-H3, which is the most variable in length and amino acid sequence. This limited set of CDR canonical structures was included in macromolecular modeling strategies for antibody structures [31,32]. The early assignments of canonical structures have been extended using an algorithm that clusters the CDRs from a set of antibody fragments with low temperature factors and low conformational energies [38]. The results are frequently updated and available online (http://dunbrack2.fccc.edu/PyIgClassify/default.aspx) from the Dunbrack Laboratory.

1.2.5. CDR-H3

One of the CDRs, CDR-H3, has a large range of lengths and amino acid sequence diversity and usually plays a primary role in the antibody–antigen interactions. The CDR-H3 conformation is quite variable in nature and canonical structures were not defined in the early cataloging efforts. In later studies, the residues in the loop nearest the framework (torso) and residues in the extended region of the loop (head) have been found to have defined conformations [39–41]. One interesting discovery by this work was that the backbone of the CDR-H3 base region can have either an 'extended' or 'kinked' conformation. The kinked conformation is a beta-bulge in the backbone of the stem region. In early studies of CDR-H3 structures, the kinked form was more prevalent than the extended one [41]. A recent study reported 16 representative Fab structures of a germline library, all having the same CDR-H3 amino acid sequence [12]. In fourteen of these structures, CDR-H3s were found in the kinked conformation, whereas in two structures CDR-H3s were in the extended conformation. This finding supports the hypothesis that the CDR-H3 conformation is controlled both by its sequence and its environment [42].

1.2.6. Antibody Modeling

The knowledge of canonical structures enabled the development of antibody modeling (Fv region) [43]. In therapeutic antibody development programs, where the number of candidates being considered far exceeds the capacity of the crystallographic structure determination process, antibody modeling has become increasingly more important. Because of this need, approaches for antibody modeling continue to evolve along with the field of protein structure prediction. Recently, antibody modeling assessment studies have been undertaken to gain insight into the quality of the results of antibody structure prediction software. These blinded studies [44,45] involved providing the antibody structure prediction software groups with the sequence of Fv regions for which structures had been determined but were not yet publicly available. Once the predictions were completed by the participants, the results were submitted to the organizers and the models were assessed and compared with the unpublished structures. In the second study [45], after the prediction of the structures of the entire Fv were completed, the participants were provided with the Fv structures without their CDR-H3s. The structures of the CDR-H3s were then predicted and submitted. This was done to assess whether more accurate structures of CDR-H3 could be predicted if the context (the Fv structural environment) was provided. The participants included Accelrys, Inc. [46], Chemical Computer Group (CCG) [47], Schrödinger [48], Jeff Gray's lab at John Hopkins University [49] Macromoltek [50], Astellas Pharma/Osaka University [51], and Prediction of ImmunoGlobulin Structure (PIGS) [52,53]. While only Accerlys, Inc. and Chemical Computer Group (CCG), and PIGS participated in the first assessment, all other aforementioned parties participated in the second assessment. In both studies, all the antibody modeling methods produced similar and reliable models for the FR, but with some exceptions in the CDRs. Each of the methods applied in these studies had different strengths and weaknesses. Overall, the second antibody assessment revealed an improved quality of the models with an incremental improvement in the accuracy of the predictions from the first assessment, but further development to improve these methods is clearly warranted [54].

1.3. Fc Region

In the 1950s, it was discovered that proteolysis of intact IgGs with papain produced large fragments about a third of the size of the intact molecule [55,56], and it was eventually discovered that one of the fragments could bind antigen and act as an inhibitor to the binding of the intact antibody. This turned out to be what we call today the Fab fragment. Another fragment approximately the same size turned out not to inhibit binding, and it was easily crystallized [57]. This crystallizable fragment is what we call the Fc. The structural features of this region of the antibody were defined in the initial structure determination of the human IgG1 Fc [58] and they have remained constant as the structures of many other Fcs have been determined (see a partial list of Fc structures in Teplyakov et al., 2013 [59]).

The three-dimensional structure of the Fc [58] revealed how the two constant domains, CH2 and CH3, of the each of the HCs interact with one another (see Figure 4). The CH3s pack tightly with each other while the CH2s have no observable protein–protein contacts with one another. Rather, the space separating the CH2s is filled in part by the carbohydrate attached at Asn297. In some structures, the two carbohydrate chains interact through hydrogen bonds, either directly or through bridging water molecules. The flexibility imparted to the CH2s contributes to their role in the interaction with C1q and the FcγRs. The Fc region of an IgG can engage with Fc gamma receptors (FcγR) and the first subcomponent of the C1 complex (C1q) to mediate antibody-dependent cellular cytotoxicity (ADCC), complement-dependent cytotoxicity (CDC), antibody-dependent cellular phagocytosis (ADCP), trogocytosis, induction of secretion of mediators, and endocytosis of opsonized particles, as well as modulation of tissue and serum half-life through interaction with the FcRn [60–62]. The Fc has been the focus of significant engineering to modulate effector function activities found on monocytes, macrophages, dendritic cells, neutrophils, T and B lymphocytes, and natural killer cells [63]. Since there is often an interchange of mAbs coming from different mammalian forms, a systematic comparison of human Fc binding to mouse, cynomolgus, and human FcγRs have been made to correlate in vitro and in vivo Fc activity [64].

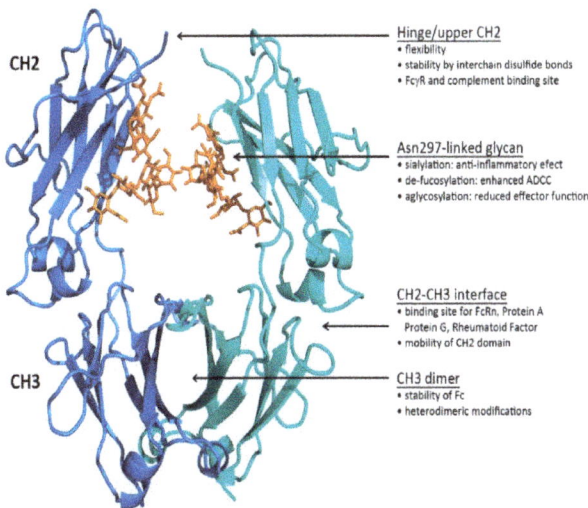

Figure 4. The structural features of the human IgG1 Fc and how they impact functionality. The Fc is represented by a ribbon image of the Fc structure (PDBid: 3ave [65]). The two heavy chains are shown in blue and cyan; the carbohydrate is represented by orange sticks.

1.3.1. The Fc CH2–CH3 Interface

The Fc CH2–CH3 interface has been recently characterized in a report of the structures of two crystal forms of the IgG2 Fc [59]. The interface is dominated by non-covalent interactions between the two domains supplemented by the presence of ordered water molecules. When the structures were compared with the structures of homologous IgG1 Fcs [65–67], it was observed that the CH2s change position relative to the CH3s. Further analysis revealed an Fc ball-and-socket joint between CH2 and CH3 that allows the CH2 domain to pivot around its Leu251 side chain, which is buried in a pocket formed by CH3 residues Met428, His429, Glu430, and His435. The movement of the CH2s is constrained by residues from both domains found at the CH2–CH3 interface and the hinge region. This Fc ball-and-socket joint is analogous to the one that is found in the Fab structures mentioned above [14,15,68], but in the Fc case, it is reversed relative to that found in Fab regions with the ball in the CH2 domain and the socket in CH3. The subset of CH2–CH3 interface residues associated with the Fc ball-and-socket are highly conserved among human IgG1, IgG2, IgG3, and IgG4 [59], indicating that it is a general structural feature that facilitates the motion of CH2 relative to CH3 in human IgGs. The positions of the amino acid residues at the interface vary as the domains change their relative orientation to one another, increasing or decreasing the gap between the domains. As the domains move, the water structure associated with the CH2–CH3 interface also adjusts. Future Fc engineering efforts can consider altering residues associated with the Fc ball and socket that could impact the flexibility of the Fc, potentially altering effector function activity.

1.3.2. The Fc CH2 Carbohydrate

The Fc CH2 carbohydrate covers a hydrophobic face of the domain and helps to fill the void between the two HC CH2s. Each of the domains has covalently bound carbohydrate with the structure described in Figure 5. This structure may vary considerably by the addition of other sugar residues, such as sialic acids, N-acetylglucosamines, and galactoses, and in some cases, the absence of fucose [69]. The presence of the glycans contributes to the biophysical stability of the protein structure [70]. Several Fc crystal structures with different glycoform variants [65,71,72] and aglycosylated forms [73,74] have been reported. In these structures, the composition of the carbohydrate dictates the separation distance between the CH2s. The composition of the carbohydrate of the Fc can substantially influence the effector functionality of the antibody as well as the pharmacokinetic profile [75].

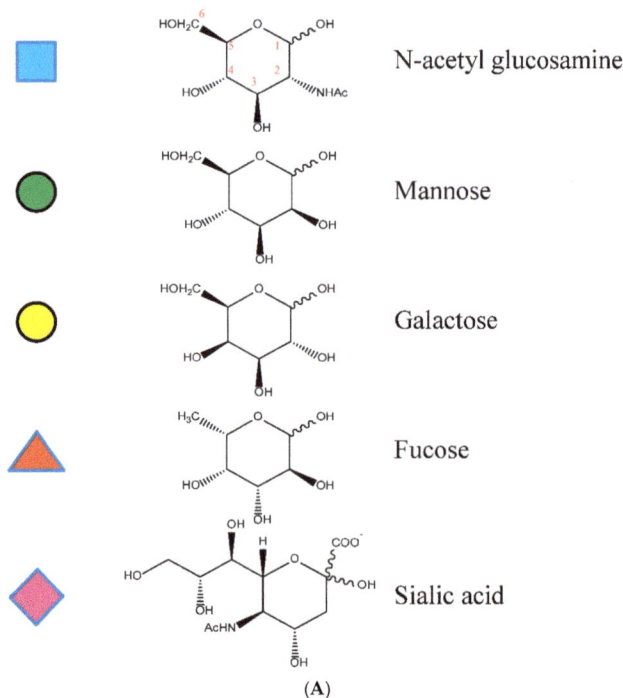

(A)

Figure 5. *Cont.*

(B)

(C)

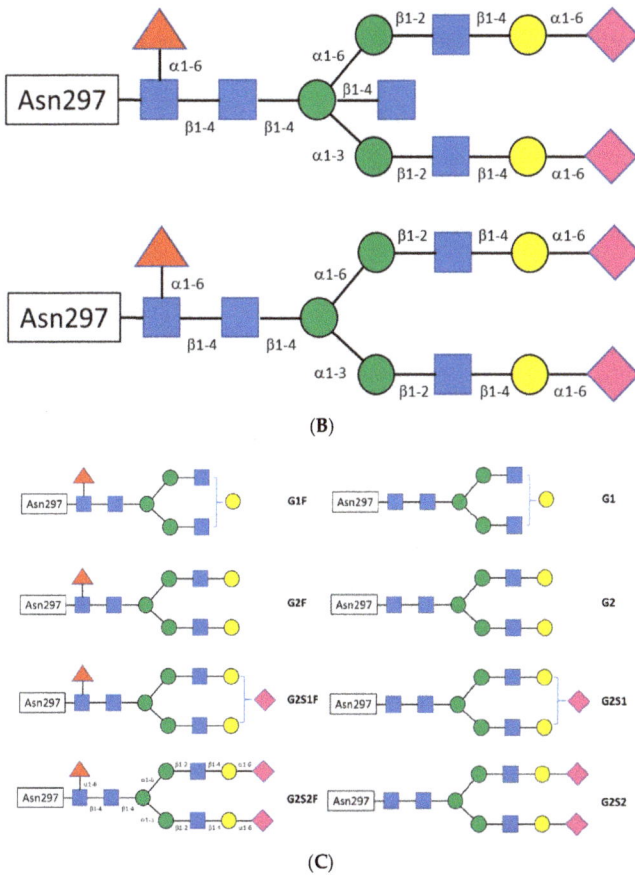

Figure 5. (**A**) Schematic representation of the most abundant recombinant N-linked oligosaccharide from human IgG Asn 297 (G2S2F) with glycosidic linkages. A similar representation of recombinant human IgG1 G2S2F is shown. The monomeric saccharides are shown as blue squares as N acetyl glucosamine; green circles as mannose; yellow circles as galactose; red squares as fucose; and purple rhombi as sialic acid or N acetyl neuraminic acid. (**B**) The glycosidic linkage numbers for representative oligosaccharides. The numbering of the glycosidic linkages are shown for oligosaccharides found in IgG molecules. The 1-4 N acetyl glucosamine can be found in human IgG structures. (**C**) Major species of N-linked oligosaccharides found in recombinant IgGs expressed in Chinese hamster ovary (CHO) cells may vary considerably by the addition of other sugar residues, such as sialic acids, N-acetylglucosamines, and galactose.

1.4. Hinge

The HC polypeptide region bridging CH1 and CH2 is called the hinge region and functionally allows the Fabs a large degree of conformational flexibility relative to the Fc. This facilitates the Fabs binding to multiple targets and allows the Fc to interact independently with other components of the immune system [76]. Structural knowledge of the IgG hinges is based upon the structures of intact mAbs, of Fcs, and of Fc:FcγR complexes. A review of structures deposited in the PDB [4] now reveals that there are 7 intact antibody structures, 87 Fc structures, and 15 FcγR complexes. There are ongoing efforts to utilize individual particle electron tomography to determine the diversity of conformational changes [77].

The antibody hinge can be divided into three regions, the upper hinge, core hinge, and lower hinge, each with a different functional role [66] (see Figure 6). On the N-terminal side, the upper hinge allows the movement and rotation of the Fabs. The central core hinge contains a variable number of cysteine residues depending on the IgG subtype that forms disulfide bonds, stabilizing the association of the HCs. On the C-terminal side is the lower hinge that allows movement of the Fc relative to the Fabs and whose amino acid residues can be involved in FcγR binding.

```
                    IgG Hinge Sequences

   Isotype      Upper           Core                      Lower
                  2             2                          2
                  1             2                          3
                  6             6                          1
      IgG1      EPKSCDKTHT      CPPCP                      APELLGGP
      IgG2      ERK            CCVECPPCP                   APPVA GP
      IgG3      ELKTPLGDTTHT   CPRCP(EPKSCDTPPPCPRCP)₃    APELLGGP
      IgG4      ESKYG....PP    CPSCP                      APEFLGGP
```

Figure 6. The hinge sequences of human IgG isotopes illustrating the upper, core, and lower hinge regions. Sequence numbers are given for the IgG1 hinge.

The hinges of Human IgG subtypes vary significantly in the number of residues and the number of possible disulfide bridges between the two heavy chains. This contributes to the overall stability of the antibody. For example, of all the IgGs, IgG4 is the only subtype that undergoes natural Fab-arm exchange producing antibody molecules that are bispecific [78]. In addition, this variability, including the differences in amino acid sequence, contributes in part to the strength of the interactions of IgGs with FcγRs.

An aspect of stability for antibodies and the hinge region is protease sensitivity. Papain [57] and other proteases [79] are used to cleave the upper hinge of IgGs, generating Fab and Fc fragments. Cleavage of the lower hinge single leads to single-clipped IgG or a double-clipped IgG with F(ab')$_2$ and Fc fragments. In humans, this cleavage can take place during inflammation, in tumor micro-environments or during bacterial infection by matrix metalloproteases, such as MMP-3, MMP-12, and MMP-7(matrix metalloproteases), and others like cathepsin G, GluV8, pepsin, and IdeS [80]. Mutations in the IgG1 [81], IgG2 [82], and IgA [83,84] hinge regions can mediate some levels of resistance to such enzymes. Such mutations can prevent hinge clipping to preserve the Fc effector function of therapeutic Abs in the inflamed tissue environment.

2. Structure-Based Antibody Engineering

Nobody is perfect, and the same applies to antibodies. Molecular engineering aims to improve the biochemical and biophysical properties of the antibodies of interest to make them good therapeutics and convenient research tools. Methodologically, there are two strategies to achieve this goal. Rational methods are based on structural knowledge derived from X-ray crystallography, Nuclear Magnetic Resonance (NMR) spectroscopy, and in silico modeling, and typically lead to the generation of a small set of variants. In contrast to rational, empirical methods are based on generating large libraries by employing phage, ribosome, or yeast display and rely on screening to select the desired variants [85]. This section of the review is focused on rational methods to engineer the antigen-binding function of the Fab arm of the antibody. The enormous progress that has been achieved in modifying the Fc-related effector function of the antibody has been reviewed recently [86–88] and will be discussed later in this review.

The availability of the three-dimensional structure of the antibody–antigen complex or even Fab alone greatly facilitates the design of the antibody variants with improved characteristics. Advances of X-ray crystallography over the last two decades coupled with the modern molecular biology and protein purification techniques have transformed structure determination into a routine procedure that requires minimal time and effort. Continual adaptations of the downstream process have included alternative purification schemes [89]. The benefits of the structural knowledge are manifold. For humanization, it helps to identify the critical positions outside of the complementarity-determining regions (CDRs) that must be preserved and positions within CDRs that may be replaced. For affinity maturation, it may point to a residue, which is otherwise unlikely to be considered as a game changer. For solubility improvement, modifications of the hydrophobic patches on the antibody surface (often not apparent in the linear sequence) are required.

In addition to crystallography, NMR and recently cryogenic electron microscopy (cryo-EM) have evolved as complementary techniques to obtain 3-D structures especially of Fab–Ag and Ab–Ag complexes. In the absence of experimental structural information, homology models are often considered as a decent alternative. However, despite the obvious development in algorithms and computer power, the quality of antibody structure prediction, particularly regarding CDR-H3, remains inadequate, and the results of antibody–antigen docking are also disappointing [90]. While homology models cannot fully substitute the experimental data, they can initiate the process for in silico design and evaluation of antibody mutants. We review such applications below.

2.1. Humanization

Historically the first and perhaps the most frequent application of antibody engineering was to reduce the immunogenicity of therapeutic antibodies of murine origin [91]. A variety of non-human species, including rodents, chicken, and rabbits, are employed today escape tolerance to obtain antibodies against human targets. All such non-human antibodies require humanization. The simplest approach was to make a chimera by combining the variable domains of non-human antibodies with human constant domains to generate molecules with 70% human content [92]. In many cases, chimeric antibodies demonstrated reduced immunogenicity but still elicited some human anti-therapeutic antibody response [93]. To further minimize immunogenicity, a CDR-grafting approach was proposed by G. Winter and coauthors [94]. The procedure involves the transfer of CDRs from a non-human (very often murine) "parental" antibody to the scaffold of a human antibody. The method was initially applied to a murine anti-hapten antibody. The CDRs from the heavy-chain variable region of the mouse antibody were substituted for the corresponding CDRs of a human anti-myeloma antibody. Following this experiment, a similar procedure produced a humanized anti-lysozyme antibody D1.3 [95], proving that CDR grafting can be used for antibodies that recognize protein antigens.

Besides CDR grafting, alternative humanization methods based on different paradigms, such as resurfacing [96], super-humanization [97], or human string content optimization [98], have been developed. All of them require the analysis of the amino acid sequence to evaluate the potential impact of the amino acid substitutions on the antibody structure and function. Typically, a relatively small number of humanized variants are produced and tested for antigen binding and functional activity. If the variants fail to meet the functional criteria, a new cycle of design, modification, and characterization is carried out to improve binding.

First, we consider CDR grafting as the principle method of antibody humanization. The procedure involves three tasks: (1) Defining the boundaries of the CDRs for grafting, (2) selecting human sequences to be utilized as framework (FR) donors, and (3) identifying residues within human FRs that may need to be replaced to maintain antibody binding. Although the tasks may seem consecutive, they are interrelated, and in practice should be carried out together.

2.1.1. CDR Definitions

Amino acid residues that constitute the CDRs were identified by Kabat [99] based on their high variability as compared to the other regions of the antibody (Figure 3). By analyzing the first crystal structures of Fabs, Chothia and Lesk [31] proposed a definition based on the conserved conformations of the antigen-binding loops named canonical structures. Accumulation of the structures of antibody–antigen complexes has led to the Martin CDR definition [100], which considers the involvement of residues in antigen binding. Symmetrical CDRs, where the N- and C-terminal residues are opposite each other in the structure, were used for the purpose of canonical structure classification [38] as implemented in the PyIgClassify database [101]. In comparison to the Martin definition, CDRs L2, H3, and H1 include an extra one, two, and three residues, respectively, at the N-terminal end of the CDRs. Universal schemes that are applicable to immunoglobulins, T-cell receptors, and major histocompatibility complex (MHC) molecules have also gained popularity [102–104]. A comparison of different CDR definitions is presented in Table 1.

Table 1. CDR definitions in Chothia numbering.

CDR	Kabat	Chothia	Martin *	PyIgClassify **	IMGT **
L1	24–34	24–34	24–34	24–34	27–32 (M − 5)
L2	50–56	50–56	50–56	49–56 (M + 1)	50–52 (M − 4)
L3	89–97	89–97	89–97	89–97	89–97
H1	31–35	26–32	26–35 (K + C)	23–35 (M + 3)	26–33 (M − 2)
H2	50–65	52–56	50–58 (K − 7)	50–58	51–57 (M − 2)
H3	95–102	95–102	95–102	93–102 (M + 2)	93–102 (M + 2)

* Martin CDRs in comparison to Kabat (K) and Chothia (C). ** PyIgClassify and IMGT CDRs in comparison to Martin (M).

For the purposes of CDR grafting, the choice of the CDR boundaries is free and not limited by the common definitions. However, two factors should be considered. First, the CDRs should be as short as possible to minimize the number of non-human residues. Second, the CDRs should include at least all residues in direct contact with the antigen. All definitions have advantages and disadvantages in terms of CDR grafting. The ImMunoGeneTics information system (IMGT) [104,105] rightfully includes residues 93 and 94 in CDR-H3 as they are very important for the CDR conformation. On the other hand, the IMGT convention excludes residues 35 and 50 from CDRs H1 and H2, respectively, although they are often involved in antigen binding. Considering all the pros and cons, the Martin definition is a good compromise (Table 1). Basically, it combines the Kabat and Chothia definitions and differs from them only in the heavy chain, where CDR-H1 includes all residues of Kabat and Chothia while CDR-H2 is seven residues shorter than that defined by Kabat. Those seven residues are in the loop between β-strands C″ and D and are never directly involved in contact with the antigen. In the light chain, there are no deviations between Kabat, Chothia, and Martin CDR definitions.

Regardless of the choice of CDRs, about 20% of the residues that bind the antigen fall outside the CDRs [106,107]. Moreover, these residues are at least as important to antigen binding as residues within the CDRs, and in some cases, they are even more important energetically. Therefore, for CDR grafting, the CDR definition is a good starting point, but the framework residues interacting with the antigen must be considered. Typically, for shorter CDRs, more FR residues, and for longer CDRs, fewer FR residues will need to be considered for back mutations.

Residue numbering schemes evolved in parallel with the CDR identification and aimed at the correct positioning of insertions and deletions in the antigen-binding loops. Since the Chothia numbering scheme [32] was based on structural considerations, it represents the best choice and is widely used in many applications. An advantage of a universal numbering versus sequential is that all structurally identical positions are numbered identically, which is convenient for alignments and comparisons. The Chothia numbering of residues is used throughout this review.

2.1.2. Human Germline Selection

The second step in the humanization process is to identify human FR donors. Initially, human antibodies of a known structure were used regardless of their homology to the non-human antibody in the so-called fixed FR approach [94,95,108]. Moreover, both VH and VL donors were often selected from a single antibody to ensure optimal pairing. However, this approach often resulted in a significant or even complete loss of affinity and was replaced with a method termed "best fit" [109], where human VH and VL sequences with the highest homology to the non-human antibody were selected. Comparison of the fixed FR and best fit strategies showed that the latter yields humanized antibodies with a higher affinity than variants obtained by the fixed FR method [110]. Another strategy of selecting human FRs as a template for humanization is by generating consensus sequences [111,112].

Regardless of the method chosen to select human FRs, there are two sources of human sequences: Mature and Germline. Mature sequences generated by the immune response carry somatic mutations [113] and therefore are potentially immunogenic. In contrast, human germline sequences are considered least immunogenic and have been recently used as FR donors almost exclusively [114].

The human repertoire consists of several dozen germline genes coding VH regions and approximately an equal number of VL genes, which are divided between κ and λ types [105]. Both heavy and light chain germlines are grouped into families according to sequence similarity. Among VH germlines, families 1, 3, and 4 are the most ubiquitous. The majority of λ VL germlines fall into families 1, 2, and 3, whereas the rest are distributed among families 4 to 10. Kappa VL germlines are almost entirely distributed over three families (1, 2, and 3) except for two genes, *IGKV4-1* and *IGKV5-2*, which represent families 4 and 5. The sequence identity within families is close to 90% while it can be as low as 50% for two germlines from different families.

Methods of human germline selection for CDR grafting are varied. One option is to base the choice on the overall sequence similarity between the non-human antibody and human germline within the variable domains. A more focused and more common approach considers sequence similarity only in the FR while neglecting the CDRs. The idea behind this is that homologous FRs provide the same scaffold for the CDRs and ensure their conformation, while the CDRs themselves are not changed at all (they are grafted). An alternative approach considers sequence similarity within the CDRs and relies on the canonical structures that are defined largely, although not exclusively, by the CDR sequence. The latter method is called super-humanization and will be discussed below.

Typically, a single germline is selected for the entire variable domain, one for VH and one for VL. However, one may apply a hybrid approach when a donor for each FR is selected independently, so that the resulting sequence will be assembled from different germlines [115]. This method has an obvious advantage of more flexibility and a potential for selecting human germlines with higher similarity score. It is believed, however, that mosaic constructs may exhibit impaired stability when compared to intact germline sequences owing to suboptimal VH-VL pairing and potential clashes in the core of the variable domain. Residues that come from different FRs may appear mutually incompatible when composed from different germlines.

A combination of sequence and structural criteria in the selection of human germlines was utilized in the humanization of mouse anti-glycoprotein VI Fab ACT017 developed for the treatment of arterial thrombosis [116]. The choice of templates for VH and VL was based upon the following four independent criteria: (1) Human germline sequences most similar to mouse germlines of the parental antibody; (2) high sequence identity and identical canonical structures of the CDRs; (3) high sequence identity and closely related CDR canonical structure; and (4) the same antibody template for both V-domains even at the cost of a less optimal template for one of the chains. Additionally, the human myeloma antibodies NEW (for VH) and REI (for VL) were selected because they are well-characterized in terms of stability and expression and they are frequently used in a fixed-FR strategy of humanization. Owing to some overlapping among the best candidates selected by applying these criteria, there were only 4 variants for each chain, resulting in a total of 16 VH-VL pairs. The 16 Fabs were expressed and evaluated for antigen binding, and only four of them showed the desired level of binding. One of the binders was based on bevacizumab, the FR donor for both VH and VL according to the selection criteria (4). The requirement of the identical canonical structure worked well only for VH and produced the best variant, which had the light chain FR from human antibody REI. Curiously, none of the variants using the VH template with the human germline most like the mouse parent retained significant binding activity.

2.1.3. VH–VL Pairing

When selecting suitable human germlines for the heavy and light chains, considerations of the VH–VL pairing is important from two points of view. Firstly, selected germlines should form a stable Fv, and secondly, the mutual orientation of the VH and VL domains should correspond to that observed in the parental antibody. While the first requirement seems obvious, the second is debatable. A completely opposite reasoning, namely that the VH–VL interface should preserve the interactions of the donor FRs, was indeed utilized (without success) in some studies [117].

The importance of maintaining the VH–VL orientation during the humanization process was demonstrated in the study of anti-lysozyme murine antibody HyHEL-10 [118]. Following humanization, the affinity dropped 10-fold. Structural analysis indicated that all interactions between antibody and antigen were conserved; however, the relative orientation of VH and VL had changed. Amino acid differences between the mouse and humanized mAbs were then mapped onto the structure. In two positions in the FR of the heavy chain, there were rather unusual residues K39 and Y47 in the parental antibody that were replaced during humanization by the conserved Q and W, respectively. A single back mutation W47Y in the humanized mAb completely recovered the affinity. The double mutant W47Y/Q39K showed a further two-fold improved affinity. The crystal structure of the final variant confirmed the VH–VL orientation to be exactly as in the parental antibody.

Early studies have established the promiscuous nature of VH and VL pairing [119–121]. The remarkable ability of the human antibody repertoire to adapt to a specific target by generating a highly diverse panel of antibodies was recently demonstrated by analyzing antibodies raised against a single protein, B-lymphocyte stimulator [122]. Over 1000 antibodies, all different in amino acid sequence, have utilized 42 functional VH, 19 λ, and 13 κ VL germlines. Analysis of the sequences revealed that a given VH sequence can pair with many light chain sequences of both λ and κ types.

Another and much broader study included over 800 different antibodies generated against 28 clinically relevant antigens and isolated from human B cells from 160 donors [123]. Nearly all possible functional germlines (45 VH, 28 λ VL, and 30 κ VL) were represented in the experimental set. The V gene usage indicated no strong bias toward any VH–VL pairing. However, the VH1-λ VL1 germline family pairings were preferentially enriched and represented a remarkable 25% of the antigen-specific selected repertoire.

Somewhat contradictory to previous observations was the conclusion from the analysis of a large dataset of paired light and heavy chains from the Kabat database (Kabat et al., 1991) that VH–VL pairing does not occur at random [124]. Apparently, germline pairing preferences do occur in human antibodies, but only for a small proportion of germlines. The VH1 family shows a strong preference for VK3. On the other hand, no correlation was found between the germline pairing and the VH–VL packing angle.

Although the total number of human germlines and hence the VH–VL pairs is quite limited, relative affinities for each possible pair have not been tabulated. A major reason is that CDR-H3 forms a significant part of the VH–VL interface and therefore affects the pairing potential. In other words, the pairing propensity of any two given germlines depends to a large extent on the sequence and conformation of CDR-H3, which is highly variable. However, for a given CDR-H3 (as is the case in humanization), the differences between various pairings may be substantial.

An interesting, albeit limited, study of VH–VL pairing has reported thermostability values of a panel of 16 Fabs that were produced by all combinations of four VH germlines and four VL (κ) germlines with a fixed CDR-H3 [12]. It was found that the melting temperatures (T_m) of the Fabs differed by more than 20 degrees. For each given light chain, the Fabs with germlines IGHV1-69 and IGHV3-23 are substantially more stable than those with germlines IGHV3-53 and IGHV5-51. Germline IGKV1-39 provides a much higher degree of stabilization than the other three light chain germlines when combined with any of the heavy chains. These results indicate that the selection of the right VH–VL pair is of prime importance during humanization.

For a given CDR-H3, the task of selecting the optimal VH–VL germline pair is reduced to the preservation of residues at the VH–VL interface. This is an additional consideration for human germline selection besides the sequence similarity to the parental antibody. Since the CDRs are grafted as they are, only a small set of FR positions should be considered. These positions may be deduced from a simple analysis of the VH–VL interface in crystal structures. The minimal set of VH–VL interface residues includes seven residues from VH and eight residues from VL (Figures 7 and 8). Besides two residues flanking CDR3, all other residues are in FR2 in both VH and VL. Most of these residues are conserved between human and mouse germlines and across human germlines. However, in those cases when they are different, either because of human vs. mouse differences or due to somatic mutations, the so-called back mutation may be required, as discussed below.

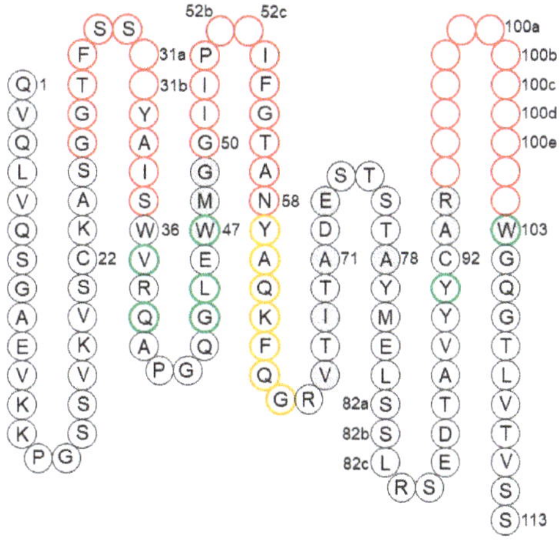

Figure 7. Collier de Perles presentation [105] of VH showing CDRs (red), Vernier zone residues (gray), and VH–VL interface residues (green). Amino acids correspond to human germline IGHV1-69*01 with the Chothia numbering of residues.

Computer programs predicting the VH–VL packing may provide some guidance in finding the best pair of germlines. Several tools have been developed recently along with the realization of the VH–VL orientation as a key parameter in antibody humanization and antibody modeling in general. A straightforward but effective approach has been implemented by Narayanan et al. [125], who used side-chain rotamer sampling for the interface residues followed by molecular mechanics energy calculations. The original main-chain conformations were from the crystal structures. A similar approach was implemented in the Rosetta Antibody modeling software [126].

A machine-learning approach to predict the VH–VL packing angle has been developed and trained on sets of interface residues taken from 567 crystal structures [127]. Rather than selecting interface residues for predicting the packing angle, a genetic algorithm was used to perform feature selection. It was designed to select a maximum of 20 interface positions that were optimal in training the neural network. Thirteen positions were identified as the most influential in determining the packing angle. The results showed an approximately normal distribution of errors with a half width at half maximum of about 2°, which is within the error observed in the crystal structures of antibodies [54].

Yet another approach for determining the VH–VL orientation [128] is also based on the identification of important residues. To describe the VH–VL orientation, six measures (five angles and a distance) were used. Correspondingly, six sets of key positions were identified, with few overlaps between them, 35 positions in total (24 in VL and 11 in VH). To consider so many positions in germline selection is impractical. Instead, the VH–VL packing orientation in the humanized variant may be predicted with one of the computational tools and compared to that of the parental antibody.

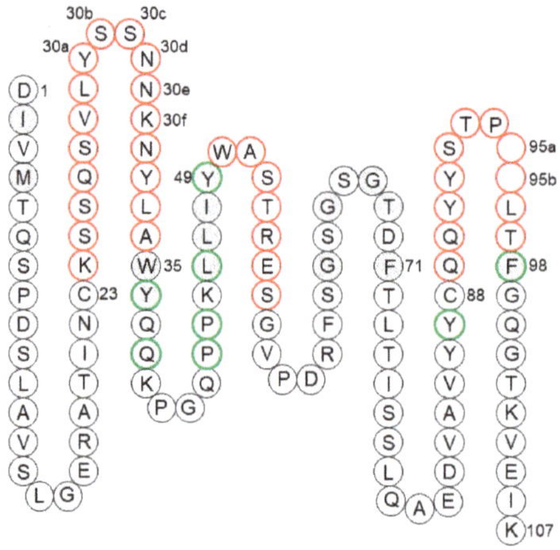

Figure 8. Collier de Perles presentation [105] of VL showing CDRs (red), Vernier zone residues (gray), and VH–VL interface residues (green). Amino acids correspond to human germline IGKV4-1*01 with the Chothia numbering of residues.

2.1.4. Back Mutations

Straightforward CDR grafting may result in reduced target binding even if the VH–VL interface residues are preserved. This problem often arises when non-human CDRs and human FRs are mutually incompatible. Therefore, any CDR grafting protocol must include a step to identify FR positions that are critical for maintaining the CDR conformation. In germline selection, these critical positions should have higher priority than the overall sequence similarity because of their direct impact on the CDR conformation. If a non-human residue in a critical position cannot be preserved because there are no such human sequences, one usually applies the so-called back mutation, i.e., a mutation of a residue in the human FR to the amino acid that occurs in the non-human parent. Such a mutation reduces the humanness score of the resulting variant, but the change should improve the binding affinity.

Foote and Winter [129] identified 30 residues underlying and in direct contact with the CDRs that potentially influence CDR conformations. These residues constitute the so-called Vernier zone. Four of them, heavy chain residues 27–30, are considered part of CDR-H1 in the Chothia, Martin, and IMGT definitions (Table 1). The remaining 26 residues are divided equally between VH and VL (Figures 7 and 8). One of the most recognized examples of a Vernier zone residue is position 71 of the heavy chain that defines the canonical structure of CDR-H2 [130]. Humanization of a few mouse antibodies, including anti-lysozyme mAb D1.3 [95], anti-acetylcholine receptor mAb 198 [131], and anti-tumor-associated glycoprotein mAb B72.3 [132], illustrates the importance of preserving the original residue in this position. However, this is not always the case. For instance, residue 71 was not a major factor in the humanization of anti-cytomegalovirus mAb 37 [133]. Similarly, substitution of Arg for Ala71 during humanization of anti-tissue factor mAb 10H10 was also well tolerated [134]. Hence, one may conclude that the importance of each critical residue depends on the involvement of different CDRs in antigen binding.

Unfortunately, it became a common practice to back mutate most of Vernier zone residues, just to reduce the possibility of a negative impact of human residues on binding [135,136]. However, this will inevitably add several 'non-human' residues to the humanized antibody. Together with CDRs, this may amount to 40% of residues in the variable domains of the antibody, which can hardly be called human. Therefore, a careful analysis of the importance of each Vernier zone position in the context of given CDRs and antibody–antigen interactions is the cornerstone of the humanization process. The availability of the crystal structure of the antibody–antigen complex greatly facilitates the design of humanized variants as it instructs on the FR positions that are indeed critical for antigen binding.

Computer modeling may to a certain extent replace the experimental structure, particularly in the regions at the periphery of the binding site where germline sequences and canonical structures dominate the landscape. The central zone around CDR-H3 remains problematic for accurate modeling, which was confirmed by the latest antibody modeling exercise [54]. Besides the limitations of antibody models, the lack of information on the CDR involvement in antigen binding often leads to an excessive number of back mutations in the humanized antibodies. To avoid such outcomes, each potentially critical position should be tested for back mutation and only those mutations that affect binding should be incorporated into the final antibody.

Back mutations may be applied not only to restore the binding affinity but also to improve the expression of the humanized variants. In the course of humanization of anti-lysozyme scFv F8, it was noticed that FR substitution of Y90F in the VH domain dramatically reduced the bacterial expression of all variants [137]. The back mutation in this position restored the expression and yielded a stable and fully functional antibody. Alternatively, there have been efforts to minimize the affinity of certain Fab domains by introducing more germline sequences [138]. This has been used to increase the potential toxicity of some binding arms.

2.1.5. Deimmunization

While some positions in FRs may require back mutation, several positions within CDRs may be converted to human germline residues when they are not involved in the interactions with antigen or they do not influence the CDR conformation. There is no need to keep non-human residues in such neutral positions. This approach was used for the humanization of three mouse antibodies targeting CD25, vascular endothelial growth factor (VEGF), and tumor necrosis factor alpha (TNFα) [139]. Successive and iterative explorations of the human germline repertoire using semi-automated computational methods allowed the selection of functional humanized mAbs with the highest level of humanness. The resulting antibodies retain the potency of the corresponding chimeric mAbs and have in vitro activity comparable to that of their respective marketed drugs, daclizumab, bevacizumab, and infliximab.

The idea of incorporating human germline residues into the CDRs is related to the finding that CDRs are likely the only segments in humanized and fully human antibodies to contain CD4+ T-cell epitopes [140]. Analysis of a set of eight humanized antibodies representing different VH and VL regions from different genomic segments and affinity maturation processes indicated that prominent CD4+ T-cell epitopes are found only in CDRs and never in FRs. The immunogenic potential of the antibodies could be reduced while retaining their binding properties by incorporating just one or two amino acid substitutions within each T-cell epitope. The approach, which is termed deimmunization, may be considered as complementary to back mutations. It was successfully applied during humanization of an anti-prostate-specific membrane antigen mAb J591 [141], a therapeutic mAb specific for the protective antigen from *Bacillus anthracis* [142], and an antibody against the αv subunit of human integrin [143].

A structure-guided deimmunization method, called EpiSweep, was developed by Parker et al. [144]. The algorithm identifies sets of mutations in potentially immunogenic peptide fragments making optimal trade-offs between structure and immunogenicity, embodied by a molecular mechanics energy function and a T-cell epitope predictor, respectively. Although the program was developed for any therapeutic protein, apparently it may be used specifically for deimmunization of antibodies.

Regarding terminology, some authors consider chimeric antibodies with human constant domains as deimmunized antibodies [145]. We use this term here for a humanized antibody that was additionally modified to enhance the human content.

2.1.6. Resurfacing

An alternative way of reducing immunogenicity risk of the humanized antibody is to replace only the surface residues in the non-human antibody with the residues present in human germlines [146]. Contrary to CDR grafting, resurfacing retains the non-exposed residues of the non-human antibody. This procedure is expected to eliminate potential B-cell epitopes while minimizing the perturbation of residues determining the antigen-binding properties of the antibody.

A systematic analysis of antibody structures was performed to determine the relative solvent accessibility distribution of residues in murine and human antibodies [96]. It appeared that residues in identical positions on the surface of human and murine variable domains are conserved with 98% fidelity across species. Thus, very few amino acid changes are needed to convert a murine Fv surface pattern to that of a human Fv surface.

The method was applied to two murine mAbs targeting CD56 and CD19 [147]. Two different procedures for selecting a human sequence were compared. For one mAb, a database of clonally derived human VL-VH sequence pairs was used while for the other, sequences for VL and VH were independently selected from the Kabat database [148]. Both resurfaced antibodies retained the affinities for their cell surface ligands.

Although most humanization projects in recent years have employed some version of the CDR-grafting method, resurfacing is still in use. For example, to reduce immunogenicity for clinical applications, mouse anti-CD34 mAb was humanized using the resurfacing approach [149]. The structural model was built using templates from the PDB to identify solvent-exposed positions for amino acid replacements with the threshold set at 30%. There were 28 solvent-accessible residues in VH and 35 in VL. Human germline sequences with the highest identity to mouse variable regions were identified, which led to amino acid substitutions in only four FR positions in VH and in five FR positions in VL. The resulting mAb retained the biological functions of the mouse mAb.

Similarly, a murine mAb, which specifically recognizes the pathogenic form of the prion protein, was resurfaced [150]. The design was based on sequence alignments and computer modeling and resulted in an scFv version bearing 13 mutations as compared to the murine parent. The deimmunized antibody demonstrated unaltered binding affinity and specificity. This is not surprising since resurfacing introduces a minimal number of mutations that are located on the surface of the molecule and are unlikely to cause conformational changes in the variable domains. Therefore, retaining affinity is virtually guaranteed, which is not the case in the CDR-grafting humanization. However, the amino acid sequence of the variable domains remains essentially non-human and may present potential epitopes for MHC class II molecules regardless of their surface exposure. Presentation of the epitope peptides to T cells may cause their activation, leading to the induction of signaling pathways [151].

2.1.7. Super-Humanization

Human FRs for CDR grafting may be selected in two different ways, by the highest sequence similarity in the FRs or within CDRs. In the second approach, the FR homology is irrelevant. This method was applied to the humanization of murine anti-CD28 antibody and was called super-humanization [97]. The donor FRs were selected from the human germline gene repertoire based on CDR canonical structures. The super-humanized antibody exhibited a 30-fold loss in affinity.

Another example involving the super-humanization of the murine anti-lysozyme mAb D1.3 was relatively successful. The affinity loss of super-humanized D1.3 was only six-fold [152]. In a final example, the application of the method to the murine mAb 1A4A1, which was raised against Venezuelan equine encephalitis virus, yielded an antibody that retained antigen-binding specificity and neutralizing activity [153]. However, given the mediocre results of the method, it has not gained popularity.

It should be noted that the term super-humanization has also been used in a different sense, particularly when human or simian antibodies contained somatic mutations in FRs and were modified to increase their humanness, as measured by, e.g., the germinality index [154]. Obviously, no CDR grafting was needed in those cases, and super-humanization simply reflected a higher human content of the engineered antibody.

2.1.8. Humanness Optimization

The humanness of the antibody can be assessed by any indicator that is able to distinguish human from non-human sequences. The human string content (HSC) score evaluates the proportion of human germline residues within a given sequence [98]. It can be calculated for a peptide in the target sequence by counting the number of residues identical to their counterparts in the most similar aligned peptide from a human germline. The validity of the HSC score was confirmed by analyzing 513 murine, 32 chimeric, 61 humanized, and 279 human antibody sequences from the IMGT database [155]. Human and humanized sequences produced significantly higher HSC scores when compared to murine and chimeric antibodies. Interestingly, the light chain scores were higher, perhaps due to relatively less diversity among light chains than among heavy chains. The HSC may be used for antibody humanization by maximizing the score rather than using a global identity measure to generate multiple diverse humanized variants. The method was successfully applied to the humanization of four antibodies with different antigen specificities [98]. The resulting variable domains differ fundamentally from those of CDR-grafted antibodies since they are immunologically more human because of being derived from several discrete germline sequences.

Because HSC optimization derives local information from multiple germlines, consideration of three-dimensional information is prudent to avoid clashes. A computational filter that screens for mutual compatibility of different fragments, called analogous contact environments (ACE), evaluates structural patches of amino acids for precedence in a database of antibody sequences [155]. For a given position, the structural precedence score measures the degree of match, weighted for distance and similarity, to the most homologous patch in the database. Averaging over all residues provides the global structural precedence for the sequence. Although a low precedence value does not necessarily mean low structural viability, a higher precedence value indicates that similar structural environments are sampled in the database, suggesting that the test sequence is more likely to behave favorably.

The humanness scores that are based on pairwise sequence identity between the sample and a set of germline human sequences may consider the average similarity [156], or the average among the top 20 sequences [157], or the highest similarity over windows of 9 residues [98,155]. In a different approach [158], the score function accounts both for local preferences and for pair correlations between residues at different positions. The method does not distinguish CDRs from FRs, which may be a plus since the latter may contain antigen-binding residues. Moreover, the relationship between the humanness score and the observed immunogenicity in patients was also considered [159].

With the growing wealth of sequence databases, statistical-inference methods could become an increasingly relevant tool, with a range of applications well beyond antibody humanization. Within a humanization protocol, the advantage of this approach over CDR grafting is that it proposes a set of candidate sequences, at increasing distance from the non-human parent toward the highest humanness score, instead of requiring the introduction of arbitrary back mutations.

2.2. Lambda to Kappa Chain Switching

Upon humanization, the type of the light chain of the parental antibody is usually not changed, i.e., if it was λ in the non-human antibody, the FR for a humanized variant is selected from the human λ repertoire. In some situations, switching the light-chain type may provide certain benefits. For instance, the production of bispecific antibodies from two mAbs might involve a purification step, which could be easily optimized if the mAbs contain light chains of different classes.

Technically, there are several significant differences between κ and λ chains that complicate the task. One is a deletion of a residue at position 10 in λ that is present in κ. Another is a different set of canonical structures for CDR L3, which are longer in λ chains and lack a conserved cis-proline at position 95. Also, λ CDR L1s differ from those of κ by being longer and fold into a helical structure.

The overall sequence similarity between κ and λ germlines is below 50%. However, the FRs sequence similarity of around 60% of identical residues is higher than between some κ sequences (e.g., IGKV4-1 and IGKV5-2, which share only 51% identity in FRs). More importantly, all eight residues at the VH–VL interface are conserved between κ and λ types (Figure 8). Therefore, light-chain type switching seems feasible, and this has been confirmed in a few reports.

As part of a bispecific antibody, the Fab arm directed against FcγRIII was humanized by CDR grafting [117]. In a first attempt, the murine λ VL was converted to a humanized λ chain, which led to a complete loss of antigen binding and extremely poor folding efficiency. Initial humanization applied a fixed-FR approach using the human myeloma protein KOL as the FR donor, which had only 51% to 54% identity to the mouse antibody. Despite several back mutations in the Vernier zone, the strategy failed. Hence, the CDRs were transplanted onto a human κ light chain using the same strategy. Humanized anti-HER2 mAb 4D5, characterized by the VH subgroup III and VL subgroup I, was selected as an FR donor. Residues in positions 46, 49, 66, and 71 in VL and 71, 73, and 78 in VH were back mutated for various reasons. This resulted in a functional Fab, yet with a 100-fold decreased antigen affinity, which was subjected to affinity maturation through random mutations in both VH and VL. The optimized Fab exhibits an affinity within a factor of two from that of the original murine antibody. This required nine mutations, six of which are in VH and three in VL. Interestingly, most of the mutations occur at the VH–VL interface. Even if the humanization strategy was not optimal, it demonstrated that switching λ to κ may be successful.

Another attempt of the λ-to-κ switch occurred during optimization of an anti-GCN4 murine scFv [160]. The CDRs were grafted onto the FR of another murine scFv, which was selected due to its high stability. In this process, the CDRs of the parental λ VL were transferred to the FR of the donor κ VL. Homology modeling of the designed variant revealed some structural inconsistencies, particularly a potential clash between CDR L1 and loop 66–71 (sometimes referred to as the DE loop or CDR4). Therefore, this loop was back mutated to the original sequence. Additionally, eight residues in VH were also back mutated. The resulting scFv was significantly more stable than the original, but lost binding by three orders of magnitude. Back mutation of seven residues at the VH–VL interface to restore the proper orientation of the domains further enhanced the stability of the construct, which still had an order of magnitude reduction in the original scFv affinity.

A successful case of λ-to-κ conversion to improve the thermodynamic properties of scFv was reported recently [161]. The heavy chain of this scFv originated from the IGHV1-69*01 germline, whereas the light chain appeared to contain a fusion of two genes, *IGLV3-19*01* and *IGLV1-44*01*, likely resulting from PCR aberration during library construction. The idea to replace the λ light chain by κ IGKV3-20 was based on the observation that this germline commonly pairs with IGHV1-69 to give highly expressed stable antibodies [162]. To guide the design process, a homology model of the converted scFv was constructed that revealed a potential clash between CDR L1 and loop 66–71. Analysis of a large set of PDB structures confirmed that this problem is typical for a λ-to-κ conversion. To facilitate CDR grafting, the DE loop from the original antibody was retained. No back mutations were necessary in VH. The resulting scFv showed increased thermostability and expression levels while retaining the binding affinity to the target. The scFv variant with the κ DE loop was less stable while also retaining binding.

These results indicate that λ and κ chains may be swapped without compromising the functional properties of the antibody. This strategy may be applied in antibody humanization or may prove useful for optimizing the biophysical properties of therapeutic candidates.

2.3. Affinity Maturation

Natural antibodies, both human and non-human, often do not possess the binding properties required for their therapeutic applications. This appears to be a consequence of the affinity ceiling that characterizes the mammalian immune system and B-cell responses [163,164]. Increasing the binding affinity is an important and almost inevitable step in the development of the lead candidate since it is related to the dose needed for treatment and the therapeutic efficacy. Different approaches, tools, and strategies are available and have been validated through the engineering of antibodies directed against various antigens. All of them can be divided into two groups according to the process of generating antibody variants. One is the rational design of the variants, followed by their expression in the system of choice. The other is the construction of a library of variants where several positions are diversified, followed by their display in a system of choice with the appropriate selection method. Owing to the large number of variants in a library covering the entire combinatorial space, the latter method is most commonly used for affinity maturation. In cases when only a few positions and a few amino acids are to be tested, perhaps the former approach may fulfill the task as it is fast and inexpensive. Whichever method is used, structure-based computational design may facilitate the process by in silico evaluation of the candidates to minimize either the library size or the number of mutants to be expressed.

A high-resolution structure of the antibody–antigen complex allows detailed analysis of the antibody–antigen interactions and greatly facilitates the design of affinity-enhanced variants. Even the structure of the Fab alone may instruct the selection of the most promising positions for mutagenesis. There are new developments of using NMR relaxation dispersion and hydroge-deuterium exchange experiments to map out regions for optimization of the affinity [165]. In the absence of any experimental information, computer modeling may fill the gap to a certain extent, and several successful examples based on theoretical models are discussed below.

First, we consider affinity maturation by structure-based rational design of the antibody variants with improved side chain packing and electrostatic interactions. A case in point is the improvement of the binding affinity of the anti-integrin antibody VLA1. Engineering increased the affinity by an order of magnitude primarily through a decrease in the dissociation rate [166]. Inspired by the crystal structure, a diverse set of single mutations (>80 variants) at the antibody–antigen interface were generated. Mutations were made to nearly every antigen-contacting residue using suggestions from computational methods. The most promising mutations were combined into a quadruple mutant with two mutations in the light chain and two in the heavy chain, and its crystal structure confirmed the predicted interactions.

A similar approach that focused on electrostatic interactions was employed to design single mutant variants with improved affinity. Selection criteria based on calculations of the improved binding electrostatics resulted in a success rate for single mutations of over 60%. By combining multiple designed mutations, the affinity of antibodies specific for various antigens was improved 10-fold for the anti-epidermal growth factor receptor antibody, cetuximab, and 140-fold for an anti-lysozyme antibody D44.1, achieving 52 and 30 pM affinity, respectively [167,168].

While antigen-contacting residues at the center of the binding interface may be an intuitive choice for mutations, many studies indicate that targeting peripheral residues may be more promising for affinity maturation. The key residues at the center of binding sites are usually hydrophobic and tightly packed and already well optimized for specific antigen interactions. In contrast, the surrounding residues are often hydrophilic and solvent exposed. Incorporating charged residues at the periphery of the interface may improve long-range interactions.

In the following example, the design strategy was based on two assumptions: (1) Mutation positions should be at the periphery of the antibody–antigen interface, and (2) substitutions should be those that frequently occur during affinity maturation in vivo. To improve the affinity of the therapeutic mAbs trastuzumab and rituximab, in silico models for a series of mutants were generated using crystal structures of the complexes, Monte Carlo-simulated annealing, and molecular dynamics simulation [169]. Single mutations at each of the 60 CDR positions to the 20 common amino acids were ranked by the total calculated binding free energy. The top 11 mutants were tested experimentally and only two of them showed improved binding. Alternatively, when only amino acids with a high usage in the binding sites of matured antibodies were considered for mutations, the success rate was 60% to 70%.

One of the most striking findings in this study was that affinity-enhancing mutations tend to cluster around positions where in vivo somatic mutations often occur. It is known that somatic hypermutation does not occur randomly within immunoglobulin V genes but is preferentially targeted to certain nucleotide positions, termed hotspots [170]. This process mainly results in the introduction of mutations that are located at or very near A/G|G|T/C|A/T and TAA sequences [171,172]. The results of the study indicate that germline hotspot sequences may point to the mutation sites in the affinity maturation process.

The combination of in silico calculations and thermodynamic analysis proved to be an effective strategy to improve the affinity of an anti-MCP-1 mAb 11K2 [173]. Amino acid substitutions were evaluated in each of the 62 CDR positions of 11K2, and all 20 amino acids were employed. Based on the crystal structure of 11K2 in complex with MCP-1, a virtual library of mutations to identify antibody variants of potentially higher affinity was generated. Each model of the mutated antibody–antigen complex was optimized by a combination of simulated annealing and molecular mechanics minimization. The variants were ranked by their electrostatic and van der Waals interaction energies and the most promising candidates were tested in vitro. Only mutations in the light chain of the antibody were effective at enhancing its affinity, suggesting that in this case, the interaction surface of the HC is not amenable to optimization. The single mutation with the highest affinity, N31R in CDR L1, yielded a variant with a five-fold higher affinity with respect to that of the wild-type antibody.

All these studies are examples of the fixed-backbone approach of computational design, where the backbones are not altered beyond energy minimization. Incorporating backbone flexibility in computational design allows conformational adjustments that may broaden the range of predicted low-energy sequences. In some cases, backbone movements are critical, for instance, when dealing with allosteric effects resulting from the changes in non-contacting residues. A comparison of different protocols for modeling backbone flexibility was performed in the affinity maturation study of the therapeutic mAb, trastuzumab [174]. An in silico approach based on the crystal structure of the trastuzumab complex with its target human epidermal growth factor receptor 2 (HER2) identified a key mutation D98W, which led to a three-fold affinity improvement of the already subnanomolar antibody.

Although the amino acid composition of protein–protein interfaces is quite diverse, there is a significant bias toward specific residues [175]. It was demonstrated that high-affinity antibodies could be obtained from restricted combinatorial libraries in which CDR positions are diversified to a combination of as few as two amino acids, Tyr and Ser [176]. Encouraged by these results, Inoue et al. [176] applied this binary code to the affinity maturation of an anti-lysozyme camelid single-domain antibody. They also used in silico screening for the selection of potential amino acid replacements. The scoring function was based on the interaction energy (IE) and electrostatic complementarity (EC) criteria [177]. When introducing mutations into CDR1 and CDR2, conserved amino acids were preserved, so that only about half of the residues were mutated to Tyr or Ser, giving a total of 512 (2^9) theoretically possible mutants. Several variants that showed improved IE and EC parameters were tested for binding. The best of them exhibited a five-fold improvement in K_D values from 2.8 to 0.5 µM. Then, based on the crystal structure of the antibody–antigen complex, two residues in CDR2 in contact with the Ag were mutated to either Arg or Asp and tested for binding. This round yielded a variant with a K_D value of 0.14 µM, i.e., 20 times lower than in the parent mAb.

The design of mutants with improved affinities relies on the 3-D structure of antibody–antigen complexes. A variety of structure-modeling tools can help in the absence of experimental data. The following examples emphasize the value of computer modeling for affinity maturation of antibodies. They emphasize the use of computational docking, the process of predicting the conformation of a complex from its separated components.

In the first example, the binding of two antibodies to the stalk region of influenza hemagglutinin was modeled by using only the structure of the target protein and compared to the known experimental structures of the complexes [178]. This study demonstrated that some of the computational docking predictions can be very accurate, but the algorithm often fails to discriminate them from inaccurate solutions.

In a second example, the binding affinity of an anti-hepatitis B virus antibody was improved. For this study, both the antibody itself and the antibody–antigen complex were modeled (by docking the 17-residue peptide) [179]. Inspection of the model instructed the design of point mutations in the putative paratope, and two mutations, Y96S in VL and D98S in VH, were predicted to have the largest drop in free energy. The double mutant indeed exhibited a 10-fold increase of affinity.

While in silico design of antibody mutants may be successful, combinatorial libraries provide a common method to improve affinity. The selection of positions for diversification and the choice of amino acids for mutations are two principal tasks in a library design. As evidenced in the literature, the best results may be achieved if these tasks are fulfilled by a structure-guided approach.

A very convincing example of a stepwise structure-guided affinity maturation procedure was reported for anti-gastrin scFv TA4 [180,181]. An impressive 454-fold improvement in affinity was obtained by a combination of walking randomization [182] and a model-based approach that was achieved without experimental 3-D information. A structural model of the antibody–antigen complex was generated by docking a seven-residue peptide representing a linear epitope into the model of the antibody. The docked complex was refined by molecular dynamics, which indicated that the peptide adopts a helical conformation. Based on this model, four positions in CDR-H3 and five positions in CDR L3 were selected for randomization. The first of the libraries, that based on CDR-H3, produced a double mutant with an almost 10-fold improved affinity. The second library based on this mutant and CDR L3 mutagenesis produced two variants with about a two-fold affinity improvement over the double mutant. Again, the 3-D model guided the selection of positions in other CDRs for further affinity improvement. This procedure yielded several variants, with the affinity in the low nanomolar range.

Analysis of the binding surface of the antibody and assessment of the relative involvement of CDRs in target binding facilitates the strategy for library design. Depending on the CDR length and the number of interactions with the antigen, central CDRs H3 and L3 may provide the best opportunities for affinity maturation. This was the case in the development of an anti-VEGF scFv isolated from a phage-displayed human antibody repertoire [183]. Two phage display libraries were constructed by diversification of CDR-H3 and CDR L3. A competitive phage-selection strategy in the presence of the parental scFv as a competitor was used to eliminate low-affinity binders. High-affinity variants were retrieved from both libraries. An optimized VL variant was designed and constructed by combining recurrent replacements found among selected variants into a single molecule, resulting in an additional affinity increase. Further affinity improvements were achieved by combining this optimized VL with the best VH variants. The final variant showed an 18-fold affinity improvement over the parental scFv and exhibited an enhanced potency to block the binding of VEGF to its receptor.

An impressive example of affinity improvement was carried out for the anti-complement protein receptors C5aR1/2 mAb [184]. The affinity of the parental antibody was improved by randomizing amino acids in CDR-H3 and CDR L3 using a phage library displaying scFv fragments. Following recombination of the two libraries and screening to identify additional synergistic increases in affinity, the best variant was selected with four mutations in CDR-H3 and two mutations in CDR L3. This variant binds its target with an affinity in the low pM range, demonstrating a gain of three orders of magnitude with respect to the parental antibody affinity.

Quite often, central CDRs, particularly CDR-H3, are too heavily involved in the interactions with the antigen, so that CDR-H2, and to a lesser extent CDRs H1 and L1, may be the focus of library design. Increasing the number of diversified positions in each library and expanding the selection to all CDRs inevitably results in a lower coverage. However, in certain cases, this approach may also be successful. Simultaneous mutagenesis of all six CDRs in a non-human primate antibody that neutralizes anthrax toxin was carried out using phage display technology [185]. The library contained 5×10^8 variants, with each variant containing an average of four mutations. The best variant selected from the library showed a 19-fold affinity improvement to 180 pM.

While selecting the positions for mutagenesis to improve binding, one may consider not only the paratope residues or surrounding residues in the Vernier zone but also positions in the core of the variable domains or even at the elbow between the variable and constant domains. Exactly this approach was realized during the development of the anti-tumor growth factor beta subunit 1 (TGF-β1) antibody metelimumab [186]. Upon conversion from the parental single-chain variable fragment (scFv) to IgG4, the binding affinity dropped by 50-fold. Following a hypothesis that this was due to decreased conformational flexibility of the IgG, insertion mutants in the elbow region were designed and screened for binding and potency. The insertion of two glycines in both the heavy chain and light chain elbow regions restored the binding affinity. The crystal structure of the mutant confirmed that the insertions provided enough flexibility for the variable domains to extend further apart than in the wild-type Fab, allowing the CDRs to make additional interactions not seen in the wild-type Fab structure.

2.4. Specificity

Conventional antibodies are monospecific and typically recognize a single antigen exclusively, owing to the binding of non-linear epitopes. Some antibodies do exhibit multi-specificity, particularly if a very similar epitope is present on more than one antigen. Typical examples include species cross-reactive antibodies that recognize orthologous proteins in different species [187,188] or antibodies that interact with different members of a conserved protein family [189,190]. The species specificity has often been a setback in assessing antibody utility as a therapeutic agent in various animal models. The same combinatorial techniques that are used to improve antibody affinity have been used to modify their cross-reactivity [191,192]. Specificity engineering also heavily relies on in silico design strategies and the availability of experimental structural information [193].

Using the crystal structure and molecular mechanics-based energy function, cross-species specificity was introduced into the antibody that inhibits cancer-associated serine protease MT-SP1 [194]. The mAb exhibits a K_D value of 12 pM towards human antigen but only 4 nM towards the mouse ortholog. There are only three residues on the protease surface that both make contact with the antibody and that are different between the human and mouse versions of the enzyme, but these residues are not critical for inhibition. Computational design was used to predict a suite of mutations that could improve the affinity to the mouse antigen. Mutations were introduced at six positions within 5 Å from these three epitope residues. Each of the selected residues, two in CDR-H1 and four in CDR-H3, was mutated in silico to all possible amino acids, and for each substitution, the change in binding energy was calculated. Most of the mutations were predicted to be neutral. Out of eight candidates tested experimentally, one variant, T98R, improved the affinity by an order of magnitude without any effect on the binding of the human ortholog.

The development of a promising therapeutic mAb targeting quiescin sulfhydryl oxidase-1 (QSOX1) was hampered by the lack of reactivity against the mouse QSOX1 ortholog [195]. To understand the molecular basis for species restriction, the crystal structures of mouse QSOX1 alone and human QSOX1 in complex with the Fab fragment were determined. Structural differences responsible for the species specificity of the antibody were identified and used for the construction of small libraries, in which up to four positions near key epitope positions were diversified. After several rounds of panning and the combination of mutations from different libraries, the affinity toward mouse QSOX1 was improved by at least four orders of magnitude, reaching the low nanomolar range and matching the affinity toward human QSOX1. The crystal structure of the re-engineered variant complexed with its mouse antigen revealed that the antibody accomplished dual-species targeting through altered VH–VL domain orientation and, most importantly, through rearrangement of the CDR-H3 backbone because of a quadruple mutation YYGS to SMDP.

In another study, a structure-based strategy was implemented to develop an anti-CD81 mAb to enable animal model testing in cynomolgus monkeys [196]. The antibody would bind tightly to human CD81 (K_D vakue of 0.9 nM) but exhibited no detectable binding to cynomolgus (cyno) CD81. The crystal structure of the scFv was determined in complex with human CD81 and used for guiding the library design. A phage-display library was constructed to diversify CDR-H2, which seemed to be a major specificity determinant. Alternating rounds of binding selections with immobilized cyno and human antigens yielded an antibody that was used as a template for the second round of selection with libraries around CDR L1 and CDR-H1. The best variant exhibited robust binding to cyno CD81 and only showed a two-fold reduction in affinity for human CD81.

Whereas species cross-reactivity or broad neutralizing potential may be beneficial for a therapeutic antibody, sometimes, the insufficient selectivity may be regarded as a liability. Although mAbs are generally very selective, close analogs of the target molecule may pose a risk of side effects. The development of anti-progesterone mAb C12G11 was hampered by its poor selectivity [197]. The mAb has a picomolar affinity for progesterone but also strongly binds 5β- and 5α-dihydroprogesterone. To reduce the cross-reaction with these analogs, a phage library randomizing five antigen-binding positions in CDR-H2 and CDR L3 was constructed. The design was based on the homology model of the antibody complemented by docking of the target molecules. Variants selected in the initial screening were further optimized by the addition of second-sphere positions to the library. The best variant demonstrates high specificity toward progesterone as compared with the 15- to 20-fold lower cross-reactivity for the analogs. The improvements are linked to a change in the canonical class of CDR L3.

Koenig et al. faced a similar task to fine-tune the specificity of an angiopoietin-2 (Ang2)/VEGF dual-action Fab [198]. This antibody utilizes overlapping CDR sites for dual antigen interaction, with affinities in the sub-nanomolar range. However, it also exhibits significant (K_D value of 4 nM) binding to Ang1, which has high sequence similarity to Ang2. An approach to specificity engineering that does not require prior knowledge of the antibody–antigen interaction was employed in this study. A large phage-displayed library of the Fab variants with all possible single mutations in all six CDRs provided information on the effect of binding for each mutation. In silico analysis identified 35 mutations predicted to decrease the affinity for Ang1 while maintaining the affinity for Ang2 and VEGF. Structural analysis showed that some of the mutations cluster near a potential Ang1/2 specificity-determining residue, while others are up to 15 Å away from the antigen-binding site and apparently influence the binding interaction indirectly. The lack of information on antibody–antigen interactions in this approach was compensated for by the size of the library.

The mechanisms of antigen recognition by antibodies vary significantly from the structurally rigid key-and-lock mechanism to the adaptable induced-fit mechanism. Correspondingly, one of the mechanisms of multispecificity lies in the plasticity of the antigen-binding site, which allows for the recognition of structurally unrelated epitopes by the same antibody [199,200]. This principle was utilized in a stepwise engineering strategy for generating dual-specific antibodies de novo, called two-in-one antibody with dual-action Fab (DAF). The first proof-of-concept DAF was targeting VEGF and HER2 [201,202]. The strategy was also successfully applied for the generation of duligotuzumab, which targets EGFR and HER3 [203].

2.5. Chemistry, Manufacturing, and Control (CMC) Considerations

Recombinant mammalian cells are the dominant production system for antibody-based therapeutics because of their ability to perform complex post-translational modifications (PTMs) that are often required for efficient secretion, drug efficacy, and stability [204]. Because of the nature of heterologous expression, there are modifications to the biologics, which include misfolding and aggregation, oxidation of methionine, deamidation of asparagine and glutamine, variable glycosylation, and proteolysis (see Table 2) [205,206]. Such unintended PTMs can pose challenges for consistent bioprocessing and can affect the molecular physicochemical properties (such as shape, size, hydrophobicity, and charge) that in turn can affect pharmacokinetic (PK) and pharmacodynamics (PD) properties. For instance, electrostatic interactions between anionic cell membranes and the predominantly positive surface charge of most antibodies can influence the blood concentration and tissue disposition kinetics in a manner that is independent of antigen recognition. Thus, charge variation can result in shifts in isoelectric point values, which can change the tissue distribution and kinetics; increases in net positive charge generally result in increased tissue retention and increased blood clearance [207–209]. Protein and peptide deamidation can occur spontaneously in vitro under relatively mild conditions that can be used to predict in vivo chemistries [210]. For antibodies and other therapeutic proteins, great effort is placed in the manufacturing process and storage conditions to minimize this form of degradation. Glycosylation control is typically controlled by the selection of the manufacturing cell line and control of cell metabolism during bioreactor conditions [211,212]. The specific glycan attachments (fucosylation, sialylation, galactosylation, high-mannose, and bisecting glycans) have great importance to the antibody properties [213,214] (See Figure 5). Simulations of the dynamic interface between the glycans and the Fc domains have been described [215]. However, it is also important to remember that under physiological temperature and pH conditions, antibody deamidation, c terminal cleavage, and glycation kinetics do occur and can affect the serum lifetime of antibodies [209,216,217].

During the process of discovery, the selection of candidate molecules should employ technologies that increase the odds of identifying potent biologics that bind the desired biologically relevant epitope. Concomitant with potency optimization, the biologics should be counter-screened to have drug-like properties early in the process. Although the germline amino acid sequences are typically left unchanged, mutagenesis of amino acids' liabilities (sites of oxidation, deamidation, clipping, glycation, glycosylation) in the CDRs should be considered. Such changes should be completed prior to the manufacturing cell line development to minimize the risks of having a less-than-robust chemistry, manufacturing, and control (CMC) process. Upstream consideration of developability metrics should reduce the frequency of failures in later downstream development stages. Optimization of stability in the early stages of discovery can reduce complications in upstream and downstream process optimization as well as increase the potential for successful drug product formulations [218,219]. Nonetheless, downstream processing can minimize some levels of oxidation via the presence of free radical scavengers, elimination of redox metal ions, addition of chelation agents, protection from light, decrease in storage temperature, and reduction of exposure to oxygen.

Table 2. Common post-translational modifications to amino acids in monoclonal antibody framework molecules.

Amino Acid Changes	Chemistry	Effect on Protein	Effect on Biology
Asn-(Gly/Ser); Asp-(Gly/Ser)	Asn deamidation, Aspartic acid isomerization	Protein degradation [220–222]; Tertiary changes to Ab structure [223]; Isoaspartic acid [224]; Aggregation [225]	Isomerization can affect IgG avidity [226]; Deamidation affects binding [227]; Deamidation affects PK [216]
Gln	Gln deamidation	Slower deamidation than Asn, heterogeneity and stability [228]	Biological activity on Fab and Fc *
Met	Oxidation	Presence of oxidized methionine affects charged state of proteins [229–231]; Methionine oxidation decreases affinity to protein A and FcRn [232]	Methionine oxidation on Fc region can modulate FcγRIIa engagement [233]; FcRn and Fcγ receptors [234]; PK [235,236]
Trp	Oxidation	Changes in Trp aromaticity [237]; color changes [238]; Effects on detergent excipients for Ab formulation [239]; Higher order structure [240]	Biological activity on Fab and Fc [241,242]
Cys	Oxidation	Cysteinylation; Hinge disulfide chemistry with Cu^{2+} ion results in hydrolysis or oxidation that can lead to cleavage of the mAb [243–245]	Cysteinylation in CDRs leads to loss of potency [246,247]; Changing disulfide patterns in IgG subtypes [248]
His	Oxidation	Oxidized histidine react with intact histidine, lysine, and free cysteine to crosslink IgG [249]. Oxidized histidine [250,251]	Biological activity on Fab and Fc *
Asp-(Pro/Gly)	Amide bond hydrolysis	Cleavage at aspartic acid under acidic conditions [252,253]; Clipping at CH2 domain leads to aggregation [254]	Biological activity on Fab and Fc *

Table 2. *Cont.*

Amino Acid Changes	Chemistry	Effect on Protein	Effect on Biology
N terminal Glu/Gln	Pyroglutamate formation	Cyclized N terminal glutamine [255]; Challenges with molecule comparability [256]	Biological activity [256]
C terminal truncation	Carboxypeptidase substrate	Human IgG is produced with C-terminal Lysines that are cleaved off in circulation. There can be changes in charge variation	C terminal lysine loss can enhance complement activation [257]
Glycation	Reducing sugar reaction with Lysines	Charge variants [218]; Structural heterogeneity [258]	Biological activity on Fab and Fc *
Glycosylation changes	Changes in glycosylation profiles	Glycan structure [67,69,259–263]; High mannose and afucosylation affect stability [264]; Sialylation [265]; Fucosylation [266]	Biological activity [267]; PK and PD [268]; Clearance [269]

* Changes to critical amino acids linked to Fab-antigen or Fc-Fc receptor binding and functions.

2.5.1. Solubility

Low solubility or high viscosity of antibody formulations at concentrations over 100 mg/mL can impede their development as products suitable for subcutaneous delivery. Antibody engineering, especially when applied at the discovery stage, may be instrumental in overcoming the challenges of the product development and pave the path to the clinic. The following examples highlight some approaches that proved to be helpful in improving the solubility of antibodies. Several studies were based on homology modeling rather than on experimental crystal structures, and in many cases, this may be enough for developability purposes. Of course, this excludes the cases when antibodies present unusual CDR in either HC or LC loops either in sequence or in length. Antigen-bound co-crystals are also very useful for identifying mutants that are more likely to retain target binding while optimizing solubility.

Structure-guided design of point mutations was carried out for the development of a therapeutic mAb candidate that was unacceptably viscous at high concentrations [270]. The idea was to test the effects of hydrophobic and electrostatic intermolecular interactions on the solution behavior of the mAb by disrupting either aggregation prone regions or clusters of charged residues. The variable region contained two hydrophobic surface patches and a negatively charged cluster. The disruption of a hydrophobic patch at the interface of VH and VL via L46K mutation in VL destabilized the mAb and abolished antigen binding. However, mutation at the preceding residue (V45K) in the same patch increased the apparent solubility and reduced viscosity without sacrificing antigen binding or thermal stability. Neutralizing the negatively charged surface patch by E60Y mutation in VL also increased apparent solubility and reduced viscosity of the mAb, whereas charge reversal at the same position (E60K/R) caused destabilization, decreased solubility, and led to difficulties in sample manipulation that precluded their viscosity measurements at high concentrations. Both V45K and E60Y mutations showed similar increases in apparent solubility. However, the viscosity profile of E60Y was considerably better than that of the V45K, providing evidence that intermolecular interactions in this mAb are electrostatically driven.

Aggregation of single-domain VH antibodies specific for Alzheimer's amyloid β-peptide was examined from the structural perspective [271–273]. These antibodies contained clusters of hydrophobic residues within the HC and LC CDR3. Inserting two or more negative charges at each edge of the CDR3 domains potently suppressed antibody aggregation without altering binding affinity. Inserting charged mutations at one edge of CDR3, either the N- or C-terminal, also prevented aggregation, but only if such mutations were located at the edge closest to the most hydrophobic portion of CDR3. In contrast, charged mutations outside of CDR3 failed to suppress aggregation. These findings demonstrate that the CDR loops can be engineered in a systematic manner to improve antibody solubility without altering binding affinity.

The case of anti-IL-13 mAb CNTO607 presents three structure-based engineering approaches to improved the solubility of a therapeutic candidate [274]. First, the isoelectric point was modified by the incorporation of charged residues in the positions remote from the binding site. A mutant with a modified pI showed a two-fold improvement in solubility while retaining full binding to IL-13. Second, the overall surface hydrophobicity was targeted by mutating residues in a hydrophobic patch found in CDR-H3. According to the crystal structure, the patch included residues F99-H100-W100a from CDR-H3 flanked by hydrophobic residues from the light chain. The triad in CDR-H3 appeared to be essential for the high affinity of the antibody to IL-13. Various mutations in these residues improved solubility but negatively impacted affinity. Mutations in CDR L3 were more promising since this CDR is less involved in antigen binding. The best variant with W91Y and M93S mutations gained a two-fold improvement in solubility and exhibited a shorter retention time on HIC, while binding to IL-13 remained in the low picomolar range. In a third approach, an N-linked glycosylation site was introduced in CDR-H2 (D53N) to shield the aggregation patch in CDR-H3. This variant indeed showed greatly improved solubility while maintaining affinity to IL-13 and proved to be the most effective route for enhancing the solubility of CNTO607. Recently, there has been a description of how optimization of the V domain framework can improve the biophysical qualities of a therapeutic antibody candidate [275].

Another example of solubility improvement through computer modeling was reported for an integrin α11-binding antibody [276]. A homology model of the parental Fv region revealed hydrophobic patches on the antigen-binding surface. A series of 97 computationally designed variants focused on the residues in the hydrophobic patches that were expressed, and their HIC retention times were measured. As intended, many of the variants reduced the overall hydrophobicity as compared to the parental antibody. Contrary to the previous study (CNTO607), replacement of aromatic residues W96, Y97, and Y98 in CDR-H3 did not cause a loss of binding, apparently because they are not in contact with integrin. Interestingly, adding charged residues in place of polar residues in the CDRs near the hydrophobic patches did not reduce the retention time.

Three-dimensional protein property descriptors were developed and evaluated for their ability to predict the hydrophobicity profiles of antibodies [276]. Analysis of recently published data for 137 clinical mAb candidates [277] indicated that the surface area of hydrophobic patches consistently correlated to the experimental HIC data across a diverse set of biotherapeutics.

A general approach to predicting aggregation-prone regions on the basis of three-dimensional structures has been realized in the algorithm termed AggScore [278]. The method uses the distribution of hydrophobic and electrostatic patches on the surface of the protein, factoring in the intensity and relative orientation of the respective surface patches into an aggregation propensity function that has been trained on a benchmark set of 31 adnectin proteins. When applied to the experimentally characterized antibodies in the clinical stage [277], AggScore accurately identified aggregation-prone regions and predicted changes in aggregation behavior upon residue mutation.

As more biotherapeutics are entering pharmaceutical pipelines, more weight is put on the early-stage developability assessment and optimization strategies. Computational methods for assessing solubility, hydrophobic interactions, and other liabilities are in high demand. Successful efforts have also been made to use rational design to reduce aggregation and improve solubility by mutating key surface residues identified from a crystal structure or a homology model.

2.5.2. Stability

After the selection based on functional properties, antibodies may be modified to improve developability or scale-up processes involving stability, expression, purification, and formulation. The stability of an antibody is influenced by a number of factors that can include: (1) core packing of individual domains that affects their intrinsic stability; (2) protein–protein interface interactions that have impact upon the HC and LC pairing; (3) burial of polar and charged residues; (4) hydrogen-bonding network for polar and charged residues; and (5) surface charge and polar residue distribution among other intra- and inter-molecular forces [279]. Potential structure-destabilizing residues may be identified based upon the crystal structure of the antibody or by molecular modeling in certain cases, and the effect of the residues on antibody stability may be tested by generating and evaluating variants harboring mutations in the identified residues. One of the ways to increase antibody stability is to raise the thermal transition midpoint (T_m) as measured by differential scanning calorimetry (DSC), differential scanning fluorimetry (DSF), or thermal transitions [280–284]. In general, the protein T_m is correlated with its stability and inversely correlated with its susceptibility to unfolding and denaturation in solution and the degradation processes that depend on the tendency of the protein to unfold [285]. A few studies have found a correlation between the ranking of the physical stability of formulations measured as thermal stability by DSC and physical stability measured by other methods [169,286–289]. Formulation studies suggest that a Fab T_m value can have implications for the long-term physical stability of a corresponding mAb. Thus, the CDR sequence selection can impact the stability of the VH–VL domain, and sequence–stability tradeoffs must be considered during the design of such libraries [290].

3. Engineering Antibody Activity

While mAbs are successful for many distinct applications, there are still limitations. First, the surface area of the IgG variable region may not bind to the small extracellular loops of transmembrane proteins, such as G-protein-coupled receptors (GPCRs). Secondly, animal models show that most administered mAbs have limited distribution into the diseased tissue. A favorable pharmacokinetic profile does require sufficient target occupancy in the diseased tissue, which ultimately requires efficient tissue penetration and retention time in the diseased tissue. Thirdly, single agent mAb efficacy can be limited because the disease phenotype can have more than one pathway that can mediate resistance. Often, the in vitro properties of the candidate antibody that probe a limited array of responses do not corroborate with in vivo profiles that can involve more complex mechanisms of action. These limitations have prompted research in generating new antibody-based therapeutics that can meet these aforementioned challenges by adopting antibody engineering approaches that can include: Binding domain engineering; avidity modulation; antibody–drug conjugation; Fc activity engineering; and bispecific antibody generation.

3.1. Binding Domain Engineering

The average surface area of an antibody epitope-containing surface is around 1600 to 2300 Å^2 [291]. Although this surface area is ideal for modulation of most protein–protein interactions, there can be target molecules with epitope surface exposures that are more restricted. For instance, there are limited examples of the obtainment of functional antibodies against GPCRs. Recent crystal structures have shown that there is limited access to epitopes due to the presence of N-terminal domain glycosylation and limited surface exposure to the extracellular membrane protein loops [292]. For applications that require smaller binding surfaces, single domain (12–15 kDa) antibodies (sdAbs, also known as VHH Abs or nanobodies) can be more suitable as targeting proteins. For such applications, VHH Abs, derived from camelid family heavy chain Abs, have smaller binding surfaces that could bind to smaller cryptic regions of GPCRs. These single domain binding proteins show promise for stabilizing active GPCR conformations and serve as chaperones for co-crystallization [293–297]. The VHH domain structure lacks the human or mouse mAb HC-LC structure (which are hydrophobic at their pairing interface), resulting in a surface that is much more hydrophilic than that of an IgG Fab region [298,299]. Therefore, camelid-derived VHH nanobodies tend to have favorable biophysical characteristics, like high solubility and low aggregation, compared to human sdAbs [300]. The smaller molecular weight can expand the range of drug-dosing modalities to include inhalation, needle free, oral, topical, and ocular delivery [298,299,301,302].

In addition, the VHH CDR3 is often longer than the IgG VH CDR3, potentially allowing it to form more favorable contacts with its binding epitope [300]. Recently, Caplacizumab (ALX-0081), an anti-von Willebrand factor humanized VHH, was launched in 2018 for the treatment of thrombotic thrombocytopenic purpura and thrombosis [303]. Since in silico analysis showed that camelid VHH sequences could be aligned to human IGKV and IGLV families based on canonical structure and sequence homology, optimization of primary sequence is possible to minimize the potential of the development of anti-drug antibodies (ADAs) [304].

Besides camelids, cartilaginous fishes produce a distinct heavy chain Ab subtype containing a single variable region immunoglobulin new antigen receptors (VNARs), These 11- to 14-kDa domains comprise two heavy chains with an antigen-binding region with no associated light chains [305] and can bind with high affinity and specificity to target molecules [306,307]. These molecules have 8 beta strands instead of the 10 beta strands found in VHH and mammalian IgGs. In addition, VNARs lack the CDR2 domains and longer hinge regions found in mammalian IgG molecules. Thus, the diversity of VNARs depends primarily on the CDR3 domain. With their small size and single domain format, VNARs are highly stable and can be produced at high levels using different expression systems [308]. Because of their size, there are recognition niches that are unique for such therapeutic sdAbs [309–311]. The development of libraries for selection has expanded the utility of generating potent molecules [312,313]. As with nanobodies, these molecules can be engineered to be more human-like and have been used to isolate binders to CXCR4, a druggable GPCR [314,315], HER2, PD1, and glypican 3 [316].

3.2. Avidity Modulation

Because each antibody has two antigen-binding sites, antibody engagement to the antigen can be multivalent when there is more than one antigen on the target surface [317]. For avidity to occur, the antigen sites must be present at a sufficient density, such that once the first Fab has bound, the second Fab can bind before the first Fab dissociates. Thus, the nature of an IgG engagement to the antigen can be more complicated than a single binding event. Rather, the functional affinity or avidity represents the accumulated strength of multiple affinities to an antigen [318,319]. In such cases, the Fab region can modulate protein–protein interactions of the antigen. Occasionally, the structural nature of the Fab region–epitope engagement can limit target neutralization via constrained avidity through steric occlusion [320]. The pharmacokinetic profile of the molecule should allow for sufficient time to allow for kinetics of avidity to occur. If not, there will be an apparent loss of potency even though there may be steep saturation curves from in vitro experiments [321].

However, Fc region clustering due to FcγR interactions can contribute to Fab target avidity [322–324]. The avidity due to Fc region–FcγR crosslinking can affect the immune cell effector function that can contribute to autoimmune disorders [325–329]. In such indications, effective FcγR blockade requires doses of intravenous immunoglobulin. However, this approach requires careful consideration of potential safety concerns related to the induction of serious acute events, such as cytokine release, platelet activation/aggregation, and complement activation [330].

Monoclonal antibodies that target the inhibitory immune checkpoint receptors, such as CTLA-4 and PD-1, stimulate antitumor immunity to treat advanced melanoma, lung cancer, and other types of human cancer [331,332]. Such agonist antibodies against the immunostimulatory receptors on T cells and antigen-presenting cells were designed to have more silent Fc regions to prevent Fc effector function but retain Fab region avidity to stimulate antitumor immunity [333]. Immune effector cells have stimulatory receptors belonging to the tumor necrosis factor (TNF) receptor superfamily (such as OX40, CD27, 4-1BB, and GITR) [334,335]. There has been much effort to develop the use of their respective ligands and agonist antibodies to activate these receptors to stimulate the proliferation and activation of T cells [336–339] to mediate anti-tumor activities [340–342].

Agonistic activities of immunomodulatory antibodies require the engagement of different types of Fc receptors and cell surface receptors. To activate downstream signaling pathways, TNF receptors undergo higher-order clustering upon binding to their respective trimeric ligands [343]. Thus, regular antibody binding may not be enough to induce the required threshold TNF receptor clustering that can occur with the binding of trimeric ligands. Instead, antibody crosslinking via Fc engagement is necessary for receptor activation in in vitro assays [344–347]. The crosslinking of IgG Fc to FcγRIIB receptors can multimerize more than one antibody molecule, which in turn can facilitate the clustering of enough TNFR for signaling pathway activation. Recent studies in mice indicated that the engagement to the inhibitory FcγRIIB receptor is critical for the agonistic activity of antibodies to a number of TNFR targets, such as CD40 [340,348], death receptor 5 (DR5) [340,349], and CD95 [350]. If such antibodies have Fc regions that can engage various activating FcγRs, effector functions, such as ADCC and ADCP, can be induced and deplete these targeted immune cells. Nonetheless, the anti-OX40 and anti-GITR antibodies may facilitate the selective elimination of intratumoral regulatory T cells in the tumor microenvironment by the effector functions of the antibody [351,352]. Such antibody-mediated killing of regulatory T cells may be more important than the antibody-mediated activation of effector T cells for the anti-tumor activities of therapeutic anti-OX40 and anti-GITR antibodies.

By design, human IgG antibodies have low binding affinities to most human Fc receptors except FcγRI [353]. To optimize the anti-tumor activity of agonist antibodies' binding to immunostimulatory TNF receptors, the Fc region of the IgG antibody was engineered to bind more strongly to the FcγRIIB receptor. In particular, Chu et al. introduced the S267E/L328F mutations on an anti-CD19 IgG1 Fc to enhance FcγRIIB-binding affinity that resulted in improved inhibition of B cell receptor-mediated activation of primary human B cells [354]. However, this Fc variant also enhanced binding to the R131 allotype of the activating FcγRIIA receptor [355]. Mimoto et al. utilized the V12 mutations (E233D/G237D/P238D/H268D/P271G/A330R) in the IgG1 Fc to selectively enhance FcγRIIB engagement without an associated increased binding to either the H131 or R131 allotype of the FcγRIIA receptor [356]. Mutations that abrogate FcγRIIIA binding can decrease the potential ADCC activity. An anti-CD137 agonistic antibody with the V12 mutations showed enhanced agonistic activity dependent on FcγRIIB engagement with less ADCC activity that was linked to FcγRIIA binding. Alternatively, FcγRII-binding Centyrins can be fused to therapeutic antibodies to bind to FcγRIIB receptor (FcγRIIB), thereby enabling the antibody multimerization that drives TNFR activation [357].

Ab agonistic activity depending on FcγRIIB engagement depends on the FcγR expression in the local microenvironment. To augment the agonism of immunostimulatory antibodies independent of FcγR engagement, White et al. recently reported that human IgG2 subtype can impart super-agonistic activity to immunostimulatory antibodies for CD40, 4-1BB, and CD28 receptors [358]. This activity is conferred by a unique configuration of disulfide bonds in the hinge region of the IgG2 subtype and is not dependent on FcγRIIB engagement. To add to the repertoire of Fc mutations that can promote antibody multimerization without the need of FcγRIIB crosslinking, Diebolder et al. reported that selective Fc mutations can facilitate hexamerization of IgG Abs upon binding targets on the cell surface [359]. Specific noncovalent interactions between Fc regions resulted in the formation of ordered antibody hexamers after antigen binding on cells. These hexamers recruit and activate C1q, the first component of complement, to trigger the complement cascade. The interactions between neighboring Fc segments could be manipulated to block, reconstitute, and enhance complement activation and killing of target cells, using all four human IgG subclasses [360]. In contrast, the E345R mutation on an anti-OX40 antibody had increased agonism by promoting the clustering of OX40 receptors without the dependence on FcγRIIB cross-linking [361]. This cross-linking to FcγRIIB can lead to a further boost of the agonism of the anti-OX40 antibody with an IgG1 Fc but not with the silent IgG2σ Fc region, which lacks binding to FcγRs. The ADCC and CDC activities of the anti-OX40 antibody with the E345R mutation were affected by the choice of IgG subtypes [362]. With so many oligomeric Ab targets, there are continuing applications of hexameric therapeutic Abs that can affect downstream signaling events [328,329,363].

Alternatively, when an IgG format does not achieve enough of an effect on a cell surface receptor, the variable regions can be transferred to an IgM format to elicit the functional activity through avidity. Clearly, the IgM's higher valency can facilitate receptor crosslinking. For instance, when anti-trail-receptor IgG did not elicit a strong response, the switch to an IgM format resulted in stronger induction of trail-receptor-induced apoptosis [364]. Antibody formats that promote crosslinking are being assessed in clinical trials [365]. Likewise, other hinge and isotype formats also affect binding to targets [366].

3.3. Antibody–Drug Conjugates

The high binding specificity of antibodies can be combined with the potent cytotoxicity of small molecule agents to generate targeted therapies with higher therapeutic indices than traditional chemotherapeutics. By delivering toxic payloads only to cells that express specified antigens, it is possible to confine toxicity to malignant tissue while theoretically minimizing collateral damage. Antibody–drug conjugates (ADCs) bearing cytotoxic moieties should ideally target antigens that are present at significantly higher amounts on tumor cells. For many ADCs, it is also beneficial to bind to internalizing receptors, which deliver the conjugate into the cell and allow the active moiety to elicit its effects.

Many types of cytotoxic agents can be conjugated to antibodies for concentration into target cells. Among these, the most common are natural products, such as the maytansinoids (derived from the macrolide maytansine of *Maytenus* plants), auristatins (derived from Dolastatin peptides of *Dolabella auricularia* sea hares), and calicheamicins (enediyne antibiotics from *Micromonospora echinospora* bacteria). The auristatins, exemplified by monomethyl auristatins E and F (MMAE, MMAF), and the maytansinoids, including DM1 and DM4, are microtubule inhibitors while calicheamicins like $\gamma1$ act by creating double-stranded DNA breaks. Due to the remarkable cytotoxicity of these agents, they must be tethered to antibodies for targeted delivery and reduction of systemic toxicity.

It was earlier recognized that the sub-picomolar potency of calicheamicins would allow for efficient tumor killing when coupled to antibodies for specific delivery. In 1993, Hinman et al. reported ADCs combining calicheamicins $\gamma1$, $\alpha2$, $\alpha3$, N-acetyl-$\gamma1$, or pseudoaglycone (PSAG) with a monoclonal antibody against the internalizing antigen, polyepithelial mucin [367]. Hydrazide analogs of the calicheamicins were prepared and conjugated to oxidized glycan residues on the antibody. A comparison of conjugate analogs revealed the importance of the rhamnose sugar in the DNA-binding region of the drug, whereas a distal amino sugar residue was more amenable to substitution or removal. Stabilization of the linker with the addition of disulfide-proximal methyl groups served to increase the therapeutic index of the ADCs.

Such calicheamicin-loaded ADCs have proven effective for the treatment of leukemia. The first approved ADC was gemtuzumab ozogamicin, which demonstrated an ablation of acute myeloid leukemia (AML) cells [368]. To form the ADC, gemtuzumab (anti-CD33) is linked to N-acetyl-γ-calicheamicin dimethyl hydrazide via non-specific lysine conjugation and a 4-(4-acetylphenoxy) butanoic acid spacer. The average drug-to-antibody ratio (DAR) is two to three, although some individual antibodies remain unconjugated and others have higher DAR values. After its approval in 2000, the ADC was voluntarily withdrawn in 2010 due to concerns over its toxicity and lack of efficacy. In 2017, the drug was re-approved after a meta-analysis and new clinical data indicated a benefit for the treatment of AML (history reviewed in [369]). Meanwhile, inotuzumab ozogamicin, a CD22 antibody conjugated to the same linker-drug moiety, demonstrated cytotoxicity against B cell lymphomas and was approved for treatment of acute lymphoblastic leukemia (ALL) in 2017 [370,371]. Clearly, calicheamicins possess satisfactory potency that, when combined with antibody specificity, allows for successful elimination of hematological malignancies.

Auristatins have also been conjugated to tumor-targeting antibodies to elicit specific and potent tumor killing. Doronina et al. attached auristatin analogs to antibodies targeting Lewis Y antigen and CD30 and compared the properties of ADCs containing acid-labile hydrazone linkers with those containing protease-sensitive dipeptide linkers [372,373]. Drug conjugation was more site specific in this case, using a maleimide group to form a covalent bond with reduced thiols from antibody cysteine residues. Since antibodies contain four relatively exposed inter-chain disulfide bonds, uniform drug loading of approximately eight auristatins per antibody was achieved. Peptide linkers, in particular, a valine–citrulline linker between the antibody and monomethyl auristatin E (MMAE), showed increased stability compared to more traditional hydrazone linkers. As a result, such linkers allowed for more specific delivery and lower systemic toxicity. Thus, optimization of conjugation and linker chemistry is important for maximization of the therapeutic index.

Subsequent work with CD30-MMAE antibodies explored the effect of drug loading on the therapeutic properties of ADCs [374]. By incubating antibodies with varying ratios of linker–drug and purifying different species using hydrophobic interaction chromatography, ADCs with defined DAR values of 2, 4, and 8 were generated. While in vitro ADC potency increased with increasing DAR, the in vivo activity was less dependent on drug loading. At equal doses, DAR 4 and DAR 8 ADCs demonstrated similar efficacy in vivo, whereas the DAR 2 species had slightly lower activity. Pharmacokinetic analysis revealed that the lower DAR species had longer half-lives and greater exposure, explaining why the DAR 8 species with high in vitro activity was not more effective in the mouse xenograft study. The intermediate DAR species progressed into clinical trials and was approved in 2011 as brentuximab vedotin for the treatment of Hodgkin lymphoma and systemic anaplastic large cell lymphoma [375].

The third main class of cytotoxic agents is the maytansinoids. In 1992, maytansine was derivatized with a disulfide group, which could be reduced and conjugated to disulfide or maleimide groups on a chemically modified HER2 antibody [376]. Chemical handles were added to the antibody via non-specific amine coupling and allowed for DAR values ranging from 1 to 6. The DAR 4 ADCs were shown to have maximal cytotoxicity in an in vitro assay. ADCs with the cleavable disulfide linker were significantly more potent than ADCs with a non-cleavable thioether linkage, presumably due to the more efficient release of the active drug within target cells. However, another study with auristatin ADCs demonstrated that non-cleavable linkers may have better therapeutic windows because of less non-specific drug release [9]. An important factor is the activity of the modified drug moiety that results from proteolytic digestion of ADCs with non-cleavable linkers.

Trastuzumab emtansine represents a successful combination of HER2 antibody, DM1 maytansinoid, and a non-cleavable linker. Lewis Phillips et al. compared trastuzumab-DM1 conjugates containing a panel of reducible disulfide linkers and a non-reducible linker [377]. The ADC with a non-cleavable linker based on thiol-maleimide conjugation to the maytansinoid and amine-succinimide conjugation to the antibody unexpectedly caused the greatest tumor inhibition in vivo. An increase in pharmacokinetic stability likely contributed to this effect, as the ADCs with reducible linkers were more likely to lose the payload before delivery to target cells. While the bystander effect does not occur when a non-cleavable linker is used due to the inability of the charged metabolite to cross membranes, lower systemic toxicity and full targeted delivery of cytotoxic payloads seem to make up for this defect. Trastuzumab-DM1 ADCs using this non-reducible linker progressed to clinical trials, and in 2013, they were approved for the treatment of HER2-positive breast cancer [378].

While the paradigm for successful ADCs has revolved around the most potent cytotoxic agents and stable linkers, other strategies have also been explored. Sacituzumab govitecan is an ADC targeting Trop-2 that contains the active metabolite SN-38 from the topoisomerase inhibitor irinotecan [379]. SN-38 is less toxic than traditional ADC payloads, having the half maximal inhibitory concentration (IC50) in the nanomolar, rather than picomolar, range. Thus, a higher DAR (7–8 compared to the more typical 3–4) is required to elicit sufficient tumor cytotoxicity. Additionally, the ADC makes use of a pH-sensitive carbonate linker, which releases drug in the lysosome of target cells but also into the circulation, with a half-life of approximately one day. This semi-stable linker is proposed to allow for the bystander effect by which molecules of SN-38 diffuse to neighboring tumor cells that may have a lower expression of Trop-2. While Trop-2 is expressed on a number of tumor types, sacituzumab govitecan has been most studied in cases of triple negative breast cancer.

In addition to natural products, toxins have also been conjugated or fused to antibodies to generate tumor-targeting immunotoxins. For example, Mansfield et al. described a disulfide-stabilized (HC R44C, LC G100C) Fv targeting CD22 that was fused to a 38-kDa truncated form of *Pseudomonas* endotoxin A (PE38) via the C-terminus of the VH domain [380,381]. The PE38 moiety contains translocation and ADP-ribosylating regions that allow it to inactivate elongation factor 2 within the cytosol of the target cell after delivery by the Fv domain. This activity leads to an inhibition of protein synthesis, induction of programmed cell death, and allowed for reduction in tumor growth in a mouse xenograft model. Subsequently, the CDR3-H3 of the Fv domain was mutated to improve CD22 binding while the PE38 domain was stabilized with the R490A mutation to reduce proteolysis [382]. The resulting molecule, moxetumomab pasudotox, was approved in 2018 for the treatment of hairy cell leukemia.

Radionuclides represent an additional class of payload that can be attached to antibodies to create antibody–radionuclide conjugates (ARCs). In one case, yttrium-90 was conjugated to the CD20 antibody ibritumomab to generate ARCs for the treatment of lymphoma [383]. The ^{90}Y isotope was used due to its generation of beta particles, which penetrate several millimeters to elicit a bystander effect, and its favorable decay half-life of 2.7 days. Conjugation was achieved using the linker-chelator tiuxetan, whose isothiocyanate group forms a stable thiourea bond with antibody amines. Notably, the ^{90}Y ARC was used for therapeutic purposes while the same antibody-linker-chelator was coupled to ^{111}In for preliminary imaging and dosimetry. Ibritumomab tiuxetan was approved in 2002 for the treatment of non-Hodgkin's lymphoma.

A similar ARC, tositumomab-^{131}I, was approved the following year for the same indication. While both ARCs target CD20, tositumomab is conjugated to iodine-131, which is a β and γ emitter with a longer physical half-life of 8 days [384]. Radiolabeling of tositumomab is achieved through oxidative iodination of aromatic residues like tyrosine and histidine, rather than through chelation. The ARC was withdrawn from the market in 2014. There is debate whether this withdrawal was due primarily to a projected decline in sales related to the complexity of administration, or follow-up studies that indicated no benefit over more traditional chemo- and immunotherapies [385]. The preceding examples demonstrate the feasibility of bringing ARCs to the market, but also highlight the complexity of generating and administering radioactive therapeutics.

3.4. Fc Activity Engineering

Although the differentiation of the antibody is often focused on the characterization of the Fab engagement to the target epitope, not all antibodies that bind to a given target have efficient effector cell function. This was demonstrated in the comparison of different anti-CD20 antibodies that had different epitopes and subsequently different levels of effector functions [386–388]. Here, the avidity of Fc presentation is critical for FcγR recognition of immune effector cells. Because mAbs depend on their Fc region to elicit certain immune reactions, engineering of this domain allows for tactical modification of activity as well as enhancement of the respective physicochemical properties. Sometimes, a simple swap by moving V regions into other IgG subtypes can result in greater efficacy [389,390]. However, there can be a greater emphasis on specific Fc mutagenesis to obtain a more selective IgG effector function [88,355,391–393]. In addition, the coupling of the Fab and Fc regions can impact the therapeutic window for the safety and efficacy of antibodies and Fc fusion proteins [394–397]. We outline several tactics to modulate Fc activity linked to FcγR for immune effector cell function and FcRn for pharmacokinetic properties. Nonetheless, it is critical to keep in mind that the Fab domain antigen binding can affect the Fc region activity via structural allostery [398–400]. Hence, evaluations of specific Fc mutations should be confirmed empirically.

3.4.1. Mutations that Modulate Effector Function

Protein and glycan engineering can modulate effector activity of antibodies to modulate ADCC, ADCP, opsonization, internalization, trogocytosis, and CDC activity. This engineering can also be applied to Fc fusions that comprise toxins, radioactive molecules, chemotherapeutic agents, or nucleic acids for targeted delivery [392].

Site-directed mutagenesis and X-ray crystal structures demonstrate that FcγRs make contact to IgG1 Fc at P232-S239, Y296-T299, and N325-332. Notwithstanding, the residues outside of this area may be linked to conformational changes that affect the Fc–FcγR complex formation. An abbreviated list of Fc modifications is shown in Table 2. Fc region residues can be mutated to increase the binding of antibodies to the activating FcγR and/or to enhance antibody effector functions. Mutations that decrease binding include G236A, S239D, F243L, T256A, K290A, R292P, S298A, Y300L, V305L, K326A, A330K, I332E, E333A, K334A, A339T, and P396L mutations (residue numbering according to the EU index) [345,391–393,401]. Experimentally, combination mutations that result in antibodies with increased ADCC or ADCP are S239D/I332E, S298A/E333A/K334A, F243L/R292P/Y300L, F243L/R292P/Y300L/P396L, F243L/R292P/Y300L/V305I/P396L, and G236A/S239D/I332E mutations on IgG1.

Fc mutations to reduce binding of the antibody to the activating FcγR and subsequently to reduce effector functions can include positions: K214T, E233P, L234V, L234A, deletion of G236, V234A, F234A, L235A, G237A, P238A, P238S, D265A, S267E, H268A, H268Q, Q268A, N297A, A327Q, P329A, D270A, Q295A, V309L, A327S, L328F, A330S, and P331S mutations on IgG1, IgG2, IgG3, or IgG4. [391,402–405]. Combinations of Fc mutations for reduced ADCC on IgG1 include: L234A/L235A; L234F/L235E/D265A; V234A/G237A; S267E/L328F; L234A/L235A/G237A/P238S/H268A/A330S/P331S; or K214T/E233P/L234V/L235A/G236-deleted/A327G/P331A/D365E/L358M. Combinations of Fc mutations for reduced ADCC on IgG2 include: H268Q/V309L/A330S/P331S or V234A/G237A /P238S/H268A/V309L/A330S/P331S. Combinations of Fc mutations for reduced ADCC on IgG4 include: F234A/L235A, S228P/F234A/L235A; S228P/F234A/L235A/G237A/P238S; or S228P/F234A/L235A/G236-deleted/G237A/P238S. Hybrid IgG2/4 Fc regions with the Fc with residues 117–260 from IgG2 and residues 261–447 from IgG4 can result in having less FcγR activity. Crystal structures and simulations of IgG1σ, IgG4σ1, and IgG4σ2 Fc variants reveal altered conformational preferences within the lower hinge and BC and FG loops relative to wild-type IgG, providing a structural rationalization for diminished Fc receptor engagement [406].

An X-ray crystal structure of the C1q-Fc region shows that complement C1q binds IgG1 at D170-K322, P329, and P331 [407,408]. To enhance CDC, Fc positions mutations can include S267E, H268F, S324T, K326A, K326W, E333A, E430S, E430F, and E430T. Combination mutations that result in antibodies with increased CDC can include K326A/E333A, K326W/E333A, H268F/S324T, S267E/H268F, S267E/S324T, E345R, and S267E/H268F/S324T [359,409,410].

The ADCC activity of antibodies can be enhanced by engineering their oligosaccharide composition. Human IgGs are N-glycosylated at Asn297 with the majority of the glycans of the well-known biantennary G0, G0F, G1, G1F, G2, or G2F forms (see Figure 5). N-linked glycosylation can be removed by using the mutation N297A on IgG1, IgG2, IgG3, or IgG4. The aglycosylated species has less FcγR activity.

Antibodies produced by non-engineered CHO cells typically have a glycan fucose content of about at least 85%. The removal of the core fucose from the biantennary complex-type oligosaccharides attached to the Fc regions enhances the ADCC of antibodies via improved FcγRIIIa binding without altering antigen binding or CDC activity. Such mAbs may be produced using different methods reported to lead to the successful expression of relatively low level fucosylated antibodies bearing the biantennary complex-type of Fc oligosaccharides, such as the control of culture osmolality [411], application of a variant CHO line Lec13 as the host cell line [412], application of a variant CHO line EB66 as the host cell line [413], application of a rat hybridoma cell line YB2/0 as the host cell line [414], introduction of small interfering RNA specifically against the α 1,6-fucosyltrasferase (*FUT8*) [415], or co-expression of β-1,4-N-acetylglucosaminyltransferase III and Golgi α-mannosidase II or a potent alpha-mannosidase I inhibitor, kifunensine [416–418]. Notwithstanding, careful monitoring of antibody glycosylation is required to control the pharmacodynamics of Abs and Fc-fusion proteins [419]. Other modifications to enhance ADCC include the introduction of bisecting N acetyl glucosamine and the removal of sialic acid residues [420].

Fc-mediated effector functions are best avoided for some applications, such as targeting cell surface antigens on immune cells or when engineering bispecific molecules to bring target diseased cells within the proximity of effector immune cells [421]. In each of these cases, it is best not to stimulate unwanted cell and tissue damage or risk undesired effector cell activation, immune cell depletion, or FcγR cross-linking that might induce cytokine release through engagement of Fc-mediated effector functions [422]. The complexity of FcγR functional properties is increased by the varying densities of activating and inhibitory receptors on the different effector cell populations [406]. Likewise, since the threshold of activation can be variable with different patients, it would be prudent for safety considerations to develop antibodies with a more silent Fc region. Thus, the development of completely silent Fc regions can be critical for biologics that do not require FcγR- or C1q-mediated effector functions [88,423,424].

3.4.2. Mutations that Alter Pharmacokinetics

The PK properties of IgG antibodies are largely governed by the neonatal Fc receptor (FcRn), a heterodimer of the FcRn α chain and β2-microglobulin. Initially, FcRn was known for its role in transferring IgG from maternal milk to the neonatal circulation via FcRn-expressing intestinal epithelial cells [425,426] (and later, for transferring human IgG from the mother to fetus via FcRn-expressing placental syncytiotrophoblasts [427]). Both of these transfer mechanisms require pH-dependent binding of IgG to FcRn, with strong binding at pH < 6.5 in the acidic intestine or endosome, and significantly weaker binding in blood (pH 7.4). While it was long recognized that an Fc-binding receptor might be responsible for IgG catabolism in adults [428], the identity of this receptor as FcRn was not confirmed until three decades later [429]. Differences in IgG-FcRn affinity with pH allow IgG to be salvaged from acidic endosomes of endothelial cells and recycled back to the blood, and thus to circulate longer than other proteins of a similar size. The IgG-binding site for FcRn was localized to the CH2-CH3 elbow, which overlaps the site for *staphylococcal* protein A binding [430–432]. With an intact binding interface at each heavy chain, a molecule of IgG can bind simultaneously to two molecules of FcRn [433].

Because FcRn is involved in lysosomal salvage and IgG serum persistence, the IgG–FcRn interaction has been engineered to modulate the PK properties of antibodies. Enhanced serum stability may be beneficial for both patients and manufacturers, as it allows for lower-level or less frequent dosing. The combination of co-crystal structures of Fc–FcRn complexes and site-directed mutagenesis may map the Fc region regions to cover L251-S254, L309-Q311, and N434-H435. Early work demonstrated that mutagenesis of IgG to disrupt FcRn binding leads to profoundly accelerated IgG clearance [432]. The mutations I253A, H310A, and Q311A in the CH2 domain, and H433A and N434A in the CH3 domain, led to two- to five-fold decrease in the β-phase half-life of mouse IgG1 Fc fragments. Subsequent work verified the feasibility of strengthening FcRn interaction at low pH, which could extend the half-lives of Fc mutants. Ghetie et al. performed random mutagenesis of mouse IgG1 residues T252, T254, and T256 coupled with phage display to isolate variants with tighter FcRn binding at pH 6.0 [434]. Their LSF mutant had 3.5-fold higher affinity for FcRn, which was driven primarily by a slower dissociation rate. The same mutant had a half-life up to 1.6-fold longer than the wild type, which translated to a 4-fold increase in exposure. These studies suggested a potential correlation between endosomal FcRn affinity and the half-life of IgG antibodies, and initiated a search for long-lived IgG mutants that outcompete endogenous antibodies for FcRn-mediated lysosomal salvage.

One historic IgG variant with altered FcRn binding and PK is the YTE (M252Y/S254T/T256E) mutant. Initial studies using human IgG1 antibodies showed that the YTE set of mutations significantly reduced antibody serum concentrations in mice, despite a 10-fold higher affinity to both mouse and human FcRn at pH 6.0 [435]. The authors attributed this unexpected result to a concomitant increase in FcRn binding at pH 7.4 that occurred for mouse, but not human, FcRn. Later work revealed that cynomolgus FcRn, like human FcRn, binds YTE 10-fold more tightly than wild-type IgG at pH 6.0, but not significantly differently at pH 7.4 [436]. When YTE antibody was administered to monkeys, its half-life was four-fold longer than that of the control. Thus, IgG PK may be improved by using engineering strategies to increase FcRn binding at low pH while maintaining weak FcRn affinity at neutral pH. More recently, motavizumab (anti-RSV) containing the YTE substitutions became the first

Fc-engineered antibody to be investigated in humans [437]. The YTE version of the antibody had a half-life of up to 100 days, or two- to four-fold longer than that of its wild-type counterpart. Although studies with intravenous administration have not indicated a higher risk of anti-drug antibodies, there is some concern that subcutaneous administration of YTE mutants could induce immunogenicity and therefore counteract any PK benefits [438].

IgG variants with altered FcRn binding not only have altered clearance; they may also have enhanced activity due to greater exposure. For instance, Zalevsky et al. developed an IgG1 LS mutant (M428L/N434S) with an 11-fold increased FcRn affinity and 3- to 5-fold increased half-life [439]. Notably, the half-life was extended for antibodies targeting a soluble antigen (vascular endothelial growth factor) and an internalizing cell-surface receptor (epidermal growth factor receptor). Thus, clearance was positively affected even in the context of an antibody with target-mediated disposition. In mouse xenograft studies, the LS antibody increased inhibition of tumor growth relative to the wild type when dosed every 10 days. Similarly, Gautam et al. showed that the LS substitutions could increase protection of rhesus macaque monkeys when incorporated into broadly neutralizing human immunodeficiency virus (HIV) antibodies [440]. Clearly, half-life extension via Fc mutation is a strategy that can be applied to a broad range of therapeutics.

Although it is tempting to oversimplify the relation between FcRn binding and PK, it must be emphasized that enhanced FcRn binding does not always translate to a longer half-life. In one informative study, Datta-Mannan et al. generated three IgG1 Fc variants with enhanced FcRn affinity at pH 6.0 and analyzed their PK profiles in cynomolgus monkeys and mice [441]. Despite up to 80-fold increases in cynomolgus FcRn affinity, the clearance of the variants in monkeys was unchanged. Furthermore, clearance was accelerated in mice even though affinity to mouse FcRn was increased almost 200-fold. As alluded to previously, the undesirable PK properties likely resulted from subtle changes in the pH dependence of FcRn binding. Borrok et al. followed up on the importance of neutral pH FcRn affinity by producing a panel of IgG1 Fc variants with variable FcRn binding at both pH 6.0 and 7.4 [442]. The authors suggest that pH 6.0 FcRn binding is directly correlated to half-life only as long as pH 7.4 binding does not also increase beyond a certain threshold. Given the sharp pH dependence required for efficient FcRn recycling, this group also proposed that half-life enhancement via Fc engineering probably cannot be improved beyond the four-fold increase already achieved.

In the Fc region, part of the CH2-CH3 domains is responsible for FcRn binding that results in recycling of antibodies for a long half-life [443–446]. mAbs with the same Fc can bind to FcRn differently, which can affect their respective PK profiles [447]. Mutations along the CH2-CH3 domains can modulate PK profiles. Single mutations that enhance the pH-sensitive binding include T250Q, M252Y, I253A, S254T, T256E, P257I, T307A, D376V, E380A, M428L, H433K, N434S, N434A, N434H, N434F, H435A, and H435R [436,439,441,442,448–454]. Combination mutations that can be made to increase the half-life of the antibody are M428L/N434S, M252Y/S254T/T256E, T250Q/M428L, N434A, and T307A/E380A/N434A. These mutations mediate pH-sensitive interactions with FcRn. In contrast, mutations that can reduce binding to FcRn, thereby decreasing the half-life of the antibody or Fc region molecules, can include: H435A, P257I/N434H, D376V/N434H, H435R, M252Y/S254T/T256E/H433K/N434F, and T308P/N434A. Although these mutations provide a guide, much of the PK profiles are determined empirically because of potential Fab–Fc non-covalent interactions [399].

A host of factors beyond FcRn interaction have been shown to affect the serum stability of antibodies. Especially for antibodies targeting distinct antigens or containing different variable regions, biological and physicochemical properties may supersede FcRn-binding properties in determining clearance. Even for antibodies of the same specificity, differences in variable region sequence may lead to altered biophysical properties like the charge and isoelectric point (pI), which can also affect PK. Igawa et al. observed that antibodies with lower pI values tended to be more stable in vivo. In a panel of four IgG4 molecules, the most acidic antibody (pI 7.2) had a half-life 2.4-fold longer than that of the most basic antibody (pI 9.2). As each antibody had the same constant regions and did not cross-react with mouse proteins, this result indicated that variable region sequences can cause differences in biophysical properties that affect serum persistence independent of FcRn. To further validate the observed trend, the variable regions of an IL-6R IgG1 antibody with a pI of 9.3 were engineered to generate more acidic variants (pI 6.9 and 5.5), with minor (two-fold) differences in antigen binding affinity. When administered to cynomolgus monkeys, the acidic variants had slower clearance than the wild-type antibody. Mechanistically, the decreasing positive charge may reduce attractive ionic interactions with negatively charged membranes and reduce pinocytotic cell uptake and degradation. Consistent with this explanation, engineering IgG variable regions to remove patches of positive charge (without greatly altering protein pI) has also been used to reduce non-specific binding and improve PK [455]. Likewise, the choice of framework mutations in the Fab can also influence the PK properties through differences in charge [456].

Glycosylation, glycation, and charges in the Fab region are also important for the PK properties of a mAb. FcγR expressed on the surface of blood and liver cells can facilitate the rapid removal of circulating Abs from circulation. Likewise, glycosylation patterns can impact both PK and PD significantly [259,260].

Biophysical liabilities, such as increased hydrophobicity and decreased Fc region thermal stability Tm values, can lead to lower levels of intracellular recycling that leads to subsequent intracellular degradation Antibodies binding to internalizing receptors and certain other antigens may undergo significant target-mediated clearance (reviewed in [457]). This saturable phenomenon causes nonlinear PK, where elimination is faster at lower antibody concentrations. As the antibody concentration surpasses that of the antigen, clearance becomes slower due to the increased contribution of FcRn-mediated catabolism. Thus, the distribution and elimination of antibodies can vary greatly in cases where antigen binding leads to active transport or degradation [209,277,451,458–460].

Extending the half-life of antigen-binding fragments and other lower molecular weight species can include strategies, such as fusion with polyethylene glycol (PEGylation), human serum albumin or an albumin-binding group, Fc region fusion, and multimerization to be above 70 kDa [461].

3.5. Bispecific Antibodies

When single component targeting is insufficient, an improved therapeutic response can require agents that can engage more than a single target linked to a single mechanism of action. There can be limitations with the use of mono-specific antibody formats in that some patients will not respond to such therapy after a period of time. Because there can be crosstalk between signaling pathways, there can be the development of resistance during the progression of diseased tissue. Thus, to regulate more than one disease-causing pathway, there are extensive efforts to use bispecific antibodies (BsAbs) to improve the therapeutic profile. BsAbs are engineered antibodies that have two domains that bind to two different antigens or to two epitopes on the same antigen.

There are strong therapeutic rationales for BsAbs: BsAb can target multiple causative agents for a disease with advantages over combination therapy using antibody mixtures; immune cell redirection via BsAb crosslinking of an effector biomolecule or effector cell to a specified target; and synergy through the coupling of multiple targets [462,463]. Likewise, the ability to bind to different ligands can exhibit an increased avidity and target residence time when both domains can bind simultaneously to their target sites [321,464]. This is because the binding of one binder forces the second tethered binding arm to stay close to its corresponding site. This 'forced proximity' favors its binding and rebinding (once dissociated) to that site. However, rebinding will also take place when the diffusion of freshly dissociated ligands is merely slowed down. Such targeting of multiple signaling pathways plays unique roles in the control of potential resistance mechanisms that are typical of the pathogenesis of various cancers. A single agent BsAb can have the advantage over a combination of mAbs by having improved compliance and less complex regulatory hurdles.

There are three approved BsAbs: Catumaxomab that can bring T cells or T lymphocytes via CD3 binding closer to cells expressing EpCAM (Trion Pharma); blinatumomab that also has a CD3-binding arm to B lymphomas with CD19(Micromet/Amgen); and Helimbra or emicizumab-kxwh

(Roche—Chugai) that mimics the cofactor VIII for patients with hemophilia A. Catumaxomab is produced using quadroma technology where the HC and LC fragment of mouse mAbs against CD3 and rat mAbs against EpCAM were secreted by fusing the respective hybridomas to form a BsAb with an intact Fc. Although quadroma molecules can be produced using a variation of the hybridoma technology, Good Manufacturing Practice (GMP) scaleup to isolate the BsAb was difficult because of the challenge of isolating the BsAb from the permutations of HC/LC fragments. This BsAb was also designed to have the Fc region that can bind to FcγR-activating receptors to permit co-localization of cells with Fc receptors, such as macrophages, dendritic cells, and natural killer (NK) cells [465,466]. While the catumaxomab Fc region can enhance activation of the patient immune system against tumor cells via T cell-mediated lysis with ADCC and ADCP, there were strong adverse effects coming from the induction of anti-drug antibodies that bound to the combination or individual mouse and rat mAb sequences. Unfortunately, the formation of anti-drug Abs against the mouse and rat mAbs led to an immune response against the BsAbs, resulting in worsening of the patient's prognosis [467].

To bypass the challenges of having mouse and rat Fc sequences, Blinatumomab was developed using the Diabody technology with a binding domain against CD19 on B cell lymphomas and CD3 binding to the surface of T cells for use in lymphoma and leukemia [468–470]. As with Catumaxomab, blinatumomab fosters the redirection of T cells to tumor cells without the constraints of the T cell receptor–major histocompatibility complex restrictions. The molecule is very potent and has to be delivered at low concentrations. Since the CMC process employed a recombinant bacterial expression of a single gene product, there was optimization of the downstream process to generate a stable drug substance free from residual impurities, such as host cell protein and DNA. However, because of the rapid clearance due to the short PK profile, this molecule is typically delivered via an infusion pump.

The third approved BsAb is Helimbra or Emicizumab-kxwh that have Fab regions that bind enzyme factor IXa (FIXa) and the substrate factor X(FX) [471]. Coagulation factor VIII (FVIII) can be added to reduce the bleeding complications of patients with hemophilia A. However, FVIII has poor PK properties. Thus, emicizumab was made to bind simultaneously the enzyme and substrate to mimic the partial function of factor VIII and restore some anticoagulant activity [472,473]. Humanized Abs that bind to FIXa and FX were put into a stabilized human IgG4 BsAb subtype with two sets of mutations. The BsAb has an S238P substitution to stabilize the hinge regions, preventing Fab-arm exchange. To generate the BsAb, a mixture of four expression vectors encoding the respective heavy and light chains of the FX and FIXa-specific Abs is used. The BsAb also had the "knobs-into-holes" mutation to promote the heterodimerization efficiency of the two heavy chains. Nonetheless, significant downstream purification efforts were required to isolate the BsAb. These difficulties in the manufacturing the BsAb were overcome by re-engineering the BsAb to have a common light chain for the anti-FX and anti-FIXa heavy chains and to modify the HC to facilitate ion exchange chromatography for BsAb purification [474].

We review different applications of how BsAbs can overcome the challenges of single target mAbs by modulating more than one pathway simultaneously, redirecting immune cells to specific targets; facilitating transport across tissue barriers; and delivering payloads to more specific targets.

Enhanced avidity has been reported for an EGFR x cMet BsAb (the use of EGFR x cMet refers to the BsAb, as compared to EGFR + cMet, which is the combination of the two parental EGFR and cMet mAbs) that is superior to the combination of EGFR mAbs and cMet mAbs (EGFR + cMet) [475,476]. This BsAb targets multiple resistance factors simultaneously by inhibiting primary or secondary mutations of the EGFR and cMet pathways [477]. Alternatively, a BsAb can target two non-overlapping epitopes of a target antigen to enhance the specificity and affinity of the therapeutic Ab. The bispecific binding can induce Her2 receptor crosslinking, which can further suppress Her2 activity [478].

There is great interest in utilizing Ab modulation of protein–protein interactions for diseases in the brain. However, the aforementioned pharmacokinetic properties of an IgG can prevent the high flux diffusion across tissue barriers. There are many researchers developing BsAbs to cross the blood–brain barrier (BBB) to target pathogenesis mediators in neurological diseases [479,480]. Couch et al. designed a bsAb that binds to transferrin receptor (TfR) and β-site APP-cleaving enzyme 1 (BACE1) to facilitate diffusion across the BBB [481,482]. TfR is highly expressed on the surface of the brain endothelium. After binding to TfR, the circulating bsAb is transported into the brain via receptor-mediated transcytosis. In the brain, BACE1 is an aspartyl protease that contributes to the pathogenesis of Alzheimer's disease. Targeting BACE1 has been a strategy for treating Alzheimer's disease [483]. The affinity between the anti-TfR Fab and TfR was selected to be weak to allow bsAb release from the endothelium and enter the brain to target the disease mediator BACE1 with the other binding arm. The safety liabilities of TfR-bispecific antibodies that cross the blood–brain barrier involves modulation of the epitopes that control the binding, internalization, and transport [481]. So far, a preclinical study showed that the BsAb could alleviate brain disease syndromes [483].

There is also interest in having molecules with asymmetric Fc regions that have different levels of engagement to FcγR and FcRn. Targeting IgG Fc region-binding selectivity of FcγRIIa versus FcγRIIb resulted in having increased ADCC activity with less other immune effector functions [484]. Asymmetric Fc with mutations in the hinge and CH2 domain can reduce binding to FcγR and C1q to decrease ADCC and CDC depletion of target cells [485]. In addition, such a strategy was used to select for Fc-silencing mutations that retain IgG1 Fc region stability to maximize CMC success. Likewise, asymmetric Fc engineering was used to have selective FcγRIIIa binding to obtain higher ADCC activity with minimal changes to IgG1 Fc stability [356].

As demonstrated in the three aforementioned BsAbs, the design was based on meeting the therapeutic hypothesis with the added challenge of meeting the CMC requirements. In the past 20 years, numerous designs can be gleaned from extensive reviews that cover many aspects of the bispecific molecule engineering, activity, and patent survey [1,486–491]. A reductionist view of bispecific agents has relied on the basic design principles of a human mAb, which include binding domains, hinge sequences, and the Fc region as shown in Figure 9. Protein engineering has extended the binding domains to include Fab, scFv, DARPins, Centyrins, Ankyrins, VHH, cytokines, enzymes, etc. The hinge has been extended to include linkers of peptide sequences from mAb sequences, subtype hinge sequences, variation of peptides having flexible linkers (Thr, Gly, Ser, Ala motifs), and rigid linkers (Pro, Arg, Phe, Thr, Glu, Gln motifs), etc. [492]. The Fc region can include the fusion of albumin-binding domains, polyethylene glycol polymers, etc. [86,397].

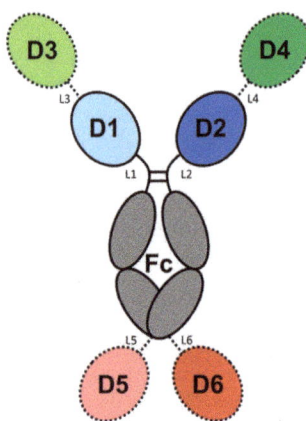

Figure 9. Schematic representation of bispecific and multispecific molecules. Domains 1–6 (D1–D6) can represent binding domains that can include Fab, scFv, DART®, VHH, and other alternative binding motifs. Linker sequences (L1–L6) can represent distinct linker regions. The Fc region can represent the IgG Fc region or be replaced with another other motif for modulation of the FcγR, FcRn, and PK profile. A standard mAb has D1 = D2, L1 = L2, and Fc = IgG Fc region.

In this view, there are five groups of BsAb formats as shown in Figure 9: Asymmetric IgG molecules with heterodimeric heavy chains (D1 ≠ D2, with L1 = L2); fusion of IgG-binding domains (combinations of D1 with or without D3, D2 with or without D4 with L1, L2, L3, and L4); fusion of binding domains to IgG molecules (combinations of D1, D2, D3, D4, D5, and/or D6 with L1, L2, L3, L4, L5, and/or L6); engineering binding domains of IgG molecules (multiple binding at D1, D2, or Fc regions); and chemically coupled IgG fragments. A normal IgG molecule has D1 = D2 and L1 = L2 with a single Fc region (Figure 1). The continuing evolving BsAb results in changes in: Valency of binding arms to control avidity; architecture via the design of binding arms and linker types to control flexibility for access and functional activity; inclusion of different binding arms that can permit the engagement of different epitopes; and pharmacokinetic control by using Fc regions or other binding arms to serum proteins. There are many variations of tethering of binding domains and PK modulation domains that employ non-Ab motifs. However, to limit the focus on BsAb, in this review, we focus on the structure–function impacts of bispecific IgG fragments and bispecific intact IgG molecules that use Fab and Fc components.

3.5.1. Bispecific Fragments

The variable region of the antibody Fab region is the smallest unit of an antibody that possesses antigen-binding capabilities. A major advantage of using fragments relative to full-length IgG is the potential for increased penetration into malignant tissue due to the decrease in size [493]. Although the absence of the Fc region abrogates FcγR binding and broadly eliminates ADCC, ADCP, and CDC, the incorporation of effector cell specificity (e.g., anti-CD3) allows for tailored effector mechanisms like T cell redirection. Similarly, FcRn-binding capability is removed. In cases where transient drug exposure is favorable (such as diagnostics), this apparent defect can lead to a desired increase in clearance. On the other hand, the half-life can be prolonged by incorporating albumin-binding functionality into one of the antigen-binding domains [494].

Fusion of Antigen-Binding Fragments

Perhaps the most obvious antigen-binding region of antibodies is the Fab region, which constitutes the light chain (VL and C_L) and heavy chain Fd (VH and C_H1). Bispecific tandem Fabs are more difficult to form than single-chain Fv and sdAb fusions, since the multiple chains may pair incorrectly to create non-functional paratopes. To address this challenge, Wu et al. made use of mutations at the LC–Fd interface that favor correct polypeptide pairing [495]. Their tandem Fabs were created using one polypeptide of linked heavy chains (VHA-C_H1A-$(G_4S)_3$-VHB-C_H1B) and two separate light chains (VLA-C_LA and VLB-C_LB). After expression in HEK cells and purification of the His-tagged protein, the EGFR × CD3 Fab was found to have similar antigen binding to the corresponding tandem scFv protein as well as better thermal stability and less aggregation. Interestingly, size had a significant impact on the ability of different bsAb formats to mediate T cell killing of EGFR-expressing cells. The 50-kDa tandem scFv had the highest potency, followed by the 100-kDa tandem Fab, and the 150-kDa bispecific IgG had the highest half-maximal effective concentration (EC50). Clearly, the size and geometry of bispecific molecules significantly impacts their ability to form a productive immunological synapse.

Fab regions can also be genetically fused to smaller binding moieties to generate bi- and trispecific molecules. For example, a human placental alkaline phosphatase (hPLAP) × BCL1 bibody (~75 kDa) was generated by fusing a BCL1 scFv to the LC C-terminus of an hPLAP Fab using a six-amino acid linker [496]. Likewise, a hPLAP × CD3 bibody was formed by fusing a CD3 scFv to the C-terminus of the hPLAP Fd using a $(G_4S)_3$ linker. More impressively, a 100-kDa tribody targeting all three antigens was generated by combining the LC (hPLAP)-scFv (BCL1) and Fd(hPLAP)-scFv (CD3) chains. While monomeric or dimeric LC contaminants were formed, the majority of antibody products had the expected composition after HEK expression. Simultaneous binding was demonstrated for each pair of tribody antigens (hPLAP × BCL1, hPLAP × CD3, BCL1 × CD3), as was T cell recognition of both tumor cell types. In addition to trispecific agents, the authors suggest that the Fab-(scFv)$_2$ format could also be used to create bispecific molecules that are bivalent for one antigen.

The VHH domain of camelid heavy chain antibodies represents another compact moiety that can be tethered to Fab fragments. Li et al. prepared 65-kDa bispecific proteins called S-Fabs by fusing a VHH domain targeting CEA to the C-terminus of the Fd chain of an anti-CD3 Fab [497]. The construct was produced in *E. coli* via transformation with the normal anti-CD3 LC and the Fd (CD3)-VHH(CEA) fusion and purified using the His tag at the LC C-terminus. Interestingly, binding of S-Fab to CEA-expressing cells was achieved despite direct fusion of the anti-CEA VHH N-terminus to the anti-CD3 Fd without a spacing linker. It is not clear whether this design choice impacted the affinity of the VHH for CEA due to steric or conformational constraints. Regardless, the S-Fab depleted CEA-bearing tumor cells in a T cell-dependent manner, demonstrating the utility of these bispecific Fab-VHH fusions.

Fusion of Single-Chain Variable Fragments

Because the Fv (VH + VL) represents the minimal structure of human antibodies containing an intact antigen-binding interface, the scFv format has been widely used to prepare small bsAb frameworks. One of the simplest ways to generate multifunctional agents is to genetically fuse two scFv domains with differing specificity, creating tandem scFvs or scFv$_2$s. Reports of such constructs date back to the 1990s. In one early study, scFvs targeting L6 tumor-associated antigen and human CD3 were genetically fused by a 27-amino acid helical linker peptide and expressed in COS cells [498]. The bispecific construct was found to co-localize T cells and L6-expressing target cells, and to elicit cytotoxic activity against the target cells. Notably, this particular scFv construct was also fused to the Fc region as a purification tag. Within a week, another report was published describing a bispecific scFv$_2$ targeting fluorescein and single-stranded DNA [499]. This protein, in contrast to the previous example, was expressed in *E. coli* and refolded from inclusion bodies. In this case, the linker between individual Fvs was based on the CBH1 peptide from *Trichoderma reesi* cellobiohydrolase I. An assessment of the binding affinity demonstrated that bacterially expressed bispecific tandem scFvs can bind their antigens with a similar affinity to the parental scFv domains. These studies established that genetic fusion of distinct Fvs is a valid approach to achieve specificity for multiple antigens.

Within the realm of cancer immunotherapy, a common approach to generate a robust anti-tumor response is to co-localize effector cells to the site of malignancy and modulate their response. Bispecific T cell-engaging (BiTE) antibodies were developed shortly after the first reports of bispecific scFV$_2$ and accomplished this feat by binding to both T cells (often via CD3ε) and target cell antigens with their scFv domains. In one early example, Mack et al. expressed BiTEs targeting CD3 and EpCAM in CHO cells and purified them via a C-terminal His tag [500]. Their 60-kDa protein used the VLA-VHA-VHB-VLB domain order, with a flexible G$_4$S or (G$_4$S)$_3$ linker joining the individual Fv fragments. In addition to binding both antigens with similar properties as the parental scFvs (as demonstrated by FACS and ELISA), nanomolar concentrations of the BiTE were sufficient to elicit T cell-mediated lysis of EpCAM-expressing cells in a ^{51}Cr release assay. A CD19 × CD3 BiTE with a similar construct design and expression/purification strategy was later generated and shown to have potent activity against CD19-positive lymphoma cells [469,470]. This molecule became blinatumomab, which was approved in 2014 for the treatment of acute lymphoblastic leukemia, thus demonstrating the aptitude of the tandem scFv/BiTE framework for treating cancer.

Similar to tandem scFvs is the diabody format, which contains two separate polypeptide chains (e.g., VHA-VLB and VHB-VLA) that associate non-covalently into a functional bispecific molecule. Use of a short linker between the VH and VL domains of a scFv prevents intrachain association between domains and instead favors dimerization with another molecule of scFv. By co-expressing two distinct VH-VL scFvs containing short linkers in *E. coli*, Hollinger et al. were able to form diabodies targeting both phenyloxazolone-bovine serum albumin and hen egg lysozyme [501]. Simultaneous binding of both antigens was shown by sandwich ELISA and surface plasmon resonance (SPR). In general, 15-residue linkers were found to promote the formation of scFv monomers, which can form intrachain associations, while shorter five-residue linkers tended to cause more dimerization for the formation of two antigen-binding sites. From here, strategic ordering of VH and VL domains to place half of each functional Fv in a different polypeptide chain allows for the formation of diabodies that can bind two distinct antigens. It is important to note that even when functional heterodimers are energetically preferred, non-functional homodimers may also form and must be removed via affinity chromatography.

Numerous diabody derivatives have been explored that attempt to mitigate the challenges associated with the non-covalent association of two separate chains. One straightforward way to stabilize proper chain pairing is to link the chains together, as was done for single-chain diabodies (scDbs) [502]. In contrast to scFv$_2$ antibodies, which place paired domains in close proximity (e.g., VHA-VLA-VHB-VLB) and have sufficiently long linkers within an Fv, the scDb design uses a staggered domain order (e.g., VHA-VLB-VHB-VLA) and a short intra-Fv linker to prevent mismatched chain pairing. Here, 58-kDa mammalian-expressed CEA × β-galactosidase scDbs successfully recruited the enzyme to the target cells, which allowed for local activation of a prodrug to cytotoxic dauromycin. Additionally, the single-chain format had superior stability in serum than the corresponding diabody based on the retention of enzyme recruitment. Thus, consolidation of the diabody framework into a single chain can simplify protein expression and prevent chain mispairing while permitting a geometry that is distinct from that of structurally distinct scFv$_2$s.

In addition to single-chain molecules, other strategies to stabilize the correctly formed diabody (especially relative to inactive homodimers) have been explored. Similar to the knob-into-hole (KiH) idea that allows for heterodimerization of half-antibodies to form intact bispecific IgG antibodies, Zhu et al. used different sets of KiH mutations to stabilize diabodies targeting HER2 and CD3 [503]. Their variant v5 (VH V37F/L45W and VL Y87A/F98M) increased bsDb purity from 72% to 92% while reducing expression yields in half. The mutations also impacted antigen binding, with HER2 affinity decreased but T cell affinity increased. The same group reported a disulfide-stabilized mutant (VH D101C and VL L46C) that increased heterodimer purity to 96%. However, this variant was difficult to produce in *E. coli*, forming insoluble aggregates and purified products that were only 65% disulfide stabilized. Shortly after, another study presented a distinct disulfide-linked diabody (VH A44C and VL G100C) that showed similar disulfide oxidation issues when expressed in *E. coli*, whereas heterodimers from *Pichia pastoris* were >90% covalently linked [504].

Dual-affinity re-targeting proteins (DARTs) are a prominent class of diabody derivatives that incorporate a C-terminal disulfide bond to stabilize the correctly formed dimer species. The first DART was described in 2010 and targeted CD16 and CD32B to recruit NK cells to act on leukemic B cells [505]. Covalent stabilization of the correct heterodimer was achieved either by appending residues LGGC at the end of each C-terminus or by adding FNRGEC to one chain and VEPKSC to the other, which mimics the sequence N-terminal to a standard IgG1κ HC-LC disulfide bond. When produced in mammalian cells, the DARTs exhibited no aggregation and were stable in serum at 37 °C for weeks and in phosphate-buffered saline (PBS) at 4 °C for months. In addition to demonstrating potent ADCC against various CD32-expressing cells with EC50 values in the pg/mL range, the DARTs were protective in a lymphoma xenograft mouse model. Rational stabilization of proper chain pairing, especially using human-derived sequences, makes the DART framework an elegant advancement in the field of bispecific fragments.

TandAbs (tandem antibodies) are a diabody-based framework that have the advantage of bivalent binding to each of two antigens. It uses the same domain orientation as scDbs but favors dimerization using a short central linker. Kipriyanov et al. first described the format in 1999, revealing that the increased valency of a CD3 × CD19 TandAb allowed for higher avidity antigen interactions as a result of slower dissociation [506]. When expressed in *E. coli*, their 57-kDa construct dimerized to form the 114-kDa tetravalent species. Similar to how diabody formation is favored using scFvs with short linkers, TandAb formation was increased using a shorter central linker (12 residues versus 27) to favor dimerization. The larger size of TandAbs was also found to increase their serum half-life relative to diabodies. In a follow-up study, TandAbs were shown to accumulate more in tumors, likely due to higher-avidity CD19 interactions [507,508]. For cases where prolonged receptor engagement is important, tetravalent TandAbs may therefore be preferred over molecules with two binding sites.

Fusion of Single-Domain Antibodies

While the scFv fragment (composed of linked VH and VL domains, ~25 kDa) is the minimal intact antigen-binding moiety of human antibodies, camelids produce heavy chain antibodies in which the antigen-binding moiety (VHH) is a single domain of just ~15 kDa. Thus, VHH domains of different specificities can be linked to create extremely compact bispecific molecules. For example, Conrath et al. created 37-kDa bispecific tandem VHH antibodies targeting lysozyme and β-lactamase [509]. These molecules used a 29-amino acid linker derived from the llama heavy chain antibody (IgG2a) hinge and were expressed in *E. coli*. While binding was maintained for both antigens, they noted a four-fold increase in the K_D value for the C-terminal VHH that might be caused by interference from the linker. This study demonstrated the potential of the tandem VHH format and paved the way for similar bispecific and multispecific molecules. Likewise, VNARs and other binding domains are amenable for similar multispecific construct design.

Another study explored whether it would be feasible to join human domain antibodies (based on VH or Vκ) to create multi-functional agents. Human dAbs against *C. albicans* secretory aspartyl proteinase 2 and mannoprotein 65 were linked by a 25-residue linker to create bispecific tandem dAbs [510]. After production in *E. coli*, mono- and bispecific dAbs were purified by protein A or protein L chromatography. Although binding parameters of the parental and bispecific dAbs were not directly compared, the bispecific molecule appeared to be more effective at clearing fungal infections. The potential advantage of using dAbs based on human VH or VL is the reduced risk of immunogenicity that may be a concern for dAbs from other animals. However, the developability properties of VHH and VNAR frameworks may be superior. Unlike the human VH/VL domains, which have hydrophobic interfaces that allow for chain pairing, VHH and VNAR domains are naturally monomeric and tend to have more hydrophilic surfaces [511].

3.5.2. Fc-Dependent Bispecific Antibodies

The Fc region of BsAb Ig formats allows for applications using Fc effector function control, such as the modulation of ADCC, CDC, half-life modulation, and heterodimerization. Ideally, BsAb could be selected via taking any one set of mAb binders and mixing them with another set of mAb binders. However, the use of four expression vectors to create BsAb from two sets of HC and LC can result in a mixture of different HC and LC combinations. To minimize the downstream efforts to isolate the BsAb, there have been two general approaches—HC and LC dimerization control.

Heavy Chain Heterodimerization

When forming IgG-like bispecific antibodies, a major challenge is that co-transfection of heavy and light chains with different specificities leads to three possible heavy chain dimers (the heterodimer and both homodimers), where the desired heterodimer makes up only 50% of the products by random chance. As a result, about half of the protein products will be contaminating homodimers that are monospecific instead of bispecific. This has been referred to as the heavy chain problem. Several engineering strategies have been applied to drive a preference for heavy chain heterodimer formation and allow for cleaner bispecific antibody preparations.

A pioneering approach to address the heavy chain problem was to engineer 'knob' and 'hole' mutations into complementary heavy chains of distinct specificities. Ridgway et al. identified residues in the IgG CH3 domain that form direct interchain contacts [512]. Amino acids at the center of this interface were rationally mutated to introduce a protruding knob with a large surface area into one heavy chain, and a sunken hole with small sidechains into the complementary heavy chain. Using this strategy, the knob mutation T366Y and hole mutation Y407T were developed. When one mutation was incorporated into an anti-CD3 heavy chain, and the other was incorporated into a CD4-IgG immunoadhesin, the co-expression of the anti-CD3 heavy chain, anti-CD3 light chain, and CD4-IgG genes allowed for preferential (>90%) formation of the corresponding hybrid molecule with CD3- and CD4-binding functionality.

Subsequently, Atwell et al. built upon this knob-into-hole approach to generate IgG variants with an even higher preference for heterodimer formation [513]. The T366W mutation was incorporated into one CH3 domain as the knob variant, and then a library of CH3 hole mutants was generated with diversity at positions 366, 368, and 407, which are in close proximity to residue 366 on the opposite CH3 domain. Variants that increased the stability of the CH3 heterodimer were selected by seven rounds of phage display. One pair of mutants (T366W in one heavy chain and T366S/L368A/Y407V in the other heavy chain) had near complete heterodimerization and produced complexes that were more stable than those produced by singly mutated variants (T366W and Y407A) as measured by guanidine- and heat-induced unfolding. Shortly after, these knob-into-hole mutations were combined with other mutations to introduce a stabilizing interchain disulfide bond in heterodimerizing CH3 domains [514]. The resulting sets of heavy chain mutations (S354C/T366W and Y349C/T366S/L368A/Y407V) were used to produce 95%-pure bispecific antibodies targeting HER3 and CD110. Light chain mispairing was precluded using a common light chain that formed functional paratopes of differing specificity when combined with either heavy chain. Finally, the conserved function of the mutant heterodimerized Fc region was demonstrated using HER2 antibodies that had similar ADCC activity whether or not they incorporated the CH3 mutations.

In addition to shape complementarity, electrostatic interactions in the Fc region have also been engineered to favor heavy chain heterodimerization. Gunasekaran et al. identified charge-mediated interactions at the CH3–CH3 interface and strategically mutated residues to cause repulsion of identical heavy chains and attraction of opposite heavy chains [515]. While single mutations in each chain (D399K and K409D) produced some preference for heterodimerization, a pair of double mutants (E356K/D399K and K392D/K409D) allowed for superior heterodimer purity (98%). To avoid the issue of light chain mispairing, a bispecific scFv-Fc format was devised using scFvs targeting CD3 and tumor-associated receptor tyrosine kinase (TARTK). The electrostatic steering mutations allowed for the production of CD3xTARTK scFv-Fc fusions that retained binding to both antigens and induced T cell killing of TARTK-bearing cells. The use of a common light chain to avert the light chain problem allowed for the development of the bispecific antibody emicizumab (factor IXa x factor X), which was approved in 2017 for the treatment of hemophilia A [474]. There have been extensions to the knob-into-hole and the electrostatic Fc heterodimerizations to introduce Fc mutations to have distinct isoelectric points so as to enable facile purification of the heterodimeric BsAb with minimum perturbations to Fc region Tm thermal stability [516]. These modifications have been applied to CD3 redirection and dual checkpoint blockade BsAbs.

Interestingly, the structural oddities of human IgG4, including a labile hinge enabled by S228 (proline in other subclasses) and weaker inter-CH3 interactions caused by R409 (lysine in other subclasses), allow it to undergo a process called Fab-arm exchange in vivo to create bispecific IgG4 antibodies [78]. Labrijn et al. drew from this natural phenomenon to create a process called controlled Fab-arm exchange (cFAE) for the formation of bispecific IgG1 antibodies [517,518]. In contrast to the other methods described, cFAE is performed in vitro using separately purified parental antibodies, rather than relying on the co-translational formation of heavy chain heterodimer during co-expression of the respective chains. A heavy chain containing the K409R mutation was found to preferentially heterodimerize with several heavy chain mutants having variation at positions 368, 370, 399, 405, or 407 when the chains were co-incubated in the presence of a reducing agent and allowed to swap half-antibodies. Mutation of F405 to leucine (which is present in rhesus monkey IgG4) largely favored the heterodimer, and the combination of F405L and K409R in opposite chains was pursued as a general bispecific platform. The pairing consistently allowed for >95% heterodimer formation when parental antibodies were combined with a reducing agent and exchanged. The rate constants have been determined for the mechanism of IgG1 BsAb formation [519].

Subsequent buffer exchange to remove reductant allows the inter-heavy chain disulfide bonds to re-oxidize, stabilizing the bispecific product. Importantly, the interaction between heavy and light chains is not disrupted during half-antibody exchange so that light chain mismatch is avoided without the need for a common light chain. The retention of Fc-mediated functions was verified by demonstrating normal levels of CDC and ADCC from CD20 antibodies containing the mutant Fc regions, as well as normal pharmacokinetics for a bispecific CD20 × EGFR bispecific antibody generated by cFAE.

The cFAE method has been extended to include other IgG1 and IgG2 parental Abs with K409R and L368E mutations to generate stable full-length BsAbs [520]. These methods can also use modulation of protein A binding to accelerate the generation of BsAbs [521,522].

Some Fab regions were selected for dual antigen recognition [203,523,524]. However, these constructs may not be capable of binding to two different epitopes simultaneously. To overcome this challenge, single chain Fv domains have been fused onto the termini of HC/LCs to generate dual action Fab molecules. The dual variable domain, DVD, combines the variable domains of two mAbs in tandem to form dual-specific IgG molecules [525–528]. However, careful selection of the binding arms and linker domains is required to bypass developability and immunogenicity concerns.

Light Chain Control

Often, in CMC development, it was found that heavy chain heterodimerization was insufficient to control LC mispairing. A method to prevent this problem is to generate BsAb that have parental Abs that have different light chains families. Kappa-lambda light chain (κλ–LC) BsAbs have the same HC and two different LCs. The production of κλ-bodies involves the co-expression of one HC and two LCs (one κ, and one λ) with different binding specificities in a single cell [529]. By using serial affinity purifications, a fully human IgG format can be prepared with scalable and more facile purification. To further control LC pairing, BsAb with common LC have been developed. The BsAb can have parental Abs that have been selected to have the common LC [530]. However, custom screening is required to find the best BsAb selection in either the common LC or κλ BsAbs.

In working towards getting any pair of mAb to prepare BsAb from four expression vectors, several methods have been explored that couple LC engineering: CrossMab technology that enforces LC domain crossover in the Fab region [531]; hydrophobic and electrostatic interactions and disulfide bonds [532–537] to minimize LC swapping for proper HC/LC pairing.

3.5.3. Considerations for Selection

The choice of what platforms are driven by the desired therapeutic product profile, ability to integrate engineering platform to discovery repertoire of pharmacophores, and access to licensed technology and platforms. Regardless of the platform, one major theme of lead selection of BsAb is to utilize multiple binding arms that span different epitopes. Concomitant to each epitope, it is critical to utilize a broad library of paratopes with different affinities and potencies for target engagement. Affinity selection on a target is not enough. Selection must be based on the pharmacology that may require inhibition of native ligands, co-receptor activation, or target-mediated agonism. Thus, it is not uncommon that high affinity is not linked to molecular efficacy [30,538].

After the selection of the binding arms, there is much to be done during lead selection by probing architecture, valency, order, specificity, and potency tuning to identify the best BsAb. An example of the therapeutic product profile driving molecular design involves the design of T cell or T lymphocyte redirection to attack tumor cells that are marked with a particular cell surface antigen protein. The inclusion of the valency of epitopes can mediate the avidity for low-density receptor cell targeting. These concepts are highlighted in an excellent review of different bispecific molecules that mediate T cell cytotoxicity to diseased tissues [539]. The review describes general strategies about how to optimize target antigen potency while preventing toxicity that can occur from engagement of normal tissue antigens. Likewise, some binding arms have to be modulated to have lower affinity that manifests in changes in the kinetic on and off rates to increase the potential therapeutic index of the molecule [138]. These classes of molecules can complement the utilization of engineered T cells with chimeric antigen receptors [540].

The engineering concepts for T cell redirection molecules include screening for: Different epitopes on target cell surface membrane proteins on the T cell and target cell; internalization of cell surface targets when bound to BsAb, and potency of the molecule engagement with T cell and target cell antigens. BsAb architecture can vary the orientation of the binders, linkers, valency, functional activity, and developability [539]. The role of the anti-CD3-binding domains in BsAb with a broader range of T cell agonism has been shown to have a potential wider therapeutic index [541]. In addition, the choice of the BsAb subtype can affect T cell redirection activity [389]. As the field evolves, more exploration into novel formats with different potency and kinetics of target engagement will expand possibilities for T cell therapeutics [539]. There is ongoing clinical development to understand how best to employ T cell redirection molecules versus chimeric antigen receptors [542].

Redirection of immune cells towards diseased cells are termed as effector cells that can comprise other cells beyond T cells and include NK cells, macrophages, neutrophils, monocytes, and granulocytes [396,543] to facilitate specific killing of tumor cells. Each of the effector cells express different types of activating receptors, and a specific population can be recruited by carefully selecting the targeted trigger receptor [544,545]. Redirected cytotoxic activity has been shown, with bispecific antibodies recruiting all effector cells, including macrophages [546]. A CD47xCD19 BsAb using an innate immune checkpoint control molecule, CD47, was used to block the macrophage suppressive signals in cancer cells [547]. Although the recruitment of T cells is far more widely used because of their proliferation ability and potent cytotoxicity, there are advantages in expanding the pool of immune cell clearance of malignant cells.

The choice of the binding arm paratopes should consider the developability of the individual arm, since the inclusion of problematic domains into the BsAb format can make that CMC process more difficult. The CMC criteria regarding ease of production, liabilities in post-translational modification, solubility, and clipping should be considered. However, regarding BsAbs with higher levels of dosing, several hurdles need to be addressed: Identification of a clear architectural format that is amenable to reasonable manufacturing costs. There has been some development of strategies to improve downstream purification using differential protein A binding and anti-lambda/anti-kappa chain affinity columns [548]. Presently, several clinical BsAb have involved cell line self-assembly using common HC or LC technologies, HC/LC association using point mutations in the framework, or controlled Fab-arm exchange [1,2].

Since BsAb engineering involves putting the different domains together, careful potency and efficacy selection is required to identify the best hits. In consideration of the therapeutic hypothesis, there are still other factors that can control the in vivo potency, which include the biodistribution, PK, dosing levels, and method of administration. Hence, the choice of scaffold, linkers, production host, and scale-up processes play critical roles in the preparations of the clinical development supply. Ideally, the sequence would minimize the presence of neoepitope to minimize the development of ADA. Besides formats that are completely derived from the antibody structure, fusion proteins can also achieve multifunctionality by combining the antigen specificity of one antibody domain with another targeting protein. The domains can be coupled together using leucine zippers to generate tetravalent-bispecific Abs. Likewise, developing a system to make BsAb with minimal human neo-epitopes can be important to generate molecules for both chronic and acute indications.

4. Evolving Applications

4.1. Multispecific Molecules

A natural extension of BsAbs to is have an increased number of binding arms to create multispecific Abs that can be effective in engaging more epitopes on a target. In Figure 9, the binding domains can involve combinations of binding domains using Fab, scFv, and VHH domains, or other scaffolds. With the opportunity of mixing in cytokines and enzymes, there is no shortage of possibilities for protein engineering. Ultimately, empirical screening will be required to determine what the best architecture of such multispecific agents should be.

There is a medical need to have broadly neutralizing antibodies against highly variable pathogens, such as infectious viruses. For instance, there have been several efforts to target clades of influenza A and B separately [549–552]. However, such efforts led to therapies that could only combat the respective class of influenza strains. Thus, a strategy was developed to prepare tethered multispecific antibodies with binding domains that could inhibit all known strains of influenza A and B [553]. A multispecific molecule has been engineered to incorporate diverse camelid single domain antibodies that recognize conserved epitopes on influenza virus hemagglutinin. For instance, the multidomain Ab has enhanced virus cross-reactivity and potency to inhibit influenza activity [553]. These binding domains engage previously mapped hemagglutinin epitopes that are close together. Because of the proximity of these epitopes, there can be steric hindrances that prevent the use of a tetravalent Fab-binding arm multispecific Ab format. Thus, the research effort focused on the discovery of binding arm-based VHH domains that have smaller paratope sizes. Such a molecule was able to bind to the different epitopes, target both influenza strains A and B, and had the ability to protect against all circulating strains of the virus with enhanced cross-reactivity and potency. The molecule was designed to have an active Fc region to utilize ADCC activity that could contribute to in vivo activity [554].

Antibody therapeutics targeting solid tumors are often limited by poor accumulation in and dispersal throughout the tumor tissues [555]. Since antibody fragments are small, they can be fused with various protein to create new molecules with novel functions. Antibody fragments, such as scFvs, can be fused with enzymes to localize such activity on a target cell. For instance, human RNase has been fused to an scFv targeting HER2 to endow cytotoxic activity on human carcinoma cell lines [556]. Permutations of multispecific fusions permit modulation of the enzyme activity desired [557]. Preclinical studies have shown how these molecules can reduce tumor volume.

Prodrug activation, by exogenously administered or endogenous enzymes, for cancer therapy is an approach to achieve better selectivity and less systemic toxicity than conventional chemotherapy. Typically, activating enzymes have short half-lives in the bloodstream. By engineering a cage or protease-sensitive peptide linker to block the activity of the enzyme or drug of interest, the trojan horse technology increases the drug or enzyme half-life and can prevent the drug or enzyme from cytotoxicity on healthy cells. Thus, the strategy is to use an antibody to deliver a pro-enzyme or pro-drug to destroy a target cell.

The EGFR-binding arm of cetuximab was engineered to be have its binding domain be unmasked by enzymes found in the tumor microenvironment [558]. In the absence of EGFR-binding properties in normal tissue, this molecule was inert in systemic circulation in animal models. However, when the mask was removed by appropriate proteases, the molecule restored the antigen binding and cell-based activities of cetuximab. Thus, this strategy to increase the therapeutic index with localized activation of the molecule has been expanded to other bispecific antibody applications, including immune cell redirection. Likewise, the Ab or BsAb can be used to deliver a pro-drug that can be released in the presence of enzymes in the tumor microenvironment [559] or by the higher reducing capacity in hypoxic tumor cells [560].

Alternatively, the binding arms of BsAb can be selected to have pH-sensitive binding that is responsive only in the diseased tissue of interest. For instance, a dual-function pH-responsive BsAb has been made that binds to the tumor-specific antigen on the cell surface but not on the proteolytically shed soluble domain of that tumor antigen [561]. In one example, antibody fragments can be fused with enzymes to develop antibody-directed enzyme therapies (ADEPTs) [562]. The enzyme can convert a non-toxic pro-drug into a cytotoxic drug when it is in close proximity to target cells so as to be a therapy for oncology. There is a report of an enzymatic tyrosinase nanoreactor based on metal-organic frameworks (MOFs) that activates the pro-drug paracetamol in cancer cells in a long-lasting manner. By generating reactive oxygen species (ROS) and depleting glutathione (GSH), the product of the enzymatic conversion of paracetamol is toxic to drug-resistant cancer cells. Tyrosinase-MOF nanoreactors cause significant cell death in the presence of paracetamol for up to three days after being internalized by cells, while free enzymes totally lose activity in a few hours. Thus, enzyme–MOF nanocomposites are envisioned to be novel persistent platforms for various biomedical applications [563]. Although the key limitation has been the immunogenicity of the enzyme, the inclusion of non-immunogenic enzymes in combination with prodrugs can generate potent molecules. ADEPT has the potential to be non-toxic to normal tissue and can therefore be combined with other modalities, including immunotherapy, for greater clinical benefit [564,565].

Antibody fragments can be fused with cytokines to generate immunocytokines [566,567]. Cytokines have been used for cancer patients but have substantial side effects and unfavorable pharmacokinetic profiles. The presence of the antibody domain allows for tissue-specific localization into malignant cells. These conjugated molecules can activate the immune systems of patients when in proximity to diseased tissues. This could prevent the systemic side effects associated with systemic administration of immune system-activating cytokines.

4.2. Intracellular Targeting

The therapeutic efficacy of anti-cancer drugs as low molecular weight chemical agents can be poor because of the inability to inhibit protein–protein interactions effectively. Because of the limitation of a small molecule drug interaction surface area, there is a great need to develop therapeutics that can control intracellular protein–protein interactions. Antibody molecules could be selected to have the specificity and potency to modulate critical cytoplasmic target molecule biology. However, antibodies cannot cross intact cellular or subcellular membranes via passive diffusion into living cells due to their size and hydrophilicity [568]. Antibody internalization into the cell can be accomplished by taking advantage of normal receptor biology: Ligand binding causes receptor activation via homo- or heterodimerization, either directly for a bivalent ligand or by causing a conformational change in the receptor for monovalent ligand and receptor-mediated endocytosis [569]. However, there is still use of the targeted receptor-mediated endocytosis machinery [570]. Manipulation of receptor-mediated endocytosis and intracellular trafficking dynamics is typically employed in the development of antibody drug conjugates. Many attempts have been made to directly deliver antibodies into intracellular compartments that include microinjection, electroporation, and protein transfection [571–573]. These are very selective therapies that cannot be generally scaled up with multifocal disease targeting [574]. Thus, there is a need for obtaining antibodies that can enter specific cells and tissues without the complications of the antibody–drug conjugate engineering.

Internalizing antibodies can be obtained via direct selection of internalizing phage antibodies by incubating phage libraries directly with the target cells [575–578]. However, the major challenge in this process is to identify the antigen bound by the internalizing antibody, which can be determined indirectly using the flow cytometry of target cells, and identifying the cognate antigen recognized by tumor-specific antibodies using immunoprecipitation and mass spectrometry [579]. Nonetheless, there are reports of tumor-specific internalizing antibodies from phage libraries that exert anti-tumor effects after systemic administration [580].

Much of this effort requires target-specific selection to identify such characteristics. Therefore, it is necessary to develop "promoter agents" that help improve tumor accumulation and penetration to improve the therapeutic index of antibody-based drugs [581]. There have been efforts to increase tumor access using tumor-penetrating ligands [582]. Conjugation with protein transduction domains, which are represented by cell-penetrating peptides (CPPs), such as the HIV-1 TAT peptide, has been extensively attempted in order to facilitate the intracellular delivery of antibodies or Fc-containing molecules formatted as single chain variable fragments (scFvs), antigen-binding fragments (Fabs), and full-length IgGs [583]. Optimization of CPPs continue to be applied to Fc-containing molecules [584].

Most antibodies that enter epithelial cells via receptor-mediated endocytosis are usually retained in endosomes and are then recycled out of the cells or are degraded in the lysosomes without being released into the cytosol [585]. Likewise, most of the CPP-conjugated antibodies inherited the intrinsic intracellular trafficking of the parent CPPs, which were either entrapped inside endocytic vesicles, translocated into the nucleus, or eventually degraded in lysosomes without efficient endosomal release into the cytosol. Molecular modifications were made to enhance release from the endosomes to allow for tumor tissue penetration [586].

Structural determinants of endosomal escape have been engineered into Ab variant with an ~three-fold improved endosomal escape efficiency [587,588]. The authors have been homing into a platform technology that enables an IgG to target cytosolic proteins via an endosomal escape mechanism. The elements of the engineering include having a domain to bind to the extracellular domain that permits endocytosis, a domain that improves endosomal escape efficiency, and a domain that can bind to the intracellular target [589]. Single domain antibodies have also been similarly modified for the knockdown of cytosolic and nuclear proteins [590]. The addition of endosomal escape protein domains and cell-penetrating peptides for efficient transfection broaden the application of inhibiting sdAbs.

A general strategy for generating intact, full-length IgG antibodies that penetrate into the cytosol of living cells is still of great interest [591]. A humanized light chain variable domain (VL) that could penetrate the cytosol of living cells was engineered for the association with various subtypes of human heavy chain variable domains (VHs). When light chains with humanized VL were co-expressed with three heavy chains (HCs), including two HCs of the clinically approved adalimumab (Humira) and bevacizumab (Avastin), all three purified IgG antibodies were internalized into the cytoplasm of living cells [589]. Although these methods are successful for delivering antibodies into the cytoplasm of cultured living cells, many issues, including cytotoxicity, loss of antibody stability, and difficulty of systemic administration, remain unresolved.

Intrabodies are Ab or Ab fragments that can be expressed intracellularly for binding to an intracellular protein [592,593]. These molecules can be created by the in-frame incorporation of intracellular, peptide-trafficking signals [594,595]. Additionally, they can be developed against a variety of target antigens that may be present at different subcellular locations, such as the cytosol, mitochondria, nucleus, and endoplasmic reticulum [596]. The interaction of these molecules with their target antigen results in the blocking or modification of molecular interactions, thereby leading to a change in the biological activity of the target proteins. Because the transport of Abs into a living cell from an extracellular environment is difficult, intrabodies can be expressed within the target cell via genetic engineering [568]. Because naturally occurring Abs are optimized to be secreted from the cell, intrabodies require special alterations, including: The use of single-chain antibodies (scFvs); modification of immunoglobulin VL domains for hyperstability [597]; selection of antibodies resistant to the more reducing intracellular environment [598]; expression as a fusion protein stable as intracellular proteins [599]; or the use of virus-like particles [600]. Several preclinical studies have demonstrated favorable results, including tumor growth inhibition and downregulation of viral envelope proteins, when such therapy candidates were used against inflammation [601], HIV [602], and hepatitis [603,604], respectively. The major challenge associated with these molecules is the absence of effective in vivo methods that can deliver the genetic material encoding the intrabody to live target cells [605].

5. Conclusions

We have provided a review of the antibody structure and function for therapeutic applications. Different examples of the engineering antibody variable domains were discussed by using rational design that is based on the experimentally derived or modeled structural information. The Fc region has been engineered to optimize effector function, clustering, and Fc receptor engagement. In general, antibody engineering of both the Fab and Fc regions is an indispensable part of the drug development process, and as such will continue to advance as more and more antibodies are generated for therapeutic use. Despite great progress in the methods of antibody engineering over the last 20 years, new approaches are in high demand. One of the remaining goals is to improve the accuracy of computational methods, which will allow for the prediction of point mutations that improve the affinity and other properties of interest. New approaches are continually being developed to create antibody-based molecules that are superior in their potency, specificity, localization, and safety. The choice of the binding domain can be tailored to engage the relevant epitopes. Likewise, engineering to change the architecture of the binding arms, Fc regions, modulatory bispecific, or multispecific domains to achieve monovalent- or avidity-driven engagement will result in more specific and potent molecules. Thus, continual process improvements to generate sufficient quantity and purity of hits will be required to facilitate comprehensive lead selection. The great diversity in antibody structure–function studies still has much room to engineer fit-for-purpose "magic bullets" with tailored PK profiles to meet different therapeutic hypotheses.

Funding: We thank William Atkins at University of Washington who helped secure funding for Dennis Goulet via NIH T32GM007750.

Conflicts of Interest: The authors declare no conflicts of interest. M.L.C., A.T., and G.L.G. are employees of Janssen Laboratories.

References

1. Kaplon, H.; Reichert, J.M. Antibodies to watch in 2019. *mAbs* **2019**, *11*, 219–238. [CrossRef] [PubMed]
2. Kaplon, H.; Reichert, J.M. Antibodies to watch in 2018. *mAbs* **2018**, *10*, 183–203. [CrossRef] [PubMed]
3. Gilliland, G.L.; Luo, J.; Vafa, O.; Almagro, J.C. Leveraging SBDD in protein therapeutic development: Antibody engineering. *Methods Mol. Biol.* **2012**, *841*, 321–349. [PubMed]
4. Berman, H.M.; Westbrook, J.; Feng, Z.; Gilliland, G.; Bhat, T.N.; Weissig, H.; Shindyalov, I.N.; Bourne, P.E. The Protein Data Bank. *Nucleic Acids Res.* **2000**, *28*, 235–242. [CrossRef] [PubMed]
5. Padlan, E.A. Structural basis for the specificity of antibody-antigen reactions and structural mechanisms for the diversification of antigen-binding specificities. *Q. Rev. Biophys.* **1977**, *10*, 35–65. [CrossRef] [PubMed]
6. Amzel, L.M.; Poljak, R.J. Three-dimensional structure of immunoglobulins. *Annu. Rev. Biochem.* **1979**, *48*, 961–997. [CrossRef]
7. Davies, D.R.; Metzger, H. Structural basis of antibody function. *Annu. Rev. Immunol.* **1983**, *1*, 87–117. [CrossRef]
8. Wilson, I.A.; Stanfield, R.L. Antibody-antigen interactions: New structures and new conformational changes. *Curr. Opin. Struct. Biol.* **1994**, *4*, 857–867. [CrossRef]
9. Poljak, R.; Amzel, L.; Avey, H.; Chen, B.; Phizackerley, R.; Saul, F. Three-dimensional structure of the Fab' fragment of a human immunoglobulin at 2,8-A resolution. *Proc. Natl. Acad. Sci. USA* **1973**, *70*, 3305–3310. [CrossRef]
10. Poljak, R.; Amzel, L.; Chen, B.; Phizackerley, R.; Saul, F. The three-dimensional structure of the Fab' fragment of a human myeloma immunoglobulin at 2.0-angstrom resolution. *Proc. Natl. Acad. Sci. USA* **1974**, *71*, 3440–3444. [CrossRef]
11. Harris, L.J.; Larson, S.B.; Hasel, K.W.; McPherson, A. Refined structure of an intact IgG2a monoclonal antibody. *Biochemistry* **1997**, *36*, 1581–1597. [CrossRef] [PubMed]
12. Teplyakov, A.; Obmolova, G.; Malia, T.J.; Luo, J.; Muzammil, S.; Sweet, R.; Almagro, J.C.; Gilliland, G.L. Structural diversity in a human antibody germline library. *mAbs* **2016**, *8*, 1045–1063. [CrossRef] [PubMed]
13. Schiffer, M.; Girling, R.L.; Ely, K.R.; Edmundson, A.B. Structure of a lambda-type Bence-Jones protein at 3.5-A resolution. *Biochemistry* **1973**, *12*, 4620–4631. [CrossRef] [PubMed]
14. Stanfield, R.L.; Zemla, A.; Wilson, I.A.; Rupp, B. Antibody elbow angles are influenced by their light chain class. *J. Mol. Biol.* **2006**, *357*, 1566–1574. [CrossRef] [PubMed]

15. Lesk, A.M.; Chothia, C. Elbow motion in the immunoglobulins involves a molecular ball-and-socket joint. *Nature* **1988**, *335*, 188–190. [CrossRef]

16. Love, R.A.; Villafranca, J.E.; Aust, R.M.; Nakamura, K.K.; Jue, R.A.; Major, J.G., Jr.; Radhakrishnan, R.; Butler, W.F. How the anti-(metal chelate) antibody CHA255 is specific for the metal ion of its antigen: X-ray structures for two Fab'/hapten complexes with different metals in the chelate. *Biochemistry* **1993**, *32*, 10950–10959. [CrossRef]

17. Wu, T.T.; Kabat, E.A. An analysis of the sequences of the variable regions of Bence Jones proteins and myeloma light chains and their implications for antibody complementarity. *J. Exp. Med.* **1970**, *132*, 211–250. [CrossRef]

18. Kabat, E.A.; Wu, T.T. Attempts to locate complementarity-determining residues in the variable positions of light and heavy chains. *Ann. N. Y. Acad. Sci.* **1971**, *190*, 382–393. [CrossRef]

19. Almagro, J.C. Identification of differences in the specificity-determining residues of antibodies that recognize antigens of different size: Implications for the rational design of antibody repertoires. *J. Mol. Recognit.* **2004**, *17*, 132–143. [CrossRef]

20. Raghunathan, G.; Smart, J.; Williams, J.; Almagro, J.C. Antigen-binding site anatomy and somatic mutations in antibodies that recognize different types of antigens. *J. Mol. Recognit.* **2012**, *25*, 103–113. [CrossRef]

21. Persson, H.; Lantto, J.; Ohlin, M. A focused antibody library for improved hapten recognition. *J. Mol. Biol.* **2006**, *357*, 607–620. [CrossRef] [PubMed]

22. Cobaugh, C.W.; Almagro, J.C.; Pogson, M.; Iverson, B.; Georgiou, G. Synthetic antibody libraries focused towards peptide ligands. *J. Mol. Biol.* **2008**, *378*, 622–633. [CrossRef] [PubMed]

23. Almagro, J.C.; Quintero-Hernandez, V.; Ortiz-Leon, M.; Velandia, A.; Smith, S.L.; Becerril, B. Design and validation of a synthetic VH repertoire with tailored diversity for protein recognition. *J. Mol. Recognit.* **2006**, *19*, 413–422. [CrossRef] [PubMed]

24. Davies, D.R.; Cohen, G.H. Interactions of protein antigens with antibodies. *Proc. Natl. Acad. Sci. USA* **1996**, *93*, 7–12. [CrossRef] [PubMed]

25. Davies, D.R.; Padlan, E.A. Twisting into shape. *Curr. Biol.* **1992**, *2*, 254–256. [CrossRef]

26. Vogt, A.D.; Pozzi, N.; Chen, Z.; di Cera, E. Essential role of conformational selection in ligand binding. *Biophys. Chem.* **2014**, *186*, 13–21. [CrossRef]

27. Foote, J.; Milstein, C. Conformational isomerism and the diversity of antibodies. *Proc. Natl. Acad. Sci. USA* **1994**, *91*, 10370–10374. [CrossRef]

28. Paul, F.; Weikl, T.R. How to Distinguish Conformational Selection and Induced Fit Based on Chemical. Relaxation Rates. *PLoS Comput. Biol.* **2016**, *12*, e1005067. [CrossRef]

29. Ma, B.; Zhao, J.; Nussinov, R. Conformational selection in amyloid-based immunotherapy: Survey of crystal structures of antibody-amyloid complexes. *Biochim. Biophys. Acta* **2016**, *1860*, 2672–2681. [CrossRef]

30. Kenakin, T.P.; Morgan, P.H. Theoretical effects of single and multiple transducer receptor coupling proteins on estimates of the relative potency of agonists. *Mol. Pharmacol.* **1989**, *35*, 214–222.

31. Chothia, C.; Lesk, A.M. Canonical structures for the hypervariable regions of immunoglobulins. *J. Mol. Biol.* **1987**, *196*, 901–917. [CrossRef]

32. Chothia, C.; Lesk, A.M.; Tramontano, A.; Levitt, M.; Smith-Gill, S.J.; Air, G.; Sheriff, S.; Padlan, E.A.; Davies, D.; Tulip, W.R.; et al. Conformations of immunoglobulin hypervariable regions. *Nature* **1989**, *342*, 877–883. [CrossRef] [PubMed]

33. Chothia, C.; Lesk, A.M.; Gherardi, E.; Tomlinson, I.M.; Walter, G.; Marks, J.D.; Llewelyn, M.B.; Winter, G. Structural repertoire of the human VH segments. *J. Mol. Biol.* **1992**, *227*, 799–817. [CrossRef]

34. Tomlinson, I.M.; Cox, J.P.; Gherardi, E.; Lesk, A.M.; Chothia, C. The structural repertoire of the human V kappa domain. *EMBO J.* **1995**, *14*, 4628–4638. [CrossRef] [PubMed]

35. Vargas-Madrazo, E.; Lara-Ochoa, F.; Almagro, J.C. Canonical structure repertoire of the antigen-binding site of immunoglobulins suggests strong geometrical restrictions associated to the mechanism of immune recognition. *J. Mol. Biol.* **1995**, *254*, 497–504. [CrossRef]

36. Collis, A.V.; Brouwer, A.P.; Martin, A.C. Analysis of the antigen combining site: Correlations between length and sequence composition of the hypervariable loops and the nature of the antigen. *J. Mol. Biol.* **2003**, *325*, 337–354. [CrossRef]

37. Al-Lazikani, B.; Lesk, A.M.; Chothia, C. Standard conformations for the canonical structures of immunoglobulins. *J. Mol. Biol.* **1997**, *273*, 927–948. [CrossRef]

38. North, B.; Lehmann, A.; Dunbrack, R.L., Jr. A new clustering of antibody CDR loop conformations. *J. Mol. Biol.* **2011**, *406*, 228–256. [CrossRef]

39. Mas, M.T.; Smith, K.C.; Yarmush, D.L.; Aisaka, K.; Fine, R.M. Modeling the anti-CEA antibody combining site by homology and conformational search. *Proteins* **1992**, *14*, 483–498. [CrossRef]

40. Shirai, H.; Kidera, A.; Nakamura, H. Structural classification of CDR-H3 in antibodies. *FEBS Lett.* **1996**, *399*, 1–8. [CrossRef]

41. Morea, V.; Tramontano, A.; Rustici, M.; Chothia, C.; Lesk, A.M. Conformations of the third hypervariable region in the VH domain of immunoglobulins. *J. Mol. Biol.* **1998**, *275*, 269–294. [CrossRef] [PubMed]

42. Weitzner, B.D.; Dunbrack, R.L., Jr.; Gray, J.J. The origin of CDR H3 structural diversity. *Structure* **2015**, *23*, 302–311. [CrossRef] [PubMed]

43. Martin, A.C.; Thornton, J.M. Structural families in loops of homologous proteins: Automatic classification, modelling and application to antibodies. *J. Mol. Biol.* **1996**, *263*, 800–815. [CrossRef] [PubMed]

44. Almagro, J.C.; Beavers, M.P.; Hernandez-Guzman, F.; Maier, J.; Shaulsky, J.; Butenhof, K.; Labute, P.; Thorsteinson, N.; Kelly, K.; Teplyakov, A.; et al. Antibody modeling assessment. *Proteins* **2011**, *79*, 3050–3066. [CrossRef] [PubMed]

45. Almagro, J.C.; Teplyakov, A.; Luo, J.; Sweet, R.W.; Kodangattil, S.; Hernandez-Guzman, F.; Gilliland, G.L. Second antibody modeling assessment (AMA-II). *Proteins* **2014**, *82*, 1553–1562. [CrossRef]

46. Fasnacht, M.; Butenhof, K.; Goupil-Lamy, A.; Hernandez-Guzman, F.; Huang, H.; Yan, L. Automated antibody structure prediction using Accelrys tools: Results and best practices. *Proteins* **2014**, *82*, 1583–1598. [CrossRef]

47. Maier, J.K.; Labute, P. Assessment of fully automated antibody homology modeling protocols in molecular operating environment. *Proteins* **2014**, *82*, 1599–1610. [CrossRef]

48. Zhu, K.; Day, T.; Warshaviak, D.; Murrett, C.; Friesner, R.; Pearlman, D. Antibody structure determination using a combination of homology modeling, energy-based refinement, and loop prediction. *Proteins* **2014**, *82*, 1646–1655. [CrossRef]

49. Weitzner, B.D.; Kuroda, D.; Marze, N.; Xu, J.; Gray, J.J. Blind prediction performance of Rosetta Antibody 3.0: Grafting, relaxation, kinematic loop modeling, and full CDR optimization. *Proteins* **2014**, *82*, 1611–1623. [CrossRef]

50. Berrondo, M.; Kaufmann, S.; Berrondo, M. Automated Aufbau of antibody structures from given sequences using Macromoltek's SmrtMolAntibody. *Proteins* **2014**, *82*, 1636–1645. [CrossRef]

51. Shirai, H.; Ikeda, K.; Yamashita, K.; Tsuchiya, Y.; Sarmiento, J.; Liang, S.; Morokata, T.; Mizuguchi, K.; Higo, J.; Standley, D.M.; et al. High-resolution modeling of antibody structures by a combination of bioinformatics, expert knowledge, and molecular simulations. *Proteins* **2014**, *82*, 1624–1635. [CrossRef] [PubMed]

52. Marcatili, P.; Rosi, A.; Tramontano, A. PIGS: Automatic prediction of antibody structures. *Bioinformatics* **2008**, *24*, 1953–1954. [CrossRef] [PubMed]

53. Lepore, R.; Olimpieri, P.P.; Messih, M.A.; Tramontano, A. PIGSPro: Prediction of immunoGlobulin structures v2. *Nucleic Acids Res.* **2017**, *45*, W17–W23. [CrossRef] [PubMed]

54. Teplyakov, A.; Luo, J.; Obmolova, G.; Malia, T.J.; Sweet, R.; Stanfield, R.L.; Kodangattil, S.; Almagro, J.C.; Gilliland, G.L. Antibody modeling assessment II. Structures and models. *Proteins* **2014**, *82*, 1563–1582. [CrossRef] [PubMed]

55. Porter, R.R. The formation of a specific inhibitor by hydrolysis of rabbit antiovalbumin. *Biochem. J.* **1950**, *46*, 479–484. [CrossRef] [PubMed]

56. Porter, R.R. A chemical study of rabbit antiovalbumin. *Biochem. J.* **1950**, *46*, 473–478. [CrossRef]

57. Porter, R.R. The hydrolysis of rabbit γ-globulin and antibodies with crystalline papain. *Biochem. J.* **1959**, *73*, 119–126. [CrossRef]

58. Deisenhofer, J. Crystallographic refinement and atomic models of a human Fc fragment and its complex with fragment B of protein A from Staphylococcus aureus at 2.9- and 2.8-A resolution. *Biochemistry* **1981**, *20*, 2361–2370. [CrossRef]

59. Teplyakov, A.; Zhao, Y.; Malia, T.J.; Obmolova, G.; Gilliland, G.L. IgG2 Fc structure and the dynamic features of the IgG CH2-CH3 interface. *Mol. Immunol.* **2013**, *56*, 131–139. [CrossRef]

60. Daeron, M. Fc receptor biology. *Annu. Rev. Immunol.* **1997**, *15*, 203–234. [CrossRef]

61. Rouard, H.; Tamasdan, S.; Moncuit, J.; Moutel, S.; Michon, J.; Fridman, W.H.; Teillaud, J.L. Fc receptors as targets for immunotherapy. *Int. Rev. Immunol.* **1997**, *16*, 147–185. [CrossRef] [PubMed]

62. Taylor, R.P.; Lindorfer, M.A. Fcgamma-receptor-mediated trogocytosis impacts mAb-based therapies: Historical precedence and recent developments. *Blood* **2015**, *125*, 762–766. [CrossRef] [PubMed]

63. Mimoto, F.; Kuramochi, T.; Katada, H.; Igawa, T.; Hattori, K. Fc Engineering to Improve the Function of Therapeutic Antibodies. *Curr. Pharm. Biotechnol.* **2016**, *17*, 1298–1314. [CrossRef] [PubMed]

64. Derebe, M.G.; Nanjunda, R.K.; Gilliland, G.L.; Lacy, E.R.; Chiu, M.L. Human IgG subclass cross-species reactivity to mouse and cynomolgus monkey Fcgamma receptors. *Immunol. Lett.* **2018**, *197*, 1–8. [CrossRef]

65. Matsumiya, S.; Yamaguchi, Y.; Saito, J.; Nagano, M.; Sasakawa, H.; Otaki, S.; Satoh, M.; Shitara, K.; Kato, K. Structural comparison of fucosylated and nonfucosylated Fc fragments of human immunoglobulin G1. *J. Mol. Biol.* **2007**, *368*, 767–779. [CrossRef]

66. Saphire, E.O.; Stanfield, R.L.; Crispin, M.D.; Parren, P.W.; Rudd, P.M.; Dwek, R.A.; Burton, D.R.; Wilson, I.A. Contrasting IgG structures reveal extreme asymmetry and flexibility. *J. Mol. Biol.* **2002**, *319*, 9–18. [CrossRef]

67. Borrok, M.J.; Jung, S.T.; Kang, T.H.; Monzingo, A.F.; Georgiou, G. Revisiting the role of glycosylation in the structure of human IgG Fc. *ACS Chem. Biol.* **2012**, *7*, 1596–1602. [CrossRef]

68. Landolfi, N.F.; Thakur, A.B.; Fu, H.; Vasquez, M.; Queen, C.; Tsurushita, N. The integrity of the ball-and-socket joint between V and C domains is essential for complete activity of a humanized antibody. *J. Immunol.* **2001**, *166*, 1748–1754. [CrossRef]

69. Jefferis, R. Glycosylation of natural and recombinant antibody molecules. *Adv. Exp. Med. Biol.* **2005**, *564*, 143–148.

70. Lee, H.S.; Qi, Y.; Im, W. Effects of N-glycosylation on protein conformation and dynamics: Protein Data Bank analysis and molecular dynamics simulation study. *Sci. Rep.* **2015**, *5*, 8926. [CrossRef]

71. Krapp, S.; Mimura, Y.; Jefferis, R.; Huber, R.; Sondermann, P. Structural analysis of human IgG-Fc glycoforms reveals a correlation between glycosylation and structural integrity. *J. Mol. Biol.* **2003**, *325*, 979–989. [CrossRef]

72. Frank, M.; Walker, R.C.; Lanzilotta, W.N.; Prestegard, J.H.; Barb, A.W. Immunoglobulin G1 Fc domain motions: Implications for Fc engineering. *J. Mol. Biol.* **2014**, *426*, 1799–1811. [CrossRef] [PubMed]

73. Sazinsky, S.L.; Ott, R.G.; Silver, N.W.; Tidor, B.; Ravetch, J.V.; Wittrup, K.D. Aglycosylated immunoglobulin G1 variants productively engage activating Fc receptors. *Proc. Natl. Acad. Sci. USA* **2008**, *105*, 20167–20172. [CrossRef] [PubMed]

74. Feige, M.J.; Groscurth, S.; Marcinowski, M.; Shimizu, Y.; Kessler, H.; Hendershot, L.M.; Buchner, J. An unfolded CH1 domain controls the assembly and secretion of IgG antibodies. *Mol. Cell* **2009**, *34*, 569–579. [CrossRef] [PubMed]

75. Kronimus, Y.; Dodel, R.; Galuska, S.P.; Neumann, S. IgG Fc N-glycosylation: Alterations in neurologic diseases and potential therapeutic target? *J. Autoimmun.* **2019**, *96*, 14–23. [CrossRef]

76. Hayashi, Y.; Miura, N.; Isobe, J.; Shinyashiki, N.; Yagihara, S. Molecular dynamics of hinge-bending motion of IgG vanishing with hydrolysis by papain. *Biophys. J.* **2000**, *79*, 1023–1029. [CrossRef]

77. Jay, J.W.; Bray, B.; Qi, Y.; Igbinigie, E.; Wu, H.; Li, J.; Ren, G. IgG Antibody 3D Structures and Dynamics. *Antibodies* **2018**, *7*, 18. [CrossRef]

78. Van der Neut Kolfschoten, M.; Schuurman, J.; Losen, M.; Bleeker, W.K.; Martinez-Martinez, P.; Vermeulen, E.; den Bleker, T.H.; Wiegman, L.; Vink, T.; Aarden, L.A.; et al. Anti-inflammatory activity of human IgG4 antibodies by dynamic Fab arm exchange. *Science* **2007**, *317*, 1554–1557. [CrossRef]

79. Turner, M.W.; Bennich, H. Subfragments from the Fc fragment of human immunoglobulin G. Isolation and physicochemical charaterization. *Biochem. J.* **1968**, *107*, 171–178.

80. Ryan, M.H.; Petrone, D.; Nemeth, J.F.; Barnathan, E.; Bjorck, L.; Jordan, R.E. Proteolysis of purified IgGs by human and bacterial enzymes in vitro and the detection of specific proteolytic fragments of endogenous IgG in rheumatoid synovial fluid. *Mol. Immunol.* **2008**, *45*, 1837–1846. [CrossRef]

81. Kinder, M.; Greenplate, A.R.; Grugan, K.D.; Soring, K.L.; Heeringa, K.A.; McCarthy, S.G.; Bannish, G.; Perpetua, M.; Lynch, F.; Jordan, R.E.; et al. Engineered protease-resistant antibodies with selectable cell-killing functions. *J. Biol. Chem.* **2013**, *288*, 30843–30854. [CrossRef] [PubMed]

82. Brezski, R.J.; Oberholtzer, A.; Strake, B.; Jordan, R.E. The in vitro resistance of IgG2 to proteolytic attack concurs with a comparative paucity of autoantibodies against peptide analogs of the IgG2 hinge. *mAbs* **2011**, *3*, 558–567. [CrossRef] [PubMed]

83. Senior, B.W.; Woof, J.M. The influences of hinge length and composition on the susceptibility of human IgA to cleavage by diverse bacterial IgA1 proteases. *J. Immunol.* **2005**, *174*, 7792–7799. [CrossRef] [PubMed]

84. Senior, B.W.; Woof, J.M. Effect of mutations in the human immunoglobulin A1 (IgA1) hinge on its susceptibility to cleavage by diverse bacterial IgA1 proteases. *Infect. Immun.* **2005**, *73*, 1515–1522. [CrossRef]

85. Almagro, J.C.; Pedraza-Escalona, M.; Arrieta, H.I.; Pérez-Tapia, S.M. Phage Display Libraries for Antibody Therapeutic Discovery and Development. *Antibodies* **2019**, *8*, 44. [CrossRef]

86. Liu, H.; Saxena, A.; Sidhu, S.S.; Wu, D. Fc Engineering for Developing Therapeutic Bispecific Antibodies and Novel Scaffolds. *Front. Immunol.* **2017**, *8*, 38. [CrossRef]

87. Fonseca, M.H.G.; Furtado, G.P.; Bezerra, M.R.L.; Pontes, L.Q.; Fernandes, C.F.C. Boosting half-life and effector functions of therapeutic antibodies by Fc-engineering: An interaction-function review. *Int. J. Biol. Macromol.* **2018**, *119*, 306–311. [CrossRef]

88. Wang, X.; Mathieu, M.; Brezski, R.J. IgG Fc engineering to modulate antibody effector functions. *Protein Cell* **2018**, *9*, 63–73. [CrossRef]

89. Kruse, T.; Schmidt, A.; Kampmann, M.; Strube, J. Integrated Clarification and Purification of Monoclonal Antibodies by Membrane Based Separation of Aqueous Two-Phase Systems. *Antibodies* **2019**, *8*, 40. [CrossRef]

90. Lensink, M.F.; Wodak, S.J. Docking, scoring, and affinity prediction in CAPRI. *Proteins* **2013**, *81*, 2082–2095. [CrossRef]

91. Klee, G.G. Human anti-mouse antibodies. *Arch. Pathol. Lab. Med.* **2000**, *124*, 921–923. [PubMed]

92. Morrison, S.L.; Johnson, M.J.; Herzenberg, L.A.; Oi, V.T. Chimeric human antibody molecules: Mouse antigen-binding domains with human constant region domains. *Proc. Natl. Acad. Sci. USA* **1984**, *81*, 6851–6855. [CrossRef] [PubMed]

93. Hwang, W.Y.; Foote, J. Immunogenicity of engineered antibodies. *Methods* **2005**, *36*, 3–10. [CrossRef] [PubMed]

94. Jones, P.T.; Dear, P.H.; Foote, J.; Neuberger, M.S.; Winter, G. Replacing the complementarity-determining regions in a human antibody with those from a mouse. *Nature* **1986**, *321*, 522–525. [CrossRef]

95. Verhoeyen, M.; Milstein, C.; Winter, G. Reshaping human antibodies: Grafting an antilysozyme activity. *Science* **1988**, *239*, 1534–1536. [CrossRef]

96. Roguska, M.A.; Pedersen, J.T.; Keddy, C.A.; Henry, A.H.; Searle, S.J.; Lambert, J.M.; Goldmacher, V.S.; Blattler, W.A.; Rees, A.R.; Guild, B.C. Humanization of murine monoclonal antibodies through variable domain resurfacing. *Proc. Natl. Acad. Sci. USA* **1994**, *91*, 969–973. [CrossRef]

97. Tan, P.; Mitchell, D.A.; Buss, T.N.; Holmes, M.A.; Anasetti, C.; Foote, J. "Superhumanized" antibodies: Reduction of immunogenic potential by complementarity-determining region grafting with human germline sequences: Application to an anti-CD28. *J. Immunol.* **2002**, *169*, 1119–1125. [CrossRef]

98. Lazar, G.A.; Desjarlais, J.R.; Jacinto, J.; Karki, S.; Hammond, P.W. A molecular immunology approach to antibody humanization and functional optimization. *Mol. Immunol.* **2007**, *44*, 1986–1998. [CrossRef]

99. Kabat, E.A.; Wu, T.T.; Bilofsky, H. Attempts to locate residues in complementarity-determining regions of antibody combining sites that make contact with antigen. *Proc. Natl. Acad. Sci. USA* **1976**, *73*, 617–619. [CrossRef]

100. Martin, A.C. Protein Sequence and Structure Analysis of Antibody Variable Domains. In *Antibody Engineering*; Kontermann, R.E., Dübel, S., Eds.; Springer: Berlin, Germany, 2014; pp. 33–51.

101. Adolf-Bryfogle, J.; Xu, Q.; North, B.; Lehmann, A.; Dunbrack, R.L., Jr. PyIgClassify: A database of antibody CDR structural classifications. *Nucleic Acids Res.* **2015**, *43*, D432–D438. [CrossRef]

102. Honegger, A.; Pluckthun, A. Yet another numbering scheme for immunoglobulin variable domains: An automatic modeling and analysis tool. *J. Mol. Biol.* **2001**, *309*, 657–670. [CrossRef] [PubMed]

103. Lefranc, M.P. IMGT, the international ImMunoGeneTics information system: A standardized approach for immunogenetics and immunoinformatics. *Immunome Res* **2005**, *1*, 3. [CrossRef] [PubMed]

104. Lefranc, M.P.; Giudicelli, V.; Kaas, Q.; Duprat, E.; Jabado-Michaloud, J.; Scaviner, D.; Ginestoux, C.; Clement, O.; Chaume, D.; Lefranc, G. IMGT, the international ImMunoGeneTics information system. *Nucleic Acids Res.* **2005**, *33*, D593–D597. [CrossRef] [PubMed]

105. Lefranc, M.P.; Ehrenmann, F.; Kossida, S.; Giudicelli, V.; Duroux, P. Use of IMGT((R)) Databases and Tools for Antibody Engineering and Humanization. *Methods Mol. Biol.* **2018**, *1827*, 35–69.

106. Kunik, V.; Ashkenazi, S.; Ofran, Y. Paratome: An online tool for systematic identification of antigen-binding regions in antibodies based on sequence or structure. *Nucleic Acids Res.* **2012**, *40*, W521–W524. [CrossRef] [PubMed]

107. Kunik, V.; Peters, B.; Ofran, Y. Structural consensus among antibodies defines the antigen binding site. *PLoS Comput. Biol.* **2012**, *8*, e1002388. [CrossRef] [PubMed]

108. Riechmann, L.; Clark, M.; Waldmann, H.; Winter, G. Reshaping human antibodies for therapy. *Nature* **1988**, *332*, 323–327. [CrossRef]

109. Queen, C.; Schneider, W.P.; Selick, H.E.; Payne, P.W.; Landolfi, N.F.; Duncan, J.F.; Avdalovic, N.M.; Levitt, M.; Junghans, R.P.; Waldmann, T.A. A humanized antibody that binds to the interleukin 2 receptor. *Proc. Natl. Acad. Sci. USA* **1989**, *86*, 10029–10033. [CrossRef]

110. Graziano, R.F.; Tempest, P.R.; White, P.; Keler, T.; Deo, Y.; Ghebremariam, H.; Coleman, K.; Pfefferkorn, L.C.; Fanger, M.W.; Guyre, P.M. Construction and characterization of a humanized anti-gamma-Ig receptor type I (Fc gamma RI) monoclonal antibody. *J. Immunol.* **1995**, *155*, 4996–5002.

111. Carter, P.; Presta, L.; Gorman, C.M.; Ridgway, J.B.; Henner, D.; Wong, W.L.; Rowland, A.M.; Kotts, C.; Carver, M.E.; Shepard, H.M. Humanization of an anti-p185HER2 antibody for human cancer therapy. *Proc. Natl. Acad. Sci. USA* **1992**, *89*, 4285–4289. [CrossRef]

112. Presta, L.G.; Lahr, S.J.; Shields, R.L.; Porter, J.P.; Gorman, C.M.; Fendly, B.M.; Jardieu, P.M. Humanization of an antibody directed against IgE. *J. Immunol.* **1993**, *151*, 2623–2632. [PubMed]

113. Neuberger, M.S.; Milstein, C. Somatic hypermutation. *Curr. Opin. Immunol.* **1995**, *7*, 248–254. [CrossRef]

114. Gorman, S.D.; Clark, M.R. Humanisation of monoclonal antibodies for therapy. *Semin. Immunol.* **1990**, *2*, 457–466. [PubMed]

115. Rother, R.P.; Wu, D. Hybrid Antibodies. U.S. Patent 8,282,924, 9 October 2012.

116. Lebozec, K.; Jandrot-Perrus, M.; Avenard, G.; Favre-Bulle, O.; Billiald, P. Design, development and characterization of ACT017, a humanized Fab that blocks platelet's glycoprotein VI function without causing bleeding risks. *mAbs* **2017**, *9*, 945–958. [CrossRef]

117. Schlapschy, M.; Fogarasi, M.; Gruber, H.; Gresch, O.; Schafer, C.; Aguib, Y.; Skerra, A. Functional humanization of an anti-CD16 Fab fragment: Obstacles of switching from murine {lambda} to human {lambda} or {kappa} light chains. *Protein Eng. Des. Sel.* **2009**, *22*, 175–188. [CrossRef]

118. Nakanishi, T.; Tsumoto, K.; Yokota, A.; Kondo, H.; Kumagai, I. Critical contribution of VH-VL interaction to reshaping of an antibody: The case of humanization of anti-lysozyme antibody, HyHEL-10. *Protein Sci.* **2008**, *17*, 261–270. [CrossRef]

119. Brezinschek, H.P.; Foster, S.J.; Dorner, T.; Brezinschek, R.I.; Lipsky, P.E. Pairing of variable heavy and variable kappa chains in individual naive and memory B cells. *J. Immunol.* **1998**, *160*, 4762–4767.

120. De Wildt, R.M.; Hoet, R.M.; van Venrooij, W.J.; Tomlinson, I.M.; Winter, G. Analysis of heavy and light chain pairings indicates that receptor editing shapes the human antibody repertoire. *J. Mol. Biol.* **1999**, *285*, 895–901. [CrossRef]

121. De Wildt, R.M.; van Venrooij, W.J.; Winter, G.; Hoet, R.M.; Tomlinson, I.M. Somatic insertions and deletions shape the human antibody repertoire. *J. Mol. Biol.* **1999**, *294*, 701–710. [CrossRef]

122. Edwards, B.M.; Barash, S.C.; Main, S.H.; Choi, G.H.; Minter, R.; Ullrich, S.; Williams, E.; Fou, L.D.; Wilton, J.; Albert, V.R.; et al. The remarkable flexibility of the human antibody repertoire; isolation of over one thousand different antibodies to a single protein, BLyS. *J. Mol. Biol.* **2003**, *334*, 103–118. [CrossRef]

123. Lloyd, C.; Lowe, D.; Edwards, B.; Welsh, F.; Dilks, T.; Hardman, C.; Vaughan, T. Modelling the human immune response: Performance of a 1011 human antibody repertoire against a broad panel of therapeutically relevant antigens. *Protein Eng. Des. Sel.* **2009**, *22*, 159–168. [CrossRef] [PubMed]

124. Jayaram, N.; Bhowmick, P.; Martin, A.C. Germline VH/VL pairing in antibodies. *Protein Eng. Des. Sel.* **2012**, *25*, 523–529. [CrossRef] [PubMed]

125. Narayanan, A.; Sellers, B.D.; Jacobson, M.P. Energy-based analysis and prediction of the orientation between light- and heavy-chain antibody variable domains. *J. Mol. Biol.* **2009**, *388*, 941–953. [CrossRef] [PubMed]

126. Sircar, A.; Kim, E.T.; Gray, J.J. RosettaAntibody: Antibody variable region homology modeling server. *Nucleic Acids Res.* **2009**, *37*, W474–W479. [CrossRef]

127. Abhinandan, K.R.; Martin, A.C. Analysis and prediction of VH/VL packing in antibodies. *Protein Eng. Des. Sel.* **2010**, *23*, 689–697. [CrossRef]

128. Dunbar, J.; Fuchs, A.; Shi, J.; Deane, C.M. ABangle: Characterising the VH-VL orientation in antibodies. *Protein Eng. Des. Sel.* **2013**, *26*, 611–620. [CrossRef]

129. Foote, J.; Winter, G. Antibody framework residues affecting the conformation of the hypervariable loops. *J. Mol. Biol.* **1992**, *224*, 487–499. [CrossRef]

130. Tramontano, A.; Chothia, C.; Lesk, A.M. Framework residue 71 is a major determinant of the position and conformation of the second hypervariable region in the VH domains of immunoglobulins. *J. Mol. Biol.* **1990**, *215*, 175–182. [CrossRef]

131. Papanastasiou, D.; Mamalaki, A.; Eliopoulos, E.; Poulas, K.; Liolitsas, C.; Tzartos, S.J. Construction and characterization of a humanized single chain Fv antibody fragment against the main immunogenic region of the acetylcholine receptor. *J. Neuroimmunol.* **1999**, *94*, 182–195. [CrossRef]

132. Xiang, J.; Sha, Y.; Jia, Z.; Prasad, L.; Delbaere, L.T. Framework residues 71 and 93 of the chimeric B72.3 antibody are major determinants of the conformation of heavy-chain hypervariable loops. *J. Mol. Biol.* **1995**, *253*, 385–390. [CrossRef]

133. Tempest, P.R.; White, P.; Buttle, M.; Carr, F.J.; Harris, W.J. Identification of framework residues required to restore antigen binding during reshaping of a monoclonal antibody against the glycoprotein gB of human cytomegalovirus. *Int. J. Biol. Macromol.* **1995**, *17*, 37–42. [CrossRef]

134. Teplyakov, A.; Obmolova, G.; Malia, T.J.; Raghunathan, G.; Martinez, C.; Fransson, J.; Edwards, W.; Connor, J.; Husovsky, M.; Beck, H.; et al. Structural insights into humanization of anti-tissue factor antibody 10H10. *mAbs* **2018**, *10*, 269–277. [CrossRef] [PubMed]

135. Shembekar, N.; Mallajosyula, V.V.; Chaudhary, P.; Upadhyay, V.; Varadarajan, R.; Gupta, S.K. Humanized antibody neutralizing 2009 pandemic H1N1 virus. *Biotechnol. J.* **2014**, *9*, 1594–1603. [CrossRef] [PubMed]

136. Jia, X.; Wang, W.; Xu, Z.; Wang, S.; Wang, T.; Wang, M.; Wu, M. A humanized anti-DLL4 antibody promotes dysfunctional angiogenesis and inhibits breast tumor growth. *Sci. Rep.* **2016**, *6*, 27985. [CrossRef] [PubMed]

137. Villani, M.E.; Morea, V.; Consalvi, V.; Chiaraluce, R.; Desiderio, A.; Benvenuto, E.; Donini, M. Humanization of a highly stable single-chain antibody by structure-based antigen-binding site grafting. *Mol. Immunol.* **2008**, *45*, 2474–2485. [CrossRef]

138. Schrade, A.; Bujotzek, A.; Spick, C.; Wagner, M.; Goerl, J.; Wezler, X.; Georges, G.; Kontermann, R.E.; Brinkmann, U. Back-to-Germline (B2G) Procedure for Antibody Devolution. *Antibodies* **2019**, *8*, 45. [CrossRef]

139. Bernett, M.J.; Karki, S.; Moore, G.L.; Leung, I.W.; Chen, H.; Pong, E.; Nguyen, D.H.; Jacinto, J.; Zalevsky, J.; Muchhal, U.S.; et al. Engineering fully human monoclonal antibodies from murine variable regions. *J. Mol. Biol.* **2010**, *396*, 1474–1490. [CrossRef]

140. Harding, F.A.; Stickler, M.M.; Razo, J.; DuBridge, R.B. The immunogenicity of humanized and fully human antibodies: Residual immunogenicity resides in the CDR regions. *mAbs* **2010**, *2*, 256–265. [CrossRef]

141. Hamilton, A.; King, S.; Liu, H.; Moy, P.; Bander, N.; Carr, F. A novel humanized antibody against prostate-specific membrane antigen also reacts with tumor vascular endothelium [abstract]. *Proc. Am. Assoc. Cancer Res.* **1998**, *39*, 440.

142. Maynard, J.A.; Maassen, C.B.; Leppla, S.H.; Brasky, K.; Patterson, J.L.; Iverson, B.L.; Georgiou, G. Protection against anthrax toxin by recombinant antibody fragments correlates with antigen affinity. *Nat. Biotechnol.* **2002**, *20*, 597–601. [CrossRef]

143. Wirth, M.; Heidenreich, A.; Gschwend, J.E.; Gil, T.; Zastrow, S.; Laniado, M.; Gerloff, J.; Zuhlsdorf, M.; Mordenti, G.; Uhl, W.; et al. A multicenter phase 1 study of EMD 525797 (DI17E6), a novel humanized monoclonal antibody targeting alphav integrins, in progressive castration-resistant prostate cancer with bone metastases after chemotherapy. *Eur. Urol.* **2014**, *65*, 897–904. [CrossRef] [PubMed]

144. Parker, A.S.; Choi, Y.; Griswold, K.E.; Bailey-Kellogg, C. Structure-guided deimmunization of therapeutic proteins. *J. Comput. Biol.* **2013**, *20*, 152–165. [CrossRef] [PubMed]

145. Jones, T.D.; Crompton, L.J.; Carr, F.J.; Baker, M.P. Deimmunization of monoclonal antibodies. *Methods Mol. Biol.* **2009**, *525*, 405–423. [PubMed]

146. Padlan, E.A. A possible procedure for reducing the immunogenicity of antibody variable domains while preserving their ligand-binding properties. *Mol. Immunol.* **1991**, *28*, 489–498. [CrossRef]

147. Roguska, M.A.; Pedersen, J.T.; Henry, A.H.; Searle, S.M.; Roja, C.M.; Avery, B.; Hoffee, M.; Cook, S.; Lambert, J.M.; Blattler, W.A.; et al. A comparison of two murine monoclonal antibodies humanized by CDR-grafting and variable domain resurfacing. *Protein Eng.* **1996**, *9*, 895–904. [CrossRef] [PubMed]

148. Kabat, E.A.; Wu, T.T.; Reid-Miller, M.; Gottesman, K. *Sequences of Proteins of Immunological Interest*; DHHS: Washington, DC, USA, 1991.

149. Fan, C.Y.; Huang, S.Y.; Chou, M.Y.; Lyu, P.C. De novo protein sequencing, humanization and in vitro effects of an antihuman CD34 mouse monoclonal antibody. *Biochem. Biophys. Rep.* **2017**, *9*, 51–60. [CrossRef] [PubMed]

150. Skrlj, N.; Vranac, T.; Popovic, M.; Serbec, V.C.; Dolinar, M. Specific binding of the pathogenic prion isoform: Development and characterization of a humanized single-chain variable antibody fragment. *PLoS ONE* **2011**, *6*, e15783. [CrossRef]

151. Bugelski, P.J.; Achuthanandam, R.; Capocasale, R.J.; Treacy, G.; Bouman-Thio, E. Monoclonal antibody-induced cytokine-release syndrome. *Expert Rev. Clin. Immunol.* **2009**, *5*, 499–521. [CrossRef]

152. Hwang, W.Y.; Almagro, J.C.; Buss, T.N.; Tan, P.; Foote, J. Use of human germline genes in a CDR homology-based approach to antibody humanization. *Methods* **2005**, *36*, 35–42. [CrossRef]

153. Hu, W.G.; Chau, D.; Wu, J.; Jager, S.; Nagata, L.P. Humanization and mammalian expression of a murine monoclonal antibody against Venezuelan equine encephalitis virus. *Vaccine* **2007**, *25*, 3210–3214. [CrossRef]

154. Pelat, T.; Bedouelle, H.; Rees, A.R.; Crennell, S.J.; Lefranc, M.P.; Thullier, P. Germline humanization of a non-human primate antibody that neutralizes the anthrax toxin, by in vitro and in silico engineering. *J. Mol. Biol.* **2008**, *384*, 1400–1407. [CrossRef] [PubMed]

155. Choi, Y.; Hua, C.; Sentman, C.L.; Ackerman, M.E.; Bailey-Kellogg, C. Antibody humanization by structure-based computational protein design. *mAbs* **2015**, *7*, 1045–1057. [CrossRef] [PubMed]

156. Abhinandan, K.R.; Martin, A.C. Analyzing the "degree of humanness" of antibody sequences. *J. Mol. Biol.* **2007**, *369*, 852–862. [CrossRef] [PubMed]

157. Gao, S.H.; Huang, K.; Tu, H.; Adler, A.S. Monoclonal antibody humanness score and its applications. *BMC Biotechnol.* **2013**, *13*, 55. [CrossRef]

158. Seeliger, D. Development of scoring functions for antibody sequence assessment and optimization. *PLoS ONE* **2013**, *8*, e1002388. [CrossRef]

159. Clavero-Alvarez, A.; di Mambro, T.; Perez-Gaviro, S.; Magnani, M.; Bruscolini, P. Humanization of Antibodies using a Statistical Inference Approach. *Sci. Rep.* **2018**, *8*, 14820. [CrossRef]

160. Worn, A.; der Maur, A.A.; Escher, D.; Honegger, A.; Barberis, A.; Pluckthun, A. Correlation between in vitro stability and in vivo performance of anti-GCN4 intrabodies as cytoplasmic inhibitors. *J. Biol. Chem.* **2000**, *275*, 2795–2803. [CrossRef]

161. Lehmann, A.; Wixted, J.H.; Shapovalov, M.V.; Roder, H.; Dunbrack, R.L., Jr.; Robinson, M.K. Stability engineering of anti-EGFR scFv antibodies by rational design of a lambda-to-kappa swap of the VL framework using a structure-guided approach. *mAbs* **2015**, *7*, 1058–1071. [CrossRef]

162. Tiller, T.; Schuster, I.; Deppe, D.; Siegers, K.; Strohner, R.; Herrmann, T.; Berenguer, M.; Poujol, D.; Stehle, J.; Stark, Y.; et al. A fully synthetic human Fab antibody library based on fixed VH/VL framework pairings with favorable biophysical properties. *mAbs* **2013**, *5*, 445–470. [CrossRef]

163. Foote, J.; Eisen, H.N. Kinetic and affinity limits on antibodies produced during immune responses. *Proc. Natl. Acad. Sci. USA* **1995**, *92*, 1254–1256. [CrossRef]

164. Batista, F.D.; Neuberger, M.S. Affinity dependence of the B cell response to antigen: A threshold, a ceiling, and the importance of off-rate. *Immunity* **1998**, *8*, 751–759. [CrossRef]

165. Yanaka, S.; Moriwaki, Y.; Tsumoto, K.; Sugase, K. Elucidation of potential sites for antibody engineering by fluctuation editing. *Sci. Rep.* **2019**, *7*, 9597. [CrossRef] [PubMed]

166. Clark, L.A.; Boriack-Sjodin, P.A.; Eldredge, J.; Fitch, C.; Friedman, B.; Hanf, K.J.; Jarpe, M.; Liparoto, S.F.; Li, Y.; Lugovskoy, A.; et al. Affinity enhancement of an in vivo matured therapeutic antibody using structure-based computational design. *Protein Sci.* **2006**, *15*, 949–960. [CrossRef]

167. Lippow, S.M.; Tidor, B. Progress in computational protein design. *Curr. Opin. Biotechnol.* **2007**, *18*, 305–311. [CrossRef]

168. Lippow, S.M.; Wittrup, K.D.; Tidor, B. Computational design of antibody-affinity improvement beyond in vivo maturation. *Nat. Biotechnol.* **2007**, *25*, 1171–1176. [CrossRef]

169. Li, B.; Zhao, L.; Wang, C.; Guo, H.; Wu, L.; Zhang, X.; Qian, W.; Wang, H.; Guo, Y. The protein-protein interface evolution acts in a similar way to antibody affinity maturation. *J. Biol. Chem.* **2010**, *285*, 3865–3871. [CrossRef]

170. Berek, C.; Milstein, C. Mutation drift and repertoire shift in the maturation of the immune response. *Immunol. Rev.* **1987**, *96*, 23–41. [CrossRef]

171. Rogozin, I.; Kondrashov, F.; Glazko, G. Use of mutation spectra analysis software. *Hum. Mutat.* **2001**, *17*, 83–102. [CrossRef]

172. Rogozin, I.B.; Pavlov, Y.I.; Bebenek, K.; Matsuda, T.; Kunkel, T.A. Somatic mutation hotspots correlate with DNA polymerase eta error spectrum. *Nat. Immunol.* **2001**, *2*, 530–536. [CrossRef]

173. Kiyoshi, M.; Caaveiro, J.M.; Miura, E.; Nagatoishi, S.; Nakakido, M.; Soga, S.; Shirai, H.; Kawabata, S.; Tsumoto, K. Affinity improvement of a therapeutic antibody by structure-based computational design: Generation of electrostatic interactions in the transition state stabilizes the antibody-antigen complex. *PLoS ONE* **2014**, *9*, e87099. [CrossRef]

174. Babor, M.; Mandell, D.J.; Kortemme, T. Assessment of flexible backbone protein design methods for sequence library prediction in the therapeutic antibody Herceptin-HER2 interface. *Protein Sci.* **2011**, *20*, 1082–1089. [CrossRef] [PubMed]

175. Lo Conte, L.; Chothia, C.; Janin, J. The atomic structure of protein-protein recognition sites. *J. Mol. Biol.* **1999**, *285*, 2177–2198. [CrossRef] [PubMed]

176. Inoue, H.; Suganami, A.; Ishida, I.; Tamura, Y.; Maeda, Y. Affinity maturation of a CDR3-grafted VHH using in silico analysis and surface plasmon resonance. *J. Biochem.* **2013**, *154*, 325–332. [CrossRef] [PubMed]

177. McCoy, A.J.; Epa, V.C.; Colman, P.M. Electrostatic complementarity at protein/protein interfaces. *J. Mol. Biol.* **1997**, *268*, 570–584. [CrossRef]

178. Pedotti, M.; Simonelli, L.; Livoti, E.; Varani, L. Computational docking of antibody-antigen complexes, opportunities and pitfalls illustrated by influenza hemagglutinin. *Int. J. Mol. Sci.* **2011**, *12*, 226–251. [CrossRef] [PubMed]

179. Wang, Z.; Li, Y.; Liang, W.; Zheng, J.; Li, S.; Hu, C.; Chen, A. A Highly Sensitive Detection System based on Proximity-dependent Hybridization with Computer-aided Affinity Maturation of a scFv Antibody. *Sci. Rep.* **2018**, *8*, 3837. [CrossRef] [PubMed]

180. Barderas, R.; Desmet, J.; Timmerman, P.; Meloen, R.; Casal, J.I. Affinity maturation of antibodies assisted by in silico modeling. *Proc. Natl. Acad. Sci. USA* **2008**, *105*, 9029–9034. [CrossRef] [PubMed]

181. Barderas, R.; Shochat, S.; Timmerman, P.; Hollestelle, M.J.; Martinez-Torrecuadrada, J.L.; Hoppener, J.W.; Altschuh, D.; Meloen, R.; Casal, J.I. Designing antibodies for the inhibition of gastrin activity in tumoral cell lines. *Int. J. Cancer* **2008**, *122*, 2351–2359. [CrossRef]

182. Yang, W.P.; Green, K.; Pinz-Sweeney, S.; Briones, A.T.; Burton, D.R.; Barbas, C.F., III. CDR walking mutagenesis for the affinity maturation of a potent human anti-HIV-1 antibody into the picomolar range. *J. Mol. Biol.* **1995**, *254*, 392–403. [CrossRef]

183. Lamdan, H.; Gavilondo, J.V.; Munoz, Y.; Pupo, A.; Huerta, V.; Musacchio, A.; Perez, L.; Ayala, M.; Rojas, G.; Balint, R.F.; et al. Affinity maturation and fine functional mapping of an antibody fragment against a novel neutralizing epitope on human vascular endothelial growth factor. *Mol. Biosyst.* **2013**, *9*, 2097–2106. [CrossRef]

184. Colley, C.S.; Popovic, B.; Sridharan, S.; Debreczeni, J.E.; Hargreaves, D.; Fung, M.; An, L.L.; Edwards, B.; Arnold, J.; England, E.; et al. Structure and characterization of a high affinity C5a monoclonal antibody that blocks binding to C5aR1 and C5aR2 receptors. *mAbs* **2018**, *10*, 104–117. [CrossRef] [PubMed]

185. Laffly, E.; Pelat, T.; Cedrone, F.; Blesa, S.; Bedouelle, H.; Thullier, P. Improvement of an antibody neutralizing the anthrax toxin by simultaneous mutagenesis of its six hypervariable loops. *J. Mol. Biol.* **2008**, *378*, 1094–1103. [CrossRef] [PubMed]

186. Lord, D.M.; Bird, J.J.; Honey, D.M.; Best, A.; Park, A.; Wei, R.R.; Qiu, H. Structure-based engineering to restore high affinity binding of an isoform-selective anti-TGFbeta1 antibody. *mAbs* **2018**, *10*, 444–452. [CrossRef] [PubMed]

187. Fuh, G.; Wu, P.; Liang, W.C.; Ultsch, M.; Lee, C.V.; Moffat, B.; Wiesmann, C. Structure-function studies of two synthetic anti-vascular endothelial growth factor Fabs and comparison with the Avastin Fab. *J. Biol. Chem.* **2006**, *281*, 6625–6631. [CrossRef] [PubMed]

188. Lee, C.V.; Hymowitz, S.G.; Wallweber, H.J.; Gordon, N.C.; Billeci, K.L.; Tsai, S.P.; Compaan, D.M.; Yin, J.; Gong, Q.; Kelley, R.F.; et al. Synthetic anti-BR3 antibodies that mimic BAFF binding and target both human and murine B cells. *Blood* **2006**, *108*, 3103–3111. [CrossRef]

189. Sanders, B.M.; Martin, L.S.; Nakagawa, P.A.; Hunter, D.A.; Miller, S.; Ullrich, S.J. Specific cross-reactivity of antibodies raised against two major stress proteins, stress 70 and chaperonin 60, in diverse species. *Environ. Toxicol. Chem.* **1994**, *13*, 1241–1249. [CrossRef]

190. Hamdani, N.; van der Velden, J. Lack of specificity of antibodies directed against human beta-adrenergic receptors. *Naunyn Schmiedebergs Arch. Pharmacol.* **2009**, *379*, 403–407. [CrossRef]

191. Liang, W.C.; Wu, X.; Peale, F.V.; Lee, C.V.; Meng, Y.G.; Gutierrez, J.; Fu, L.; Malik, A.K.; Gerber, H.P.; Ferrara, N.; et al. Cross-species vascular endothelial growth factor (VEGF)-blocking antibodies completely

inhibit the growth of human tumor xenografts and measure the contribution of stromal VEGF. *J. Biol. Chem.* **2006**, *281*, 951–961. [CrossRef]

192. Garcia-Rodriguez, C.; Levy, R.; Arndt, J.W.; Forsyth, C.M.; Razai, A.; Lou, J.; Geren, I.; Stevens, R.C.; Marks, J.D. Molecular evolution of antibody cross-reactivity for two subtypes of type A botulinum neurotoxin. *Nat. Biotechnol.* **2007**, *25*, 107–116. [CrossRef]

193. Joachimiak, L.A.; Kortemme, T.; Stoddard, B.L.; Baker, D. Computational design of a new hydrogen bond network and at least a 300-fold specificity switch at a protein-protein interface. *J. Mol. Biol.* **2006**, *361*, 195–208. [CrossRef]

194. Farady, C.J.; Sellers, B.D.; Jacobson, M.P.; Craik, C.S. Improving the species cross-reactivity of an antibody using computational design. *Bioorg. Med. Chem. Lett.* **2009**, *19*, 3744–3747. [CrossRef] [PubMed]

195. Grossman, I.; Ilani, T.; Fleishman, S.J.; Fass, D. Overcoming a species-specificity barrier in development of an inhibitory antibody targeting a modulator of tumor stroma. *Protein Eng. Des. Sel.* **2016**, *29*, 135–147. [CrossRef] [PubMed]

196. Nelson, B.; Adams, J.; Kuglstatter, A.; Li, Z.; Harris, S.F.; Liu, Y.; Bohini, S.; Ma, H.; Klumpp, K.; Gao, J.; et al. Structure-Guided Combinatorial Engineering Facilitates Affinity and Specificity Optimization of Anti-CD81 Antibodies. *J. Mol. Biol.* **2018**, *430*, 2139–2152. [CrossRef] [PubMed]

197. Dubreuil, O.; Bossus, M.; Graille, M.; Bilous, M.; Savatier, A.; Jolivet, M.; Menez, A.; Stura, E.; Ducancel, F. Fine tuning of the specificity of an anti-progesterone antibody by first and second sphere residue engineering. *J. Biol. Chem.* **2005**, *280*, 24880–24887. [CrossRef]

198. Koenig, P.; Sanowar, S.; Lee, C.V.; Fuh, G. Tuning the specificity of a Two-in-One Fab against three angiogenic antigens by fully utilizing the information of deep mutational scanning. *mAbs* **2017**, *9*, 959–967. [CrossRef]

199. James, L.C.; Roversi, P.; Tawfik, D.S. Antibody multispecificity mediated by conformational diversity. *Science* **2003**, *299*, 1362–1367. [CrossRef]

200. James, L.C.; Tawfik, D.S. The specificity of cross-reactivity: Promiscuous antibody binding involves specific hydrogen bonds rather than nonspecific hydrophobic stickiness. *Protein Sci.* **2003**, *12*, 2183–2193. [CrossRef]

201. Bostrom, J.; Lee, C.V.; Haber, L.; Fuh, G. Improving antibody binding affinity and specificity for therapeutic development. *Methods Mol. Biol.* **2009**, *525*, 353–376.

202. Bostrom, J.; Yu, S.F.; Kan, D.; Appleton, B.A.; Lee, C.V.; Billeci, K.; Man, W.; Peale, F.; Ross, S.; Wiesmann, C.; et al. Variants of the antibody herceptin that interact with HER2 and VEGF at the antigen binding site. *Science* **2009**, *323*, 1610–1614. [CrossRef]

203. Schaefer, G.; Haber, L.; Crocker, L.M.; Shia, S.; Shao, L.; Dowbenko, D.; Totpal, K.; Wong, A.; Lee, C.V.; Stawicki, S.; et al. A two-in-one antibody against HER3 and EGFR has superior inhibitory activity compared with monospecific antibodies. *Cancer Cell* **2011**, *20*, 472–486. [CrossRef]

204. Jenkins, N.; Murphy, L.; Tyther, R. Post-translational modifications of recombinant proteins: Significance for biopharmaceuticals. *Mol. Biotechnol.* **2008**, *39*, 113–118. [CrossRef] [PubMed]

205. Yang, R.; Jain, T.; Lynaugh, H.; Nobrega, R.P.; Lu, X.; Boland, T.; Burnina, I.; Sun, T.; Caffry, I.; Brown, M.; et al. Rapid assessment of oxidation via middle-down LCMS correlates with methionine side-chain solvent-accessible surface area for 121 clinical stage monoclonal antibodies. *mAbs* **2017**, *9*, 646–653. [CrossRef] [PubMed]

206. Lu, X.; Nobrega, R.P.; Lynaugh, H.; Jain, T.; Barlow, K.; Boland, T.; Sivasubramanian, A.; Vasquez, M.; Xu, Y. Deamidation and isomerization liability analysis of 131 clinical-stage antibodies. *mAbs* **2019**, *11*, 45–57. [CrossRef] [PubMed]

207. Vlasak, J.; Ionescu, R. Heterogeneity of monoclonal antibodies revealed by charge-sensitive methods. *Curr. Pharm. Biotechnol.* **2008**, *9*, 468–481. [CrossRef] [PubMed]

208. Boswell, C.A.; Tesar, D.B.; Mukhyala, K.; Theil, F.P.; Fielder, P.J.; Khawli, L.A. Effects of charge on antibody tissue distribution and pharmacokinetics. *Bioconjug. Chem.* **2010**, *21*, 2153–2163. [CrossRef] [PubMed]

209. Bumbaca, D.; Boswell, C.A.; Fielder, P.J.; Khawli, L.A. Physiochemical and biochemical factors influencing the pharmacokinetics of antibody therapeutics. *AAPS J.* **2012**, *14*, 554–558. [CrossRef]

210. Yang, N.; Tang, Q.; Hu, P.; Lewis, M.J. Use of In Vitro Systems to Model In Vivo Degradation of Therapeutic Monoclonal Antibodies. *Anal. Chem.* **2018**, *90*, 7896–7902. [CrossRef]

211. Jimenez del Val, I.; Nagy, J.M.; Kontoravdi, C. A dynamic mathematical model for monoclonal antibody N-linked glycosylation and nucleotide sugar donor transport within a maturing Golgi apparatus. *Biotechnol. Prog.* **2011**, *27*, 1730–1743. [CrossRef]

212. Loebrich, S.; Clark, E.; Ladd, K.; Takahashi, S.; Brousseau, A.; Kitchener, S.; Herbst, R.; Ryll, T. Comprehensive manipulation of glycosylation profiles across development scales. *mAbs* **2019**, *11*, 335–349. [CrossRef]

213. Wang, Q.; Chung, C.Y.; Chough, S.; Betenbaugh, M.J. Antibody glycoengineering strategies in mammalian cells. *Biotechnol. Bioeng.* **2018**, *115*, 1378–1393. [CrossRef]

214. Liu, L.; Stadheim, A.; Hamuro, L.; Pittman, T.; Wang, W.; Zha, D.; Hochman, J.; Prueksaritanont, T. Pharmacokinetics of IgG1 monoclonal antibodies produced in humanized Pichia pastoris with specific glycoforms: A comparative study with CHO produced materials. *Biologicals* **2011**, *39*, 205–210. [CrossRef] [PubMed]

215. Yanaka, S.; Yogo, R.; Inoue, R.; Sugiyama, M.; Itoh, S.G.; Okumura, H.; Miyanoiri, H.; Yagi, H.; Satoh, T.; Yamaguchi, T.; et al. Dynamic Views of the Fc Region of Immunoglobulin G Provided by Experimental and Computational Observations. *Antibodies* **2019**, *8*, 39. [CrossRef]

216. Liu, Y.D.; van Enk, J.Z.; Flynn, G.C. Human antibody Fc deamidation in vivo. *Biologicals* **2009**, *37*, 313–322. [CrossRef] [PubMed]

217. Yin, S.; Pastuskovas, C.V.; Khawli, L.A.; Stults, J.T. Characterization of therapeutic monoclonal antibodies reveals differences between in vitro and in vivo time-course studies. *Pharm. Res.* **2013**, *30*, 167–178. [CrossRef]

218. Chung, S.; Tian, J.; Tan, Z.; Chen, J.; Lee, J.; Borys, M.; Li, Z.J. Industrial bioprocessing perspectives on managing therapeutic protein charge variant profiles. *Biotechnol. Bioeng.* **2018**, *115*, 1646–1665. [CrossRef] [PubMed]

219. Chung, S.; Tian, J.; Tan, Z.; Chen, J.; Zhang, N.; Huang, Y.; Vandermark, E.; Lee, J.; Borys, M.; Li, Z.J. Modulating cell culture oxidative stress reduces protein glycation and acidic charge variant formation. *mAbs* **2019**, *11*, 205–2016. [CrossRef]

220. Geiger, T.; Clarke, S. Deamidation, isomerization, and racemization at asparaginyl and aspartyl residues in peptides. Succinimide-linked reactions that contribute to protein degradation. *J. Biol. Chem.* **1987**, *262*, 785–794.

221. Stephenson, R.C.; Clarke, S. Succinimide formation from aspartyl and asparaginyl peptides as a model for the spontaneous degradation of proteins. *J. Biol. Chem.* **1989**, *264*, 6164–6170.

222. Aswad, D.W. *Deamidation and Isoaspartate Formation in Peptides and Proteins*; Aswad, D.W., Ed.; CRC Series in Analytical Biotechnology; CRC Press: Boca Raton, FL, USA, 1994.

223. Phillips, J.J.; Buchanan, A.; Andrews, J.; Chodorge, M.; Sridharan, S.; Mitchell, L.; Burmeister, N.; Kippen, A.D.; Vaughan, T.J.; Higazi, D.R.; et al. Rate of Asparagine Deamidation in a Monoclonal Antibody Correlating with Hydrogen Exchange Rate at Adjacent Downstream Residues. *Anal. Chem.* **2017**, *89*, 2361–2368. [CrossRef]

224. Ni, W.; Dai, S.; Karger, B.L.; Zhou, Z.S. Analysis of isoaspartic Acid by selective proteolysis with Asp-N and electron transfer dissociation mass spectrometry. *Anal. Chem.* **2010**, *82*, 7485–7491. [CrossRef]

225. Alam, M.E.; Barnett, G.V.; Slaney, T.R.; Starr, C.G.; Das, T.K.; Tessier, P.M. Deamidation Can Compromise Antibody Colloidal Stability and Enhance Aggregation in a pH-Dependent Manner. *Mol. Pharm.* **2019**, *16*, 1939–1949. [CrossRef]

226. Rehder, D.S.; Chelius, D.; McAuley, A.; Dillon, T.M.; Xiao, G.; Crouse-Zeineddini, J.; Vardanyan, L.; Perico, N.; Mukku, V.; Brems, D.N.; et al. Isomerization of a single aspartyl residue of anti-epidermal growth factor receptor immunoglobulin gamma2 antibody highlights the role avidity plays in antibody activity. *Biochemistry* **2008**, *47*, 2518–2530. [CrossRef] [PubMed]

227. Qiu, H.; Wei, R.; Jaworski, J.; Boudanova, E.; Hughes, H.; VanPatten, S.; Lund, A.; Day, J.; Zhou, Y.; McSherry, T.; et al. Engineering an anti-CD52 antibody for enhanced deamidation stability. *mAbs* **2019**, *11*, 1266–1275. [CrossRef] [PubMed]

228. Liu, H.; Gaza-Bulseco, G.; Chumsae, C. Glutamine deamidation of a recombinant monoclonal antibody. *Rapid Commun. Mass Spectrom.* **2008**, *22*, 4081–4088. [CrossRef]

229. Schechter, Y. Selective oxidation and reduction of methionine residues in peptides and proteins by oxygen exchange between sulfoxide and sulfide. *J. Biol. Chem.* **1986**, *261*, 66–70.

230. Schechter, Y.; Burstein, Y.; Patchornik, A. Proceedings: Selective oxidation of methionine residues in proteins. *Isr. J. Med. Sci.* **1975**, *11*, 1171. [CrossRef]

231. Brot, N.; Weissbach, H. Biochemistry and physiological role of methionine sulfoxide residues in proteins. *Arch. Biochem. Biophys.* **1983**, *223*, 271–281. [CrossRef]

232. Pan, H.; Chen, K.; Chu, L.; Kinderman, F.; Apostol, I.; Huang, G. Methionine oxidation in human IgG2 Fc decreases binding affinities to protein A and FcRn. *Protein Sci.* **2009**, *18*, 424–433. [CrossRef]

233. Cymer, F.; Thomann, M.; Wegele, H.; Avenal, C.; Schlothauer, T.; Gygax, D.; Beck, H. Oxidation of M252 but not M428 in hu-IgG1 is responsible for decreased binding to and activation of hu-FcgammaRIIa (His131). *Biologicals* **2017**, *50*, 125–128. [CrossRef]

234. Bertolotti-Ciarlet, A.; Wang, W.; Lownes, R.; Pristatsky, P.; Fang, Y.; McKelvey, T.; Li, Y.; Li, Y.; Drummond, J.; Prueksaritanont, T.; et al. Impact of methionine oxidation on the binding of human IgG1 to Fc Rn and Fc gamma receptors. *Mol. Immunol.* **2009**, *46*, 1878–1882. [CrossRef]

235. Wang, W.; Vlasak, J.; Li, Y.; Pristatsky, P.; Fang, Y.; Pittman, T.; Roman, J.; Wang, Y.; Prueksaritanont, T.; Ionescu, R. Impact of methionine oxidation in human IgG1 Fc on serum half-life of monoclonal antibodies. *Mol. Immunol.* **2011**, *48*, 860–866. [CrossRef] [PubMed]

236. Stracke, J.; Emrich, T.; Rueger, P.; Schlothauer, T.; Kling, L.; Knaupp, A.; Hertenberger, H.; Wolfert, A.; Spick, C.; Lau, W.; et al. A novel approach to investigate the effect of methionine oxidation on pharmacokinetic properties of therapeutic antibodies. *mAbs* **2014**, *6*, 1229–1242. [CrossRef] [PubMed]

237. Creed, D. The photophysics and photochemistry of the near-uv absorbing amino acids–i. Tryptophan and its simple derivatives. *Photochem. Photobiol.* **1984**, *39*, 537–562. [CrossRef]

238. Li, Y.; Polozova, A.; Gruia, F.; Feng, J. Characterization of the degradation products of a color-changed monoclonal antibody: Tryptophan-derived chromophores. *Anal. Chem.* **2014**, *86*, 6850–6857. [CrossRef]

239. Lam, X.M.; Lai, W.G.; Chan, E.K.; Ling, V.; Hsu, C.C. Site-specific tryptophan oxidation induced by autocatalytic reaction of polysorbate 20 in protein formulation. *Pharm. Res.* **2011**, *28*, 2543–2555. [CrossRef]

240. Barnett, G.V.; Balakrishnan, G.; Chennamsetty, N.; Hoffman, L.; Bongers, J.; Tao, L.; Huang, Y.; Slaney, T.; Das, T.K.; Leone, A.; et al. Probing the Tryptophan Environment in Therapeutic Proteins: Implications for Higher Order Structure on Tryptophan Oxidation. *J. Pharm. Sci.* **2019**, *108*, 1944–1952. [CrossRef]

241. Wei, Z.; Feng, J.; Lin, H.Y.; Mullapudi, S.; Bishop, E.; Tous, G.I.; Casas-Finet, J.; Hakki, F.; Strouse, R.; Schenerman, M.A. Identification of a single tryptophan residue as critical for binding activity in a humanized monoclonal antibody against respiratory syncytial virus. *Anal. Chem.* **2007**, *79*, 2797–2805. [CrossRef]

242. Pavon, J.A.; Xiao, L.; Li, X.; Zhao, J.; Aldredge, D.; Dank, E.; Fridman, A.; Liu, Y.H. Selective Tryptophan Oxidation of Monoclonal Antibodies: Oxidative Stress and Modeling Prediction. *Anal. Chem.* **2019**, *91*, 2192–2200. [CrossRef]

243. Glover, Z.K.; Basa, L.; Moore, B.; Laurence, J.S.; Sreedhara, A. Metal ion interactions with mAbs: Part 1. *mAbs* **2015**, *7*, 901–911. [CrossRef]

244. Zhu, F.; Glover, M.S.; Shi, H.; Trinidad, J.C.; Clemmer, D.E. Populations of metal-glycan structures influence MS fragmentation patterns. *J. Am. Soc. Mass Spectrom.* **2015**, *26*, 25–35. [CrossRef]

245. Moritz, B.; Stracke, J.O. Assessment of disulfide and hinge modifications in monoclonal antibodies. *Electrophoresis* **2017**, *38*, 769–785. [CrossRef] [PubMed]

246. McSherry, T.; McSherry, J.; Ozaeta, P.; Longenecker, K.; Ramsay, C.; Fishpaugh, J.; Allen, S. Cysteinylation of a monoclonal antibody leads to its inactivation. *mAbs* **2016**, *8*, 718–725. [CrossRef] [PubMed]

247. Wust, C.J. Interference with antibody neutralization by coenzyme and reducing agents. *Ann. N. Y. Acad. Sci.* **1963**, *103*, 849–857. [CrossRef] [PubMed]

248. Liu, Y.D.; Chen, X.; Enk, J.Z.; Plant, M.; Dillon, T.M.; Flynn, G.C. Human IgG2 antibody disulfide rearrangement in vivo. *J. Biol. Chem.* **2008**, *283*, 29266–29272. [CrossRef] [PubMed]

249. Xu, C.F.; Chen, Y.; Yi, L.; Brantley, T.; Stanley, B.; Sosic, Z.; Zang, L. Discovery and Characterization of Histidine Oxidation Initiated Cross-links in an IgG1 Monoclonal Antibody. *Anal. Chem.* **2017**, *89*, 7915–7923. [CrossRef]

250. Amano, M.; Kobayashi, N.; Yabuta, M.; Uchiyama, S.; Fukui, K. Detection of histidine oxidation in a monoclonal immunoglobulin gamma (IgG) 1 antibody. *Anal. Chem.* **2014**, *86*, 7536–7543. [CrossRef]

251. Bane, J.; Mozziconacci, O.; Yi, L.; Wang, Y.J.; Sreedhara, A.; Schoneich, C. Photo-oxidation of IgG1 and Model Peptides: Detection and Analysis of Triply Oxidized His and Trp Side Chain Cleavage Products. *Pharm. Res.* **2017**, *34*, 229–242. [CrossRef]

252. Inglis, A.S. Cleavage at aspartic acid. *Methods Enzymol.* **1983**, *91*, 324–332.

253. Oliyai, C.; Borchardt, R.T. Chemical pathways of peptide degradation. IV. Pathways, kinetics, and mechanism of degradation of an aspartyl residue in a model hexapeptide. *Pharm. Res.* **1993**, *10*, 95–102. [CrossRef]

254. Kameoka, D.; Ueda, T.; Imoto, T. Effect of the conformational stability of the CH2 domain on the aggregation and peptide cleavage of a humanized IgG. *Appl. Biochem. Biotechnol.* **2011**, *164*, 642–654. [CrossRef]

255. Dick, L.W., Jr.; Kim, C.; Qiu, D.; Cheng, K.C. Determination of the origin of the N-terminal pyro-glutamate variation in monoclonal antibodies using model peptides. *Biotechnol. Bioeng.* **2007**, *97*, 544–553. [CrossRef]

256. Chelius, D.; Jing, K.; Lueras, A.; Rehder, D.S.; Dillon, T.M.; Vizel, A.; Rajan, R.S.; Li, T.; Treuheit, M.J.; Bondarenko, P.V. Formation of pyroglutamic acid from N-terminal glutamic acid in immunoglobulin gamma antibodies. *Anal. Chem.* **2006**, *78*, 2370–2376. [CrossRef] [PubMed]

257. Van den Bremer, E.T.; Beurskens, F.J.; Voorhorst, M.; Engelberts, P.J.; de Jong, R.N.; van der Boom, B.G.; Cook, E.M.; Lindorfer, M.A.; Taylor, R.P.; van Berkel, P.H.; et al. Human IgG is produced in a pro-form that requires clipping of C-terminal lysines for maximal complement activation. *mAbs* **2015**, *7*, 672–680. [CrossRef] [PubMed]

258. Yuk, I.H.; Zhang, B.; Yang, Y.; Dutina, G.; Leach, K.D.; Vijayasankaran, N.; Shen, A.Y.; Andersen, D.C.; Snedecor, B.R.; Joly, J.C. Controlling glycation of recombinant antibody in fed-batch cell cultures. *Biotechnol. Bioeng.* **2011**, *108*, 2600–2610. [CrossRef]

259. Jefferis, R. Glycosylation as a strategy to improve antibody-based therapeutics. *Nat. Rev. Drug Discov.* **2009**, *8*, 226–234. [CrossRef]

260. Jefferis, R. Glycosylation of antibody therapeutics: Optimisation for purpose. *Methods Mol. Biol.* **2009**, *483*, 223–238.

261. Jefferis, R. Glycosylation of recombinant antibody therapeutics. *Biotechnol. Prog.* **2005**, *21*, 11–16. [CrossRef]

262. Jefferis, R. The glycosylation of antibody molecules: Functional significance. *Glycoconj. J.* **1993**, *10*, 358–361.

263. Shibata-Koyama, M.; Iida, S.; Okazaki, A.; Mori, K.; Kitajima-Miyama, K.; Saitou, S.; Kakita, S.; Kanda, Y.; Shitara, K.; Kato, K.; et al. The N-linked oligosaccharide at Fc gamma RIIIa Asn-45: An inhibitory element for high Fc gamma RIIIa binding affinity to IgG glycoforms lacking core fucosylation. *Glycobiology* **2009**, *19*, 126–134. [CrossRef]

264. Zheng, K.; Yarmarkovich, M.; Bantog, C.; Bayer, R.; Patapoff, T.W. Influence of glycosylation pattern on the molecular properties of monoclonal antibodies. *mAbs* **2014**, *6*, 649–658. [CrossRef]

265. Zhong, X.; Ma, W.; Meade, C.L.; Tam, A.S.; Llewellyn, E.; Cornell, R.; Cote, K.; Scarcelli, J.J.; Marshall, J.K.; Tzvetkova, B.; et al. Transient CHO expression platform for robust antibody production and its enhanced N-glycan sialylation on therapeutic glycoproteins. *Biotechnol. Prog.* **2019**, *35*, e2724. [CrossRef]

266. Yamane-Ohnuki, N.; Satoh, M. Production of therapeutic antibodies with controlled fucosylation. *mAbs* **2009**, *1*, 230–236. [CrossRef] [PubMed]

267. Jefferis, R. Recombinant antibody therapeutics: The impact of glycosylation on mechanisms of action. *Trends Pharmacol. Sci.* **2009**, *30*, 356–362. [CrossRef] [PubMed]

268. Huhn, C.; Selman, M.H.; Ruhaak, L.R.; Deelder, A.M.; Wuhrer, M. IgG glycosylation analysis. *Proteomics* **2009**, *9*, 882–913. [CrossRef] [PubMed]

269. Liu, Y.D.; Flynn, G.C. Effect of high mannose glycan pairing on IgG antibody clearance. *Biologicals* **2016**, *44*, 163–169. [CrossRef] [PubMed]

270. Nichols, P.; Li, L.; Kumar, S.; Buck, P.M.; Singh, S.K.; Goswami, S.; Balthazor, B.; Conley, T.R.; Sek, D.; Allen, M.J. Rational design of viscosity reducing mutants of a monoclonal antibody: Hydrophobic versus electrostatic inter-molecular interactions. *mAbs* **2015**, *7*, 212–230. [CrossRef]

271. Perchiacca, J.M.; Tessier, P.M. Engineering aggregation-resistant antibodies. *Annu. Rev. Chem. Biomol. Eng.* **2012**, *3*, 263–286. [CrossRef]

272. Perchiacca, J.M.; Ladiwala, A.R.; Bhattacharya, M.; Tessier, P.M. Aggregation-resistant domain antibodies engineered with charged mutations near the edges of the complementarity-determining regions. *Protein Eng. Des. Sel.* **2012**, *25*, 591–601. [CrossRef]

273. Perchiacca, J.M.; Ladiwala, A.R.; Bhattacharya, M.; Tessier, P.M. Structure-based design of conformation- and sequence-specific antibodies against amyloid beta. *Proc. Natl. Acad. Sci. USA* **2012**, *109*, 84–89. [CrossRef]

274. Wu, S.J.; Luo, J.; O'Neil, K.T.; Kang, J.; Lacy, E.R.; Canziani, G.; Baker, A.; Huang, M.; Tang, Q.M.; Raju, T.S.; et al. Structure-based engineering of a monoclonal antibody for improved solubility. *Protein Eng. Des. Sel.* **2010**, *23*, 643–651. [CrossRef]

275. Douillard, P.; Freissmuth, M.; Antoine, G.; Thiele, M.; Fleischander, D.; Matthiessen, P.; Voelkel, D.; Kerschbaumer, R.J.; Scheiflinger, F.; Sabarth, N. Optimization of an Antibody Light Chain Framework Enhances Expression, Biophysical Properties and Pharmacokinetics. *Antibodies* **2019**, *8*, 46. [CrossRef] [PubMed]

276. Jetha, A.; Thorsteinson, N.; Jmeian, Y.; Jeganathan, A.; Giblin, P.; Fransson, J. Homology modeling and structure-based design improve hydrophobic interaction chromatography behavior of integrin binding antibodies. *mAbs* **2018**, *10*, 890–900. [CrossRef] [PubMed]

277. Jain, T.; Sun, T.; Durand, S.; Hall, A.; Houston, N.R.; Nett, J.H.; Sharkey, B.; Bobrowicz, B.; Caffry, I.; Yu, Y.; et al. Biophysical properties of the clinical-stage antibody landscape. *Proc. Natl. Acad. Sci. USA* **2017**, *114*, 944–949. [CrossRef] [PubMed]

278. Sankar, K.; Krystek, S.R., Jr.; Carl, S.M.; Day, T.; Maier, J.K.X. AggScore: Prediction of aggregation-prone regions in proteins based on the distribution of surface patches. *Proteins* **2018**, *86*, 1147–1156. [CrossRef]

279. Worn, A.; Pluckthun, A. Stability engineering of antibody single-chain Fv fragments. *J. Mol. Biol.* **2001**, *305*, 989–1010. [CrossRef] [PubMed]

280. Weiss, W.F.T.; Young, T.M.; Roberts, C.J. Principles, approaches, and challenges for predicting protein aggregation rates and shelf life. *J. Pharm. Sci.* **2009**, *98*, 1246–1277. [CrossRef] [PubMed]

281. He, X. Thermostability of biological systems: Fundamentals, challenges, and quantification. *Open Biomed. Eng. J.* **2011**, *5*, 47–73. [CrossRef]

282. He, F.; Woods, C.E.; Trilisky, E.; Bower, K.M.; Litowski, J.R.; Kerwin, B.A.; Becker, G.W.; Narhi, L.O.; Razinkov, V.I. Screening of monoclonal antibody formulations based on high-throughput thermostability and viscosity measurements: Design of experiment and statistical analysis. *J. Pharm. Sci.* **2011**, *100*, 1330–1340. [CrossRef]

283. Thiagarajan, G.; Semple, A.; James, J.K.; Cheung, J.K.; Shameem, M. A comparison of biophysical characterization techniques in predicting monoclonal antibody stability. *mAbs* **2016**, *8*, 1088–1097. [CrossRef]

284. Schermeyer, M.T.; Woll, A.K.; Kokke, B.; Eppink, M.; Hubbuch, J. Characterization of highly concentrated antibody solution—A toolbox for the description of protein long-term solution stability. *mAbs* **2017**, *9*, 1169–1185. [CrossRef]

285. Remmele, R.L.; Gombotz, W.R. Differential scanning calorimetry: A practical tool for elucidating stability of liquid biopharmaceuticals. *Biopharm* **2000**, *13*, 36–46.

286. Maa, Y.F.; Hsu, C.C. Aggregation of recombinant human growth hormone induced by phenolic compounds. *Int. J. Pharm.* **1996**, *140*, 155–168. [CrossRef]

287. Remmele, R.L.; Nightlinger, N.S.; Srinivasan, S.; Gombotz, W.R. Interleukin-1 receptor (IL-1R) liquid formulation development using differential scanning calorimetry. *Pharm. Res.* **1997**, *15*, 200–208. [CrossRef] [PubMed]

288. Gupta, S.; Kaisheva, E. Development of a multidose formulation for a humanized monoclonal antibody using experimental design techniques. *AAPS PharmSci* **2003**, *5*, E8. [CrossRef]

289. Bedu-Addo, F.K.; Johnson, C.; Jeyarajah, S.; Henderson, I.; Advant, S.J. Use of biophysical characterization in preformulation development of a heavy-chain fragment of botulinum serotype B: Evaluation of suitable purification process conditions. *Pharm. Res.* **2004**, *21*, 1353–1361. [CrossRef]

290. Henry, K.A.; Kim, D.Y.; Kandalaft, H.; Lowden, M.J.; Yang, Q.; Schrag, J.D.; Hussack, G.; MacKenzie, C.R.; Tanha, J. Stability-Diversity Tradeoffs Impose Fundamental Constraints on Selection of Synthetic Human VH/VL Single-Domain Antibodies from In Vitro Display Libraries. *Front. Immunol.* **2017**, *8*, 1759. [CrossRef]

291. Ramaraj, T.; Angel, T.; Dratz, E.A.; Jesaitis, A.J.; Mumey, B. Antigen-antibody interface properties: Composition, residue interactions, and features of 53 non-redundant structures. *Biochim. Biophys. Acta* **2012**, *1824*, 520–532. [CrossRef]

292. Katritch, V.; Cherezov, V.; Stevens, R.C. Structure-function of the G protein-coupled receptor superfamily. *Annu. Rev. Pharmacol. Toxicol.* **2013**, *53*, 531–556. [CrossRef]

293. Abskharon, R.N.; Soror, S.H.; Pardon, E.; El Hassan, H.; Legname, G.; Steyaert, J.; Wohlkonig, A. Combining in-situ proteolysis and microseed matrix screening to promote crystallization of PrPc-nanobody complexes. *Protein Eng. Des. Sel.* **2011**, *24*, 737–741. [CrossRef]

294. Domanska, K.; Vanderhaegen, S.; Srinivasan, V.; Pardon, E.; Dupeux, F.; Marquez, J.A.; Giorgetti, S.; Stoppini, M.; Wyns, L.; Bellotti, V.; et al. Atomic structure of a nanobody-trapped domain-swapped dimer of an amyloidogenic beta2-microglobulin variant. *Proc. Natl. Acad. Sci. USA* **2011**, *108*, 1314–1319. [CrossRef]

295. Rasmussen, S.G.; Choi, H.J.; Fung, J.J.; Pardon, E.; Casarosa, P.; Chae, P.S.; Devree, B.T.; Rosenbaum, D.M.; Thian, F.S.; Kobilka, T.S.; et al. Structure of a nanobody-stabilized active state of the beta (2) adrenoceptor. *Nature* **2011**, *469*, 175–180. [CrossRef] [PubMed]

296. Steyaert, J.; Kobilka, B.K. Nanobody stabilization of G protein-coupled receptor conformational states. *Curr. Opin. Struct. Biol.* **2011**, *21*, 567–572. [CrossRef] [PubMed]

297. Park, Y.J.; Pardon, E.; Wu, M.; Steyaert, J.; Hol, W.G. Crystal structure of a heterodimer of editosome interaction proteins in complex with two copies of a cross-reacting nanobody. *Nucleic Acids Res.* **2012**, *40*, 1828–1840. [CrossRef] [PubMed]

298. Hassanzadeh-Ghassabeh, G.; Devoogdt, N.; de Pauw, P.; Vincke, C.; Muyldermans, S. Nanobodies and their potential applications. *Nanomedicine* **2013**, *8*, 1013–1026. [CrossRef]

299. Muyldermans, S. Nanobodies: Natural single-domain antibodies. *Annu. Rev. Biochem.* **2013**, *82*, 775–797. [CrossRef]

300. Sheridan, C. Ablynx's nanobody fragments go places antibodies cannot. *Nat. Biotechnol.* **2017**, *35*, 1115–1117. [CrossRef]

301. Hu, Y.; Liu, C.; Muyldermans, S. Nanobody-Based Delivery Systems for Diagnosis and Targeted Tumor Therapy. *Front. Immunol.* **2017**, *8*, 1442. [CrossRef]

302. Arezumand, R.; Alibakhshi, A.; Ranjbari, J.; Ramazani, A.; Muyldermans, S. Nanobodies As Novel Agents for Targeting Angiogenesis in Solid Cancers. *Front. Immunol.* **2017**, *8*, 1746. [CrossRef]

303. Peyvandi, F.; Scully, M.; Hovinga, J.A.K.; Cataland, S.; Knobl, P.; Wu, H.; Artoni, A.; Westwood, J.P.; Taleghani, M.M.; Jilma, B.; et al. Caplacizumab for Acquired Thrombotic Thrombocytopenic Purpura. *N. Engl. J. Med.* **2016**, *374*, 511–522. [CrossRef]

304. Klarenbeek, A.; El Mazouari, K.; Desmyter, A.; Blanchetot, C.; Hultberg, A.; de Jonge, N.; Roovers, R.C.; Cambillau, C.; Spinelli, S.; Del-Favero, J.; et al. Camelid Ig V genes reveal significant human homology not seen in therapeutic target genes, providing for a powerful therapeutic antibody platform. *mAbs* **2015**, *7*, 693–706. [CrossRef]

305. Konning, D.; Zielonka, S.; Grzeschik, J.; Empting, M.; Valldorf, B.; Krah, S.; Schroter, C.; Sellmann, C.; Hock, B.; Kolmar, H. Camelid and shark single domain antibodies: Structural features and therapeutic potential. *Curr. Opin. Struct. Biol.* **2017**, *45*, 10–16. [CrossRef] [PubMed]

306. Goodchild, S.A.; Dooley, H.; Schoepp, R.J.; Flajnik, M.; Lonsdale, S.G. Isolation and characterisation of Ebolavirus-specific recombinant antibody fragments from murine and shark immune libraries. *Mol. Immunol.* **2011**, *48*, 2027–2037. [CrossRef] [PubMed]

307. Walsh, R.; Nuttall, S.; Revill, P.; Colledge, D.; Cabuang, L.; Soppe, S.; Dolezal, O.; Griffiths, K.; Bartholomeusz, A.; Locarnini, S. Targeting the hepatitis B virus precore antigen with a novel IgNAR single variable domain intrabody. *Virology* **2011**, *411*, 132–141. [CrossRef] [PubMed]

308. Nuttall, S.D. Overview and discovery of IgNARs and generation of VNARs. *Methods Mol. Biol.* **2012**, *911*, 27–36.

309. Kovalenko, O.V.; Olland, A.; Piche-Nicholas, N.; Godbole, A.; King, D.; Svenson, K.; Calabro, V.; Muller, M.R.; Barelle, C.J.; Somers, W.; et al. Atypical antigen recognition mode of a shark immunoglobulin new antigen receptor (IgNAR) variable domain characterized by humanization and structural analysis. *J. Biol. Chem.* **2013**, *288*, 17408–17419. [CrossRef]

310. Kovaleva, M.; Ferguson, L.; Steven, J.; Porter, A.; Barelle, C. Shark variable new antigen receptor biologics—A novel technology platform for therapeutic drug development. *Expert Opin. Biol. Ther.* **2014**, *14*, 1527–1539. [CrossRef]

311. Zielonka, S.; Empting, M.; Grzeschik, J.; Konning, D.; Barelle, C.J.; Kolmar, H. Structural insights and biomedical potential of IgNAR scaffolds from sharks. *mAbs* **2015**, *7*, 15–25. [CrossRef]

312. Grzeschik, J.; Yanakieva, D.; Roth, L.; Krah, S.; Hinz, S.C.; Elter, A.; Zollmann, T.; Schwall, G.; Zielonka, S.; Kolmar, H. Yeast Surface Display in Combination with Fluorescence-activated Cell Sorting Enables the Rapid Isolation of Antibody Fragments Derived from Immunized Chickens. *Biotechnol. J.* **2019**, *14*, e1800466. [CrossRef]

313. Grzeschik, J.; Konning, D.; Hinz, S.C.; Krah, S.; Schroter, C.; Empting, M.; Kolmar, H.; Zielonka, S. Generation of Semi-Synthetic Shark IgNAR Single-Domain Antibody Libraries. *Methods Mol. Biol.* **2018**, *1701*, 147–167.

314. Streltsov, V.A.; Varghese, J.N.; Carmichael, J.A.; Irving, R.A.; Hudson, P.J.; Nuttall, S.D. Structural evidence for evolution of shark Ig new antigen receptor variable domain antibodies from a cell-surface receptor. *Proc. Natl. Acad. Sci. USA* **2004**, *101*, 12444–12449. [CrossRef]

315. Griffiths, K.; Dolezal, O.; Cao, B.; Nilsson, S.K.; See, H.B.; Pfleger, K.D.; Roche, M.; Gorry, P.R.; Pow, A.; Viduka, K.; et al. I-bodies, Human Single Domain Antibodies That Antagonize Chemokine Receptor CXCR4. *J. Biol. Chem.* **2016**, *291*, 12641–12657. [CrossRef] [PubMed]

316. Feng, M.; Bian, H.; Wu, X.; Fu, T.; Fu, Y.; Hong, J.; Fleming, B.D.; Flajnik, M.F.; Ho, M. Construction and next-generation sequencing analysis of a large phage-displayed VNAR single-domain antibody library from six naive nurse sharks. *Antib. Ther.* **2019**, *2*, 1–11. [CrossRef] [PubMed]

317. Murphy, K.M.; Weaver, C. *Janeway's Immunobiology*, 9th ed.; Garland Science, Taylor and Science Group: New York, NY, USA, 2017.

318. Kitov, P.I.; Bundle, D.R. On the nature of the multivalency effect: A thermodynamic model. *J. Am. Chem. Soc.* **2003**, *125*, 16271–16284. [CrossRef]

319. Vorup-Jensen, T. On the roles of polyvalent binding in immune recognition: Perspectives in the nanoscience of immunology and the immune response to nanomedicines. *Adv. Drug Deliv. Rev.* **2012**, *64*, 1759–1781. [CrossRef] [PubMed]

320. Klein, J.S.; Gnanapragasam, P.N.; Galimidi, R.P.; Foglesong, C.P.; West, A.P., Jr.; Bjorkman, P.J. Examination of the contributions of size and avidity to the neutralization mechanisms of the anti-HIV antibodies b12 and 4E10. *Proc. Natl. Acad. Sci. USA* **2009**, *106*, 7385–7390. [CrossRef]

321. Vauquelin, G.; Charlton, S.J. Exploring avidity: Understanding the potential gains in functional affinity and target residence time of bivalent and heterobivalent ligands. *Br. J. Pharmacol.* **2013**, *168*, 1771–1785. [CrossRef]

322. Nesspor, T.C.; Raju, T.S.; Chin, C.N.; Vafa, O.; Brezski, R.J. Avidity confers FcgammaR binding and immune effector function to aglycosylated immunoglobulin G1. *J. Mol. Recognit.* **2012**, *25*, 147–154. [CrossRef]

323. Loyau, J.; Malinge, P.; Daubeuf, B.; Shang, L.; Elson, G.; Kosco-Vilbois, M.; Fischer, N.; Rousseau, F. Maximizing the potency of an anti-TLR4 monoclonal antibody by exploiting proximity to Fcgamma receptors. *mAbs* **2014**, *6*, 1621–1630. [CrossRef]

324. Read, T.; Olkhov, R.V.; Williamson, E.D.; Shaw, A.M. Label-free Fab and Fc affinity/avidity profiling of the antibody complex half-life for polyclonal and monoclonal efficacy screening. *Anal. Bioanal. Chem.* **2015**, *407*, 7349–7357. [CrossRef]

325. Bruhns, P. Properties of mouse and human IgG receptors and their contribution to disease models. *Blood* **2012**, *119*, 5640–5649. [CrossRef]

326. Jain, A.; Olsen, H.S.; Vyzasatya, R.; Burch, E.; Sakoda, Y.; Merigeon, E.Y.; Cai, L.; Lu, C.; Tan, M.; Tamada, K.; et al. Fully recombinant IgG2a Fc multimers (stradomers) effectively treat collagen-induced arthritis and prevent idiopathic thrombocytopenic purpura in mice. *Arthritis Res. Ther.* **2012**, *14*, R192. [CrossRef] [PubMed]

327. Ortiz, D.F.; Lansing, J.C.; Rutitzky, L.; Kurtagic, E.; Prod'homme, T.; Choudhury, A.; Washburn, N.; Bhatnagar, N.; Beneduce, C.; Holte, K.; et al. Elucidating the interplay between IgG-Fc valency and FcgammaR activation for the design of immune complex inhibitors. *Sci. Transl. Med.* **2016**, *8*, 365ra158. [CrossRef] [PubMed]

328. Qureshi, O.S.; Rowley, T.F.; Junker, F.; Peters, S.J.; Crilly, S.; Compson, J.; Eddleston, A.; Bjorkelund, H.; Greenslade, K.; Parkinson, M.; et al. Multivalent Fcgamma-receptor engagement by a hexameric Fc-fusion protein triggers Fcgamma-receptor internalisation and modulation of Fcgamma-receptor functions. *Sci. Rep.* **2017**, *7*, 17049. [CrossRef] [PubMed]

329. Spirig, R.; Campbell, I.K.; Koernig, S.; Chen, C.G.; Lewis, B.J.B.; Butcher, R.; Muir, I.; Taylor, S.; Chia, J.; Leong, D.; et al. rIgG1 Fc Hexamer Inhibits Antibody-Mediated Autoimmune Disease via Effects on Complement and FcgammaRs. *J. Immunol.* **2018**, *200*, 2542–2553. [CrossRef] [PubMed]

330. Warncke, M.; Calzascia, T.; Coulot, M.; Balke, N.; Touil, R.; Kolbinger, F.; Heusser, C. Different adaptations of IgG effector function in human and nonhuman primates and implications for therapeutic antibody treatment. *J. Immunol.* **2012**, *188*, 4405–4411. [CrossRef] [PubMed]

331. Mellman, I.; Coukos, G.; Dranoff, G. Cancer immunotherapy comes of age. *Nature* **2011**, *480*, 480–489. [CrossRef]

332. Chen, L.; Flies, D.B. Molecular mechanisms of T cell co-stimulation and co-inhibition. *Nat. Rev. Immunol.* **2013**, *13*, 227–242. [CrossRef]

333. Schaer, D.A.; Hirschhorn-Cymerman, D.; Wolchok, J.D. Targeting tumor-necrosis factor receptor pathways for tumor immunotherapy. *J. Immunother. Cancer* **2014**, *2*, 7. [CrossRef]

334. Wajant, H. Principles of antibody-mediated TNF receptor activation. *Cell Death Differ.* **2015**, *22*, 1727–1741. [CrossRef]

335. Dostert, C.; Grusdat, M.; Letellier, E.; Brenner, D. The TNF Family of Ligands and Receptors: Communication Modules in the Immune System and Beyond. *Physiol. Rev.* **2019**, *99*, 115–160. [CrossRef]

336. Pollok, K.E.; Kim, Y.J.; Zhou, Z.; Hurtado, J.; Kim, K.K.; Pickard, R.T.; Kwon, B.S. Inducible T cell antigen 4-1BB. Analysis of expression and function. *J. Immunol.* **1993**, *150*, 771–781. [PubMed]

337. Gramaglia, I.; Weinberg, A.D.; Lemon, M.; Croft, M. Ox-40 ligand: A potent costimulatory molecule for sustaining primary CD4 T cell responses. *J. Immunol.* **1998**, *161*, 6510–6517. [PubMed]

338. Kanamaru, F.; Youngnak, P.; Hashiguchi, M.; Nishioka, T.; Takahashi, T.; Sakaguchi, S.; Ishikawa, I.; Azuma, M. Costimulation via glucocorticoid-induced TNF receptor in both conventional and CD25+ regulatory CD4+ T cells. *J. Immunol.* **2004**, *172*, 7306–7314. [CrossRef] [PubMed]

339. Ramakrishna, V.; Sundarapandiyan, K.; Zhao, B.; Bylesjo, M.; Marsh, H.C.; Keler, T. Characterization of the human T cell response to in vitro CD27 costimulation with varlilumab. *J. Immunother. Cancer* **2015**, *3*, 37. [CrossRef]

340. Wilson, N.S.; Yang, B.; Yang, A.; Loeser, S.; Marsters, S.; Lawrence, D.; Li, Y.; Pitti, R.; Totpal, K.; Yee, S.; et al. An Fcgamma receptor-dependent mechanism drives antibody-mediated target-receptor signaling in cancer cells. *Cancer Cell* **2011**, *19*, 101–113. [CrossRef]

341. He, L.Z.; Prostak, N.; Thomas, L.J.; Vitale, L.; Weidlick, J.; Crocker, A.; Pilsmaker, C.D.; Round, S.M.; Tutt, A.; Glennie, M.J.; et al. Agonist anti-human CD27 monoclonal antibody induces T cell activation and tumor immunity in human CD27-transgenic mice. *J. Immunol.* **2013**, *191*, 4174–4183. [CrossRef]

342. Mangsbo, S.M.; Broos, S.; Fletcher, E.; Veitonmaki, N.; Furebring, C.; Dahlen, E.; Norlen, P.; Lindstedt, M.; Totterman, T.H.; Ellmark, P. The human agonistic CD40 antibody ADC-1013 eradicates bladder tumors and generates T-cell-dependent tumor immunity. *Clin. Cancer Res.* **2015**, *21*, 1115–1126. [CrossRef]

343. Vanamee, E.S.; Faustman, D.L. Structural principles of tumor necrosis factor superfamily signaling. *Sci. Signal.* **2018**, *11*, eaao4910. [CrossRef]

344. Morris, N.P.; Peters, C.; Montler, R.; Hu, H.M.; Curti, B.D.; Urba, W.J.; Weinberg, A.D. Development and characterization of recombinant human Fc:OX40L fusion protein linked via a coiled-coil trimerization domain. *Mol. Immunol.* **2007**, *44*, 3112–3121. [CrossRef]

345. Stavenhagen, J.B.; Gorlatov, S.; Tuaillon, N.; Rankin, C.T.; Li, H.; Burke, S.; Huang, L.; Johnson, S.; Koenig, S.; Bonvini, E. Enhancing the potency of therapeutic monoclonal antibodies via Fc optimization. *Adv. Enzym. Regul.* **2008**, *48*, 152–164. [CrossRef]

346. Kim, J.M.; Ashkenazi, A. Fcgamma receptors enable anticancer action of proapoptotic and immune-modulatory antibodies. *J. Exp. Med.* **2013**, *210*, 1647–1651. [CrossRef] [PubMed]

347. Furness, A.J.; Vargas, F.A.; Peggs, K.S.; Quezada, S.A. Impact of tumour microenvironment and Fc receptors on the activity of immunomodulatory antibodies. *Trends Immunol.* **2014**, *35*, 290–298. [CrossRef] [PubMed]

348. White, A.L.; Chan, H.T.; Roghanian, A.; French, R.R.; Mockridge, C.I.; Tutt, A.L.; Dixon, S.V.; Ajona, D.; Verbeek, J.S.; Al-Shamkhani, A.; et al. Interaction with FcgammaRIIB is critical for the agonistic activity of anti-CD40 monoclonal antibody. *J. Immunol.* **2011**, *187*, 1754–1763. [CrossRef] [PubMed]

349. Li, F.; Ravetch, J.V. Apoptotic and antitumor activity of death receptor antibodies require inhibitory Fcgamma receptor engagement. *Proc. Natl. Acad. Sci. USA* **2012**, *109*, 10966–10971. [CrossRef] [PubMed]

350. Xu, Y.; Szalai, A.J.; Zhou, T.; Zinn, K.R.; Chaudhuri, T.R.; Li, X.; Koopman, W.J.; Kimberly, R.P. Fc gamma Rs modulate cytotoxicity of anti-Fas antibodies: Implications for agonistic antibody-based therapeutics. *J. Immunol.* **2003**, *171*, 562–568. [CrossRef] [PubMed]

351. Bulliard, Y.; Jolicoeur, R.; Zhang, J.; Dranoff, G.; Wilson, N.S.; Brogdon, J.L. OX40 engagement depletes intratumoral Tregs via activating FcgammaRs, leading to antitumor efficacy. *Immunol. Cell Biol.* **2014**, *92*, 475–480. [CrossRef] [PubMed]

352. Bulliard, Y.; Jolicoeur, R.; Windman, M.; Rue, S.M.; Ettenberg, S.; Knee, D.A.; Wilson, N.S.; Dranoff, G.; Brogdon, J.L. Activating Fc gamma receptors contribute to the antitumor activities of immunoregulatory receptor-targeting antibodies. *J. Exp. Med.* **2013**, *210*, 1685–1693. [CrossRef]

353. Guilliams, M.; Bruhns, P.; Saeys, Y.; Hammad, H.; Lambrecht, B.N. The function of Fcgamma receptors in dendritic cells and macrophages. *Nat. Rev. Immunol.* **2014**, *14*, 94–108. [CrossRef]

354. Chu, S.Y.; Vostiar, I.; Karki, S.; Moore, G.L.; Lazar, G.A.; Pong, E.; Joyce, P.F.; Szymkowski, D.E.; Desjarlais, J.R. Inhibition of B cell receptor-mediated activation of primary human B cells by coengagement of CD19 and FcgammaRIIb with Fc-engineered antibodies. *Mol. Immunol.* **2008**, *45*, 3926–3933. [CrossRef]

355. Mimoto, F.; Katada, H.; Kadono, S.; Igawa, T.; Kuramochi, T.; Muraoka, M.; Wada, Y.; Haraya, K.; Miyazaki, T.; Hattori, K. Engineered antibody Fc variant with selectively enhanced FcgammaRIIb binding over both FcgammaRIIa(R131) and FcgammaRIIa(H131). *Protein Eng. Des. Sel.* **2013**, *26*, 589–598. [CrossRef]

356. Mimoto, F.; Igawa, T.; Kuramochi, T.; Katada, H.; Kadono, S.; Kamikawa, T.; Shida-Kawazoe, M.; Hattori, K. Novel asymmetrically engineered antibody Fc variant with superior FcgammaR binding affinity and specificity compared with afucosylated Fc variant. *mAbs* **2013**, *5*, 229–236. [CrossRef] [PubMed]

357. Zhang, D.; Whitaker, B.; Derebe, M.G.; Chiu, M.L. FcgammaRII-binding Centyrins mediate agonism and antibody-dependent cellular phagocytosis when fused to an anti-OX40 antibody. *mAbs* **2018**, *10*, 463–475. [CrossRef] [PubMed]

358. White, A.L.; Chan, H.T.; French, R.R.; Willoughby, J.; Mockridge, C.I.; Roghanian, A.; Penfold, C.A.; Booth, S.G.; Dodhy, A.; Polak, M.E.; et al. Conformation of the human immunoglobulin g2 hinge imparts superagonistic properties to immunostimulatory anticancer antibodies. *Cancer Cell* **2015**, *27*, 138–148. [CrossRef] [PubMed]

359. Diebolder, C.A.; Beurskens, F.J.; de Jong, R.N.; Koning, R.I.; Strumane, K.; Lindorfer, M.A.; Voorhorst, M.; Ugurlar, D.; Rosati, S.; Heck, A.J.; et al. Complement is activated by IgG hexamers assembled at the cell surface. *Science* **2014**, *343*, 1260–1263. [CrossRef] [PubMed]

360. De Jong, R.N.; Beurskens, F.J.; Verploegen, S.; Strumane, K.; van Kampen, M.D.; Voorhorst, M.; Horstman, W.; Engelberts, P.J.; Oostindie, S.C.; Wang, G.; et al. A Novel Platform for the Potentiation of Therapeutic Antibodies Based on Antigen-Dependent Formation of IgG Hexamers at the Cell Surface. *PLoS Biol.* **2016**, *14*, e1002344. [CrossRef] [PubMed]

361. Zhang, D.; Goldberg, M.V.; Chiu, M.L. Fc Engineering Approaches to Enhance the Agonism and Effector Functions of an Anti-OX40 Antibody. *J. Biol. Chem.* **2016**, *291*, 27134–27146. [CrossRef]

362. Zhang, D.; Armstrong, A.A.; Tam, S.H.; McCarthy, S.G.; Luo, J.; Gilliland, G.L.; Chiu, M.L. Functional optimization of agonistic antibodies to OX40 receptor with novel Fc mutations to promote antibody multimerization. *mAbs* **2017**, *9*, 1129–1142. [CrossRef]

363. Rowley, T.F.; Peters, S.J.; Aylott, M.; Griffin, R.; Davies, N.L.; Healy, L.J.; Cutler, R.M.; Eddleston, A.; Pither, T.L.; Sopp, J.M.; et al. Engineered hexavalent Fc proteins with enhanced Fc-gamma receptor avidity provide insights into immune-complex interactions. *Commun. Biol.* **2018**, *1*, 146. [CrossRef]

364. Piao, X.; Ozawa, T.; Hamana, H.; Shitaoka, K.; Jin, A.; Kishi, H.; Muraguchi, A. TRAIL-receptor 1 IgM antibodies strongly induce apoptosis in human cancer cells in vitro and in vivo. *Oncoimmunology* **2016**, *5*, e1131380. [CrossRef]

365. Dubuisson, A.; Micheau, O. Antibodies and Derivatives Targeting DR4 and DR5 for Cancer Therapy. *Antibodies* **2017**, *6*, 16. [CrossRef]

366. Roux, K.H.; Strelets, L.; Brekke, O.H.; Sandlie, I.; Michaelsen, T.E. Comparisons of the ability of human IgG3 hinge mutants, IgM, IgE, and IgA2, to form small immune complexes: A role for flexibility and geometry. *J. Immunol.* **1998**, *161*, 4083–4090. [PubMed]

367. Hinman, L.M.; Hamann, P.R.; Wallace, R.; Menendez, A.T.; Durr, F.E.; Upeslacis, J. Preparation and characterization of monoclonal antibody conjugates of the calicheamicins: A novel and potent family of antitumor antibiotics. *Cancer Res.* **1993**, *53*, 3336–3342. [PubMed]

368. Sievers, E.L.; Appelbaum, F.R.; Spielberger, R.T.; Forman, S.J.; Flowers, D.; Smith, F.O.; Shannon-Dorcy, K.; Berger, M.S.; Bernstein, I.D. Selective ablation of acute myeloid leukemia using antibody-targeted chemotherapy: A phase I study of an anti-CD33 calicheamicin immunoconjugate. *Blood* **1999**, *93*, 3678–3684. [CrossRef] [PubMed]

369. Egan, P.C.; Reagan, J.L. The return of gemtuzumab ozogamicin: A humanized anti-CD33 monoclonal antibody-drug conjugate for the treatment of newly diagnosed acute myeloid leukemia. *Onco Targets Ther.* **2018**, *11*, 8265–8272. [CrossRef] [PubMed]

370. DiJoseph, J.F.; Goad, M.E.; Dougher, M.M.; Boghaert, E.R.; Kunz, A.; Hamann, P.R.; Damle, N.K. Potent and specific antitumor efficacy of CMC-544, a CD22-targeted immunoconjugate of calicheamicin, against systemically disseminated B-cell lymphoma. *Clin. Cancer Res.* **2004**, *10*, 8620–8629. [CrossRef] [PubMed]

371. DiJoseph, J.F.; Armellino, D.C.; Boghaert, E.R.; Khandke, K.; Dougher, M.M.; Sridharan, L.; Kunz, A.; Hamann, P.R.; Gorovits, B.; Udata, C.; et al. Antibody-targeted chemotherapy with CMC-544: A CD22-targeted immunoconjugate of calicheamicin for the treatment of B-lymphoid malignancies. *Blood* **2004**, *103*, 1807–1814. [CrossRef] [PubMed]

372. Doronina, S.O.; Toki, B.E.; Torgov, M.Y.; Mendelsohn, B.A.; Cerveny, C.G.; Chace, D.F.; DeBlanc, R.L.; Gearing, R.P.; Bovee, T.D.; Siegall, C.B.; et al. Development of potent monoclonal antibody auristatin conjugates for cancer therapy. *Nat. Biotechnol.* **2003**, *21*, 778–784. [CrossRef]

373. Doronina, S.O.; Mendelsohn, B.A.; Bovee, T.D.; Cerveny, C.G.; Alley, S.C.; Meyer, D.L.; Oflazoglu, E.; Toki, B.E.; Sanderson, R.J.; Zabinski, R.F.; et al. Enhanced activity of monomethylauristatin F through monoclonal antibody delivery: Effects of linker technology on efficacy and toxicity. *Bioconjug. Chem.* **2006**, *17*, 114–124. [CrossRef]

374. Hamblett, K.J.; Senter, P.D.; Chace, D.F.; Sun, M.M.; Lenox, J.; Cerveny, C.G.; Kissler, K.M.; Bernhardt, S.X.; Kopcha, A.K.; Zabinski, R.F.; et al. Effects of drug loading on the antitumor activity of a monoclonal antibody drug conjugate. *Clin. Cancer Res.* **2004**, *10*, 7063–7070. [CrossRef]

375. Scott, L.J. Brentuximab Vedotin: A Review in CD30-Positive Hodgkin Lymphoma. *Drugs* **2017**, *77*, 435–445. [CrossRef]

376. Chari, R.V.; Martell, B.A.; Gross, J.L.; Cook, S.B.; Shah, S.A.; Blattler, W.A.; McKenzie, S.J.; Goldmacher, V.S. Immunoconjugates containing novel maytansinoids: Promising anticancer drugs. *Cancer Res.* **1992**, *52*, 127–131. [PubMed]

377. Lewis Phillips, G.D.; Li, G.; Dugger, D.L.; Crocker, L.M.; Parsons, K.L.; Mai, E.; Blattler, W.A.; Lambert, J.M.; Chari, R.V.; Lutz, R.J.; et al. Targeting HER2-positive breast cancer with trastuzumab-DM1, an antibody-cytotoxic drug conjugate. *Cancer Res.* **2008**, *68*, 9280–9290. [CrossRef] [PubMed]

378. Verma, S.; Miles, D.; Gianni, L.; Krop, I.E.; Welslau, M.; Baselga, J.; Pegram, M.; Oh, D.Y.; Dieras, V.; Guardino, E.; et al. Trastuzumab emtansine for HER2-positive advanced breast cancer. *N. Engl. J. Med.* **2012**, *367*, 1783–1791. [CrossRef] [PubMed]

379. Cardillo, T.M.; Govindan, S.V.; Sharkey, R.M.; Trisal, P.; Arrojo, R.; Liu, D.; Rossi, E.A.; Chang, C.H.; Goldenberg, D.M. Sacituzumab Govitecan (IMMU-132), an Anti-Trop-2/SN-38 Antibody-Drug Conjugate: Characterization and Efficacy in Pancreatic, Gastric, and Other Cancers. *Bioconjug. Chem.* **2015**, *26*, 919–931. [CrossRef]

380. Mansfield, E.; Chiron, M.F.; Amlot, P.; Pastan, I.; FitzGerald, D.J. Recombinant RFB4 single-chain immunotoxin that is cytotoxic towards CD22-positive cells. *Biochem. Soc. Trans.* **1997**, *25*, 709–714. [CrossRef]

381. Mansfield, E.; Amlot, P.; Pastan, I.; FitzGerald, D.J. Recombinant RFB4 immunotoxins exhibit potent cytotoxic activity for CD22-bearing cells and tumors. *Blood* **1997**, *90*, 2020–2026. [CrossRef]

382. Bang, S.; Nagata, S.; Onda, M.; Kreitman, R.J.; Pastan, I. HA22 (R490A) is a recombinant immunotoxin with increased antitumor activity without an increase in animal toxicity. *Clin. Cancer Res.* **2005**, *11*, 1545–1550. [CrossRef]

383. Knox, S.J.; Goris, M.L.; Trisler, K.; Negrin, R.; Davis, T.; Liles, T.M.; Grillo-Lopez, A.; Chinn, P.; Varns, C.; Ning, S.C.; et al. Yttrium-90-labeled anti-CD20 monoclonal antibody therapy of recurrent B-cell lymphoma. *Clin. Cancer Res.* **1996**, *2*, 457–470.

384. Kaminski, M.S.; Zasadny, K.R.; Francis, I.R.; Milik, A.W.; Ross, C.W.; Moon, S.D.; Crawford, S.M.; Burgess, J.M.; Petry, N.A.; Butchko, G.M.; et al. Radioimmunotherapy of B-cell lymphoma with [131I]anti-B1(anti-CD20) antibody. *N. Engl. J. Med.* **1993**, *329*, 459–465. [CrossRef]

385. Prasad, V. The withdrawal of drugs for commercial reasons: The incomplete story of tositumomab. *JAMA Intern. Med.* **2014**, *174*, 1887–1888. [CrossRef]

386. Teeling, J.L.; Mackus, W.J.; Wiegman, L.J.; van den Brakel, J.H.; Beers, S.A.; French, R.R.; van Meerten, T.; Ebeling, S.; Vink, T.; Slootstra, J.W.; et al. The biological activity of human CD20 monoclonal antibodies is linked to unique epitopes on CD20. *J. Immunol.* **2006**, *177*, 362–371. [CrossRef] [PubMed]

387. Boross, P.; Leusen, J.H. Mechanisms of action of CD20 antibodies. *Am. J. Cancer Res.* **2012**, *2*, 676–690. [PubMed]

388. Klein, C.; Lammens, A.; Schafer, W.; Georges, G.; Schwaiger, M.; Mossner, E.; Hopfner, K.P.; Umana, P.; Niederfellner, G. Epitope interactions of monoclonal antibodies targeting CD20 and their relationship to functional properties. *mAbs* **2013**, *5*, 22–33. [CrossRef] [PubMed]

389. Kapelski, S.; Cleiren, E.; Attar, R.M.; Philippar, U.; Hasler, J.; Chiu, M.L. Influence of the bispecific antibody IgG subclass on T cell redirection. *mAbs* **2019**, *11*, 1012–1024. [CrossRef]

390. Konitzer, J.D.; Sieron, A.; Wacker, A.; Enenkel, B. Reformatting Rituximab into Human IgG2 and IgG4 Isotypes Dramatically Improves Apoptosis Induction In Vitro. *PLoS ONE* **2015**, *10*, e0145633. [CrossRef]

391. Shields, R.L.; Namenuk, A.K.; Hong, K.; Meng, Y.G.; Rae, J.; Briggs, J.; Xie, D.; Lai, J.; Stadlen, A.; Li, B.; et al. High resolution mapping of the binding site on human IgG1 for Fc gamma RI, Fc gamma RII, Fc gamma RIII, and FcRn and design of IgG1 variants with improved binding to the Fc gamma R. *J. Biol. Chem.* **2001**, *276*, 6591–6604. [CrossRef]

392. Lazar, G.A.; Dang, W.; Karki, S.; Vafa, O.; Peng, J.S.; Hyun, L.; Chan, C.; Chung, H.S.; Eivazi, A.; Yoder, S.C.; et al. Engineered antibody Fc variants with enhanced effector function. *Proc. Natl. Acad. Sci. USA* **2006**, *103*, 4005–4010. [CrossRef]

393. Stavenhagen, J.B.; Gorlatov, S.; Tuaillon, N.; Rankin, C.T.; Li, H.; Burke, S.; Huang, L.; Vijh, S.; Johnson, S.; Bonvini, E.; et al. Fc optimization of therapeutic antibodies enhances their ability to kill tumor cells in vitro and controls tumor expansion in vivo via low-affinity activating Fcgamma receptors. *Cancer Res.* **2007**, *67*, 8882–8890. [CrossRef]

394. Reddy, M.P.; Kinney, C.A.; Chaikin, M.A.; Payne, A.; Fishman-Lobell, J.; Tsui, P.; Monte, P.R.D.; Doyle, M.L.; Brigham-Burke, M.R.; Anderson, D.; et al. Elimination of Fc receptor-dependent effector functions of a modified IgG4 monoclonal antibody to human CD4. *J. Immunol.* **2000**, *164*, 1925–1933. [CrossRef]

395. Strohl, W.R. Optimization of Fc-mediated effector functions of monoclonal antibodies. *Curr. Opin. Biotechnol.* **2009**, *20*, 685–691. [CrossRef]

396. Strohl, W.R. Current progress in innovative engineered antibodies. *Protein Cell* **2018**, *9*, 86–120. [CrossRef]

397. Czajkowsky, D.M.; Hu, J.; Shao, Z.; Pless, R.J. Fc-fusion proteins: New developments and future perspectives. *EMBO Mol. Med.* **2012**, *4*, 1015–1028. [CrossRef]

398. Yogo, R.; Yamaguchi, Y.; Watanabe, H.; Yagi, H.; Satoh, T.; Nakanishi, M.; Onitsuka, M.; Omasa, T.; Shimada, M.; Maruno, T.; et al. The Fab portion of immunoglobulin G contributes to its binding to Fcγ receptor III. *Sci. Rep.* **2019**, *9*, 1–10. [CrossRef]

399. Kilar, F.; Zavodszky, P. Non-covalent interactions between Fab and Fc regions in immunoglobulin G molecules. *Eur. J. Biochem.* **1987**, *162*, 57–81. [CrossRef] [PubMed]

400. Zhao, J.; Nussinov, R.; Ma, B. Antigen binding allosterically promotes Fc receptor recognition. *mAbs* **2019**, *11*, 58–74. [CrossRef] [PubMed]

401. Richards, J.O.; Karki, S.; Lazar, G.A.; Chen, H.; Dang, W.; Desjarlais, J.R. Optimization of antibody binding to FcgammaRIIa enhances macrophage phagocytosis of tumor cells. *Mol. Cancer Ther.* **2008**, *7*, 2517–2527. [CrossRef]

402. Bolt, S.; Routledge, E.; Lloyd, I.; Chatenoud, L.; Pope, H.; Gorman, S.D.; Clark, M.; Waldmann, H. The generation of a humanized, non-mitogenic CD3 monoclonal antibody which retains in vitro immunosuppressive properties. *Eur. J. Immunol.* **1993**, *23*, 403–411. [CrossRef]

403. Alegre, M.L.; Peterson, L.J.; Xu, D.; Sattar, H.A.; Jeyarajah, D.R.; Kowalkowski, K.; Thistlethwaite, J.R.; Zivin, R.A.; Jolliffe, L.; Bluestone, J.A. A non-activating "humanized" anti-CD3 monoclonal antibody retains immunosuppressive properties in vivo. *Transplantation* **1994**, *57*, 1537–1543. [CrossRef]

404. Cole, M.S.; Stellrecht, K.E.; Shi, J.D.; Homola, M.; Hsu, D.H.; Anasetti, C.; Vasquez, M.; Tso, J.Y. HuM291, a humanized anti-CD3 antibody, is immunosuppressive to T cells while exhibiting reduced mitogenicity in vitro. *Transplantation* **1999**, *68*, 563–571. [CrossRef]

405. Xu, D.; Alegre, M.L.; Varga, S.S.; Rothermel, A.L.; Collins, A.M.; Pulito, V.L.; Hanna, L.S.; Dolan, K.P.; Parren, P.W.; Bluestone, J.A.; et al. In vitro characterization of five humanized OKT3 effector function variant antibodies. *Cell. Immunol.* **2000**, *200*, 16–26. [CrossRef]

406. Tam, S.H.; McCarthy, S.G.; Armstrong, A.A.; Somani, S.; Wu, S.J.; Liu, X.S.; Gervais, A.; Ernst, R.; Saro, D.; Decker, R.; et al. Functional, Biophysical, and Structural Characterization of Human IgG1 and IgG4 Fc Variants with Ablated Immune Functionality. *Antibodies* **2017**, *6*, 12. [CrossRef]

407. Lee, C.H.; Romain, G.; Yan, W.; Watanabe, M.; Charab, W.; Todorova, B.; Lee, J.; Triplett, K.; Donkor, M.; Lungu, O.I.; et al. IgG Fc domains that bind C1q but not effector Fcgamma receptors delineate the importance of complement-mediated effector functions. *Nat. Immunol.* **2017**, *18*, 889–898. [CrossRef] [PubMed]

408. Schneider, S.; Zacharias, M. Atomic resolution model of the antibody Fc interaction with the complement C1q component. *Mol. Immunol.* **2012**, *51*, 66–72. [CrossRef] [PubMed]

409. Idusogie, E.E.; Wong, P.Y.; Presta, L.G.; Gazzano-Santoro, H.; Totpal, K.; Ultsch, M.; Mulkerrin, M.G. Engineered antibodies with increased activity to recruit complement. *J. Immunol.* **2001**, *166*, 2571–2575. [CrossRef] [PubMed]

410. Moore, G.L.; Chen, H.; Karki, S.; Lazar, G.A. Engineered Fc variant antibodies with enhanced ability to recruit complement and mediate effector functions. *mAbs* **2010**, *2*, 181–189. [CrossRef] [PubMed]

411. Konno, Y.; Kobayashi, Y.; Takahashi, K.; Takahashi, E.; Sakae, S.; Wakitani, M.; Yamano, K.; Suzawa, T.; Yano, K.; Ohta, T.; et al. Fucose content of monoclonal antibodies can be controlled by culture medium osmolality for high antibody-dependent cellular cytotoxicity. *Cytotechnology* **2012**, *64*, 249–265. [CrossRef]

412. Shields, R.L.; Lai, J.; Keck, R.; O'Connell, L.Y.; Hong, K.; Meng, Y.G.; Weikert, S.H.; Presta, L.G. Lack of fucose on human IgG1 N-linked oligosaccharide improves binding to human Fcgamma RIII and antibody-dependent cellular toxicity. *J. Biol. Chem.* **2002**, *277*, 26733–26740. [CrossRef]

413. Olivier, S.; Jacoby, M.; Brillon, C.; Bouletreau, S.; Mollet, T.; Nerriere, O.; Angel, A.; Danet, S.; Souttou, B.; Guehenneux, F.; et al. EB66 cell line, a duck embryonic stem cell-derived substrate for the industrial production of therapeutic monoclonal antibodies with enhanced ADCC activity. *mAbs* **2010**, *2*, 405–415. [CrossRef]

414. Shinkawa, T.; Nakamura, K.; Yamane, N.; Shoji-Hosaka, E.; Kanda, Y.; Sakurada, M.; Uchida, K.; Anazawa, H.; Satoh, M.; Yamasaki, M.; et al. The absence of fucose but not the presence of galactose or bisecting N-acetylglucosamine of human IgG1 complex-type oligosaccharides shows the critical role of enhancing antibody-dependent cellular cytotoxicity. *J. Biol. Chem.* **2003**, *278*, 3466–3473. [CrossRef]

415. Mori, K.; Kuni-Kamochi, R.; Yamane-Ohnuki, N.; Wakitani, M.; Yamano, K.; Imai, H.; Kanda, Y.; Niwa, R.; Iida, S.; Uchida, K.; et al. Engineering Chinese hamster ovary cells to maximize effector function of produced antibodies using FUT8 siRNA. *Biotechnol. Bioeng.* **2004**, *88*, 901–908. [CrossRef]

416. Ferrara, C.; Stuart, F.; Sondermann, P.; Brunker, P.; Umana, P. The carbohydrate at FcgammaRIIIa Asn-162. An element required for high affinity binding to non-fucosylated IgG glycoforms. *J. Biol. Chem.* **2006**, *281*, 5032–5036. [CrossRef] [PubMed]

417. Ferrara, C.; Brunker, P.; Suter, T.; Moser, S.; Puntener, U.; Umana, P. Modulation of therapeutic antibody effector functions by glycosylation engineering: Influence of Golgi enzyme localization domain and co-expression of heterologous beta1, 4-N-acetylglucosaminyltransferase III and Golgi alpha-mannosidase II. *Biotechnol. Bioeng.* **2006**, *93*, 851–861. [CrossRef] [PubMed]

418. Zhou, Q.; Shankara, S.; Roy, A.; Qiu, H.W.; Estes, S.; McVie-Wylie, A.; Culm-Merdek, K.; Park, A.; Pan, C. Edmunds, Development of a simple and rapid method for producing non-fucosylated oligomannose containing antibodies with increased effector function. *Biotechnol. Bioeng.* **2007**, *99*, 652–665. [CrossRef] [PubMed]

419. Liu, L. Antibody glycosylation and its impact on the pharmacokinetics and pharmacodynamics of monoclonal antibodies and Fc-fusion proteins. *J. Pharm. Sci.* **2015**, *104*, 1866–1884. [CrossRef]

420. Chames, P.; Van Regenmortel, M.; Weiss, E.; Baty, D. Therapeutic antibodies: Successes, limitations and hopes for the future. *Br. J. Pharmacol.* **2009**, *157*, 220–233. [CrossRef]

421. Pollreisz, A.; Assinger, A.; Hacker, S.; Hoetzenecker, K.; Schmid, W.; Lang, G.; Wolfsberger, M.; Steinlechner, B.; Bielek, E.; Lalla, E.; et al. Intravenous immunoglobulins induce CD32-mediated platelet aggregation in vitro. *Br. J. Dermatol.* **2008**, *159*, 578–584. [CrossRef]

422. Labrijn, A.F.; Aalberse, R.C.; Schuurman, J. When binding is enough: Nonactivating antibody formats. *Curr. Opin. Immunol.* **2008**, *20*, 479–485. [CrossRef]

423. Vafa, O.; Gilliland, G.L.; Brezski, R.J.; Strake, B.; Wilkinson, T.; Lacy, E.R.; Scallon, B.; Teplyakov, A.; Malia, T.J.; Strohl, W.R. An engineered Fc variant of an IgG eliminates all immune effector functions via structural perturbations. *Methods* **2014**, *65*, 114–126. [CrossRef]

424. Kinder, M.; Greenplate, A.R.; Strohl, W.R.; Jordan, R.E.; Brezski, R.J. An Fc engineering approach that modulates antibody-dependent cytokine release without altering cell-killing functions. *mAbs* **2015**, *7*, 494–504. [CrossRef]

425. Jones, E.A.; Waldmann, T.A. The mechanism of intestinal uptake and transcellular transport of IgG in the neonatal rat. *J. Clin. Investig.* **1972**, *51*, 2916–2927. [CrossRef]

426. Waldmann, T.A.; Jones, E.A. The role of cell-surface receptors in the transport and catabolism of immunoglobulins. *Ciba Found. Symp.* **1972**, *9*, 5–23. [PubMed]

427. Leach, J.L.; Sedmak, D.D.; Osborne, J.M.; Rahill, B.; Lairmore, M.D.; Anderson, C.L. Isolation from human placenta of the IgG transporter, FcRn, and localization to the syncytiotrophoblast: Implications for maternal-fetal antibody transport. *J. Immunol.* **1996**, *157*, 3317–3322. [PubMed]

428. Brambell, F.W.; Hemmings, W.A.; Morris, I.G. A Theoretical Model of Gamma-Globulin Catabolism. *Nature* **1964**, *203*, 1352–1354. [CrossRef] [PubMed]

429. Junghans, R.P.; Anderson, C.L. The protection receptor for IgG catabolism is the beta2-microglobulin-containing neonatal intestinal transport receptor. *Proc. Natl. Acad. Sci. USA* **1996**, *93*, 5512–5516. [CrossRef] [PubMed]

430. Burmeister, W.P.; Huber, A.H.; Bjorkman, P.J. Crystal structure of the complex of rat neonatal Fc receptor with Fc. *Nature* **1994**, *372*, 379–383. [CrossRef] [PubMed]

431. Burmeister, W.P.; Gastinel, L.N.; Simister, N.E.; Blum, M.L.; Bjorkman, P.J. Crystal structure at 2.2 A resolution of the MHC-related neonatal Fc receptor. *Nature* **1994**, *372*, 336–343. [CrossRef] [PubMed]

432. Kim, J.K.; Tsen, M.F.; Ghetie, V.; Ward, E.S. Identifying amino acid residues that influence plasma clearance of murine IgG1 fragments by site-directed mutagenesis. *Eur. J. Immunol.* **1994**, *24*, 542–548. [CrossRef]

433. Huber, A.H.; RKelley, F.; Gastinel, L.N.; Bjorkman, P.J. Crystallization and stoichiometry of binding of a complex between a rat intestinal Fc receptor and Fc. *J. Mol. Biol.* **1993**, *230*, 1077–1083. [CrossRef]

434. Ghetie, V.; Popov, S.; Borvak, J.; Radu, C.; Matesoi, D.; Medesan, C.; Ober, R.J.; Ward, E.S. Increasing the serum persistence of an IgG fragment by random mutagenesis. *Nat. Biotechnol.* **1997**, *15*, 637–640. [CrossRef]

435. Dall'Acqua, W.F.; Woods, R.M.; Ward, E.S.; Palaszynski, S.R.; Patel, N.K.; Brewah, Y.A.; Wu, H.; Kiener, P.A.; Langermann, S. Increasing the affinity of a human IgG1 for the neonatal Fc receptor: Biological consequences. *J. Immunol.* **2002**, *169*, 5171–5180.

436. Dall'Acqua, W.F.; Kiener, P.A.; Wu, H. Properties of human IgG1s engineered for enhanced binding to the neonatal Fc receptor (FcRn). *J. Biol. Chem.* **2006**, *281*, 23514–23524. [CrossRef] [PubMed]

437. Robbie, G.J.; Criste, R.; Dall'acqua, W.F.; Jensen, K.; Patel, N.K.; Losonsky, G.A.; Griffin, M.P. A novel investigational Fc-modified humanized monoclonal antibody, motavizumab-YTE, has an extended half-life in healthy adults. *Antimicrob. Agents Chemother.* **2013**, *57*, 6147–6153. [CrossRef] [PubMed]

438. Rosenberg, Y.J.; Lewis, G.K.; Montefiori, D.C.; LaBranche, C.C.; Lewis, M.G.; Urban, L.A.; Lees, J.P.; Mao, L.; Jiang, X. Introduction of the YTE mutation into the non-immunogenic HIV bnAb PGT121 induces anti-drug antibodies in macaques. *PLoS ONE* **2019**, *14*, e0212649. [CrossRef] [PubMed]

439. Zalevsky, J.; Chamberlain, A.K.; Horton, H.M.; Karki, S.; Leung, I.W.; Sproule, T.J.; Lazar, G.A.; Roopenian, D.C.; Desjarlais, J.R. Enhanced antibody half-life improves in vivo activity. *Nat. Biotechnol.* **2010**, *28*, 157–159. [CrossRef] [PubMed]

440. Gautam, R.; Nishimura, Y.; Gaughan, N.; Gazumyan, A.; Schoofs, T.; Buckler-White, A.; Seaman, M.S.; Swihart, B.J.; Follmann, D.A.; Nussenzweig, M.C.; et al. A single injection of crystallizable fragment domain-modified antibodies elicits durable protection from SHIV infection. *Nat. Med.* **2018**, *24*, 610–616. [CrossRef] [PubMed]

441. Datta-Mannan, A.; Witcher, D.R.; Tang, Y.; Watkins, J.; Jiang, W.; Wroblewski, V.J. Humanized IgG1 variants with differential binding properties to the neonatal Fc receptor: Relationship to pharmacokinetics in mice and primates. *Drug Metab. Dispos.* **2007**, *35*, 86–94. [CrossRef]

442. Borrok, M.J.; Wu, Y.; Beyaz, N.; Yu, X.Q.; Oganesyan, V.; Dall'Acqua, W.F.; Tsui, P. pH-dependent binding engineering reveals an FcRn affinity threshold that governs IgG recycling. *J. Biol. Chem.* **2015**, *290*, 4282–4290. [CrossRef]

443. Roopenian, D.C.; Sun, V.Z. Clinical ramifications of the MHC family Fc receptor FcRn. *J. Clin. Immunol.* **2010**, *30*, 790–797. [CrossRef]

444. Roopenian, D.C.; Christianson, G.J.; Sproule, T.J. Human FcRn transgenic mice for pharmacokinetic evaluation of therapeutic antibodies. *Methods Mol. Biol.* **2010**, *602*, 93–104.

445. Roopenian, D.C.; Akilesh, S. FcRn: The neonatal Fc receptor comes of age. *Nat. Rev. Immunol.* **2007**, *7*, 715–725. [CrossRef]

446. Rath, T.; Kuo, T.T.; Baker, K.; Qiao, S.W.; Kobayashi, K.; Yoshida, M.; Roopenian, D.; Fiebiger, E.; Lencer, W.I.; Blumberg, R.S. The immunologic functions of the neonatal Fc receptor for IgG. *J. Clin. Immunol.* **2013**, *33* (Suppl. 1), S9–S17. [CrossRef] [PubMed]

447. Wang, W.; Lu, P.; Fang, Y.; Hamuro, L.; Pittman, T.; Carr, B.; Hochman, J.; Prueksaritanont, T. Monoclonal antibodies with identical Fc sequences can bind to FcRn differentially with pharmacokinetic consequences. *Drug Metab. Dispos.* **2011**, *39*, 1469–1477. [CrossRef] [PubMed]

448. Kim, J.K.; Firan, M.; Radu, C.G.; Kim, C.H.; Ghetie, V.; Ward, E.S. Mapping the site on human IgG for binding of the MHC class I-related receptor, FcRn. *Eur. J. Immunol.* **1999**, *29*, 2819–2825. [CrossRef]

449. Vaccaro, C.; Zhou, J.; Ober, R.J.; Ward, E.S. Engineering the Fc region of immunoglobulin G to modulate in vivo antibody levels. *Nat. Biotechnol.* **2005**, *23*, 1283–1288. [CrossRef]

450. Hinton, P.R.; Xiong, J.M.; Johlfs, M.G.; Tang, M.T.; Keller, S.; Tsurushita, N. An engineered human IgG1 antibody with longer serum half-life. *J. Immunol.* **2006**, *176*, 346–356. [CrossRef]

451. Datta-Mannan, A.; Chow, C.K.; Dickinson, C.; Driver, D.; Lu, J.; Witcher, D.R.; Wroblewski, V.J. FcRn affinity-pharmacokinetic relationship of five human IgG4 antibodies engineered for improved in vitro FcRn binding properties in cynomolgus monkeys. *Drug Metab. Dispos.* **2012**, *40*, 1545–1555. [CrossRef]

452. Datta-Mannan, A.; Witcher, D.R.; Lu, J.; Wroblewski, V.J. Influence of improved FcRn binding on the subcutaneous bioavailability of monoclonal antibodies in cynomolgus monkeys. *mAbs* **2012**, *4*, 267–273. [CrossRef]

453. Yeung, Y.A.; Wu, X.; Reyes, A.E.; Vernes, J.M.; Lien, S.; Lowe, J.; Maia, M.; Forrest, W.F.; Meng, Y.G.; Damico, L.A.; et al. A therapeutic anti-VEGF antibody with increased potency independent of pharmacokinetic half-life. *Cancer Res.* **2010**, *70*, 3269–3277. [CrossRef]

454. Oganesyan, V.; Damschroder, M.M.; Cook, K.E.; Li, Q.; Gao, C.; Wu, H.; Dall'Acqua, W.F. Structural insights into neonatal Fc receptor-based recycling mechanisms. *J. Biol. Chem.* **2014**, *289*, 7812–7824. [CrossRef]

455. Igawa, T.; Tsunoda, H.; Tachibana, T.; Maeda, A.; Mimoto, F.; Moriyama, C.; Nanami, M.; Sekimori, Y.; Nabuchi, Y.; Aso, Y.; et al. Reduced elimination of IgG antibodies by engineering the variable region. *Protein Eng. Des. Sel.* **2010**, *23*, 385–392. [CrossRef]

456. Li, B.; Tesar, D.; Boswell, C.A.; Cahaya, H.S.; Wong, A.; Zhang, J.; Meng, Y.G.; Eigenbrot, C.; Pantua, H.; Diao, J.; et al. Framework selection can influence pharmacokinetics of a humanized therapeutic antibody through differences in molecule charge. *mAbs* **2014**, *6*, 1255–1264. [CrossRef]

457. Liu, L. Pharmacokinetics of monoclonal antibodies and Fc-fusion proteins. *Protein Cell* **2018**, *9*, 15–32. [CrossRef]

458. Igawa, T.; Tsunoda, H.; Kuramochi, T.; Sampei, Z.; Ishii, S.; Hattori, K. Engineering the variable region of therapeutic IgG antibodies. *mAbs* **2011**, *3*, 243–252. [CrossRef]

459. Datta-Mannan, A.; Thangaraju, A.; Leung, D.; Tang, Y.; Witcher, D.R.; Lu, J.; Wroblewski, V.J. Balancing charge in the complementarity-determining regions of humanized mAbs without affecting pI reduces non-specific binding and improves the pharmacokinetics. *mAbs* **2015**, *7*, 483–493. [CrossRef]

460. Datta-Mannan, A.; Lu, J.; Witcher, D.R.; Leung, D.; Tang, Y.; Wroblewski, V.J. The interplay of non-specific binding, target-mediated clearance and FcRn interactions on the pharmacokinetics of humanized antibodies. *mAbs* **2015**, *7*, 1084–1093. [CrossRef]

461. Fan, G.; Wang, Z.; Hao, M.; Li, J. Bispecific antibodies and their applications. *J. Hematol. Oncol.* **2015**, *8*, 130. [CrossRef]

462. Suresh, T.; Lee, L.X.; Joshi, J.; Barta, S.K. New antibody approaches to lymphoma therapy. *J. Hematol. Oncol.* **2014**, *7*, 58. [CrossRef]

463. Sedykh, S.E.; Prinz, V.V.; Buneva, V.N.; Nevinsky, G.A. Bispecific antibodies: Design, therapy, perspectives. *Drug Des. Dev. Ther.* **2018**, *12*, 195–208. [CrossRef]

464. Vauquelin, G.; Bricca, G.; van Liefde, I. Avidity and positive allosteric modulation/cooperativity act hand in hand to increase the residence time of bivalent receptor ligands. *Fundam. Clin. Pharmacol.* **2014**, *28*, 530–543. [CrossRef]

465. Ruf, P.; Lindhofer, H. Induction of a long-lasting antitumor immunity by a trifunctional bispecific antibody. *Blood* **2001**, *98*, 2526–2534. [CrossRef]

466. Fossati, M.; Buzzonetti, A.; Monego, G.; Catzola, V.; Scambia, G.; Fattorossi, A.; Battaglia, A. Immunological changes in the ascites of cancer patients after intraperitoneal administration of the bispecific antibody catumaxomab (anti-EpCAMxanti-CD3). *Gynecol. Oncol.* **2015**, *138*, 343–351. [CrossRef]

467. Ott, M.G.; Marme, F.; Moldenhauer, G.; Lindhofer, H.; Hennig, M.; Spannagl, R.; Essing, M.M.; Linke, R.; Seimetz, D. Humoral response to catumaxomab correlates with clinical outcome: Results of the pivotal phase II/III study in patients with malignant ascites. *Int. J. Cancer* **2012**, *130*, 2195–2203. [CrossRef]

468. Loffler, A.; Kufer, P.; Lutterbuse, R.; Zettl, F.; Daniel, P.T.; Schwenkenbecher, J.M.; Riethmuller, G.; Dorken, B.; Bargou, R.C. A recombinant bispecific single-chain antibody, CD19 x CD3, induces rapid and high lymphoma-directed cytotoxicity by unstimulated T lymphocytes. *Blood* **2000**, *95*, 2098–2103. [CrossRef]

469. Goebeler, M.E.; Knop, S.; Viardot, A.; Kufer, P.; Topp, M.S.; Einsele, H.; Noppeney, R.; Hess, G.; Kallert, S.; Mackensen, A.; et al. Bispecific T-Cell Engager (BiTE) Antibody Construct Blinatumomab for the Treatment of Patients with Relapsed/Refractory Non-Hodgkin Lymphoma: Final Results from a Phase I Study. *J. Clin. Oncol.* **2016**, *34*, 1104–1111. [CrossRef]

470. Goebeler, M.E.; Bargou, R. Blinatumomab: A CD19/CD3 bispecific T cell engager (BiTE) with unique anti-tumor efficacy. *Leuk Lymphoma* **2016**, *57*, 1021–1032. [CrossRef]

471. Kitazawa, T.; Igawa, T.; Sampei, Z.; Muto, A.; Kojima, T.; Soeda, T.; Yoshihashi, K.; Okuyama-Nishida, Y.; Saito, H.; Tsunoda, H.; et al. A bispecific antibody to factors IXa and X restores factor VIII hemostatic activity in a hemophilia A model. *Nat. Med.* **2012**, *18*, 1570–1574. [CrossRef]

472. Lenting, P.J.; Denis, C.V.; Christophe, O.D. Emicizumab, a bispecific antibody recognizing coagulation factors IX and X: How does it actually compare to factor VIII? *Blood* **2017**, *130*, 2463–2468. [CrossRef]

473. Knight, T.; Callaghan, M.U. The role of emicizumab, a bispecific factor IXa- and factor X-directed antibody, for the prevention of bleeding episodes in patients with hemophilia A. *Ther. Adv. Hematol.* **2018**, *9*, 319–334. [CrossRef]

474. Sampei, Z.; Igawa, T.; Soeda, T.; Okuyama-Nishida, Y.; Moriyama, C.; Wakabayashi, T.; Tanaka, E.; Muto, A.; Kojima, T.; Kitazawa, T.; et al. Identification and multidimensional optimization of an asymmetric bispecific IgG antibody mimicking the function of factor VIII cofactor activity. *PLoS ONE* **2013**, *8*, e57479. [CrossRef]

475. Jarantow, S.W.; Bushey, B.S.; Pardinas, J.R.; Boakye, K.; Lacy, E.R.; Sanders, R.; Sepulveda, M.A.; Moores, S.L.; Chiu, M.L. Impact of Cell-surface Antigen Expression on Target Engagement and Function of an Epidermal Growth Factor Receptor x c-MET Bispecific Antibody. *J. Biol. Chem.* **2015**, *290*, 24689–24704. [CrossRef]

476. Zheng, S.; Moores, S.; Jarantow, S.; Pardinas, J.; Chiu, M.; Zhou, H.; Wang, W. Cross-arm binding efficiency of an EGFR x c-Met bispecific antibody. *mAbs* **2016**, *8*, 551–561. [CrossRef]

477. Moores, S.L.; Chiu, M.L.; Bushey, B.S.; Chevalier, K.; Luistro, L.; Dorn, K.; Brezski, R.J.; Haytko, P.; Kelly, T.; Wu, S.J.; et al. A Novel Bispecific Antibody Targeting EGFR and cMet Is Effective against EGFR Inhibitor-Resistant Lung Tumors. *Cancer Res.* **2016**, *76*, 3942–3953. [CrossRef]

478. Yang, F.; Wen, W.; Qin, W. Bispecific Antibodies as a Development Platform for New Concepts and Treatment Strategies. *Int. J. Mol. Sci.* **2017**, *18*, 48. [CrossRef]

479. Farrington, G.K.; Caram-Salas, N.; Haqqani, A.S.; Brunette, E.; Eldredge, J.; Pepinsky, B.; Antognetti, G.; Baumann, E.; Ding, W.; Garber, E.; et al. A novel platform for engineering blood-brain barrier-crossing bispecific biologics. *FASEB J.* **2014**, *28*, 4764–4778. [CrossRef]

480. Stanimirovic, D.; Kemmerich, K.; Haqqani, A.S.; Farrington, G.K. Engineering and pharmacology of blood-brain barrier-permeable bispecific antibodies. *Adv. Pharmacol.* **2014**, *71*, 301–335.

481. Couch, J.A.; Yu, Y.J.; Zhang, Y.; Tarrant, J.M.; Fuji, R.N.; Meilandt, W.J.; Solanoy, H.; Tong, R.K.; Hoyte, K.; Luk, W.; et al. Addressing safety liabilities of TfR bispecific antibodies that cross the blood-brain barrier. *Sci. Transl. Med.* **2013**, *5*, 1–12. [CrossRef]

482. Yu, Y.J.; Atwal, J.K.; Zhang, Y.; Tong, R.K.; Wildsmith, K.R.; Tan, C.; Bien-Ly, N.; Hersom, M.; Maloney, J.A.; Meilandt, W.J.; et al. Therapeutic bispecific antibodies cross the blood-brain barrier in nonhuman primates. *Sci. Transl. Med.* **2014**, *6*, 261ra154. [CrossRef]

483. Atwal, J.K.; Chen, Y.; Chiu, C.; Mortensen, D.L.; Meilandt, W.J.; Liu, Y.; Heise, C.E.; Hoyte, K.; Luk, W.; Lu, Y.; et al. A therapeutic antibody targeting BACE1 inhibits amyloid-beta production in vivo. *Sci. Transl. Med.* **2011**, *3*, 84ra43. [CrossRef]

484. Liu, Z.; Gunasekaran, K.; Wang, W.; Razinkov, V.; Sekirov, L.; Leng, E.; Sweet, H.; Foltz, I.; Howard, M.; Rousseau, A.M.; et al. Asymmetrical Fc engineering greatly enhances antibody-dependent cellular cytotoxicity (ADCC) effector function and stability of the modified antibodies. *J. Biol. Chem.* **2014**, *289*, 3571–3590. [CrossRef]

485. Escobar-Carbrera, E.; Lario, P.I.; Shrag, J.; Durocher, Y.; Dixit, S.B. Asymmetric Fc Engineering for Bispecific Antibodies with Reduced Effector Function. *Antibodies* **2017**, *6*, 7. [CrossRef]

486. Riethmuller, G. Symmetry breaking: Bispecific antibodies, the beginnings, and 50 years on. *Cancer Immun.* **2012**, *12*, 12. [PubMed]

487. Spiess, C.; Zhai, Q.; Carter, P.J. Alternative molecular formats and therapeutic applications for bispecific antibodies. *Mol. Immunol.* **2015**, *67 Pt A*, 95–106. [CrossRef]

488. Ha, J.H.; Kim, J.E.; Kim, Y.S. Immunoglobulin Fc Heterodimer Platform Technology: From Design to Applications in Therapeutic Antibodies and Proteins. *Front. Immunol.* **2016**, *7*, 394. [CrossRef] [PubMed]

489. Godar, M.; de Haard, H.; Blanchetot, C.; Rasser, J. Therapeutic bispecific antibody formats: A patent applications review (1994–2017). *Expert Opin. Ther. Pat.* **2018**, *28*, 251–276. [CrossRef] [PubMed]

490. Labrijn, A.F.; Janmaat, M.L.; Reichert, J.M.; Parren, P. Bispecific antibodies: A mechanistic review of the pipeline. *Nat. Rev. Drug Discov.* **2019**, *18*, 585–608. [CrossRef] [PubMed]

491. Wang, Q.; Chen, Y.; Park, J.; Liu, X.; Hu, Y.; Wang, T.; McFarland, K.; Betenbaugh, M.J. Design and Production of Bispecific Antibodies. *Antibodies* **2019**, *8*, 43. [CrossRef]

492. Reddy Chichili, V.P.; Kumar, V.; Sivaraman, J. Linkers in the structural biology of protein-protein interactions. *Protein Sci.* **2013**, *22*, 153–167. [CrossRef]

493. Xenaki, K.T.; Oliveira, S.; van Bergen, P.M.P. En henegouwen. Antibody or antibody fragments: Implications for molecular imaging and targeted therapy of solid tumors. *Front. Immunol.* **2017**, *8*, 1287. [CrossRef]

494. Dave, E.; Adams, R.; Zaccheo, O.; Carrington, B.; Compson, J.E.; Dugdale, S.; Airey, M.; Malcolm, S.; Hailu, H.; Wild, G.; et al. Fab-dsFv: A bispecific antibody format with extended serum half-life through albumin binding. *mAbs* **2016**, *8*, 1319–1335. [CrossRef]

495. Wu, X.; Sereno, A.J.; Huang, F.; Lewis, S.M.; Lieu, R.L.; Weldon, C.; Torres, C.; Fine, C.; Batt, M.A.; Fitchett, J.R.; et al. Fab-based bispecific antibody formats with robust biophysical properties and biological activity. *mAbs* **2015**, *7*, 470–482. [CrossRef]

496. Schoonjans, R.; Willems, A.; Schoonooghe, S.; Fiers, W.; Grooten, J.; Mertens, N. Fab chains as an efficient heterodimerization scaffold for the production of recombinant bispecific and trispecific antibody derivatives. *J. Immunol.* **2000**, *165*, 7050–7057. [CrossRef] [PubMed]

497. Li, L.; He, P.; Zhou, C.; Jing, L.; Dong, B.; Chen, S.Y.; Zhang, N.; Liu, Y.; Maio, J.; Wang, Z.; et al. A Novel Bispecific Antibody, S-Fab, Induces Potent Cancer Cell Killing. *J. Immunother.* **2015**, *38*, 350–356. [CrossRef] [PubMed]

498. Hayden, M.S.; Linsley, P.S.; Gayle, M.A.; Bajorath, J.; Brady, W.A.; Norris, N.A.; Fell, H.P.; Ledbetter, J.A.; Gilliland, L.K. Single-chain mono- and bispecific antibody derivatives with novel biological properties and antitumour activity from a COS cell transient expression system. *Ther. Immunol.* **1994**, *1*, 3–15. [PubMed]

499. Mallender, W.D.; Voss, E.W., Jr. Construction, expression, and activity of a bivalent bispecific single-chain antibody. *J. Biol. Chem.* **1994**, *269*, 199–206. [PubMed]

500. Mack, M.; Riethmuller, G.; Kufer, P. A small bispecific antibody construct expressed as a functional single-chain molecule with high tumor cell cytotoxicity. *Proc. Natl. Acad. Sci. USA* **1995**, *92*, 7021–7025. [CrossRef] [PubMed]

501. Holliger, P.; Prospero, T.; Winter, G. "Diabodies": Small bivalent and bispecific antibody fragments. *Proc. Natl. Acad. Sci. USA* **1993**, *90*, 6444–6448. [CrossRef]

502. Brusselbach, S.; Korn, T.; Volkel, T.; Muller, R.; Kontermann, R.E. Enzyme recruitment and tumor cell killing in vitro by a secreted bispecific single chain diabody. *Tumor Target.* **1999**, *4*, 115–123.

503. Zhu, Z.; Presta, L.G.; Zapata, G.; Carter, P. Remodeling domain interfaces to enhance heterodimer formation. *Protein Sci.* **1997**, *6*, 781–788. [CrossRef]

504. FitzGerald, K.; Holliger, P.; Winter, G. Improved tumour targeting by disulphide stabilized diabodies expressed in Pichia pastoris. *Protein Eng.* **1997**, *10*, 1221–1225. [CrossRef]

505. Johnson, S.; Burke, S.; Huang, L.; Gorlatov, S.; Li, H.; Wang, W.; Zhang, W.; Tuaillon, N.; Rainey, J.; Barat, B.; et al. Effector cell recruitment with novel Fv-based dual-affinity re-targeting protein leads to potent tumor cytolysis and in vivo B-cell depletion. *J. Mol. Biol.* **2010**, *399*, 436–449. [CrossRef]

506. Kipriyanov, S.M.; Moldenhauer, G.; Schuhmacher, J.; Cochlovius, B.; von der Lieth, C.W.; Matys, E.R.; Little, M. Bispecific tandem diabody for tumor therapy with improved antigen binding and pharmacokinetics. *J. Mol. Biol.* **1999**, *293*, 41–56. [CrossRef] [PubMed]

507. Cochlovius, B.; Kipriyanov, S.M.; Stassar, M.J.; Schuhmacher, J.; Benner, A.; Moldenhauer, G.; Little, M. Cure of Burkitt's lymphoma in severe combined immunodeficiency mice by T cells, tetravalent CD3 × CD19 tandem diabody, and CD28 costimulation. *Cancer Res.* **2000**, *60*, 4336–4341. [PubMed]

508. Cochlovius, B.; Kipriyanov, S.M.; Stassar, M.J.; Christ, O.; Schuhmacher, J.; Strauss, G.; Moldenhauer, G.; Little, M. Treatment of human B cell lymphoma xenografts with a CD3 × CD19 diabody and T cells. *J. Immunol.* **2000**, *165*, 888–895. [CrossRef] [PubMed]

509. Els Conrath, K.; Lauwereys, M.; Wyns, L.; Muyldermans, S. Camel single-domain antibodies as modular building units in bispecific and bivalent antibody constructs. *J. Biol. Chem.* **2001**, *276*, 7346–7350. [CrossRef] [PubMed]

510. De Bernardis, F.; Liu, H.; O'Mahony, R.; la Valle, R.; Bartollino, S.; Sandini, S.; Grant, S.; Brewis, N.; Tomlinson, I.; Basset, R.C.; et al. Human domain antibodies against virulence traits of Candida albicans inhibit fungus adherence to vaginal epithelium and protect against experimental vaginal candidiasis. *J. Infect. Dis.* **2007**, *195*, 149–157. [CrossRef]

511. Bannas, P.; Hambach, J.; Koch-Nolte, F. Nanobodies and Nanobody-Based Human Heavy Chain Antibodies as Antitumor Therapeutics. *Front. Immunol.* **2017**, *8*, 1603. [CrossRef]

512. Ridgway, J.B.; Presta, L.G.; Carter, P. 'Knobs-into-holes' engineering of antibody CH3 domains for heavy chain heterodimerization. *Protein Eng.* **1996**, *9*, 617–621. [CrossRef]

513. Atwell, S.; Ridgway, J.B.; Wells, J.A.; Carter, P. Stable heterodimers from remodeling the domain interface of a homodimer using a phage display library. *J. Mol. Biol.* **1997**, *270*, 26–35. [CrossRef]

514. Merchant, A.M.; Zhu, Z.; Yuan, J.Q.; Goddard, A.; Adams, C.W.; Presta, L.G.; Carter, P. An efficient route to human bispecific IgG. *Nat. Biotechnol.* **1998**, *16*, 677–681. [CrossRef]

515. Gunasekaran, K.; Pentony, M.; Shen, M.; Garrett, L.; Forte, C.; Woodward, A.; Ng, S.B.; Born, T.; Retter, M.; Manchulenko, K.; et al. Enhancing antibody Fc heterodimer formation through electrostatic steering effects: Applications to bispecific molecules and monovalent IgG. *J. Biol. Chem.* **2010**, *285*, 19637–19646. [CrossRef]

516. Moore, G.L.; Bernett, M.J.; Rashid, R.; Pong, E.W.; Nguyen, D.T.; Jacinto, J.; Eivazi, A.; Nisthal, A.; Diaz, J.E.; Chu, S.Y.; et al. A robust heterodimeric Fc platform engineered for efficient development of bispecific antibodies of multiple formats. *Methods* **2019**, *154*, 38–50. [CrossRef] [PubMed]

517. Labrijn, A.F.; Meesters, J.I.; Priem, P.; de Jong, R.N.; van den Bremer, E.T.; van Kampen, M.D.; Gerritsen, A.F.; Schuurman, J.; Parren, P.W. Controlled Fab-arm exchange for the generation of stable bispecific IgG1. *Nat. Protoc.* **2014**, *9*, 2450–2463. [CrossRef] [PubMed]

518. Labrijn, A.F.; Meesters, J.I.; de Goeij, B.E.; van den Bremer, E.T.; Neijssen, J.; van Kampen, M.D.; Strumane, K.; Verploegen, S.; Kundu, A.; Gramer, M.J.; et al. Efficient generation of stable bispecific IgG1 by controlled Fab-arm exchange. *Proc. Natl. Acad. Sci. USA* **2013**, *110*, 5145–5150. [CrossRef] [PubMed]

519. Goulet, D.R.; Orcutt, S.J.; Zwolak, A.; Rispens, T.; Labrijn, A.F.; de Jong, R.N.; Atkins, W.M.; Chiu, M.L. Kinetic mechanism of controlled Fab-arm exchange for the formation of bispecific immunoglobulin G1 antibodies. *J. Biol. Chem.* **2018**, *293*, 651–661. [CrossRef] [PubMed]

520. Strop, P.; Ho, W.H.; Boustany, L.M.; Abdiche, Y.N.; Lindquist, K.C.; Farias, S.E.; Rickert, M.; Appah, C.T.; Pascua, E.; Radcliffe, T.; et al. Generating bispecific human IgG1 and IgG2 antibodies from any antibody pair. *J. Mol. Biol.* **2012**, *420*, 204–219. [CrossRef]

521. Zwolak, A.; Leettola, C.N.; Tam, S.H.; Goulet, D.R.; Derebe, M.G.; Pardinas, J.R.; Zheng, S.; Decker, R.; Emmell, E.; Chiu, M.L. Rapid Purification of Human Bispecific Antibodies via Selective Modulation of Protein a Binding. *Sci. Rep.* **2017**, *7*, 15521. [CrossRef]

522. Zwolak, A.; Armstrong, A.A.; Tam, S.H.; Pardinas, J.R.; Goulet, D.R.; Zheng, S.; Brosnan, K.; Emmell, E.; Luo, J.; Gilliland, G.L.; et al. Modulation of protein A binding allows single-step purification of mouse bispecific antibodies that retain FcRn binding. *mAbs* **2017**, *9*, 1306–1316. [CrossRef]

523. Eigenbrot, C.; Fuh, G. Two-in-One antibodies with dual action Fabs. *Curr. Opin. Chem. Biol.* **2013**, *17*, 400–405. [CrossRef]

524. Lee, C.V.; Koenig, P.; Fuh, G. A two-in-one antibody engineered from a humanized interleukin 4 antibody through mutation in heavy chain complementarity-determining regions. *mAbs* **2014**, *6*, 622–627. [CrossRef]

525. Wu, C.; Ying, H.; Grinnell, C.; Bryant, S.; Miller, R.; Clabbers, A.; Bose, S.; McCarthy, D.; Zhu, R.R.; Santora, L.; et al. Simultaneous targeting of multiple disease mediators by a dual-variable-domain immunoglobulin. *Nat. Biotechnol.* **2007**, *25*, 1290–1297. [CrossRef]

526. Wu, C.; Ying, H.; Bose, S.; Miller, R.; Medina, L.; Santora, L.; Ghayur, T. Molecular construction and optimization of anti-human IL-1alpha/beta dual variable domain immunoglobulin (DVD-Ig) molecules. *mAbs* **2009**, *1*, 339–347. [CrossRef]

527. Lacy, S.E.; Wu, C.; Ambrosi, D.J.; Hsieh, C.M.; Bose, S.; Miller, R.; Conlon, D.M.; Tarcsa, E.; Chari, R.; Ghayur, T.; et al. Generation and characterization of ABT-981, a dual variable domain immunoglobulin (DVD-Ig(TM)) molecule that specifically and potently neutralizes both IL-1alpha and IL-1beta. *mAbs* **2015**, *7*, 605–619. [CrossRef]

528. Jakob, C.G.; Edalji, R.; Judge, R.A.; DiGiammarino, E.; Li, Y.; Gu, J.; Ghayur, T. Structure reveals function of the dual variable domain immunoglobulin (DVD-Ig) molecule. *mAbs* **2013**, *5*, 358–363. [CrossRef]

529. Fischer, N.; Elson, G.; Magistrelli, G.; Dheilly, E.; Fouque, N.; Laurendon, A.; Gueneau, F.; Ravn, U.; Depoisier, J.F.; Moine, V.; et al. Exploiting light chains for the scalable generation and platform purification of native human bispecific IgG. *Nat. Commun.* **2015**, *6*, 6113. [CrossRef]

530. Krah, S.; Sellmann, C.; Rhiel, L.; Schroter, C.; Dickgiesser, S.; Beck, J.; Zielonka, S.; Toleikis, L.; Hock, B.; Kolmar, H.; et al. Engineering bispecific antibodies with defined chain pairing. *New Biotechnol.* **2017**, *39*, 167–173. [CrossRef]

531. Klein, C.; Schaefer, W.; Regula, J.T.; Dumontet, C.; Brinkmann, U.; Bacac, M.; Umana, P. Engineering therapeutic bispecific antibodies using CrossMab technology. *Methods* **2019**, *154*, 21–31. [CrossRef]

532. Bonisch, M.; Sellmann, C.; Maresch, D.; Halbig, C.; Becker, S.; Toleikis, L.; Hock, B.; Ruker, F. Novel CH1:CL interfaces that enhance correct light chain pairing in heterodimeric bispecific antibodies. *Protein Eng. Des. Sel.* **2017**, *30*, 685–696. [CrossRef]

533. Dillon, M.; Yin, Y.; Zhou, J.; McCarty, L.; Ellerman, D.; Slaga, D.; Junttila, T.T.; Han, G.; Sandoval, W.; Ovacik, M.A.; et al. Efficient production of bispecific IgG of different isotypes and species of origin in single mammalian cells. *mAbs* **2017**, *9*, 213–230. [CrossRef]

534. Lewis, S.M.; Wu, X.; Pustilnik, A.; Sereno, A.; Huang, F.; Rick, H.L.; Guntas, G.; Leaver-Fay, A.; Smith, E.M.; Ho, C.; et al. Generation of bispecific IgG antibodies by structure-based design of an orthogonal Fab interface. *Nat. Biotechnol.* **2014**, *32*, 191–198. [CrossRef]

535. Liu, Z.; Leng, E.C.; Gunasekaran, K.; Pentony, M.; Shen, M.; Howard, M.; Stoops, J.; Manchulenko, K.; Razinkov, V.; Liu, H.; et al. A novel antibody engineering strategy for making monovalent bispecific heterodimeric IgG antibodies by electrostatic steering mechanism. *J. Biol. Chem.* **2015**, *290*, 7535–7562. [CrossRef]

536. Woods, R.J.; Xie, M.H.; von Kreudenstein, T.S.; Ng, G.Y.; Dixit, S.B. LC-MS characterization and purity assessment of a prototype bispecific antibody. *mAbs* **2013**, *5*, 711–722. [CrossRef]

537. Von Kreudenstein, T.S.; Escobar-Carrera, E.; Lario, P.I.; D'Angelo, I.; Brault, K.; Kelly, J.; Durocher, Y.; Baardsnes, J.; Woods, R.J.; Xie, M.H.; et al. Improving biophysical properties of a bispecific antibody scaffold to aid developability: Quality by molecular design. *mAbs* **2013**, *5*, 646–654. [CrossRef]

538. Stephenson, R.P. A Modification of Receptor Theory. *Br. J. Pharmacol.* **1956**, *11*, 379–393. [CrossRef]

539. Ellerman, D. Bispecific T-cell engagers: Towards understanding variables influencing the in vitro potency and tumor selectivity and their modulation to enhance their efficacy and safety. *Methods* **2019**, *154*, 102–117. [CrossRef]

540. Zhukovsky, E.A.; Morse, R.J.; Maus, M.V. Bispecific antibodies and CARs: Generalized immunotherapeutics harnessing T cell redirection. *Curr. Opin. Immunol.* **2016**, *40*, 24–35. [CrossRef]

541. Trinklein, N.D.; Pham, D.; Schellenberger, U.; Buelow, B.; Boudreau, A.; Choudhry, P.; Clarke, S.C.; Dang, K.; Harris, K.E.; Iyer, S.; et al. Efficient tumor killing and minimal cytokine release with novel T-cell agonist bispecific antibodies. *mAbs* **2019**, *11*, 639–652. [CrossRef]

542. Strohl, W.R.; Naso, M.F. Bispecific T-Cell Redirection versus Chimeric Antigen. Receptor (CAR)-T Cells as Approaches to Kill. Cancer Cells. *Antibodies* **2019**, *8*, 41. [CrossRef]

543. Suurs, F.V.; Lub-de Hooge, M.N.; de Vries, E.G.E.; de Groot, D.J.A. A review of bispecific antibodies and antibody constructs in oncology and clinical challenges. *Pharmacol. Ther.* **2019**, *201*, 103–119. [CrossRef]

544. Kellner, C.; Peipp, M.; Valerius, T. Effector Cell Recruitment by Bispecific Antibodies. In *Bispecific Antibodies*; Kontermann, R.E., Ed.; Springer: Heidelberg/Berlin, Germany, 2011; pp. 217–241.

545. Peipp, M.; Wesch, D.; Oberg, H.H.; Lutz, S.; Muskulus, A.; van de Winkel, J.G.J.; Parren, P.; Burger, R.; Humpe, A.; Kabelitz, D.; et al. CD20-Specific Immunoligands Engaging NKG2D Enhance gammadelta T Cell-Mediated Lysis of Lymphoma Cells. *Scand. J. Immunol.* **2017**, *86*, 196–206. [CrossRef]

546. Li, B.; Xu, L.; Pi, C.; Yin, Y.; Xie, K.; Tao, F.; Li, R.; Gu, H.; Fang, J. CD89-mediated recruitment of macrophages via a bispecific antibody enhances anti-tumor efficacy. *Oncoimmunology* **2017**, *7*, e1380142. [CrossRef]

547. Dheilly, E.; Moine, V.; Broyer, L.; Salgado-Pires, S.; Johnson, Z.; Papaioannou, A.; Cons, L.; Calloud, S.; Majocchi, S.; Nelson, R.; et al. Selective Blockade of the Ubiquitous Checkpoint Receptor CD47 Is Enabled by Dual-Targeting Bispecific Antibodies. *Mol. Ther.* **2017**, *25*, 523–533. [CrossRef] [PubMed]

548. Brinkmann, U.; Kontermann, R.E. The making of bispecific antibodies. *mAbs* **2017**, *9*, 182–212. [CrossRef] [PubMed]

549. Dreyfus, C.; Laursen, N.S.; Kwaks, T.; Zuijdgeest, D.; Khayat, R.; Ekiert, D.C.; Lee, J.H.; Metlagel, Z.; Bujny, M.V.; Jongeneelen, M.; et al. Highly conserved protective epitopes on influenza B viruses. *Science* **2012**, *337*, 1343–1348. [CrossRef] [PubMed]

550. Ekiert, D.C.; Bhabha, G.; Elsliger, M.A.; Friesen, R.H.; Jongeneelen, M.; Throsby, M.; Goudsmit, J.; Wilson, I.A. Antibody recognition of a highly conserved influenza virus epitope. *Science* **2009**, *324*, 246–251. [CrossRef]

551. Laursen, N.S.; Wilson, I.A. Broadly neutralizing antibodies against influenza viruses. *Antivir. Res.* **2013**, *98*, 476–483. [CrossRef]

552. Tan, G.S.; Lee, P.S.; Hoffman, R.M.; Mazel-Sanchez, B.; Krammer, F.; Leon, P.E.; Ward, A.B.; Wilson, I.A.; Palese, P. Characterization of a broadly neutralizing monoclonal antibody that targets the fusion domain of group 2 influenza A virus hemagglutinin. *J. Virol.* **2014**, *88*, 13580–13592. [CrossRef]

553. Laursen, N.S.; Friesen, R.H.E.; Zhu, X.; Jongeneelen, M.; Blokland, S.; Vermond, J.; van Eijgen, A.; Tang, C.; van Diepen, H.; Obmolova, G.; et al. Universal protection against influenza infection by a multidomain antibody to influenza hemagglutinin. *Science* **2018**, *362*, 598–602. [CrossRef]

554. DiLillo, D.J.; Tan, G.S.; Palese, P.; Ravetch, J.V. Broadly neutralizing hemagglutinin stalk-specific antibodies require FcgammaR interactions for protection against influenza virus in vivo. *Nat. Med.* **2014**, *20*, 143–151. [CrossRef]

555. Thurber, G.M.; Schmidt, M.M.; Wittrup, K.D. Antibody tumor penetration: Transport opposed by systemic and antigen-mediated clearance. *Adv. Drug Deliv. Rev.* **2008**, *60*, 1421–1434. [CrossRef]

556. Balandin, T.G.; Edelweiss, E.; Andronova, N.V.; Treshalina, E.M.; Sapozhnikov, A.M.; Deyev, S.M. Antitumor activity and toxicity of anti-HER2 immunoRNase scFv 4D5-dibarnase in mice bearing human breast cancer xenografts. *Investig. New Drugs* **2011**, *29*, 22–32. [CrossRef]

557. Deyev, S.M.; Waibel, R.; Lebedenko, E.N.; Schubiger, A.P.; Pluckthun, A. Design of multivalent complexes using the barnase*barstar module. *Nat. Biotechnol.* **2003**, *21*, 1486–1492. [CrossRef]

558. Desnoyers, L.R.; Vasiljeva, O.; Richardson, J.H.; Yang, A.; Menendez, E.E.; Liang, T.W.; Wong, C.; Bessette, P.H.; Kamath, K.; Moore, S.J.; et al. Tumor-specific activation of an EGFR-targeting probody enhances therapeutic index. *Sci. Transl. Med.* **2013**, *5*, 207ra144. [CrossRef]

559. Tietze, L.F.; Major, F.; Schuberth, I. Antitumor agents: Development of highly potent glycosidic duocarmycin analogues for selective cancer therapy. *Angew. Chem. Int. Ed. Eng.* **2006**, *45*, 6574–6577. [CrossRef]

560. Jin, W.; Trzupek, J.D.; Rayl, T.J.; Broward, M.A.; Vielhauer, G.A.; Weir, S.J.; Hwang, I.; Boger, D.L. A unique class of duocarmycin and CC-1065 analogues subject to reductive activation. *J. Am. Chem. Soc.* **2007**, *129*, 15391–15397. [CrossRef]

561. Bogen, J.P.; Hinz, S.C.; Grzeschik, J.; Ebening, A.; Krah, S.; Zielonka, S.; Kolmar, H. Dual Function pH Responsive Bispecific Antibodies for Tumor Targeting and Antigen Depletion in Plasma. *Front. Immunol.* **2019**, *10*, 1892. [CrossRef]

562. Andrady, C.; Sharma, S.K.; Chester, K.A. Antibody-enzyme fusion proteins for cancer therapy. *Immunotherapy* **2011**, *3*, 193–211. [CrossRef]

563. Lian, X.; Huang, Y.; Zhu, Y.; Fang, Y.; Zhao, R.; Joseph, E.; Li, J.; Pellois, J.P.; Zhou, H.C. Enzyme-MOF Nanoreactor Activates Nontoxic Paracetamol for Cancer Therapy. *Angew. Chem. Int. Ed. Eng.* **2018**, *57*, 5725–5730. [CrossRef]

564. Sharma, S.K.; Bagshawe, K.D. Antibody Directed Enzyme Prodrug Therapy (ADEPT): Trials and tribulations. *Adv. Drug Deliv. Rev.* **2017**, *118*, 2–7. [CrossRef]

565. Sharma, S.K.; Bagshawe, K.D. Translating antibody directed enzyme prodrug therapy (ADEPT) and prospects for combination. *Expert Opin. Biol. Ther.* **2017**, *17*, 1–13. [CrossRef]

566. Kiefer, J.D.; Neri, D. Immunocytokines and bispecific antibodies: Two complementary strategies for the selective activation of immune cells at the tumor site. *Immunol. Rev.* **2016**, *270*, 178–192. [CrossRef]

567. Hutmacher, C.; Neri, D. Antibody-cytokine fusion proteins: Biopharmaceuticals with immunomodulatory properties for cancer therapy. *Adv. Drug Deliv. Rev.* **2018**, *141*, 67–91. [CrossRef]

568. Marschall, A.L.; Zhang, C.; Frenzel, A.; Schirrmann, T.; Hust, M.; Perez, F.; Dubel, S. Delivery of antibodies to the cytosol: Debunking the myths. *mAbs* **2014**, *6*, 943–956. [CrossRef]

569. Ullrich, A.; Schlessinger, J. Signal transduction by receptors with tyrosine kinase activity. *Cell* **1990**, *61*, 203–212. [CrossRef]

570. Ritchie, M.; Tchistiakova, L.; Scott, N. Implications of receptor-mediated endocytosis and intracellular trafficking dynamics in the development of antibody drug conjugates. *mAbs* **2013**, *5*, 13–21. [CrossRef]

571. Weill, C.O.; Biri, S.; Erbacher, P. Cationic lipid-mediated intracellular delivery of antibodies into live cells. *Biotechniques* **2008**, *44*, Pvii–Pxi. [CrossRef]

572. Weill, C.O.; Biri, S.; Adib, A.; Erbacher, P. A practical approach for intracellular protein delivery. *Cytotechnology* **2008**, *56*, 41–48. [CrossRef]

573. Freund, G.; Sibler, A.P.; Desplancq, D.; Oulad-Abdelghani, M.; Vigneron, M.; Gannon, J.; van Regenmortel, M.H.; Weiss, E. Targeting endogenous nuclear antigens by electrotransfer of monoclonal antibodies in living cells. *mAbs* **2013**, *5*, 518–522. [CrossRef]

574. Fowler, D.A.; Filla, M.B.; Little, C.D.; Rongish, B.J.; Larsson, H.C.E. Live tissue antibody injection: A novel method for imaging ECM in limb buds and other tissues. *Methods Cell. Biol.* **2018**, *143*, 41–56.

575. Becerril, B.; Poul, M.A.; Marks, J.D. Toward selection of internalizing antibodies from phage libraries. *Biochem. Biophys. Res. Commun.* **1999**, *255*, 386–393. [CrossRef]

576. Poul, M.A.; Becerril, B.; Nielsen, U.B.; Morisson, P.; Marks, J.D. Selection of tumor-specific internalizing human antibodies from phage libraries. *J. Mol. Biol.* **2000**, *301*, 1149–1161. [CrossRef]

577. Zhou, Y.; Zhao, L.; Marks, J.D. Selection and characterization of cell binding and internalizing phage antibodies. *Arch. Biochem. Biophys.* **2012**, *526*, 107–113. [CrossRef]

578. Zhou, Y.; Marks, J.D. Discovery of internalizing antibodies to tumor antigens from phage libraries. *Methods Enzymol.* **2012**, *502*, 43–66.

579. Goenaga, A.L.; Zhou, Y.; Legay, C.; Bougherara, H.; Huang, L.; Liu, B.; Drummond, D.C.; Kirpotin, D.B.; Auclair, C.; Marks, J.D.; et al. Identification and characterization of tumor antigens by using antibody phage display and intrabody strategies. *Mol. Immunol.* **2007**, *44*, 3777–3788. [CrossRef]

580. Shin, S.M.; Choi, D.K.; Jung, K.; Bae, J.; Kim, J.S.; Park, S.W.; Song, K.H.; Kim, Y.S. Antibody targeting intracellular oncogenic Ras mutants exerts anti-tumour effects after systemic administration. *Nat. Commun.* **2017**, *8*, 15090. [CrossRef]

581. Beyer, I.; van Rensburg, R.; Strauss, R.; Li, Z.; Wang, H.; Persson, J.; Yumul, R.; Feng, Q.; Song, H.; Bartek, J.; et al. Epithelial junction opener JO-1 improves monoclonal antibody therapy of cancer. *Cancer Res.* **2011**, *71*, 7080–7090. [CrossRef]

582. Sugahara, K.N.; Teesalu, T.; Karmali, P.P.; Kotamraju, V.R.; Agemy, L.; Greenwald, D.R.; Ruoslahti, E. Coadministration of a tumor-penetrating peptide enhances the efficacy of cancer drugs. *Science* **2010**, *328*, 1031–1035. [CrossRef]

583. Avignolo, C.; Bagnasco, L.; Biasotti, B.; Melchiori, A.; Tomati, V.; Bauer, I.; Salis, A.; Chiossone, L.; Mingari, M.C.; Orecchia, P.; et al. Internalization via Antennapedia protein transduction domain of an scFv antibody toward c-Myc protein. *FASEB J.* **2008**, *22*, 1237–1245. [CrossRef]

584. Montrose, K.; Yang, Y.; Sun, X.; Wiles, S.; Krissansen, G.W. Xentry, a new class of cell-penetrating peptide uniquely equipped for delivery of drugs. *Sci. Rep.* **2013**, *3*, 1661. [CrossRef]

585. Halaby, R. Role of lysosomes in cancer therapy. *Res. Rep. Biol.* **2015**, *6*, 147–155. [CrossRef]

586. Shin, T.H.; Sung, E.S.; Kim, Y.J.; Kim, K.S.; Kim, S.H.; Kim, S.K.; Lee, Y.D.; Kim, Y.S. Enhancement of the tumor penetration of monoclonal antibody by fusion of a neuropilin-targeting peptide improves the antitumor efficacy. *Mol. Cancer Ther.* **2014**, *13*, 651–661. [CrossRef]

587. Kim, Y.J.; Bae, J.; Shin, T.H.; Kang, S.H.; Jeong, M.; Han, Y.; Park, J.H.; Kim, S.K.; Kim, Y.S. Immunoglobulin Fc-fused, neuropilin-1-specific peptide shows efficient tumor tissue penetration and inhibits tumor growth via anti-angiogenesis. *J. Control. Release* **2015**, *216*, 56–68. [CrossRef] [PubMed]

588. Kim, J.S.; Choi, D.K.; Shin, J.Y.; Shin, S.M.; Park, S.W.; Cho, H.S.; Kim, Y.S. Endosomal acidic pH-induced conformational changes of a cytosol-penetrating antibody mediate endosomal escape. *J. Control. Release* **2016**, *235*, 165–175. [CrossRef] [PubMed]

589. Kim, J.S.; Park, J.Y.; Shin, S.M.; Park, S.W.; Jun, S.Y.; Hong, J.S.; Choi, D.K.; Kim, Y.S. Engineering of a tumor cell-specific, cytosol-penetrating antibody with high endosomal escape efficacy. *Biochem. Biophys. Res. Commun.* **2018**, *503*, 2510–2516. [CrossRef] [PubMed]

590. Boldicke, T. Single domain antibodies for the knockdown of cytosolic and nuclear proteins. *Protein Sci.* **2017**, *26*, 925–945. [CrossRef] [PubMed]

591. Choi, D.K.; Bae, J.; Shin, S.M.; Shin, J.Y.; Kim, S.; Kim, Y.S. A general strategy for generating intact, full-length IgG antibodies that penetrate into the cytosol of living cells. *mAbs* **2014**, *6*, 1402–1414. [CrossRef] [PubMed]

592. Cardinale, A.; Biocca, S. The potential of intracellular antibodies for therapeutic targeting of protein-misfolding diseases. *Trends Mol. Med.* **2008**, *14*, 373–380. [CrossRef] [PubMed]

593. Cardinale, A.; Biocca, S. Combating protein misfolding and aggregation by intracellular antibodies. *Curr. Mol. Med.* **2008**, *8*, 2–11.

594. Cohen, P.A.; Mani, J.C.; Lane, D.P. Characterization of a new intrabody directed against the N-terminal region of human p53. *Oncogene* **1998**, *17*, 2445–2456. [CrossRef]

595. Cohen, P.A. Intrabodies. Targeting scFv expression to eukaryotic intracellular compartments. *Methods Mol. Biol.* **2002**, *178*, 367–378.

596. Wheeler, Y.Y.; Kute, T.E.; Willingham, M.C.; Chen, S.Y.; Sane, D.C. Intrabody-based strategies for inhibition of vascular endothelial growth factor receptor-2: Effects on apoptosis, cell growth, and angiogenesis. *FASEB J.* **2003**, *17*, 1733–1735. [CrossRef]

597. Tanaka, T.; Lobato, M.N.; Rabbitts, T.H. Single domain intracellular antibodies: A minimal fragment for direct in vivo selection of antigen-specific intrabodies. *J. Mol. Biol.* **2003**, *331*, 1109–1120. [CrossRef]

598. Auf der Maur, A.; Escher, D.; Barberis, A. Antigen-independent selection of stable intracellular single-chain antibodies. *FEBS Lett.* **2001**, *508*, 407–412. [CrossRef]

599. Shaki-Loewenstein, S.; Zfania, R.; Hyland, S.; Wels, W.S.; Benhar, I. A universal strategy for stable intracellular antibodies. *J. Immunol. Methods* **2005**, *303*, 19–39. [CrossRef] [PubMed]

600. Abraham, A.; Natraj, U.; Karande, A.A.; Gulati, A.; Murthy, M.R.; Murugesan, S.; Mukunda, P.; Savithri, H.S. Intracellular delivery of antibodies by chimeric Sesbania mosaic virus (SeMV) virus like particles. *Sci. Rep.* **2016**, *6*, 21803. [CrossRef] [PubMed]

601. Strebe, N.; Guse, A.; Schungel, M.; Schirrmann, T.; Hafner, M.; Jostock, T.; Hust, M.; Muller, W.; Dubel, S. Functional knockdown of VCAM-1 at the posttranslational level with ER retained antibodies. *J. Immunol. Methods* **2009**, *341*, 30–40. [CrossRef] [PubMed]

602. Nordin, M.A.C.; Teow, S.-Y. Review of Current Cell-Penetrating Antibody Developments for HIV-1 Therapy. *Molecules* **2018**, *23*, 335. [CrossRef]

603. Serruys, B.; van Houtte, F.; Verbrugghe, P.; Leroux-Roels, G.; Vanlandschoot, P. Llama-derived single-domain intrabodies inhibit secretion of hepatitis B virions in mice. *Hepatology* **2009**, *49*, 39–49. [CrossRef]

604. Serruys, B.; van Houtte, F.; Farhoudi-Moghadam, A.; Leroux-Roels, G.; Vanlandschoot, P. Production, characterization and in vitro testing of HBcAg-specific VHH intrabodies. *J. Gen. Virol.* **2010**, *91*, 643–652. [CrossRef]

605. Behar, G.; Chames, P.; Teulon, I.; Cornillon, A.; Alshoukr, F.; Roquet, F.; Pugniere, M.; Teillaud, J.L.; Gruaz-Guyon, A.; Pelegrin, A.; et al. Llama single-domain antibodies directed against nonconventional epitopes of tumor-associated carcinoembryonic antigen absent from nonspecific cross-reacting antigen. *FEBS J.* **2009**, *276*, 3881–3893. [CrossRef]

antibodies

MDPI

Article

Dynamic Views of the Fc Region of Immunoglobulin G Provided by Experimental and Computational Observations

Saeko Yanaka [1,2,3], Rina Yogo [1,2], Rintaro Inoue [5], Masaaki Sugiyama [5], Satoru G. Itoh [1,4], Hisashi Okumura [1,4], Yohei Miyanoiri [6], Hirokazu Yagi [2], Tadashi Satoh [2], Takumi Yamaguchi [2,7] and Koichi Kato [1,2,3,*]

[1]   Exploratory Research Center on Life and Living Systems (ExCELLS) and Institute for Molecular Science (IMS), National Institutes of Natural Sciences, 5-1 Higashiyama, Myodaiji, Okazaki 444-8787, Japan
[2]   Graduate School of Pharmaceutical Sciences, Nagoya City University, 3-1 Tanabe-dori, Mizuho-ku, Nagoya, Aichi 467-8603, Japan
[3]   Department of Functional Molecular Science, SOKENDAI (The Graduate University for Advanced Studies), Okazaki, Aichi 444-8787, Japan
[4]   Department of Structural Molecular Science, SOKENDAI (The Graduate University for Advanced Studies), Okazaki, Aichi 444-8585, Japan
[5]   Institute for Integrated Radiation and Nuclear Science, Kyoto University, 2-1010 Asashiro-Nishi, Kumatori, Osaka 590-0494, Japan
[6]   Institute for Protein Research, Osaka University, 3-2 Yamadaoka, Suita, Osaka 565-0871, Japan
[7]   School of Materials Science, Japan Advanced Institute of Science and Technology (JAIST), 1-1 Asahidai, Nomi 923-1292, Japan
*   Correspondence: kkatonmr@ims.ac.jp

Received: 28 April 2019; Accepted: 12 June 2019; Published: 1 July 2019

Abstract: The Fc portion of immunoglobulin G (IgG) is a horseshoe-shaped homodimer, which interacts with various effector proteins, including Fcγ receptors (FcγRs). These interactions are critically dependent on the pair of N-glycans packed between the two C_H2 domains. Fucosylation of these N-glycans negatively affects human IgG1-FcγRIIIa interaction. The IgG1-Fc crystal structures mostly exhibit asymmetric quaternary conformations with divergent orientations of C_H2 with respect to C_H3. We aimed to provide dynamic views of IgG1-Fc by performing long-timescale molecular dynamics (MD) simulations, which were experimentally validated by small-angle X-ray scattering and nuclear magnetic resonance spectroscopy. Our simulation results indicated that the dynamic conformational ensembles of Fc encompass most of the previously reported crystal structures determined in both free and complex forms, although the major Fc conformers in solution exhibited almost symmetric, stouter quaternary structures, unlike the crystal structures. Furthermore, the MD simulations suggested that the N-glycans restrict the motional freedom of C_H2 and endow quaternary-structure plasticity through multiple intramolecular interaction networks. Moreover, the fucosylation of these N-glycans restricts the conformational freedom of the proximal tyrosine residue of functional importance, thereby precluding its interaction with FcγRIIIa. The dynamic views of Fc will provide opportunities to control the IgG interactions for developing therapeutic antibodies.

Keywords: Immunoglobulin G; Fc; conformational dynamics; molecular dynamics simulation; small-angle X-ray scattering; nuclear magnetic resonance; N-glycan; core fucosylation

1. Introduction

Antibodies play pivotal roles in the immune system as multifunctional glycoproteins, coupling between antigen recognition and effector functions, as typified by immunoglobulin G (IgG). The Fab region of each IgG binds to its specific antigen while the Fc region interacts with effector proteins, including the complement component C1q and a series of Fcγ receptors (FcγRs) [1–4]. Further, the Fc portion of IgG interacts with various proteins other than the effector proteins, such as staphylococcal protein A and streptococcal protein G [5]. This versatile functionality of IgG and the other classes of antibodies is believed to be attributed to their structural flexibility and plasticity [6,7].

Antibodies have modular structures, in which the Ig-fold domains as building blocks are connected through flexible linkers. In IgG, the two identical light chains are each divided into the V_L and C_L domains, whilst the two identical heavy chains are each composed of the V_H, C_H1, C_H2, and C_H3 domains. The C_H1 and C_H2 domains are separated by the hinge region, which possesses significant degrees of freedom for internal flexibility.

Each Ig domain is structurally characterized by nine or seven β strands (in the V and C domains, respectively) connected by turn or loop segments. The hypervariable loops in the V_L and V_H domains are directly involved in antigen binding. In particular, the third hypervariable loop in the V_H domain (H3 loop) is extremely divergent in amino acid sequence and it exhibits dynamic conformational multiplicity [8–10]. Antigen binding generally renders the hypervariable loops less mobile. The conformational plasticity of the hypervariable loops is the key property for the antigen recognition mechanism. Kinetic data have shown that the antibody combining site undergoes conformational changes upon interacting with the antigen, thereby involving conformational selection as well as induced-fit processes [11]. Antigen binding often induces non-local conformational rearrangements of the V_L and V_H domains, thereby altering their orientation relative to each other and with respect to the C_L and C_H1 domains [4,12]. It has been speculated that such allosteric effects are involved in functional cooperativity between antigen binding and interaction with the effector proteins [4].

Dynamic views of antibody structures thus offer deep insights into their functional mechanisms. Dynamic structures of antibodies in solution have recently been characterized by sophisticated experimental techniques, including nuclear magnetic resonance (NMR) spectroscopy [13,14], solution scattering [15–19], cryo-electron microscopy [7,20], and high-speed atomic force microscopy [21]. In addition, computational approaches offer powerful tools for providing dynamic views of antibody structures [22]. These techniques highlight considerable variability in the spatial arrangements of the two Fab arms with respect to the Fc stem in IgG, which is provided by the conformational freedom of the hinge region, facilitating IgG's bivalent binding to antigens with various spatial arrangements, such as bacterial surfaces.

By contrast to the extensively characterized Fab and hinge dynamics, the dynamic properties of the IgG-Fc structure in solution remain to be fully understood, although the crystal structures of Fc exhibit significant conformational variability [23–25]. In addition, the IgG-Fc possesses a conserved glycosylation site at Asn297 in each C_H2 domain, at which a bi-antennary complex-type oligosaccharide is present with heterogeneity in sequence and conformation [26,27]. This N-glycosylation is critically important for IgG interactions with the effector proteins and the consequent effector functions [3,28,29]. Deglycosylation of IgG-Fc impairs its binding to C1q and FcγRs. In contrast, the removal of the core fucose residue from the N-glycan of human IgG1-Fc results in enhancement of its interaction with FcγRIIIa, thereby dramatically improving the antibody-dependent cell-mediated cytotoxicity (ADCC) [30–32]. Therefore, in this study, we have attempted to provide dynamic views of the Fc region of human IgG1 by computational approaches with experimental validation to discuss their functional significance. In particular, we will shed light on intramolecular interaction networks involving the N-glycans, providing mechanistic insights into the improved efficacy of the therapeutic antibodies by the glycan remodeling.

2. Materials and Methods

2.1. Molecular Dynamics Simulation

The starting structures were built on the basis of the crystal structure Protein Data Bank (PDB) entry 3AVE, 5IW3, and 2DTS, supplemented with crystallographically unobserved segments in each structure as follows: In 3AVE and 2DTS, N-terminal segments (T224–E233 in chain A and T224–G236 in chain B), C-terminal segments (P445–K447), and the terminal galactose residues of the α1-6Man branches; and in 5IW3, N-terminal segments (T224–L235), C-terminal segments (S444–K447), and the core fucose residue. The crystallographically unobserved galactose residues of 3AVE and 2DTS were modeled by superposition of their GlcNAc residues of the α1-6Man branches and the corresponding Gal-β1,4-GlcNAc disaccharide moieties of an sFcγRIIIa-bound Fc crystal structure (3AY4). The core fucose residues of 5IW3 were modeled by superposition of its reducing terminal GlcNAc residues and the corresponding Fuc-α1,6-GlcNAc disaccharide moieties of 3AVE. In the 5IW3-based model, E379 and M381 were substituted in silico with aspartate and leucine, respectively, by using PyMOL (https://www.pymol.org). The AMBER14 [33] program package was used with the force fields AMBER ff14SB [34] and GLYCAM06 [35] for proteins and glycans, respectively, along with the TIP3P water model [36]. The IgG1-Fc models derived from 3AVE, 5IW3, and 2DTS were placed in boxes containing 28,135, 24,372, and 28,243 water molecules, respectively. After the preparation of the initial conformations, we performed equilibrium simulations for 4 ns with NPT ensemble with periodic boundary conditions. The system was heated from 5 to 310 K for 100 ps at a constant pressure of 1 atm. Eight simulations were then performed with different velocities. Each simulation time period was 400 ns. All production simulations were done at 300 K with the weak-coupling algorithm in the NPT ensemble. The bonds involving hydrogen atoms were constrained by the SHAKE algorithm [37]. The electrostatic interactions were treated with the particle mesh Ewald method and the cutoff distance for the non-bonded interactions was 8.0 Å. The first 80 ns were removed considering the time needed for the initial structure to reach equilibrium. Molecular dynamics (MD) simulations for the Fc fragments without the hinge and without the N-glycans were performed using the same protocol and initial models based on 3AVE (with 19,410 water molecules) and 5IW3 (with 24,551 water molecules), respectively. Ensemble models were created from 25,600 conformers extracted every 100 ps from the 320 ns simulation results of the eight production runs. The root mean square fluctuation (RMSF) for each amino acid Cα atom of IgG1-Fc was calculated from 3,200 conformers extracted from each of the eight production runs, which were superimposed by the Cα atoms, yielding an average structure. The RMSF was derived on the average structure.

2.2. Sample Preparation

For the small-angle X-ray scattering (SAXS) measurements, the Fc fragment with core fucosylation was cleaved from commercially available IgG1 antibodies (Chugai Pharmaceutical, Tokyo, Japan). For the NMR measurements, the fucosyl Fc fragment was cleaved from IgG1 metabolically labeled with [CO, α, β, γ, ε1, ε2-$^{13}C_6$; β2, δ1, δ2-$^2H_3$; ^{15}N] tyrosine (Taiyo Nippon Sanso, Tokyo, Japan). The metabolic isotope labeling, proteolytic digestion, and purification of the Fc fragments were performed according to previously reported protocols [13,38].

2.3. Small-Angle X-Ray Scattering

SAXS experiments were performed using Fc dissolved in 50 mM Tris-HCl, pH.8.0, 150 mM NaCl at a protein concentration of 5.0 mg/mL with NANOPIX (Rigaku, Tokyo, Japan) at 25 °C. X-rays from a high-brilliance point-focused X-ray generator (MicroMAX-007HF, Rigaku, Tokyo, Japan) were focused with a confocal mirror (OptiSAXS) and collimated with a confocal multilayer mirror and two pinholes collimation system with the lower parasitic scattering, "ClearPinhole", supplied for the X-rays with the flux of 2.0×10^8 cps at the sample position (high flux mode). The scattered X-rays were detected using a two-dimensional semiconductor detector (HyPix-6000, Rigaku, Tokyo, Japan) having the spatial resolution of 100 μm. By measuring SAXS profiles in two sample-to-detector distance (SDD) conditions, 1320 mm and 300 mm, the wide Q-range (0.015 Å^{-1}–0.5 Å^{-1}) were covered. In addition, for removal of unfavorable aggregates from the sample solution, the laboratory-based SEC-SAXS System (LA-SSS) was employed to measure the SAXS profile in the lower-Q range (0.015 Å^{-1}–0.08 Å^{-1}, SDD = 1320 mm).

The theoretical SAXS profiles were independently calculated from the atomic coordinates of the ensemble model containing 25,600 conformers. Finally, an MD-derived SAXS profile was obtained by averaging the 25,600 calculating SAXS profiles. The χ^2 values for the-goodness-of-fit is defined as follows:

$$\chi^2 = \frac{1}{N-1} \sum_j \left[\frac{I_{\text{exp}}(q_j) - cI_{\text{calc}}(q_j)}{\sigma(q_j)} \right]^2$$

where N, $I_{\text{exp}}(q)$, $\sigma(q)$, $I_{\text{calc}}(q)$, c is the number of experimental data points, the experimental scattering intensity, its error, the calculated scattering intensity and the scaling factor, respectively.

2.4. NMR Measurement

Two-dimensional heteronuclear single-quantum correlation nuclear Overhauser effect spectroscopy (HSQC-NOESY) spectra were acquired for fucosylated IgG1-Fc labeled with [CO, α, β, γ, ε1, ε2-$^{13}C_6$; β2, δ1, δ2-$^2H_3$; ^{15}N] tyrosine and dissolved in 5 mM sodium phosphate buffer, pH 6.0, containing 50 mM NaCl and 10% D_2O at a protein concentration of 10 mg/mL by using an AVANCEIII 950 spectrometer (Bruker BioSpin) equipped with a TCI cryogenic probe at 300 K. The NMR spectral data were recorded at a proton observation frequency of 950.3 MHz with 128(t_1) × 2048(t_2) complex points and 512 scans per t_1 increment, with a mixing time of 300 ms.

3. Results and Discussion

3.1. Overall Conformation of IgG-Fc

The Fc portion of IgG has a horseshoe-shape, harboring a pair of N-glycans packed between the two C_H2 domains, which consequently make no direct contacts, while the C_H3 domains extensively contact each other (Figure 1a). This domain arrangement renders the C_H2 domains more mobile than the C_H3 domains as implicated by the crystallographic data [23–25]. The B factors of the C_H2 domains are generally higher than those of the C_H3 domains [39]. Moreover, the C_H2 domains exhibit divergent orientations in crystal structures in free and liganded states and in various glycoforms. Indeed, the great majority of the IgG1-Fc crystal structures deposited in Protein Data Bank (PDB) exhibit asymmetric quaternary structures even in uncomplexed states, with few exceptions, for example, 5IW3 with a crystallographic two-fold axis. However, these conformational deformations might be, at least partially, ascribed to non-physiological crystal contacts. Frank et al. have performed a 200 ns MD simulation of human IgG1-Fc with fully galactosylated glycans and demonstrated that the C_H2 domains showed significant degrees of motional freedom [24]. In general, MD simulation results depend on the calculation protocol, including the initial structure and simulation time as well as the force field.

We performed long-timescale MD simulations in explicit water, using our determined crystal structure of human IgG1-Fc (3AVE) [40] as the initial model. We attempt to deal with a major glycoform of Fc, in which two complex-type N-glycans are mono-galactosylated at the α1-6Man branch. The crystal structure was supplemented with models of the hinge and the C-terminal regions along with the non-reducing terminal galactose residues in the α1-6Man branches, because these parts gave no interpretable electron density in this crystal structure (Figure 1a). From the MD trajectories (2.56 μs in total for each Fc glycoform), 25,600 conformers were extracted to create an ensemble model reproducing possible conformational spaces of the Fc glycoproteins.

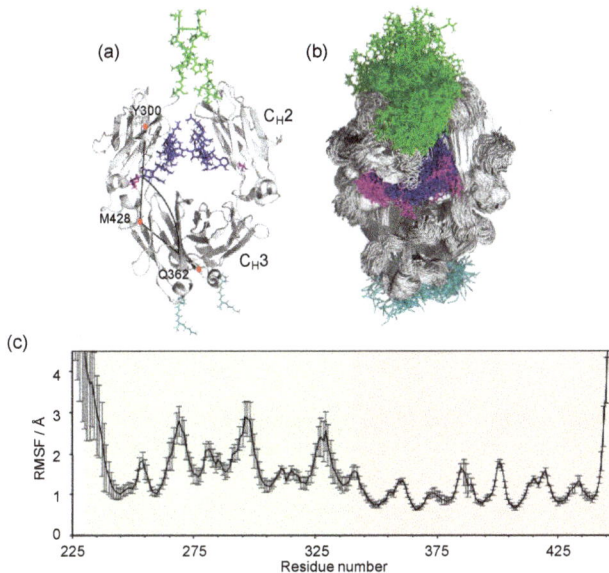

Figure 1. MD simulation of IgG1-Fc. (**a**) The starting structure of the MD simulation, based on the crystal structure of fucosyl IgG1-Fc (3AVE) supplemented with the hinge (green; T224–E233 in chain A and T224–G236 in chain B) and C-terminal (cyan; P445–K447) segments along with the terminal galactose residues (magenta) of the α1-6Man branches. The N-glycans are colored blue except for the terminal galactose. The intra-chain domain-orientation angle between C_H2 and C_H3 defined by Cα atoms of Y300, M428, and Q362 are shown in chain A. (**b**) The superposition of 256 structures extracted every 100 ns from the MD trajectory. The structures were visualized by PyMOL (https://www.pymol.org). (**c**) The RMSF for each amino-acid Cα atom of IgG1-Fc, which was calculated as described in Materials and Methods. White, hinge; light green, C_H2; light orange, C_H3.

For experimental validation of the simulation results, we measured SAXS of the Fc region, which has been applied for the characterization of Fc structures in solution [41–43]. The SAXS profile computed from the Fc ensemble model was in good agreement with the experimentally obtained SAXS profile (Figure 2). The MD-derived SAXS profile reproduced the experimental data ($x^2 = 6.8$) better than did a profile computed from the crystal structure used for building the initial model ($x^2 = 13.6$). The smallest x^2 value, 6.8, was achieved with the ensemble model created from the total 2.56 μs MD simulation, while the x^2 value calculated for each of the eight production runs were larger (range 7.3–32.2), suggesting that a shorter MD simulation was not enough for exploring the Fc conformational space.

Although our simulation results confirmed that, besides the hinge and the C-terminal segments, the C_H2 domains exhibit considerable motion in comparison with the C_H3 domains (Figure 1b,c), the major conformers exhibited almost symmetric structures with C_H2-C_H3 angles (approximately 90 degrees) that were more acute in comparison with that of the uncomplexed Fc crystal structures, including those used as the initial structures in the simulations (Figure 3). We performed an additional MD simulation by using a two-fold symmetrical Fc model (based on the crystal structure 5IW3) as the initial model. Despite the remarkably different starting model, the MD result showed a similar tendency: The MD-derived conformers still exhibited symmetrical but significantly stouter quaternary structures (Figure 3). These results were quantitatively consistent with the previously reported MD simulation of the fully galactosylated Fc [24], suggesting that the Fc packed in crystal lattices is apt to adopt asymmetric slim quaternary structures as compared with those in solution.

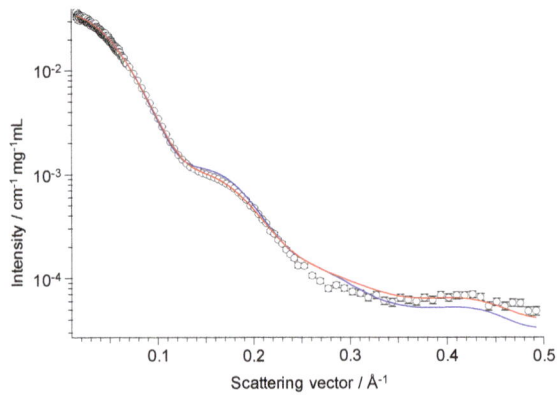

Figure 2. SAXS-based experimental validation of the MD-derived ensemble model. SAXS profile of fucosyl IgG1-Fc (open circle) shown with theoretical profiles computed from the MD-derived ensemble model (red) and the crystal structure (3AVE) (blue).

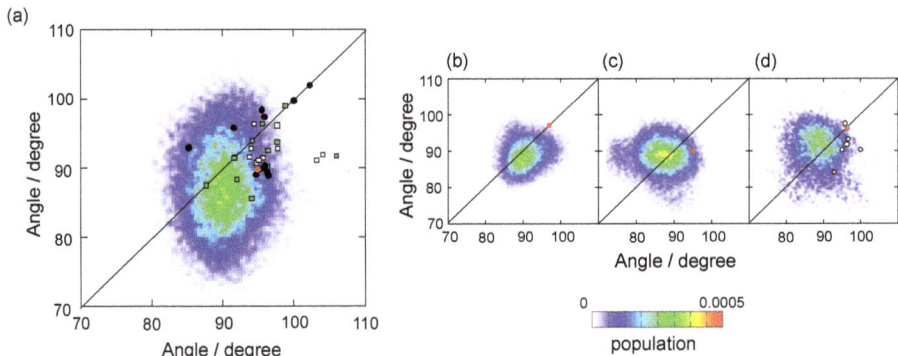

Figure 3. Distribution of intra-chain domain-orientation angles between C_H2 and C_H3 for the ensemble models of IgG1-Fc and various crystal structures of IgG1-Fc. The angles between the C_H2 and C_H3 domains of chain A and chain B were plotted on the X axis and Y axis, respectively, for the ensemble models derived from MD simulations starting from the initial structures based on (**a**) asymmetric crystal structure (3AVE) supplemented with the crystallographically unobserved N- and C-terminal segments, (**b**) symmetric crystal structure (5IW3) supplemented with crystallographically unobserved N- and C-terminal segments, (**c**) crystallographically observed parts of 3AVE, and (**d**) 5WI3 supplemented with the crystallographically unobserved N- and C-terminal segments with deletion of the N-glycans. In A, B, and C, the N-glycans of each initial structure were modeled to have the core fucose residue and the terminal galactose residue of the α1-6Man branch. The angles observed in the crystal structures are represented as circles for uncomplexed Fc structures (red, the starting structures used for the corresponding MD simulations; black, Fc with native N-glycans; white, Fc with enzymatically trimmed N-glycans; yellow, aglycosylated Fc), rectangles for complexed Fc structures (white, complex with sFcγRs; green, complex with other ligands).

The experimentally validated conformational ensembles of IgG1-Fc encompassed most of the previously reported crystal structures because of the variability of the C_H2 domains with respect to the C_H3 domains (Figure 3). The conformational ensemble included the crystal structures of human IgG1-Fc in complex with Fc-binding proteins such as protein A domains, though they are minor conformational species. The Cα RMSD was 1.1 Å between the protein A B domain-bound Fc crystal structure (IL6X) [44] and its most resembling conformer found in the MD-derived ensemble.

These findings imply that interactions of Fc with these proteins involve conformational selection mechanisms. However, the asymmetrically deformed Fc structures bound to sFcγRIIa were obviously far from the MD-derived conformational cluster and very rarely found in the corresponding ensemble model: The Cα RMSD was 3.2 Å between the sFcγRIIa-bound Fc crystal structure (3RY6) [45] and its most resembling conformer found in the MD-derived ensemble (excluding the Asn286–Gln295 segments because of inconsistent interpretation of their electron densities). This finding suggested that induced-fit mechanisms are involved in the binding process.

The two C_H2 domains are tethered at their N-termini through the disulfide-linked hinge region and they bracket the pair of N-linked oligosaccharides. Therefore, it was highly plausible that the hinge and the glycans critically affect the conformational space of the Fc region. Indeed, the Fc crystal structures of IgG1-Fc with different glycoforms showed different quaternary conformations [23]. We performed MD simulations of IgG1-Fc without the hinge region and that of IgG1-Fc without the N-glycans. Elimination of these parts resulted in greater degrees of motional freedom of the C_H2 domains with increases in the population of extremely asymmetric quaternary conformations. This is seemingly inconsistent with a crystal structure of aglycosyl Fc reported by Borrok et al., which adopted a more closed conformation [41]. However, in that study, they suggested that aglycosyl Fc assumes a more open C_H2 orientation based on their SAXS observation, which was apparently consistent with our MD simulation results. The asymmetrically distorted quaternary conformation was found in the previously reported MD simulation of a hinge-truncated Fc [24], which is also qualitatively consistent with our data.

It has been reported that either disulfide cleavage at the hinge or deglycosylation of Fc impairs the interactions of IgG1 with the effector proteins [2,3]. In addition to the local conformational perturbations suggested by NMR studies [14,46,47], the increased mobility of the C_H2 domains may negatively contribute to the affinities of the IgG-Fc for the effector proteins, at least partially, due to the increase in the conformational entropic penalty upon their binding.

3.2. Intramolecular Interaction Networks of N-Glycans

In general, carbohydrate chains are conformationally dynamic compared to polypeptide chains and, therefore, yield ambiguous electron densities in the crystal structures of glycoproteins [48,49]. However, this is not true for the N-glycans of IgG-Fc because they are packed between the two C_H2 domains and, therefore, are restricted in terms of internal motion, which has been confirmed by the previous and present MD simulations [24]. Consequently, in many crystal structures, the Fc N-glycans have been visualized except for the non-reducing terminal galactose residues in the α1-3Man branches, which are projected to the inner space of the horseshoe-structure and therefore are considerably mobile [14,28].

The crystal structures indicated that the core part and the α1-6Man branch of the N-glycans make extensive contacts with the amino acid residues located in the inner surface of the C_H2 domain, while the interactions between the two glycans are quite limited (Figure 4) [28]. The crystallographic data of the sFcγRIIIa-bound Fc showed rearrangements of the interaction network, creating new contact pairs with concomitant loss of a number of contact pairs, resulting in the disappearance of the intramolecular glycan-glycan interactions [40,50]. All the contact pairs between the N-glycan and the C_H2 amino acid residues in both the free and the complexed forms were found in the MD-derived ensemble model. Moreover, the ensemble model included more contact pairs not only between glycan and amino acid residues but also between the two glycans, which were not observed in the Fc crystal structures, demonstrating dynamic behaviors of the N-glycans within the inner space of the Fc. This is consistent with the previously reported NMR observation for the mobility of the terminal galactose residue of the α1-6Man branch [27] and its missing electron densities in many crystal structures exemplified by 3AVE and 2DTS [40]. All these data suggest that the pair of N-glycans not only restrict motional freedom of the C_H2 domains but also endow quaternary structure plasticity through multiple intramolecular interaction networks. It is conceivable that the intramolecular N-glycan interaction networks critically depend on the Fc glycoform. Lee and Im reported MD simulations of human IgG1-Fc glycoforms exhibiting a series of sequentially truncated high-mannose-type glycans. In this extreme case, the N-glycans dynamically interconverted between C_H2-bound and unbound forms and the glycan truncation affected the Fc quaternary conformational dynamics [51].

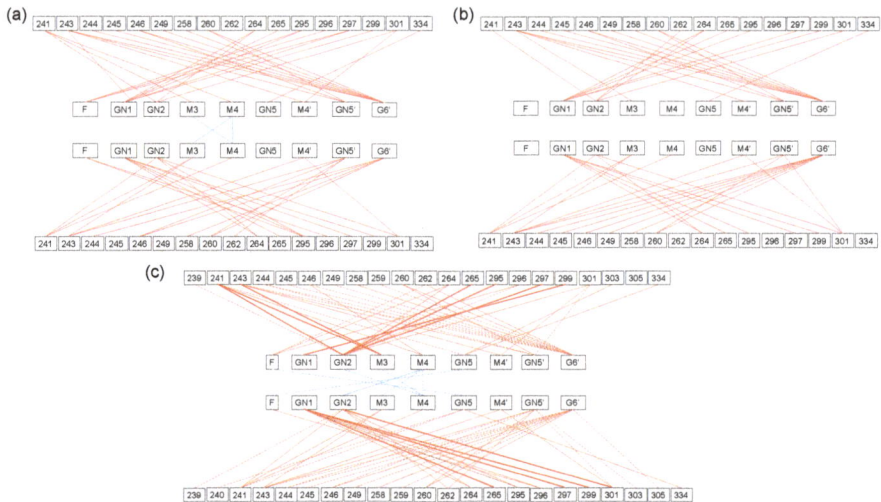

Figure 4. Intramolecular interaction networks of the IgG1-Fc glycans. A pair of contact residues between the N-glycan and the polypeptide chain or between the two N-glycans found within 4 Å is connected by a line segment for the crystal structures of (**a**) Fc alone (3AVE) and (**b**) sFcγRIIIa-bound Fc (5XJE). The terminal galactose residues of α1-6Man branch in 3AVE were modeled as described in the Materials and Methods. The contact pairs involving either of these galactose residues are conserved in the crystal structure 5IW3, which gave electron densities of the terminal galactose residues. (**c**) Pairs of contact residues found within 4 Å in the ensemble model derived from the MD simulation are connected by different types of line segments (red for carbohydrate-protein contact and cyan for carbohydrate-carbohydrate contact) according to incidence as follows: More than 24,000 pairs (thick solid line), 24,000 to 16,000 pairs (thin solid line), and 16,000 to 8,000 pairs (dashed line).

3.3. Effects of Fc Defucosylation of its Dynamic Conformation and FcγR Interaction

The major forms of IgG1-Fc N-glycans share the core fucose residue, which negatively contributes to binding to FcγRIIIa, and thereby impairs ADCC activity [30–32]. While our crystallographic data have indicated that the core fucosylation does not affect the overall conformation of IgG1-Fc, our NMR data have shown defucosylation-induced microenvironmental changes surrounding this fucose residue. This is best exemplified by Tyr296, which is crucially involved in the interaction with FcγRs, including FcγRIIIa [40,52]. We performed a long-timescale MD simulation starting from a model based on a crystal structure of non-fucosylated IgG1-Fc (2DTS) for comparing the local conformation of this tyrosine residue between the fucosylated and non-fucosylated glycoforms (Figure 5). In either of the crystal structures, where free IgG1-Fc in fucosylated or non-fucosylated form was used as the initial structure in the simulation, the Tyr296 side chain adopts a semi-outward conformation with χ1 dihedral angles of −96 and −91 degrees, respectively [40]. Remarkably, these conformations were not the major ones during our MD simulation. In the fucosyl form, the most populated conformer of the side chain of Tyr296 stays in a "flipped-in" state with a χ1 angle of approximately 80 degrees, making contacts with the core fucose. This inward conformation has been experimentally confirmed based on the observation of nuclear Overhauser effect (NOE) connectivities between the core fucose and the Tyr296 side chain (Figure 5c). By contrast, in the non-fucosyl form, the conformational state of this tyrosine is more divergent with a significantly increased outward conformation with a χ1 angle of approximately 180 degrees. The conformational multiplicity of Tyr296 in the non-fucosyl form was also indicated by our NMR data. Interestingly, the crystallographic data indicated that Tyr296 is involved in the interaction with FcγRIIIa in a flipped-out state with a χ1 angle of 189 and 200 degrees in the fucosylated and non-fucosylated glycoforms, respectively [50,53]. These data indicated that the Tyr296 side chain is stabilized in the inward conformation through interaction with the core fucose and, on defucosylation, undergoes a conformational population shift with an increased outward conformation, which is favorable for FcγRIIIa binding. It is possible that, in the fucosylated IgG1, the flipping-out of this tyrosine side chain can be a rate-limiting step in its interaction with FcγRIIIa. This view is consistent with the previously reported kinetic data indicating that the core fucosylation of IgG1 primarily affects its association phase of FcγRIIIa binding [54].

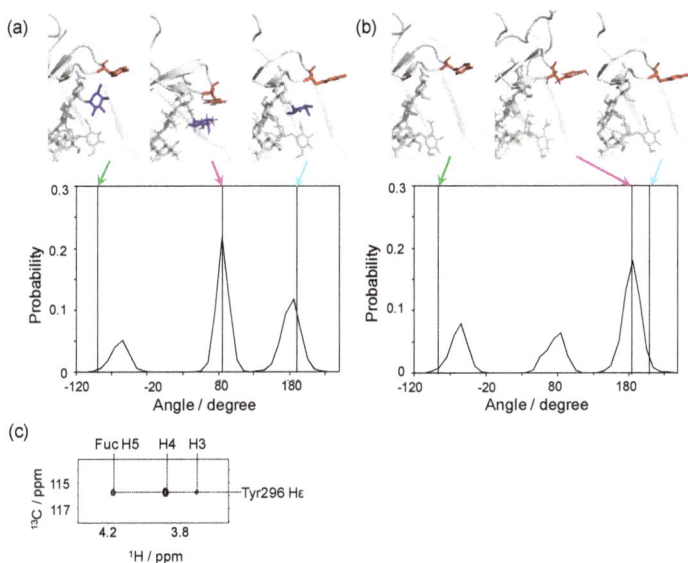

Figure 5. Conformational dynamics of the side chain of Tyr296 of IgG1-Fc depending on the core fucosylation. Distributions of χ1 dihedral angles of Tyr296 in the ensemble models derived from MD simulations are plotted for (**a**) fucosylated IgG1-Fc and (**b**) non-fucosylated IgG1-Fc. The typical conformational snapshots of derived from the major conformational states (magenta arrows) in the simulation trajectory are shown along with the crystal structures used for building the starting models (green arrows; A, 3AVE; B, 2DTS) and those of sFcγRIIIa-bound Fc (cyan arrows; A, 5XJE; B, 3AY4). (**c**) 2D HSQC-NOESY spectrum of IgG1-Fc labeled with [CO, α, β, γ, ε1, ε2-^{13}C$_6$; β2, δ1, δ2-^{2}H$_3$; ^{15}N] tyrosine.

4. Conclusions

In this study, the conformational spaces in IgG1-Fc glycoproteins were investigated by performing long-timescale MD simulations in explicit water with validation by experiments in solution. The MD-derived conformational ensembles included most of the crystallographic snapshots of IgG1-Fc thus far reported, whilst its conformational space was restricted by the hinge disulfides and the Asn297 glycans. We presumed that during the evolutionary process, the freedom of the Fc quaternary conformation was optimally restricted to strike a balance between the increase in adaptability to a variety of binding partners and the decrease in a conformational entropic penalty upon their interactions. From the viewpoint of antibody engineering, the conformational plasticity of Fc can be targeted to control its interactions with specific binding partners. As in the case of the removal of the core fucose, engineering strategies have been developed to control local conformational dynamics around the specific target binding sites of the Fc region, through amino acid substitutions and/or N-glycan remodeling [30–32,55,56]. In a complementary approach, a detailed understanding of the quaternary conformational dynamics of the Fc region, through concerted theoretical and experimental approaches will open up opportunities for developing novel therapeutic antibodies by allosteric control of their interactions with effector molecules.

Author Contributions: S.Y., R.Y., and K.K. conceived and designed the study; R.Y. and H.Y. carried out sample preparation; S.Y., R.Y., S.G.I., H.O., T.Y., and T.S. performed MD simulations and analyses; R.I. and M.S. performed SAXS experiments and analyses; S.Y. and Y.M. performed NMR experiments and analyses; S.Y., R.Y., and K.K. mainly drafted the manuscript.

Funding: This work was supported in part by the Research Fellowship for Young Scientists, the Grants-in-Aid for Scientific Research (Grant Numbers, JP17H06414 to H.Y., JP17H05893, JP18K14892 to S.Y., JP18H05203, JP19H01017 to K.K., JP19J15602 to R.Y., JP17K07361 to R.I., JP18H05229 and JP18H03681 to M.S.) from the Ministry of Education, Culture, Sports, Science and Technology (MEXT), Japan, the Cooperative Research Program of Institute for Protein Research, Osaka University, NMRCR-16-05,-17-05,-18-05, Joint Research by Exploratory Research Center on Life and Living Systems (ExCELLS) (ExCELLS program No, 18-301), and the project for Construction of the basis for the advanced materials science and analytical study by the innovative use of quantum beam and nuclear sciences in Institute for Integrated Radiation and Nuclear Science, Kyoto University.

Acknowledgments: We thank Kumiko Hattori and Kiyomi Senda (Nagoya City University) for their help with the preparation of the recombinant proteins. We thank Tsutomu Terauchi (Taiyo Nippon Sanso) for his help in the stable isotope labeling. The computations were performed at the Research Center for Computational Science, IMS and Research Center for Advanced Computing Infrastructure, JAIST. The funders had no role in the design of the study; in the collection, analyses, or interpretation of data; in the writing of the manuscript, or in the decision to publish the results.

Conflicts of Interest: The authors declare no conflict of interest.

Abbreviations

ADCC	Antibody-dependent cell-mediated cytotoxicity
FcγR	Fcγ receptor
Fuc	Fucose
Gal	Galactose
GlcNAc	*N*- Acetylglucosamine
HSQC	Heteronuclear single-quantum correlation
IgG	Immunoglobulin G
Man	Mannose
MD	Molecular dynamics
NMR	Nuclear magnetic resonance
NOE	Nuclear Overhauser effect
NOESY	Nuclear Overhauser effect spectroscopy
PDB	Protein Data Bank
RMSD	Root mean square deviation
RMSF	Root mean square fluctuation
SAXS	Small-angle X-ray scattering

References

1. Dorrington, K.J.; Klein, M.H. Binding sites for Fcγ receptors on immunoglobulin G and factors influencing their expression. *Mol. Immunol.* **1982**, *19*, 1215–1221. [CrossRef]
2. Burton, D.R. Immunoglobulin G: Functional sites. *Mol. Immunol.* **1985**, *22*, 161–206. [CrossRef]
3. Jefferis, R.; Lund, J.; Pound, J.D. IgG-Fc-mediated effector functions: Molecular definition of interaction sites for effector ligands and the role of glycosylation. *Immunol. Rev.* **1998**, *163*, 59–76. [CrossRef] [PubMed]
4. Yang, D.; Kroe-Barrett, R.; Singh, S.; Roberts, C.J.; Laue, T.M. IgG cooperativity—Is there allostery? Implications for antibody functions and therapeutic antibody development. *mAbs* **2017**, *9*, 1231–1252. [CrossRef] [PubMed]

5. Roberts, G.C.K.; Lian, L.Y.; Barsukov, I.L.; Derrick, J.P.; Kato, K.; Arata, Y. Interactions of bacterial cell-surface proteins with antibodies: A versatile set of protein-protein interactions. *Tech. Prot. Chem.* **1995**, *6*, 409–416.
6. Nezlin, R. Internal movements in immunoglobulin molecules. *Adv. Immunol.* **1990**, *48*, 1–40. [PubMed]
7. Jay, J.W.; Bray, B.; Qi, Y.; Igbinigie, E.; Wu, H.; Li, J.; Ren, G. IgG antibody 3D structures and dynamics. *Antibodies* **2018**, *7*, 18. [CrossRef]
8. Arata, Y.; Kato, K.; Takahashi, H.; Shimada, I. Nuclear-magnetic-resonance study of antibodies—A multinuclear approach. *Methods Enzymol.* **1994**, *239*, 440–464.
9. Nakasako, M.; Oka, T.; Mashumo, M.; Takahashi, H.; Shimada, I.; Yamaguchi, Y.; Kato, K.; Arata, Y. Conformational dynamics of complementarity-determining region H3 of an anti-dansyl Fv fragment in the presence of its hapten. *J. Mol. Biol.* **2005**, *351*, 627–640. [CrossRef]
10. Fernández-Quintero, M.L.; Loeffler, J.R.; Kraml, J.; Kahler, U.; Kamenik, A.S.; Liedl, K.R. Characterizing the diversity of the CDR-H3 loop conformational ensembles in relationship to antibody binding properties. *Front. Immunol.* **2018**, *9*, 3065. [CrossRef]
11. Foote, J.; Milstein, C. Conformational isomerism and the diversity of antibodies. *Proc. Natl. Acad. Sci. USA* **1994**, *91*, 10370–10374. [CrossRef] [PubMed]
12. Sela-Culang, I.; Kunik, V.; Ofran, Y. The structural basis of antibody-antigen recognition. *Front. Immunol.* **2013**, *4*, 302. [CrossRef] [PubMed]
13. Yanaka, S.; Yagi, H.; Yogo, R.; Yagi-Utsumi, M.; Kato, K. Stable isotope labeling approaches for NMR characterization of glycoproteins using eukaryotic expression systems. *J. Biomol. NMR* **2018**, *71*, 193–202. [CrossRef] [PubMed]
14. Kato, K.; Yamaguchi, Y.; Arata, Y. Stable-isotope-assisted NMR approaches to glycoproteins using immunoglobulin G as a model system. *Prog. Nucl. Magn. Reson. Spectrosc.* **2010**, *56*, 346–359. [CrossRef] [PubMed]
15. Clark, N.J.; Zhang, H.L.; Krueger, S.; Lee, H.J.; Ketchem, R.R.; Kerwin, B.; Kanapuram, S.R.; Treuheit, M.J.; McAuley, A.; Curtis, J.E.; et al. Small-angle neutron scattering study of a monoclonal antibody using free-energy constraints. *J. Phys. Chem. B* **2013**, *117*, 14029–14038. [CrossRef] [PubMed]
16. Eryilmaz, E.; Janda, A.; Kim, J.; Cordero, R.J.; Cowburn, D.; Casadevall, A. Global structures of IgG isotypes expressing identical variable regions. *Mol. Immunol.* **2013**, *56*, 588–598. [CrossRef] [PubMed]
17. Inouye, H.; Houde, D.; Temel, D.B.; Makowski, L. Utility of solution X-ray scattering for the development of antibody biopharmaceuticals. *J. Pharm. Sci.* **2016**, *105*, 3278–3289. [CrossRef]
18. Tian, X.; Vestergaard, B.; Thorolfsson, M.; Yang, Z.; Rasmussen, H.B.; Langkilde, A.E. In-depth analysis of subclass-specific conformational preferences of IgG antibodies. *IUCrJ* **2015**, *2*, 9–18. [CrossRef]
19. Castellanos, M.M.; Howell, S.C.; Gallagher, D.T.; Curtis, J.E. Characterization of the NISTmAb reference material using small-angle scattering and molecular simulation. *Anal. Bioanal. Chem.* **2018**, *410*, 2141–2159. [CrossRef]
20. Ugurlar, D.; Howes, S.C.; de Kreuk, B.J.; Koning, R.I.; de Jong, R.N.; Beurskens, F.J.; Schuurman, J.; Koster, A.J.; Sharp, T.H.; Parren, P.W.H.I.; et al. Structures of C1-IgG1 provide insights into how danger pattern recognition activates complement. *Science* **2018**, *359*, 794–797. [CrossRef]
21. Preiner, J.; Kodera, N.; Tang, J.L.; Ebner, A.; Brameshuber, M.; Blaas, D.; Gelbmann, N.; Gruber, H.J.; Ando, T.; Hinterdorfer, P.; et al. IgGs are made for walking on bacterial and viral surfaces. *Nat. Commun.* **2014**, *5*. [CrossRef] [PubMed]
22. Brandt, J.P.; Patapoff, T.W.; Aragon, S.R. Construction, md simulation, and hydrodynamic validation of an all-atom model of a monoclonal IgG antibody. *Biophys. J.* **2010**, *99*, 905–913. [CrossRef]
23. Krapp, S.; Mimura, Y.; Jefferis, R.; Huber, R.; Sondermann, P. Structural analysis of human IgG-Fc glycoforms reveals a correlation between glycosylation and structural integrity. *J. Mol. Biol.* **2003**, *325*, 979–989. [CrossRef]
24. Frank, M.; Walker, R.C.; Lanzilotta, W.N.; Prestegard, J.H.; Barb, A.W. Immunoglobulin G1 Fc domain motions: Implications for Fc engineering. *J. Mol. Biol.* **2014**, *426*, 1799–1811. [CrossRef] [PubMed]
25. Caaveiro, J.M.; Kiyoshi, M.; Tsumoto, K. Structural analysis of Fc/FcγR complexes: A blueprint for antibody design. *Immunol. Rev.* **2015**, *268*, 201–221. [CrossRef]
26. Jefferis, R. Glycoforms of human IgG in health and disease. *Trends Glycosci. Glycotechnol.* **2009**, *21*, 105–117.
27. Barb, A.W.; Prestegard, J.H. NMR analysis demonstrates immunoglobulin G N-glycans are accessible and dynamic. *Nat. Chem. Biol.* **2011**, *7*, 147–153. [CrossRef]

28. Yamaguchi, Y.; Takahashi, N.; Kato, K. *Antibody Structures*; Elsevier: Oxford, UK, 2007; Volume 3.

29. Dekkers, G.; Treffers, L.; Plomp, R.; Bentlage, A.E.H.; de Boer, M.; Koeleman, C.A.M.; Lissenberg-Thunnissen, S.N.; Visser, R.; Brouwer, M.; Mok, J.Y.; et al. Decoding the human immunoglobulin G-Glycan repertoire reveals a spectrum of Fc-receptor- and complement-mediated-effector activities. *Front. Immunol.* **2017**, *8*. [CrossRef]

30. Niwa, R.; Shoji-Hosaka, E.; Sakurada, M.; Shinkawa, T.; Uchida, K.; Nakamura, K.; Matsushima, K.; Ueda, R.; Hanai, N.; Shitara, K.; et al. Defucosylated chimeric anti-CC chemokine receptor 4 IgG1 with enhanced antibody-dependent cellular cytotoxicity shows potent therapeutic activity to T-cell leukemia and lymphoma. *Cancer Res.* **2004**, *64*, 2127–2133. [CrossRef]

31. Yamane-Ohnuki, N.; Satoh, M. Production of therapeutic antibodies with controlled fucosylation. *mAbs* **2009**, *1*, 230–236. [CrossRef]

32. Ferrara, C.; Stuart, F.; Sondermann, P.; Brunker, P.; Umana, P. The carbohydrate at FcγRIIIa Asn-162. An element required for high affinity binding to non-fucosylated IgG glycoforms. *J. Biol. Chem.* **2006**, *281*, 5032–5036. [CrossRef] [PubMed]

33. Case, D.A.; Babin, V.; Berryman, J.T. *Amber14*; University of California: San Francisco, CA, USA, 2014.

34. Maier, J.A.; Martinez, C.; Kasavajhala, K.; Wickstrom, L.; Hauser, K.E.; Simmerling, C. Ff14sb: Improving the accuracy of protein side chain and backbone parameters from ff99sb. *J. Chem. Theory Comput.* **2015**, *11*, 3696–3713. [CrossRef] [PubMed]

35. Kirschner, K.N.; Yongye, A.B.; Tschampel, S.M.; González-Outeiriño, J.; Daniels, C.R.; Foley, B.L.; Woods, R.J. Glycam06: A generalizable biomolecular force field. Carbohydrates. *J. Comput. Chem.* **2008**, *29*, 622–655. [CrossRef] [PubMed]

36. Jorgensen, W.L.; Chandrasekhar, J.; Madura, J.D.; Impey, R.W.; Klein, M.L. Comparison of simple potential functions for simulating liquid water. *J. Chem. Phys.* **1983**, *79*, 926–935. [CrossRef]

37. Ryckaert, J.-P.; Ciccotti, G.; Berendsen, H.J.C. Numerical integration of the cartesian equations of motion of a system with constraints: Molecular dynamics of n-alkanes. *J. Comput. Phys.* **1977**, *23*, 327–341. [CrossRef]

38. Kato, K.; Yanaka, S.; Yagi, H. *Technical Basis for Nuclear Magnetic Resonance Approach for Glycoproteins*; Springer: Singapore, 2018; pp. 415–438.

39. Deisenhofer, J. Crystallographic refinement and atomic models of a human Fc fragment and its complex with fragment B of protein A from *Staphylococcus aureus* at 2.9- and 2.8- resolution. *Biochemistry* **1981**, *20*, 2361–2370. [CrossRef] [PubMed]

40. Matsumiya, S.; Yamaguchi, Y.; Saito, J.; Nagano, M.; Sasakawa, H.; Otaki, S.; Satoh, M.; Shitara, K.; Kato, K. Structural comparison of fucosylated and nonfucosylated Fc fragments of human immunoglobulin G1. *J. Mol. Biol.* **2007**, *368*, 767–779. [CrossRef] [PubMed]

41. Borrok, M.J.; Jung, S.T.; Kang, T.H.; Monzingo, A.F.; Georgiou, G. Revisiting the role of glycosylation in the structure of human IgG Fc. *ACS Chem. Biol.* **2012**, *7*, 1596–1602. [CrossRef] [PubMed]

42. Remesh, S.G.; Armstrong, A.A.; Mahan, A.D.; Luo, J.; Hammel, M. Conformational plasticity of the immunoglobulin Fc domain in solution. *Structure* **2018**, *26*, 1007–1014.e2. [CrossRef]

43. Yageta, S.; Imamura, H.; Shibuya, R.; Honda, S. C_H2 domain orientation of human immunoglobulin G in solution: Structural comparison of glycosylated and aglycosylated Fc regions using small-angle X-ray scattering. *mAbs* **2019**, *11*, 453–462. [CrossRef]

44. Idusogie, E.E.; Presta, L.G.; Gazzano-Santoro, H.; Totpal, K.; Wong, P.Y.; Ultsch, M.; Meng, Y.G.; Mulkerrin, M.G. Mapping of the C1q binding site on rituxan, a chimeric antibody with a human IgG1 Fc. *J. Immunol.* **2000**, *164*, 4178–4184. [CrossRef] [PubMed]

45. Ramsland, P.A.; Farrugia, W.; Bradford, T.M.; Sardjono, C.T.; Esparon, S.; Trist, H.M.; Powell, M.S.; Tan, P.S.; Cendron, A.C.; Wines, B.D.; et al. Structural basis for FcγRIIa recognition of human IgG and formation of inflammatory signaling complexes. *J. Immunol.* **2011**, *187*, 3208–3217. [CrossRef] [PubMed]

46. Yamaguchi, Y.; Nishimura, M.; Nagano, M.; Yagi, H.; Sasakawa, H.; Uchida, K.; Shitara, K.; Kato, K. Glycoform-dependent conformational alteration of the Fc region of human immunoglobulin G1 as revealed by NMR spectroscopy. *Biochim. Biophys. Acta* **2006**, *1760*, 693–700. [CrossRef] [PubMed]

47. Subedi, G.P.; Barb, A.W. The structural role of antibody N-glycosylation in receptor interactions. *Structure* **2015**, *23*, 1573–1583. [CrossRef] [PubMed]

48. Wormald, M.R.; Petrescu, A.J.; Pao, Y.L.; Glithero, A.; Elliott, T.; Dwek, R.A. Conformational studies of oligosaccharides and glycopeptides: Complementarity of NMR, X-ray crystallography, and molecular modelling. *Chem. Rev.* **2002**, *102*, 371–386. [CrossRef] [PubMed]

49. Kamiya, Y.; Satoh, T.; Kato, K. Recent advances in glycoprotein production for structural biology: Toward tailored design of glycoforms. *Curr. Opin. Struct. Biol.* **2014**, *26*, 44–53. [CrossRef] [PubMed]

50. Sakae, Y.; Satoh, T.; Yagi, H.; Yanaka, S.; Yamaguchi, T.; Isoda, Y.; Iida, S.; Okamoto, Y.; Kato, K. Conformational effects of N-Glycan core fucosylation of immunoglobulin G Fc region on its interaction with Fcγ receptor IIIa. *Sci. Rep.* **2017**, *7*. [CrossRef] [PubMed]

51. Lee, H.S.; Im, W. Effects of N-Glycan composition on structure and dynamics of IgG1 Fc and their implications for antibody engineering. *Sci. Rep.* **2017**, *7*. [CrossRef]

52. Isoda, Y.; Yagi, H.; Satoh, T.; Shibata-Koyama, M.; Masuda, K.; Satoh, M.; Kato, K.; Iida, S. Importance of the side chain at position 296 of antibody Fc in interactions with FcγRIIIa and other Fcγ receptors. *PLoS ONE* **2015**, *10*, e0140120. [CrossRef]

53. Mizushima, T.; Yagi, H.; Takemoto, E.; Shibata-Koyama, M.; Isoda, Y.; Iida, S.; Masuda, K.; Satoh, M.; Kato, K. Structural basis for improved efficacy of therapeutic antibodies on defucosylation of their Fc glycans. *Genes Cells Devoted Mol. Cell. Mech.* **2011**, *16*, 1071–1080. [CrossRef]

54. Okazaki, A.; Shoji-Hosaka, E.; Nakamura, K.; Wakitani, M.; Uchida, K.; Kakita, S.; Tsumoto, K.; Kumagai, I.; Shitara, K. Fucose depletion from human IgG1 oligosaccharide enhances binding enthalpy and association rate between IgG1 and FcγRIIIa. *J. Mol. Biol.* **2004**, *336*, 1239–1249. [CrossRef] [PubMed]

55. Mimoto, F.; Igawa, T.; Kuramochi, T.; Katada, H.; Kadono, S.; Kamikawa, T.; Shida-Kawazoe, M.; Hattori, K. Novel asymmetrically engineered antibody Fc variant with superior FcγR binding affinity and specificity compared with afucosylated Fc variant. *mAbs* **2013**, *5*, 229–236. [CrossRef] [PubMed]

56. Richards, J.O.; Karki, S.; Lazar, G.A.; Chen, H.; Dang, W.; Desjarlais, J.R. Optimization of antibody binding to FcγRIIa enhances macrophage phagocytosis of tumor cells. *Mol. Cancer Ther.* **2008**, *7*, 2517–2527. [CrossRef] [PubMed]

antibodies

MDPI

Review

Design and Production of Bispecific Antibodies

Qiong Wang, Yiqun Chen, Jaeyoung Park, Xiao Liu, Yifeng Hu, Tiexin Wang, Kevin McFarland and Michael J. Betenbaugh *

Department of Chemical and Biomolecular Engineering, Johns Hopkins University, Baltimore, MD 21218, USA
* Correspondence: beten@jhu.edu

Received: 12 June 2019; Accepted: 31 July 2019; Published: 2 August 2019

Abstract: With the current biotherapeutic market dominated by antibody molecules, bispecific antibodies represent a key component of the next-generation of antibody therapy. Bispecific antibodies can target two different antigens at the same time, such as simultaneously binding tumor cell receptors and recruiting cytotoxic immune cells. Structural diversity has been fast-growing in the bispecific antibody field, creating a plethora of novel bispecific antibody scaffolds, which provide great functional variety. Two common formats of bispecific antibodies on the market are the single-chain variable fragment (scFv)-based (no Fc fragment) antibody and the full-length IgG-like asymmetric antibody. Unlike the conventional monoclonal antibodies, great production challenges with respect to the quantity, quality, and stability of bispecific antibodies have hampered their wider clinical application and acceptance. In this review, we focus on these two major bispecific types and describe recent advances in the design, production, and quality of these molecules, which will enable this important class of biologics to reach their therapeutic potential.

Keywords: single-chain variable fragment (scFv); bispecific antibody; quadroma technology; knobs-into-holes; CrossMAb; bispecific T-cell engager (BiTE)

1. Introduction

Over recent decades, immunotherapies, including checkpoint inhibitors, adoptive cell transfer, monoclonal antibodies, and vaccine treatments, have become efficient and highly specific treatments to fight cancer by boosting a patient's immune system. These treatments can specifically target tumor cells and the tumor microenvironment with less cytotoxicity and fewer side effects [1]. Through three decades of development and exploration, therapeutic monoclonal antibodies have become the most widely used and approved immunotherapy method in clinical practice to treat various malignant tumors [1]. These antibodies are designed to bind to specific targets found on cancer cells and destroy them by activating the patient's immune system. More recently, bispecific antibodies represent a valuable alternative antibody platform in immunotherapy treatment. These bispecifics work by binding to two different antigen sites and can provide more robust and tailored immunogenic targeting than what is possible with natural antibodies.

An antibody produced against a single epitope of an antigen is called a monoclonal antibody (mAb) produced by a single plasma cell type, while polyclonal antibodies bind to multiple epitopes of an antigen or multiple antigens and are typically produced by multiple plasma cells [2]. Bispecific antibodies are engineered artificial antibodies capable of recognizing two epitopes of an antigen or two antigens. Human immunoglobulin G (IgG) is the most common type of antibody found in human serum and is further broken down into four subclasses, IgG_{1-4}. These subclasses differ in their constant regions, particularly the γ-chain sequences and disulfide bond patterns, but share the same basic structure [3]. As illustrated in Figure 1a, the basic structure of IgG is composed of two light chains (LC) and two heavy chains (HC) to form a complex quaternary Y-shaped structure with three independent protein moieties connected through a flexible hinge region [4]. These moieties are symmetrical with two identical fragment antigen-specific binding (Fab) regions and one fragment crystallizable (Fc) region [3]. Antibodies bind to specific antigens through the Fab domain formed by hypervariable regions of heavy and light chains. Natural Abs can bind to natural and artificial antigens with high affinity and specificity with a remarkable diversity of 10^8–10^{10} different variants for each antigen-binding site [5,6]. The Fc region of the antibody binds to receptors or other proteins of the host immune system such as Fcγ receptors (FcγRs), C1q, and neonatal Fc receptor (FcRn) to initiate distinct effector functions. Small differences in the amino acid sequence and glycosylation pattern on the Fc domain can highly impact key attributes such as IgG thermal stability, FcγR-binding effectiveness, and serum half-life [7].

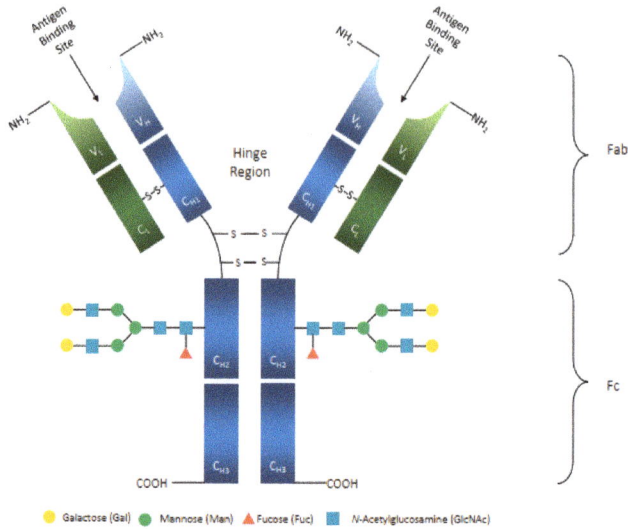

(a)

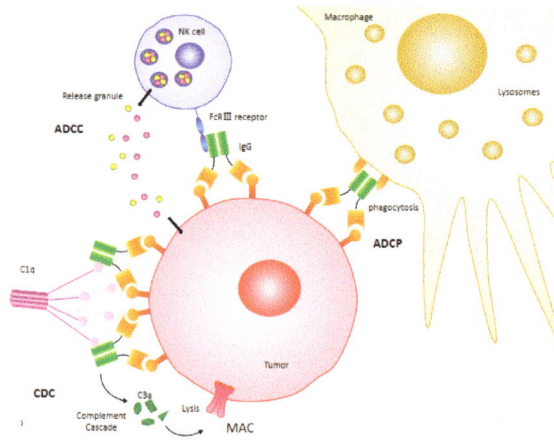

(b)

Figure 1. (**a**) The illustration of human IgG$_1$ structure with N-glycan attached at Asn297 site in the CH2 of the Fc region. Light chains (L) are highlighted in green, and heavy chains (H) are highlighted in blue. C: constant domain; V: variable domain; H: heavy chain; L: light chain; S-S: disulfide bond; Fab: Fragment antigen-binding domain; Fc: Fragment crystallizable domain. Fc regions which bind effector molecules and cells. (**b**) The schematic diagram of antibody-dependent cellular cytotoxicity (ADCC), complement-dependent cytotoxicity (CDC) and antibody-dependent cell-mediated phagocytosis (ADCP) mechanisms in cancer treatment. For ADCC, natural killer (NK) cells recognize the Fc region mediated by surface FcγRIIIa receptors, causing apoptosis of the antibody-coated tumor cells. Once activated, NK cells release cytotoxic granules containing perforin and granzymes to induce apoptosis in targeted cells [8]. Another antibody-induced pathway is the classical pathway in the complement system, often called complement-dependent cytotoxicity. CDC is initiated when the C1q complex interacts with antibodies bound to the pathogen surface, facilitating the lysis of cells by forming the membrane attack complex (MAC) which induces lethal colloid-osmotic swelling [9–11]. The third major effector mechanism is ADCP. ADCP is a potent mechanism by which IgG-opsonized tumor cells activate the FcγRIIa and FcγRI expressed on the surface of macrophages to induce phagocytosis, resulting in engulfment and degradation of the target cell through acidification of the phagosome and the fusion with lysosomes [12].

Cancer cells are abnormal cells differentiated from normal healthy cells, but they may be difficult for the body's immune system to detect. Monoclonal antibodies have been successful as cancer therapeutics by targeting surface antigens over-expressed or expressed uniquely, on tumor cells [13]. The efficacy of antibody-based cancer immunotherapy involves two main factors as discussed below: (1) Blocking or binding factors activate cell death or inhibit activation of signal pathways used by cancer cells to grow and survive, resulting in tumor cell death [14,15]. For example, trastuzumab targets HER2 receptors in breast and stomach cancer cells in order to inhibit their proliferation and survival [16], and cetuximab inhibits epidermal growth factor receptor (EGFR) in colorectal and lung cancer [17,18]; and (2) Immune effector functions that engage the Fc region of antibodies via Fc receptors (FcR) on immune cells. The mechanisms of antibody-dependent cellular cytotoxicity (ADCC), complement-dependent cytotoxicity (CDC) and antibody-dependent cellular phagocytosis (ADCP) are illustrated in Figure 1b. All three mechanisms (ADCC, CDC, and ADCP) can induce target cell death and aid in the efficacy of the treatment.

Even with the success of various anti-tumor therapeutic antibody drugs, however, there also have been significant limits to conventional antibody therapy. One major drawback of antibody therapy is low tumor penetration and retention rate. Many therapeutic mAbs directed against tumor-specific antigens largely remain in the circulation with typically only 20% of the administered dose interacting with the surface proteins of solid tumors [19]. This disproportionality reflects the challenges of achieving effective penetration and retention within solid tumor tissue [19]. Furthermore, most mAbs serve to prevent the binding of growth factors to their receptors but fail to induce apoptosis of tumor due to an insufficient immune response from patients, especially the re-activation of T-cells to destroy tumors. For the effector functions (mainly ADCC, CDC, and ADCP) triggered by mAbs, the exact effectiveness depends on specific antibodies. Numerous studies have demonstrated that the amino acid sequence of the CH2 and CH3 domain and the Fc conserved glycan profile both impact the antibody ADCC and CDC activities [11]. Moreover, the unsatisfactory performance of natural antibody treatment also comes from the extensive cross-talk among some signaling pathways in cancer cells and nearby cells contributing to relapses in mAb treatment, which works by blocking signaling pathways and the induction of apoptosis [20,21].

Bispecific antibodies were proposed three decades ago and have been extensively investigated to overcome the limitation of natural mAbs, which can only bind a single epitope [22]. Bispecific antibodies can target two different antigens at the same time [5], such as simultaneously binding tumor cell receptors and recruiting cytotoxic immune cells. This enhanced functionality may potentially result in fewer side effects and fewer injections. Furthermore, from a biopharmaceutical manufacturer's perspective, fewer clinical trials and reduced production costs can be accomplished by making a single molecule instead of two [23]. With more than 100 bispecifics in clinical trials [5], bispecific antibodies are under development to cover a broad spectrum of applications including diagnosis, imaging, prophylaxis, and therapy, with the majority of drug candidates focusing on cancer therapy [22].

Through decades of exploration and development of bispecific antibodies and their derivatives, there are two common formats of bispecific antibodies on the market: the single-chain variable fragment (scFv)-based (no Fc fragment) antibody and the full-length IgG-based antibody. Unlike the conventional monoclonal antibodies, great production challenges with respect to the quantity, quality, and stability of bispecific antibodies have hampered their wider clinical application and acceptance [24]. Meanwhile, advanced design strategies around phage display screening, antibody linker engineering, quadroma technology [25], knobs-into-holes technology [26], common light chain [27], CrossMAb technology [28], and protein engineering have all been extensively investigated, and make up the principal knowledge base of this fast-growing and diverse field [22,29–31]. Therefore, this review will focus on the design and manufacture of these two major bispecific molecule types and describe the recent developments in the therapeutic potential and opportunities in bispecific antibody production capacity and quality achieved by employing a range of operational strategies.

2. Strategies to Improve Bispecific Antibody Production and Quality

2.1. Single-Chain Variable Fragment (scFv) Antibodies

Single-chain variable fragments (scFvs) are minimalist forms of a functional antibody, generated by fusing variable domains of the IgG heavy chain (VH) and light chain (VL) through a flexible polypeptide linker [32]. ScFv molecules have a molecular weight in the range of 25 kDa, with a single antigen-binding site that is comprised of components from each arm of the antibody [22]. Several important considerations in developing scFv antibodies are the antibody fragment types, the linker type, and production capability. More recently, another exciting area for using scFv technology is the chimeric antigen receptor (CAR) T-cell approach for adoptive cell transfer immunotherapy [33]. Given the scope and volume of this review, the use of scFvs for CAR-Ts will not be included here [34].

2.1.1. Antibody Fragment Types

Currently, there are three main bispecific antibody fragment formats: bispecific T-cell engager (BiTE), dual-affinity re-targeting proteins (DARTs) and Tandem diabodies (TandAbs), as depicted in Figure 2a.

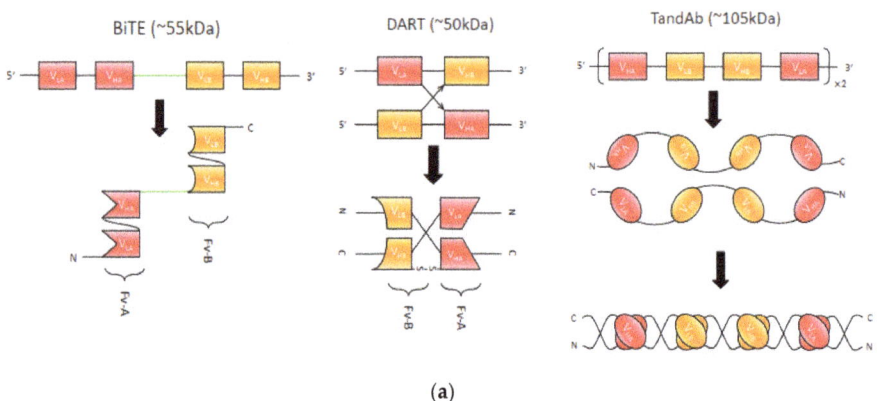

(a)

Figure 2. *Cont.*

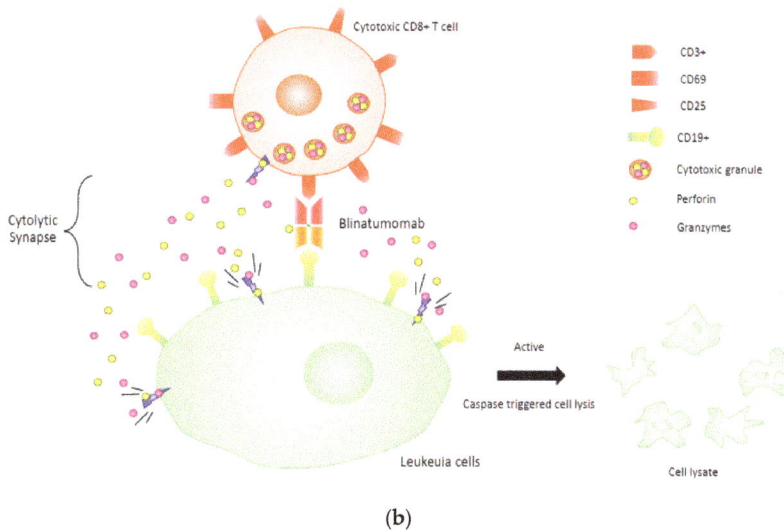

(b)

Figure 2. (a) The construction of three main bispecific antibody fragment molecules. (b) The mechanism of Blinatumomab treatment. Blinatumomab as a bispecific antibody can simultaneously bind to CD3+ T-cells and CD19+ leukemia cells and has been approved for the treatment of B-cell precursor acute lymphoblastic leukemia (ALL).

BiTE molecules have been extensively applied in cancer immunotherapy for re-targeting of T-cells to tumor cells or tumor-associated cells in the tumor microenvironment. They employ scFv fragments from two different monoclonal antibodies connected by a peptide linker, enabling them to retain each antibody's binding activity when assembled [35]. The short flexible linker connecting the two scFvs enables free rotation of the two arms, which is vital for flexible interaction with targeted receptors on two opposing cell membranes (cytotoxic T-cell and tumor cell) and the subsequent induction of T-cell activation [36].

One of the most successful BiTE drugs is blinatumomab (Blincyto®, DrugBank entry DB09052), which has been approved by the FDA for the treatment of B-cell precursor acute lymphoblastic leukemia (ALL). Blinatumomab is comprised of an anti-CD 19 scFv in the VL-VH orientation linked through a short glycine/serine (GGGGS) linker to an anti-CD3 scFv in the VH-VL orientation [30]. The mechanism of blinatumomab has been illustrated in Figure 2b. Due to its small size, blinatumomab can reach in close proximity to T-cell and target cell membranes, but this feature also leads to the rapid clearance from circulation with a short elimination half-life (mean ± SD) of 1.25 ± 0.63 h [37], which is presumed to be eliminated renally [38]. As a result, BiTE requires continuous dosing at a high concentration (15–28 μg per day) to recruit and activate a large amount of suboptimal T cells to achieve half-maximal target cell lysis [36]. Therefore, this antibody is administered as a 4-week continuous intravenous (IV) infusion to maintain sufficient therapeutic serum concentration [37], which increases costs by having to produce more clinical-grade antibodies [39]. The single polypeptide chain structure that enhances BiTE antibody-antigen recognition, however, comes at the cost of increased aggregation and decreasing protein stability [24,40].

Partly in response to these issues, researchers developed a potential alternative—dual-affinity re-targeting proteins (DARTs). As shown in Figure 2a, a DART is composed of two Fv fragments, with two unique antigen-binding sites formed when two Fv fragments heterodimerize [41]. Specifically, Fv1 consists of a VH from antibody A and a VL from antibody B, while Fv2 is made from a VH from antibody B and VL from antibody A. Unlike BiTE antibodies which are connected by a polypeptide linker, this combination allows DART to mimic natural interaction within an IgG molecule. Adding

211

another cysteine residue to the end of each heavy-chain improves stability by forming a C-terminal disulfide bridge (see Figure 2a) [41]. Compared to a BiTE, DART molecules are able to retain potency for both in vitro and in vivo administration as well but can be produced at scale with lower aggregation rates [21,42]. A recent comparison by Moore et al. of the in vitro ability of CD19xCD3 DART and BiTE molecules to kill B-cell lymphoma found that DART molecules outperformed BiTE molecules consistently. In this study, both DART and BiTE molecules were derived from the same parental antibodies (mouse anti-human CD3 and CD19 mAbs), with DART molecules performing better in maximal B-cell lysis, requiring less concentration for half-maximal B-cell lysis, and in molecular markers of T-cell activation [40].

Currently, DART and BiTE proteins can also be further engineered to integrate better with patient immune systems. For example, several BiTE molecules can be linked to the IgG Fc domain to generate BiTE-Fc fusion drugs compatible with once-weekly dosing for treatments [43–45]. Anti-CD19x CD3 BiTE-Fc fusion protein binds with high affinity to human and non-human primate (NHP) CD19 as well as CD3 with a serum half-life of 210 h following a single intravenous administration of a 5 µg/kg dose in NHP, without obvious signs of toxicity in clinical and laboratory animal studies [45]. An anti-BCMA BiTE-Fc fusion protein for the treatment of multiple myeloma has a serum half-life of 112 h following a single 15 µg/kg dose in NHP [44]. In addition, BiTE molecule can also be linked to human albumin to extend serum half-life. A comparison of anti-CD33 x CD3 BiTE-Fc and BiTE-albumin fusion proteins reveals that the Fc-based BiTE antibody constructs provided a similar survival advantage when administered every four or five days as the canonical BiTE when administered on a daily basis in the mouse model. Alternatively, the albumin fusion-based BiTE was less efficacious when administered every four days than the daily administered canonical BiTE [43]. DART proteins can also be fused with the Fc region of an IgG, creating a DART-Fc construct which can significantly extend the serum half-life when compared to the DART protein alone [46,47]. One group designed HIVxCD3 DART and DART-Fc and evaluated their killing activity on mononuclear cell cultures isolated from HIV-infected participants [46]. Their results showed that both DART formats reduced cell-to-cell virus spreading in resting or activated CD4 T-cell cultures [46]. Additionally, though HIVxCD3 DART-Fc performed similarly in killing activity to HIVxCD3 DART, the DART-Fc extended the DART in vivo half-life from less than 10 to 70.2 h [46].

The small size of scFvs contributes to a high renal clearance rate in comparison to natural antibodies. One solution to the size issue is to generate Tandem diabodies (TandAbs), which are shown in detail in Figure 2a. These tetravalent bispecific antibodies provide two binding sites for each antigen to maintain the avidity of a natural bivalent antibody [48,49]. Moreover, TandAbs have a molecular weight (approximately 105 kDa) exceeding the first-pass renal clearance threshold, thus offering a longer half-life compared to smaller antibody constructs [48,50]. Two TandAb format drugs are in clinical trials—AFM13 (CD30xCD16) for NK cell recruitment and AFM11 (CD19xCD3) for T-cell recruitment [51].

2.1.2. Linker Engineering

The linker region between light and heavy-chain domains plays a significant role in stabilizing the antibody and is therefore an important target for optimizing scFvs [52]. The scFv can be assembled with a single polypeptide chain in the form of VH-linker-VL or VL-linker-VH, where the linker bridges the gap between C and N termini of the respective domains. Studies investigating the orientation of the heavy and light-chain domains imply that both orientations can be favorable in different cases and that linker design may impact biophysical properties [53–55]. Two essential considerations in linker design are amino acid composition and sequence length. Firstly, the amino acid composition is critical in designing a viable and flexible linker peptide; for instance, a hydrophilic sequence is indispensable to avoid intercalation of the peptide within or between the V domains during protein folding [56]. Currently, the most commonly used amino acid sequence motif is $(G4S)_n$ (G: glycine, S: serine; G4S is four glycine residues and one serine residue). Glycine and serine are preferred because their short side chains grant conformational flexibility and minimal immunogenicity, while serine additionally improves solubility [31,57]. Besides the conventional Gly-Ser linker, other designs include charged residues such as glutamic acid and lysine to enhance solubility [58], while high-throughput selection methods such as phage display also facilitate the design and generation of linkers that are specifically optimized for certain antibodies [59].

In addition to composition, the length of the linker between heavy and light chains of the Fv domain is also critical in assembling the correct conformation of the scFv. It has been reported that the linker should be able to span 3.5 nm (35 Å) between the C terminus of one V domain and the N-terminus of the other V domain without affecting the native Fv conformation [31]. The length of the linker exerts a significant impact on multimer formation of the antibodies, with studies showing that a linker length longer than 12 amino acid residues allows sufficient distance between heavy and light-chain domains to associate and form monomers [60–62]. Shorter linkers connecting VH and VL can prevent the direct association of the two domains, resulting in an increased possibility for pairings between heavy and light chains of different scFv molecules, forming dimers, trimers or higher order oligomers [40,63]. Therefore, by properly designing the linker length, one can effectively promote the formation of scFv molecules that are designed as diabodies (in particular, the TandAbs molecule), or whose multivalent forms are desired over their monovalent forms. For example, pharmacokinetic studies suggest that for particular antibodies, such as CC49, the formation of dimer or tetramers are favorable, improving tumor targeting compared to the monomeric form [64,65], while maintaining efficient in vivo tumor localization and in blood [66]. Therefore, it is critical to design linker length to achieve desired scFv conformation or distribution of multivalent forms.

The design of linkers for optimal bispecific scFv configurations necessitates careful consideration of the above principles. Initial attempts to construct bispecific scFvs focused on designing linkers that directly connects two monovalent single-chain antibodies. Examples of these designs include the linker CBH1 composed of 24 amino acids [35] and the 205C linker which has 25 amino acid residues [67]. Factors such as the amino acid composition of these inter-chain linkers can have an impact on the function of single-chain bispecific antibodies [68]. More recent designs of linkers emphasize more on linker length and achieving the desired antibody conformations and domain associations, which can be exemplified by BiTE, DART, and TandAb, as shown in Figure 2a. For successful construction of bispecific scFv, it is important to control linker length to avoid or minimize non-cognate pairing between heavy and light chains of heterologous antibodies, since appropriate VH-VL association is critical in antibody affinity and specificity that ensures proper functioning [29,69]. The BiTE design places long linkers between heavy and light chains of homologous domains to ensure association, and short linkers between heterologous heavy-chain fragments (the GGGGS linker) to form the connection between the two Fvs [70]. The DART molecules are dual chains that bind to each other to form functional dimers, where the linkers between VHA and VLB or VHB and VLA needs to be as short as five amino acids to prevent undesired non-homologous pairing [40]. Moreover, the positioning of the disulfide bond is another key feature of DART molecules, which holds the molecule together in the correct orientation. The linker design for TandAb is similar to DART, for which the linkers between adjacent domains are six amino acids (GGSGGS) so that two identical chains are likely to bind and form large dimers that are favorable in the aspects of in vivo half-life [21,48,71].

2.1.3. Stability Engineering of scFv Antibodies

Stability of scFv molecule is a critical factor because it is believed that there is a direct correlation between the stability and biological activity [72,73], and stable scFv molecules can be considered as building blocks for functional bispecific antibodies. A number of different approaches through changing expression environment and introducing helper molecules (e.g., chaperons) to improve the scFv solubility and stability are discussed in the following bispecific scFv expression and production (Section 2.1.4). In this section, we focus on the approaches that achieve optimized protein stability through direct modification of scFv frameworks. Two types of commonly used methods to engineer scFv structure include loop grafting and mutagenesis. The loop grafting approach may be favorable for the generation of therapeutic scFv because the process achieves both stabilization and humanization in one step, by grafting the antigen-specific complementarity determining regions (CDRs) onto frameworks with suitable biophysical properties including stability [74–78]. For example, Borras et al. [79] reported a successful attempt to humanize and stabilize rabbit variable domains by grafting CDRs from 15 different rabbit monoclonal antibodies onto a human scFv scaffold, resulting in similar affinity but significantly improved biophysical properties. Alternatively, for the mutagenesis approaches, enhanced stability is achieved by either optimizing structure by rational site-specific mutation [80–84] or directed protein evolution (i.e., inducing random mutagenesis followed by positive selection steps) [74,85–91]. Compared to the laborious directed evolution method which requires iterative steps to reach an optimum, site-specific mutagenesis approaches are relatively easy to implement with the well-established techniques [84]. Rational designs of site-specific mutation are generally knowledge-based and different mutations can be combined and introduced simultaneously with the assumption that mutations have cumulative effects on improving stability [84]. For example, one of the most common mutation-based optimization methods is the consensus sequence approach, in which the most frequent amino acid at any position in homologous Fv domain is assumed to contribute to the stability considering molecular evolution and selection, and a mutation toward this collective consensus sequence is expected to have a positive effect on stability [77,81,82]. Besides the consensus approach, other methods alter amino acid residues to achieve certain goals such as creation of inter-domain disulfide bonds [84,92–94], creation of intramolecular hydrogen bonds [83,95] and optimization of hydrophobicity [77,83,96]. As a successful case of utilizing these mutation-based approaches, Miller et al. [72] used a combination of sequence-based statistical analyses (residue frequency analysis and consensus methods) and structure-based design approach to identify the target residues for mutations in VH and VL sequences. Then, the scFv mutants were screened through a high-throughput antigen-binding assay with thermal challenges [72]. The isolated stability-engineered scFv variant was able to produce in suspension Chinese hamster ovary cells with high yield (21.5 mg/L), purity and biological activity [72].

2.1.4. Bispecific scFv Antibody Expression and Production

Appropriate host platforms are determinant to the efficient expression and production of scFv antibodies, and there exist several different viable platforms for scFv expression including bacteria, yeast, mammalian cells, insect cells, plant and cell-free systems [97–100]. Given that bispecific scFvs are composed of two or more scFv molecules, the various expression hosts for the bispecific scFvs may vary from those used for the production of scFv single molecules. The "best" expression system for bispecific scFv proteins is yet to be determined because differences in size, amino acid sequence, and conformation of the recombinant protein make it difficult to conclude a universal expression system that optimizes the yield and quality of the protein, which can be affected by many factors such as solubility and stability [99]. However, several studies listed in Table 1 have reported successful expression of bispecific scFv and its fusion molecules using bacterial and mammalian systems.

Table 1. Examples of the expression of bispecific antibody fragment molecules in various hosts.

Platform	Species	Molecule	Yield	Purification	Type	Reference
Bacteria	TG1 *E. coli*	anti-HER2/*neu* × anti-CD16	around 3.7 mg/L	Ni-NTA column	BiTE	[101]
Bacteria	*E. coli*	anti-TCR × anti-fluorescein	1 mg/L	Fluorescein affinity chromatography	Tandem bispecific scFv molecule linked by 212 and 205 c′ linkers	[67]
Bacteria	*E. coli, periplasmic*	anti-EpCAM × anti-CD3	12–15 mg/L	Ni-NTA column	BiTE	[102–104]
Bacteria	*E. coli BL21(DE3)*	anti-HER2 × anti-CD3	3 mg/L	Ni-NTA column	BiTE	[105]
Mammalian	CHO-K1	anti-CD123 × anti-CD3	2–5 mg/L	Protein G chromatography	BiTE-Fc	[106]
Mammalian	CHO cell	anti-P-cadherin × anti-CD3	1300 mg/L	Protein A chromatography	DART-Fc	[107]
Mammalian	CHO-S	anti-CD19 × anti-CD3		SEC	DART & BiTE	[40]
Mammalian	CHO	anti-CD33 × anti-CD3		IMAC + SEC	TandAbs	[48]
Mammalian	CHO	anti-CD19 × anti-CD3	>200 mg/L	Ni-NTA column	TandAbs	[71]
Mammalian	CHO cell	anti-EpCAM × ani-CD3		IMAC + gel filtration + CEX	BiTE	[108]
Mammalian	HEK 293	anti-EpCAM × ani-CD3		IMAC	BiTE	[109]
Mammalian	CHO-S	anti-GD2/DOTA-metal complex	5–10 mg/L	Protein A chromatography	IgG-ScFv	[110]

SEC: size exclusive chromatography, IMAC: immobilized metal affinity chromatography, CEX: cation exchange chromatography.

E. coli is one of the most widely used hosts for scFv expression. Some of the major advantages of using *E. coli* include its rapid growth, cost efficiency, high heterologous protein productivity, well-understood genetics as well as easy genetic manipulation [111–113]. Unlike the glycosylated whole antibody protein, scFv molecules are much easier to produce in bacteria. However, challenges still remain for harnessing this high-yield expression system, one of which is insufficient protein solubility. It was reported by multiple studies that proteins produced from the *E. coli* expression system result in misfolding and inclusion body [114–117]. This inefficiency in producing soluble scFv is known to be caused by the lack of chaperon and post-translational machinery and the reducing environment of *E. coli* cytoplasm which prevents disulfide bonds to be formed [97,118], and for scFv molecules, formation of intra-domain disulfide bonds is essential for the key structure known as the "immunoglobulin fold" [119,120]. Therefore, successful expression of functional scFv molecules from *E. coli* systems usually requires additional procedures or modifications. For example, subsequent protein refolding and recovery steps can be integrated into the process, including solubilization treatment with agents such as urea and guanidine hydrochloride, and a step to refold solubilized protein by removing solubilization agents by methods such as dialysis [121]. Gruber et al. (Table 1, [67]) reported the production of bispecific scFv in *E. coli* with these refolding steps. The solubility of scFv molecules can also be improved by secreting them into the bacterial periplasm that has an oxidizing environment, through genetically attaching the secretory signal sequence to the N-terminus of scFv sequence [7,122–124]. A number of studies have reported the periplasmic expression of BiTE type molecules in *E. coli* [4,103,104]. Besides the above methods to tackle solubility issue, there are other approaches exist that may in the future be applied to facilitate the production of bispecific scFvs in bacterial platforms. For example, expression of scFv as a fusion protein with solubility enhancing tags such as MBP, NusA, and TRx can promote and facilitate correct protein folding [38,97,125,126], despite that these tags need to be removed afterward to allow normal antibody usage. Furthermore, studies have shown that co-expression of molecular chaperons such as Skp, OmpH, HlpA and FkpA, and folding modulators/catalysts such as disulfide bond metabolizing enzymes can effectively tackle protein aggregation and misfolding problems, and the "cocktails" approach has been an increasingly common expression strategy that involves the simultaneous usage of various chaperons or folding catalysts [32,127–130].

Single-chain Fv molecules are also expressed through other platforms to exploit particular advantages that are not granted by bacterial hosts. Mammalian cells represent the most widely used production platform for therapeutic proteins and a promising expression vehicle for bispecific scFvs due to their advanced protein folding pathways and post-translational modifications [99,131]. Mammalian cells allow for stable expression and robust production of soluble recombinant proteins. For example, Vendel et al. [114] and Jain et al. [132] managed to express bioactive scFv molecules via Chinese Hamster Ovary (CHO) cells while the same proteins expressed in bacteria shows less activity or even a significantly different secondary structure.

Besides bacterial and mammalian cells, other expression systems have demonstrated advantages in various studies involving the expression of scFvs and may be potential candidates of large-scale production platforms for bispecific scFvs in the future. Yeast as a eukaryotic microorganism is not only capable of producing correctly folded and fully functional proteins, but can also survive and grow rapidly in simple media [99]. Another organism of interest is insect cell, which allows the utilization of the baculovirus-mediated gene expression system [99]. The advantage of using baculovirus expression vector system (BEVS) is the high gene expression level achieved though the polyhedron gene (polh) promoter in virus-infected insect cells [96,133]. Production of recombinant protein from plant is believed to be a desired method for large-scale protein production considering factors including scalability, cost efficiency, and safety [134]. Growing transiently expressed or stable transgenic plants followed by protein extraction from leaf tissue allows the production of scFv molecules [4,135,136]. Besides the host-based expression systems, cell-free protein synthesis system allows high-throughput protein library generation due to the efficiency and flexibility this approach offers [137,138], and the expression of scFv molecules can be achieved without the time-consuming steps of expression vector generation and transformation.

2.2. Full-Size IgG-like Asymmetric Bispecific Antibody

Although IgG-like asymmetric bispecific antibodies have some properties that are similar to natural monoclonal antibodies, they are engineered molecules that have not been generated by typical B-cells [139]. As a result, these differences lead to significant production challenges. One of the greatest challenges for asymmetric IgG-like bispecific antibodies manufacturing is ensuring the correct assembly of antibody fragments, which is a prerequisite for bispecific antibody large-scale production. Random assembly of four distinctive polypeptide chains (two different heavy and two different light chains) results in 16 combinations (10 different molecular configurations), among which only two represent the desirable asymmetric heterodimeric bispecific antibody (12.5% of the statistical probability) [22,140,141]. The remaining pairings result in non-functional or monospecific molecules [22]. So not only the quality but also the quantity of bispecific antibodies generated from *E. coli* and mammalian cells can be greatly improved by optimizing the correct assembly of bispecific antibodies. Production examples of several IgG-like bispecific antibody molecules are summarized in Table 2. There are mainly two problems that must be solved to produce the desired IgG-like bispecific antibody—the heterodimerization of two different heavy chains and the discrimination between the two light-chain/heavy-chain interactions [142]. Judicious genetic and cellular engineering strategies, such as quadroma technology, knobs-into-holes, common heavy chain, and common light-chain strategies, CrossMab and co-culture methods, have been implemented to produce optimized Y-shape IgG-like bispecific antibodies. We will describe each of these important strategies briefly in the following sections.

Table 2. Examples of the expression of IgG-like bispecific antibody molecules in various hosts.

Platform	Name	Target	Heavy-Chain Engineering	Heavy/Light-Chain Engineering	Yield	Purification	Note	Reference
CHO-DG44 cells	MCLA-128	Human epidermal growth factor receptors (HER2 and HER3)	knobs-into-holes	common light chain	0.6–1.2 g/L	Protein A + IEC	stable expression	[27]
HEK293F suspension cells	Ang-2-VEGF-A CrossMab	angiopoietin-2 (Ang-2) and vascular endothelial growth factor A (VEGF-A)	knobs-into-holes	CrossMab (CH1-CL)	0.03 g/L	Protein A + SEC	transient expression	[142]
cell-free system (*E. coli* extract)	ScFv-KiH, BiTE-KiH	CD3, EpCAM, HER2	knobs-into-holes		0.2–0.4 g/L	Protein A	in vitro	[143]
HEK293	M315-14D2 (scFv-Fc)	mouse NKG2D and mouse p55TNFR	Electrostatic Steering Effects		0.1 g/L	protein A	transient expression	[144]
Expi 293 cells *	10E8V2.0/iMab	human CD4 and HIV-1	knobs-into-holes	CrossMab (CH1-CL)		Protein A + SEC	transient expression	[145]
HEK293F suspension cells	CD20-243 CrossMab	CD20 and HLA-DR	knobs-into-holes	CrossMab (CH1-CL)		Protein A + SEC	transient expression	[146]
CHO-K1 suspension cell culture	anti-FGFR1/βKL	FGFR1/βKL	knobs-into-holes	Co-culture	0.35 g/L	Protein A + IEC	stable expression	[147]
E. coli K-12 W3110 suspension cell	Anti-Her2/CD3	Her2/CD3	knobs-into-holes	Co-culture	4.8 g/L	Protein A + HIC	stable expression	[148]
E. coli K-12 W3110 suspension cell	Anti-CD19/CD3	CD19/CD3	knobs-into-holes	Co-culture	1 g/L.	Protein A + HIC	stable expression	[148]

* Expi293 cells are developed for high-yield transient expression purpose by Gibco company, which is based on suspension-adapted human embryonic kidney (HEK) cells. IEC: ion-exchange chromatography, SEC: size-exclusion chromatography, HIC: hydrophobic interaction chromatography.

Antibodies **2019**, *8*, 43

2.2.1. Quadroma (or Hybrid-Hybridoma) Technology

Initially, a bispecific antibody was generated by the somatic fusion of two hybridomas, as illustrated in Figure 3a. [149]. Each hybridoma cell expresses a unique monoclonal antibody with predefined specificity. Then, the two antibody-expressing cells are fused and the resulting hybrid-hybridoma cell expresses the immunoglobulin heavy and light chains from both parents [149], where assembly allows the formation of both parental and hybrid immunoglobulins. The quadroma technology represents the foundation of bispecific antibody production, but also suffers from low production yields and high product heterogeneity. [4]. The random assembly of two different heavy and two different light chains can theoretically result in 10 different molecular configurations and only one of those is functional bispecific antibody [22]. The real percentage of functional bispecific antibody by a quadroma cell line is unpredictable and a laborious process is required to isolate the bispecific antibody from the side products [150,151]. Later, a chimeric quadroma technology was developed by fusing a murine and a rat hybridoma cell line [139]. The content of chimeric mouse/rat bsAb was significantly enriched due to preferential species-restricted heavy/light-chain pairing in contrast to the random pairing in conventional mouse/mouse or rat/rat quadromas [152,153]. Furthermore, rat heavy chains did not bind to protein A for purification, while the mouse heavy chains in bsAbs can be eluted at pH 5.8 while the full-size parental mouse Ab can be eluted at pH 3.5 [153]. This feature provided an easy and simple purification process through protein A and ion-exchange chromatography to isolate the desired bispecific component. With the improvements of quadroma technology, Catumaxomab (anti-EpCAM x anti-CD3) was the first approved IgG-like bispecific antibody in Europe in 2009 for the intraperitoneal treatment of patients with malignant ascites [154]. Catumaxomab is generated via quadroma technology and composed of mouse IgG2a and rat IgG2b [154]. As a trifunctional antibody, one Fab antigen-binding site binds T-cells via CD3 receptor, the other site binds tumor cells via the tumor antigen epithelial cell adhesion molecule (EpCAM) and the Fc region provides a third binding site to recruit and activate immune effector cells via binding to FcγRI, IIa and III receptors [154]. Nevertheless, Catumaxomab cannot bind to the inhibitory Fcγ IIb receptor. Immunogenicity is another concern—human anti-mouse or anti-rat antibody response are sometimes observed in patients with catumaxomab treatment [25,51].

(a)

(b)

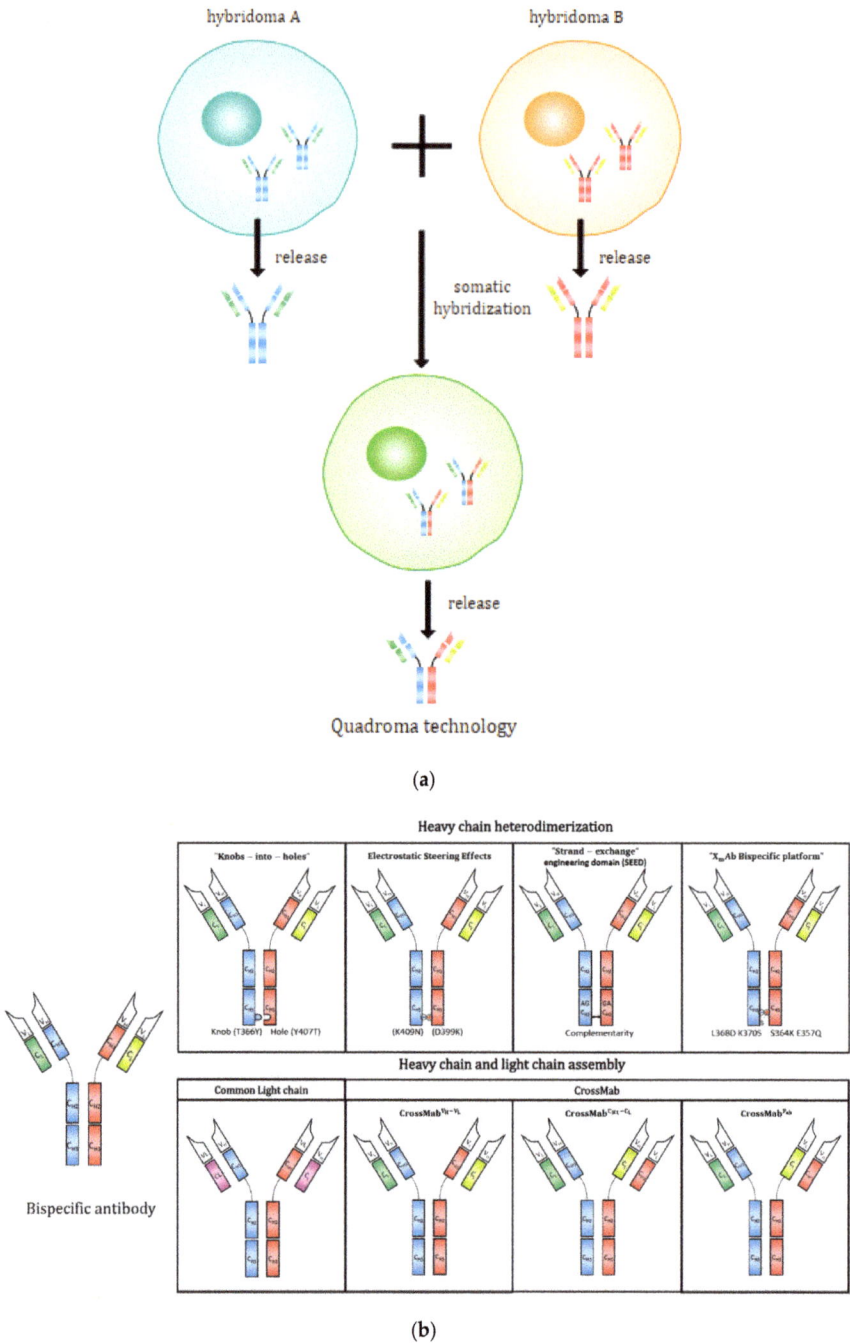

Figure 3. The strategies for improving IgG-like bispecific antibody product quality. (**a**) The illustration of quadroma technology. (**b**) A summary of the heavy and light-chain genetic and protein engineering strategies to achieve homogeneous asymmetric heterodimeric bispecific antibody product.

2.2.2. Heavy-Chain Assembly

Fc heterodimerization is a particularly important design to reduce the number of possible combinations of different forms while exclusively producing asymmetric antibodies by eliminating the formation of normal monoclonal antibodies. Heterodimeric heavy chains are achieved by combining two complementary but not identical heavy chains that result in a single heavy-chain combination. Each heavy chain can then bind to different light chains, resulting in four possibilities: one bispecific molecule, one non-functional combination, and two monospecific molecules [22]. Using this approach, the possible antibody combinations are thus substantially reduced from 10 different molecules to just the four remaining molecules [22]. The dimerization of Fc is achieved by CH3 domain of Fc (the last domain of the constant region) interfacing with each other [155]. Different technologies can be applied to engineer the CH3 domain so that two different Fc domains can be properly linked to one another, as shown in Figure 3b.

Knobs-into-holes technology, which involves engineering CH3 domains to create either a "knob" or a "hole" in each heavy chain to promote Fc heterodimerization [143], has been extensively applied for Fc engineering. The knobs-into-holes model was first proposed by Francis Crick to pack amino acid backbones of coiled-coil alpha-helix domains of proteins [156]. Ridgway et al. applied the knobs-into-holes as a novel design strategy to engineer heavy chains of Fc domains rendering them able to form heterodimers [26]. A small amino acid in a CH3 domain was replaced with larger one (T366Y) to make a knob variant, and a large amino acid in another CH3 domain was replaced with smaller one (Y407T) to produce a hole so that the two engineered domains can fit into one another favoring the heterodimerization [26]. Furthermore, additional mutation sites including S354C and T366W in a CH3 domain were found to generate knobs while Y349C, T366S, L368A, and Y407V were examined in the other CH3 domain for holes while L351C was used to form disulfide bonds and further enhance the heterodimerization [157,158]. The engineering sites were identified and examined according to three criteria: (1) The distances between alpha-carbons should be around 5.0–6.8 Å, which is the average distance found in naturally formed disulfide bonds, but can reach up to 7.6 Å; (2) the pairings of amino acid residues should be distinct from those on each natural CH3 interface; (3) the formation of disulfide bonds between cysteine residues should be favorable conformationally [158]. As a result, the heavy chain (HC) heterodimerization was further improved up to 95% under co-expression conditions, making it feasible for scalable production [143,159]. The knobs-into-holes heterodimerization not only solves the heavy-chain problem via the correct heterodimeric pairing of bispecific antibodies but also renders them conformationally stable [160] and allows for antibody purification by protein A [22]. Zhang et al. demonstrated the stability of knobs-into-holes heterodimers in comparison to holes-holes homodimer variants, further supporting the knobs-into-holes heterodimerization as a rational design strategy [160]. The heteromeric heavy chains produced functional bispecific antibodies and also retained Fc-mediated effector functions, such as ADCC. Compared to an *E. coli* host system producing unglycosylated antibodies, a mammalian host expression systems can produce glycosylated, effector-function competent heterodimeric antibodies. One study revealed that afucosylation of half the asymmetric anti-CD20 antibody by knobs-into-holes technology from CHO cells is sufficient to produce ADCC-enhancement similar to that observed for a fully afucosylated symmetric wild-type anti-CD20 antibody [161].

Alternatively, strand-exchange engineered domain (SEED) heterodimerization represents another steric mutation-based design strategy which utilizes complementarity of alternating sequences derived from IgG and IgA CH3 domains also known as AG SEED CH3 and GA SEED CH3. The IgG and IgA CH3 derivatives generate complementary sequences so that the two complementary heavy-chain heterodimers are assembled while excluding the assembly of homodimers lacking complementarity [162]. According to Muda et al., Fab-SEED fusions retained desirable binding affinity and characteristics comparable to other antibodies including favorable pharmacokinetics and stability [163].

In addition to the steric mutations mentioned earlier, electrostatic steering interactions have also been widely used to promote the formation of heavy-chain heterodimers by substituting a residue in a CH3 domain and another residue in the other CH3 domain by a negatively charged aspartic acid or glutamic acid residue and a positively charged lysine residue respectively. Then, the charge pair substitution favors the assembly of heterodimers while inhibiting the formation of homodimers via electrostatic repulsion. Gunasekaran et al. first demonstrated the Fc heterodimerization of antibodies using the electrostatic steering effects as applied to the production of bispecific antibodies [144]. In their work, novel engineering strategy was applied to support favorable opposite charge interactions between heterodimers and also to induce unfavorable repulsive charge interactions between homodimers at the same time by replacing K409 and D399 in different CH3 domains with aspartate and lysine, respectively, in order to suppress the formation of homodimers [144].

Indeed, site-specific mutations can significantly improve both the quality and quantity of bispecific antibodies by circumventing the heavy-chain problem. However, the steric mutations and the introduction of charge pairs can reduce the thermostability of bispecific antibodies. Moore et al. reported an efficient method called XmAb bispecific platform, which leads to enhanced thermostability that combines charge interactions, conformational aspects of CH3 domains, and hydrogen bonding [159]. The novel Fc mutations include a side chain swap of native IgG1 to S364K and K370S heterodimer to form a hydrogen bond in between followed by L368D/K370S substitutions to drive salt bridge interactions. The engineered Fc sites were specified and selected based on the minimum exposure area in order to ensure near net-isovolumetric substitutions without interfering the receptor binding or generating extra potential N-linked glycosylation sites [159]. Additionally, due to the engineered structure and charge pair mutations, the formation of homodimers was disfavored by driving the steric hindrance and charge repulsion between the sites [159]. In addition, the sites were examined and engineered to modulate different isoelectric points (pI) between the two CH3 sites. This is of a particular interest to improve the robustness of the heterodimeric Fc platform at scale because an engineered pI of heterodimers significantly different from those of homodimers can facilitate and ease the purification process of heterodimeric bispecific antibodies from non-bispecific antibodies via standard ion-exchange chromatography, whose performance is independent of the variable domains and format of bispecific antibodies.

In summary, bispecific antibody heavy-chain heterodimerization, especially within the CH3 region, represents a rapidly emerging approach including multiple design strategies, such as steric mutation, electrostatic steering interactions and charge difference of heavy chains to facilitate purification. These approaches are often applied together to achieve bispecific antibody heavy-chain heterodimerization with minimum homodimer formation. However, an alternative strategy is to generate a common heavy chain with one lambda and one kappa light chains without any modification, which is called a kappa-lambda (κλ) body [164]. The co-expression of a heavy chain and one κ and one λ in CHO or HEK293 cells generated both monospecific and bispecific antibodies [140,164]. It has been reported that the expression of light chains is a determinant for the bispecific antibody specificity and affinity [164]. Columns specific for kappa- and lambda-monospecific antibodies isolation were then adopted followed by Protein A purification, although only around 50% of the final product is κλ body with the rest mainly including kappa-kappa and lambda-lambda antibody side products [140,164]. Interestingly, another research group reported that codon de-optimization of the lambda chain sequence increased the κλ body yield two-fold and enhanced the relative distribution of bispecific antibodies in a low kappa chain expressing κλ body cell line [165].

2.2.3. Heavy Chain and Light-Chain Assembly

While deliberate modifications of Fc CH3 domains enable correct heavy-chain heterodimerization, using two different light chains still results in a low yield of desired bispecific antibodies (the generation of four different combinations, with only one being bispecific). Advanced approaches have, therefore, been developed to allow the correct pairing of light chain and heavy chain to resolve the improper heavy chain and light-chain interaction problem, such as the common light-chain method and CrossMab to swap the VH and VL Fab fragments partially. These strategies are often applied in combination with Fc-modified heavy chains, as shown in Figure 3b.

First, a common light-chain strategy was applied to assemble IgG-like bispecific antibodies which can be combined together with the knobs-into-holes approach [158]. The mechanism of a common light-chain strategy is based on the fact that antibodies discovered from phase display screening against diverse antigens often share the same VL domain, reflecting the very limited size of the L chain repertoire in the phage library [22]. One of the great advantages of the common light-chain format is that it allows the use of methods that simplify the antibody engineering and the purification process in industrial production [166]. For example, based on computational prediction, one Fc variant pair dubbed "DEKK" consisted of substitutions L351D and L368E in one heavy-chain combined with L351K and T366K in the other drove efficient heterodimerization of the antibody heavy chains [27]. Additionally, using a common light chain, the bispecific antibody MCLA-128, targeting human EGF receptors 2 and 3, was produced and purified with a standard CHO cell culture platform and a routine purification protocol under Good Manufacturing Practice (GMP) conditions [27]. More recently, a full-length bispecific IgG-like bsAb was approved in 2017 was emicizumab (Hemlibra®) for the treatment of Hemophilia A patients [167]. Engineered on the structure of humanized IgG4, emicizumab mimics the function of activated FVIII to restore the FVIII binding to factor IX (FIX) and factor X (FX), which is missing in Hemophilia A patients [168,169]. Large-scale manufacturing of emicizumab was achieved by a combination of three antibody engineering strategies-a common light chain to assemble heavy and light chain, changing the charges of two different heavy chains to facilitate antibody purification, and the application of electrostatic steering of two different heavy chains to promote expression of heavy chains in cells [169]. Currently, numerous common light chain and common heavy-chain discovery platforms have been developed to enable the effective generation of antibodies for bsAb assembly. These include but are not limited to transgenic mice with a fixed single light chain [27] as well as screening phage display libraries with common heavy chain (as described above) [170–172]. Therefore, the application of a common light chain is becoming increasingly popular in this field in order to overcome the stability, yield, and immunogenicity problems of bispecific antibodies. However, this approach may lower flexibility in antibody engineering, which limits antibody optimization in some cases [173]. Furthermore, the screening process for common light chain requires animal immunization and/or phage display, which may be problematic due to time and development costs [166].

Different from the common light chain approach, CrossMab represents one of the most widely utilized generic approaches to solve the light-chain problem by exchanging the sequences of heavy and light-chain domains of Fab fragments. Crossmab technology allows for the generation of various bispecific antibodies including bi-, tri- and tetra-valent antibodies, as well as other novel Fab-based antibody derivatives [174]. Three different formats typically proposed are displayed in Figure 3b [142]. The first format involves simply replacing the entire Fab-arm of a heavy chain with a cognate light chain (CrossMab Fab) of one half of the bispecific antibody, and the "crossover" still retains the binding affinity while favoring the assembly of the engineered Fab fragment. The second format involves the swapping of VH of a Fab domain with its corresponding VL domain (CrossMab VH-VL) so that the molecular architectures of the heavy chain and light-chain interfaces in both arms of a bispecific antibody are not identical to prevent the light-chain mispairing. Likewise, for the third format, CH1 and CL of one arm of the bispecific antibody are also interchanged for the correct assembly between heavy and light chains (CrossMab CH1-CL) [142]. CrossMabCH1-CL yields no theoretical side products while CrossMabFab can result in the formation of a non-functional monovalent antibody due to the interaction between "VL-CL" of first IgG and "VH-CH1" of the second IgG [142,175]. CrossMabVH-VL can lead to the development of a Bence-Jones-like side product; to be more specific, successful CrossMabVH-VL should result in a pairing between "VH-CL" and "VL-CH1" while a Bence–Jones-like side product has a "VL-CL" chain paired with a crossed "VL-CH1" chain, meaning that two light-chain domains are assembled to one another. The Bence–Jones-like antibody can theoretically be prevented by making the two constant CH1 and CL electrostatically repulsive to one another [174].

The crossover design has been shown effective in target binding affinity and potency such as anti-tumor activity. Vanucizumab is one of the products that first utilized the CrossMabCH1-CL approach and was designed to target vascular endothelial growth factor (VEGF)-A and angiopoietin-2 (Ang-2) [142,176]. The design was optimized for clinical trial by using the original non-mutated bevacizumab targeting VEGF-A while applying the CrossMabCH1-CL mutation to LC06-bearing antigen-binding site targeting Ang-2. Furthermore, disulfide-stabilized knobs-into-holes mutations were introduced to ensure the correct heavy-chain assembly. As a result, vanucizumab exhibited high potency against patient-derived human tumors as well as several mouse tumors and was able to suppress micro-metastatic growth through Ang-2 inhibition without any side effect from anti-VEGF activity on physiologic vessels [176–178]. The technology has also been utilized to generate bispecific heterodimeric antibodies for many different purposes. For example, knobs-into-holes and CrossMabCH1-CL technology were used to produce a bispecific antibody targeting CD20 and HLA-DR as reported by Zhao et al. [146], and also to generate one of the broadest and potent HIV-1 neutralizing antibodies by Huang et al. [145]. The optimized CrossMab approach can be very effective and thus a powerful design strategy for improving selective light-chain pairing especially when used in combination with additional design approaches such as knobs-into-holes. The crossover design has been shown its effectiveness in achieving proper pairing of light chains for correct target binding affinity and consequently high yields.

2.2.4. Co-Culture Method

Alternatively, to solve the light-chain mispairing issue and retain the natural antibody architecture, Spiess et al. proposed to produce bispecific antibodies by combining two distinct half-antibodies, expressed from two different cell lines in vitro [148] as depicted in Figure 4. Half-antibodies are then purified and mixed with 1:1 molar ratio in vitro to generate functional bispecific antibodies. While this half-antibody method can be effective, the method is attended with some inherent challenges. Using two separate cell lines means that two culture vessels, harvests, and purification processes must be performed before combining in vitro, potentially increasing costs and the risk of contamination [147]. The co-culture method was first demonstrated in *E. coli* in which cells were transformed with plasmids containing different half-antibodies genes (A and B) containing knobs-into-holes to prevent self-dimerization of the heavy chains prior to association with light chains [148]. After culturing, the cells are lysed to harvest half-antibodies. Since the processes mentioned above are identical for both cell lines, a co-culture strategy can be applied to lower the risk and cost. Both *E. coli* cell lines containing plasmid for half-antibody A and plasmid for half-antibody B are inoculated into the same vessel with same cell numbers. Using comparable cell numbers for both cell lines is a way to ensure having the same amount of antibodies A and B produced at the end. After culturing and processing, functional bispecific antibodies are successfully detected and harvested at the end. This method has been proven to be simple for the design and production of a wide range of stable antibodies [148].

(a)

(b)

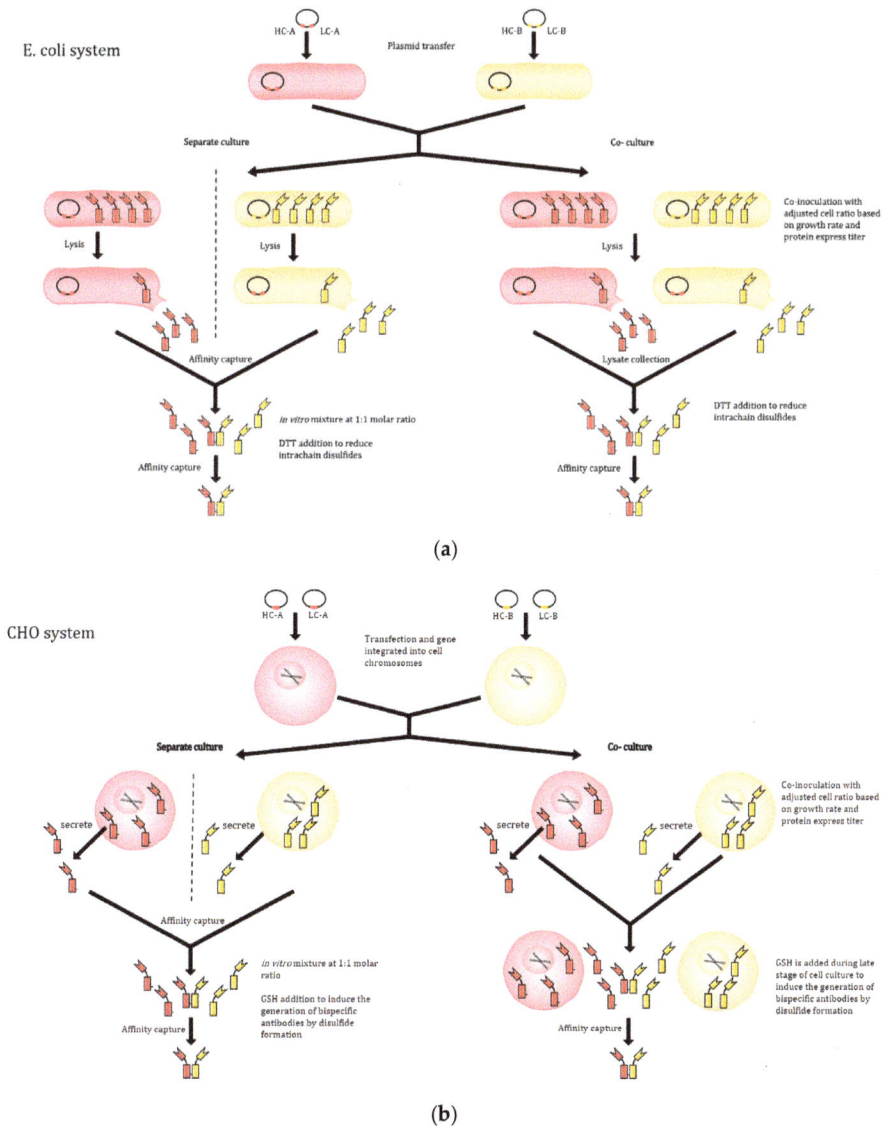

Figure 4. The illustration of the co-culture method to product IgG-like bispecific antibody in *E. coli* and CHO systems. (**a**) *E. coli* production vessels are constructed by transferring the cells with individual plasmids containing both heavy and light-chain gene for a half-antibodies. Cell lysis is used to harvest half-antibodies then assembled in vitro. Both separate culture and co-culture system can be used. (**b**) Similar to the *E. coli* system, CHO cells are co-transfected separate plasmids containing heavy or light-chain gene for a half-antibodies. Secreted antibodies are assembled by GSH induction. GSH: reduced glutathione.

Recently, the co-culture methodology has been shown to work for CHO cells as well [147]. Although similar in approach, there are some differences between the methods used, mainly in plasmids' designs, inoculation, and harvest. For CHO cells, heavy chains and light chains are introduced into the cells on separate DNA plasmid to avoid non-cognate heavy chain and light-chain pairing and ensure a sufficient antibody titer. It is also critical to adjust the ratio of the two CHO cell lines for each half-antibody production based on antibody titers and cell growth rate to maximize production of half-antibodies A and B with a 1:1 molar ratio. Prior to harvest, reduced glutathione (GSH) was added in order to enhance bispecific antibody productivity, since CHO cells can still generate a minimum number of knob or hole homodimers of the desired half-antibody [143,160]. The addition of GSH as a reducing agent can help prevent the homodimerization of the half-antibodies and formation of disulfide. In the study, they also determined a 0.04 to 40 L scalable range for bispecific antibody production using co-culture [147]. With the simplicity of antibody design, relatively low risk and low cost, compatibility with current technology like controlled Fab-arm exchange (cFAE) and SEED, this methodology provides a simple and effective tool to produce a wide range of bispecific antibodies via co-culture of half-antibody secreting CHO cells.

2.2.5. Expression and Production of IgG-like Bispecific Antibodies

Mammalian cells are the predominant workhorses for IgG production in industry, and the production platform is widely scalable for high titers of antibodies in order to meet clinical and commercial demands. However, the production of bispecific antibodies is more complex and typically requires at least two plasmids for heterodimerized heavy chains and one plasmid for a common light chain or two light-chain plasmids if two different light chains are used. Notably, expressing HC and LC on separate plasmids is recommended because the manipulation of the plasmid ratio is an easy and efficient approach to optimize protein assembly for desired products [179]. Subsequently, a laborious and time-consuming process is typically needed to select the most desirable clonal cell lines from a heterogeneous stable transfectant pool for large-scale antibody production [180]. CHO cell is well-known for its high protein productivity, low contamination rates, and human immunological compatibility [181]. The expression levels for antibodies via stable CHO cells can reach >3 g/L and sometimes >5 g/L and beyond and be successfully scaled up in bioreactors to large volumes [182]. Nevertheless, the yield of bispecific IgG-like antibody from CHO cells is lower about 1–3 g/L and often even lower [27].

Compared to stable transfections, transient transfection can deliver results in a few days without integrating recombinant DNA into the host genome. Human embryonic kidney (HEK293) and HEK-based Expi 293 cells are human cells for transient expression, which has been used early in bsAb development. However, due to the multiple number of plasmids required for the production of bispecific antibodies, transient expression of IgG can sometimes be difficult to scale and result in relatively low titers compared to stable cell lines. As an alternative, Rajendra et al. designed a single plasmid vector containing all the engineered light chains and heavy chains for both transient and stable expression of CHO cells. The CHO cell pools transiently transfected with two plasmid vectors of which one heavy chain and a light-chain pairing of the bispecific antibody were harbored on individual plasmids yielded 0.09–0.15 g/L with 73–92% correctly paired bispecific antibody as determined by mass spectrometry. The cell lines transiently transfected with a single plasmid vector containing all the components for bispecific antibody generated similar yields of correctly paired bispecific antibody [183]. However, a stable CHO pool with a single plasmid expression resulted in a higher titer ranging from 0.6 up to 2.2 g/L while the percentage of correct pairing ranged from 74 to 98% [183]. Their results indicated that a single plasmid system could be comparable to multiple plasmid system in terms of titers and it may facilitate the generation of stable CHO cells [183].

Indeed, for in vivo assembly, the efficient co-expression of engineered heavy and light chains relies deeply on the selection of stably expressing cell clones. Plus, culture conditions such as temperature also affect the half-antibody expression and aggregate formation [184]. In contrast, following downstream Protein A purification of CH3 mutants, in vitro assembly (separate expression of two different heavy-chain types followed by mixing them at appropriate redox conditions to induce assembly) have demonstrated their capability in generating high-quality molecules. Such is the case for Duobody technology [185]. With the Fab-arm exchange (FAE) process occurring in IgG4 in vivo and in vitro, researchers engineered FAE-associated IgG4-specific mutation pairs in IgG1 and generated stable IgG1 bispecific antibody with high yield and stability via in vitro assembly [186,187]. In addition, amenable bispecific antibody production can also be achieved by cell-free expression systems using *E. coli* based extracts. The flexibility of this system enables manipulation of the knob: hole plasmid ratio to achieve the most efficient HC heterodimer assembly, which can achieve protein yield at g/L scale within hours [143]. Nonetheless, these methods may include additional obstacles such as increased costs of production. As a result, co-expression in stable cells remains the predominant approach for IgG-like bispecific antibody production.

3. Conclusions and Future Thoughts

This review has focused on the design, production, and quality of bispecific antibodies. A key challenge is how to produce uniform bispecific antibody with high quality and limited or negligible side products and impurities. For scFv-type bispecifics, the protein stability and tissue penetration ability vary and depend on different types of scFv antibody. Furthermore, with multiple host options to choose from, the determination of the most suitable system depends on the specific scFv antibody size, amino acid sequence, protein conformation, solubility, stability, purification, and scalability. For IgG-like full-size bispecific antibody, the production of pure heterodimer is achieved by complete heavy chain and light-chain heterodimerizations. Knobs-into holes method is an efficient means with which to associate different heavy chains. The common light chain and CrossMab technology are also useful approaches for varying light chain and heavy-chain assembly. More recently, co-culture and cell-free systems are also emerging as complementary production platforms to generate bispecific antibodies readily.

Advanced protein and production engineering technologies in the antibody field have boosted the development of bispecific antibodies and their derivatives, which represent one of the fastest-growing next-generation of antibody therapeutics [188]. Diversity has been obtained in the bispecific antibody structure design both in the scFv- and IgG-like formats or by using a combination of both. Furthermore, the addition of small molecules such as aptamers, affibodies, and synthetic drugs can further expand their applicability, creating a plethora of novel bispecific antibody-related products [4]. Bispecific antibodies have found wide applicability to immunotherapy for cancer treatment, and these diverse molecules have the potential to treat other diseases, such as infections, acquired immune deficiency syndrome (AIDS) and genetic diseases [21] as well as serving for medical diagnosis purposes. Looking forward, with continuous efforts to improve their design, production, and purification on an industrial scale, bispecific antibodies will represent an increasing share of the therapeutics in the market with the capacity to reach their full potential as a complementary approach to the conventional therapy in the next decade.

Author Contributions: Q.W. and M.J.B. wrote and edited the paper. Y.C. wrote the scFv bispecific antibody section partly. J.P. and Y.H. wrote the IgG-like bispecific antibody section partly. X.L. designed the figures. T.W. prepared the tables. K.M. edited the paper.

Funding: This research was supported by the National Science Foundation (grant no. 1512265).

Conflicts of Interest: The authors declare that there is no conflict of interest regarding the publication of this article.

References

1. Kimiz-Gebologlu, I.; Gulce-Iz, S.; Biray-Avci, C. Monoclonal antibodies in cancer immunotherapy. *Mol. Biol. Rep.* **2018**, *45*, 2935–2940. [CrossRef] [PubMed]

2. Howard, G.C.; Bethell, D.R. *Basic Methods in Antibody Production and Characterization*; CRC Press: Boca Raton, FL, USA, 2000.

3. Wang, Q.; Chung, C.Y.; Chough, S.; Betenbaugh, M.J. Antibody glycoengineering strategies in mammalian cells. *Biotechnol. Bioeng.* **2018**, *115*, 1378–1393. [CrossRef] [PubMed]

4. Liu, H.; Saxena, A.; Sidhu, S.S.; Wu, D. Fc Engineering for Developing Therapeutic Bispecific Antibodies and Novel Scaffolds. *Front. Immunol.* **2017**, *8*, 38. [CrossRef] [PubMed]

5. Sedykh, S.E.; Prinz, V.V.; Buneva, V.N.; Nevinsky, G.A. Bispecific antibodies: Design, therapy, perspectives. *Drug Des. Dev. Ther.* **2018**, *12*, 195–208. [CrossRef]
6. Sela-Culang, I.; Kunik, V.; Ofran, Y. The structural basis of antibody-antigen recognition. *Front. Immunol.* **2013**, *4*, 302. [CrossRef]
7. Kiyoshi, M.; Tsumoto, K.; Ishii-Watabe, A.; Caaveiro, J.M. Glycosylation of IgG-Fc: A molecular perspective. *Int. Immunol.* **2017**, *29*, 311–317. [CrossRef] [PubMed]
8. Wang, W.; Erbe, A.K.; Hank, J.A.; Morris, Z.S.; Sondel, P.M. NK Cell-Mediated Antibody-Dependent Cellular Cytotoxicity in Cancer Immunotherapy. *Front. Immunol.* **2015**, *6*, 368. [CrossRef] [PubMed]
9. Zhou, X.; Hu, W.; Qin, X. The Role of Complement in the Mechanism of Action of Rituximab for B-Cell Lymphoma: Implications for Therapy. *Oncologist* **2008**, *13*, 954–966. [CrossRef] [PubMed]
10. Beck, A.; Wurch, T.; Bailly, C.; Corvaia, N. Strategies and challenges for the next generation of therapeutic antibodies. *Nat. Rev. Immunol.* **2010**, *10*, 345–352. [CrossRef]
11. Jefferis, R. Glycosylation as a strategy to improve antibody-based therapeutics. *Nat. Rev. Drug Discov.* **2009**, *8*, 226–234. [CrossRef]
12. Kamen, L.A.; Kho, E.; Ordonia, B.; Langsdorf, C.; Chung, S. A method for determining antibody-dependent cellular phagocytosis. *J. Immunol.* **2017**, *198*, 157.17. [CrossRef]
13. Kontermann, R.E. Dual targeting strategies with bispecific antibodies. *MAbs* **2012**, *4*, 182–197. [CrossRef]
14. Hanahan, D.; Weinberg, R.A. Hallmarks of Cancer: The Next Generation. *Cell* **2011**, *144*, 646–674. [CrossRef]
15. Witsch, E.; Sela, M.; Yarden, Y. Roles for growth factors in cancer progression. *Physiology (Bethesda)* **2010**, *25*, 85–101. [CrossRef]
16. Tai, W.; Mahato, R.; Cheng, K. The role of HER2 in cancer therapy and targeted drug delivery. *J. Control. Release* **2010**, *146*, 264–275. [CrossRef]
17. Pabla, B.; Bissonnette, M.; Konda, V.J. Colon cancer and the epidermal growth factor receptor: Current treatment paradigms, the importance of diet, and the role of chemoprevention. *World J. Clin. Oncol.* **2015**, *6*, 133. [CrossRef]
18. Ponz-Sarvise, M.; Rodriguez, J.; Viudez, A.; Chopitea, A.; Calvo, A.; Garcia-Foncillas, J.; Gil-Bazo, I. Epidermal growth factor receptor inhibitors in colorectal cancer treatment: What's new? *World J. Gastroenterol.* **2007**, *13*, 5877. [CrossRef]
19. Beckman, R.A.; Weiner, L.M.; Davis, H.M. Antibody constructs in cancer therapy. *Cancer* **2007**, *109*, 170–179. [CrossRef]
20. Speirs, C.K.; Hwang, M.; Kim, S.; Li, W.; Chang, S.; Varki, V.; Mitchell, L.; Schleicher, S.; Lu, B. Harnessing the cell death pathway for targeted cancer treatment. *Am. J. Cancer Res.* **2011**, *1*, 43–61.
21. Yang, F.; Wen, W.; Qin, W. Bispecific Antibodies as a Development Platform for New Concepts and Treatment Strategies. *Int. J. Mol. Sci.* **2016**, *18*. [CrossRef]
22. Brinkmann, U.; Kontermann, R.E. The making of bispecific antibodies. *MAbs* **2017**, *9*, 182–212. [CrossRef]
23. Fraser, L. Engineering Bispecific Antibodies is Challenging, Creating Unwanted Side Products. CrossMAb Technology, Designed as a Simple Platform for Complex Molecules, Solves those Problems. Available online: https://www.roche.com/research_and_development/what_we_are_working_on/research_technologies/bispecific-antibodies.htm (accessed on 17 May 2019).
24. Rader, C. DARTs take aim at BiTEs. *Blood* **2011**, *117*, 4403–4404. [CrossRef]
25. Zhang, X.; Yang, Y.; Fan, D.; Xiong, D. The development of bispecific antibodies and their applications in tumor immune escape. *Exp. Hematol. Oncol.* **2017**, *6*, 12. [CrossRef]
26. Ridgway, J.B.; Presta, L.G.; Carter, P. 'Knobs-into-holes' engineering of antibody CH3 domains for heavy chain heterodimerization. *Protein Eng.* **1996**, *9*, 617–621. [CrossRef]
27. De Nardis, C.; Hendriks, L.J.A.; Poirier, E.; Arvinte, T.; Gros, P.; Bakker, A.B.H.; de Kruif, J. A new approach for generating bispecific antibodies based on a common light chain format and the stable architecture of human immunoglobulin G1. *J. Biol. Chem.* **2017**. [CrossRef]
28. Klein, C.; Schaefer, W.; Regula, J.T.; Dumontet, C.; Brinkmann, U.; Bacac, M.; Umaña, P. Engineering therapeutic bispecific antibodies using CrossMab technology. *Methods* **2019**, *154*, 21–31. [CrossRef]
29. Carter, P. Bispecific human IgG by design. *J. Immunol. Methods* **2001**, *248*, 7–15. [CrossRef]
30. Huehls, A.M.; Coupet, T.A.; Sentman, C.L. Bispecific T-cell engagers for cancer immunotherapy. *Immunol. Cell Biol.* **2015**, *93*, 290–296. [CrossRef]

31. Huston, J.S.; Mudgett-Hunter, M.; Tai, M.S.; McCartney, J.; Warren, F.; Haber, E.; Oppermann, H. Protein engineering of single-chain Fv analogs and fusion proteins. *Methods Enzymol.* **1991**, *203*, 46–88.

32. Ahmad, Z.A.; Yeap, S.K.; Ali, A.M.; Ho, W.Y.; Alitheen, N.B.; Hamid, M. scFv antibody: Principles and clinical application. *Clin. Dev. Immunol.* **2012**, *2012*, 980250. [CrossRef]

33. Slaney, C.Y.; Wang, P.; Darcy, P.K.; Kershaw, M.H. CARs versus BiTEs: A Comparison between T Cell–Redirection Strategies for Cancer Treatment. *Cancer Discov.* **2018**, *8*, 924–934. [CrossRef]

34. Strohl, W.R.; Naso, M. Bispecific T-Cell Redirection versus Chimeric Antigen Receptor (CAR)-T Cells as Approaches to Kill Cancer Cells. *Antibodies* **2019**, *8*, 41. [CrossRef]

35. Mallender, W.D.; Voss, E.W., Jr. Construction, expression, and activity of a bivalent bispecific single-chain antibody. *J. Biol. Chem.* **1994**, *269*, 199–206.

36. Wolf, E.; Hofmeister, R.; Kufer, P.; Schlereth, B.; Baeuerle, P.A. BiTEs: Bispecific antibody constructs with unique anti-tumor activity. *Drug Discov. Today* **2005**, *10*, 1237–1244. [CrossRef]

37. Portell, C.A.; Wenzell, C.M.; Advani, A.S. Clinical and pharmacologic aspects of blinatumomab in the treatment of B-cell acute lymphoblastic leukemia. *Clin. Pharmacol. Adv. Appl.* **2013**, *5*, 5–11. [CrossRef]

38. Klinger, M.; Brandl, C.; Zugmaier, G.; Hijazi, Y.; Bargou, R.C.; Topp, M.S.; Gokbuget, N.; Neumann, S.; Goebeler, M.; Viardot, A.; et al. Immunopharmacologic response of patients with B-lineage acute lymphoblastic leukemia to continuous infusion of T cell-engaging CD19/CD3-bispecific BiTE antibody blinatumomab. *Blood* **2012**, *119*, 6226–6233. [CrossRef]

39. Ghaderi, D.; Zhang, M.; Hurtado-Ziola, N.; Varki, A. Production platforms for biotherapeutic glycoproteins. Occurrence, impact, and challenges of non-human sialylation. *Biotechnol. Genet. Eng. Rev.* **2012**, *28*, 147–175. [CrossRef]

40. Moore, P.A.; Zhang, W.; Rainey, G.J.; Burke, S.; Li, H.; Huang, L.; Gorlatov, S.; Veri, M.C.; Aggarwal, S.; Yang, Y.; et al. Application of dual affinity retargeting molecules to achieve optimal redirected T-cell killing of B-cell lymphoma. *Blood* **2011**, *117*, 4542–4551. [CrossRef]

41. Holliger, P.; Prospero, T.; Winter, G. "Diabodies": Small bivalent and bispecific antibody fragments. *Proc. Natl. Acad. Sci. USA* **1993**, *90*, 6444–6448. [CrossRef]

42. Walseng, E.; Nelson, C.G.; Qi, J.; Nanna, A.R.; Roush, W.R.; Goswami, R.K.; Sinha, S.C.; Burke, T.R., Jr.; Rader, C. Chemically Programmed Bispecific Antibodies in Diabody Format. *J. Biol. Chem.* **2016**, *291*, 19661–19673. [CrossRef]

43. Arvedson, T.L.; Balazs, M.; Bogner, P.; Black, K.; Graham, K.; Henn, A.; Friedrich, M.; Hoffmann, P.; Kischel, R.; Kufer, P.; et al. Abstract 55: Generation of half-life extended anti-CD33 BiTE® antibody constructs compatible with once-weekly dosing. *Cancer Res.* **2017**, *77*, 55. [CrossRef]

44. Goyos, A.; Li, C.M.; Deegen, P.; Bogner, P.; Thomas, O.; Matthias, K.; Wahl, J.; Goldstein, R.; Coxon, A.; Balazs, M. Generation of half-life extended anti-BCMA Bite® antibody construct compatible with once-weekly dosing for treatment of multiple myeloma (MM). *Am. Soc. Hematol.* **2017**, *130*, 5389.

45. Lorenczewski, G.; Friedrich, M.; Kischel, R.; Dahlhoff, C.; Anlahr, J.; Balazs, M.; Rock, D.; Boyle, M.C.; Goldstein, R.; Coxon, A.; et al. Generation of a Half-Life Extended Anti-CD19 BiTE® Antibody Construct Compatible with Once-Weekly Dosing for Treatment of CD19-Positive Malignancies. *Blood* **2017**, *130*, 2815.

46. Sloan, D.D.; Lam, C.Y.; Irrinki, A.; Liu, L.; Tsai, A.; Pace, C.S.; Kaur, J.; Murry, J.P.; Balakrishnan, M.; Moore, P.A.; et al. Targeting HIV Reservoir in Infected CD4 T Cells by Dual-Affinity Re-targeting Molecules (DARTs) that Bind HIV Envelope and Recruit Cytotoxic T Cells. *PLoS Pathog.* **2015**, *11*, e1005233. [CrossRef]

47. McDonagh, C.F.; Huhalov, A.; Harms, B.D.; Adams, S.; Paragas, V.; Oyama, S.; Zhang, B.; Luus, L.; Overland, R.; Nguyen, S.; et al. Antitumor activity of a novel bispecific antibody that targets the ErbB2/ErbB3 oncogenic unit and inhibits heregulin-induced activation of ErbB3. *Mol. Cancer Ther.* **2012**, *11*, 582–593. [CrossRef]

48. Reusch, U.; Harrington, K.H.; Gudgeon, C.J.; Fucek, I.; Ellwanger, K.; Weichel, M.; Knackmuss, S.H.; Zhukovsky, E.A.; Fox, J.A.; Kunkel, L.A.; et al. Characterization of CD33/CD3 Tetravalent Bispecific Tandem Diabodies (TandAbs) for the Treatment of Acute Myeloid Leukemia. *Clin. Cancer Res. Off. J. Am. Assoc. Cancer Res.* **2016**, *22*, 5829–5838. [CrossRef]

49. Reusch, U.; Burkhardt, C.; Fucek, I.; Le Gall, F.; Le Gall, M.; Hoffmann, K.; Knackmuss, S.H.; Kiprijanov, S.; Little, M.; Zhukovsky, E.A. A novel tetravalent bispecific TandAb (CD30/CD16A) efficiently recruits NK cells for the lysis of CD30+ tumor cells. *MAbs* **2014**, *6*, 728–739. [CrossRef]

50. Compte, M.; Alvarez-Cienfuegos, A.; Nunez-Prado, N.; Sainz-Pastor, N.; Blanco-Toribio, A.; Pescador, N.; Sanz, L.; Alvarez-Vallina, L. Functional comparison of single-chain and two-chain anti-CD3-based bispecific antibodies in gene immunotherapy applications. *Oncoimmunology* **2014**, *3*, e28810. [CrossRef]

51. Kontermann, R.E.; Brinkmann, U. Bispecific antibodies. *Drug Discov. Today* **2015**, *20*, 838–847. [CrossRef]

52. Pantoliano, M.W.; Bird, R.E.; Johnson, S.; Asel, E.D.; Dodd, S.W.; Wood, J.F.; Hardman, K.D. Conformational stability, folding, and ligand-binding affinity of single-chain Fv immunoglobulin fragments expressed in *Escherichia coli*. *Biochemistry* **1991**, *30*, 10117–10125. [CrossRef]

53. Dorai, H.; McCartney, J.E.; Hudziak, R.M.; Tai, M.-S.; Laminet, A.A.; Houston, L.L.; Huston, J.S.; Oppermann, H. Mammalian Cell Expression of Single–Chain Fv (sFv) Antibody Proteins and Their C–terminal Fusions with Interleukin–2 and Other Effector Domains. *Biol. Technol.* **1994**, *12*, 890–897. [CrossRef]

54. Desplancq, D.; King, D.J.; Lawson, A.D.; Mountain, A. Multimerization behaviour of single chain Fv variants for the tumour-binding antibody B72.3. *Protein Eng.* **1994**, *7*, 1027–1033. [CrossRef]

55. Long, N.E.; Sullivan, B.J.; Ding, H.; Doll, S.; Ryan, M.A.; Hitchcock, C.L.; Martin, E.W.; Kumar, K.; Tweedle, M.F.; Magliery, T.J. Linker engineering in anti-TAG-72 antibody fragments optimizes biophysical properties, serum half-life, and high-specificity tumor imaging. *J. Biol. Chem.* **2018**, *293*, 9030–9040. [CrossRef]

56. Argos, P. An investigation of oligopeptides linking domains in protein tertiary structures and possible candidates for general gene fusion. *J. Mol. Biol.* **1990**, *211*, 943–958. [CrossRef]

57. Ekerljung, L.; Wallberg, H.; Sohrabian, A.; Andersson, K.; Friedman, M.; Frejd, F.Y.; Stahl, S.; Gedda, L. Generation and evaluation of bispecific affibody molecules for simultaneous targeting of EGFR and HER2. *Bioconjugate Chem.* **2012**, *23*, 1802–1811. [CrossRef]

58. Whitlow, M.; Bell, B.A.; Feng, S.L.; Filpula, D.; Hardman, K.D.; Hubert, S.L.; Rollence, M.L.; Wood, J.F.; Schott, M.E.; Milenic, D.E.; et al. An improved linker for single-chain Fv with reduced aggregation and enhanced proteolytic stability. *Protein Eng.* **1993**, *6*, 989–995. [CrossRef]

59. Tang, Y.; Jiang, N.; Parakh, C.; Hilvert, D. Selection of Linkers for a Catalytic Single-chain Antibody Using Phage Display Technology. *J. Biol. Chem.* **1996**, *271*, 15682–15686. [CrossRef]

60. Atwell, J.L.; Breheney, K.A.; Lawrence, L.J.; McCoy, A.J.; Kortt, A.A.; Hudson, P.J. scFv multimers of the anti-neuraminidase antibody NC10: Length of the linker between VH and VL domains dictates precisely the transition between diabodies and triabodies. *Protein Eng.* **1999**, *12*, 597–604. [CrossRef]

61. Dolezal, O.; Pearce, L.A.; Lawrence, L.J.; McCoy, A.J.; Hudson, P.J.; Kortt, A.A. ScFv multimers of the anti-neuraminidase antibody NC10: Shortening of the linker in single-chain Fv fragment assembled in VL to VH orientation drives the formation of dimers, trimers, tetramers and higher molecular mass multimers. *Protein Eng. Des. Sel.* **2000**, *13*, 565–574. [CrossRef]

62. Alfthan, K.; Takkinen, K.; Sizmann, D.; Söderlund, H.; Teeri, T.T. Properties of a single-chain antibody containing different linker peptides. *Protein Eng. Des. Sel.* **1995**, *8*, 725–731. [CrossRef]

63. Gil, D.; Schrum, A.G. Strategies to stabilize compact folding and minimize aggregation of antibody-based fragments. *Adv. Biosci. Biotechnol. (Print)* **2013**, *4*, 73–84. [CrossRef]

64. Pavlinkova, G.; Beresford, G.W.; Booth, B.J.M.; Batra, S.K.; Colcher, D. Pharmacokinetics and Biodistribution of Engineered Single-Chain Antibody Constructs of MAb CC49 in Colon Carcinoma Xenografts. *J. Nucl. Med.* **1999**, *40*, 1536–1546.

65. Goel, A.; Colcher, D.; Baranowska-Kortylewicz, J.; Augustine, S.; Booth, B.J.M.; Pavlikova, G.; Batra, S.K. Genetically Engineered Tetravalent Single-Chain Fv of the Pancarcinoma Monoclonal Antibody CC49: Improved Biodistribution and Potential for Therapeutic Application. *Cancer Res.* **2000**, *60*, 6964–6971.

66. Goel, A.; Baranowska-Kortylewicz, J.; Hinrichs, S.H.; Wisecarver, J.; Pavlinkova, G.; Augustine, S.; Colcher, D.; Booth, B.J.M.; Batra, S.K. 99mTc-Labeled Divalent and Tetravalent CC49 Single-Chain Fv's: Novel Imaging Agents for Rapid In Vivo Localization of Human Colon Carcinoma. *J. Nucl. Med.* **2001**, *42*, 1519–1527.

67. Gruber, M.; Schodin, B.A.; Wilson, E.R.; Kranz, D.M. Efficient tumor cell lysis mediated by a bispecific single chain antibody expressed in Escherichia coli. *J. Immunol.* **1994**, *152*, 5368–5374.

68. Hao, C.H.; Han, Q.H.; Shan, Z.J.; Hu, J.T.; Zhang, N.; Zhang, X.P. Effects of different interchain linkers on biological activity of an anti-prostate cancer single-chain bispecific antibody. *Theor. Biol. Med Model.* **2015**, *12*, 14. [CrossRef]

69. Nakanishi, T.; Tsumoto, K.; Yokota, A.; Kondo, H.; Kumagai, I. Critical contribution of VH-VL interaction to reshaping of an antibody: The case of humanization of anti-lysozyme antibody, HyHEL-10. *Protein Sci. Publ. Protein Soc.* **2008**, *17*, 261–270. [CrossRef]

70. Nagorsen, D.; Kufer, P.; Baeuerle, P.A.; Bargou, R. Blinatumomab: A historical perspective. *Pharmacol. Ther.* **2012**, *136*, 334–342. [CrossRef]

71. Reusch, U.; Duell, J.; Ellwanger, K.; Herbrecht, C.; Knackmuss, S.H.; Fucek, I.; Eser, M.; McAleese, F.; Molkenthin, V.; Gall, F.L.; et al. A tetravalent bispecific TandAb (CD19/CD3), AFM11, efficiently recruits T cells for the potent lysis of CD19(+) tumor cells. *MAbs* **2015**, *7*, 584–604. [CrossRef]

72. Miller, B.R.; Demarest, S.J.; Lugovskoy, A.; Huang, F.; Wu, X.; Snyder, W.B.; Croner, L.J.; Wang, N.; Amatucci, A.; Michaelson, J.S.; et al. Stability engineering of scFvs for the development of bispecific and multivalent antibodies. *Protein Eng. Des. Sel.* **2010**, *23*, 549–557. [CrossRef]

73. Wörn, A.; Auf der Maur, A.; Escher, D.; Honegger, A.; Barberis, A.; Plückthun, A. Correlation between in Vitro Stability and in Vivo Performance of Anti-GCN4 Intrabodies as Cytoplasmic Inhibitors. *J. Biol. Chem.* **2000**, *275*, 2795–2803. [CrossRef]

74. Jung, S.; Honegger, A.; Plückthun, A. Selection for improved protein stability by phage display. *J. Mol. Biol.* **1999**, *294*, 163–180. [CrossRef]

75. Jung, S.; Plückthun, A. Improving in vivo folding and stability of a single-chain Fv antibody fragment by loop grafting. *Protein Eng. Des. Sel.* **1997**, *10*, 959–966. [CrossRef]

76. Willuda, J.; Honegger, A.; Waibel, R.; Schubiger, P.A.; Stahel, R.; Zangemeister-Wittke, U.; Plückthun, A. High Thermal Stability Is Essential for Tumor Targeting of Antibody Fragments. *Eng. Humaniz. Anti Epithel. Glycoprotein 2 (Epithel. Cell Adhes. Mol.) Single Chain Fv Fragm.* **1999**, *59*, 5758–5767.

77. Ewert, S.; Honegger, A.; Plückthun, A. Stability improvement of antibodies for extracellular and intracellular applications: CDR grafting to stable frameworks and structure-based framework engineering. *Methods* **2004**, *34*, 184–199. [CrossRef]

78. Jones, P.T.; Dear, P.H.; Foote, J.; Neuberger, M.S.; Winter, G. Replacing the complementarity-determining regions in a human antibody with those from a mouse. *Nature* **1986**, *321*, 522–525. [CrossRef]

79. Borras, L.; Gunde, T.; Tietz, J.; Bauer, U.; Hulmann-Cottier, V.; Grimshaw, J.P.A.; Urech, D.M. Generic approach for the generation of stable humanized single-chain Fv fragments from rabbit monoclonal antibodies. *J. Biol. Chem.* **2010**, *285*, 9054–9066. [CrossRef]

80. Wörn, A.; Plückthun, A. Mutual Stabilization of VL and VH in Single-Chain Antibody Fragments, Investigated with Mutants Engineered for Stability. *Biochemistry* **1998**, *37*, 13120–13127. [CrossRef]

81. Steipe, B. Consensus-based engineering of protein stability: From intrabodies to thermostable enzymes. *Methods Enzymol.* **2004**, *388*, 176–186. [CrossRef]

82. Steipe, B.; Schiller, B.; Plückthun, A.; Steinbacher, S. Sequence Statistics Reliably Predict Stabilizing Mutations in a Protein Domain. *J. Mol. Biol.* **1994**, *240*, 188–192. [CrossRef]

83. Ewert, S.; Honegger, A.; Plückthun, A. Structure-Based Improvement of the Biophysical Properties of Immunoglobulin VH Domains with a Generalizable Approach. *Biochemistry* **2003**, *42*, 1517–1528. [CrossRef]

84. Monsellier, E.; Bedouelle, H. Improving the Stability of an Antibody Variable Fragment by a Combination of Knowledge-based Approaches: Validation and Mechanisms. *J. Mol. Biol.* **2006**, *362*, 580–593. [CrossRef]

85. Jermutus, L.; Honegger, A.; Schwesinger, F.; Hanes, J.; Plückthun, A. Tailoring *in vitro* evolution for protein affinity or stability. *Proc. Natl. Acad. Sci. USA* **2001**, *98*, 75–80. [CrossRef]

86. Proba, K.; Wörn, A.; Honegger, A.; Plückthun, A. Antibody scFv fragments without disulfide bonds, made by molecular evolution11Edited by I. A. Wilson. *J. Mol. Biol.* **1998**, *275*, 245–253. [CrossRef]

87. Demarest, S.J.; Chen, G.; Kimmel, B.E.; Gustafson, D.; Wu, J.; Salbato, J.; Poland, J.; Elia, M.; Tan, X.; Wong, K.; et al. Engineering stability into Escherichia coli secreted Fabs leads to increased functional expression. *Protein Eng. Des. Sel.* **2006**, *19*, 325–336. [CrossRef]

88. Jespers, L.; Schon, O.; Famm, K.; Winter, G. Aggregation-resistant domain antibodies selected on phage by heat denaturation. *Nat. Biotechnol.* **2004**, *22*, 1161. [CrossRef]

89. Martineau, P.; Jones, P.; Winter, G. Expression of an antibody fragment at high levels in the bacterial cytoplasm11Edited by J. Karn. *J. Mol. Biol.* **1998**, *280*, 117–127. [CrossRef]

90. Graff, C.P.; Chester, K.; Begent, R.; Wittrup, K.D. Directed evolution of an anti-carcinoembryonic antigen scFv with a 4-day monovalent dissociation half-time at 37 °C. *Protein Eng. Des. Sel.* **2004**, *17*, 293–304. [CrossRef]

91. Brockmann, E.-C.; Cooper, M.; Strömsten, N.; Vehniäinen, M.; Saviranta, P. Selecting for antibody scFv fragments with improved stability using phage display with denaturation under reducing conditions. *J. Immunol. Methods* **2005**, *296*, 159–170. [CrossRef]

92. Glockshuber, R.; Malia, M.; Pfitzinger, I.; Plueckthun, A. A comparison of strategies to stabilize immunoglobulin Fv-fragments. *Biochemistry* **1990**, *29*, 1362–1367. [CrossRef]

93. Reiter, Y.; Brinkmann, U.; Lee, B.; Pastan, I. Engineering antibody Fv fragments for cancer detection and therapy: Disulfide-stabilized Fv fragments. *Nat. Biotechnol.* **1996**, *14*, 1239–1245. [CrossRef]

94. Zhao, J.X.; Yang, L.; Gu, Z.N.; Chen, H.Q.; Tian, F.W.; Chen, Y.Q.; Zhang, H.; Chen, W. Stabilization of the single-chain fragment variable by an interdomain disulfide bond and its effect on antibody affinity. *Int. J. Mol. Sci.* **2010**, *12*, 1–11. [CrossRef]

95. Pokkuluri, P.R.; Gu, M.; Cai, X.; Raffen, R.; Stevens, F.J.; Schiffer, M. Factors contributing to decreased protein stability when aspartic acid residues are in β-sheet regions. *Protein Sci.* **2002**, *11*, 1687–1694. [CrossRef]

96. Kaufmann, M.; Lindner, P.; Honegger, A.; Blank, K.; Tschopp, M.; Capitani, G.; Plückthun, A.; Grütter, M.G. Crystal Structure of the Anti-His Tag Antibody 3D5 Single-chain Fragment Complexed to its Antigen. *J. Mol. Biol.* **2002**, *318*, 135–147. [CrossRef]

97. Vaks, L.; Benhar, I. Production of stabilized scFv antibody fragments in the E. coli bacterial cytoplasm. *Methods Mol. Biol.* **2014**, *1060*, 171–184. [CrossRef]

98. Miller, K.D.; Weaver-Feldhaus, J.; Gray, S.A.; Siegel, R.W.; Feldhaus, M.J. Production, purification, and characterization of human scFv antibodies expressed in Saccharomyces cerevisiae, Pichia pastoris, and Escherichia coli. *Protein Expr. Purif.* **2005**, *42*, 255–267. [CrossRef]

99. Verma, R.; Boleti, E.; George, A.J. Antibody engineering: Comparison of bacterial, yeast, insect and mammalian expression systems. *J. Immunol. Methods* **1998**, *216*, 165–181. [CrossRef]

100. Galeffi, P.; Lombardi, A.; Pietraforte, I.; Novelli, F.; Di Donato, M.; Sperandei, M.; Tornambé, A.; Fraioli, R.; Martayan, A.; Natali, P.G.; et al. Functional expression of a single-chain antibody to ErbB-2 in plants and cell-free systems. *J. Transl. Med.* **2006**, *4*, 39. [CrossRef]

101. McCall, A.M.; Adams, G.P.; Amoroso, A.R.; Nielsen, U.B.; Zhang, L.; Horak, E.; Simmons, H.; Schier, R.; Marks, J.D.; Weiner, L.M. Isolation and characterization of an anti-CD16 single-chain Fv fragment and construction of an anti-HER2/neu/anti-CD16 bispecific scFv that triggers CD16-dependent tumor cytolysis. *Mol. Immunol.* **1999**, *36*, 433–445. [CrossRef]

102. Offner, S.; Hofmeister, R.; Romaniuk, A.; Kufer, P.; Baeuerle, P.A. Induction of regular cytolytic T cell synapses by bispecific single-chain antibody constructs on MHC class I-negative tumor cells. *Mol. Immunol.* **2006**, *43*, 763–771. [CrossRef]

103. Mack, M.; Riethmüller, G.; Kufer, P. A small bispecific antibody construct expressed as a functional single-chain molecule with high tumor cell cytotoxicity. *Proc. Natl. Acad. Sci. USA* **1995**, *92*, 7021–7025. [CrossRef]

104. Mack, M.; Gruber, R.; Schmidt, S.; Riethmüller, G.; Kufer, P. Biologic properties of a bispecific single-chain antibody directed against 17-1A (EpCAM) and CD3: Tumor cell-dependent T cell stimulation and cytotoxic activity. *J. Immunol.* **1997**, *158*, 3965–3970.

105. Lin, L.; Li, L.; Zhou, C.; Li, J.; Liu, J.; Shu, R.; Dong, B.; Li, Q.; Wang, Z. A HER2 bispecific antibody can be efficiently expressed in Escherichia coli with potent cytotoxicity. *Oncol. Lett.* **2018**, *16*, 1259–1266. [CrossRef]

106. Kuo, S.R.; Wong, L.; Liu, J.S. Engineering a CD123xCD3 bispecific scFv immunofusion for the treatment of leukemia and elimination of leukemia stem cells. *Protein Eng. Des. Sel. PEDS* **2012**, *25*, 561–569. [CrossRef]

107. Root, R.A.; Cao, W.; Li, B.; LaPan, P.; Meade, C.; Sanford, J.; Jin, M.; O'Sullivan, C.; Cummins, E.; Lambert, M.; et al. Development of PF-06671008, a Highly Potent Anti-P-cadherin/Anti-CD3 Bispecific DART Molecule with Extended Half-Life for the Treatment of Cancer. *Antibodies* **2016**, *5*, 6. [CrossRef]

108. Brischwein, K.; Schlereth, B.; Guller, B.; Steiger, C.; Wolf, A.; Lutterbuese, R.; Offner, S.; Locher, M.; Urbig, T.; Raum, T.; et al. MT110: A novel bispecific single-chain antibody construct with high efficacy in eradicating established tumors. *Mol. Immunol.* **2006**, *43*, 1129–1143. [CrossRef]

109. Stadler, C.R.; Bahr-Mahmud, H.; Plum, L.M.; Schmoldt, K.; Kolsch, A.C.; Tureci, O.; Sahin, U. Characterization of the first-in-class T-cell-engaging bispecific single-chain antibody for targeted immunotherapy of solid tumors expressing the oncofetal protein claudin 6. *Oncoimmunology* **2016**, *5*, e1091555. [CrossRef]

110. Cheal, S.M.; Xu, H.; Guo, H.F.; Zanzonico, P.B.; Larson, S.M.; Cheung, N.K. Preclinical evaluation of multistep targeting of diasialoganglioside GD2 using an IgG-scFv bispecific antibody with high affinity for GD2 and DOTA metal complex. *Mol. Cancer Ther.* **2014**, *13*, 1803–1812. [CrossRef]

111. Rosano, G.L.; Ceccarelli, E.A. Recombinant protein expression in Escherichia coli: Advances and challenges. *Front. Microbiol.* **2014**, *5*. [CrossRef]

112. Arbabi-Ghahroudi, M.; Tanha, J.; MacKenzie, R. Prokaryotic expression of antibodies. *Cancer Metastasis Rev.* **2005**, *24*, 501–519. [CrossRef]

113. Power, B.E.; Hudson, P.J. Synthesis of high avidity antibody fragments (scFv multimers) for cancer imaging. *J. Immunol. Methods* **2000**, *242*, 193–204. [CrossRef]

114. Vendel, M.C.; Favis, M.; Snyder, W.B.; Huang, F.; Capili, A.D.; Dong, J.; Glaser, S.M.; Miller, B.R.; Demarest, S.J. Secretion from bacterial versus mammalian cells yields a recombinant scFv with variable folding properties. *Arch. Biochem. Biophys.* **2012**, *526*, 188–193. [CrossRef]

115. Joosten, V.; Lokman, C.; van den Hondel, C.A.; Punt, P.J. The production of antibody fragments and antibody fusion proteins by yeasts and filamentous fungi. *Microb. Cell Fact.* **2003**, *2*, 1. [CrossRef]

116. Geng, S.; Chang, H.; Qin, W.; Li, Y.; Feng, J.; Shen, B. Overexpression, effective renaturation, and bioactivity of novel single-chain antibodies against TNF-alpha. *Prep. Biochem. Biotechnol.* **2008**, *38*, 74–86. [CrossRef]

117. Liu, M.; Wang, X.; Yin, C.; Zhang, Z.; Lin, Q.; Zhen, Y.; Huang, H. A novel bivalent single-chain variable fragment (scFV) inhibits the action of tumour necrosis factor alpha. *Biotechnol. Appl. Biochem.* **2008**, *50*, 173–179. [CrossRef]

118. Carrio, M.M.; Cubarsi, R.; Villaverde, A. Fine architecture of bacterial inclusion bodies. *FEBS Lett.* **2000**, *471*, 7–11. [CrossRef]

119. Berg, J.M.; Tymoczko, J.L.; Stryer, L. The Immunoglobulin Fold Consists of a Beta-Sandwich Framework with Hypervariable Loops. 2002. Available online: https://www.ncbi.nlm.nih.gov/books/NBK22461/ (accessed on 14 March 2019).

120. Guglielmi, L.; Martineau, P. Expression of single-chain Fv fragments in E. coli cytoplasm. *Methods Mol. Biol.* **2009**, *562*, 215–224. [CrossRef]

121. Singh, A.; Upadhyay, V.; Upadhyay, A.K.; Singh, S.M.; Panda, A.K. Protein recovery from inclusion bodies of Escherichia coli using mild solubilization process. *Microb. Cell Fact.* **2015**, *14*, 41. [CrossRef]

122. Kipriyanov, S.M. High-level periplasmic expression and purification of scFvs. *Methods Mol. Biol.* **2009**, *562*, 205–214. [CrossRef]

123. Chi, W.-J.; Kim, H.; Yoo, H.; Kim, Y.P.; Hong, S.-K. Periplasmic expression, purification, and characterization of an anti-epidermal growth factor receptor antibody fragment in Escherichia coli. *Biotechnol. Bioprocess Eng.* **2016**, *21*, 321–330. [CrossRef]

124. Dewi, K.S.; Retnoningrum, D.S.; Riani, C.; Fuad, A.M. Construction and Periplasmic Expression of the Anti-EGFRvIII ScFv Antibody Gene in Escherichia coli. *Sci. Pharm.* **2016**, *84*, 141–152. [CrossRef]

125. Yang, H.; Zhong, Y.; Wang, J.; Zhang, Q.; Li, X.; Ling, S.; Wang, S.; Wang, R. Screening of a ScFv Antibody With High Affinity for Application in Human IFN-gamma Immunoassay. *Front. Microbiol.* **2018**, *9*, 261. [CrossRef] [PubMed]

126. Nogi, T.; Sangawa, T.; Tabata, S.; Nagae, M.; Tamura-Kawakami, K.; Beppu, A.; Hattori, M.; Yasui, N.; Takagi, J. Novel affinity tag system using structurally defined antibody-tag interaction: Application to single-step protein purification. *Protein Sci.* **2008**, *17*, 2120–2126. [CrossRef]

127. Wang, R.; Xiang, S.; Feng, Y.; Srinivas, S.; Srinivas, S.; Lin, M.; Wang, S. Engineering production of functional scFv antibody in E. coli by co-expressing the molecule chaperone Skp. *Front. Cell. Infect. Microbiol.* **2013**, *3*, 72. [CrossRef] [PubMed]

128. de Marco, A.; De Marco, V. Bacteria co-transformed with recombinant proteins and chaperones cloned in independent plasmids are suitable for expression tuning. *J. Biotechnol.* **2004**, *109*, 45–52. [CrossRef] [PubMed]

129. Sonoda, H.; Kumada, Y.; Katsuda, T.; Yamaji, H. Functional expression of single-chain Fv antibody in the cytoplasm of Escherichia coli by thioredoxin fusion and co-expression of molecular chaperones. *Protein Expr. Purif.* **2010**, *70*, 248–253. [CrossRef]

130. Sonoda, H.; Kumada, Y.; Katsuda, T.; Yamaji, H. Effects of cytoplasmic and periplasmic chaperones on secretory production of single-chain Fv antibody in Escherichia coli. *J. Biosci. Bioeng.* **2011**, *111*, 465–470. [CrossRef]

131. Jäger, V.; Büssow, K.; Wagner, A.; Weber, S.; Hust, M.; Frenzel, A.; Schirrmann, T. High level transient production of recombinant antibodies and antibody fusion proteins in HEK293 cells. *BMC Biotechnol.* **2013**, *13*, 52. [CrossRef]

132. Jain, S.; Aresu, L.; Comazzi, S.; Shi, J.; Worrall, E.; Clayton, J.; Humphries, W.; Hemmington, S.; Davis, P.; Murray, E.; et al. The Development of a Recombinant scFv Monoclonal Antibody Targeting Canine CD20 for Use in Comparative Medicine. *PloS ONE* **2016**, *11*, e0148366. [CrossRef]

133. Chambers, A.C.; Aksular, M.; Graves, L.P.; Irons, S.L.; Possee, R.D.; King, L.A. Overview of the Baculovirus Expression System. *Curr. Protoc. Protein Sci.* **2018**, *91*, 5.4.1–5.4.6. [CrossRef]

134. Ma, J.K.; Drake, P.M.; Christou, P. The production of recombinant pharmaceutical proteins in plants. *Nat. Rev. Genet.* **2003**, *4*, 794–805. [CrossRef]

135. Rech, E.; Vianna, G.; Murad, A.; Cunha, N.; Lacorte, C.; Araujo, A.; Brigido, M.; Michael, W.; Fontes, A.; Barry, O.; et al. Recombinant proteins in plants. *BMC Proc.* **2014**, *8*, 1. [CrossRef]

136. Makvandi-Nejad, S.; McLean, M.D.; Hirama, T.; Almquist, K.C.; Mackenzie, C.R.; Hall, J.C. Transgenic tobacco plants expressing a dimeric single-chain variable fragment (scfv) antibody against Salmonella enterica serotype Paratyphi B. *Transgenic Res.* **2005**, *14*, 785–792. [CrossRef]

137. Stech, M.; Hust, M.; Schulze, C.; Dübel, S.; Kubick, S. Cell-free eukaryotic systems for the production, engineering, and modification of scFv antibody fragments. *Eng. Life Sci.* **2014**, *14*, 387–398. [CrossRef]

138. Carlson, E.D.; Gan, R.; Hodgman, C.E.; Jewett, M.C. Cell-free protein synthesis: Applications come of age. *Biotechnol. Adv.* **2012**, *30*, 1185–1194. [CrossRef]

139. Spasevska, I. *An Outlook on Bispecific Antibodies: Methods of Production and Therapeutic Benefits*; BioSciences Master Reviews: Lyon, France, 2013.

140. Krah, S.; Kolmar, H.; Becker, S.; Zielonka, S. Engineering IgG-Like Bispecific Antibodies—An Overview. *Antibodies* **2018**, *7*, 28. [CrossRef]

141. Efficient Protein A Chromatography for Bispecific Antibodies. Available online: https://www.gelifesciences. com/en/us/solutions/bioprocessing/knowledge-center/purifying-bispecific-antibodies-in-a-single-step (accessed on 8 July 2019).

142. Schaefer, W.; Regula, J.T.; Bähner, M.; Schanzer, J.; Croasdale, R.; Dürr, H.; Gassner, C.; Georges, G.; Kettenberger, H.; Imhof-Jung, S.; et al. Immunoglobulin domain crossover as a generic approach for the production of bispecific IgG antibodies. *Proc. Natl. Acad. Sci. USA* **2011**, *108*, 11187. [CrossRef]

143. Xu, Y.; Lee, J.; Tran, C.; Heibeck, T.H.; Wang, W.D.; Yang, J.; Stafford, R.L.; Steiner, A.R.; Sato, A.K.; Hallam, T.J.; et al. Production of bispecific antibodies in "knobs-into-holes" using a cell-free expression system. *MAbs* **2015**, *7*, 231–242. [CrossRef]

144. Gunasekaran, K.; Pentony, M.; Shen, M.; Garrett, L.; Forte, C.; Woodward, A.; Ng, S.B.; Born, T.; Retter, M.; Manchulenko, K.; et al. Enhancing antibody Fc heterodimer formation through electrostatic steering effects: applications to bispecific molecules and monovalent IgG. *J. Biol. Chem.* **2010**, *285*, 19637–19646. [CrossRef]

145. Huang, Y.; Yu, J.; Lanzi, A.; Yao, X.; Andrews, C.D.; Tsai, L.; Gajjar, M.R.; Sun, M.; Seaman, M.S.; Padte, N.N.; et al. Engineered Bispecific Antibodies with Exquisite HIV-1-Neutralizing Activity. *Cell* **2016**, *165*, 1621–1631. [CrossRef]

146. Zhao, L.; Xie, F.; Tong, X.; Li, H.; Chen, Y.; Qian, W.; Duan, S.; Zheng, J.; Zhao, Z.; Li, B.; et al. Combating non-Hodgkin lymphoma by targeting both CD20 and HLA-DR through CD20-243 CrossMab. *MABs* **2014**, *6*, 740–748. [CrossRef]

147. Shatz, W.; Ng, D.; Dutina, G.; Wong, A.W.; Dunshee, D.R.; Sonoda, J.; Shen, A.; Scheer, J.M. An efficient route to bispecific antibody production using single-reactor mammalian co-culture. *MAbs* **2016**, *8*, 1487–1497. [CrossRef]

148. Spiess, C.; Merchant, M.; Huang, A.; Zheng, Z.; Yang, N.Y.; Peng, J.; Ellerman, D.; Shatz, W.; Reilly, D.; Yansura, D.G.; et al. Bispecific antibodies with natural architecture produced by co-culture of bacteria expressing two distinct half-antibodies. *Nat. Biotechnol.* **2013**, *31*, 753–758. [CrossRef]

149. Suresh, M.R.; Cuello, A.C.; Milstein, C. Bispecific monoclonal antibodies from hybrid hybridomas. *Methods Enzymol.* **1986**, *121*, 210–228.

150. Kroesen, B.J.; Helfrich, W.; Molema, G.; de Leij, L. Bispecific antibodies for treatment of cancer in experimental animal models and man. *Adv. Drug Deliv. Rev.* **1998**, *31*, 105–129. [CrossRef]

151. Tustian, A.D.; Endicott, C.; Adams, B.; Mattila, J.; Bak, H. Development of purification processes for fully human bispecific antibodies based upon modification of protein A binding avidity. *MAbs* **2016**, *8*, 828–838. [CrossRef]

152. Krishnamurthy, A.; Jimeno, A. Bispecific antibodies for cancer therapy: A review. *Pharmacol. Ther.* **2018**, *185*, 122–134. [CrossRef]

153. Lindhofer, H.; Mocikat, R.; Steipe, B.; Thierfelder, S. Preferential species-restricted heavy/light chain pairing in rat/mouse quadromas. Implications for a single-step purification of bispecific antibodies. *J. Immunol. (Baltimore)* **1995**, *155*, 219–225.

154. Linke, R.; Klein, A.; Seimetz, D. Catumaxomab: Clinical development and future directions. *MAbs* **2010**, *2*, 129–136. [CrossRef]

155. Dall'Acqua, W.; Simon, A.L.; Mulkerrin, M.G.; Carter, P. Contribution of domain interface residues to the stability of antibody CH3 domain homodimers. *Biochemistry* **1998**, *37*, 9266–9273. [CrossRef]

156. Crick, F.H.C. The packing of α-helices: Simple coiled-coils. *Acta Crystallogr.* **1953**, *6*, 689–697. [CrossRef]

157. Atwell, S.; Ridgway, J.B.; Wells, J.A.; Carter, P. Stable heterodimers from remodeling the domain interface of a homodimer using a phage display library. *J. Mol. Biol.* **1997**, *270*, 26–35. [CrossRef]

158. Merchant, A.M.; Zhu, Z.; Yuan, J.Q.; Goddard, A.; Adams, C.W.; Presta, L.G.; Carter, P. An efficient route to human bispecific IgG. *Nat. Biotechnol.* **1998**, *16*, 677–681. [CrossRef]

159. Moore, G.L.; Bernett, M.J.; Rashid, R.; Pong, E.W.; Nguyen, D.-H.T.; Jacinto, J.; Eivazi, A.; Nisthal, A.; Diaz, J.E.; Chu, S.Y.; et al. A robust heterodimeric Fc platform engineered for efficient development of bispecific antibodies of multiple formats. *Methods* **2019**, *154*, 38–50. [CrossRef]

160. Zhang, H.M.; Li, C.; Lei, M.; Lundin, V.; Lee, H.Y.; Ninonuevo, M.; Lin, K.; Han, G.; Sandoval, W.; Lei, D.; et al. Structural and Functional Characterization of a Hole–Hole Homodimer Variant in a "Knob-Into-Hole" Bispecific Antibody. *Anal. Chem.* **2017**, *89*, 13494–13501. [CrossRef]

161. Shatz, W.; Chung, S.; Li, B.; Marshall, B.; Tejada, M.; Phung, W.; Sandoval, W.; Kelley, R.F.; Scheer, J.M. Knobs-into-holes antibody production in mammalian cell lines reveals that asymmetric afucosylation is sufficient for full antibody-dependent cellular cytotoxicity. *MAbs* **2013**, *5*, 872–881. [CrossRef]

162. Davis, J.H.; Aperlo, C.; Li, Y.; Kurosawa, E.; Lan, Y.; Lo, K.M.; Huston, J.S. SEEDbodies: Fusion proteins based on strand-exchange engineered domain (SEED) CH3 heterodimers in an Fc analogue platform for asymmetric binders or immunofusions and bispecific antibodies. *Protein Eng. Des. Sel. PEDS* **2010**, *23*, 195–202. [CrossRef]

163. Muda, M.; Gross, A.W.; Dawson, J.P.; He, C.; Kurosawa, E.; Schweickhardt, R.; Dugas, M.; Soloviev, M.; Bernhardt, A.; Fischer, D.; et al. Therapeutic assessment of SEED: A new engineered antibody platform designed to generate mono- and bispecific antibodies. *Protein Eng. Des. Sel. PEDS* **2011**, *24*, 447–454. [CrossRef]

164. Fischer, N.; Elson, G.; Magistrelli, G.; Dheilly, E.; Fouque, N.; Laurendon, A.; Gueneau, F.; Ravn, U.; Depoisier, J.-F.; Moine, V.; et al. Exploiting light chains for the scalable generation and platform purification of native human bispecific IgG. *Nat. Commun.* **2015**, *6*, 6113. [CrossRef]

165. Magistrelli, G.; Poitevin, Y.; Schlosser, F.; Pontini, G.; Malinge, P.; Josserand, S.; Corbier, M.; Fischer, N. Optimizing assembly and production of native bispecific antibodies by codon de-optimization. *MAbs* **2017**, *9*, 231–239. [CrossRef]

166. Shiraiwa, H.; Narita, A.; Kamata-Sakurai, M.; Ishiguro, T.; Sano, Y.; Hironiwa, N.; Tsushima, T.; Segawa, H.; Tsunenari, T.; Ikeda, Y.; et al. Engineering a bispecific antibody with a common light chain: Identification and optimization of an anti-CD3 epsilon and anti-GPC3 bispecific antibody, ERY974. *Methods* **2019**, *154*, 10–20. [CrossRef]

167. Dimasi, N.; Fleming, R.; Wu, H.; Gao, C. Molecular engineering strategies and methods for the expression and purification of IgG1-based bispecific bivalent antibodies. *Methods* **2019**, *154*, 77–86. [CrossRef]

168. Adamkewicz, J.I.; Chen, D.C.; Paz-Priel, I. Effects and Interferences of Emicizumab, a Humanised Bispecific Antibody Mimicking Activated Factor VIII Cofactor Function, on Coagulation Assays. *Thromb. Haemost.* **2019**, *119*, 1084–1093. [CrossRef]

169. ART-Ig®(Bispecific Antibody Manufacturing Technology). Available online: https://www.chugai-pharm.co.jp/english/ir/rd/technologies_popup3.html (accessed on 4 July 2019).

170. Husain, B.; Ellerman, D. Expanding the Boundaries of Biotherapeutics with Bispecific Antibodies. *BioDrugs* **2018**, *32*, 441–464. [CrossRef]

171. Jackman, J.; Chen, Y.; Huang, A.; Moffat, B.; Scheer, J.M.; Leong, S.R.; Lee, W.P.; Zhang, J.; Sharma, N.; Lu, Y.; et al. Development of a two-part strategy to identify a therapeutic human bispecific antibody that inhibits IgE receptor signaling. *J. Biol. Chem.* **2010**, *285*, 20850–20859. [CrossRef]

172. Van Blarcom, T.; Lindquist, K.; Melton, Z.; Cheung, W.L.; Wagstrom, C.; McDonough, D.; Valle Oseguera, C.; Ding, S.; Rossi, A.; Potluri, S.; et al. Productive common light chain libraries yield diverse panels of high affinity bispecific antibodies. *MAbs* **2018**, *10*, 256–268. [CrossRef]

173. Dillon, M.; Yin, Y.; Zhou, J.; McCarty, L.; Ellerman, D.; Slaga, D.; Junttila, T.T.; Han, G.; Sandoval, W.; Ovacik, M.A. Efficient production of bispecific IgG of different isotypes and species of origin in single mammalian cells. *MAbs* **2017**, *9*, 213–230. [CrossRef]

174. Klein, C.; Schaefer, W.; Regula, J.T. The use of CrossMab technology for the generation of bi- and multispecific antibodies. *MAbs* **2016**, *8*, 1010–1020. [CrossRef]

175. Regula, J.T.; Imhof-Jung, S.; Mølhøj, M.; Benz, J.; Ehler, A.; Bujotzek, A.; Schaefer, W.; Klein, C. Variable heavy-variable light domain and Fab-arm CrossMabs with charged residue exchanges to enforce correct light chain assembly. *Protein Eng. Des. Sel. PEDS* **2018**, *31*, 289–299. [CrossRef]

176. Thomas, M.; Kienast, Y.; Scheuer, W.; Bahner, M.; Kaluza, K.; Gassner, C.; Herting, F.; Brinkmann, U.; Seeber, S.; Kavlie, A.; et al. A novel angiopoietin-2 selective fully human antibody with potent anti-tumoral and anti-angiogenic efficacy and superior side effect profile compared to Pan-Angiopoietin-1/-2 inhibitors. *PLoS ONE* **2013**, *8*, e54923. [CrossRef]

177. Kienast, Y.; Klein, C.; Scheuer, W.; Raemsch, R.; Lorenzon, E.; Bernicke, D.; Herting, F.; Yu, S.; The, H.H.; Martarello, L.; et al. Ang-2-VEGF-A CrossMab, a novel bispecific human IgG1 antibody blocking VEGF-A and Ang-2 functions simultaneously, mediates potent antitumor, antiangiogenic, and antimetastatic efficacy. *Clin. Cancer Res.* **2013**, *19*, 6730–6740. [CrossRef]

178. Gassner, C.; Lipsmeier, F.; Metzger, P.; Beck, H.; Schnueriger, A.; Regula, J.T.; Moelleken, J. Development and validation of a novel SPR-based assay principle for bispecific molecules. *J. Pharm. Biomed. Anal.* **2015**, *102*, 144–149. [CrossRef]

179. Bratt, J.; Linderholm, A.; Monroe, B.; Chamow, S.M. *Therapeutic IgG-Like Bispecific Antibodies: Modular Versatility and Manufacturing Challenges*; Part 2; BioProcess International: Boston, MA, USA, 2018.

180. Priola, J.J.; Calzadilla, N.; Baumann, M.; Borth, N.; Tate, C.G.; Betenbaugh, M.J. High-throughput screening and selection of mammalian cells for enhanced protein production. *Biotechnol. J.* **2016**, *11*, 853–865. [CrossRef]

181. Hossler, P.; Khattak, S.F.; Li, Z.J. Optimal and consistent protein glycosylation in mammalian cell culture. *Glycobiology* **2009**, *19*, 936–949. [CrossRef]

182. Ye, J.; Alvin, K.; Latif, H.; Hsu, A.; Parikh, V.; Whitmer, T.; Tellers, M.; de la Cruz Edmonds, M.C.; Ly, J.; Salmon, P.; et al. Rapid protein production using CHO stable transfection pools. *Biotechnol. Prog.* **2010**, *26*, 1431–1437. [CrossRef]

183. Rajendra, Y.; Peery, R.B.; Hougland, M.D.; Barnard, G.C.; Wu, X.; Fitchett, J.R.; Bacica, M.; Demarest, S.J. Transient and stable CHO expression, purification and characterization of novel hetero-dimeric bispecific IgG antibodies. *Biotechnol. Prog.* **2017**, *33*, 469–477. [CrossRef]

184. Gomez, N.; Wieczorek, A.; Lu, F.; Bruno, R.; Diaz, L.; Agrawal, N.J.; Daris, K. Culture temperature modulates half antibody and aggregate formation in a Chinese hamster ovary cell line expressing a bispecific antibody. *Biotechnol. Bioeng.* **2018**, *115*, 2930–2940. [CrossRef]

185. Ha, J.-H.; Kim, J.-E.; Kim, Y.-S. Immunoglobulin Fc heterodimer platform technology: From design to applications in therapeutic antibodies and proteins. *Front. Immunol.* **2016**, *7*, 394. [CrossRef]

186. Labrijn, A.F.; Meesters, J.I.; de Goeij, B.E.C.G.; van den Bremer, E.T.J.; Neijssen, J.; van Kampen, M.D.; Strumane, K.; Verploegen, S.; Kundu, A.; Gramer, M.J.; et al. Efficient generation of stable bispecific IgG1 by controlled Fab-arm exchange. *Proc. Natl. Acad. Sci. USA* **2013**, *110*, 5145. [CrossRef]

187. Labrijn, A.F.; Meesters, J.I.; Priem, P.; de Jong, R.N.; van den Bremer, E.T.J.; van Kampen, M.D.; Gerritsen, A.F.; Schuurman, J.; Parren, P.W.H.I. Controlled Fab-arm exchange for the generation of stable bispecific IgG1. *Nat. Protoc.* **2014**, *9*, 2450. [CrossRef]

188. Igawa, T. Next Generation Antibody Therapeutics Using Bispecific Antibody Technology. *Yakugaku Zasshi J. Pharm. Soc. Jpn.* **2017**, *137*, 831–836. [CrossRef]

antibodies

MDPI

Review

Bispecific T-Cell Redirection versus Chimeric Antigen Receptor (CAR)-T Cells as Approaches to Kill Cancer Cells

William R. Strohl [1,*] and Michael Naso [2]

[1] BiStro Biotech Consulting, LLC, 1086 Tullo Farm Rd., Bridgewater, NJ 08807, USA
[2] Century Therapeutics, 3675 Market St., Philadelphia, PA 19104, USA
* Correspondence: wrstrohl@gmail.com; Tel.: +1-908-745-8576

Received: 27 May 2019; Accepted: 24 June 2019; Published: 3 July 2019

Abstract: The concepts for T-cell redirecting bispecific antibodies (TRBAs) and chimeric antigen receptor (CAR)-T cells are both at least 30 years old but both platforms are just now coming into age. Two TRBAs and two CAR-T cell products have been approved by major regulatory agencies within the last ten years for the treatment of hematological cancers and an additional 53 TRBAs and 246 CAR cell constructs are in clinical trials today. Two major groups of TRBAs include small, short-half-life bispecific antibodies that include bispecific T-cell engagers (BiTE®s) which require continuous dosing and larger, mostly IgG-like bispecific antibodies with extended pharmacokinetics that can be dosed infrequently. Most CAR-T cells today are autologous, although significant strides are being made to develop off-the-shelf, allogeneic CAR-based products. CAR-Ts form a cytolytic synapse with target cells that is very different from the classical immune synapse both physically and mechanistically, whereas the TRBA-induced synapse is similar to the classic immune synapse. Both TRBAs and CAR-T cells are highly efficacious in clinical trials but both also present safety concerns, particularly with cytokine release syndrome and neurotoxicity. New formats and dosing paradigms for TRBAs and CAR-T cells are being developed in efforts to maximize efficacy and minimize toxicity, as well as to optimize use with both solid and hematologic tumors, both of which present significant challenges such as target heterogeneity and the immunosuppressive tumor microenvironment.

Keywords: chimeric antigen receptor; bispecific antibody; T-cell redirection; immune synapse; CD3ε, T cells; NK cells; tumor cell killing; tumor microenvironment

1. Introduction and History

1.1. Historical Context for Immunotherapy

While the concept of immunotherapy goes back to ancient Greek times, the first significant use of prospective immunotherapy was by William B. Coley, in the late nineteenth century [1]. In the early 1880s, an immigrant patient named Fred Stein had a neck tumor that had re-emerged after each attempt to remove it by surgery. Finally, after one surgical procedure, Stein developed an erysipelas infection (*Streptococcus pyogenes*), leading his attending physicians to assume that he would succumb to the infection and die. Stein, however, recovered not only from the infection but also from the cancer. Years later, upon researching Stein's case, cancer physician William Coley became convinced that the bacterial infection led to a response against the tumor [2,3]. Coley then systematically treated some of his own cancer patients with live bacteria in efforts to stimulate their immune response against the tumors [1–3]. These studies yielded variable results but with some clear clinical successes, particularly against sarcomas. Later, Coley used heat-killed pathogens to stimulate the immune system, now known as "Coley's vaccine" [2,3]. While Coley's ground-breaking results were hailed by a few, the concept of immune stimulation to treat cancers was not widely accepted and was even scorned by the American Cancer Society for many years [3]. Then, in the late 1990s, over a century after Coley's initial observations, Bruce Beutler and his colleagues demonstrated that bacterial lipopolysaccharides could agonize toll like receptors (TLRs) [4], which in turn could activate the immune system against cancer [5]. This century-long story has continued to evolve and now has become a major focus in cancer therapy. This review describes two fundamental T-cell-based strategies, as well as variations on those central themes, to harness the power of the immune system to eradicate tumors.

One of the major mechanisms by which cancer cells evade the immune system is via down regulation and loss of their major histocompatibility complex class I (MHC-I) molecules (aka human leukocyte antigens (HLAs)) [6]. Normally, MHC-I-positive tumor cells would be targeted by T-cells with T cell receptors (TCRs) recognizing tumor-specific peptides displayed by the MHC class-I molecules. The recognition and binding of cancer cell surface peptide-loaded MHCs (pMHCs) by TCRs results in the formation of a cytolytic synapse between the T-cell and cancer cell, leading to the directed massive release of cytotoxic proteins such as perforin and granzymes [7], as well as clonal T-cell activation and proliferation [8]. Optimal activation of the T-cells in this and other synaptic interactions requires two signals, the TCR-MHC interaction, known as "Signal 1" and a costimulatory signal ("Signal 2") through one of several costimulatory receptors on T cells (e.g., CD28, CD137, OX40, CD27, ICOS, GITR) and their cognate ligands (e.g., CD80/86, CD137L, OX40L, CD70, ICOS-L, GITR-L) on the targeted cells or professional antigen-presenting cell (APC) [9]. A third signal, production of immunostimulatory cytokines, helps to drive T cell differentiation and expansion [10]. The MHC-I loss-based mechanism of tumor escape is further complicated by the fact that tumor-cell specific neo-antigens are often "minimal" or difficult to discriminate because they may only be single residue different from their wild-type allele, low affinity or not presented well by MHC-I complex [11].

Loss or downregulation of the MHC-I molecules and absence of strong tumor antigens in cancer cells allow those cells to escape recognition and killing by tumor-infiltrating T-cells which are key components of anti-tumor immunological response [6,12,13]. Additionally, loss of costimulatory molecules (e.g., CD86, CD54) [14], overproduction of checkpoint inhibitory molecules (e.g., PD-1, CTLA4) [15] and tumor production of the tryptophan degrading-enzyme indoleamine 2,3-dioxygenase (IDO), which eliminates tryptophan, a key amino acid required for T-cell proliferation [16], are other examples of mechanisms utilized by tumors to evade cytotoxic T cells.

Today, various therapeutic strategies seek to harness the killing power of T-cells in a TCR functionality-independent manner, bypassing the limitation of HLA-restricted antigen recognition. Two of the most important TCR function-independent T-cell-based therapeutic strategies employed today are T-cell redirecting bispecific antibodies (TRBAs) and chimeric antigen receptor (CAR)-T cells. With TRBAs, the epsilon (ε) domain of cluster of differentiation 3 (CD3), a component of the TCR complex, is targeted with one combining (i.e., binding) domain, while a second binding domain (hence, "bispecific" antibody) binds a tumor cell surface antigen (Figure 1B). These TRBAs function to bring the T-cells and targeted cells into close proximity to form a cytolytic synapse resulting in tumor cell death [17]. In the case of chimeric antigen receptor (CAR)-T cells, a cancer cell surface antigen-targeting antibody fragment, fused to T-cell activating intracellular domains, is expressed as a neo-receptor on the surface of the T-cells (Figure 1D). These tumor antigen-recognizing, "armed" T-cells then will identify, bind and kill the targeted cancer cells. Both of these strategies rely on antibodies to replace the function of the TCR, making them independent of the TCR and its cognate MHC-I/peptide recognition and both can be employed to recognize and target tumor-specific antigens outside the realm of MHC-I-displayed neo-antigen peptides.

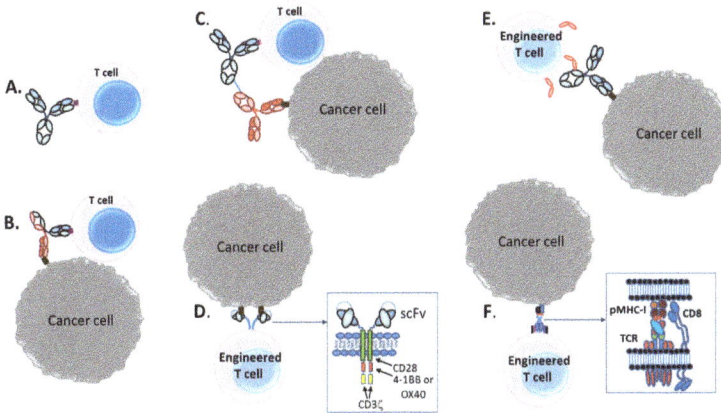

Figure 1. Examples of T-cell based therapeutics in clinical development. (**A**) Inhibition of checkpoint receptors such as PD-1 and CTLA-4 to improve T-cell activity [18]; (**B**) T-cell redirection with bispecific antibodies (TRBAs) in which one binding arm recognizes a tumor antigen and the other binding arm recognizes CD3ε on T-cells [17,19,20]; (**C**) Autologous T-cells activated ex vivo, combined with bispecific antibody conjugates recognizing tumor antigen with one mAb and CD3ε on T cells with the other mAb, followed by re-administration to the patient to kill tumors [21]; (**D**) Genetically engineered autologous chimeric antigen receptor (CAR)-T cells in which an antibody, typically a single chain variable fragment (scFv), fused to intracellular T-cell activation domains such as CD28, 4-1BB, OX40 and CD3ζ, replace the function of the T-cell receptor (TCR), making the T-cells killers of specific antigen-bearing cells [22–24]; (**E**) Autologous or allogeneic T-cells or NK cells genetically engineered with FcγRIIIa (CD16a), which, when administered with an anti-tumor monoclonal antibody (mAb) such as the anti-CD20 mAb, rituximab, binds to the Fc of the antibodies and functionally redirects the T-or NK-cells to the tumor to kill the cancer cells [25]; (**F**). Autologous T cells with engineered TCRs.

Both forms of therapeutic approaches, that is, redirection of T-cells by TRBAs to kill tumor cells and the generation of autologous CAR-T cells from patient T-cells, offer great hope today as next generation antitumor biotherapies [26,27]. These approaches also are being adopted as potential antiviral therapies as well [28–30]. As of 20 June 2019, two therapeutics have been approved by major regulatory agencies for each of these T-cell based approaches. In all, there are at least 289 unique T/NK-cell redirected therapeutic candidates, including 61 different TRBAs, 225 unique CAR-Ts and three T/NK cells transduced with CD16a currently being tested in over 320 unique clinical trials (Table 1). Moreover, additional therapeutic approaches utilizing concepts based on these two major T-cell based therapeutic strategies also are being tested in clinical trials (Figure 1).

Table 1. Overview of T-cell and NK-cell redirected therapeutics in clinical trials *.

Type	Clinical Stage			Total
	Phase I/II	Phase III	Approved	
T-cell or NK cell-redirecting bispecific Abs	59	0	2 **	61
Autologous CAR-T, CAR-NK, CAR-NKT cells	207	2	2	211
Allogeneic CAR-T, CAR-NK, CAR-NKT cells	14	0	0	14
Allogeneic NK or Autologous T cells engineered with Fc RIIIa for binding therapeutic antibodies	3	0	0	3
Total of CAR-T, T-cell or NK cell redirected killing of tumor cells	283	2	4	289

* BiStro Biotech Consulting LLC database, locked 20 June 2019. Data obtained from Clinicaltrials.gov, literature papers, company websites, analyst reports and other sources. ** One of these, Removab®, was voluntarily discontinued in 2017 by the sponsor.

1.2. Brief History of T-Cell Redirecting Bispecific Antibodies

Two fundamental discoveries from the mid-1970s ultimately led to the concept of the TCR function-independent (i.e., as defined by not requiring the recognition and binding of TCR α/β to pMHC) T-cell based therapeutic approaches that are now amongst the most promising paradigms for treating at least some forms of cancer. The first of these is the well-known, Nobel Prize-winning, discovery by Köhler and Milstein [31] of the methods for making and characterizing monoclonal antibodies from hybridomas. The second was the fundamental observation that activated cytotoxic T lymphocytes (CTLs) could function as serial killers of targeted cancer cells [32,33], via formation of an immunological synapse with the targeted cells [34], followed by degranulation and release of cytolytic proteins such as perforin and granzymes [7]. These and other early studies ultimately led to both the development of TRBAs to engage and redirect T-cells to induce serial killing of antigen-specific, targeted cancer cells [20] and to the genetic engineering of autologous T-cells to empower them with cancer cell surface antigen-specific targeting antibody-based receptors (i.e., CARs) fused to T-cell activating domains [35–37].

Within ten years of the initial isolation of monoclonal antibodies (mAbs) from immunized mice [31], the first bispecific antibodies were generated using a variety of approaches, including hybrid-hybridomas [38,39], chemical conjugation of both full-length IgGs and of Fabs [40–42], formation of bispecific F(ab')$_2$ antibodies using reduction and oxidation of sulfhydryl bonds processes [43] and recombinant approaches to make bispecific antibody fragments [44,45] based on single chain variable fragments (scFvs) [46,47].

Perhaps underappreciated today in the tsunami of T-cell redirecting bispecific antibodies and CAR-T cells, the first concepts and practice of redirecting T cells through binding of one antibody recombining (or binding) site to T cell surface markers to kill tumor cells bound by the other recombining site were laid out in several papers in the 1985–1986 time frame [41–43]. The use of the CD3 component of the TCR as the T cell target for redirection was first described shortly thereafter, in 1987 [48].

Figure 2 lays out a brief history of T-cell redirected bispecific antibodies and CAR-T therapeutics. Several key advances in the 1990s laid the foundation for the wide variety of T-cell redirecting bispecific antibody formats used for clinical stage candidate antibodies today. The first clinical trial in which T-cell redirecting bispecific antibodies were dosed was in 1990, when patients with glioblastoma were treated with an anti-CD3 IgG chemically coupled to an anti-glioma antigen IgG [49]. This was closely followed by the generation [50] and use in clinical studies [51] of an anti-CD19 × anti-CD3 bispecific IgG-like rat/mouse hybrid bispecific antibody for treatment of B-cell lymphomas. This antibody was the first IgG-like T cell redirecting bispecific antibody targeting malignant B cells to be studied in clinical trials [51].

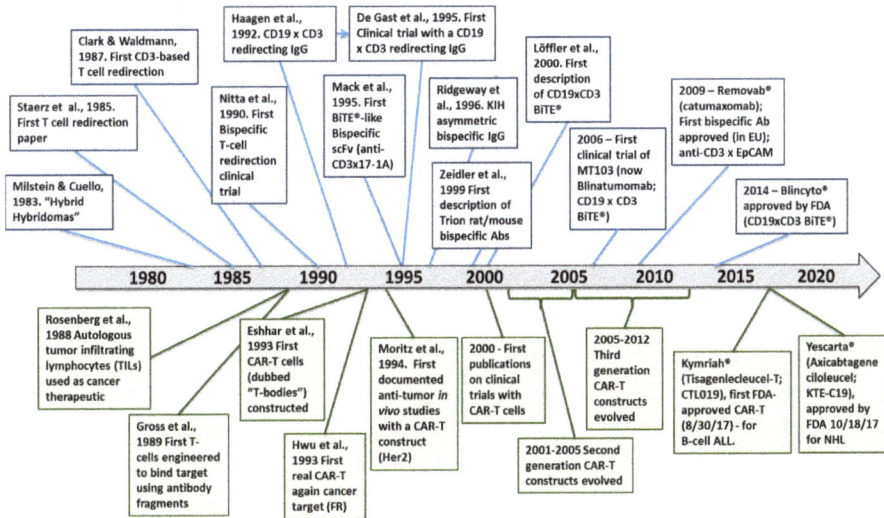

Figure 2. Key milestones in the history of T-cell redirected bispecific antibodies (TRBAs; top) and chimeric antigen receptor (CAR)-T cell (bottom) therapeutics. Specific references cited for TRBAs are: Milstein and Cuello, 1983 [38], Staerz et al., 1985 [41], Clark and Waldmann, 1987 [48], Nitta et al., 1990 [49], Haagen et al., 1992 [50], De Gast et al., 1995 [51], Mack et al., 1995 [45], Ridgeway et al., 1996 [52], Zeidler et al., 1999 [53] and Löffler et al., 2000 [54]. Specific references for CAR-T development cited are: Rosenberg et al., 1988 [55], Gross et al., 1989 [35], Eshhar et al., 1993 [37], Hwu et al., 1993 [56] and Moritz et al., 1994 [57].

Another significant advance in the late 1980s and 1990s was the discovery of methods to generate single chain variable fragment (scFv) antibody constructs by linking the two domains of an Fv, the variable heavy (V_H) and variable light (V_L) domains together using a short flexible linker [46,47], followed by the fusion of two scFvs together via a peptide linker to generate the first bispecific T cell engager (BiTE®)-like antibody [45] Figure 2). The first BiTE® targeted the tumor antigen 17-1A on the target cell with one scFv and CD3 on the T-cell with the other scFv arm [45]. The first description of an anti-CD19 × CD3 BiTE® was in 2000 [54]. The final significant advance in the 1990s was the generation of the now well-known asymmetric, heterodimeric Fc platform, "knobs-into-holes" (KIH), by scientists at Genentech [52,58,59] (Figure 2). This platform became the prototype for an entire generation of IgG-like asymmetric bispecific antibodies modified in the C_H3 domain to allow for heterodimeric antibody formation [60]. After engineering a production cell line with two heavy chains, one with a "knob" or protruding amino acid residue, mutation in the interface region of the C_H3 domain and the other with a compensating "hole" or small amino acid residue mutation and two light chains, the resultant heterodimer could be formed in four possible HC–LC pairings, in which the desired format is only one of the antibody molecules [59]. This technology was subsequently improved with the use of common LCs to eliminate the "light chain issue," that is, pairing of the light chains with the correct Fc half [59]. Interestingly, though, in the decade following the Merchant et al. paper [59], very few advances were made in the engineering of bispecific antibodies and most of the activity was focused on just two clinical candidates. Starting in about the 2007–2009 timeframe, however, the interest in developing new bispecific antibody platforms and using these to make TRBAs and other bispecific antibody therapeutics literally exploded, resulting in the development of more than one hundred different new platforms [60–62].

The first TRBA and bispecific antibody of any kind, to be approved by a major regulatory agency for commercial use was catumaxomab (trade name Removab®), a hybrid mouse-rat IgG-like bispecific antibody targeting CD3ε on T-cells with one arm and the cancer antigen, epithelial cell adhesion molecule (EpCAM), with the other arm [53,63]. Catumaxomab, which appears to have first entered clinical trials around the 2001–2002 timeframe [64], was approved in 2009 by the European Medicines Agency (EMA) as a therapy to treat malignant ascites [65]. However, due to its high immunogenicity rates in humans (being a fully rodent antibody), narrow and rare approved indication (i.e., malignant ascites) and subsequently poor sales, Removab® was not actively marketed past 2014 and was voluntarily discontinued by its sponsor in 2017. Removab® was never approved by the United States Food and Drug Administration (US-FDA).

The second T-cell redirecting antibody to be approved for therapeutic use was blinatumomab (trade name Blincyto®), a fragment-based bispecific antibody called a BiTE®, in which two single chain variable fragments (scFvs), one targeting the B cell antigen, CD19 and the other CD3ε, were linked together with a short, five residue $(G_4S)_1$ linker, that is,: $((V_LCD19\text{-}(GGGGS)_3\text{-}V_HCD19)\text{-}GGGGS\text{-}(V_HCD3\text{-}(GGS)_4GG\text{-}V_LCD3\epsilon))$ [66]. Blinatumomab, which was first known as Micromet MT103 (aka MedImmune MEDI-538), first entered clinical trials in 2006. Blincyto® was approved by the US-FDA in 2014 for treatment of Philadelphia chromosome-negative, B-cell acute lymphoblastic leukemia (ALL), making it the second TRBA to be approved for therapeutic use [67].

From just three TRBAs being studied in clinical trials in the 2008 timeframe (catumaxomab [63–65]), blinatumomab [68] and ertumaxomab, a rat/mouse TRBA targeting HER2 [69], there are now 59 unique clinical candidate CD3ε-binding TRBAs either approved by a regulatory agency or being studied in clinical trials today, with another two redirecting NK cells, totaling 61 TRBAs (Table 1).

1.3. Brief History of CAR-T Cells

It was clear from studies in the late 1970s that CTLs were capable of serial killing of targeted cancer cells [32,33]. This concept logically led to the idea of utilizing the power of autologous tumor infiltrating lymphocytes (TILs) to treat the tumor from which they were derived [55,70]. For this approach, TILs were harvested from human tumors, expanded ex vivo for four to eight weeks and then were re-administered intravenously along with a dose of interleukin-2 (IL-2) to help stimulate the re-administered lymphocytes [55]. This treatment resulted in regression of metastatic tumors in 60% of patients treated. While these results were preliminary, they clearly demonstrated the potential use of tumor-specific, expanded and activated autologous T-cells in cancer therapy [55].

The use of autologous TILs as therapeutics, however, still suffered from the lack of robust tumor targeting and the ability to control which cells were targeted. The first successful engineering of T-cells with a known and specific artificial binding capability was reported in 1989 (Figure 2), when Gross et al. [35] fused the V_H and V_L chains of an anti-2,4,6-trinitrophenol (TNP) antibody onto either the Cα-chain or Cβ-chain (i.e., V_H-Cα/V_L-Cβ and vice versa) of the TCR to generate artificial, chimeric TCRs. T-cells engineered in this manner were capable of killing TNP-coated target cells in a non-MHC-restricted manner [35]. While this engineered cell construct itself was not a CAR-T cell as we think of it today, it led directly to the formation of first-generation CAR-T cells.

The first CARs, which targeted the hapten TNP, consisted of an scFv (V_L-linker-V_H) fused directly to the human Fc receptor γ-chain, replete with its short extracellular domain, transmembrane domain and the immunoreceptor tyrosine activation motifs (ITAMs) [37] (Figure 3). The CD3ζ chain, which is highly similar in sequence and function to the γ-chain, also was used as the intracellular signaling domain in the fusion [37]. These first CAR-T cells were dubbed "T-bodies" denoting the construction of T cells with CARs made up of antibodies [71]. While the concept of CAR-T cells has been around since the early 1990s [37], only in the past decade have the technologies advanced to the point required to turn this into a viable "manufacturable" process. Thus, analogous to the TRBA approach, CAR-Ts are conceptually old but functionally still relatively young and developing [72].

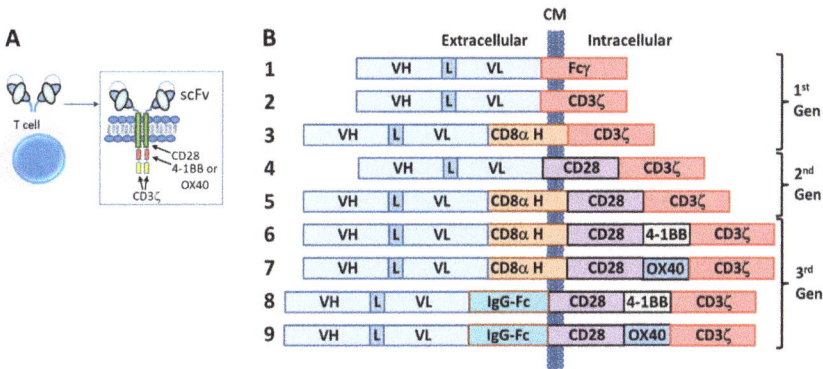

Figure 3. Generations of CAR-T cell therapeutics. Generations of CAR-T cell therapeutics as described in the text. (**A**) Generalized drawing of a CAR-T showing the fusion of the scFv to the transmembrane domain and intracellular activation domains. (**B**) Drawing depicting examples of first generation (1–3), second generation (4, 5) and third generation (6–9) CAR-T constructs as described in the text.

One of the earliest "real" cancer targets for CAR-T cell engineering was a cell surface folate binding protein (later determined to be folate receptor (FR)), implicated as a target in ovarian cancer. A first-generation anti-FR CAR was constructed by fusion of an scFv derived from the anti-FR antibody, MOv18, with Fcγ-chain, as described above [73] (Figure 3). T-cells transduced with this CAR (named *Mov-γ*) killed FR$^+$ IGROV-1 ovarian adenocarcinoma cells in vitro [73] and increased the survival of mice implanted with IGROV-1 ovarian adenocarcinoma cells [56]. This construct then was used in one of the early clinical trials of autologous CAR-Ts to treat ovarian cancer patients [74]. Due to the limited first-generation design, treatment with the *Mov-γ* CAR-Ts resulted in the lack of CAR-T persistence, poor trafficking to the tumor site and no reduction in tumor burden for any patient [74]. In the same period, Moritz et al. [57] carried out the first preclinical in vivo studies with a CAR-T cell line targeting HER2 (Figure 2).

The first two reports of clinical trials using CAR-T cells were published in the year 2000 (Figure 2). Mitsuyasu et al. [75] described the treatment of HIV-infected patients with a CAR comprised of the extracellular and transmembrane domains of human CD4 fused with the intracellular domain of the CD3ζ, resulting in a few patients having a transient drop in viral titer. Additionally, Junghans et al. [76] reported the results of a clinical trial in which cancer patients were treated with a CAR against carcinoembryonic antigen (CEA). Additional early clinical efforts using CAR-T cells have been reviewed by Eshhar [77].

Other first-generation CARs incorporated the inert transmembrane domain from CD8 between the scFv and the intracellular signaling γ-chain or CD3ζ chain [78,79] (Figure 3). All of these first-generation CAR-T constructs suffered from the fact that, while they could engage and kill targeted cells in vitro and in in vivo rodent models, they lacked the ability to persist in vivo [22,74]. This is most likely due to the absence of a costimulatory signal (i.e., signal 2), because tumor cells rarely express a costimulatory receptor ligand (e.g., B7, OX40L) [80]. Additionally, the lack of the costimulatory signal can render the T-cells anergic [81] and potentially susceptible to apoptosis [82]. Thus, it was quickly realized that additional signaling would be required to construct biologically active CAR-T cells that would persist in vivo.

Second generation CAR-T cells were designed by adding to the γ-chain or CD3ζ CAR constructs a cytoplasmic signaling domain from a costimulatory receptor, such as CD28 [83–85], 4-1BB (CD137) [84] or OX40 (CD134) [84] (Figure 3). These constructs typically resulted in improved production of activating cytokines such as IL-2 and IFN-γ, increased antigen-dependent proliferation in vitro and upregulated apoptotic factors such as Bcl-X$_L$ [83–85]. Nevertheless, even with second generation CARs, it appeared that T-cell activation was still not complete [80]. Thus, a series of third generation CARs was designed and these are starting to be incorporated into clinical trials today. Third generation CARs combine internal domains for CD28 plus intracellular signaling domains from either OX40 (CD134) [80,86] or 4-1BB (CD137) [86,87] (Figure 3), resulting in cytolytic T cells fortified with both proliferation and survival signals that enhance both their cell killing activity and their persistence in circulation.

Subsequently, it was demonstrated that a longer and more flexible "hinge" region (i.e., extracellular spacer such as regions from IgG-Fc or CD8α) was required for optimal CAR activity [88] (Figure 3) and, over the years since then, significant efforts have been made to optimize both the length and the structural characteristics of the extracellular spacer [89–91].

It might seem obvious that the addition of more T-cell activating signals to CARs would result in more robust tumor cell killing. Although certain studies have shown this to be the case, it is still not clear that "more is better" in every case. Various in vitro and in vivo studies have described both improvement and limitations in engineered T-cell function dependent on the design of the CAR [92–95]. T cell exhaustion and anergy, as well as the often negative influence on the T-cell by the tumor microenvironment, involve a carefully orchestrated series of signals within the T-cell that are poorly understood and not easily accommodated by CAR engineering, as of yet [96]. Similarly, the fine tuning of the molecular architecture of the CAR is also recognized as an area that needs to be improved, as the complicated physiochemical nature of the complete T-cell receptor complex is starting to be revealed [97].

For fourth generation CAR-T cells, new functions have been added beyond the target binding and T-cell activating signals. These most recent approaches include functions such as an inducible caspase-based suicide mechanism to eliminate the CAR-T cells on demand [98], expression and secretion of T-cell activating cytokines [99], the incorporation of trafficking receptors such as CCR2 to help the T-cell home to tumor microenvironments [100] or the use of virus-specific T cells that recognize viral antigens which can be used as "vaccines" to increase the persistence of the CAR-T construct [101–103].

Two CAR-T based therapeutics have thus far been approved by major regulatory agencies. Kymriah® (Tisagenlecleucel-T; also known as CTL019), the first CAR-T to be approved for therapeutic use, was approved on August 30, 2017 by the US-FDA for treatment of B-cell ALL [104]. Yescarta® (Axicabtagene ciloleucel; also known as KTE-C19), was approved by the US-FDA on October 18, 2017 for treatment of diffuse large B-cell lymphoma (DLBCL) [105]. There are now at least 223 additional recombinant CAR cell-based candidates being studied in clinical trials today (Table 1).

2. T-Cell Synapse and Killing Target Cells

2.1. Introduction to Immunological Synapse

The immunological or immune, synapse is a central mechanism of action for lymphocytes to communicate via cell-cell interaction with antigen-presenting cells (APCs), antigen-specific targeted cells and other lymphocytes. In normal T cell biology, small (~5 μm diameter) circulating naïve CD8+ T cells find an antigen-presenting cell (APC) and form a synapse with the APC via interaction of clustered TCRs on the surface of T cells with the neo- or non-self peptide antigen-loaded MHC molecules on the surface of the APC [8]. This interaction results in differentiation and activation of the CD8+ T-cells over the next 4–5 days into "armed" antigen-specific killer T cells loaded with granules full of the cytolytic proteins, granzyme and perforin [8]. These primed antigen-specific T cells expand and proliferate, increase in diameter to ~10 μm, induce a more sophisticated cytoskeletal system to "load" the cytolytic granules in proximity to the cell membrane [106] and express additional receptors of activation and response [8,107]. Upon locating the cells expressing the non-self or neo- antigen, typically either neoplastic or infected cells, the T-cells form the cytolytic synapse with the target cells and release their cytolytic toxins to kill those cells [107]. Additionally, T-cell membrane lytic factors such as FasL also can act in the synapse to induce apoptosis in the targeted cells [107].

The immune synapse, also known as the supramolecular activation cluster (SMAC), is responsible for initiating and completing the cell-cell response between APCs and T cells [8]. The SMAC is formed in three concentric rings, similar to a "bullseye" (Figure 4), with the central SMAC (cSMAC) forming the center ring, encircled by the peripheral SMAC (pSMAC) and the distal SMAC (dSMAC). Each ring has its own special function and structure [8]. The cSMAC contains a concentration of TCRs and the costimulatory molecule CD28 and is responsible for the key T-cell activation signaling events that accompany synapse formation, the pSMAC contains a series of adhesion molecules such as LFA-1 that stabilize the cell-cell interaction and the dSMAC is comprised of filamentous actin that helps to exert a mechanical force on the synapse [8].

There are multiple forms of the immune synapse, each with its own special function. The classical immune synapse, as exemplified by naïve CD4$^+$ T cells interacting with APCs, is an antigen recognition synapse. CD8$^+$ cells and NK cells can form stimulatory synapses leading to cytokine secretion or alternatively, inhibitory synapses [108]. CD8$^+$ T cells and NK cells also can form a cytolytic synapse with target cells leading to killing, which is the basis on which T-cell and NK cell redirected therapies are based.

Cytolytic synapses are very similar to the classic immune synapse but with additional activities to drive target cell killing. These include actin and microtubule guided localization of the lytic proteins [106], signals directing the secretion of cytotoxic proteins such as perforin and granzymes [109] and use of the mechanical forces of the dSMAC to enhance perforin activity and focus the cytotoxic killing in a directional, polarized manner [8,107].

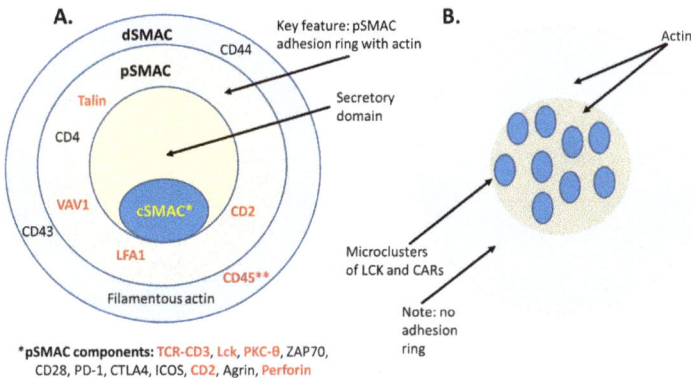

Figure 4. Classical immune synapse as compared with a bispecific T-cell engager (BiTE®)-induced synapse and a CAR-T synapse. (**A**) Diagrammatic representation of the immune synapse, adapted and modified from Huppa and Davis [110] and Watanabe et al. [111]. The classical immune synapse forms as a "bullseye" with the center central supramolecular activation cluster (cSMAC) surrounded by the peripheral SMAC (pSMAC) adhesion ring and the distal dSMAC ring. CD3, PKC-θ, perforin, CD28, CTLA4 and Agrin are found in the cSMAC. Additionally, Lck initially accumulates in the cSMAC and then distributes more broadly [110]. A key feature of the immune synapse is exclusion of CD45 from the cSMAC (noted by **). The pSMAC ring includes Talin, LFA1, VAV1 and CD4. LFA-1 is a key synapse stabilizing force in the pSMAC. The dSMAC markers are CD43, CD44, CD45 and filamentous actin. Offnexr et al. [13] compared the synapses formed by an anti-EpCAM × CD3 BiTE® TRBA to those formed by MHC-Her2-peptide/TCR. The markers denoted in red were positioned similarly in both the normal peptide-loaded major histocompatibility complex (pMHC)/TCR synapse and the BiTE-induced synapse [13]. CD45 was found to be excluded from both the BiTE®-induced synapse and the control pMHC/TCR synapse [13]. (**B**) A diagrammatic representation of the synapse formed by CAR-T cells, adopted and modified from Davenport et al. [112]. They described the CAR-T/target cell synapse as disorganized, with multifocal clusters containing LCK, no apparent LFA-1 stabilization and the absence of the adhesion ring that helps to define the classical immune synapse [112,113].

2.2. Normal TCR-pMHC Synapses vs. CAR-T and TRBA-Induced Synapses

The delicately orchestrated events associated with a TCR complex-MHC interaction signaling into the T cell has been an area of research for some time. The TCR complex is a complicated structure of a TCR α/β or δ/γ heterodimer that, analogous to an antibody, precisely recognizes peptide-MHC (pMHC) complexes. However, the intracellular signals associated with this interaction come from the other members of this complex, namely CD3ε, δ, γ and ζ. These chains are specifically associated with the TCR α/β (δ/γ) heterodimer at the cell surface through ionic bonds made between the transmembrane and hinge/stalk domains [114] (Figure 5). Although previously thought to be just a clustering-driven event that drives downstream signaling through phosphorylation of key residues within ITAMs, it is now recognized that important structural changes during this interaction drive the strength and duration of the downstream events [114]. In addition, the co-receptors CD4 and CD8, both dimers themselves, are required to interact and specifically bind to either class I (CD8) or class II (CD4) MHC in the context of TCR α/β complexes (TCR δ/γ complexes do not require CD4/CD8 co-receptor engagement to function). The intracellular domains of CD4 and CD8 also associate with the Src kinase LCK to provide additional signaling, the function of which is not completely understood [115]. In total, this complicated structure has evolved to control one of the most complex cellular activities within mammals and other organisms. The goals of TRBA and CARs has always been to mimic this complexity as much as possible.

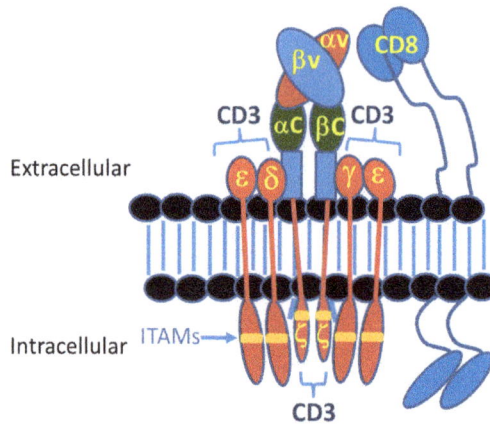

Figure 5. Diagram of the T-cell receptor (TCR) complex. Normal TCR-pMHC interactions are in the range of 1–100 μM, with the inherent avidity of clustered TCRs providing the required attraction [11]. Affinities of the MHC to the presented peptide have a profound influence on the ability of natural T cells to kill and eradicate tumors. When the peptide-MHC affinity was found to be <10 nM as determined in in vitro assays, it was shown that T cells recognizing those pMHC complexes were able to cause tumor rejection [116]. On the other hand, when the peptide-MHC affinity was >100 nM, the tumors relapsed, indicating that the T cells were not capable of killing those tumors cells [116]. Both CAR-T cells and TRBAs function independently of this parameter.

The T cell/target cell synapse is driven by a delicate balance between affinity and receptor-target density, with regard to natural TCR-pMHC interactions, TRBA and CARs. The consequence of these interactions can be influenced by natural regulatory receptors, such as CD45 isoforms, which can naturally down-regulate the signals emanating from the TCR or CARs [117]. Embedded in the cell membrane, CD45, which is a complex, highly differentially spliced molecule of varying extracellular size, can interfere with these synapse-based interactions and prevent downstream signaling [118]. Low-affinity interactions, typical of TCR-pMHC interactions, are very susceptible to the effects of CD45 isoforms, serving as a natural safety mechanism to prevent undesired T cell activation [117]. However, higher affinity interactions or multiple interactions between the TCR and pMHC can overcome these effects [117]. Similar considerations must also be taken into account when designing TRBA and CARs.

The spacing between T cells and APCs during synapse formation has been measured in the range of 5–25 nm [8] and the normal spacing in synapses formed between T cell TCRs and peptide-loaded MHC (pMHC) complexes has been shown to be about 13 nm [119] (Figure 6). Experimentally forced longer distances between the cell membranes decreased the TCR activation and response [119], which is a key issue for both TRBAs and CAR-Ts, as described later.

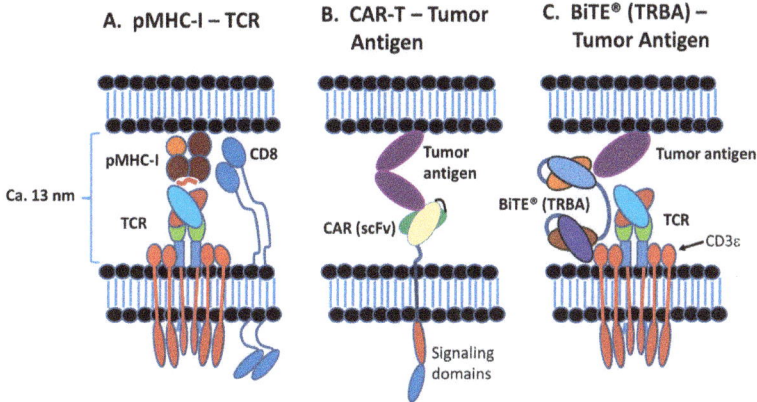

Figure 6. Diagram of the T-cell receptor complex. Diagram of the molecules driving synapse formation in pMHC-1/TCR T-cell/APC interactions, in CAR-T/target cell interactions and in TRBA-induced T-cell/target cell interactions. (**A**) Classic TCR/pMHC-1 type of interaction with a membrane to membrane spacing in the range of 13 nm [119]. (**B**) scFv-based CAR-T cell binding to tumor antigen on target cell. (**C**) BiTE® binding to CD3ε on T-cell and to tumor antigen on target cell to bring the cells into close proximity to form the cytolytic synapse.

For natural CTLs, it has been shown that as few as 1–3 peptide-MHC/TCR interactions are required to trigger a cytolytic killing event [107,120,121]. In those cases, however, the elaborate SMAC complex is neither required nor fully formed [121]. Additionally, it has recently been demonstrated with NK cells, another cytolytic lymphocyte, that NK cell lines only produce about 200 perforin-positive granules and a single degranulation event at the cytolytic synapse results in only about 20 granules being released, only about 2–4 of which are actually required to kill a target cell [122]. Thus, the machinery for cytolytic synapse-based killing is exquisitely potent and sensitive to activation.

How do the synapses induced by TRBAs or those formed between CAR-T cells with targeted cells, compare with natural synapses? A study by Baeuerle and colleagues demonstrated that the synapse formed between T-cells and EpCAM⁺ cells, brought together by an anti-EpCAM × CD3ε BiTE® TRBA, was highly similar in its structure and function to normal cytolytic T-cell synapses, including formation of the concentric rings and presence of many of the same protein markers, such as LCK, PKC-θ, LFA-1, VAV1, Talin, CD3, perforin and CD2 [13] (Figure 4). Furthermore, it was demonstrated separately that the TRBA-induced synapse also possessed classical synapse hallmarks [13], including target clustering, ZAP70 translocation and exclusion of the negative regulatory protein, CD45, from the cSMAC [15].

While synapses formed by the function of TRBAs appear to be highly similar to normal MHC/TCR mediated synapses, CAR-T synapses appear to be significantly different from normal T-cell synapses [113]. CAR-T synapses are not highly organized as SMACs but rather, they are disorganized, patchy signaling clusters lacking a defined structure [112]. Additionally, CAR-T synapses do not require LFA-1 for stabilization and do not form the characteristic pSMAC. Thus, it is clear that immune synapses formed by CAR-T cells with their target cells are structurally distinct from both classical immune synapses and those formed by TRBAs (Figure 4). These structural differences between CAR-T synapses and classical pMHC/TCR synapses also result in functional differences [113]. CAR-T cells yield faster proximal signaling and recruit lysosomes to the immune synapse faster than classical synapses, suggesting that they are able to mount a more rapid killer response than TCR-mediated killing [112]. Additionally, they have a significantly faster off-rate, that is, dissolution of the synapse and detachment from the target cell, than found with TCR-driven T cell interactions [112,113,123]. In a time course comparison of TCR-mediated killing versus CAR-T killing after synapse formation, CAR-T signal strength was both greater and ramped up faster than TCR signaling, perforin and granzyme release were faster (peak release within two minutes of initiation for CAR-T vs three min for TCR) and detachment was significantly faster (at five min for CAR-T vs seven min for TCR) [123]. Thus, CAR-Ts appear to kill target cells faster and then move on faster than CTLs. Additionally, it has been demonstrated in vitro that a CAR-T cell can kill a target antigen-positive tumor cell within about 25 minutes of initially recognizing the target antigen [124]. Recently, Xiong et al. [125] demonstrated that the strength of the CAR-T synapse, as measured by quantification of actin and lytic granules, was more predictive of killing effectiveness than either cytokine production of a 4-h killing assay, demonstrating how important the CAR-T synapse is to CAR-T function, even if it is structured significantly different from traditional immune synapses [112].

As noted above, cytolytic TCR/pMHC synapses, TRBA-induced synapses and CAR-T-target cell synapses all result in death of the target cells, typically by perforin and granzyme-induced apoptosis [109], although with CAR-T cells, the FAS/FAS-L axis is also involved [123]. As expected from the differences in structures (Figure 6), however, the TCR-independent synapses formed by TRBAs and CAR-Ts have some unique features as compared with TCR-pMHC synapses. A few of these differences will be highlighted below.

First, both TRBAs and CAR-T cells function independently of TCR/pMHC and thus, do not require expression of MHC receptors on target cells. This was demonstrated in a study in which BiTE®s were shown to kill EpCAM-positive, MHC Class-I-negative cell lines, indicating that BiTE®s could function to kill target cells in the total absence of the MHC T cell recognition molecules [13]. Second, the affinities of TCRs for pMHCs are typically in the range of 1–100 µM [11], whereas affinities of CARs or TRBAs to their targets are typically below 100 nM and often below 10 nM.

Third, T-cell activation due to interaction via TCR/pMHC is in part governed by the expression of pMHC on the target cells, which are typically found at very low copy number [126]. Targets for TRBAs and CAR-T cells, on the other hand, typically number in the 1000 s to 10,000 s and sometimes even higher. Nevertheless, TRBAs have been shown to elicit T-cell killing even with very low antigen densities on target cells, such as only 200 target molecules/cell [15]. Thus, even at low target densities, at least in some circumstances, TRBAs can still be effective killing agents.

CAR-T cells also appear to be able to kill targeted cells when antigen densities reach levels as low as 200 target molecules/cell [127,128]. In studies directly comparing the activity of TRBAs and analogous CAR-T cells, it appears that the CAR-T cells are more effective at killing targeted cells at low antigen density than was the analogous BiTE®s [127,128]. Interestingly, similar to the hierarchical threshold found with TCR signaling [129], the antigen density required to trigger CAR-T killing (100–200 targets/cell) was significantly lower than the antigen density required to trigger CAR-T cytokine release (~5000 targets/cell) [111,128]. Thus, it appears that, similar to TCR responses, there are distinct thresholds in CAR-T cell activation for killing, proliferation and cytokine release, with a full response requiring a density of at least 5000 antigens/cell [128].

On the other hand, in what appears to be very different from TRBA or pMHC/TCR-mediated synapse formation, CAR-T cells may directly contribute to target cell antigen loss resulting in low antigen density. In a very recent study, it was demonstrated that CAR-T cells decreased antigen density on target cells through the mechanism of trogocytosis, a process by which targeted antigens are transferred to the CAR-Ts [130]. This process not only decreased antigen density on the target cells but also once the CAR-T cells obtain the target antigen, they themselves can become targets of CAR-T fratricide, potentially contributing to lack of persistence [130].

Fourth, it was calculated that, for the most potent BiTE® with a fM IC50, "double-digit" (i.e., <100) TRBA-driven cell-cell interactions were required to drive synapse formation [131]. Similarly, with CAR-T cells, it has been estimated that as low as 100 or less CAR-antigen interactions are required to drive synapse formation between CAR-T cells and their target cells [111]. These estimates both are at least a log greater than the number of interactions required than the minimal number of TCR-pMHC interactions required to initiate a synaptic killing event [107,120,121].

Finally, T-cell activation via TCR-pMHC interaction is enhanced by recruitment of CD8 [115] and incorporates "signal 2" costimulatory pathways [9,18]. Conversely, T-cells are negatively regulated by checkpoint interactions such as PD-1/PD-L1 and CTLA4/CD80-86 [18,132]. Both T-cells engaged by TRBAs and CAR-T cells can function independently of signal 2. Nevertheless, the effector function and targeted killing by T-cells engaged by TRBAs can be enhanced by costimulatory molecules such as CD28 and CD137 (4-1-BB) [133]. For CAR-T cells, the CARs themselves are designed with intracellular costimulatory signaling domains from CD28, OX40, ICOS and/or 4-1BB, providing the costimulatory signal upon binding of the CAR to the targeted cells [94,125,134].

Similar to the regulation of CTLs, T-cells engaged by TRBAs [15,135–137] and CAR-T cells [138] can be subjected to inhibition by checkpoint pathways such as PD-1/PD-L1 and CTLA4/CD80-86. With this in mind, clinical trials are currently underway combining the treatment of B-cell lymphomas or leukemias with the anti-CD19 BiTE®, Blincyto®, with anti-PD-1 [139,140] or anti-PD-1 and anti-CTLA4 antibodies [141].

Similarly, anti-PD-1 and/or anti-CTLA4 checkpoint inhibitor co-therapy also is being tested clinically with CAR-T therapeutics in efforts to relieve checkpoint inhibition of the CAR-T cells [142]. In some cases, fourth generation CAR-T cells are being engineered to express antibodies or antibody fragments that can function in an autocrine/paracrine manner to block checkpoint inhibitors such as PD-1 or PD-L1 [143–145] or engineer into the CAR-T cells dominant-negative PD-1 that negates the PD-1/PD-L1 inhibitory pathway [138]. In both of these cases, the inhibitory effects of PD-1 were blocked, increasing the effector functions and persistence of the CAR-T cells. In addition, clinical trials with CAR-T cells engineered by knocking out their endogenous PD-1 gene are being run to test their hypothesized improved efficacy [146].

As noted above, while there are significant differences in the mechanisms by which T-cells are activated in the CTL (TCR/pMHC), TRBA or CAR-T cell paradigms, there are also many similarities. These include the formation of synapses by redirected T cells with target cells, ability to function as serial killers, their mechanism of killing, (e.g., directed release of cytolytic proteins such as perforin and granzymes to kill the targeted cells), their regulation via costimulatory and checkpoint inhibitory pathways and their ability to proliferate and secrete cytokines [17].

3. T-Cell Redirecting Bispecific Antibodies (TRBAs)

3.1. Introduction

Bispecific antibodies are antibodies that have two different types of combining regions (variable domain-based binding sites), which makes them capable of binding two different antigens simultaneously. Of the approximately 858 antibodies either currently in clinical trials or approved by a major regulatory agency (WR Strohl, BiStro Biotech Consulting Antibody and CAR-T Database, last updated 20 June 2019), there are currently 122 unique clinical stage bispecific antibodies, 59 of which are CD3ε-binding, T-cell redirecting bispecific antibodies and two (GT Biopharma GTB-3550 [147] and Affimed AFM13 [148] of which are CD16a (FcγRIIIa) NK-cell redirecting antibodies (Table 2).

Fundamentally, there are two major types of bispecific antibodies used as TRBAs, bispecific antibody fragments (e.g., Figure 7A–D) and IgG-like asymmetric heterobispecific antibodies (e.g., Figure 7E–H). There are many variations on these two themes, some of which are absolutely critical to the unique function of the antibodies. A sampling of these platforms is shown in Figure 7. Additional details on these various platforms can be found in various recent reviews [60–62,149–152]. This section will describe a few of these platforms briefly and how structure can be very important to the function of TRBAs.

Table 2. Bispecific antibody T- and NK-cell redirecting antibody formats represented in clinical trials *,**.

Bispecific T- or NK-Cell Redirecting Antibody Format ***	Clinical Stage			Total
	Phase I/II	Phase III	Approved	
Short half-life bivalent fragments (e.g., BiTE®s, DART®s, ImTACs, other bivalent fragments)	15	0	1	16
Half-life extended bivalent fragments (e.g., DART®-Fc, Extended half-life BiTE®s, TriTAC)	11	0	0	11
Asymmetric bivalent IgG-like (e.g., Trion, BEAT, Xencor H/A platform, Duobodies, other asymmetric platforms)	21	0	1 ****	22
Roche TCB 2:1, Chugai ART-Ig®-scFv and Teneobio 2:1 platforms (two binding sites for target cell, one for CD3)	4	0	0	4
ADAPTIR®and TandAb platforms (tetravalent platforms)	4	0	0	4
Chemically conjugated IgGs (tetravalent; two IgGs)	4	0	0	4
Total	59	0	2	61

Abbreviations: ADAPTIR, modular protein technology; ART-Ig®, asymmetric re-engineering technology–immunoglobulin; BEAT, bispecific engagement by antibodies based on the T cell receptor; BiTE®, bispecific T cell engager; DART®, dual affinity retargeting (antibody); ImmTAC, immune-mobilizing monoclonal TCR against cancer; TandAb, tandem diabody; TCB, T-cell bispecific; TriTAC, Trispecific T cell activating construct. * BiStro Biotech Consulting LLC database, locked 20 June 2019. Data obtained from Clinicaltrials.gov, literature papers, company websites, analyst reports and other sources. ** Out of a total known 122 bispecific antibodies being studied in clinical trials as of 20 June 2019. *** The platforms and abbreviations are described in the text and in Figure 7. **** Voluntarily removed from marketing in 2017.

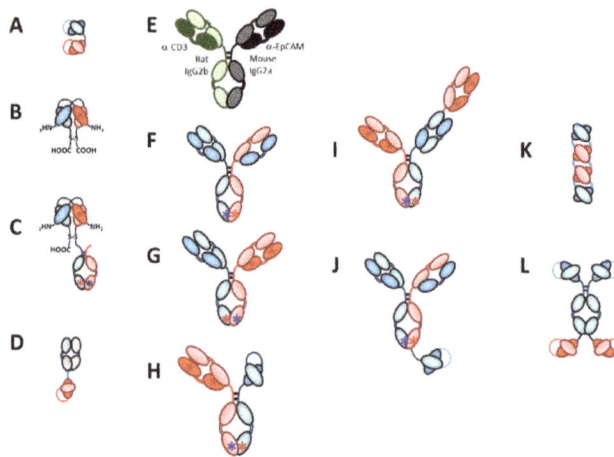

Figure 7. Examples of bispecific antibody platforms used to make clinical stage TRBAs. (**A**) Bispecific T-cell engager (BiTE®) [45]; (**B**) dual affinity retargeting (DART®) antibody [153]; (**C**) DART®-Fc for elongated half-life in vivo [154]; (**D**) TCR fused to scFv called immune-mobilizing monoclonal TCRs against cancer (ImmTAC) [155]; (**E**) mouse/rat hybrid IgG [53]; (**F**) Asymmetric IgG with common LC [59]; (**G**) Asymmetric IgG with different light chains [156]; (**H**) Asymmetric IgG-like molecule with a Fab arm and an scFv arm to eliminate light chain resorting [157,158]; (**I**) Asymmetric IgG using cross-mab technology for LC fidelity and extra Fab arm to make a 2:1 (target cell antigen:CD3ε) antibody call "TCBs," for "T-cell Bispecifics" [159,160]; (**J**) Chugai's asymmetric IgG using "Asymmetric Re-engineering Technology–Immunoglobulin" (ART-Ig®) platform [161,162] technology, with an scFv fused to one HC to make an ART-Ig®-scFv 2:1 (target cell antigen:CDε) antibody; (**K**) tetravalent, bispecific tandem diabody (TandAb) [163]; (**L**) tetravalent, bispecific ADAPTIR™ platform with two different scFvs fused to each Fc [164].

3.2. Bispecific Bivalent Antibody Fragments Used to Make TRBAs

Most of the bispecific antibody fragments in development today are based in some manner on scFv antibodies, discovered independently by two research groups in 1988 [46,47], which are comprised of a V_H domain linked to a V_L domain via a short, flexible linker. It is notable that Huston et al. [47] utilized the $(GGGGS)_3$ flexible linker to fuse V_H to V_L; this "gly-ser" linker or variations thereof, is one of the most widely used linkers in bispecific antibodies today. From both an historical perspective as well as a therapeutic perspective, the most significant types of bispecific antibody fragments are tandem scFvs [165], bispecific T cell engagers (BiTE®s [166]), dual affinity retargeting (DART®s [153]) antibodies, diabodies [44] and tandem domain antibodies [167]. Additional recently described fragment-based bispecific antibody constructs, such as HSA-antibody fragment fusions [168], bispecific killer engagers (BiKEs) [169–171], trispecific killer engagers (TriKEs) [169–172] and dock-and-lock Fabs [173] have also been developed as bispecific antibody fragment platforms. There are several recent reviews that describe these bispecific antibody fragment platforms in detail [60–62,149,152,174].

As previously mentioned, the anti-CD19 × anti-CD3 BiTE®, blinatumomab, has been approved for commercial use under the trade name of Blincyto®. Including blinatumomab, there are currently 27 bivalent, bispecific antibody fragments being tested as TRBAs in clinical trials. Of these, 11 are short-half-life BiTE®s or bispecific scFv-based molecules similar to BiTE®s, six are next-generation half-life extended BiTE®s (i.e., BiTE®-Fcs), one is a short half-life DART® construct, three are long half-life DART®-Fc molecules, three are immune-mobilizing monoclonal TCRs against cancer (ImmTACs) [155], one is a trispecific killer engager (TriKE) [147] and two are a Trispecific T cell Activating Constructs (TriTACs) [175].

3.3. Bispecific Bivalent Asymmetric IgG-Like Antibodies Used to Make TRBAs

One of the most widely utilized approaches for making bispecific antibodies today to be used as TRBAs is the generation of asymmetric heterobispecific IgGs containing two different types of heavy chains. Two important components are required to make asymmetric IgG-like bispecific antibodies that can be developed and manufactured in a consistent manner. The first is the preferred formation and/or isolation of the asymmetric heterodimerized heavy chains (HCs) over the parental IgG homodimeric antibodies [60]. The second is the proper pairing of the light chains (LCs) of each arm with the cognate HC [176].

3.3.1. Asymmetric Pairing of HCs

It is well established that the primary driver for HC dimerization is the high affinity (ca. 10 pM [177,178]) interaction between the C_H3-C_H3 domains [60,156,177,178]. The interactions between the C_H3-C_H3 domains, which bury over 2400Å^2 of surface area [178], are driven by a strong central hydrophobic core surrounded by a series of charged residues that provide electrostatic interactions between the two Fc domains [60]. Essentially, three basic strategies have been used to promote asymmetric heterodimerization over formation of the parental homodimers: (i) the first based on physical/spatial interactions, for example, adding a protrusion to one heavy chain and a corresponding cleft in the other (e.g., "knobs-into-holes") [52–54]; (ii) the second depends on alteration of specific amino acid interactions at the C_H3-C_H3 interface [60,156,179] and (iii) the third focuses on charge, that is, changing charged amino acid residues to generate a repulsion of homodimers and a corresponding charge attraction between the heterodimer pairs [178,180,181].

Several different platforms have been used to make asymmetric heterobispecific IgG-like antibodies to be used as TRBAs. The first was the three-way fusion of a mouse B-cell, a rat B-cell and a myeloma cell to form a quadroma cell line (Triomab® technology) [53] (Figure 7E). In this platform, the LCs naturally sort to the proper heavy chains due to species specificity. The downside, of course, is that any antibody made using this approach will be highly immunogenic, as mentioned previously was one of the problems with catumaxomab. This platform also was used to generate the first asymmetric heterodimeric IgG-like TRBA to be clinically tested in 1995 [51].

All of the rest of the asymmetric bispecific IgG-like platforms depend on engineering of the Fc to promote either formation of or purification of, the heterodimeric IgG over the parental homodimeric IgGs. The first *bona fide* bispecific IgG-like antibody platform was the knobs-into holes (KIH) platform [52–54]. Other platforms, which rely on charge attraction/repulsion, include the electrostatic steering (ES) platform [178], ES plus hinge mutations [180], the Oncomed IgG2-based ES "Bimab" technology [182] and the Chugai "Asymmetric Re-engineering Technology—Immunoglobulin" (Art-Ig®) platform [161,162]. Other asymmetric heterobispecific IgG approaches include the modification of specific amino acid pairings in the interface of the C_H3 domains, such as the Duobody® approach [156], the Zymeworks azymetric platform based largely on hydrophobic interactions [179], the BEAT (bispecific engagement by antibodies based on the T cell receptor) platform from Glenmark [157], the Xencor H/A platform [158,183] and the Merus Biclonics® Platform [184]. Additional platforms used to make asymmetric heterobispecific IgG-like antibodies include Regeneron's modified protein A-binding platform [185,186], the "strand exchange engineered domain" (SEED), which consists of alternating sequences derived from IgG and IgA C_H3 domains resulting in asymmetric but complementary pairs, AG and GA, in a manner that only the heterodimeric protein would bind and fold into an active Fc [187,188] and NovImmune's kλ-antibody platform, which incorporates common HCs paired with a lambda LC in one Fab arm and a kappa LC in the other Fab arm [189]. In this case, essentially all of the binding activity rests with the LCs [189], exactly opposite of what would be the case using heterologous HCs with a common LC. While there are several additional examples of platforms recently developed to generate asymmetric heterobispecific IgG-like antibodies for example, References [19,20,60–62,152,190,191], these will not be further described here.

3.3.2. LC Issue for Asymmetric Heterobispecific IgG-Like Antibodies

Whether by generating hybrid-hybridomas by the fusion of two different hybridomas or by genetic engineering, the introduction into a cell line of four antibody genes encoding two different heavy chains (HCs) and two different light chains (LCs), will result in the formation of ten different potential combinations, only one of which is the desired bispecific antibody, due to promiscuous heavy chain (HC)- light chain (LC) pairing [192]. This LC pairing problem can be reduced to four possible pairings if there is a forced pairing of the two different heavy chains to make an asymmetric, heterologous Fc [59]. Thus, even with the high efficiency formation of the heterodimeric Fc via generation of asymmetric Fcs, there still needs to be a solution for the light chain independent distribution issue.

Multiple solutions have been found to alleviate the light chain pairing issue in asymmetric bispecific IgG-like antibodies. One approach, which is essentially an "avoidance" strategy, is the formation of asymmetric IgG-like antibodies with a Fab arm on one side and an scFv on the other half (Figure 7H), such as Xencor's Xmab H/A platform [158,183] or Glenmark's BEAT platform [157]. A second solution is the "common LC" approach, in which both Fabs of the asymmetric IgG-like bispecific antibody possess identical LCs [59,161,184,185,193,194]. Another strategy is to generate differences in the HC-LC interactions by switching out C_H and C_L domains on one half of the antibody to generate a "CrossMab" (CM) [195–197]. The result of this switch is the pairing of one normal heavy and light chain on one side of the bispecific antibody and a pairing of V_H-C_L-hinge-C_H2-C_H3 with V_L-C_H1 on the other half (CM^{CH1-CL}; [195]). A similar strategy to CrossMab would be to mutate certain sequences in the interfaces of the HC and LC in one and/or the other Fab arm to ensure proper pairing [198–201].

A final method to control LC distribution is via separate upstream production of the two parental antibodies with post-Protein A recombination of the two antibody halves, typically by reduction and re-oxidation processes [156,180,187], leading to the asymmetric heterodimeric bispecific antibody. Two platforms are built around this concept, the Duobody® platform [156] and the SEEDbody platform [187]. These methods depend on the fact that the heavy chains can be separated via reduction of the interchain disulfide bonds while the HC-LC interactions and disulfide bonds remain stable [156,187]. A variation on this theme is the production of bispecific antibodies in two cocultured strains of *Escherichia coli*, each strain containing a half antibody (HC+LC). After growth of the strains and production of the antibody halves, the antibodies were reduced and re-oxidized to form the heterodimeric IgGs [202]. A mammalian coculture protocol for generating asymmetric heterodimeric bispecific antibodies also has been devised [200], combining aspects of the Duobody® [156] and the bacterial [202] approaches.

Excluding catumaxomab, which had been approved for malignant ascites in Europe but now discontinued, there are currently 21 known asymmetric, bivalent heterobispecific TRBAs in clinical trials that utilize the designs mentioned above, all of which are designed to redirect T cells, via CD33ε binding, to kill targeted cells (Table 2).

3.3.3. Trivalent, Bispecific Antibody Platforms

There are several platforms that have been developed recently to provide two binding arms for target cells and a single binding arm for CD3ε, resulting in trivalent but bispecific, antibodies. The concept behind these antibody formats is to provide better binding to the target cell through avidity, while only providing a single binding arm for CD3ε [159,160], because it is known that providing two binding arms for CD3ε may result in non-specific and undesired T-cell activation [203].

One trivalent, bispecific antibody format is an asymmetric heterodimeric IgG, constructed using the KIH technology, with a single anti-CD3ε Fab arm appended to one of the HCs as described earlier (Figure 7I). LC fidelity is maintained through use of CrossMab technology [159,160]. These 2:1 (target cell antigen:CD3ε) TRBAs have been dubbed by Roche as "TCBs," for "T-cell Bispecifics." There are currently two clinical stage TCBs that incorporate the Fab as an extra appendage, including cibisatamab (aka RG7802, RO6958688, CEA TCB) [159,204], which has two binding arms for CEA and a single binding arm for CD3ε and RG6026 (aka RO7082859) [205], which has two binding arms for CD20 and one for CD3ε [160]. Two additional TCBs, one targeting BCMA [206] and another targeting a carboxyl-terminal fragment of HER2 expressed in about half of HER2-positive tumors [207], have been reported but are not yet in clinical trials.

Another trivalent, bispecific antibody is ERY974, which is a silenced IgG1 asymmetric mAb with an anti-CD3 scFv fused to the C-terminal sequence of one of the heavy chains (Figure 7J) [162]. ERY974, which has two binding sites for glypican-3 to optimize avidity but only a single anti-CD3 arm to reduce the chance of non-specific T cell activation [162], is in Phase I clinical trials [208].

Other trivalent, bispecific antibody platforms providing two binding sites for target cells and a single binding site for T-cell CD3ε are "Asymmetric Tandem Trimerbody for T cell Activation and Cancer Killing" (ATTACK) [209] and the new trivalent, IgG-shaped tri-Fab format [210]. These platforms are not yet represented in the clinic.

3.3.4. Tetravalent Bispecific Antibody Platforms

It is generally considered that bivalent targeting of CD3ε can lead to non-specific T cell activation and release of cytokines [203,211], which is not desirable in a T cell redirection platform. Thus, as described in the previous sections, the vast majority of both fragment and IgG-like TRBAs are bivalent, with one antigen combining site binding to the target cell and the other antigen combining site binding to CD3ε on T cells. Having stated that, however, a few clinical-stage platforms stand out as antitheses to that trend. The first of these is the Affimed tandem diabody (TandAb) platform, which is a bispecific, tetravalent tandem diabody with a molecular weight of approximately 114 kDa [163]. Clinical stage TandAbs include AMV564, Aphivena's anti-CD33 × CD3 TRBA for myelodysplastic syndromes [212] and AFM13, Affimed's NK-cell redirected anti-CD30 × CD16a candidate for Hodgkin's lymphoma [148].

The Aptevo ADAPTIR^TM platform is a tetravalent, bispecific antibody consisting of two identical scFvs binding to target A fused to the hinges of an Fc and two identical scFvs binding to target B fused via a short linker to the C-terminal sequences of that Fc (see Figure 7L) [164]. The anti-prostate specific membrane antigen (PSMA) × anti-CD3 tetravalent, bispecific ADAPTIR^TM TRBA molecule

MOR209 (also called ES414) [164] is currently being tested in Phase 1 clinical trials [213] for the potential treatment of prostate cancer. In preclinical studies, even though it possesses two CD3ε combining regions, MOR209 did not appear to activate T cells indiscriminately and it induced lower levels of pro-inflammatory cytokines than some other platforms [164]. Thus, TRBA geometry may play a significant role in whether binding to two CD3ε s on the T-cell surface causes non-specific T-cell activation or not.

Other tetravalent, bispecific platforms not currently represented in clinical trials include IgG-scFv fusions [214], dual variable domain-immunoglobulins (DVD-Ig) [215,216] and Fabs-in-tandem immunoglobulins (FIT-Ig) [217]. IgG-scFv fusions, which are IgGs with scFvs fused to either the C- or N-termini of each HC or LC [214,218–220], often suffer from instability due to the unfolding and aggregation of the scFvs, requiring additional modifications to achieve stable, manufacturable candidates [219,220]. DVD-Igs are tetravalent, bispecific antibodies of about 200 kDa comprised of an IgG to which an extra Fv is appended to the N-terminus [215,216]. The Abbvie team that developed the DVD-Ig, however, has moved to the half-DVD platform for their TRBA constructs to reduce non-specific T-cell activation [203]. Finally, the Epimab Biotherapeutics "Fabs-in-tandem immunoglobulins" (FIT-Ig) platform is similar in some respects to the DVD-Ig, except that an entire Fab is appended to the N-termini of each HC [217]. There are no TRBAs in the clinic currently from any of these platforms.

3.4. Factors Affecting TRBA Potency

Factors affecting the potency of TRBAs include location of the epitope on the target antigen, size of target antigen, affinity of the TRBA arms to target antigen and CD3ε, valency of the TRBA on the target antigen, antibody size and geometry, antigen density on the target cell [15,17,221], effector-to-target ratios and TRBA concentration [222]. Jiang et al. [222] used a variety of parameters to build a model of the target cell-biologic-effector cell complex (TBE complex) to demonstrate the sensitivities to killing potency with a variety of parameters. One key result that emerged was that TRBA concentration appears to be critical, with a TRBA concentration greater than the K_D of the lower affinity binding arm resulting in decreased TBE complex formation due to a shift toward monovalent binding [222]. On the other hand, too high a concentration of TRBAs can result in separate coating of both antigen-positive cells and CD3ε -positive T cells, resulting in poorer killing [223].

While the hierarchy of factors affecting TRBA potency is not entirely understood [17,222], a few factors have become clear over the past several years. First, the epitope for antibody binding to the target antigen is critical, with the best epitopes being membrane proximal [15,17,221,224], especially with larger target antigens [17]. It was demonstrated that targeting a membrane proximal epitope also causes exclusion of the negative regulatory protein CD45 in the synapse, increasing the potency of the T-cell response and killing [15]. This observation that membrane proximal epitopes on the target antigen should provide the greatest TRBA potency has been supported with TRBA-based studies on the cancer targets P-cadherin [225] and ROR1 [226]. Second, the size of the antigen, which can effectively increase the distance within the synapse between the T-cell and target cell, also can affect potency, with much larger targets resulting in lower TRBA potencies [15,221].

It has been shown multiple times that increase in affinity of a TRBA to the target antigen can significantly increase potency [222,225]. Another approach to increase binding of a TRBA to target cells is to increase the avidity, that is, to have more binding arms on the target cell. As described in Section 3.3.3., there are two TRBA formats that provide two binding arms for the target cell with only a single binding arm for CD3ε, the three-Fab TCB format designed by Roche scientists [159,160] and the ART-Ig®-scFv format designed by Chugai scientists [162]. The key to understanding whether a 2:1 construct shows better potency than a 1:1 construct is to have each type of construct made for the same target. Bacac et al. [160] compared their 2:1 anti-CD20 × CD3ε versus a 1:1 construct made similarly and demonstrated that the 2:1 format had a 10–100× greater potency in vitro than a similar 1:1 construct. Additionally, the 2:1 format outperformed the 1:1 format in ex vivo assays and showed very potent activity in in vivo animal models [160]. Moreover, Bacac et al. [227] demonstrated that pretreatment with the anti-CD20 mAb, obinutuzumab, prior to treatment with the anti-CD20 × CD3ε 2:1 TCB, RG6026, resulted in significantly lower cytokine release, which may translate into a clinical benefit.

Considering that the anti-CD20 × CD3ε 2:1 TCB format appears to be more potent than various 1:1 anti-CD20 × CD3ε formats [160], it might be interesting to see how other multiple-tumor-target-binding × single CD3ε binding formats might behave. To that end, IGM Biosciences has an anti-CD20 × CD3ε IgM pentameric antibody in preclinical studies that has 10 binding arms for CD20 and one for CD3ε [228]. It will be interesting to follow this highly avid antibody (on the target side) to see how it ultimately compares with the 1:1 and 2:1 formats.

Significantly, antibody format and size, from small antibodies such as BiTE®s (ca. 24 kDa) to much larger formats such as the asymmetric IgGs (ca. 150 kDa), appear to have a lower differential effect on potency than the distance of the epitope to the membrane or affinity to the target antigen [17]. Additionally, binding geometry, which would include target epitope, antibody size and format, binding angle and perhaps other local factors, can influence the potency of the TRBA [17]. In an unpublished study carried out at Janssen R&D, several versions of a bispecific antibody (CD3ε arm)/centyrin (tumor antigen arm) combination were made and tested in vitro for tumor cell killing activity. As shown in Figure 8, the position of the tumor antigen binding arm on the molecule, that is, the geometry/distance of binding between the CD3ε arm and the tumor antigen arm, made a ca. 100-fold difference in in vitro killing activity.

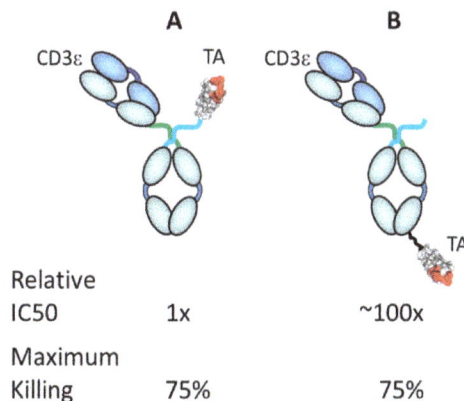

Figure 8. Effect of geometry on in vitro activity of an example TRBA. Cartoon of two antibody-centyrin fusions targeting CD3ε with the Fab arm and a solid tumor antigen with the centyrin (~12 kDa engineered FN3 domain). (**A**). Centyrin fused to the hinge, making it close to the anti-CD3 Fab arm; (**B**). Centyrin fused to the C-terminus of the heavy chain, making it distal from the Fab arm. All components, molecule sizes and in vitro killing assay conditions (E:T 5:1, 24 h assay) were identical. Thus, only the architecture and geometry of the two TRBAs are different, resulting in a ca. 100-fold difference in potency. This simple example demonstrates how important the geometry of the binding arms can be in the design of future TRBAs. Data presented were derived from experiments provided Steve Jacobs, Janssen R&D. These data were previously presented at PEGS Boston, April 2018 with permission.

Another area that has not yet been fully investigated with respect to T cell redirection is the role of Fc functionality. The Triomab® platform on which Removab® was designed has a highly active Fc domain that interacts with human FcγRs to increase the immune response [229,230]. It is generally accepted that the presence of an active Fc in a T cell redirecting bispecific antibody would increase the likelihood of pro-inflammatory cytokine release by T cells and other effector cells in the tumor microenvironment [229,230]. The release of these pro-inflammatory cytokines is thought to be part of the therapeutic mechanism of action of these antibodies [231], so while it is desired, it also needs to be

Antibodies **2019**, *8*, 41

controlled [232]. On the other hand, most of the current fragment-Fc, asymmetric IgG or appended IgG platforms have used muted or silenced Fcs so as not to over-stimulate the immune system via interactions with myeloid effector cells. Even with the absence of Fc activity, many treatments with T cell redirecting bispecific antibodies are accompanied by a cytokine release syndrome (CRS) that needs to be addressed as part of the therapeutic paradigm [232]. Thus, it seems likely that many, if not most, T cell redirecting antibodies made now and in the future, will likely continue to avoid Fc activity to limit the potential for immune-mediated toxicities.

4. Ex Vivo T-Cell–Bispecific Antibody Approaches

4.1. T Cells Armed Ex Vivo with Bispecific Antibody Conjugates

The tetravalent bispecific TRBAs described in Section 3.3.4. are not the only tetravalent platforms being tested. Based on protein conjugation approaches of Nisonoff and Rivers [233] and the ability to generate monoclonal antibodies via hybridoma technologies [31], several groups in the 1980s–1990s generated heterobispecific antibodies using conjugation methodologies [40,234,235]. The first clinical trial ever run with a TRBA was with an anti-CD3 mAb chemically conjugated to an anti-glioma antigen antibody [49] (Figure 2). More recently, chemical conjugation methods have been used to generate bispecific antibodies from existing antibodies for clinical studies. One method that has been widely used is to react the anti-CD3 antibody, OKT3, with Traut's reagent (2-iminothiolane HCl) and to treat the second antibody targeting cancer cells with sulfosuccinimidyl 4-(*N*-maleimidomethyl) cyclohexane-1-carboxylate (sulfo-SMCC) [236]. Mixed together at equimolar concentrations, these form 1:1 heterobispecific conjugates between the anti-CD3ε mAb, OKT3 and the targeting IgGs [236–238] (see Figure 5F). Based on pioneering efforts by Lawrence Lum and his colleagues at the Karmanos Cancer Institute (KCI), these and similar methods have been used to generate conjugated bispecific IgGs of several anti-tumor antibodies. These include clinical candidates such as anti-HER2 trastuzumab × OKT3 (Her2bi) [239] (clinical trial NCT03406858 [240]), anti-EGFR cetuximab × OKT3 (EGFR-bi) [237] (clinical trial NCT02620865 [241]), anti-GD2 3F8 × OKT3 (GD2bi) [238] (clinical trial NCT02173093 [242]) and anti-CD20 rituximab × OKT3 (called CD20bi) [243]. Note that all of these approaches utilize long-existing antibodies (e.g., trastuzumab, rituximab and cetuximab, antibody 3F8) which are chemically conjugated with the "original" anti-CD3ε antibody, OKT3 [244].

All four of these heterobispecific conjugates have been, or are currently being, evaluated in clinical trials by mixing them ex vivo with patient-derived leukapheresed T cells to "arm" and activate those T cells, followed by the administration of the armed and activated T cells autologously to the patients from which they were isolated [21]. Other than various academic efforts, very little has been done to advance chemically coupled bispecific antibodies. Newer technologies, however, may improve the coupling procedures to allow for greater efficiency and the potential for manufacturability. One such method recently described to generated chemically coupled bispecific antibodies with high efficiency was the use of sortase enzyme, which recognized the LPxTG peptide sequence, in conjunction with click chemistry [245]. This approach allowed for the high efficiency formation of a bispecific antibody comprised of two fully active, heterologous anti-influenza IgGs using the mild conditions of an aqueous environment at room temperature [245].

4.2. Cytokine-Induced Killer Cells

Another ex vivo approach combines the power of a subset of activated killer cells with the redirection provided by a bispecific antibody. In this case, $CD3^+CD56^+$ natural killer (NK) cells, often described as NKT cells [246], are isolated from a patient's PBMCs (i.e., autologous), activated ex vivo using anti-CD3 Mabs and cytokines, typically IL-1, IL-2 and IFN-γ and then combined with a bispecific antibody with one arm binding the CD3ε on the surface of the cytokine-induced killers (CIKs) and the other arm targeting a cancer cell surface protein (e.g., CD19, CD20, HER2, EGFR and so forth). This preparation is then administered to the patient to redirect the autologous CIKs to the targeted tumor. While several examples of this approach have been tried in clinical trials, currently the most advanced therapy is an anti-MUC1 × anti-CD3ε bispecific antibody that is dosed in conjunction with autologous CIKs at the Fuda Hospital in Guangzhou, China in collaboration with Benhealth Pharmaceutical Co., Ltd. (clinical trial NCT03554395 [247]).

CIKs are a particularly useful cell phenotype for immune-oncology uses. They possess properties of both T-cells and NK cells, the latter of which can kill target cells in an MHC-independent manner and are primed to kill upon target engagement without the need of additional signaling and stimulus [246]. Key molecular drivers of their cytotoxic activities are the NK receptors NKG2D, NKp30 and NKp46 [246]. In addition, their direct cytotoxic killing through cytotoxic granule secretion is accompanied by robust cytokine secretion of IFN-γ, TNF-α and IL-2 that further support the overall activated immune state in the tumor microenvironment [246].

5. Other Examples of Immune Cell Redirection

5.1. Early Immune Cell Redirection Efforts

As far back as ca. 2000, several preclinical and clinical efforts were made to use bispecific antibody approaches to redirect CD64 (FcγRI)-bearing immune cells (monocytes, macrophages, activated neutrophils) to kill cancer cells [248–250]. One of these, MDX-H210, an anti-CD64 × anti-HER2 bispecific antibody, reached Phase 2 clinical trials [248] but ultimately was discontinued. There are currently no CD64-based bispecific antibody redirection programs in clinical trials. Similarly, there were efforts to utilize bispecific antibody approaches by targeting the IgA receptor, CD89, to redirect activated neutrophils to kill cancer cells [251–254] and HIV [255]. As yet, none of these efforts has yet resulted in a clinical development candidate.

5.2. NK Cell Redirection, BiKEs and TriKEs

Immune cell redirection is not necessarily limited to T cells. There have been efforts since the 1990s to redirect other types of immune cells using targets such as NKG2D [256] and CD16a [257,258] or other surface markers on natural killer (NK) immune cells. NK cells are perhaps the most efficient immune cell killing machines available. There is a long history of therapeutic antibodies using NK cell-mediated antibody-dependent cellular cytotoxicity (ADCC) as a major mechanism for killing targeted cancer cells, including rituximab (Rituxan®), trastuzumab (Herceptin®) and cetuximab (Erbitux®), targeting CD20, Her2 and EGFR, respectively [259]. The composition and release of NK lytic granules filled with perforin and granzymes are comparable to those of the cytotoxic T cell subset but without as strong cytokine release [260]. Thus, there has been a recent push to use bispecific antibody technology to redirect CD16a$^+$ (FcγRIIIa) NK cells to kill cancer cells [169–172,261]. One of these molecules, AFM13, an anti-CD30 × anti-CD16a TandAb [262], is currently being studied in Phase 2 clinical trials [263] for the treatment of Hodgkin's Lymphoma.

One approach that has been revisited recently is to redirect natural killer (NK) cells to the cancer cells using scFv antibodies targeting CD16a (FcγRIIIa) fused via a short linker to another scFv targeting receptors on the cancer cells (e.g., CD19), similar to a BiTE® format. At least a few groups are renaming these bispecific antibodies "BiKEs," for "bispecific killer cell engager" [170,171]. In some cases, a third scFv against another cancer target (e.g., CD22) is added to increase the targeting ability [173]. These constructs are called "TriKEs," for "trispecific killer engagers." Another variation on this theme is to add immune cell-stimulatory cytokines, such as IL-15, in some "TRiKE"–like constructs, resulting in not only broader targeting with the two targeting scFvs but also NK and T cell activation via the activity of the stimulatory cytokine [264]. Other than AFM13 mentioned above, the only other NK cell redirecting antibody approaching clinical trials is the anti-CD33 × CD16a × modified IL-15 TriKE, OXS-3550 (aka GTB-3550; 161533), which is registered with ClinicalTrials.Gov but not yet recruiting [147].

5.3. Combining Engineered Cells with mAb Therapy

Recently, a few companies, namely Unum and Nantkwest, have generated recombinant, autologous or allogeneic NKT or T-cells expressing on their surface the high affinity allele (158V) of the CD16a receptor (FcγRIIIa) [25,265] normally expressed on NK cells. These cells are then used in conjunction with existing monoclonal antibody therapeutics that naturally bind FcγRIIIa via their Fc functionality [265]. The signaling domain for these constructs is the natural ITAM found in FcγRIIIa [265], different from most CAR-Ts today, which utilize CD3ζ as the primary signaling domain.

There are currently three of these types of constructs in clinical trials. The first is Nantkwest's haNK™, which is a CD16a (158V)/IL-2 expressing NK92 cell line (i.e., allogeneic) currently in Phase I clinical trials by itself [266] and, soon, in conjunction with the anti-PD-L1 Mab, avelumab [267] (NCT03853317; not yet recruiting). Unum Therapeutics has two clinical stage candidates, both of which are derived from patient's T cells (i.e., autologous approach). The first of these is ACTR707 [268], which is an autologous T cell construct expressing the high affinity allele (158V) of CD16a fused with the CD3ζ activation domain, dosed in combination with rituximab to target CD20$^+$ B cells, currently in Phase I [269]. The second is ACTR087 [270], an autologous T cell construct expressing the high affinity variant (158V) of CD16a fused with 4-1BB and CD3ζ activation domains. This construct is being dosed with either rituximab [271] or with Seattle Genetics' anti-BCMA antibody, SEA-BCMA [272], a humanized, non-fucosylated antibody with significantly increased affinity to CD16a over a normal IgG1 [273]. In all of these cases, the antibodies and recombinant cell are administered separately, so there is no need for formulating for combination therapy.

The potential issue with this approach is that the CD16a-positive T-cells will interact with all IgGs in circulation that can naturally interact with CD16a, including serum IgG1 and IgG3 isotypes which, when combined, make up approximately 10 mg/mL of CD16a-interacting "circulation matrix" IgG. A therapeutic IgG1, such as rituximab, typically is present in circulation at concentrations below 100 μg/mL, so it would comprise about 1% or even less, of the total molecules of antibody vying for binding to the CD16a on the surface of the engineered therapeutic T cells. On the other hand, antibodies engineered to possess a 10-fold (or more) higher affinity to CD16a, such as those with either low or no fucose glycans attached at N297 in the C_H2 domain [273], might be better candidates to be used in conjunction with the CD16a-modified NK/T-cell therapeutics. It is also possible for auto-antibodies present in the matrix of IgGs to interact with the CD16a engineered T cells and drive autoimmunity. Although typically these auto-antibodies would not cause problems as a result of the natural check-point systems in place with NK cell that express CD16a, the engineered T-cells may not be governed by those same checks [274].

5.4. Engineered T or NK Cells with Recombinant Target-Specific TCRs

Another approach to targeting tumor or viral antigens is to utilize the natural TCR machinery of αβT cells but with enhancements to increase the ability of the T cell to kill targeted cells. As differentiated from all of the other approaches in this paper, this strategy is both TCR and pMHC-dependent. Also, in this case, the antigens are typically intracellular and are displayed as neoantigen peptides on MHC I [275]. Different from CARs, the engineered TCRs replace the natural αV-βV of the TCR, they signal through the ITAMs of the TCR complex and their target is the pMHC on the surface of the targeted cells (see Figure 1F). The potential upside of this approach is that the regulatory apparatus of the TCR is in place. The downsides, however, include the possible loss of target with MHC loss or down-regulation on cancer cells as noted earlier [6,12], HLA restriction and the requirement for HLA matching and the effects of a repressive tumor environment, including T cell checkpoints [275]. Nevertheless, there are at least 20 clinical candidates in which autologous T cells from patients have been collected, engineered ex vivo with modified TCRs and then re-administered to the patients [275]. Cancer targets for these clinical stage–engineered TCR T cells include: preferentially expressed antigen in melanoma (PRAME), New York esophageal squamous cell carcinoma 1 (NY-ESO), melanoma-associated antigen (MAGE) A4, MAGE A3/A6, MAGE A10 and alpha-fetoprotein (AFP) [276].

The candidates noted above are all autologous. A different approach using engineered TCRs is the introduction of the TCR machinery in the stable NK-92 cell line to produce an allogeneic cell line that could be used as an off-the-shelf NK-TCR product [276,277]. While this approach is very attractive, it is still in the early stages of development.

6. Chimeric Antigen Receptor (CAR)-T and NK Cells

6.1. Introduction

There are at least 225 unique, documented CAR constructs currently being tested in clinical trials. Over 90% of the current clinical stage CAR-based cell therapies are autologous CAR-T cell products generated from αβ-T cells (Table 3). Of these, 176 are single CAR constructs per cell, dosed alone. Another 30 clinical candidate CARs either have multiple CARs targeting different tumor antigens per cell or are multiple CAR-T cell lines targeting different tumor antigens and mixed for dosing. One clinical stage autologous CAR construct is made using NKT cells. Finally, 14 CAR constructs in clinical trials are allogeneic, 11 of which are from T cells and 3 from NK or NKT cells (Table 3).

Table 3. CAR-T formats *.

CAR-T Format	Clinical Stage			Total
	Phase I/II	Phase III	Approved	
Autologous CAR-T **—Single CAR	176	2	2	180
Autologous CAR-T **—Multiple CARs for different targets or multiple CAR-Ts dosed in combination	30	0	0	30
Autologous CAR-NKT	1	0	0	1
Allogeneic CAR-T	11	0	0	11
Allogeneic CAR-NK or -NKT	3	0	0	3
Total	221	2	2	225

* BiStro Biotech Consulting LLC database, locked 20 June 2019. Data obtained from Clinicaltrials.gov, literature papers, company websites, analyst reports and other sources. ** As far as can be ascertained from public documents, all of these appear to be based on αβ-T cells.

As mentioned previously, two CAR-T cell products, Kymriah® (Tisagenlecleucel-T; aka CTL019, CART-19) [104] and Yescarta® (Axicabtagene ciloleucel; aka KTE-C19) [105], have been approved by the US-FDA, both in 2017. Both of the approved CAR-T products target CD19, with Kymriah® approved for treatment of B-cell ALL and, more recently, DLBCL and Yescarta® approved for treatment of DLBCL. The most advanced clinical candidates not yet approved are bluebird bb2121 [278] and JCAR017 [lisocabtagene maraleucel or liso-cel] [279], both in Phase III.

Nearly half (106/225) of the clinical stage CAR-T clinical candidates have originated in China, with the remainder originating in the US (96) or rest-of-the-world (23) (WR Strohl, BiStro Biotech Consulting Antibody and CAR-T Database, last updated 20 June 2019). Interestingly, over half (125/225, ~56%) of current CAR-based clinical trials are sponsored and run by academic groups, government institutes or hospitals without apparent industry funding; the remainder (100/235, ~44%) are either sponsored by biotechnology or pharmaceutical companies or are collaborations between medical or academic institutions and industry (WR Strohl, BiStro Biotech Consulting Antibody and CAR-T Database, last updated 20 June 2019). The fact that over half of the clinical trials are investigator-driven is significant, as it demonstrates the leading role that academic (e.g., University of Pennsylvania), medical/institutional (e.g., MD Anderson Cancer Center, Fred Hutchinson Cancer Research Center, Memorial Sloan Kettering Cancer Center) and government (e.g., National Institutes of Health) investigators have taken in moving CAR-T therapies forward from an idea to a new and powerful mode of cancer therapy.

6.2. Autologous CARs

We have briefly described the first, second and third generation of CAR-T cells (Section 1.3; Figures 2 and 3) and there are many reviews of the autologous CAR-T process [22–24,280–283], so we will only summarize the details of a few leading CAR-Ts here. Most of the current clinical stage CAR-based cell therapies are autologous CAR-T cell products (Table 3), that is, personalized products that are derived from a patient, activated (e.g., typically using CD3/CD28 beads), genetically modified and expanded ex vivo and then reinfused back into that same patient [282,283]. Vectors used to transduce the T cells include lentivirus [284], replication-defective gamma retrovirus [285] or transposons such as PiggyBac [286] or Sleeping Beauty [287]. Table 4 compares the salient features of the two approved autologous CAR-Ts targeting CD19 as well as the two Phase III autologous CAR-T therapeutic candidates, JCAR-017 [279]—targeting CD19 for treatment of NHL and bb2121 [278]—targeting BCMA for treatment of multiple myeloma. Key features of these CAR-Ts are the source of the cells (patient PBMCs vs defined CD4/CD8 populations), parameters for ex vivo stimulation, vector used to transduce the T cells with the CAR (lentivirus or retrovirus), hinge and transmembrane domains, costimulatory domains and the common use of CD3ζ signaling domain (Table 4). Where known, production from leukapheresis to re-administration to patient took from 10 to 28 days, depending on the candidate and all of these candidates required lymphodepletion with fludarabine and cyclophosphamide treatment, typically two-to-three days prior to administration of the CAR-T therapeutic.

Although the currently approved products are non-standardized mixtures of CD4 and CD8 T cells, studies are on-going to determine the optimal proportion of each cell type [288]. In addition, since these products are not nearly 100% CAR positive, as gene modification efficiencies vary from vector to vector and T cell pool-to-T-cell pool [280], these CAR-T cell products are typically dosed in terms of a standardized number of CAR$^+$ cells (Table 4). One further step in the direction of standardization and product homogeneity is JCAR017 (lisocabtagene maraleucel or liso-cel), which is dosed using a fixed ratio of CD4$^+$ and CD8$^+$ T cells [279,289].

Table 4. Brief comparison of autologous CAR-T constructs.

Property	CAR-T Constructs			
	Kymriah® (Tisagenlecleucel-T; CTL019)	Yescarta® (Axicabtagene Ciloleucel; KTE-C19)	Lisocaptagene Maraleucel (Liso-cel, JCAR-017)	bb2121
Sponsor	Novartis	Gilead (Kite)	Celgene	Celgene/bluebird
KEGG Number #	D11386	D11144	Na	Na
Clinical stage	Approved by USFDA	Approved by USFDA	Phase III (NCT03575351)	Phase III (NCT03651128)
Base cost (US)	$475,000 for B-ALL; $373,000 for R/R DLBCL	$373,000	Na	Na
Indication	B-ALL, R/R DLBCL	R/R DLBCL; PMBCL	R/R DLBCL; CLL	MM
T-cell source	Patient PBMCs; autologous; unspecified	Patient PBMCs; autologous; unspecified	Patient CD4 and CD8 T cells 1:1 ratio; autologous	Patient PBMCs; autologous
Vector	Lentivirus	Retrovirus	Lentivirus	Lentivirus
Antibody	Anti-CD19 mouse scFv FMC63	Anti-CD19 mouse scFv FMC63	Anti-CD19 mouse scFv FMC63	Anti-BCMA
Costimulatory domain	4-1BB	CD28	4-1BB	4-1BB
Signaling domain	CD3ε	CD3ε	CD3ε	CD3ε
Hinge and transmembrane	CD8α	IgG1 Fc	IgG4 Fc spacer; CD28tm	CD8α
Other Markers	Nk	nk	EGFRt	nk
Ex vivo activation	CD3, CD28	CD3, IL-2	nk	CD3, CD28
Lymphodepletion	Yes	yes	yes	yes
Time from leukapheresis to infusion	21–28 days	17 days	nk	10 days

Table 4. *Cont.*

	CAR-T Constructs			
Property	Kymriah® (Tisagenlecleucel-T; CTL019)	Yescarta® (Axicabtagene Ciloleucel; KTE-C19)	Lisocaptagene Maraleucel (Liso-cel, JCAR-017)	bb2121
Dose	0.2 to 5×10^6 CAR-positive viable T cells/kg	$0.4–2 \times 10^6$ anti-CD19 CAR-positive viable T cells/kg	5×10^7 CD8$^+$ and 5×10^7 CD4$^+$ CAR-positive (not weight based)	$50–800 \times 10^6$ CAR-positive T cells (not weight based)

Abbreviations: B-ALL, B-cell acute lymphoblastic leukemia; BCMA, B-cell maturation antigen; DLBCL, diffuse large B-cell lymphoma; CAR, chimeric antigen receptor; CD, cluster of differentiation; CLL, chronic lymphocytic leukemia; MM, multiple myeloma; Na, not applicable; nk, not known to these authors; PBMC, peripheral blood mononuclear cells; PMBCL, primary mediastinal large B-cell lymphoma; R/R, relapsed or refractory. #: Number. References used to build this table: [278–282,290–295].

6.3. Allogeneic CARs

There are at least 14 allogeneic CAR-NK, CAR-NKT or CAR-T cells in clinical trials (Table 3). These are all constructed uniquely and have a wide range of properties based on the host cell (NK, NKT or T cell), the genes knocked out and how the CARs are made. For allogeneic CAR-Ts, either primary cells from a healthy subject or stable cell lines from a source other than the treated patient are used.

These allogeneic cell products will not naturally be histocompatible and may either be rejected by the patient's immune system or cause graft versus host disease (GVHD) in the case of T cell products. To prevent rejection, the major source of histocompatibility mismatch, the class I HLA surface proteins, can be deleted by genetic engineering [20,296,297]. To prevent GVHD, cells that are naturally devoid of a TCR (e.g., NK cells) or T cells in which the TCR has been deleted by genetic engineering can be used [298]. These cells or cell lines are then further engineered to express a CAR, as with the autologous products described in Section 1.3 and above but can now be given to any patient, representing a truly universal cell therapy. First generation allogeneic cell products were derived from NK cells and lacked the potential to elicit GVHD [299]. However, such products would eventually be rejected by the adaptive immune system (T and B cells), limiting their utility.

The current wave of allogeneic CAR-T cells is perhaps best described as "allogeneic, transiently engrafted T-cell therapeutics." To give a general sense of how these might be made, the process developed by Cellectis is briefly reviewed here. Allogeneic cells are derived from health donors, are typically engineered to remove the TCR function (for example, by knockout of *TRAC*, eliminating TCR-α) and the pan-lymphocyte maker, CD52 (target of alemtuzumab) but are not made HLA Class I deficient (Figure 9) [300,301]. Prior to treatment, the patients are lymphodepleted using alemtuzumab and then the allogeneic cells, which are resistant to alemtuzumab via CD52 knockout, are dosed. Over time, the patients' immune systems will naturally be restored and with that, the HLA$^+$ allogeneic cells will be eliminated [300]. The manufacturing method for BCMA-targeted, allogeneic CAR-T cells was described as an 18-day process including apheresis from donors, activation, transduction, TALEN gene editing, expansion, removal of TCRα$^+$ cells and finally storage [300].

An additional level of safety has been incorporated into the Cellectis allogeneic CAR-T cells. The RQR8 polypeptide, which contains two anti-CD20 rituximab epitopes and a single anti-CD34 Qbend10 epitope [302] has been incorporated into the hinge portion of the CAR, which can be used as a target for rituximab killing of the allogeneic CAR-T (Figure 9) [301,303]. While this first generation of allogeneic CAR-Ts is perhaps not a truly "off-the-shelf" product manufactured from stable cell lines like a biologic drug, it is a significant step in that direction.

Next generation allogeneic products utilizing state-of-the-art genetic engineering technologies in which both the class I HLA and TCRs have been deleted are approaching the clinic. Because these cells will not likely be recognized by the immune system, it will be imperative to include safety switches or even multiple orthogonal kill switches, in such products. It remains to be seen whether these gene edited allogeneic products will be as effective as autologous products and whether they represent a commercially viable source of engineered cell products [304].

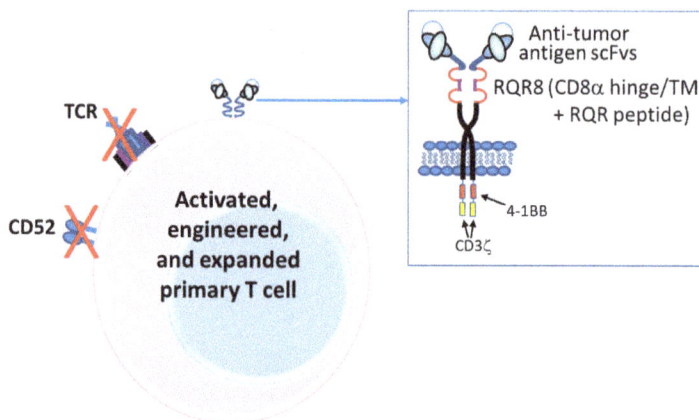

Figure 9. Drawing depicting the construction of an example allogeneic CAR-T cell. The CAR contains, from N-to-C terminus, anti-BCMA scFv, 136 amino acid residue RQR peptide, CD8α hinge and transmembrane (TM) domain, 4-1BB costimulatory domain and CDζ signaling domain in a cell in which TCR- α and CD52 have been knocked out using gene editing technology [301].

6.4. Alternative Cell Types for CAR Expression

T-cells are not the only cell type that is being explored for engineered immune cell therapies. Natural killer cells (NK), natural killer T (NKT) cells and even macrophages expressing CARs are all being tested in the clinic [305]. Like T cells, NK and NKT cells share many of the killer functions of T cells but have differences that can be either an advantage or disadvantage, depending on the situation. NK cells are derived from the same immunological precursor cells as T cells and, therefore, have many of the same activities [306,307]. The most obvious difference is that NK cells do not possess a TCR. As a result, it is possible to use NK cells in an allogeneic setting, as the risk of GVHD associated with a mismatched donor TCR does not exist [299]. However, NK cells, unlike T cells, do not robustly expand and do not persist in vivo. In scenarios where the tumor target antigen is expressed at a low level on normal tissues, this could be an advantage, as the product would have a limited life span and penetrance *in vivo*. However, the current dogma is that persistence is important to maintain good responses and long-term remissions, which is not afforded by NK cell-based products [299,306,307]. NKT cells, on the other hand, share properties of both NK cell and T cells. Although NKT cells will not persist like T cells, they do undergo expansion in vivo [308]. Invariant NKT (iNKT) cells—A subclass of NKT cells—Express a TCR that does not recognize classical MHC I presented antigen but, instead, recognizes non-classical MHC I presented antigens that are mainly confined to infectious agents. Therefore, iNKTs are not expected to elicit a GVH response, similar to true NK cells.

Gamma/delta (γ/δ) T cells are very similar in surface marker expression and function to NKT cells in that the γ/δ TCR does not recognize classic MHC I-peptide complexes and are also thought to be useful in an allogeneic setting [309]. There do not appear to be any γ/δ-CAR-T cells currently in clinical trials, although there is one registered clinical trial for collecting γ/δ-CAR-T cells as a feasibility study for constructing CD33-CD28 γ/δ-CAR-T cells [310]. The company involved with that trial, TC Biopharm, also claims to have a non-recombinant allogeneic γ/δ-T cell product in Phase I clinical trials for AML (no NCT provided). Additionally, other companies such as Gammadelta Therapeutics and Gadeta are also working with γ/δ-T cell therapeutics and Lava Therapeutics is making TRBAs that recognize γ/δ-T cells instead of αβ-T cells, so it is expected that TRBA and/or CAR-T products utilizing the unique biology of γ/δ-T cells will soon be tested in clinical studies. CAR constructs employed with NK, NKT, iNKT and γ/δ T-cells utilize the same signaling molecules as αβ TCR T-cell CARs [311].

Macrophages represent the newest cell type being engineered for cell therapies. Unlike lymphocyte-derived cell lineages, monocyte-derived lineage cells like macrophages function mainly as professional antigen presenters and in the removal of invaders through phagocytosis [312]. Engineering macrophages with CARs that signal through a FcγRIIIa (CD16a) receptor intracellular domain imparts on the cell the ability to recognize, phagocytose and then present tumor cell antigens to other T cells to further amplify the immune response. Although not efficient killers, engineered macrophages have the ability to broaden the immune response to the tumor cells. Carisma Therapeutics is on track to reach the clinic with their first CAR-macrophage in early 2020.

A key shortcoming of many of these alternative cell products is the quantity of each cell type present in normal, let alone compromised, patients. The ability to generate a sufficient amount of product for each patient will have to be significantly improved [313].

6.5. CAR Designs

6.5.1. scFvs

The first CAR-T cells were nearly universally constructed using mouse-derived scFvs [76,281]. In fact, both of the approved CAR-T constructs, Kymriah® (Tisagenlecleucel-T; aka CTL019, CART-19) [104] and Yescarta® (Axicabtagene ciloleucel; aka KTE-C19) [105], as well as the Phase III candidate, JCAR017 (Table 4) and many of the other 86 CD19-targeted CARs in clinical studies, utilize the same mouse-derived anti-CD19 scFv, FMC63 [281]. One of the first CARs to be made with a humanized scFv was a CAR-T targeting HER2, which utilized an scFv derived from trastuzumab [314]. Only in the past several years have the antibodies for incorporation into CARs been constructed using humanized or human scFvs, a practice that should grow significantly.

The selection and engineering of antibodies for scFvs is actually much more important than many CAR-T engineers have historically perceived. Many of today's CARs were constructed with antibodies that were retrieved from the freezer, the literature or from academic colleagues. Most of those antibodies were not designed or engineered for stability, lack of aggregation or optimized folding as scFvs or even, in many cases as noted above, for humanization. The reformatting of an IgG antibody into an scFv, although currently reduced to practice, can result in a less stable domain compared to the antibody from which it was derived [315]. As noted earlier, scFvs are notorious for their instability, unfolding, domain swapping and aggregation [220]. The consequence of this loss of stability can have functional and toxic consequences, as a poorly stable binding domains can aggregate and lose binding capability when on the cells surface or cause non-target-specific, tonic signaling [316]. Thus, scFv stability, as well as human-ness, epitope targeting and affinity, are key issues that need to be addressed in the design of future scFv-derived CARs.

6.5.2. Domain Antibodies and Alternative Scaffolds

As with the extensive protein engineering that has helped evolve the TRBAs, creative formatting of CARs and their uses in CAR-T cells has helped expand the capabilities of functional control of CAR T-cell products going into clinical development [317]. Most current CARs utilize scFv-formatted antibody fragments for tumor targeting. Recently, alternatives to scFv formats have been explored, some making it into full clinical development. Humanized llama or camelid V_{HH} fragments or single domain human antibody fragments, that can be engineered to possess all of the binding selectivity and specificity of full V_H/V_L dual-domain antibody fragments, are becoming more frequently used for tumor targeting [318,319]. V_{HH} domains possess the added benefit of simplicity afforded by the single domain, as compared to the two domains separated by a linker of the more typical scFv.

In addition to single domains, others have been pursuing alternative scaffolds based on fibronectin repeats (e.g., Adnectins or Centyrins) [320] and ankyrin repeats (DARPINs) [321,322]. Similar in their simplicity, these alternative scaffolds possess many of the binding properties of antibodies and antibody fragments. Human and humanized versions of these scaffolds have lessened some of the anxiety associated with their use in clinical settings, although with the exception of Centyrins [323], none have been tested in the clinic. The potential to use additional scaffolds, like Centyrins or Anticalins, is only limited to the imagination, as clinical experiences with other protein-based scaffolds are starting to demonstrate their additional potential use in CAR-T programs.

6.5.3. Multiple CAR Designs

In their simplest forms, CARs target only one tumor associated antigen. However, multiple targeting domains can be linked together on one CAR to generate multi-specific CARs. Typically separated by the common G_4S linker that is used for scFv engineering, "beads-on-a-string" designs linking scFv or single domain binding elements have produced CARs that can be activated to kill target cells that express more than one antigen or bind to different epitopes on the same antigen [324,325]. For example, Janssen R&D and Nanjing Legend Biotech have a dual-BCMA-epitope-binding CAR-T product, LCAR-B38M, in Phase 2 clinical development for treatment of multiple myeloma [326,327]. Perhaps more exciting are multiple CARs on the same cell in which each CAR possesses only one signaling domain. CARs that provide signal 1 through a CD3ζ ITAM are not very effective killers and require a second signal provided by the intracellular domains of 4-1BB or CD28. These signaling domains can be split between two CARs, each with a different binding specificity, producing logic-based CAR T-cells. These so-called "and" (target A and target B) CARs only produce robust target cell killing if both target antigens are engaged on the surface of the target cell [328–330]. This strategy and other related logic-based designs, for example, the "but not" CAR (target A but not target B) and the syn/notch CAR, are being explored to address the complex nature of solid tumors and to generate safer products [331,332].

The sophisticated nature of the TCR complex has been shown to be a key mechanism involved in the many biological facets of T cell biology. Attempts to mimic this complex structure using CARs have proven to be difficult [112]. More recently, additional protein engineering designs have attempted to improve on the original CAR concepts. The modular nature of the various Ig-domains that encompass the TCR α/β, CD3δε and CD3γε domains and CD4/8 chains has been open to interesting engineering. The biotechnology company, TCR2, is exploring CD3ε-scFv fusions (Figure 10). Utilizing their TruC platform, the tumor specific binding scFv is fused to the extracellular domain of CD3ε, allowing for complex activation upon scFv-target binding, bypassing TCRα/β-MHC interactions [333]. Similarly, Triumvira has fused two scFvs in place of the MHC I binding head domain of CD8. The more membrane proximal scFv binds specifically to one of the CD3 extracellular domains and the more distal scFv binds to tumor antigens (Figure 10). In Triumvira's so-called TAC technology, this design essentially locks down the tumor antigen binding to the entire TCR complex structure as a result of the multiple ionic associations of all the players, while simultaneously including the CD8 activities as well [334]. One final spin on this theme is Eureka's Artemis platform that replaces TCR αV and βV regions with antibody V_H/V_L regions, similar to the very first pre-CAR-like constructs made in 1989 [35]. In this format, antibody specificity is dialed into the TCR complex and can either operate independent of MHC presentation, or, if the antibody variable regions were raised against pMHC, retain the class I restriction, mimicking a true TCR [335]. All of the variations have one primary goal in mind, that is, to more closely resemble the intricate control associated with the natural TCR complex. Future studies will need to be performed to confirm the degree to which these and most certainly others, have achieved that goal.

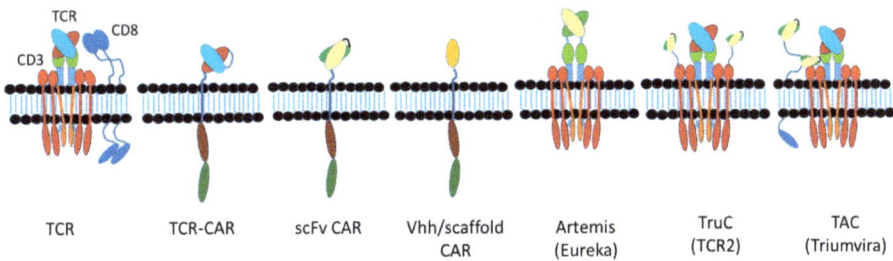

Figure 10. Schematic representation of various CAR formats. Left, natural organization of a CD8$^+$ TCR expressed on cytotoxic T cells driving MHC-restricted CD8$^+$ T cell effects. TCR-CARs are formed by fusing the TCR α/β variable domains to a second-generation CAR scaffold. The Eureka Therapeutics Artemis platform fuses antibody variable domains to TCR α/β constant regions. TCR2 platform fuses scFvs to CD3ϵ. Triumvira platform fusing two scFv chains to CD8 stalk, hinge and intracellular domains-one scFv binds to CD3ϵ engaging the rest of the TCR complex, while the second is free to interact with tumor antigens.

6.6. Additional Enhancements for Tuning CAR-T Cells

In the past few years, CAR-T cells have been made with additional enhancements, either to modulate the immune system, to help the T-cell home to the tumor or to increase the safety of the CAR-T therapeutic [336–338]. As noted in Section 1.3, second or third generation CAR-Ts (see Figure 3) to which these enhancements are made are sometimes called "fourth generation CAR-Ts" [338–340]. This section describes a few of these "add-ons," some of which have already become incorporated into clinical candidate CARs.

6.6.1. Safety Switches

As mentioned in Section 6.3, when delivering CAR-T cells that may have potential safety issues, for example, allogeneic CAR-Ts, highly persistent CAR-Ts, CAR-Ts expressing cytokines or CAR-Ts with the capability of generating a strong CRS, it is important for safety reasons to have a method for either killing the CAR-T cells or turning off their function. There have been multiple approaches to this issue, including a variety of different types of kill switches, as well as the application of adapter molecules that link the CAR-T cells to the tumor antigens. Several different types of kill switches have been developed for CAR-T cells, a few of which will be described here. One of the oldest switches is the use of herpes simplex virus thymidine kinase (HSV-TK), which is sensitive to the drug gangcyclovir [341]. Treatment of cells containing HSV-TK with gangcyclovir, however, is quite slow, taking about 72 h to be fully effective and the viral gene product itself can lead to immunogenicity via MHC I presentation, so this approach is not considered to be particularly attractive [342]. A more recent and now more widely used "kill switch" is CaspaCIDE®, which couples an inducible caspase-9 (iCasp9) with a small molecule inducer known as AP1903 [98,342,343]. The effect of AP1903 treatment in vivo on cells containing iCasp9 is quick, with 90% loss of iCasp9$^+$ cells within 30 min and full 100% effect within 24 h [342]. The iCasp9 kill switch system has been used in CAR-Ts cells since at least 2010 [344] and it is being employed more widely every year since. There are currently at least 11 unique CAR-T cell constructs being tested in clinical trials today that employ the iCasp9/AP1903 system as a safety switch (WR Strohl, BiStro Biotech Consulting Antibody and CAR-T Database, last updated 20 June 2019). A second small molecule kill switch based on caspase-9 is the very recently developed rapamycin-induced caspase-9 dimerization safety switch (iRC9) [345].

Conversely, a different approach to regulate the activity and thus, the safety, of CAR-T cells is to provide the CAR-T system with an inducible "only on" switch, which would rely on the constant presence of a theoretically innocuous small molecule to keep the expression of the CAR gene cassette going. One of these inducible "on" systems is the rimiducid-inducible MyD88/CD24-CAR (iMC-CAR) system [345], which give "on demand" rimiducid-dependent co-stimulation, which enhances proliferation and activation. The CAR, however, is not fully activated until binding of the targeting scFv to the tumor cell surface antigen triggers CD3ε signaling as well [345]. Similarly, there are multiple versions of a tetracycline (doxycycline)-inducible (tet-inducible) CAR, in which CAR expression is completely dependent on the presence of the tetracycline [346,347]. The tet-inducible system, however, appears to be leaky with a significant background expression level in the absence of the inducer [347].

Another effective and popular approach to kill CAR-T cells, when required, is the use of commercially available, approved antibodies targeting tumor targets such as EGFR and CD20. As mentioned in Section 6.3 (see also Figure 9), the RQR8 epitope was constructed as a target for the anti-CD20 mAb rituximab [302] and has been employed in allogeneic cells as a cell-surface kill switch [301,303]. Additionally, a truncated version of epidermal growth factor receptor (EGFRt), which encompasses residues 334–668 of mature EGFR (domains III and IV), has been fused to the GM-CSF leader peptide to make a surface expressed "tag" for CAR-T cells [348–350]. This EGFRt tag is recognized by the approved drug cetuximab (Erbitux®), which can then be used as a kill switch to eliminate cells expressing the tag [348–350]. There are currently at least 12 unique CAR-T cell constructs in clinical trials that have incorporated the EGFRt tag (WR Strohl, BiStro Biotech Consulting Antibody and CAR-T Database, last updated 20 June 2019). This tag can be used not only as a kill switch but also as a marker for sorting and tracking EGFRt-positive cells [348,349].

6.6.2. Adapters

A completely orthogonal way to control the activity of CAR-T cells is to have CAR-T cells that recognize a molecule which is fused to the targeting antibody, that is, an "adapter molecule," rather than to the targeted antigen itself. In this manner, CAR-T activity only occurs when the CAR-T, the adapter and the targeted antigen are present together [351,352]. Many adapter systems have been described in the literature, including those that use FITC, GCN4 or biotin as the adapter molecule. In these cases, the adapter is conjugated or fused to a tumor antigen-binding scFv (or other binding moiety) and the CAR recognizes the adapter molecule. In this manner, the activity of the CAR-T is controlled by availability of the adapter molecule. A potential significant upside for this kind of molecule could be for use in the allogeneic setting. If one could construct a stable CAR+ cell line that targets an adapter, then this could potentially be a universal, allogeneic CAR-T that could be made specific for new targets simply by changing out the scFv that is coupled to the adapter [351,352]. A good example of this is the peptide-specific switchable CAR (sCAR), recently developed, which specifically recognizes a 14-amino acid residue peptide derived from yeast transcription factor GCN4 [353]. The peptide was fused to a Fab targeting CD19 to target B cells and then fused to a CD20 Fab to demonstrate that it too would work similarly [353]. This sCAR platform is an example of how a universal adapter CAR-T system might work.

In a sense, the NK and T cells expressing CD16a [25,265,268,270], described in Section 5.3, would be similar to universal adapter CARs because the only change needed to make a new "drug" would be a different, already approved (or even one in development), targeting antibody [351].

6.6.3. Homing Receptors

At least three critical factors are required to construct a CAR-T cell to treat solid tumors. The first is having the proper tumor antigen target, the second is efficient homing of the CAR-T cells to the tumors and then penetration of the targeting killer cells into the tumors and, finally, the third is a successful defense against the immunosuppressive environment within the tumor microenvironments [354]. These next two sections will cover the latter two of these requirements.

T cells traffic to different tissues based on a wide variety of different homing signals [337,355]. For example, the receptor/ligand pair CXCR4/CXCL12 will help to target T cells to the gut, lung and bone marrow, whereas CCR4/CCL17 will help to home T cells to skin, lung and heart [355]. It is the combination of the proper tissue-based signals and their cognate homing receptors that help home T cells to the proper tissues under the appropriate conditions. While this is a simplified view of very complex biology that requires cell-cell interactions, chemokine gradients and immune signal-mediated upregulation of chemokine receptors, it nevertheless illustrates the importance of specific chemokines and chemokine receptors for homing T cells to target tissues. Most autologous CAR-T cell products currently suffer from the fact that they are very heterogenous, are derived from peripheral blood rather than tumor tissues and are deficient in many or all of the homing receptors required to traffic them to cancerous tissues. To improve the ability of T cells to home to and then penetrate, solid tumors, the proper receptors need to be present; if they are naturally lacking, then they will need to be engineered into the cells.

There are now several examples of attempts to clone chemokine receptors into T cells to improve the ability of those cells to traffic to and penetrate, solid tumor tissues. One of the keys to this approach is understanding which tumors overexpress which chemokines, so that the proper match can be made for each tumor. Chheda et al. [356] demonstrated that the knockout of CXCR3 and leukotriene B4 receptor (BLT1) in CTLs abrogated their ability to home to tumors in mice, which correlated with a loss of efficacy, demonstrating the criticality of those two chemoattractant receptors in CTL migration to the tumors. In a separate study, cultured NK cells were found to have lost the ability to express CXCR2, which led to a defect in trafficking to RCC tumors [357]. When enforced CXCR2 expression was reinstalled into those cells via genetic manipulation, trafficking to the RCC tumors as restored, with a concomitant improvement in tumor cell killing [357].

In an early attempt to manipulate CAR-T cells to home to tumors, Moon et al. [100] transduced mesothelin-targeting CAR-Ts with CCR2, which resulted in more than a 12-fold increase in trafficking of those CAR-T cells to the mesothelin-expressing tumors in a mouse model [100]. Similarly, Siddiqui et al. [358] demonstrated that transduction of T cells with CXCR1 significantly improved the homing of those T cells to tumors expressing CX3CL1 in mice, with concomitant improvement in tumor suppression. In a separate study, both mouse and human pancreatic cancers were demonstrated to over-express the chemokine, CCL22 [359]. Transduction of T cells with CCR4, the receptor for CCL22, led to improved interaction of the T cells with dendritic cells (DCs), increased T cell activation and improved T cell tumor penetration in a mouse model [359], suggesting that addition of this chemokine receptor to CAR-T cells intended for treatment of pancreatic cancer might be fruitful.

Perhaps one of the most critical findings was the apparent requirement for the expression of CXCR3 for extravasation of T cells from the vasculature into the tumor in a mouse model [360]. Neither CCR2 nor CCR5 could fulfil that role, indicating the likely requirement for transduction of multiple chemokine receptors into CAR-Ts intended to treat solid tumors, each with its own function along the path from circulation to tumor tissue. Additionally, these studies exemplify the need to understand which chemokine attractants are expressed in which tumors under what conditions. This information will help to design and construct "smart" CAR-Ts for targeting solid tumors.

6.6.4. Counteracting PD-1/PD-L1-Based Immunosuppression

Not only are trafficking to the site of a tumor and tumor penetration critical to the success of CAR-Ts for solid tumors but also dealing with the typically strong immunosuppressive tumor microenvironment (TME) [111,337,338]. The immune-repressive environments especially found in the TME of solid tumors is clearly a potential issue for causing immune suppression of both CAR-T cells and TRBAs. It is known that PD-1 interaction with PD-L1 has the ability to override the activating signals from CD28, providing a dominant immunosuppressive effect in the TME [361]. To this point, it has also been demonstrated that anti-PD-1 antibodies can rescue the CD28 function in animal models [138]. Thus, exogenous anti-checkpoint target antibodies are being tested in clinical trials to counteract the suppressive immune environment for either TRBA therapy [139–141] or CAR-T therapy [142].

With CAR-T constructs, there are additional opportunities to address the PD-1/PD-L1-based immunosuppression by modification of the CAR-T cell itself. There are multiple approaches to this concept. First, there are several reports of CAR-Ts in which either PD-1 or PD-L1 were knocked out [146,362–365] or modified to generate of a dominant-negative PD-1 lacking the signaling domain [138]. In a slightly different twist, Liu et al. [366] described the construction of a "PD1CD28" CAR-T cell in which the extracellular domain of PD-1 was fused to the intracellular activating domain of CD28. With these cells, the presence of PD-1 in the TME would act as a stimulatory signal through the CD28 signaling domain [366]. Finally, CAR-T cells have been engineered not only with the CARs but also with the ability to express and secrete into the TME either anti-PD-1 antibodies [144,145], anti-PD-L1 antibodies [143], anti-CTLA4 antibodies [145] or decoys to block PD-1/PD-L1 interaction such as the C_H3-PD-1 fusion [367].

In a different twist to dealing with the PD-1/PD-L1 immunosuppressive axis, Xie et al. [364] generated CAR-T cells that target PD-L1 in the tumor microenvironment. In this way, they not only were targeting PD-L1-positive cancer cells but also stromal cells in the TME that help provide for the immunosuppressive environment. In targeting PD-L1, Xie et al. [364] found that CAR-T cells themselves produced a low level of PD-L1, so knock-outs of endogenous CAR-T PD-L1 were made that were significantly better than wild-types. The PD-L1 CAR-T cells reduced tumor growth and increased survival in an animal model. A clinical candidate CAR-T cell targeting PD-L1 has been registered but not yet recruiting patients, for the treatment of non-small cell lung cancer (NSCLC) [368].

6.6.5. Cytokine-Expressing CAR-Ts

Another approach to dealing with the immunosuppressive environment of solid tumors is the generation of CAR-T cells that produce T cell stimulatory cytokines themselves [111,338]. A variety of cytokine-expressing and secreting CARs have been made and tested preclinically [340]. These constructs have been given names such as "armored CARs" [340] or TRUCKs (i.e., T cells redirected for universal cytokine killing) [99,339,369]. Koneru et al. [370] reported the generation of CAR-T cells targeting MUC-16^{ecto} which also expressed and secreted IL-12 to the TME, which now has been taken into Phase I clinical trials [370] where it is being delivered directly into the tumor [371].

7. Targets for Clinical Stage TRBAs and CAR-T Cells

As shown in Table 1, there are currently 289 unique TRBAs and CARs being tested in clinical trials today, targeting a total of 53 unique targets (BiStro Biotech Database, last updated 20 June 2019). Currently, the 61 unique TRBAs in clinical trials target 31 different antigens, with seven candidates targeting BCMA, six candidates targeting CD33, five candidates each targeting CD20 and CD123 and four candidates targeting PSMA (Table 5). A total of 18 different targets are currently being targeted each by a single known TRBA clinical candidate. Of the 61 clinical stage TRBAs, 31 (~51%) primarily target heme malignancies, 25 (~41%) primarily target non-central nervous system (non-CNS) based solid tumors, four target neurological tumors and one targets human immunodeficiency virus (HIV). There are only two clinical-stage TRBAs currently targeting CD19, blinatumomab, which has been approved under the trade name Blincyto® for treatment of B cell acute lymphoblastic leukemia (B-ALL) and AMG 562, a half-life extended BiTE® construct in Phase I clinical trials [372].

On the other hand, of the 225 clinical candidate CARs, 88 (39%), including the two approved CAR-T therapeutics, primarily target CD19 (of these, 11 target CD19 plus at least one additional B-cell target). Other tumor antigens most frequently targeted by CAR constructs include BCMA (26 candidates targeting), mesothelin (12), GD2 (10), CD123 (8), CD22 (8) and HER2 (6) (Table 5). Approximately 67% of all clinical stage CAR constructs target hematological cancers, largely driven by the vast number of CD19 targeted clinical candidates. Altogether, CAR constructs are being studied in clinical trials against 43 known unique targets.

Table 5. Targets of clinical stage T-cell redirected therapeutics, TRBAs and CARs *,**.

Primary Target	Primary Indications	Therapeutic Format			Total
		TRBAs	CAR-T/NKs	rCells Expressing FcγRIIIa	
CD19	B-cell cancer (NHL, etc.)	2	88	0	90
BCMA	MM	7	26	1	34
CD123	AML	5	8	0	13
Mesothelin	Solid tumors	1	12	0	13
GD2	Solid and neurological tumors	2	10	0	12
CD20	B-cell cancer (NHL, etc.)	5	4	2	11
CD33	AML	6	4	0	10
HER2	Solid tumors	3	6	0	9
CD22	B-cell cancer (NHL, etc.)	0	8	0	8
CD30	HL	1	5		6
PSMA	Solid tumor (prostate)	4	2	0	6
EGFRvIII	Neurological tumors	2	4	0	6
EGFR	Solid tumors	1	3	0	4
CD38	MM	2	2	0	4
EpCAM	Solid tumors	2	2	0	4
PSCA	Solid tumor (prostate)	1	3	0	4
CEA (CEACAM5)	Solid tumors	2	1	0	3
HIV	Virus	1	1	0	2
Glypican-3	Solid tumors	1	1	0	2
Flt3	AML	1	1	0	2
NKG2D ligands	Solid tumors	0	2	0	2
Claudin 18.2	Solid tumors	0	2	0	2
DLL3	SCLC	1	1	0	2
CS1 (SLAMF7)	MM	0	2	0	2
MUC16	Solid tumors	1	1	0	2
Lewis-Y	Solid tumors	0	2	0	2
cMet	Solid tumors	0	2	0	2
Others with single candidate	Mostly solid tumors	10	16	0	26
Undisclosed/other	Unknown	0	6	0	6
Total	–	61	225	3	289

* BiStro Biotech Consulting LLC database, locked 20 June 2019. Data obtained from Clinicaltrials.gov, literature papers, company websites, analyst reports and other sources. ** Abbreviations: AML, adult acute myeloid leukemia; BCMA, B-cell maturation antigen; CAR, chimeric antigen receptor; CD, cluster of differentiation; CEA, carcinoembryonic antigen; EGFR, epidermal growth factor receptor; EpCAM, epithelial cellular adhesion molecule; GD2, disialoganglioside antigen; HER2, human epidermal growth factor receptor; HL, Hodgkin's lymphoma; MM, multiple myeloma; NHL, non-Hodgkin lymphoma; NK, natural killer (cells); NKG2D, natural killer group 2D; PD-L1, programmed death ligand-1; PSCA, prostate stem cell antigen; PSMA, prostate-specific membrane antigen; rCells, recombinant NK or T-cells; SCLC, small cell lung cancer; TRBA, T-cell redirecting bispecific antibody.

As mentioned in Section 5.3, there are three clinical candidate recombinant NK or T cells that express FcγRIIIa to be redirected by therapeutic antibodies to tumors. Two of these candidates target CD20 for NHL and one targets BCMA for MM. As noted previously, a clinical trial of recombinant FcγRIIIa+-NK-92 cells with the anti-PD-L1 mAb, avelumab [267], is already planned, which could potentially extend this approach to solid tumors. If this combined cell/targeting mAb approach yields significant efficacy in the clinic, this number could rise quickly, because the cell products themselves would not need to be changed. The existing cell product candidates would just need to be paired in clinical trials with different approved therapeutic antibodies, such as trastuzumab for HER2-positive tumors, daratumumab for CD38-positive multiple myeloma (this would work for T cells but not for NK-92 cells, which are CD38-positive) and atezolizumab, avelumab or durvalumab for PD-L1-positive tumors and so forth.

An interesting twist on choice of targets is the recent report of using CAR-T cells to target the tumor micro-environment instead of the cancer cells themselves [364]. They generated CAR-T cells targeting the tumor TME-specific fibronectin splice variant, EIIIB [364]. Dosing of the EIIIB CAR-T cells helped to drive immune response to the tumor, suppressing tumor growth [364]. This strategy is not yet in clinical trials but seems promising.

As mentioned above in Section 6.6.3., one of the three critical issues for building successful CAR-Ts is finding the right tumor antigen targets. Due to the potency and toxicity of CAR-Ts and TRBAs to the target cells, it is critical to have either tumor-specific targets or targets that are vastly over-expressed in tumors as compared with normal tissues. The search for truly tumor-specific antigens has been ongoing for decades. Although some antigens have been discovered that are very tumor selective (e.g., MUC1, EGFRvIII, CEA, GD2 ganglioside, PSCA), essentially no antigens have been discovered that are absolutely restricted to tumor cells [373]. As a result, strategies to overcome the potential toxicity associated with killing normal cells that express these targets, even at low levels, are required.

In addition to the paucity of truly tumor-specific antigens, there is significant tumor antigen heterogeneity, that is, not all cells within the tumor will possess the targeted antigen, which makes the antigen selection bar even higher [374,375]. For example, antigen presence on cells within a tumor, for example, HER2 in NSCLC tumors, may be present on only 40% of the tumor cells [375]. Moreover, the copy number of tumor surface antigens can vary significantly from cell-to-cell within a tumor, as well. It is common for many cancers, including lung cancer, renal cell carcinoma, breast cancer, AML and CLL, amongst others, to have significant subclonal populations within the tumor [376]. Several genomic studies have revealed the extreme heterogeneity within tumors and even a wide range of heterogeneity amongst patients with a single type of tumor and heterogeneity amongst different types of tumors [377,378]. A mathematical model based on genomic analysis of tumor heterogeneity has even been developed called "MATH" (mutant-allele tumor heterogeneity) score, which has been used to help identify the extent of heterogeneity in various tumor types [378,379]. The problem with tumor heterogeneity, no matter how carefully measured, is that it generally works against any tumor-targeting approach, including TRBAs and CAR-Ts.

Even efficient killing of antigen-positive cells will only eliminate part of the tumor. Thus, it is important to note that there are examples of both CAR-T cells [380] and BiTE®s [381] demonstrating bystander killing of antigen-negative cells that were in direct contact with antigen-positive cells. In the case of BiTE®-induced bystander killing, FAS and ICAM-1 were both upregulated on the antigen-negative cells, which helped contribute to the bystander killing process that took place over a matter of hours after initial contact of the BiTE® with the antigen-positive tumor cells [381]. It is known that IFN-γ can upregulate Fas (CD95) on the surface of cancer cells [382]. Since both CAR-T cells [380] and TRBA-induced T-cell killers [381] both induce the production of IFN-γ as part of their activation and killing process, it is likely that Fas-mediated apoptosis of bystander cells may be more prevalent than shown in just these few studies. Additional studies need to be carried out to determine the extent to which bystander killing can help T-cell redirected strategies to eradicate tumors.

For both reasons mentioned above, that is, the need for greater tumor specificity and tumor antigen heterogeneity, one of the key approaches is to build CAR-T cells that have the ability to target multiple antigens, which appears to be part of a trend in tumor targeting going forward [375]. In our most recent analysis, we count 30 "multiple" targeting CARs out of 225 total unique CARs (Table 3), which is ca. 13% of all CARs being tested in clinical trials today.

Highly engineered CAR-T cell products in the future will not only have the ability to regulate what combinations of targets are engaged to drive an activation response but also to control through signal domain optimization the strength of that activation [329–332]. In any case, using either TRBAs or CAR-T as treatment options, it will be necessary to determine the presence of tumor antigens before treatment and to score for the change in tumor antigens after treatment [373].

8. Cytokine Release Syndrome (CRS) and Its Effect on Treatment

Cytokine release syndrome (CRS) is a significant concern for both CAR-T and TRBA mediated therapies [383]. Virtually every CAR-T and TRBA candidate tested in the clinic thus far has had at least some patients experiencing CRS adverse events. For Blincyto®, the only currently approved TRBA on the market, CRS occurs in only about 7–15% of patients depending on indication but in some cases, it was quite severe [384]. More severe cases of CRS can present clinical signs resembling severe inflammatory syndromes such as hemophagocytic lymphohistiocytosis (HLH) or macrophage activation syndrome (MAS) [383].

CRS is a major concern for CAR-T cell products. For example, treatment of 101 patients for refractory aggressive B-cell NHL with axicabtagene ciloleucel (KTE-C19) resulted in grade ≥3 CRS in 13% of the patients, whereas treatment of 51 patients with tisagenlecleucel for relapsed or refractory DLBCL led to grade ≥3 CRS in 26% of the patients [385]. As seen in virtually all CD19 CAR-T trials, the rapid expansion and activation upon CAR-T cell product target engagement results in the expression of toxic cytokines including IL-6, TNF-α and IFN-γ [386]. The release of these cytokines can be lethal, unless appropriately managed in the hospital setting. Current management includes immunosuppressive corticosteroid infusion and/or inclusion of an IL-6R blocking antibody (Actemra®) [386]. Most recently, the prominent role of IL-1 produced by monocyte/macrophages during CAR-T therapy has been elucidated as a major driver of CRS; potential treatments with IL-1 blocking antibodies are being tested in the clinic [387]. Very recently, it was demonstrated that GM-CSF stimulation of monocytes/macrophages may play a significant role in driving CRS [388,389]. Deletion of the GM-CSF gene in CAR-T cells eliminated CRS and actually improved CAR-T functionality in mouse models [388,389]. Similarly, GM-CSF function could be managed using an anti-GM-CSF mAb, such as lenzilumab [388,389], or one of the other clinical-stage anti-GM-CSF mAbs (e.g., otilimab, namilumab, TJ003234). These promising results suggest that as we learn more about the mechanisms by which CRS occurs, improved management will quickly follow.

It is important to note that for CAR-T cells, CRS is driven by the expansion and activation of the CAR-T cells and that in the early clinical trials in B-ALL, the CD19 target was highly and widely expressed in the total B cell compartment [383]. This may have resulted in an over-activation paradigm that may not be seen in more selectively targeted CAR-T cell products.

CRS often occurs with a short time window (e.g., 0–72 h) after dosing. Two of the key predictors or correlates, of severe CRS with either TRBA or CAR-T therapy are tumor burden and therapeutic dose [383]. For T-cell engaging bispecific antibodies, many investigators are now using step-up dosing (low dose preceding before regular dose) [390] or, in the case of the CD20-TCB RG6026, pre-dosing with the anti-CD20 mAb, obinutuzumab [227]. These types of dosing protocols appear to help lower the incidence of CRS. For CAR-T cell therapeutics, investigators have tried split-dosing to help manage CRS [391,392].

Neurotoxicity is also a significant AE that can occur with both TRBA and CAR-T therapy. Neurological toxicities occur in about two-thirds of all patients taking Blincyto®, typically within the first two weeks of therapy [384]. Pretreatment with dexamethasone or concomitant treatment with corticosteroids seemed to help reduce the neurological adverse events [393]. In clinical trials, neurotoxicity associated with CAR-T therapy (CAR-T-related encephalopathy syndrome or CRES) occurred in about 28% of patients treated with axicabtagene ciloleucel (KTE-019) and in 13% of patients treated with tisagenlecleucel (CTL019) [385].

9. Comparison of TRBAs and CAR-Ts Therapeutic Approaches

9.1. General Comparison of TRBAs with CAR-T Cells

Table 6 shows the overall comparisons of TRBAs with autologous and allogeneic CAR-T cells. Since there is such diversity amongst both TRBAs (e.g., short vs log half-life; bivalent vs trivalent and tetravalent; geometry) and CAR-Ts (affinity of antibodies used; use of different activation domains; methods for production and activation ex vivo, T cell types included, etc.), some the comparisons are necessarily generalized. Several other recent reviews also have compared many of the salient features of TRBAs and CAR-Ts [27,223,394,395].

Fundamentally, it is generally considered that CAR-T cells are more potent and efficacious than current TRBAs [17,127] but this comparison is largely made on Kymriah® and Yescarta® CAR-T therapies versus the only currently approved TRBA, Blincyto®. It may turn out that new TRBA therapeutics in development may leapfrog Blincyto® in efficacy and be much more competitive with the CAR-T therapies while maintaining the virtues of TRBAs. When it comes to cost, availability, convenience, ability to control dosing, ability to re-dose over time and the requirement with CAR-Ts for lymphodepletion, however, TRBAs have a significant "usefulness" advantage over CAR-T cells.

From a competitive marketing stance, Yescarta® brought in $264M (US dollars) in sales in 2018, largely due to the need for better treatment for DLBCL [396]. Kymriah®, which was the first CAR-T to be approved, was limited to 2018 sales of $28M (US dollars), hampered by both manufacturing concerns and approval for only B-ALL until May 2018, when it was also approved for DLBCL [397]. Blincyto®, which is currently only approved for B-ALL, brought in $230M (US dollars) in 2018 [398], which demonstrates that at least in B-ALL, it is competitive market-wise with the CAR-T thus far. A significant caveat to this, however, is that Blincyto® has been approved since 2014, so it has had a longer period to build its market share.

Table 6. Comparison of TRBAs and CAR-T cell platforms as therapeutics.

Properties	Therapeutic Approach		
	T cell Redirection with TRBAs	Autologous CAR-T or NK Cells	Allogeneic CAR-T or NK Cells (Projected)
Currently approved and marketed (as of 20 June 2019)	1; Blincyto® (anti-CD19 × CD3 BiTE®)	2; Kymriah® and Yescarta®, both CD19-targeting autologous CAR-Ts	None
Current indications covered	R/R B-ALL	DLBCL, R/R NHL, B-ALL	None
Structure	Bispecific antibodies that bind both a tumor antigen and CD3ε on T cells	T cells engineered with synthetic gene construct encoding scFv fused to linker and activation domains	T cells engineered with synthetic gene construct encoding scFv fused to linker and activation domains
Source and homogeneity of T cell component	Endogenous T cells; No homogeneity (i.e., all CD3+ T cells may be engaged)	Expanded and activated endogenous T cells; homogeneity depends on process used	Could be homogeneous CD8+ T-cells, depending on cell type and approach
Antibody	Short half-life vs long half-life formats	Currently, mostly scFvs; possible unfolding, aggregation, tonic signaling; need for better binding constructs	Currently mostly scFvs—possible unfolding, aggregation, tonic signaling; need for better binding constructs
T-cell signaling domain(s)	CD3ζ	CD3ζ + 4-1BB (or OX40) and/or CD28	CD3ζ + 4-1BB (or OX40) and/or CD28
PD-1 inhibition of CD28 activity	Likely significant issue; may need to co-dose with PD-1 inhibitor	Use of 4-1-BB signaling domain should alleviate	Use of 4-1-BB signaling domain should alleviate
Drug-like properties	"Off-the-shelf" drug	Must be engineered from patient's T cells (2–4 week process)	Depends on cell type and construct
Dosing	Multiple dosing; short half-life formats may require continuous dosing via pumps	Single dose	Single dose; multiple dose potentially available if engineered to eliminate HLA
Route of administration	IV; possible subcutaneous for future candidates	IV only	IV only
Long-term persistence and memory	Short half-life – only as long as continuously infused; long half-life – typically measured in weeks	Yes, but variable; longer persistence correlated with activity	Unknown but likely to be similar to autologous T-cells
Immune synapse	Normal and concentric; normal detachment	Abnormal and multifocal; fast detachment	Expected to be similar to autologous CAR-T cells
T cell signals at synapse	Signals 1, 3	Signals 1, 2 (sometimes), 3	Expected to be similar to autologous CAR-T cells
Killing mechanisms	Perforin and granzyme; Secondary: cytokine modulation of TME [123]	Perforin and granzyme; Fas/FasL axis; Secondary: cytokine modulation of TME [123]	Expected to be similar to autologous CAR-T cells
Serial killing	Yes, similar to CTLs	Yes, faster than TRBAs an CTLs	Expected to be the same as autologous T cells
	None; related to dosing and half-life	Yes, in responders	Unknown but expected
Bystander killing of antigen-negative cells	Demonstrated, as long as antigen-negative cells were in direct contact with antigen-positive cells [381]	Demonstrated, as long as antigen-negative cells were in direct contact with antigen-positive cells [380]	Unknown but expected based on CAR-T results
Toxicity	CRS, neurotoxicity	Higher CRS and neurotoxicity than TRBAs	Unknown but expected

Table 6. *Cont.*

Properties	Therapeutic Approach		
	T cell Redirection with TRBAs	Autologous CAR-T or NK Cells	Allogeneic CAR-T or NK Cells (Projected)
Ability to attack solid tumors	To be determined; early data are mixed but not encouraging	To be determined; early data are mixed but not encouraging	Potential based on TIL correlation data
Trafficking	Passive	Active but limited; can be engineered to match tumor needs	Active; possible to engineered to match tumor needs
Trafficking into CNS	Not demonstrated; Unlikely if BBB is intact [395]	Demonstrated trafficking into CNS [399]	Unknown but expected based on CAR-T results
Need for lymphodepletion prior to treatment	No	Yes	Yes
Technical risk	Moderate; many platforms are working well	High but may be manageable	Currently very high
Need for "kill switch" or turn-off methodology	No but nice to have, especially for long half-life formats	Moderate; nice to have	Very high; must have for safety
Accessibility	High–off-the-shelf biologic drug	Only available at specific medical centers thus far; 2–4 week process time before therapy	Projected to eventually have availability similar to biologic drugs
Cost of goods	Relatively low; Antibody-like or slightly higher depending on type of TRBA platform	Very high (more than a $75,000 process)	Projected to be low to medium once cell manufacturing process is established
Cost to patient/payers	Medium ($89,000/course; $178,000 for predicted two course therapy) *	Very high ($373,000 for treatment of DLBCL; $475,000 for Kymriah® treatment of B-cell ALL) **	Projected as medium to high, depending on cell type and construct

Abbreviations: BBB, blood-brain barrier; B-ALL, B-cell acute lymphoblastic leukemia; BiTE®, bispecific T-cell engager; CAR-T cell, chimeric antigen receptor-T cell; CD, cluster of differentiation; CNS, central nervous system; CRS, cytokine release syndrome; DLBCL, diffuse large B-cell lymphoma; IV, intravenous; NHL, non-Hodgkin lymphoma; NK, natural killer (cells); R/R, relapsed or refractory; PD-1, programmed cell death protein 1; scFv, single chain, Fragment variable (antibodies); TIL, tumor-infiltrating lymphocytes; TRBA, T-cell redirecting bispecific antibody. * Quote is for the cost of two-course treatment with Blincyto® [400]. ** Quotes for cost of CAR-T therapies [401].

The overall therapeutic strategy behind both CAR T-cells and TRBA is the same, that is, harness the cytotoxic function of T (or NK, NKT) cells to target and kill tumor cells, as well as to overcome the escape mechanisms utilized by the tumor cells. However, these two technologies differ significantly. TRBAs typically have a significantly higher affinity on the tumor antigen arm than on the T-cell arm, so once dosed, they will largely distribute to the site of the tumor antigen and coat the antigen-positive tumor cells. These coated cells then become targets for T cells that enter the tumor. Thus, TRBAs have a significant dependency on both the ability of cytotoxic T-cells to enter the tumors and for those cells to become activated upon binding. It has been shown, however, that TRBAs can even cause T-regulatory cells to kill tumors cells [402], so the forced formation of the synapse also serves to activate the T-cells. The activation signals for TRBAs are limited to CD3ζ and there is no provision for signal 2, at least in the current generation of TRBAs. Thus, TRBAs are potentially limited in the TME by the suppressive effects of PD-1/PD-L1 and similar regulatory immune checkpoints.

CAR-T cells, on the other hand already are armed with the antibodies and seek out tumor cells as a "ready-made killing machine." Additionally, autologous CAR-T cells are activated ex vivo, typically by CD3, CD28, and/or cytokine treatment, providing a stimulant for killing tumor cells immediately upon dosing. Finally, when CAR-T cells bind to the target cells, the costimulatory domains activate the CAR-Ts further, giving them a significant edge over TRBAs in terms of killing potential.

As described in Section 4.1, another approach that falls in between CAR-Ts and TRBAs is the use of autologous T cells, activated and armed with a TRBA *ex vivo*, followed by reintroduction to the patient [21]. This approach is similar to CAR-Ts in the sense that autologous T-cells are used and activated ex vivo but more similar to TRBAs in the sense that the "drug" is a bispecific antibody conjugate. Also, in this case, there is no recombinant T-cell activation domain, so the activation rests solely with CD3ζ, thus missing one of the key components of a CAR-T strategy. This is probably one of the key reasons why the ex vivo T-cell/TRBA approach has garnered only limited interest.

As mentioned in Section 2.2, the potency of the CAR-T cells may come from the unusual multifocal synapses they form with the target cells, allowing for faster killing rates and quicker time to release [112,113]. Additionally, CAR-T cells possess their own ability to provide signal 2 in the intracellular activation domains (e.g., CD27, 4-1BB, OX40). Finally, as mentioned in Section 6.6.4., one potential checkpoint issue with natural T cells and bispecific antibodies is the ability of PD-1/PD-L1 interaction to override the CD28 activation signal [361]. In CAR-T cells possessing the 4-1BB signaling domain, which is becoming an increasing number due to the activation and proliferative signaling provided by 4-1BB as well as its ability to promote persistence and reduce T-cell exhaustion [138,403,404], this should not be an issue. Third generation CAR-Ts can include both CD28 and 4-1BB costimulatory domains (or other combinations; see Figure 3), so these newer CAR-T constructs should better promote both T-cell function, proliferation and persistence [134,404].

In both CAR T-cells and TRBA therapeutics, healthy and active T cells are a prerequisite. In most cancer patients, especially patients with hematological malignancies, multiple lines of therapy previously administered can severely impact the number and health of circulating T cells [313]. The impact of the patients T cell status will certainly limit the effectiveness of both CAR T-cells and TRBAs. However, during the process of generating a CAR T-cell product, the patients T-cells are removed and manipulated ex vivo during their genetic manipulation to express the CAR, allowing for recovery and expansion prior to dosing [313]. As a result, the impact of the pretreatment can be mitigated to some extent and less chance is factored into the overall success of the CAR T-cell approach, compared to TRBA approach. This will be even less of an issue for truly allogeneic CAR T-cell products in which healthy, highly characterized cells are used as the starting material.

Currently, TRBAs do have an advantage with regard to ease of manufacturing, consistent product and dose control. This is especially significant in situations in which multiple dosing is required or preferred or for tumor target antigens that are also expressed on normal tissues. The ability to titer the dose and control the exposure of the agent in vivo has a significant advantage when tight control of the therapeutic agent is required to limit on-target, off-tumor toxicity. In contrast, currently used autologous CAR T-cell therapeutics are usually only dosed once (or at most, twice), owing to the challenges associated with current manufacturing technologies and the products themselves are a heterogeneous mixture of different T-cell subtypes with varying CAR expression.

9.2. Clinical Comparison of TRBAs vs. CAR-T Cells

T-cell redirecting therapeutics, including both TRBAs and autologous CAR-Ts, have revolutionized the treatment of hematological malignancies over the past half-dozen years or so. These forms of therapy, as described in this review, have significant promise for use and success in many forms of cancer. As shown in Tables 4 and 6, two autologous CAR-T cell therapies, Kymriah® and Yescarta®, were approved by the US-FDA in 2017, just a few months apart. Kymriah® has now been approved for treatment of B-cell ALL (August 2017) and R/R DLBCL (May 2018), whereas Yescarta® has been approved for R/R DLBCL and PMBCL (Oct 2017). Similarly, Blincyto® was approved in 2014 for Philadelphia-negative chromosome B-cell ALL and has more recently (July 2017) tacked on approval for Philadelphia-chromosome-positive B-cell ALL. The CD19-targeting CAR, Lisocabtagene maraleucel (Liso-cell; JCAR017) and BCMA-targeting bb2121 are currently in Phase 3 clinical trials for DLBCL and multiple myeloma, respectively, so it would be expected that they may be the next T-cell redirected biologics in line for consideration for marketing approval.

Table 7 provides a glimpse at a few example clinical trial outcomes into the current status of these lead T-cell redirecting biologics. Key results that have enthused clinicians and patients are objective response rates (ORR) of up to ~80% and complete responses (CRs) above 50% for CAR-T treatments of the aggressive B-cell lymphoma, DLBCL. Median duration time for responses in DLBCL have been over 10 months for the CAR-Ts and 7 months for Blincyto® (Table 7). Some of these responses in DLBCL have been durable, while others have faded over time. Kymriah® has been impressive for the treatment of B-cell ALL, with CRs above 80%, as compared with Blincyto®, which at least for Philadelphia chromosome-negative B-ALL, CRs were above 30% (Table 7).

Table 7. Examples of clinical data with T-cell redirecting biologics.

Property	T-Cell Redirecting Biologic Drug							
	Kymriah®	Blincyto®	Kymriah®	Yescarta®	Liso-cel	Blincyto®	bb2121	AMG420
Sponsor	Novartis	Amgen	Novartis	Gilead (Kite)	Juno/ Celgene	Amgen	Bluebird/ Celgene	Amgen
Format	CAR-T; 4-1BB CS	TRBA (BiTE®)	CAR-T; 4-1BB CS	CAR-T; CD28 CS	CAR-T; 4-1BB CS	TRBA (BiTE®)	CAR-T; 4-1BB CS	TRBA (BiTE®)
Trial	EL	TW	JU	ZU	TC	Phase 1	Phase 1	Phase 1
Target	CD19	CD19	CD19	CD19	CD19	CD19	BCMA	BCMA
Indication	B-ALL	B-ALL (PCN)	DLBCL	DLBCL	DLBCL	DLBCL	MM	MM
# Number of Patients	63	271	93	101	73	11	33	42
ORR	ND	ND	52%	83%	80%	55%	85%	31%
CR/CR *	83%	34%	40%	58%	59%	36%	45%	17%
PR	20%	ND	12%	25%	21%	18%	39%	10%
Median response duration time	NR	7.7 mo	11.7 mo	11.1 mo	10.2 mo NR	13.3 mo	11.8 mo	NR
Grade 3+ AEs	ND	ND	89%	98%	16%	90%	ND	ND
CRS incidence	77%	15%	58%	58%	37%	ND	76%	38%
Grade 3+ CRS	ND	nk	22%	11%	1%	ND	6%	ND
Neurotoxicity	ND	65%	21%	64%	25%	71%	ND	ND
Grade 3+ Neurotoxicity	ND	13%	12%	32%	15%	20%	ND	0%
Elimination half-life	21.7 d RP; 2.7 d NRP	NA	91.3 d RP; 15.4 d NRP	ND	ND	NA	ND	NA
References	[290]	[384,405]	[290,406]	[291,407]	[408]	[393,409]	[295,410]	[411,412]

* Abbreviations: B-ALL, B-cell acute lymphoblastic leukemia; BCMA, B-cell maturation antigen; CD, cluster of differentiation; BiTE®, bispecific T-cell engager (short half-life, continuously infused); d, days; CAR-T, chimeric antigen receptor T-cell; CS, co-stimulation domain; DLBCL, diffuse large B-cell lymphoma; EL, ELIANA trial; JU, JULIET trial; Liso-cel, Lisocabtagene maraleucel (JCAR017); MM, multiple myeloma; mo, months; NA, not applicable; ND, no data; NR, not reached (during testing period covered); NRP, non-responding patients; PCN, Philadelphia chromosome-negative; RP, responding patients; TC, TRANSCEND-CORE; TRBA, T-cell redirecting bispecific antibody; TW, Tower trial; ZU, ZUMA-1 trial. #: Number.

It is now well known that persistence of the CAR-Ts is highly critical for their success and is correlated with responses [413]. This is borne out by data provided in clinical trials with Kymriah®. For B-cell ALL patients who responded to the treatment, the geometric mean half-life for the CAR-T cells was ca. 22 days; for non-responders, the half-life was only 2.7 days [290]. The time to the last detectable CAR-Ts was 170 days for responders versus only about 29 days for non-responders [290]. Similarly, for DLBCL responding patients treated with Kymriah®, the half-life was calculated to be about 91 days for responders versus 15.4 days for non-responder patients. In this case, times to the last detectable CAR-Ts in circulation were 289 days for responders versus 57 days for non-responders [290]. There are two take-aways from these data: first, the persistence of the same CAR-T was significantly different between the two different patient populations, indicating that disease-specific factors play a role in persistence of autologous CAR-T cells; second, it is clear that longer CAR-T persistence in both disease types was correlated with responders. As we learn more about what differentiates responders from non-responders, we may be able to influence the level of persistence of CAR-Ts in patients to improve response rates.

While the data for anti-BCMA treatment of MM by CAR-Ts and TRBAs is in its infancy, a few results recently cited from early clinical trials give hope for future MM treatment paradigms. The very high ORR (85%) and CR (45%) associated with bb2121 treatment of MM (Table 7) is very encouraging, as are the data with the anti-BCMA TRBA, AMG420, which are from a Phase I ascending dose paradigm (i.e., several of the patients dosed were dosed at below therapeutic levels, which skews the numbers downward). At the projected optimal dose of 400 μg/day, there were four CRs [411].

The real goals of clinical treatment are durable, complete responses. Currently, both TRBAs and CAR-Ts have high initial response rates but have a significant relapse rate. Thus, both increasing the number of patients with CRs and extending the durability through CAR-T persistence and improved design of TRBAs will be critical to clinical improvements. While it appears thus far that CAR-T cell therapy is more potent that TRBA therapy, this analysis comes with huge caveats, including the very small sample size, the fact that the only TRBA included in the analysis is Blincyto®, which due to its short half-life must be dosed continuously and the overall toxicity profile, which appears to be higher for CAR-Ts than for TRBAs. Moreover, it is still too early to pass judgement on the various forms of T-cell redirected therapies, as so many new variables have been tested over the past few years that we really are just now beginning to learn the critical quality attributes for each type of therapeutic.

It has been suggested that for TRBAs to compete with future CAR-Ts, CRs will need to be >50% with durations longer than a year, with progression-free survival and overall survival of 12 and 18 months, respectively and cure rates of at least 30%. Those kinds of data, however, particularly extending to treatments beyond B-ALL and DLBCL, will make virtually any TRBA or CAR-T highly competitive.

There are many reasons for failure of either a TRBA or CAR-T. Target antigen loss accounts for about a third of all TRBA and CAR-T failures. For example, in six clinical trials using CD19 CAR-T cells to treat B-ALL, the relapse rates ranged from 29–57% [414]. Of those relapses, 7–25% were due to loss of the targeted antigen, CD19, on the tumor cells [414]. Related to antigen loss is antigen down-regulation, which may change the copy number of the tumor-associated antigen from high to low and perhaps even as low or lower than copy number on normal tissues, which presents a significant challenge with respect to therapeutic window. Another major reason for treatment failure is modification of the tumor antigen on the surface of cells due to mutations, many of which have lost the targeted epitope [415] and possibly also alternative expression of splice variants. Antigen down-regulation and modification are thought to make up a second third of treatment failures. The final major factor resulting in failures is treatment-related toxicities. While this is being controlled by treatment with drugs to limit CRS and by changes to dosing paradigm, it is still a significant factor in failure.

9.3. Future Improvements

9.3.1. TRBAs

There are several improvements that may be made to increase TRBA efficacy and reduce toxicity, many of which have been described in this paper. Some of these include greater emphasis on targeting epitopes close to the membrane that allow the greatest level of T-cell activation [221,224,226], greater emphasis on geometry of both binding arms that allows for strongest synapse formation and killing potency (c.f., Figure 8 [17]) and improvement in understanding and applying affinity to both the CD3ε and tumor antigen arms. With respect to affinity, it has been often considered that the affinity between the CD3ε arm should be lower than the affinity of the second arm to the tumor antigen. This allows for distribution to the tumor and coating of the tumor antigen-positive tumor cells [416]. Too high of an affinity on the CD3ε arm of a TRBA can potentially lead to distribution into T-cell rich tissues such as lymph node and spleen rather than into tumor tissues [416] and may potentially lead to toxicity [417]. One rule of thumb, which may or may not bear out with time, is to have at least a ten-fold higher affinity to the tumor antigen than to the CD3ε. This makes sense, since the goal is to have the TRBA distribute to the tumor, coat antigen-positive tumor cells and attract T-cells to those TRBA-coated tumor cells. Another approach to increasing the apparent affinity to the tumor versus T-cells is to increase the number of tumor antigen binding arms while keeping only a single CD3ε binding arm [227,228]. The 2:1 TCB platform recently described by Roche does exactly that, with a concomitant improvement in activity [227]. Intriguing designs for TRBAs in the future may include multiple tumor antigen binding arms with only a single CD3ε binding arm, such as the IGM Biosciences IGM-based TRBA, which has 10 binding arms for the tumor antigen and a single binding arm for CD3ε) [228].

Getting the right TRBA dose is a significant issue. Because TRBAs are so potent and even low doses can lead to significant CRS in some cases, the FDA carried out a study on clinical stage TRBAs concerning first-in-human dosing and concluded that initial doses needed to follow careful MABEL (minimum anticipated biological effect level) calculations to ensure safe dose escalation in the clinic [418]. Additional improvements in TRBAs may be more related to dosing paradigms, including step-up dosing when necessary to prevent or reduce the chance of CRS [390]. Another issue with TRBA dosing is the potential to overdose, which can lead to separate coating of T-cells and tumor cells in a manner that they will not interact, which would lead to a lack of efficacy [223]. Additionally, it is well known that continuous stimulation of T cells can lead to T-cell anergy, so TRBA dosing paradigms may be designed to provide oscillations in serum concentrations, which may decrease T-cell anergy [27]. Finally, since TRBAs do not provide a signal 2, combinations of TRBAs and either checkpoint inhibitors or possibly activators may increase the efficacy and therapeutic window.

The most important improvement in TRBAs would be to develop a TRBA format that either significantly limits or eliminates CRS, since CRS appears to be dose-limiting in most cases. A very recent publication highlighted the development of a new TRBA format that appears, at least in vitro and in animal experiments, to limit CRS significantly [417]. This TRBA consisted of a Fab arm binding to CD3ε on one side and two domain antibodies binding the tumor antigen in place of the other Fab arm [417]. Of critical importance, though, was the fact that the correlation of high potency and low CRS was largely due to the properties of the CD3ε-binding arm, F2B, which binds the CD3δε heterodimer at about 34 nM but does not measurably bind the CD3γε heterodimer [417]. Well-studied anti-CD3ε mAbs such as OKT3 bind both the CD3δε and CD3γε heterodimers at relatively high affinity, indicating that the mAb F2B epitope must be unique, which may also offer a unique signaling pattern [417]. Unfortunately, the CD3δε-specific F2B mAb does not cross-react with cynomolgus monkey CD3ε [417], so that will make preclinical toxicology more difficult to assess. Hopefully, additional TRBA platforms with improved binding arms, geometry, and/or affinity/avidity will be found that improve potency with concomitant low CRS.

9.3.2. CARs

For CAR-based cell therapy designs, there are so many potential improvements that could be made, it is difficult to define which are the most important. Clearly, if a stable, off-the-shelf allogeneic CAR-T or CAR-NKT cell line with a defined PK/PD profile could be established that rivals the clinical results demonstrated by the approved autologous CAR-Ts, Kymriah® and Yescarta®, that would be perhaps the most significant advance. This type of CAR would allow for broader distribution, immediate use as a therapeutic and hopefully, a less costly therapy. Additionally, such a cell line could be engineered in all the ways mentioned in this paper, including making it PD-1 or PD-L1-negative [146,362–365], engineering in the ability to produce antibodies in the local TME environment [143–145], adding in cytokine expression to help activation and proliferation [99,339,369,370], kill switches to control any adverse circumstances [98,301–303,341–344,348–350] and cloning in chemokine receptors to help in trafficking the cells to tumors [100,337,355,358,360].

As shown in Tables 4 and 7, the different CAR-T therapeutics are constructed with different spacers, transmembrane domains and costimulatory domains. Virtually all CAR-T constructs today utilized the CD3ζ signaling domain, so that is a constant. Additionally, as mentioned previously, affinity, epitope and type of antibody used may have a significant impact on activity. A signaling analysis recently carried out suggested that co-stimulation with CD28 resulted in faster and higher downstream protein phosphorylation which correlated with effector T-cell function [419]. On the other hand, co-stimulation with 4-1BB was more correlated with genes associated with T-cell memory and as previously noted, also correlates with sustained activity and persistence [134,138,403,404]. The importance of one or the other profile may change with cell types, cancer types and indications, so it is too early to state broadly that one costimulatory form is preferable over another.

The appropriate affinity for a CAR has been a topic of considerable discussion. Early CARs were constructed with whatever antibodies were already available, many of them of mouse origin, as previously discussed. As CAR design has become more sophisticated, both the affinity [404,420] and target antigen epitope [421] of the CAR have become more critically designed. It has been noted that higher affinity CARs may interfere with serial killing and persistence, as well as potentially promoting T cell exhaustion or even activation-induced cell death [111]. In some cases, lower affinity CARs were correlated with greater cancer cell killing and tumor clearance [422] and may help to promote a faster off-rate from the tumor cell [112,113], which itself correlates with increased cytokine and chemokine expression [423]. Lower CAR affinity, however, can result in the requirement for higher target numbers to achieve activation, which is potentially great for differentiating normal tissue with low antigen expression from antigen-over-expressing cancer tissue but can limit tumor killing when target antigen is decreased by either lower expression [111,128] or through trogocytosis [130]. In fact, the strategy of lowering the CAR affinity to reduce on-target but off-tumor toxicity (i.e., binding to tissues with lower target antigen numbers (has been used to limit toxicities associated with targeting solid tumors) [420]. Thus, a key to improvement of future CARS will be the balancing of CAR affinity with the application of activation and signaling domains. A recent publication demonstrated that lowering the affinity of the CAR, coupled with the combination of both CD28 and 4-1BB signaling, produced a CAR with balanced and potentially optimal affinity/signaling properties [404].

As has been shown with TRBAs (Section 3.4) [15,17,221,224,226], it is also believed that membrane proximal epitopes on targeted antigens provide the best activity for CARs [421]. More distal epitopes might not be sufficient to generate a strong immunological synapse [224,424]. However, other considerations are equally important and not as implied based on sequence of the target antigen. For example, the "neighborhood" of the target antigen and how it can interfere with epitope access, as well as, its structural conformation can also interfere with the interaction of the TRBA interaction [425]. Thus, understanding the activity of the CAR in the presence of primary target cells, which may possess the most realistic target neighborhood, may be of significant benefit over cell lines into which the target antigen has been cloned for expression.

Changes in CAR construction not involving either the target-binding scFv or the signaling and/or costimulatory domains also can have a huge impact on CAR function and safety. A recent result suggests that minor changes can have enormous impacts on the performance and safety of a CAR-T cell. Ying et al. [426] generated and compared a series of modified CAR-T cells based on CTL019 (Kymriah®; Tisagenlecleucel-T) in which they altered the length of the CD8α transmembrane and hinge used. They found that one of their constructs, CD19-BBz(86), which possessed a 10 amino acid residue longer extracellular hinge as well as a 4 amino acid longer intracellular domain, was far superior to CTL019 in CRS and other functions [426]. This construct was recently taken into a Phase 1 clinical trial [427], demonstrating significantly superior safety (very low CRS and neurotoxicity) over most other clinical stage CAR-Ts and slower proliferation, while retaining high functionality (54.5% CRS obtained with B-cell lymphoma patients) [426]. This study demonstrates how even minor changes to "seemingly non-critical components" of the CAR can yield significantly improved CAR-Ts.

Once the CAR T-cell product is dosed, the expansion and persistence are completely under the control of the biology and state of the patient and the CAR T-cells. This loss of control is often the driver behind the toxic CRS and neurotoxicity observed in CAR T-cell therapeutics and has been forced CAR T-cell developers to design safety switch technologies and other management strategies to control these products once dosed [428,429]. Control of PK/PD and persistence of CAR-T products would be a significant advance. It is well recognized that CAR-T persistence is correlated with response [395]. This is exemplified by data from clinical trials with Kymriah®, in which the investigators were able to identify a clear correlation between patient population and pharmacokinetic properties of the

CAR-T [290]. There were two distinct populations of patients, that is, responding patients in which the geometric mean half-life of the CAR-T was over 20 days and the time to last measurable CAR-T cells was 170 days and non-responding patients, in which the geometric mean of the CAR-T in circulation was less than three days and the CAR-Ts were gone within a month [290]. Understanding what biological signals or circumstances that differentiate these two populations might allow us to turn non-responders into responders, which could make the CAR-T therapeutics significantly more effective than they already are.

Similar to TRBAs, CAR-Ts suffer from the influences of two main biological facts: The potential negative influences of the tumor microenvironment of the T-cells, especially for solid tumors; and our ability to identify truly tumor specific target antigens. There are many negative influences on the biological responses to T-cells including the upregulation of check-point molecules like PD-1/PD-L1 and PD-L2, expression of suppressive cytokines like VEGF and TGF-β, the infiltration of myeloid-derived suppressive cells, tumor associated macrophages, cancer associated fibroblasts and regulatory T-cells, poor vascularization and hypoxic conditions [425]. Although we are just beginning to understand some of the factors driving these barriers, our ability to manipulate them will be key to overcoming them. With so many different pathways contributing to this physiochemical barrier, it is difficult to envision that a single agent therapeutic will be able to overcome them all and allow for the immune cells to have an effect. Consequently, TRBAs will no doubt have the most difficult time as they are only designed to address one aspect of the therapeutic strategy of localizing and activating T-cells near the tumor cells. The simplicity of these agents loses some of its appeal when one considers the need to now address many additional pathways that will need to be manipulated in order to achieve therapeutic responses. CAR T-cells, on the other hand, have built-in components that can be manipulated to address many, if not all, of these barriers. As previously discussed, the generation of a CAR T-cell involves genetic engineering to insert the CAR construct into the T-cell genome. It is entirely feasible to add additional modifications, even in a single genetic manipulation step, to address some of these barriers. For example, CAR T-cells can be engineered to express antibody fragments that will block the action of suppressive cytokines or check-point molecules. They can be stimulated ex vivo to upregulate chemokine receptors or other receptors that drive tumor localization [430].

Efforts to generate allogeneic CAR T-cell products will undoubtedly be highly engineered to not only address the allo-reactivity but also many of these barriers. Future allogeneic CAR-based cell products that are expected to be more homogenous and well characterized will be a significant improvement over current autologous CAR products with regard to understanding dose-response relationships. However, it still remains to be seen if even with these highly controlled allogeneic products they will be more predictable once dosed.

10. Summary and Future State

This is an exciting time for T-cell redirected therapeutics, both for protein-based bispecifics and cell-based CAR therapeutics. Both formats have shown significant glimpses of promise but also many shortcomings, several of which may be able to be addressed with engineering. With so many different types of TRBAs (Table 2) and CAR-Ts (Tables 3 and 4) being studied in clinical trials today, it will be years before we understand the optimal structures and/or constructs with critical quality attributes that best address each particular type of cancer or indication. The differences between the two platforms, CAR-Ts and TRBAs, today is enormous, with ease of use, availability and costs favoring TRBAs where sheer potency driving CRs and duration of response currently favoring CAR-Ts. If allogeneic CAR-Ts become more widely available in easily engineered formats, this could significantly change the outlook, as the "off-the-shelf" CAR-Ts will then start to take on more of the characteristics of regular biologic drugs, that is, dosing and redosing, no need to wait for processing, lower costs and availability. This review has documented many or the ways in which the molecules and/or cells can be engineered to potentially increase the therapeutic window. Many of these engineered modifications or addons will be much easier to accomplish by engineering cells rather than protein biologics, so in the long-run, there does appear to be a potentially higher upside potential for CAR-Ts than TRBAs but that remains to be seen. The key to both formats is the ability to increase the therapeutic window by significantly decreasing both CRS-related and neuro-related toxicity, while maintaining or even increasing potency.

Author Contributions: W.R.S. and M.N. wrote the manuscript together, designed the figures and edited the manuscript.

Funding: This research received no external funding.

Conflicts of Interest: The authors declare the following potential conflicts of interest: W.R.S. and M.N. both were recent employees at Janssen R&D, Johnson & Johnson and hold stock in that company. W.R.S. currently is a consultant for several small biotechnology companies and is a member of the Board of Directors of IGM Biosciences. M.N. currently is an employee of Century Therapeutics, Philadelphia, PA.

Abbreviations

AML	adult acute myeloid leukemia
APC	antigen presenting cell
ART-Ig	asymmetric re-engineering technology—immunoglobulin
ATTACK	asymmetric tandem trimerbody for T cell activation and cancer killing
B-ALL	B cell acute lymphoblastic leukemia
BBB	blood-brain barrier
BEAT	bispecific engagement by antibodies based on the T cell receptor
BiKE	bispecific killer engager
BiTE	bispecific T-cell engager
CAR	chimeric antigen receptor
CD	cluster of differentiation
CIKs	cytokine-induced killers
CLL	chronic lymphocytic leukemia
CNS	central nervous system
CR	complete response
CRS	cytokine released syndrome
CTL	cytotoxic T lymphocytes
DART	dual affinity retargeting (antibody)
DLBCL	diffuse large B cell lymphoma
DVD-Ig	dual variable domain immunoglobulins
EGFRt	truncated version of epidermal growth factor receptor
EMA	European Medicines Agency
EpCAM	epithelial cell adhesion molecule
Fc	fragment, crystallizable
FIT-Ig	Fabs-in-tandem immunoglobulins
GVH, GVHD	graft-versus-host (disease)
HC	heavy chain
HLA	human leukocyte antigen
HSV-TK	herpes simplex virus thymidine kinase
iCasp9	inducible caspase-9
Ig	immunoglobulin
ImmTAC	immune-mobilizing monoclonal TCR against cancer
iNKT	invariant NKT (cells)
ITAM	immunoreceptor tyrosine activation motif
KIH	knobs-into-holes
LC	light chain
MATH	mutant-allele tumor heterogeneity
MHC	major histocompatibility complex
MM	multiple myeloma
NK	natural killer (cell)

NKG2D	natural killer group 2D
NKT	natural killer T (cell)
NSCLC	non-small cell lung cancer
OR	objective response
PBMCs	peripheral blood mononuclear cells
pMHC	peptide-MHC complex
PR	partial response
R/R	relapsed/refractory
sCAR	switchable chimeric antigen receptor
scFv	single chain, fragment variable
SEED	strand exchange engineered domain
SMAC	supramolecular activation cluster
TandAb	tandem diabody
TBE	target cell-biologic-effector cell (complex)
TCB	T-cell bispecifics
TCR	T-cell receptor
TILs	tumor infiltrating lymphocytes
TME	tumor microenvironment
TRBA	T-cell redirecting bispecific antibody
TriKE	trispecific killer engager
TITAC	trispecific T cell activating construct
US-FDA	United States Food and Drug Administration

References

1. Coley, W.B. Contribution to the knowledge of sarcoma. *Ann. Surg.* **1891**, *14*, 199–220. [CrossRef] [PubMed]
2. Hoption Cann, S.A.; Van Netten, J.J.; Van Netten, C. Dr William Coley and tumor regression: A place in history or in the future. *Postgrad. Med.* **2003**, *79*, 672–680.
3. Vernon, L.F. William Bradley Coley, MD and the phenomenon of spontaneous regression. *Immunotargets Ther.* **2018**, *7*, 29–34. [CrossRef] [PubMed]
4. Poltorak, A.; He, X.; Smirnova, I.; Liu, M.-Y.; Van Huffel, C.; Du, X.; Birdwell, D.; Alejos, E.; Silva, M.; Galanos, C.; et al. Defective LPS signaling in C3H/HeJ and C57BL/10ScCr mice: Mutations in Tlr4 gene. *Science* **1998**, *282*, 2085–2088. [CrossRef] [PubMed]
5. Reisser, D.; Pance, A.; Jeannin, J.F. Mechanisms of the antitumoral effect of lipid A. *Bioessays* **2002**, *24*, 284–289. [CrossRef] [PubMed]
6. Aptsiauri, N.; Ruiz-Cabello, F.; Garrido, F. The transition from HLA-I positive to HLA-I negative primary tumors: The road to escape from T-cell responses. *Curr. Opin. Immunol.* **2018**, *51*, 123–132. [CrossRef]
7. Isaaz, S.; Baetz, K.; Olsen, K.; Podack, E.; Griffiths, G.M. Serial killing by cytotoxic T lymphocytes: T cell receptor triggers degranulation, re-filling of the lytic granules and secretion of lytic proteins via a non-granule pathway. *Eur. J. Immunol.* **1995**, *25*, 1071–1079. [CrossRef]
8. De La Roche, M.; Asano, Y.; Griffiths, G.M. Origins of the cytolytic synapse. *Nat. Rev. Immunol.* **2016**, *16*, 421–432. [CrossRef]
9. Han, X.; Vesely, M.D. Stimulating T cells against cancer with agonist immunostimulatory monoclonal antibodies. *Int. Rev. Cell Mol. Biol.* **2019**, *342*, 1–25.
10. Kim, M.T.; Harty, J.T. Impact of inflammatory cytokines on effector and memory CD8+ T cells. *Front. Immunol.* **2014**, *5*, 295. [CrossRef]
11. Kammertoens, T.; Blankenstein, T. It's the peptide-MHC affinity, stupid. *Cancer Cell* **2013**, *23*, 429–431. [CrossRef] [PubMed]
12. Bubenik, J. Tumour MHC class I downregulation and immunotherapy (Review). *Oncol. Rep.* **2003**, *10*, 2005–2008. [CrossRef] [PubMed]
13. Offner, S.; Hofmeister, R.; Romaniuk, A.; Kufer, P.; Baeuerle, P.A. Induction of regular cytolytic T cell synapses by bispecific single-chain antibody constructs on MHC class I-negative tumor cells. *Mol. Immunol.* **2006**, *43*, 763–771. [CrossRef] [PubMed]

14. Stopeck, A.T.; Gessner, A.; Miller, T.P.; Hersh, E.M.; Johnson, C.S.; Cui, H.; Frutiger, Y.; Grogan, T.M. Loss of B7.2 (CD86) and intracellular adhesion molecule 1 (CD54) expression is associated with decreased tumor-infiltrating T lymphocytes in diffuse B-cell large-cell lymphoma. *Clin. Cancer Res.* **2000**, *6*, 3904–3909. [PubMed]

15. Li, J.; Stagg, N.J.; Johnston, J.; Harris, M.J.; Menzies, S.A.; DiCara, D.; Clark, V.; Hristopoulos, M.; Cook, R.; Slaga, D.; et al. Membrane-proximal epitope facilitates efficient T cell synapse formation by anti-FcRH5/CD3 and is a requirement for myeloma cell killing. *Cancer Cell* **2017**, *31*, 383–395. [CrossRef] [PubMed]

16. Uyttenhove, C.; Pilotte, L.; Theate, I.; Stroobant, V.; Colau, D.; Parmentier, N.; Boon, T.; Van Den Eynde, B.J. Evidence for atumoral immune resistance mechanism based on tryptophan degradation by indoleamine 2,3-dioxygenase. *Nat. Med.* **2003**, *9*, 1269–1274. [CrossRef] [PubMed]

17. Ellerman, D. Bispecific T-cell engagers: Towards understanding variables influencing the in vitro potency and tumor selectivity and their modulation to enhance their efficacy and safety. *Methods* **2019**, *154*, 102–117. [CrossRef]

18. Chen, D.; Mellman, I. Oncology meets immunology: The cancer-immunity cycle. *Immunity* **2013**, *39*, 1–10. [CrossRef]

19. Lum, L.G.; Thakur, A. Targeting T cells with bispecific antibodies for cancer therapy. *BioDrugs* **2011**, *25*, 365–379. [CrossRef]

20. Clynes, R.A.; Desjarlais, J.R. Redirected T cell cytotoxicity in cancer therapy. *Annu. Rev. Med.* **2018**, *70*, 437–450. [CrossRef]

21. Lum, L.G.; Thakur, A.; Al-Kadhimi, Z.; Colvin, G.A.; Cummings, F.J.; Legare, R.D.; Dizon, D.S.; Kouttab, N.; Maizei, A.; Colaiace, W.; et al. Targeted T-cell therapy in stage IV breast cancer: A phase I clinical trial. *Clin. Cancer Res.* **2015**, *21*, 2305–2314. [CrossRef] [PubMed]

22. Park, J.H.; Brentjens, R.J. Adoptive immunotherapy for B-cell malignancies with autologous chimeric antigen receptor modified tumor targeted T cells. *Discov. Med.* **2010**, *9*, 277–288. [PubMed]

23. Chmielewski, M.; Hombach, A.A.; Abken, H. Antigen-specific T-cell activation independently of the MHC: Chimeric antigen receptor-redirected T cells. *Front. Immunol.* **2013**, *4*, 371. [CrossRef] [PubMed]

24. June, C.H.; Sadelain, M. Chimeric antigen receptor therapy. *N. Engl. J. Med.* **2018**, *379*, 64–73. [CrossRef] [PubMed]

25. Jochems, C.; Hodge, J.W.; Fantini, M.; Fujii, R.; Morillon, Y.M., 2nd; Greiner, J.W.; Padget, M.R.; Tritsch, S.R.; Tsang, K.Y.; Campbell, K.S.; et al. An NK cell line (haNK) expressing high levels of granzyme and engineered to express the high affinity CD16 allele. *Oncotarget* **2016**, *7*, 86359–86373. [CrossRef] [PubMed]

26. Satta, A.; Mezzanzanica, D.; Turatti, F.; Canevari, S.; Figini, M. Redirection of T-cell effector functions for cancer therapy: Bispecific antibodies and chimeric antigen receptors. *Future Oncol.* **2013**, *9*, 527–539. [CrossRef]

27. Zhukovsky, E.A.; Morse, R.J.; Maus, M.V. Bispecific antibodies and CARs: Generalized immunotherapeutics harnessing T cell redirection. *Curr. Opin. Immunol.* **2016**, *40*, 24–35. [CrossRef]

28. Sahu, G.K.; Sango, K.; Selliah, N.; Ma, Q.; Skowron, G.; Junghans, R.P. Anti-HIV designer T cells progressively eradicate a latently infected cell line by sequentially inducing HIV reactivation then killing the newly gp120-positive cells. *Virology* **2013**, *446*, 268–275. [CrossRef]

29. Liu, B.; Zou, F.; Lu, L.; Chen, C.; He, D.; Zhang, X.; Tang, X.; Liu, C.; Li, L.; Zhang, H. Chimeric antigen receptor T cells guided by the single-chain Fv of a broadly neutralizing antibody specifically and effectively eradicate virus reactivated from latency in CD4+ T lymphocytes isolated from HIV-1-infected individuals receiving suppressive combined antiretroviral therapy. *J. Virol.* **2016**, *90*, 9712–9724.

30. Hale, M.; Mesojednik, T.; Romano Ibarra, G.S.; Sahni, J.; Bernard, A.; Sommer, K.; Scharenberg, A.M.; Rawlings, D.J.; Wagner, T.A. Engineering HIV-resistant, anti-HIV chimeric antigen receptor T cells. *Mol. Ther.* **2017**, *25*, 570–579. [CrossRef]

31. Köhler, G.; Milstein, C. Continuous cultures of fused cells secreting antibody of predefined specificity. *Nature* **1975**, *256*, 495–497. [CrossRef] [PubMed]

32. Martz, E. Multiple target cell killing by the cytolytic T lymphocyte and the mechanism of cytotoxicity. *Transplantation* **1976**, *21*, 5–11. [CrossRef] [PubMed]

33. Rothstein, T.L.; Mage, M.; Jones, G.; McHugh, L.L. Cytotoxic T lymphocyte sequential killing of immobilized allogeneic tumor target cells measured by time-lapse microcinematography. *J. Immunol.* **1978**, *121*, 1652–1656. [PubMed]

34. Grakoui, A.; Bromley, S.K.; Sumen, C.; Davis, M.M.; Shaw, A.S.; Allen, P.M.; Dustin, M.L. The immunological synapse: A molecular machine controlling T cell activation. *Science* **1999**, *285*, 221–227. [CrossRef] [PubMed]

35. Gross, G.; Waks, T.; Eshhar, Z. Expression of immunoglobulin-T-cell receptor chimeric molecules as functional receptors with antibody-type specificity. *Proc. Natl. Acad. Sci. USA* **1989**, *86*, 10024–10028. [CrossRef] [PubMed]

36. Eshhar, Z.; Gross, G. Chimeric T cell receptor which incorporates the anti-tumour specificity of a monoclonal antibody with the cytolytic activity of T cells: A model system for immunotherapeutical approach. *Br. J. Cancer Suppl.* **1990**, *10*, 27–29. [PubMed]

37. Eshhar, Z.; Waks, T.; Gross, G.; Schindler, D.G. Specific activation and targeting of cytotoxic lymphocytes through chimeric single chains consisting of antibody binding domains and the gamma or zeta subunits of the immunoglobulin and T-cell receptors. *Proc. Natl. Acad. Sci. USA* **1993**, *90*, 720–724. [CrossRef] [PubMed]

38. Milstein, C.; Cuello, A.C. Hybrid hybridomas and their use in immunohistochemistry. *Nature* **1983**, *305*, 537–540. [CrossRef]

39. Jantscheff, P.; Winkler, L.; Karawajew, L.; Kaiser, G.; Böttger, V.; Micheel, B. Hybrid hybridomas producing bispecific antibodies to CEA and peroxidase isolated by a combination of HAT medium selection and fluorescence activated cell sorting. *J. Immunol. Methods* **1993**, *163*, 91–97. [CrossRef]

40. Brennan, M.; Davison, P.F.; Paulus, H. Preparation of bispecific antibodies by chemical recombination of monoclonal immunoglobulin G1 fragments. *Science* **1985**, *229*, 81–83. [CrossRef]

41. Staerz, U.D.; Kanagaw, O.; Bevan, M.J. Hybrid antibodies can target sites for attack by T cells. *Nature* **1985**, *314*, 628–631. [CrossRef] [PubMed]

42. Perez, P.; Hoffman, R.W.; Shaw, S.; Bluestone, J.A.; Segal, D.M. Specific targeting of cytotoxic T cells by anti-T3 linked to anti-target cell antibody. *Nature* **1985**, *316*, 354–356. [CrossRef] [PubMed]

43. Staerz, U.D.; Bevan, M.J. Hybrid hybridoma producing a bispecific monoclonal antibody that can focus effector T-cell activity. *Proc. Natl. Acad. Sci. USA* **1986**, *83*, 1453–1457. [CrossRef] [PubMed]

44. Holliger, P.; Prospero, T.; Winter, G. "Diabodies": Small bivalent and bispecific antibody fragments. *Proc. Natl. Acad. Sci. USA* **1993**, *90*, 6444–6448. [CrossRef] [PubMed]

45. Mack, M.; Riethmüller, G.; Kufer, P. A small bispecific antibody construct expressed as a functional single-chain molecule with high tumor cell cytotoxicity. *Proc. Natl. Acad. Sci. USA* **1995**, *92*, 7021–7025. [CrossRef] [PubMed]

46. Bird, R.E.; Hardman, K.D.; Jacobson, J.W.; Johnson, S.; Kaufman, B.M.; Lee, S.M.; Lee, T.; Pope, S.H.; Riordan, G.S.; Whitlow, M. Single-chain antigen-binding proteins. *Science* **1988**, *242*, 423–426. [CrossRef] [PubMed]

47. Huston, J.S.; Levinson, D.; Mudgett-Hunter, M.; Tai, M.S.; Novotný, J.; Margolies, M.N.; Ridge, R.J.; Bruccoleri, R.E.; Haber, E.; Crea, R.; et al. Protein engineering of antibody binding sites: Recovery of specific activity in an anti-digoxin single-chain Fv analogue produced in *Escherichia coli*. *Proc. Natl. Acad. Sci. USA* **1988**, *85*, 5879–5883. [CrossRef]

48. Clark, M.; Waldmann, H. T-cell killing of target cells induced by hybrid antibodies: Comparison of two bispecific monoclonal antibodies. *J. Natl. Cancer Inst.* **1987**, *79*, 1393–1401.

49. Nitta, T.; Sato, K.; Yagita, H.; Okumura, K.; Ishii, S. Preliminary trial of specific targeting therapy against malignant glioma. *Lancet* **1990**, *335*, 368–371. [CrossRef]

50. Haagen, I.A.; Van De Griend, R.; Clark, M.; Geerars, A.; Bast, B.; De Gast, B. Killing of human leukaemia/lymphoma B cells by activated cytotoxic T lymphocytes in the presence of a bispecific monoclonal antibody (alpha CD3/alpha CD19). *Clin. Exp. Immunol.* **1992**, *90*, 368–375. [CrossRef]

51. De Gast, G.C.; Van Houten, A.A.; Haagen, I.A.; Klein, S.; De Weger, R.A.; Van Dijk, A.; Phillips, J.; Clark, M.; Bast, B.J. Clinical experience with CD3 × CD19 bispecific antibodies in patients with B cell malignancies. *J. Hematother.* **1995**, *4*, 433–437. [CrossRef] [PubMed]

52. Ridgeway, J.B.; Presta, L.G.; Carter, P. 'Knobs-into-holes' engineering of antibody C_H3 domains for heavy chain heterodimerization. *Protein Eng.* **1996**, *9*, 617–621. [CrossRef]

53. Zeidler, R.; Reisbach, G.; Wollenberg, B.; Lang, S.; Chaubel, S.; Schmitt, B.; Lindhofer, H. Simultaneous activation of T cells and accessory cells by a new class of intact bispecific antibody results in efficient tumor cell killing. *J. Immunol.* **1999**, *163*, 1246–1252. [PubMed]

54. Löffler, A.; Kufer, P.; Lutterbüse, R.; Zettl, F.; Daniel, P.T.; Schwenkenbecher, J.M.; Riethmüller, G.; Dörken, B.; Bargou, R.C. A recombinant bispecific single-chain antibody, CD19 × CD3, induces rapid and high lymphoma-directed cytotoxicity by unstimulated T lymphocytes. *Blood* **2000**, *95*, 2098–2103. [PubMed]

55. Rosenberg, S.A.; Packard, B.S.; Aebersold, P.M.; Solomon, D.; Topalian, S.L.; Toy, S.T.; Simon, P.; Lotze, M.T.; Yang, J.C.; Seipp, C.A.; et al. Use of tumor infiltrating lymphocytes and interleukin-2 in the immunotherapy of patients with metastatic melanoma. Preliminary report. *N. Engl. J. Med.* **1988**, *319*, 1676–1680. [CrossRef] [PubMed]

56. Hwu, P.; Shafer, G.E.; Treisman, J.; Schindler, G.; Gross, G.; Cowherd, R.; Rosenberg, S.A.; Eshhar, Z. Lysis of ovarian cancer cells by human lymphocytes redirected with a chimeric gene composed of an antibody variable region and the Fc receptor gamma chain. *J. Exp. Med.* **1993**, *178*, 361–366. [CrossRef] [PubMed]

57. Moritz, D.; Wels, W.; Mattern, J.; Groner, B. Cytotoxic T lymphocytes with a grafted recognition specificity for ERBB2-expressing tumor cells. *Proc. Natl. Acad. Sci. USA* **1994**, *91*, 4318–4322. [CrossRef]

58. Atwell, S.; Ridgway, J.B.; Wells, J.A.; Carter, P. Stable heterodimers from remodeling the domain interface of a homodimer using a phage display library. *J. Mol. Biol.* **1997**, *270*, 26–35. [CrossRef]

59. Merchant, A.M.; Zhu, Z.; Yuan, J.Q.; Goddard, A.; Adams, C.W.; Presta, L.G.; Carter, P. An efficient route to human bispecific IgG. *Nat. Biotechnol.* **1998**, *16*, 677–681. [CrossRef]

60. Ha, J.H.; Kim, J.E.; Kim, Y.S. Immunoglobulin Fc heterodimer platform technology: From design to applications in therapeutic antibodies and proteins. *Front. Immunol.* **2016**, *7*, 394. [CrossRef]

61. Brinkmann, U.; Kontermann, R.E. The making of bispecific antibodies. *MAbs* **2017**, *9*, 182–212. [CrossRef] [PubMed]

62. Koch, J.; Tesar, M. Recombinant antibodies to arm cytotoxic lymphocytes in cancer immunotherapy. *Transfus. Med. Hemother.* **2017**, *44*, 337–350. [CrossRef] [PubMed]

63. Sebastian, M.; Kiewe, P.; Schuette, W.; Brust, D.; Peschel, C.; Schneller, F.; Rühle, K.H.; Nilius, G.; Ewert, R.; Lodziewski, S.; et al. Treatment of malignant pleural effusion with the trifunctional antibody catumaxomab (Removab) (anti-EpCAM × Anti-CD3): Results of a phase 1/2 study. *J. Immunother.* **2009**, *32*, 195–202. [CrossRef] [PubMed]

64. Burges, A.; Wimberger, P.; Kümper, C.; Gorbounova, V.; Sommer, H.; Schmalfeldt, B.; Pfisterer, J.; Lichinitser, M.; Makhson, A.; Moiseyenko, V.; et al. Effective relief of malignant ascites in patients with advanced ovarian cancer by a trifunctional anti-EpCAM × anti-CD3 antibody: A phase I/II study. *Clin. Cancer Res.* **2007**, *13*, 3899–3905. [CrossRef] [PubMed]

65. Sebastian, M. Review of catumaxomab in the treatment of malignant ascites. *Cancer Manag. Res.* **2010**, *2*, 283–286. [CrossRef]

66. Mølhøj, M.; Crommer, S.; Brischwein, K.; Rau, D.; Sriskandarajah, M.; Hoffmann, P.; Kufer, P.; Hofmeister, R.; Baeuerle, P.A. CD19-/CD3-bispecific antibody of the BiTE class is far superior to tandem diabody with respect to redirected tumor cell lysis. *Mol. Immunol.* **2007**, *44*, 1935–1943. [CrossRef] [PubMed]

67. Przepiorka, D.; Ko, C.W.; Deisseroth, A.; Yancey, C.L.; Candau-Chacon, R.; Chiu, H.J.; Gehrke, B.J.; Gomez-Broughton, C.; Kane, R.C.; Kirshner, S.; et al. FDA approval blinatumomab. *Clin. Cancer Res.* **2015**, *21*, 4035–4039. [CrossRef]

68. Bargou, R.; Leo, E.; Zugmaier, G.; Klinger, M.; Goebeler, M.; Knop, S.; Noppeney, R.; Viardot, A.; Hess, G.; Schuler, M.; et al. Tumor regression in cancer patients by very low doses of a T cell-engaging antibody. *Science* **2008**, *321*, 974–977. [CrossRef]

69. Kiewe, P.; Thiel, E. Ertumaxomab: A trifunctional antibody for breast cancer treatment. *Expert Opin. Investig. Drugs* **2008**, *17*, 1553–1558. [CrossRef]

70. Topalian, S.L.; Solomon, D.; Avis, F.P.; Chang, A.E.; Freerksen, D.L.; Linehan, W.M.; Lotze, M.T.; Robertson, C.N.; Seipp, C.A.; Simon, P.; et al. Immunotherapy of patients with advanced cancer using tumor-infiltrating lymphocytes and recombinant interleukin-2: A pilot study. *J. Clin. Oncol.* **1988**, *6*, 839–853. [CrossRef]

71. Eshhar, Z.; Waks, T.; Gross, G. The emergence of T-bodies/CAR T cells. *Cancer J.* **2014**, *20*, 123–126. [CrossRef] [PubMed]

72. Lim, W.A.; June, C.H. The principles of engineering immune cells to treat cancer. *Cell* **2017**, *168*, 724–740. [CrossRef] [PubMed]

73. Hwu, P.; Yang, J.C.; Cowherd, R.; Treisman, J.; Shafer, G.E.; Eshhar, Z.; Rosenberg, S.A. In vivo antitumor activity of T cells redirected with chimeric antibody/T-cell receptor genes. *Cancer Res.* **1995**, *55*, 3369–3373. [PubMed]

74. Kershaw, M.H.; Westwood, J.A.; Parker, L.L.; Wang, G.; Eshhar, Z.; Mavroukakis, S.A.; White, D.E.; Wunderlich, J.R.; Canevari, S.; Rogers-Freezer, L. A phase I study on adoptive immunotherapy using gene-modified T cells for ovarian cancer. *Clin. Cancer Res.* **2006**, *12*, 6106–6115. [CrossRef] [PubMed]

75. Mitsuyasu, R.T.; Anton, P.A.; Deeks, S.G.; Scadden, D.T.; Connick, E.; Downs, M.T.; Bakker, A.; Roberts, M.R.; June, C.H.; Jalali, S.; et al. Prolonged survival and tissue trafficking following adoptive transfer of CD4zeta gene-modified autologous CD4(+) and CD8(+) T cells in human immunodeficiency virus-infected subjects. *Blood* **2000**, *96*, 785–793. [PubMed]

76. Junghans, R.P.; Safar, M.; Huberman, M.S. Preclinical and phase I data of anti-CEA "designer T cell" therapy for cancer: A new immunotherapeutic modality. *Proc. Am. Assoc. Cancer Res.* **2000**, *41*, 543.

77. Eshhar, Z. The T-body approach: Redirecting T cells with antibody specificity. *Handb. Exp. Pharmacol.* **2008**, *181*, 329–342.

78. Brentjens, R.J.; Latouche, J.B.; Santos, E.; Marti, F.; Gong, M.C.; Lyddane, C.; King, P.D.; Larson, S.; Weiss, M.; Riviere, I.; et al. Eradication of systemic B-cell tumors by genetically targeted human T lymphocytes co-stimulated by CD80 and interleukin-15. *Nat. Med.* **2003**, *9*, 279–286. [CrossRef]

79. Cooper, L.J.; Topp, M.S.; Serrano, L.M.; Gonzalez, S.; Chang, W.C.; Naranjo, A.; Wright, C.; Popplewell, L.; Raubitschek, A.; Forman, S.J.; et al. T-cell clones can be rendered specific for CD19: Toward the selective augmentation of the graft-versus-B-lineage leukemia effect. *Blood* **2003**, *101*, 1637–1644. [CrossRef]

80. Pulè, M.A.; Straathof, K.C.; Dotti, G.; Heslop, H.E.; Rooney, C.M.; Brenner, M.K. A chimeric T cell antigen receptor that augments cytokine release and supports clonal expansion of primary human T cells. *Mol. Ther.* **2005**, *12*, 933–941. [CrossRef]

81. Gimmi, C.D.; Freeman, G.J.; Gribben, J.G.; Gray, G.; Nadler, L.M. Human T-cell clonal anergy is induced by antigen presentation in the absence of B7 co-stimulation. *Proc. Natl. Acad. Sci. USA* **1993**, *90*, 6586–6590. [CrossRef] [PubMed]

82. Jenkins, M.K.; Chen, C.A.; Jung, G.; Mueller, D.L.; Schwartz, R.H. Inhibition of antigen-specific proliferation of type 1 murine T cell clones after stimulation with immobilized anti-CD3 monoclonal antibody. *J. Immunol.* **1990**, *144*, 16–22.

83. Hombach, A.; Sent, D.; Schneider, C.; Heuser, C.; Koch, D.; Pohl, C.; Seliger, B.; Abken, H. T-cell activation by recombinant receptors: CD28 co-stimulation is required for interleukin 2 secretion and receptor-mediated T-cell proliferation but does not affect receptor-mediated target cell lysis. *Cancer Res.* **2001**, *61*, 1976–1982. [PubMed]

84. Finney, H.M.; Akbar, A.N.; Lawson, A.D. Activation of resting human primary T cells with chimeric receptors: Co-stimulation from CD28, inducible costimulator, CD134 and CD137 in series with signals from the TCR zeta chain. *J. Immunol.* **2004**, *172*, 104–113. [CrossRef]

85. Kowolik, C.M.; Topp, M.S.; Gonzalez, S.; Pfeiffer, T.; Olivares, S.; Gonzalez, N.; Smith, D.D.; Forman, S.J.; Jensen, M.C.; Cooper, L.J. CD28 co-stimulation provided through a CD19-specific chimeric antigen receptor enhances in vivo persistence and antitumor efficacy of adoptively transferred T cells. *Cancer Res.* **2006**, *66*, 10995–11004. [CrossRef]

86. Hombach, A.A.; Abken, H. Costimulation by chimeric antigen receptors revisited the T cell antitumor response benefits from combined CD28-OX40 signalling. *Int. J. Cancer* **2011**, *129*, 2935–2944. [CrossRef] [PubMed]

87. Milone, M.C.; Fish, J.D.; Carpenito, C.; Carroll, R.G.; Binder, G.K.; Teachey, D.; Samanta, M.; Lakhal, M.; Gloss, B.; Danet-Desnoyers, G.; et al. Chimeric receptors containing CD137 signal transduction domains mediate enhanced survival of T cells and increased antileukemic efficacy *in vivo*. *Mol. Ther.* **2009**, *17*, 1453–1464. [CrossRef]

88. Moritz, D.; Groner, B. A spacer region between the single chain antibody and the CD3 zeta-chain domain of chimeric T cell receptor components is required for efficient ligand binding and signaling activity. *Gene Ther.* **1995**, *2*, 539–546.

89. Guest, R.D.; Hawkins, R.E.; Kirillova, N.; Kirillova, N.; Cheadle, E.J.; Arnold, J.; O'Neill, A.; Irlam, J.; Chester, K.A.; Kemshead, J.T.; et al. The role of extracellular spacer regions in the optimal design of chimeric immune receptors: Evaluation of four different scFvs and antigens. *J. Immunother.* **2005**, *28*, 203–211. [CrossRef]

90. Hudecek, M.; Sommermeyer, D.; Kosasih, P.L.; Silva-Benedict, A.; Liu, L.; Rader, C.; Jensen, M.C.; Riddell, S.R. The nonsignaling extracellular spacer domain of chimeric antigen receptors is decisive for in vivo antitumor activity. *Cancer Immunol. Res.* **2015**, *3*, 125–135. [CrossRef]

91. Qin, L.; Lai, Y.; Zhao, R.; Wei, X.; Weng, J.; Lai, P.; Li, B.; Lin, S.; Wang, S.; Wu, Q.; et al. Incorporation of a hinge domain improves the expansion of chimeric antigen receptor T cells. *J. Hematol. Oncol.* **2017**, *10*, 68. [CrossRef] [PubMed]

92. Zhao, Z.; Condomines, M.; Van Der Stegen, S.J.C.; Perna, F.; Kloss, C.C.; Gunset, G.; Plotkin, J.; Sadelain, M. Structural design of engineered co-stimulation determines tumor rejection kinetics and persistence of CAR T cells. *Cancer Cell* **2015**, *28*, 415–428. [CrossRef] [PubMed]

93. Yu, Z.; Prinzing, B.; Cao, F.; Gottschalk, S.; Krenciute, G. Optimizing EphA2-CAR T cells for the adoptive immunotherapy of glioma. *Mol. Ther. Methods Clin. Dev.* **2018**, *9*, 70–80.

94. Guedan, S.; Posey, A.D., Jr.; Shaw, C.; Wing, A.; Da, T.; Patel, P.R.; McGettigan, S.E.; Casado-Medrano, V.; Kawalekar, O.U.; Uribe-Herranz, M. Enhancing CAR T cell persistence through ICOS and 4-1BB co-stimulation. *JCI Insight* **2018**, *3*, 96976. [CrossRef] [PubMed]

95. Pang, Y.; Hou, X.; Yang, C.; Liu, Y.; Jiang, G. Advances on chimeric antigen receptor-modified T-cell therapy for oncotherapy. *Mol. Cancer* **2018**, *17*, 91. [CrossRef] [PubMed]

96. Kasakovski, D.; Xu, L.; Li, Y. T cell senescence and CAR-T cell exhaustion in hematological malignancies. *J. Hematol. Oncol.* **2018**, *11*, 91. [CrossRef]

97. Ramello, M.C.; Benzaïd, I.; Kuenzi, B.M.; Lienlaf-Moreno, M.; Kandell, W.M.; Santiago, D.N.; Pabón-Saldaña, M.; Darville, L.; Fang, B.; Rix, U. An immunoproteomic approach to characterize the CAR interactome and signalosome. *Sci. Signal.* **2019**, *12*, eaap9777. [CrossRef]

98. Di Stasi, A.; Tey, S.K.; Dotti, G.; Fujita, Y.; Kennedy-Nasser, A.; Martinez, C.; Straathof, K.; Liu, E.; Durett, A.G.; Grilley, B.; et al. Inducible apoptosis as a safety switch for adoptive cell therapy. *N. Engl. J. Med.* **2011**, *365*, 1673–1683. [CrossRef]

99. Chmielewski, M.; Abken, H. CAR T cells transform to trucks: Chimeric antigen receptor–redirected T cells engineered to deliver inducible IL-12 modulate the tumour stroma to combat cancer. *Cancer Immunol. Immunother.* **2012**, *61*, 1269–1277. [CrossRef]

100. Moon, E.K.; Carpenito, C.; Sun, J.; Wang, L.C.; Kapoor, V.; Predina, J.; Powel, D.J., Jr.; Riley, J.L.; June, C.H.; Albelda, S.M. Expression of a functional CCR2 receptor enhances tumor localization and tumor eradication by retargeted human T cells expressing a mesothelin-specific chimeric antibody receptor. *Clin. Cancer Res.* **2011**, *17*, 4719–4730. [CrossRef]

101. Rossig, C.; Bollard, C.M.; Nuchtern, J.G.; Rooney, C.M.; Brenner, M.K. Epstein-Barr virus-specific human T lymphocytes expressing antitumor chimeric T-cell receptors: Potential for improved immunotherapy. *Blood* **2002**, *99*, 2009–2016. [CrossRef] [PubMed]

102. Pulè, M.A.; Savoldo, B.; Myers, G.D.; Rossig, C.; Russell, H.V.; Dotti, G.; Huls, M.H.; Liu, E.; Gee, A.P.; Mei, Z.; et al. Virus-specific T cells engineered to coexpress tumor-specific receptors: Persistence and antitumor activity in individuals with neuroblastoma. *Nat. Med.* **2008**, *14*, 1264–1270. [CrossRef] [PubMed]

103. Rossig, C.; Pulè, M.; Altvater, B.; Saiagh, S.; Wright, G.; Ghorashian, S.; Clifton-Hadley, L.; Champion, K.; Sattar, Z.; Popova, B.; et al. Vaccination to improve the persistence of CD19CAR gene-modified T cells in relapsed pediatric acute lymphoblastic leukemia. *Leukemia* **2017**, *31*, 1087–1095. [CrossRef] [PubMed]

104. Geyer, M.B. First CAR to pass the road test. Tisagenlecleucel's drive to FDA approval. *Clin. Cancer Res.* **2019**, *25*, 1133–1135. [CrossRef] [PubMed]

105. Bouchkouj, N.; Kasamon, Y.L.; De Claro, R.A.; George, B.; Lin, X.; Lee, S.; Blumenthal, G.M.; Bryan, W.; McKee, A.E.; Pazdur, R. FDA approval summary: Axicabtagene ciloleucel for relapsed or refractory large B-cell lymphoma. *Clin. Cancer Res.* **2019**, *25*, 1702–1708. [CrossRef] [PubMed]

106. Stinchcombe, J.C.; Majorovits, E.; Bossi, G.; Fuller, S.; Griffiths, G.M. Centrosome polarization delivers secretory granules to the immunological synapse. *Nature* **2006**, *443*, 462–465. [CrossRef] [PubMed]

107. Kabanova, A.; Zurli, V.; Baldari, C.T. Signals controlling lytic granule polarization at the cytotoxic immune synapse. *Front. Immunol.* **2018**, *9*, 307. [CrossRef] [PubMed]

108. Liu, D.; Tian, S.; Zhang, K.; Xiong, W.; Lubaki, N.M.; Chen, Z.; Han, W. Chimeric antigen receptor (CAR)-modified natural killer cell-based immunotherapy and immunological synapse formation in cancer and HIV. *Protein Cell* **2017**, *8*, 861–877. [CrossRef] [PubMed]

109. Stinchcombe, J.C.; Bossi, G.; Booth, S.; Griffiths, G.M. The immunological synapse of CTL contains a secretory domain and membrane bridges. *Immunity* **2001**, *15*, 751–761. [CrossRef]

110. Huppa, J.B.; Davis, M.M. T-cell-antigen recognition and the immunological synapse. *Nat. Rev.* **2003**, *3*, 973–983. [CrossRef]

111. Watanabe, K.; Kuramitsu, S.; Posey, A.D., Jr.; June, C.H. Expanding the therapeutics window for CAR-T cell therapy in solid tumors: The knowns and unknowns of CAR-T cell biology. *Front. Immunol.* **2018**, *9*, 2486. [CrossRef] [PubMed]

112. Davenport, A.J.; Cross, R.S.; Watson, K.A.; Liao, Y.; Shi, W.; Prince, H.M.; Beavis, P.A.; Trapani, J.A.; Kershaw, M.H.; Ritchie, D.S.; et al. Chimeric antigen receptor T cells form nonclassical and potent immune synapses driving rapid cytotoxicity. *Proc. Natl. Acad. Sci. USA* **2018**, *115*, E2068–E2076. [CrossRef] [PubMed]

113. Davenport, A.J.; Jenkins, M.R. Programming a serial killer: CAR T cells form non-classical immune synapses. *Oncoscience* **2018**, *5*, 69–70. [PubMed]

114. Wucherpfennig, K.W.; Gagnon, E.; Call, M.J.; Huseby, E.S.; Call, M.E. Structural biology of the T-cell receptor: Insights into receptor assembly, ligand recognition and initiation of signaling. *Cold Spring Harb. Perspect. Biol.* **2009**, *2*, a005140. [CrossRef] [PubMed]

115. Artyomov, M.N.; Lis, M.; Devadas, S.; Davis, M.M.; Chakraborty, A.K. CD4 and CD8 binding to MHC molecules primarily acts to enhance Lck delivery. *Proc. Natl. Acad. Sci. USA* **2010**, *107*, 16916–16921. [CrossRef]

116. Engels, B.; Engelhard, V.H.; Sidney, J.; Sette, A.; Binder, D.C.; Liu, R.B.; Kranz, D.M.; Meredith, S.C.; Rowley, D.A.; Schreiber, H. Relapse or eradication of cancer is predicted by peptide-major histocompatibility complex affinity. *Cancer Cell* **2013**, *23*, 516–526. [CrossRef] [PubMed]

117. Furlan, G.; Minowa, T.; Hanagata, N.; Kataoka-Hamai, C.; Kaizuka, Y. Phosphatase CD45 both positively and negatively regulates T cell receptor phosphorylation in reconstituted membrane protein clusters. *J. Biol. Chem.* **2014**, *289*, 28514–28525. [CrossRef]

118. Penninger, J.M.; Irie-Sasaki, J.; Sasaki, T.; Oliveira-dos-Santos, A.J. CD45: New jobs for an old acquaintance. *Nat. Immunol.* **2001**, *2*, 389–396. [CrossRef]

119. Choudhuri, K.; Wiseman, D.; Brown, M.H.; Gould, K.; Van Der Merwe, P.A. T-cell receptor triggering is critically dependent on the dimensions of its peptide-MHC ligand. *Nature* **2005**, *436*, 578–582. [CrossRef]

120. Sykulev, Y.; Joo, M.; Vturina, I.; Tsomides, T.J.; Eisen, H.N. Evidence that a single peptide-MHC complex on a target cell can elicit a cytolytic T cell response. *Immunity* **1996**, *4*, 565–571. [CrossRef]

121. Purbhoo, M.A.; Irvine, D.J.; Huppa, J.B.; Davis, M.M. T cell killing does not require the formation of a stable mature immunological synapse. *Nat. Immunol.* **2004**, *5*, 524–530. [CrossRef] [PubMed]

122. Gwalani, L.A.; Orange, J.S. Single degranulations in NK cells can mediate target cell killing. *J. Immunol.* **2018**, *200*, 3231–3243. [CrossRef] [PubMed]

123. Benmebarek, M.R.; Karches, C.H.; Cadilha, B.L.; Lesch, S.; Endres, S.; Kobold, S. Killing mechanisms of chimeric antigen receptor (CAR) T cells. *Int. J. Mol. Sci.* **2019**, *20*, 1283. [CrossRef] [PubMed]

124. Cazaux, M.; Grandjean, C.L.; Lemaître, F.; Garcia, Z.; Beck, R.J.; Milo, I.; Postat, J.; Beltman, J.B.; Cheadle, E.J.; Bousso, J. Single-cell imaging of CAR T cell activity in vivo reveals extensive functional and anatomical heterogeneity. *J. Exp. Med.* **2019**, *216*, 1038. [CrossRef] [PubMed]

125. Xiong, W.; Chen, Y.; Kang, X.; Chen, Z.; Zheng, P.; Hsu, Y.H.; Jang, J.H.; Qin, L.; Liu, H.; Dotti, G.; et al. Immunological synapse predicts effectiveness of chimeric antigen receptor cells. *Mol. Ther.* **2018**, *26*, 963–975. [CrossRef] [PubMed]

126. Purbhoo, M.A.; Sutton, D.H.; Brewer, J.E.; Mullings, R.E.; Hill, M.E.; Mahon, T.M.; Karbach, J.; Jäger, E.; Cameron, B.J.; Lissin, N.; et al. Quantifying and imaging NY-ESO-1/LAGE-1-derived epitopes on tumzr cells using high affinity T cell receptors. *J. Immunol.* **2006**, *176*, 7308–7316. [CrossRef]

127. Stone, J.D.; Aggen, D.H.; Schietinger, A.; Schreiber, H.; Kranz, D.M. A sensitivity scale for targeting T cells with chimeric antigen receptors (CARs) and bispecific T-cell Engagers (BiTEs). *Oncoimmunology* **2012**, *1*, 863–873. [CrossRef] [PubMed]

128. Watanabe, K.; Terakura, S.; Martens, A.C.; Van Meerten, T.; Uchiyama, S.; Imai, M.; Sakemura, R.; Goto, T.; Hanajiri, R.; Imahashi, N.; et al. Target antigen density governs the efficacy of anti-CD20-CD28-CD3 ζ chimeric antigen receptor-modified effector CD8+ T cells. *J. Immunol.* **2015**, *194*, 911–920. [CrossRef]

129. Au-Yeung, B.B.; Zikherman, J.; Mueller, J.L.; Ashouri, J.F.; Matloubian, M.; Cheng, D.A.; Chen, Y.; Shokat, K.M.; Weiss, A. A sharp T-cell antigen receptor signaling threshold for T-cell proliferation. *Proc. Natl. Acad. Sci. USA* **2014**, *111*, E3679–E3688. [CrossRef]

130. Hamieh, M.; Dobrin, A.; Cabriolu, A.; Van Der Stegen, S.J.C.; Giavridis, T.; Mansilla-Soto, J.; Eyquem, J.; Zhao, Z.; Whitlock, B.M.; Miele, M.M.; et al. CAR T cell trogocytosis and cooperative killing regulate tumour antigen escape. *Nature* **2019**, *568*, 112–116. [CrossRef]

131. Wolf, E.; Hofmeister, R.; Kufer, P.; Schlereth, B.; Baeuerle, P.A. BiTEs: Bispecific antibody constructs with unique anti-tumor activity. *Drug Discov. Today* **2005**, *10*, 1237–1244. [CrossRef]

132. Parry, R.V.; Chemnitz, J.M.; Frauwirth, K.A.; Lanfranco, A.R.; Braunstein, I.; Kobayashi, S.V.; Linsley, P.S.; Thompson, C.B.; Riley, J.L. CTLA-4 and PD-1 receptors inhibit T-cell activation by distinct mechanisms. *Mol. Cell. Biol.* **2005**, *25*, 9543–9553. [CrossRef] [PubMed]

133. Velasquez, M.P.; Szoor, A.; Vaidya, A.; Thakkar, A.; Nguyen, P.; Wu, M.-F.; Liu, H.; Gottschalk, S. CD28 and 41BB co-stimulation enhances the effector function of CD19-specific engager T cells. *Cancer Immunol. Res.* **2017**, *5*, 860–870. [CrossRef] [PubMed]

134. Quintarelli, C.; Orlando, D.; Boffa, I.; Guercio, M.; Polito, V.A.; Petretto, A.; Lavarello, C.; Sinibaldi, M.; Weber, G.; Del Bufalo, F.; et al. Choice of costimulatory domains and of cytokines determines CAR T-cell activity in neuroblastoma. *Oncoimmunology* **2018**, *7*, e1433518. [CrossRef] [PubMed]

135. Feucht, J.; Kayser, S.; Gorodezki, D.; Hamieh, M.; Döring, M.; Blaeschke, F.; Schlegel, P.; Bösmüller, H.; Quintanilla-Fend, L.; Ebinger, M.; et al. T-cell responses against CD19+ pediatric acute lymphoblastic leukemia mediated by bispecific T-cell engager (BiTE) are regulated contrarily by PD-L1 and CD80/CD86 on leukemic blasts. *Oncotarget* **2016**, *7*, 76902–76919. [CrossRef] [PubMed]

136. Herrmann, M.; Krupka, C.; Deiser, K.; Brauchle, B.; Marcinek, A.; Ogrinc Wagner, A.; Rataj, F.; Mocikat, R.; Metzeler, K.H.; Spiekermann, K.; et al. Bifunctional PD-1 × αCD3 × αCD33 fusion protein reverses adaptive immune escape in acute myeloid leukemia. *Blood* **2018**, *132*, 2484–2494. [CrossRef]

137. Kobold, S.; Pantelyushin, S.; Rataj, F.; Berg, J.V. Rationale for combining bispecific T cell activating antibodies with checkpoint blockade for cancer therapy. *Front. Oncol.* **2018**, *8*, 285. [CrossRef]

138. Cherkassky, L.; Morello, A.; Villena-Vargas, J.; Feng, Y.; Dimitrov, D.S.; Jones, D.R.; Sadelain, M.; Adusumilli, P.S. Human CAR T cells with cell-intrinsic PD-1 checkpoint blockade resist tumor-mediated inhibition. *J. Clin. Investig.* **2016**, *8*, 3130–3144. [CrossRef]

139. Blinatumomab Relapsed/Refractory Acute Leukemia or Lymphoma—Clinical Trial: NCT03605589. Available online: https://clinicaltrials.gov/ct2/show/NCT03605589?term=NCT03605589&rank=1 (accessed on 4 April 2019).

140. Safety and Efficacy of Blinatumomab—KEYNOTE-348 Clinical Trial: NCT03340766. Available online: https://clinicaltrials.gov/ct2/show/NCT03340766?term=NCT03340766&rank=1 (accessed on 4 April 2019).

141. Blinatumomab in CD19+ Precursor B-Lymphoblastic Leukemia. Clinical Trial: NCT02879695. Available online: https://clinicaltrials.gov/ct2/show/NCT02879695?term=NCT02879695&rank=1 (accessed on 4 April 2019).

142. CART-EGFR-vIII+ Pembrolizumab Clinical Trial: NCT03726515. Available online: https://clinicaltrials.gov/ct2/show/NCT03726515?term=NCT03726515&rank=1 (accessed on 4 April 2019).

143. Suarez, E.R.; Chang, D.-K.; Sun, J.; Sui, J.; Freeman, G.J.; Signoretti, S.; Zhu, Q.; Marasco, W.A. Chimeric antigen receptor T cells secreting anti-PD-L1 antibodies more effectively regress renal cell carcinoma in a humanized mouse model. *Oncotarget* **2016**, *7*, 34341–34355. [CrossRef]

144. Rafiq, S.; Yeku, O.O.; Jackson, H.J.; Purdon, T.J.; Van Leeuwen, D.G.; Drakes, D.J.; Song, M.; Miele, M.M.; Li, Z.; Wang, P.; et al. Targeted delivery of a PD-1-blocking scFv by CAR-T cells enhances anti-tumor efficacy *in vivo*. *Nat. Biotechnol.* **2018**, *36*, 847–856. [CrossRef]

145. CTLA-4 and PD-1 Expressing EGFR-CAR-T Cells Clinical Trial: NCT03182816. Available online: https://clinicaltrials.gov/ct2/show/NCT03182816?term=NCT03182816&rank=1 (accessed on 4 April 2019).

146. PD-1 Gene-Knocked Out Mesothelin-Directed CAR-T Cells. Clinical Trial: NCT03747965. Available online: https://clinicaltrials.gov/ct2/show/NCT03747965?term=NCT03747965&rank=1 (accessed on 4 April 2019).

147. CD16/IL-15/CD33 Tri-Specific Killer Engagers (TriKEs) Clinical Trial: NCT03214666. Available online: https://clinicaltrials.gov/ct2/show/NCT03214666?term=NCT03214666&rank=1 (accessed on 10 April 2019).

148. AFM13 in Relapsed/Refractory Cutaneous Lymphomas Clinical Trial: NCT03192202. Available online: https://clinicaltrials.gov/ct2/show/NCT03192202?term=NCT03192202&rank=1 (accessed on 10 April 2019).

149. Wang, Q.; Chen, Y.; Park, J.; Liu, X.; Hu, Y.; Wang, T.; McFarland, K.; Betenbaugh, M.J. Design and biomanufacturing of bispecific antibodies. *Antibodies* **2019**. under review.

150. Husain, B.; Ellerman, D. Expanding the boundaries of biotherapeutics with bispecific antibodies. *BioDrugs* **2018**, *32*, 441–464. [CrossRef] [PubMed]

151. Bhatta, P.; Humphreys, D.F. Relative contribution of framework and CDR regions in antibody variable domains to multimerisation of Fv- and scFv-containing bispecific antibodies. *Antibodies* **2018**, *7*, 35. [CrossRef]

152. Labrijn, A.F.; Janmaat, M.L.; Reichert, J.M.; Parren, P.W.H.I. Bispecific antibodies: A mechanistic review of the pipeline. *Nat. Rev. Drug Disc.* **2019**, *1*. [CrossRef] [PubMed]

153. Johnson, S.; Burke, S.; Huang, L.; Gorlatov, S.; Li, H.; Wang, W.; Zhang, W.; Tuaillon, N.; Rainey, J.; Barat, B.; et al. Effector cell recruitment with novel Fv-based dual-affinity re-targeting protein leads to potent tumor cytolysis and in vivo B-cell depletion. *J. Mol. Biol.* **2010**, *399*, 436–449. [CrossRef] [PubMed]

154. Liu, L.; Lam, C.K.; Long, V.; Widjaja, L.; Yang, Y.; Li, H.; Jin, L.; Burke, S.; Gorlatov, S.; Brown, J.; et al. MGD011, a CD19 × CD3 dual-affinity retargeting bi-specific molecule incorporating extended circulating half-life for the treatment of B-cell malignancies. *Clin. Cancer Res.* **2017**, *23*, 1506–1518. [CrossRef]

155. Bossi, G.; Buisson, S.; Oates, J.; Jakobsen, B.K.; Hassan, N.J. ImmTAC-redirected tumour cell killing induces and potentiates antigen cross-presentation by dendritic cells. *Cancer Immunol. Immunother.* **2014**, *63*, 437–448. [CrossRef] [PubMed]

156. Labrijn, A.F.; Meesters, J.I.; De Goeij, B.E.; Van Den Bremer, E.T.; Neijssen, J.; Van Kampen, M.D.; Strumane, K.; Verploegen, S.; Kundu, A.; Gramer, M.J.; et al. Efficient generation of stable bispecific IgG1 by controlled Fab-arm exchange. *Proc. Natl. Acad. Sci. USA* **2013**, *110*, 5145–5250. [CrossRef]

157. Skegro, D.; Stutz, C.; Ollier, R.; Svensson, E.; Wassmann, P.; Bourquin, F.; Monney, T.; Gn, S.; Blein, S. Immunoglobulin domain interface exchange as a platform technology for the generation of Fc heterodimers and bispecific antibodies. *J. Biol. Chem.* **2017**, *292*, 9745–9759. [CrossRef]

158. Moore, G.L.; Bautista, C.; Pong, E.; Nguyen, D.H.; Jacinto, J.; Eivazi, A.; Muchhal, U.S.; Karki, S.; Chu, S.Y.; Lazar, G.A. A novel bispecific antibody format enables simultaneous bivalent and monovalent co-engagement of distinct target antigens. *MAbs* **2011**, *3*, 546–557. [CrossRef]

159. Bacac, M.; Klein, C.; Umaña, P. CEA TCB: A novel head-to-tail 2:1 T cell bispecific antibody for treatment of CEA-positive solid tumors. *Oncoimmunol.* **2016**, *5*, e1203498. [CrossRef] [PubMed]

160. Bacac, M.; Umaña, P.; Herter, S.; Colombetti, S.; Sam, J.; Le Clech, M.; Freimoser-Grundschober, A.; Richard, M.; Nicolini, V.; Gerdes, C.; et al. CD20 Tcb (RG6026), a novel "2:1" T cell bispecific antibody for the treatment of B cell malignancies. *Blood* **2016**, *128*, 1836.

161. Sampei, Z.; Igawa, T.; Soeda, T.; Okuyama-Nishida, Y.; Moriyama, C.; Wakabayashi, T.; Tanaka, E.; Muto, A.; Kojima, T.; Kitazawa, T.; et al. Identification and multidimensional optimization of an asymmetric bispecific IgG antibody mimicking the function of factor VIII cofactor activity. *PLoS ONE* **2013**, *8*, e57479. [CrossRef] [PubMed]

162. Shiraiwa, H.; Narita, A.; Kamata-Sakurai, M.; Ishiguro, T.; Sano, Y.; Hironiwa, N.; Tsushima, T.; Segawa, H.; Tsunenari, T.; Ikeda, Y.; et al. Engineering a bispecific antibody with a common light chain: Identification and optimization of an anti-CD3 epsilon and anti-GPC3 bispecific antibody, ERY974. *Methods* **2019**, *154*, 10–20. [CrossRef] [PubMed]

163. Kipriyanov, S.M.; Moldenhauer, G.; Schuhmacher, J.; Cochlovius, B.; Von Der Lieth, C.-W.; Matys, E.R.; Little, M. Bispecific tandem diabody for tumor therapy with improved antigen binding and pharmacokinetics. *J. Mol. Biol.* **1999**, *293*, 41–56. [CrossRef] [PubMed]

164. Hernandez-Hoyos, G.; Sewell, T.; Bader, R.; Bannink, J.; Chenault, R.A.; Daugherty, M.; Dasovich, M.; Fang, H.; Gottschalk, R.; Kumer, J.; et al. MOR209/ES414, a novel bispecific antibody targeting PSMA for the treatment of metastatic castration-resistant prostate cancer. *Mol. Cancer Ther.* **2016**, *15*, 2155–2165. [CrossRef] [PubMed]

165. Madrenas, J.; Chau, L.A.; Teft, W.A.; Wu, P.W.; Jussif, J.; Kasaian, M.; Carreno, B.M.; Ling, V. Conversion of CTLA-4 from inhibitor to activator of T cells with a bispecific tandem single-chain Fv ligand. *J. Immunol.* **2004**, *172*, 5948–5956. [CrossRef] [PubMed]

166. Stieglmaier, J.; Benjamin, J.; Nagorsen, D. Utilizing the BiTE (bispecific T-cell engager) platform for immunotherapy of cancer. *Expert Opin. Biol. Ther.* **2015**, *15*, 1093–1099. [CrossRef] [PubMed]

167. Conrath, K.E.; Lauwereys, M.; Wyns, L.; Muyldermans, S. Camel single-domain antibodies as modular building units in bispecific and bivalent antibody constructs. *J. Biol. Chem.* **2001**, *276*, 7346–7350. [CrossRef]

168. Müller, D.; Karle, A.; Meissburger, B.; Höfig, I.; Stork, R.; Kontermann, R.E. Improved pharmacokinetics of recombinant bispecific antibody molecules by fusion to human serum albumin. *J. Biol. Chem.* **2007**, *282*, 12650–12660. [CrossRef]

169. Gleason, M.K.; Verneris, M.R.; Todhunter, D.A.; Zhang, B.; McCullar, V.; Zhou, S.X.; Panoskaltsis-Mortari, A.; Weiner, L.M.; Vallera, D.A.; Miller, J.S. Bispecific and trispecific killer cell engagers directly activate human NK cells through CD16 signaling and induce cytotoxicity and cytokine production. *Mol. Cancer Ther.* **2012**, *11*, 2674–2684. [CrossRef] [PubMed]

170. Tay, S.S.; Carol, H.; Biro, M. TriKEs and BiKEs join CARs on the cancer immunotherapy highway. *Hum. Vaccines Immunother.* **2016**, *12*, 2790–2796. [CrossRef] [PubMed]

171. Felices, M.; Lenvik, T.R.; Davis, Z.B.; Miller, J.S.; Vallera, D.A. Generation of BiKEs and TriKEs to improve NK cell-mediated targeting of tumor cells. *Meth. Mol. Biol.* **2016**, *1441*, 333–346.

172. Schmohl, J.U.; Felices, M.; Taras, E.; Miller, J.S.; Vallera, D.A. Enhanced ADCC and NK cell activation of an anticarcinoma bispecific antibody by genetic insertion of a modified IL-15 cross-linker. *Mol. Ther.* **2016**, *24*, 1312–1322. [CrossRef] [PubMed]

173. Schoffelen, R.; Boerman, O.C.; Goldenberg, D.M.; Sharkey, R.M.; Van Herpen, C.M.; Franssen, G.M.; McBride, W.J.; Chang, C.H.; Rossi, E.A.; Van Der Graaf, W.T.; et al. Development of an imaging-guided CEA-pretargeted radionuclide treatment of advanced colorectal cancer: First clinical results. *Br. J. Cancer* **2013**, *109*, 934–942. [CrossRef] [PubMed]

174. Bates, A.; Power, C.A. David vs. Goliath: The structure, function, and clinical prospects of antibody fragments. *Antibodies* **2019**, *8*, 28. [CrossRef]

175. Austin, R.; Aaron, W.; Baeuerle, P.; Barath, M.; Jones, A.; Jones, S.D.; Law, C.-L.; Kwant, K.; Lemon, B.; Muchnik, A.; et al. HPN536, a T cell-engaging, mesothelin/CD3-specific TriTAC for the treatment of solid tumors (abstract). *Cancer Res.* **2018**, *78*, 1781.

176. Yang, F.; Wen, W.; Qin, W. Bispecific antibodies as a development platform for new concepts and treatment strategies. *Int. J. Mol. Sci.* **2017**, *18*, 48. [CrossRef]

177. Ellerson, J.R.; Yasmeen, D.; Painter, R.H.; Dorrington, K.J. Structure and function of immunoglobulin domains. III. Isolation and characterization of a fragment corresponding to the Cgamma2 homology region of human immunoglobin G1. *J. Immunol.* **1976**, *116*, 510–517.

178. Gunasekaran, K.; Pentony, M.; Shen, M.; Garrett, L.; Forte, C.; Woodward, A.; Ng, S.B.; Born, T.; Retter, M.; Manchulenko, K.; et al. Enhancing antibody Fc heterodimer formation through electrostatic steering effects: Applications to bispecific molecules and monovalent IgG. *J. Biol. Chem.* **2010**, *285*, 19637–19646. [CrossRef]

179. Von Kreudenstein, T.S.; Escobar-Carbrera, E.; Lario, P.I.; D'Angelo, I.; Brault, K.; Kelly, J.; Durocher, Y.; Baardsnes, J.; Woods, R.J.; Xie, M.H.; et al. Improving biophysical properties of a bispecific antibody scaffold to aid developability: Quality by molecular design. *MAbs* **2013**, *5*, 646–654. [CrossRef] [PubMed]

180. Strop, P.; Ho, W.H.; Boustany, L.M.; Abdiche, Y.N.; Lindquist, K.C.; Farias, S.E.; Rickert, M.; Appah, C.T.; Pascua, E.; Radcliffe, T.; et al. Generating bispecific human IgG1 and IgG2 antibodies from any antibody pair. *J. Mol. Biol.* **2012**, *420*, 204–219. [CrossRef] [PubMed]

181. Klein, C.; Sustmann, C.; Thomas, M.; Stubenrauch, K.; Croasdale, R.; Schanzer, J.; Brinkmann, U.; Kettenberger, H.; Regula, J.T.; Schaefer, W. Progress in overcoming the chain association issue in bispecific heterodimeric IgG antibodies. *MAbs* **2012**, *4*, 653–663. [CrossRef] [PubMed]

182. Navicixizumabum, Proposed INN: List 114. WHO Drug Information 2015. Volume 29, pp. 550–551. Available online: https://www.who.int/medicines/publications/druginformation/issues/PL_114.pdf?ua=1 (accessed on 19 April 2019).

183. Moore, G.L.; Bernett, M.J.; Rashid, R.; Pong, E.W.; Nguyen, D.T.; Jacinto, J.; Eivazi, A.; Nisthal, A.; Diaz, J.E.; Chu, S.Y.; et al. A robust heterodimeric Fc platform engineered for efficient development of bispecific antibodies of multiple formats. *Methods* **2019**, *154*, 38–50. [CrossRef] [PubMed]

184. De Nardis, C.; Hendriks, L.J.A.; Poirier, E.; Arvinte, T.; Gros, P.; Bakker, A.B.H.; De Kruif, J. A new approach for generating bispecific antibodies based on a common light chain format and the stable architecture of human immunoglobulin G_1. *J. Biol. Chem.* **2017**, *292*, 14706–14717. [CrossRef] [PubMed]

185. Smith, E.J.; Olson, K.; Haber, L.J.; Varghese, B.; Duramad, P.; Tustian, A.D.; Oyejide, A.; Kirshner, J.R.; Canova, L.; Menon, J.; et al. A novel, native-format bispecific antibody triggering T-cell killing of B-cells is robustly active in mouse tumor models and cynomolgus monkeys. *Sci. Rep.* **2015**, *5*, 17943. [CrossRef] [PubMed]

186. Tustian, A.D.; Endicott, C.; Adams, B.; Mattila, J.; Bak, H. Development of purification processes for fully human bispecific antibodies based upon modification of protein A binding avidity. *MAbs* **2016**, *8*, 828–838. [CrossRef] [PubMed]

187. Davis, J.H.; Aperlo, C.; Li, Y.; Kurosawa, E.; Lan, Y.; Lo, K.-M.; Huston, J.S. SEEDbodies: Fusion proteins based on strand-exchange engineered domain (SEED) C_H3 heterodimers in an Fc analogue platform for asymmetric binders or immunofusions and bispecific antibodies. *Protein Eng. Des. Sel.* **2010**, *23*, 195–202. [CrossRef]

188. Muda, M.; Gross, A.W.; Dawson, J.P.; He, C.; Kurosawa, E.; Schweickhardt, R.; Dugas, M.; Soloviev, M.; Bernhardt, A.; Fischer, D.; et al. Therapeutic assessment of SEED: A new engineered antibody platform designed to generate mono- and bispecific antibodies. *Protein Eng. Des. Sel.* **2011**, *24*, 447–454. [CrossRef]

189. Fischer, N.; Elson, G.; Magistrelli, G.; Dheilly, E.; Fouque, N.; Laurendon, A.; Gueneau, F.; Ravn, U.; Depoisier, J.F.; Moine, V.; et al. Exploiting light chains for the scalable generation and platform purification of native human bispecific IgG. *Nat. Commun.* **2015**, *6*, 6113. [CrossRef]

190. Choi, H.J.; Kim, Y.J.; Choi, D.K.; Kim, Y.S. Engineering of immunoglobulin Fc heterodimers using yeast surface-displayed combinatorial Fc library screening. *PLoS ONE* **2015**, *10*, e0145349. [CrossRef] [PubMed]

191. Leaver-Fay, A.; Froning, K.J.; Atwell, S.; Aldaz, H.; Pustilnik, A.; Lu, F.; Huang, F.; Yuan, R.; Hassanali, S.; Chamberlain, A.K.; et al. Computationally designed bispecific antibodies using negative state repertoires. *Structure* **2016**, *24*, 641–651. [CrossRef] [PubMed]

192. Suresh, M.R.; Cuello, A.C.; Milstein, C. Bispecific monoclonal antibodies from hybrid hybridomas. *Meth. Enzymol.* **1986**, *121*, 210–228. [PubMed]

193. Van Blarcom, T.; Lindquist, K.; Melton, Z.; Cheung, W.L.; Wagstrom, C.; McDonough, D.; Valle Oseguera, C.; Ding, S.; Rossi, A.; Potluri, S.; et al. Productive common light chain libraries yield diverse panels of high affinity bispecific antibodies. *MAbs* **2018**, *10*, 256–268. [CrossRef] [PubMed]

194. Krah, S.; Sellmann, C.; Rhiel, L.; Schröter, C.; Dickgiesser, S.; Beck, J.; Zielonka, S.; Toleikis, L.; Hock, B.; Kolmar, H.; et al. Engineering bispecific antibodies with defined chain pairing. *Nat. Biotechnol.* **2017**, *39*, 167–173. [CrossRef]

195. Schaefer, W.; Regula, J.T.; Bähner, M.; Schanzer, J.; Croasdale, R.; Dürr, H.; Gassner, C.; Georges, G.; Kettenberger, H.; Imhof-Jung, S.; et al. Immunoglobulin domain crossover as a generic approach for the production of bispecific IgG antibodies. *Proc. Natl. Acad. Sci. USA* **2011**, *108*, 11187–11192. [CrossRef] [PubMed]

196. Fenn, S.; Schiller, C.B.; Griese, J.J.; Duerr, H.; Imhof-Jung, S.; Gassner, C.; Moelleken, J.; Regula, J.T.; Schaefer, W.; Thomas, M.; et al. Crystal structure of an anti-Ang2 CrossFab demonstrates complete structural and functional integrity of the variable domain. *PLoS ONE* **2013**, *8*, e61953. [CrossRef]

197. Klein, C.; Schaefer, W.; Regula, J.T. The use of CrossMab technology for the generation of bi- and multispecific antibodies. *MAbs* **2016**, *8*, 1010–1020. [CrossRef]

198. Lewis, S.M.; Wu, X.; Pustilnik, A.; Sereno, A.; Huang, F.; Rick, H.L.; Guntas, G.; Leaver-Fay, A.; Smith, E.M.; Ho, C.; et al. Generation of bispecific IgG antibodies by structure-based design of an orthogonal Fab interface. *Nat. Biotechnol.* **2014**, *32*, 191–198. [CrossRef]

199. Golay, J.; Choblet, S.; Iwaszkiewicz, J.; Cérutti, P.; Ozil, A.; Loisel, S.; Pugnière, M.; Ubiali, G.; Zoete, V.; Michielin, O.; et al. Design and validation of a novel generic platform for the production of tetravalent IgG1-like bispecific antibodies. *J. Immunol.* **2016**, *196*, 3199–3211. [CrossRef]

200. Dillon, M.; Yin, Y.; Zhou, J.; McCarty, L.; Ellerman, D.; Slaga, D.; Junttila, T.T.; Han, G.; Sandoval, W.; Ovacik, M.A.; et al. Efficient production of bispecific IgG of different isotypes and species of origin in single mammalian cells. *MAbs* **2017**, *9*, 213–230. [CrossRef] [PubMed]

201. Corper, A.L.; Urosev, D.; Tom-Yew, S.A.L.; Bleile, D.W.B.; Von Kreudenstein, T.S.; Dixit, S.; Lario, P.I. Engineered Immunoglobulin Heavy Chain-Light Chain Pairs and Uses Thereof. U.S. Patent 2014/0200331, 17 July 2014.

202. Spiess, C.; Merchant, M.; Huang, A.; Zheng, Z.; Yang, N.Y.; Peng, J.; Ellerman, D.; Shatz, W.; Reilly, D.; Yansura, D.G.; et al. Bispecific antibodies with natural architecture produced by co-culture of bacteria expressing two distinct half-antibodies. *Nat. Biotechnol.* **2013**, *31*, 753–758. [CrossRef] [PubMed]

203. Bardwell, P.D.; Staron, M.M.; Liu, J.; Tao, Q.; Scesney, S.; Bukofzer, G.; Rodriguez, L.E.; Choi, C.H.; Wang, J.; Chang, Q.; et al. Potent and conditional redirected T cell killing of tumor cells using half DVD-Ig. *Protein Cell* **2018**, *9*, 121–129. [CrossRef] [PubMed]

204. Safety, Pharmacokinetics and Therapeutic Activity of RO6958688. Clinical Trial NCT02650713. Available online: https://clinicaltrials.gov/ct2/show/NCT02650713?term=NCT02650713&rank=1 (accessed on 16 April 2019).

205. Dose Escalation Study with RO7082859. Clinical Trial NCT03075696. Available online: https://clinicaltrials.gov/ct2/show/NCT03075696?term=NCT03075696&rank=1 (accessed on 16 April 2019).

206. Vu, M.D.; Moser, S.; Delon, C.; Latzko, M.; Gianotti, R.; Lüoend, R.; Friang, C.; Murr, R.; Duerner, L.; Weinzierl, T.; et al. A new class of T-cell bispecific antibodies for the treatment of multiple myeloma, binding to B cell maturation antigen and CD3 and showing potent, specific antitumor activity in myeloma cells and long duration of action in cynomolgus monkeys. *Blood* **2015**, *126*, 2998.

207. Rius Ruiz, I.; Vicario, R.; Morancho, B.; Morales, C.B.; Arenas, E.J.; Herter, S.; Freimoser-Grundschober, A.; Somandin, J.; Sam, J.; Ast, O.; et al. p95HER2–T cell bispecific antibody for breast cancer treatment. *Sci. Transl. Med.* **2018**, *10*, eaat1445. [CrossRef] [PubMed]

208. Study of ERY974 in Patients with Advanced Solid Tumors. Clinical Trial NCT02748837. Available online: https://clinicaltrials.gov/ct2/show/NCT02748837?term=NCT02748837&rank=1 (accessed on 16 April 2019).

209. Harwood, S.L.; Alvarez-Cienfuegos, A.; Nunez-Prado, N.; Compte, M.; Hernandez-Perez, S.; Merino, N.; Bonet, J.; Navarro, R.; Van Bergen En Henegouwen, P.M.P.; Lykkemark, S.; et al. ATTACK, a novel bispecific T cell-recruiting antibody with trivalent EGFR binding and monovalent CD3 binding for cancer immunotherapy. *Oncoimmunology* **2017**, *7*, e1377874. [CrossRef] [PubMed]

210. Dickopf, S.; Lauer, M.E.; Ringler, P.; Spick, C.; Kern, P.; Brinkmann, U. Highly flexible, IgG-shaped, trivalent antibodies effectively target tumor cells and induce T cell-mediated killing. *Biol. Chem.* **2019**, *400*, 343–350. [CrossRef]

211. Shiheido, H.; Chen, C.; Hikida, M.; Watanabe, T.; Shimizu, J. Modulation of the human T cell response by a novel non-mitogenic anti-CD3 antibody. *PLoS ONE* **2014**, *9*, e94324. [CrossRef]

212. Phase 1 Study of AMV564 in Patients with Myelodysplastic Syndromes. Clinical Trial NCT03516591. Available online: https://clinicaltrials.gov/ct2/show/NCT03516591?term=NCT03516591&rank=1 (accessed on 18 April 2019).

213. Study of ES414 in Metastatic Castration-Resistant Prostate Cancer. Clinical Trial NCT02262910. Available online: https://clinicaltrials.gov/ct2/show/NCT02262910?term=NCT02262910&rank=1 (accessed on 18 April 2019).

214. Coloma, M.J.; Morrison, S.L. Design and production of novel tetravalent bispecific antibodies. *Nat. Biotechnol.* **1997**, *15*, 159–163. [CrossRef]

215. Wu, C.; Ting, H.; Grinnell, C.; Bryant, S.; Miller, R.; Clabbers, A.; Bose, S.; McCarthy, D.; Zhu, R.R.; Santora, L.; et al. Simultaneous targeting of multiple disease mediators by a dual-variable-domain immunoglobulin. *Nat. Biotechnol.* **2007**, *25*, 1290–1297. [CrossRef]

216. Wu, C.; Ying, H.; Bose, S.; Miller, R.; Medina, L.; Santora, L.; Ghayur, T. Molecular construction and optimization of anti-human IL-1alpha/beta dual variable domain immunoglobulin (DVD-Ig) molecules. *MAbs* **2009**, *1*, 339–347. [CrossRef] [PubMed]

217. Gong, S.; Ren, F.; Wu, D.; Wu, X.; Wu, C. Fabs-in-tandem immunoglobulin is a novel and versatile bispecific design for engaging multiple therapeutic targets. *MAbs* **2017**, *9*, 1118–1128. [CrossRef]

218. Lu, D.; Zhu, Z. Construction and production of an IgG-like tetravalent bispecific antibody, IgG-single-chain Fv fusion. *Meth. Mol. Biol.* **2014**, *1060*, 185–213.

219. Dong, J.; Sereno, A.; Snyder, W.B.; Miller, B.R.; Tamraz, S.; Doern, A.; Favis, M.; Wu, X.; Tran, H.; Langley, E.; et al. Stable IgG-like bispecific antibodies directed toward the type I insulin-like growth factor receptor demonstrate enhanced ligand blockade and anti-tumor activity. *J. Biol. Chem.* **2011**, *286*, 4703–4717. [CrossRef] [PubMed]

220. Miller, B.R.; Demarest, S.J.; Lugovskoy, A.; Huang, F.; Wu, X.; Snyder, W.B.; Croner, L.J.; Wang, N.; Amatucci, A.; Michaelson, J.S.; et al. Stability engineering of scFvs for the development of bispecific and multivalent antibodies. *Protein Eng. Des. Sel.* **2010**, *23*, 549–557. [CrossRef] [PubMed]

221. Bluemel, C.; Hausmann, S.; Fluhr, P.; Sriskandarajah, M.; Stallcup, W.B.; Baeuerle, P.A.; Kufer, P. Epitope distance to the target cell membrane and antigen size determine the potency of T cell-mediated lysis by BiTE antibodies specific for a large melanoma surface antigen. *Cancer Immunol. Immunother.* **2010**, *59*, 1197–1209. [CrossRef]

222. Jiang, X.; Chen, X.; Carpenter, T.J.; Wang, J.; Zhou, R.; Davis, H.M.; Heald, D.L.; Wang, W. Development of a target cell-biologics-effector cell (TBE) complex-based cell killing model to characterize target cell depletion by T cell redirecting bispecific agents. *MAbs* **2018**, *10*, 876–889. [CrossRef] [PubMed]

223. Slaney, C.Y.; Wang, P.; Darcy, P.K.; Kershaw, M.H. CARs versus BiTEs: A comparison between T cell-redirection strategies for cancer treatment. *Cancer Discov.* **2018**, *8*, 924–934. [CrossRef]

224. James, S.E.; Greenberg, P.D.; Jensen, M.C.; Lin, Y.; Wang, J.; Till, B.G.; Raubitschek, A.A.; Forman, S.J.; Press, O.W. Antigen sensitivity of CD22-specific chimeric TCR is modulated by target epitope distance from the cell membrane. *J. Immunol.* **2008**, *180*, 7028–7038. [CrossRef]

225. Root, A.; Cao, W.; Li, B.; LaPan, P.; Meade, C.; Sanford, J.; Jin, M.; O'Sullivan, C.; Cummins, E.; Lambert, M.; et al. PF-06671008, a highly potent anti-P-cadherin/anti-CD3 bispecific DART molecule with extended half-life for the treatment of cancer. *Antibodies* **2016**, *5*, 6. [CrossRef]

226. Qi, J.; Li, X.; Peng, H.; Cook, E.M.; Dadashian, E.L.; Wiestner, A.; Park, H.; Rader, C. Potent and selective antitumor activity of a T cell-engaging bispecific antibody targeting a membrane-proximal epitope of ROR1. *Proc. Natl. Acad. Sci. USA* **2018**, *115*, E5467–E5476. [CrossRef] [PubMed]

227. Bacac, M.; Colombetti, S.; Herter, S.; Sam, J.; Perro, M.; Chen, S.; Bianchi, R.; Richard, M.; Schoenle, A.; Nicolini, V.; et al. CD20-TCB with obinutuzumab pretreatment as next-generation treatment of hematologic malignancies. *Clin. Cancer Res.* **2018**, *24*, 4785–4797. [CrossRef] [PubMed]

228. IGM Biosciences Anti-CD20 × CD3 IgM. Available online: http://igmbio.com/pipeline/cd20-x-dc3/ (accessed on 18 April 2019).

229. Chelius, D.; Ruf, P.; Gruber, P.; Plöscher, M.; Liedtke, R.; Gansberger, E.; Hess, J.; Wasiliu, M.; Lindhofer, H. Structural and functional characterization of the trifunctional antibody catumaxomab. *MAbs* **2010**, *2*, 309–319. [CrossRef] [PubMed]

230. Heiss, M.M.; Murawa, P.; Koralewski, P.; Kutarska, E.; Kolesnik, O.O.; Ivanchenko, V.V.; Dudnichenko, A.S.; Aleknaviciene, B.; Razbadauskas, A.; Gore, M.; et al. The trifunctional antibody catumaxomab for the treatment of malignant ascites due to epithelial cancer: Results of a prospective randomized phase II/III trial. *Int. J. Cancer* **2010**, *127*, 2209–2221. [CrossRef] [PubMed]

231. Linke, R.; Klein, A.; Seimetz, D. Catumaxomab: Clinical development and future directions. *MAbs* **2010**, *2*, 129–136. [CrossRef]

232. Lee, K.J.; Chow, V.; Weissman, A.; Tulpule, S.; Aldoss, I.; Akhtari, M. Clinical use of blinatumomab for B-cell acute lymphoblastic leukemia in adults. *Ther. Clin. Risk Manag.* **2016**, *12*, 1301–1310.

233. Nisonoff, A.; Rivers, M.M. Recombination of a mixture of univalent antibody fragments of different specificity. *Arch. Biochem. Biophys.* **1961**, *93*, 460–462. [CrossRef]

234. Karpovsky, B.; Titus, J.A.; Stephany, D.A.; Segal, D.M. Production of target-specific effector cells using hetero-cross-linked aggregates containing anti-target cell and anti-Fc gamma receptor antibodies. *J. Exp. Med.* **1984**, *160*, 1686–1701. [CrossRef]

235. Glennie, M.J.; McBride, H.M.; Worth, A.T.; Stevenson, G.T. Preparation and performance of bispecific F(ab' gamma)2 antibody containing thioether-linked Fab' gamma fragments. *J. Immunol.* **1987**, *139*, 2367–2375.

236. Sen, M.; Wankowski, D.M.; Garlie, N.K.; Siebenlist, R.E.; Van Epps, D.; LeFever, A.V.; Lum, L.G. Use of anti-CD3 × anti-HER2/neu bispecific antibody for redirecting cytotoxicity of activated T cells toward HER2/neu+ tumors. *J. Hematother. Stem Cell Res.* **2001**, *10*, 247–260. [CrossRef]

237. Reusch, U.; Sundaram, M.; Davol, P.A.; Olson, S.D.; Davis, J.B.; Demel, K.; Nissim, J.; Rathore, R.; Liu, P.Y.; Lum, L.G. Anti-CD3 × anti-epidermal growth factor receptor (EGFR) bispecific antibody redirects T-cell cytolytic activity to EGFR-positive cancers in vitro and in an animal model. *Clin. Cancer Res.* **2006**, *12*, 183–190. [CrossRef] [PubMed]

238. Yankelevich, M.; Kondadasula, S.V.; Thakur, A.; Buck, S.; Cheung, N.K.; Lum, L.G. Anti-CD3 × anti-GD2 bispecific antibody redirects T-cell cytolytic activity to neuroblastoma targets. *Pediatr. Blood Cancer* **2012**, *59*, 1198–1205. [CrossRef] [PubMed]

239. Ma, J.; Han, H.; Liu, D.; Li, W.; Feng, H.; Xue, X.; Wu, X.; Niu, G.; Zhang, G.; Zhao, Y.; et al. HER2 as a promising target for cytotoxicity T cells in human melanoma therapy. *PLoS ONE* **2013**, *8*, e73261.

240. HER2Bi-Armed Activated T Cells for Castration Resistant Prostate Cancer. Clinical Trial NCT03406858. Available online: https://clinicaltrials.gov/ct2/show/NCT03406858?term=NCT03406858&rank=1 (accessed on 20 June 2019).

241. Bispecific Antibody Armed Activated T-Cells. Clinical Trial NCT02620865. Available online: https://clinicaltrials.gov/ct2/show/NCT02620865?term=NCT02620865&rank=1 (accessed on 20 April 2019).

242. Activated T Cells Armed with GD2 Bispecific Antibody. Clinical Trial NCT02173093. Available online: https://clinicaltrials.gov/ct2/show/NCT02173093?term=NCT02173093&rank=1 (accessed on 20 April 2019).

243. Thakur, A.; Sorenson, C.; Norkina, O.; Schalk, D.; Ratanatharathorn, V.; Lum, L.G. Activated T cells from umbilical cord blood armed with anti-CD3 × anti-CD20 bispecific antibody mediate specific cytotoxicity against CD20⁺ targets with minimal allogeneic reactivity: A strategy for providing antitumor effects after cord blood transplants. *Transfusion* **2012**, *52*, 63–75. [CrossRef] [PubMed]

244. Kung, P.; Goldstein, G.; Reinherz, E.L.; Schlossman, S.F. Monoclonal antibodies defining distinctive human T cell surface antigens. *Science* **1979**, *206*, 347–349. [CrossRef] [PubMed]

245. Wagner, K.; Kwakkenbos, M.J.; Claassen, Y.B.; Maijoor, K.; Böhne, M.; Van Der Sluijs, K.F.; Witte, M.D.; Van Zoelen, D.J.; Cornelissen, L.A.; Beaumont, T.; et al. Bispecific antibody generated with sortase and click chemistry has broad antiinfluenza virus activity. *Proc. Natl. Acad. Sci. USA* **2014**, *111*, 16820–16825. [CrossRef] [PubMed]

246. Gao, X.; Mi, Y.; Guo, N.; Xu, H.; Xu, L.; Gou, X.; Jin, W. Cytokine-induced killer cells as pharmacological tools for cancer immunotherapy. *Front. Immunol.* **2017**, *8*, 774. [CrossRef]

247. Study of Activated Cytokine-Induced Killer. Clinical Trial NCT03554395. Available online: https://clinicaltrials.gov/ct2/show/NCT03554395?term=NCT03554395&rank=1 (accessed on 20 April 2019).

248. James, N.D.; Atherton, P.J.; Jones, J.; Howie, A.J.; Tchekmedyian, S.; Curnow, R.T. A phase II study of the bispecific antibody MDX-H210 (anti-HER2 × CD64) with GM-CSF in HER2⁺ advanced prostate cancer. *Br. J. Cancer* **2001**, *85*, 152–156. [CrossRef]

249. Repp, R.; Van Ojik, H.H.; Valerius, T.; Groenewegen, G.; Wieland, G.; Oetzel, C.; Stockmeyer, B.; Becker, W.; Eisenhut, M.; Steininger, H.; et al. Phase I clinical trial of the bispecific antibody MDX-H210 (anti-FcgammaRI × anti-HER-2/neu) in combination with Filgrastim (G-CSF) for treatment of advanced breast cancer. *Br. J. Cancer* **2003**, *89*, 2234–2243. [CrossRef]

250. Balaian, L.; Ball, E.D. Inhibition of acute myeloid leukemia cell growth by mono-specific and bi-specific anti-CD33 × anti-CD64 antibodies. *Leuk. Res.* **2004**, *28*, 821–829. [CrossRef]

251. Stockmeyer, B.; Dechant, M.; Van Egmond, M.; Tutt, A.L.; Sundarapandiyan, K.; Graziano, R.F.; Repp, R.; Kalden, J.R.; Gramatzki, M.; Glennie, M.J.; et al. Triggering Fc alpha-receptor I (CD89) recruits neutrophils as effector cells for CD20-directed antibody therapy. *J. Immunol.* **2000**, *165*, 5954–5961. [CrossRef] [PubMed]

252. Tacken, P.J.; Hartshorn, K.L.; White, M.R.; Van Kooten, C.; Van De Winkel, J.G.; Reid, K.B.; Batenburg, J.J. Effective targeting of pathogens to neutrophils via chimeric surfactant protein D/anti-CD89 protein. *J. Immunol.* **2004**, *172*, 4934–4940. [CrossRef] [PubMed]

253. Guettinger, Y.; Barbin, K.; Peipp, M.; Bruenke, J.; Dechant, M.; Horner, H.; Thierschmidt, D.; Valerius, T.; Repp, R.; Fey, G.H.; et al. A recombinant bispecific single-chain fragment variable specific for HLA class II and Fc alpha RI (CD89) recruits polymorphonuclear neutrophils for efficient lysis of malignant B lymphoid cells. *J. Immunol.* **2010**, *184*, 1210–1217. [CrossRef] [PubMed]

254. Boross, P.; Lohse, S.; Nederend, M.; Jansen, J.H.; Van Tetering, G.; Dechant, M.; Peipp, M.; Royle, L.; Liew, L.P.; Boon, L.; et al. IgA EGFR antibodies mediate tumour killing in vivo. *EMBO Mol. Med.* **2013**, *5*, 1213–1226. [CrossRef] [PubMed]

255. Yu, X.; Duval, M.; Gawron, M.; Posner, M.R.; Cavacini, L.A. Overcoming the constraints of anti-HIV/CD89 bispecific antibodies that limit viral inhibition. *J. Immunol. Res.* **2016**, *2016*, 1–5. [CrossRef]

256. Germain, C.; Campigna, E.; Salhi, I.; Morisseau, S.; Navarro-Teulon, I.; Mach, J.P.; Pèlegrin, A.; Robert, B. Redirecting NK cells mediated tumor cell lysis by a new recombinant bifunctional protein. *Protein Eng. Des. Sel.* **2008**, *21*, 665–672. [CrossRef] [PubMed]

257. Silla, L.M.; Chen, J.; Zhong, R.K.; Whiteside, T.L.; Ball, E.D. Potentiation of lysis of leukaemia cells by a bispecific antibody to CD33 and CD16 (Fc gamma RIII) expressed by human natural killer (NK) cells. *Br. J. Haematol.* **1995**, *89*, 712–718. [CrossRef]

258. Hartmann, F.; Renner, C.; Jung, W.; Da Costa, L.; Tembrink, S.; Held, G.; Sek, A.; König, J.; Bauer, S.; Kloft, M.; et al. Anti-CD16/CD30 bispecific antibody treatment for Hodgkin's disease: Role of infusion schedule and costimulation with cytokines. *Clin. Cancer Res.* **2001**, *7*, 1873–1881.

259. Lo Nigro, C.; Macagno, M.; Sangiolo, D.; Bertolaccini, L.; Aglietta, M.; Merlano, M.C. NK-mediated antibody-dependent cell-mediated cytotoxicity in solid tumors: Biological evidence and clinical perspectives. *Ann. Transl. Med.* **2019**, *7*, 105. [CrossRef]

260. Chiang, S.C.; Theorell, J.; Entesarian, M.; Meeths, M.; Mastafa, M.; Al-Herz, W.; Frisk, P.; Gilmour, K.C.; Ifversen, M.; Langenskiöld, C.; et al. Comparison of primary human cytotoxic T-cell and natural killer cell responses reveal similar molecular requirements for lytic granule exocytosis but differences in cytokine production. *Blood* **2013**, *121*, 1345–1356. [CrossRef]

261. Li, W.; Yang, H.; Dimitrov, D.S. Identification of high-affinity anti-CD16A allotype-independent human antibody domains. *Exp. Mol. Pathol.* **2016**, *101*, 281–289. [CrossRef] [PubMed]

262. Wu, J.; Fu, J.; Zhang, M.; Liu, D. AFM13: A first-in-class tetravalent bispecific anti-CD30/CD16A antibody for NK cell-mediated immunotherapy. *J. Hematol. Oncol.* **2015**, *8*, 96. [CrossRef] [PubMed]

263. GHSG-AFM13 an Open-Label, Multicenter Phase II Trial. Clinical Trial NCT02321592. Available online: https://clinicaltrials.gov/ct2/show/NCT02321592?term=NCT02321592&rank=1 (accessed on 21 April 2019).

264. Vallera, D.A.; Felices, M.; McElmurry, R.; McCullar, V.; Zhou, X.; Schmohl, J.U.; Zhang, B.; Lenvik, A.J.; Panoskaltsis-Mortari, A.; Verneris, M.R.; et al. IL15 trispecific killer engagers (TriKE) make natural killer cells specific to CD33+ targets while also inducing persistence, in vivo expansion and enhanced function. *Clin. Cancer Res.* **2016**, *22*, 3440–3450. [CrossRef] [PubMed]

265. Jochems, C.; Hodge, J.W.; Fantini, M.; Tsang, K.Y.; Vandeveer, A.J.; Gulley, J.L.; Schlom, J. ADCC employing an NK cell line (haNK) expressing the high affinity CD16 allele with avelumab, an anti-PD-L1 antibody. *Int. J. Cancer* **2017**, *141*, 583–593. [CrossRef] [PubMed]

266. Quilt-3.028: Study of HANK™. Clinical Trial NCT03027128. Available online: https://clinicaltrials.gov/ct2/show/NCT03027128?term=NCT03027128&rank=1 (accessed on 21 April 2019).

267. Evaluate Efficacy of Avelumab, HaNK and N-803. Clinical Trial NCT03853317. Available online: https://clinicaltrials.gov/ct2/show/NCT03853317?term=NCT03853317&rank=1 (accessed on 21 April 2019).

268. Motz, G.; Whiteman, K.; Shin, J.; Pai, T.; Judge, C.; Barnitz, A.; Hemphill, J.; Kim, J.; Ranger, A.; Huet, H.; et al. ACTR707: A novel T-cell therapy for the treatment of relapsed or refractory CD20+ B cell lymphoma in combination with rituximab. *Mol. Cancer Ther.* **2018**, *17*, B105. [CrossRef]

269. Study of ACTR707 for Relapsed or Refractory B Cell Lymphoma. Clinical Trial NCT03189836. Available online: https://clinicaltrials.gov/ct2/show/NCT03189836?term=NCT03189836&rank=1 (accessed on 21 April 2019).

270. Akard, L.P.; Jaglowski, S.; Devine, S.M.; McKinney, M.S.; Vasconcelles, M.; Huet, H.; Ettenberg, S.; Ranger, A.; Abramson, J.S. ACTR087, autologous T lymphocytes expressing antibody coupled T-cell receptors (ACTR), induces complete responses in patients with relapsed or refractory CD20-positive B-cell lymphoma, in combination with rituximab. *Blood* **2017**, *130*, 580.

271. Study of ACTR087 for Relapsed or Refractory B Cell Lymphoma. Clinical Trial NCT02776813. Available online: https://clinicaltrials.gov/ct2/show/NCT02776813?term=NCT02776813&rank=1 (accessed on 21 April 2019).

272. Study of ACTR087 for Relapsed or Refractory Multiple Myeloma. Clinical Trial NCT03266692. Available online: https://clinicaltrials.gov/ct2/show/NCT03266692?term=NCT03266692&rank=1 (accessed on 21 April 2019).

273. Van Epps, H.; Anderson, M.; Yu, C.; Klussman, K.; Westendorf, L.; Carosino, C.; Manlove, L.; Cochran, J.; Neale, J.; Benjamin, D.; et al. SEA-BCMA: A highly active enhanced antibody for multiple myeloma. *Cancer Res.* **2018**, *78*, 3833.

274. Fujio, K.; Okamura, T.; Sumitomo, S.; Yamamoto, K. Regulatory T cell-mediated control of autoantibody-induced inflammation. *Front. Immunol.* **2012**, *3*, 28. [CrossRef] [PubMed]

275. Getts, D.; Hofmeister, R.; Quintás-Cardama, A. Synthetic T cell receptor-based lymphocytes for cancer therapy. *Adv. Drug Deliv. Rev.* **2019**. [CrossRef]

276. Mensali, N.; Dillard, P.; Hebeisen, M.; Lorenz, S.; Theodossiou, T.; Myhre, M.R.; Fåne, A.; Gaudernack, G.; Kvalheim, G.; Myklebust, J.H.; et al. NK cells specifically TCR-dressed to kill cancer cells. *EBioMedicine* **2019**, *40*, 106–117. [CrossRef]

277. Walseng, E.; Köksal, H.; Sektioglu, I.M.; Fåne, A.; Skorstad, G.; Kvalheim, G.; Gaudernack, G.; Inderberg, E.M.; Wälchli, S. A TCR-based chimeric antigen receptor. *Sci. Rep.* **2017**, *7*, 10713. [CrossRef] [PubMed]

278. Efficacy and Safety Study of Bb2121. Clinical Trial NCT03651128. Available online: https://clinicaltrials.gov/ct2/show/NCT03651128?term=NCT03651128&rank=1 (accessed on 21 April 2019).

279. A Study to Compare Efficacy and Safety of JCAR017 to Standard of Care. Clinical Trial NCT03575351. Available online: https://clinicaltrials.gov/ct2/show/NCT03575351?term=NCT03575351&rank=1 (accessed on 21 April 2019).

280. Hartmann, J.; Schüßler-Lenz, M.; Bondanza, A.; Buchholz, C.J. Clinical development of CAR T cells—Challenges and opportunities in translating innovative treatment concepts. *EMBO Mol. Med.* **2017**, *9*, 1183–1197. [CrossRef] [PubMed]

281. Salmikangas, P.; Kinsella, N.; Chamberlain, P. Chimeric antigen receptor T-cells (CAR T-cells) for cancer immunotherapy—Moving target for industry? *Pharm. Res.* **2018**, *35*, 152. [CrossRef] [PubMed]

282. Makita, S.; Imaizumi, K.; Kurosawa, S.; Tobinai, K. Chimeric antigen receptor T-cell therapy for B-cell non-Hodgkin lymphoma: Opportunities and challenges. *Drugs Context* **2019**, *8*, 212567. [CrossRef] [PubMed]

283. Johnson, L.A.; June, C.H. Driving gene-engineered T cell immunotherapy of cancer. *Cell Research* **2017**, *27*, 38–58. [CrossRef] [PubMed]

284. Milone, M.C.; O'Doherty, U. Clinical use of lentiviral vectors. *Leukemia* **2018**, *32*, 1529–1541. [CrossRef] [PubMed]

285. Murad, J.M.; Baumeister, S.H.; Werner, L.; Daley, H.; Trébéden-Negre, H.; Reder, J.; Sentman, C.L.; Gilham, D.; Lehmann, F.; Snykers, S.; et al. Manufacturing development and clinical production of NKG2D chimeric antigen receptor-expressing T cells for autologous adoptive cell therapy. *Cytotherapy* **2018**, *20*, 952–963. [CrossRef]

286. He, J.; Zhang, Z.; Lv, S.; Liu, X.; Cui, L.; Jiang, D.; Zhang, Q.; Li, L.; Qin, W.; Jin, H. Engineered CAR T cells targeting mesothelin by piggyBac transposon system for the treatment of pancreatic cancer. *Cell. Immunol.* **2018**, *329*, 31–40. [CrossRef]

287. Kebriaei, P.; Singh, H.; Huls, M.H.; Figliola, M.J.; Bassett, R.; Olivares, S.; Jena, B.; Dawson, M.J.; Kumaresan, P.R.; Su, S.; et al. Phase I trials using Sleeping Beauty to generate CD19-specific CAR T cells. *J. Clin. Investig.* **2016**, *126*, 3363–3376. [CrossRef]

288. Gardner, R.A.; Finney, O.; Annesley, C.; Brakke, H.; Summers, C.; Leger, K.; Bleakley, M.; Brown, C.; Mgebroff, S.; Kelly-Spratt, K.S.; et al. Intent-to-treat leukemia remission by CD19 CAR T cells of defined formulation and dose in children and young adults. *Blood* **2017**, *129*, 3322–3331.

289. Chow, V.A.; Shadman, M.; Gopal, A.K. Translating anti-CD19 CAR T-cell therapy into clinical practice for relapsed/refractory diffuse large B-Cell lymphoma. *Blood* **2018**, *132*, 777–781. [CrossRef] [PubMed]

290. Kymriah European Public Assessment Reports (EPAR) Product Information. Available online: https://www.ema.europa.eu/en/documents/product-information/kymriah-epar-product-information_en.pdf (accessed on 21 April 2019).

291. Yescarta European Public Assessment Reports (EPAR) Product Information. Available online: https://www.ema.europa.eu/en/documents/product-information/yescarta-epar-product-information_en.pdf (accessed on 21 April 2019).

292. Friedman, K.M.; Garrett, T.E.; Evans, J.W.; Horton, H.M.; Latimer, H.J.; Seidel, S.L.; Horvath, C.J.; Morgan, R.A. Effective targeting of multiple B-cell maturation antigen-expressing hematological malignances by anti-B-cell maturation antigen chimeric antigen receptor T cells. *Hum. Gene Ther.* **2018**, *29*, 585–601. [CrossRef] [PubMed]

293. Bb2121 Website. Available online: https://www.researchoncology.com/translational-research/bb2121-car-t/ (accessed on 21 April 2019).

294. Ramsborg, C.G.; Guptill, P.; Weber, C.; Christin, B.; Larson, R.P.; Lewis, K.; Mallaney, M.; Bowen, M.; Higham, E.; Albertson, T. JCAR017 is a defined composition CAR T cell product with product and process controls that deliver precise doses of CD4 and CD8 CAR T cell to patients with NHL. *Blood* **2017**, *130*, 4471.

295. Raje, N.; Berdeja, J.; Lin, Y.; Siegel, D.; Jagannath, S.; Madduri, D.; Liedtke, M.; Rosenblatt, J.; Maus, M.V.; Turka, A.; et al. Anti-BCMA CAR T-cell therapy bb2121 in relapsed or refractory multiple myeloma. *N. Engl. J. Med.* **2019**, *380*, 1726–1737. [CrossRef] [PubMed]

296. Torikai, H.; Reik, A.; Soldner, F.; Warren, E.H.; Yuen, C.; Zhou, Y.; Crossland, D.L.; Huls, H.; Littman, N.; Zhang, Z.; et al. Toward eliminating HLA class I expression to generate universal cells from allogeneic donors. *Blood* **2013**, *122*, 1341–1349. [CrossRef] [PubMed]

297. Xu, H.; Wang, B.; Ono, M.; Yoshida, Y.; Kaneko, S.; Hotta, A. Targeted disruption of HLA genes via CRISPR-Cas9 generates iPSCs with enhanced immune compatibility. *Cell Stem Cell* **2019**, *24*, 566–578. [CrossRef] [PubMed]

298. Torikai, H.; Reik, A.; Yuen, C.; Zhou, Y.; Kellar, S.; Huls, H.; Warren, E.E., III; Tykodi, S.S.; Gregory, P.D.; Holmes, M.C.; et al. HLA and TCR knockout by zinc finger nucleases: Toward "off-the-shelf" allogeneic T-cell therapy for CD19⁺ malignancies. *Blood* **2010**, *116*, 3766.

299. Mehta, R.S.; Rezvani, K. Chimeric antigen receptor expressing natural killer cells for the immunotherapy of cancer. *Front. Immunol.* **2018**, *9*, 283. [CrossRef]

300. Poirot, L.; Philip, B.; Schiffer-Mannioui, C.; Le Clerre, D.; Chion-Sotinel, I.; Derniame, S.; Potrel, P.; Bas, C.; Lemaire, L.; Galetto, R.; et al. Multiplex genome-edited T-cell manufacturing platform for "off-the-shelf" adoptive T-cell immmunotherapies. *Cancer Res.* **2015**, *75*, 3853–3864. [CrossRef]

301. Sommer, C.; Boldajipour, B.; Kuo, T.C.; Bentley, T.; Sutton, J.; Chen, A.; Geng, T.; Dong, H.; Galetto, R.; Valton, J.; et al. Preclinical evaluation of allogeneic CAR T cells targeting BCMA for the treatment of multiple myeloma. *Mol. Ther.* **2019**, *27*, 1126–1138. [CrossRef]

302. Philip, B.; Kokalaki, E.; Mekkaoui, L.; Thomas, S.; Straathof, K.; Flutter, B.; Marin, V.; Marafioti, T.; Chakraverty, R.; Linch, D.; et al. A highly compact epitope based marker/suicide gene for easier and safer T-cell therapy. *Blood* **2014**, *124*, 1277–1287. [CrossRef] [PubMed]

303. Valton, J.; Guyot, V.; Boldajipour, B.; Sommer, C.; Pertel, T.; Juillerat, A.; Duclert, A.; Sasu, B.J.; Duchateau, P.; Poirot, L. A versatile safeguard for chimeric antigen receptor T-cell immunotherapies. *Sci. Rep.* **2018**, *8*, 8972. [CrossRef] [PubMed]

304. Gautron, A.S.; Juillerat, A.; Guyot, V.; Filhol, J.-M.; Dessez, E.; Duclert, A.; Duchateau, P.; Poirot, L. Fine and predictable tuning of TALEN gene editing targeting for improved T cell adoptive immunotherapy. *Mol. Ther. Nucleic Acids* **2017**, *9*, 312–321. [CrossRef] [PubMed]

305. Harrer, D.C.; Dörrie, J.; Schaft, N. Chimeric antigen receptors in different cell types: New vehicles join the race. *Hum. Gene Ther.* **2018**, *29*, 547–558. [CrossRef] [PubMed]

306. Narni-Mancinelli, E.; Vivier, E.; Kerdiles, Y.M. The 'T-cell-ness' of NK cells: Unexpected similarities between NK cells and T cells. *Internat. Immunol.* **2011**, *23*, 427–431. [CrossRef]

307. Abel, A.M.; Yang, C.; Thakar, M.S.; Malarkannan, S. Natural killer cells: Development, maturation and clinical utilization. *Front. Immunol.* **2018**, *9*, 1869. [CrossRef]

308. Wolf, B.J.; Choi, J.E.; Exley, M.A. Novel approaches to exploiting invariant NKT cells in cancer immunotherapy. *Front. Immunol.* **2018**, *9*, 384. [CrossRef]

309. Fisher, J.; Anderson, J. Engineering approaches in human gamma delta T cells for cancer immunotherapy. *Front. Immunol.* **2018**, *9*, 1409. [CrossRef]

310. Gamma Delta T Cells in AML. Clinical Trial NCT03885076. Available online: https://clinicaltrials.gov/ct2/show/NCT03885076?term=NCT03885076&rank=1 (accessed on 21 April 2019).

311. Capsomidis, A.; Benthall, G.; Van Acker, H.H.; Fisher, J.; Kramer, A.M.; Abeln, Z.; Majani, Y.; Gileadi, T.; Wallace, R.; Gustafsson, K.; et al. Chimeric antigen receptor-engineered human gamma delta T cells: Enhanced cytotoxicity with retention of cross presentation. *Mol. Ther.* **2018**, *26*, 354–365. [CrossRef]

312. Morrissey, M.A.; Williamson, A.P.; Steinbach, A.M.; Roberts, E.W.; Kern, N.; Headley, M.B.; Vale, R.D. Chimeric antigen receptors that trigger phagocytosis. *eLife* **2018**, *7*, e36688. [CrossRef]

313. Allen, E.S.; Stroncek, D.F.; Ren, J.; Eder, A.F.; West, K.A.; Fry, T.J.; Lee, D.W.; Mackall, C.L.; Conry-Cantilena, C. Autologous lymphapheresis for the production of chimeric antigen receptor (CAR) T Cells. *Transfusion* **2017**, *57*, 1133–1141. [CrossRef] [PubMed]

314. Zhao, Y.; Wang, Q.J.; Yang, S.; Kochenderfer, J.N.; Zheng, Z.; Zhong, X.; Sadelain, M.; Eshhar, Z.; Rosenberg, S.A.; Morgan, R.A. A herceptin-based chimeric antigen receptor with modified signaling domains leads to enhanced survival of transduced T lymphocytes and antitumor activity. *J. Immunol.* **2009**, *183*, 5563–5574. [CrossRef] [PubMed]

315. Watanabe, N.; Bajgain, P.; Sukumaran, S.; Ansari, S.; Heslop, H.E.; Rooney, C.M.; Brenner, M.K.; Leen, A.M.; Vera, J.F. Fine-tuning the CAR spacer improves T-cell potency. *Oncoimmunol.* **2016**, *5*, e1253656. [CrossRef] [PubMed]

316. Ajina, A.; Maher, J. Strategies to address chimeric antigen receptor tonic signaling. *Mol. Cancer Ther.* **2018**, *17*, 1795–1815. [CrossRef] [PubMed]

317. Kulemzin, S.V.; Kuznetsova, V.V.; Mamonkin, M.; Taranin, A.V.; Gorchakov, A.A. Engineering chimeric antigen receptors. *Acta Naturae* **2017**, *9*, 6–14. [CrossRef]

318. Bannas, P.; Hambach, J.; Koch-Nolte, F. Nanobodies and nanobody-based human heavy chain antibodies as antitumor therapeutics. *Front. Immunol.* **2017**, *8*, 1603. [CrossRef] [PubMed]

319. Rahbarizadeh, F.; Ahmadvand, D.; Moghimi, S.M. CAR T-cell bioengineering: Single variable domain of heavy chain antibody targeted CARs. *Adv. Drug Deliv. Rev.* **2019**. [CrossRef]

320. Chin, C.-N.; Lee, J.; McCabe, T.; Mooney, J.; Naso, M.; Strohl, W.R. Chimeric Antigen Receptors Comprising BCMA-Specific Fibronectin Type III Domains and Uses Thereof. Patent WO/2018/052828, 22 March 2018.

321. Han, X.; Cinay, G.E.; Zhao, Y.; Guo, Y.; Zhang, X.; Wang, P. Adnectin-based design of chimeric antigen receptor for T cell engineering. *Mol. Ther.* **2017**, *25*, 2466–2476. [CrossRef]

322. Hammill, J.A.; VanSeggelen, H.; Helsen, C.W.; Denisova, G.F.; Evelegh, C.; Tantalo, D.G.; Bassett, J.D.; Bramson, J.L. Designed ankyrin repeat proteins are effective targeting elements for chimeric antigen receptors. *J. Immunother. Cancer* **2015**, *3*, 55. [CrossRef]

323. Hermanson, D.L.; Barnett, B.E.; Rengarajan, S.; Codde, R.; Wang, X.; Tan, Y.; Martin, C.E.; Smith, J.B.; He, J.; Mathur, R.; et al. A novel BCMA-specific, centyrin-based CAR-T product for the treatment of multiple myeloma. *Blood* **2016**, *128*, 2127.

324. Martyniszyn, A.; Krahl, A.-C.; André, M.C.; Hombach, A.A.; Abken, H. CD20-CD19 bispecific CAR T cells for the treatment of B-cell malignancies. *Hum. Gene Ther.* **2017**, *28*, 1147–1157. [CrossRef] [PubMed]

325. De Munter, S.; Ingels, J.; Goetgeluk, G.; Bonte, S.; Pille, M.; Weening, K.; Kerre, T.; Abken, H.; Vandekerckhove, B. Nanobody based dual specific CARs. *Int. J. Mol. Sci.* **2018**, *19*, 403. [CrossRef] [PubMed]

326. Zhao, W.H.; Liu, J.; Wang, B.Y.; Chen, Y.X.; Cao, X.M.; Yang, Y.; Zhang, Y.L.; Wang, F.X.; Zhang, P.Y.; Lei, B.; et al. A phase 1, open-label study of LCAR-B38M, a chimeric antigen receptor T cell therapy directed against B cell maturation antigen, in patients with relapsed or refractory multiple myeloma. *J. Hematol. Oncol.* **2018**, *11*, 141. [CrossRef] [PubMed]

327. A Study of LCAR-B38M CAR-T Cells. Clinical Trial NCT03758417. Available online: https://clinicaltrials.gov/ct2/show/NCT03758417?term=bcma+car+janssen&rank=1 (accessed on 10 May 2019).

328. Wilkie, S.; Van Schalkwyk, M.C.; Hobbs, S.; Davies, D.M.; Van Der Stegen, S.J.; Pereira, A.C.; Burbridge, S.E.; Box, C.; Eccles, S.A.; Maher, J. Dual targeting of ErbB2 and MUC1 in breast cancer using chimeric antigen receptors engineered to provide complementary signaling. *Clin. Immunol.* **2012**, *32*, 1059–1070. [CrossRef] [PubMed]

329. Kloss, C.C.; Condomines, M.; Cartellieri, M.; Bachmann, M.; Sadelain, M. Combinatorial antigen recognition with balanced signaling promotes selective tumor eradication by engineered T cells. *Nat. Biotechnol.* **2013**, *31*, 71–75. [CrossRef] [PubMed]

330. Lanitis, E.; Poussin, M.; Klattenhoff, A.W.; Song, D.; Sandaltzopoulos, R.; June, C.H.; Powell, D.J., Jr. Chimeric antigen receptor T cells with dissociated signaling domains exhibit focused antitumor activity with reduced potential for toxicity *in vivo*. *Cancer Immunol. Res.* **2013**, *1*, 43–53. [CrossRef] [PubMed]

331. Roybal, K.T.; Williams, J.Z.; Morsut, L.; Rupp, L.J.; Kolinko, I.; Choe, J.H.; Walker, W.J.; McNally, K.A.; Lim, W.A. Engineering T cells with customized therapeutic response programs using synthetic notch receptors. *Cell* **2016**, *167*, 419–432. [CrossRef] [PubMed]

332. Ebert, L.M.; Yu, W.; Gargett, T.; Brown, M.P. Logic-gated approaches to extend the utility of chimeric antigen receptor T-cell technology. *Biochem. Soc. Trans.* **2018**, *46*, 391–401. [CrossRef] [PubMed]

333. Patel, E.; Ding, J.; Thorausch, N.; Krishnamurthy, J.; Choudhary, R.; Weiler, S.; Le, B.; Tavares, P.; Zieba, A.; Quinn, J.; et al. Abstract 3589: Preclinical evaluation of mesothelin-specific T cell receptor (TCR) fusion constructs (TRuC™s) utilizing the signaling power of the complete TCR complex: A new opportunity for solid tumor therapy. *Cancer Res.* **2018**, *78*, 3589.

334. Helsen, C.W.; Hammill, J.A.; Lau, V.W.C.; Mwawasi, K.A.; Afsahi, A.; Bezverbnaya, K.; Newhook, L.; Hayes, D.L.; Aarts, C.; Bojovic, B.; et al. The chimeric TAC receptor co-opts the T cell receptor yielding robust anti-tumor activity without toxicity. *Nat. Commun.* **2018**, *9*, 3049. [CrossRef]

335. Xu, Y.; Yang, Z.; Horan, L.H.; Zhang, P.; Liu, L.; Zimdahl, B.; Green, S.; Lu, J.; Morales, J.F.; Barrett, D.M.; et al. A novel antibody-TCR (AbTCR) platform combines Fab-based antigen recognition with gamma/delta-TCR signaling to facilitate T-cell cytotoxicity with low cytokine release. *Cell Discov.* **2018**, *4*, 62. [CrossRef] [PubMed]

336. D'Aloia, M.M.; Zizzari, I.G.; Sacchetti, B.; Pierelli, L.; Alimandi, M. CAR-T cells: The long and winding road to solid tumors. *Cell Death Dis.* **2018**, *9*, 282. [CrossRef] [PubMed]

337. Zhang, E.; Gu, J.; Xu, H. Prospects for chimeric antigen receptor-modified T cell therapy for solid tumors. *Mol. Cancer* **2018**, *17*, 7. [CrossRef] [PubMed]

338. Knochelmann, H.M.; Smith, A.S.; Dwyer, C.J.; Wyatt, M.M.; Mehrotra, S.; Paulos, C.M. CAR T Cells in solid tumors: Blueprints for building effective therapies. *Front. Immunol.* **2018**, *9*, 1740. [CrossRef] [PubMed]

339. Chmielewski, M.; Abken, H. TRUCKs: The fourth generation of CARs. *Expert Opin. Biol. Ther.* **2015**, *15*, 1145–1154. [CrossRef] [PubMed]

340. Martinez, M.; Moon, E.K. CAR T cells for solid tumors: New strategies for finding, infiltrating and surviving in the tumor microenvironment. *Front. Immunol.* **2019**, *10*, 128. [CrossRef] [PubMed]

341. Ciceri, F.; Bonini, C.; Stanghellini, M.T.; Bondanza, A.; Traversari, C.; Salomoni, M.; Turchetto, L.; Colombi, S.; Bernardi, M.; Peccatori, J.; et al. Infusion of suicide-gene-engineered donor lymphocytes after family haploidentical haemopoietic stem-cell transplantation for leukaemia (the TK007 trial): A non-randomised phase I-II study. *Lancet Oncol.* **2009**, *10*, 489–500. [CrossRef]

342. Gargett, T.; Brown, M.P. The inducible caspase-9 suicide gene system as a "safety switch" to limit on-target, off-tumor toxicities of chimeric antigen receptor T cells. *Front. Pharmacol.* **2014**, *5*, 235. [CrossRef]

343. Straathof, K.C.; Pulè, M.A.; Yotnda, P.; Dotti, G.; Vanin, E.F.; Brenner, M.K.; Heslop, H.E.; Spencer, D.M.; Rooney, C.M. An inducible caspase 9 safety switch for T-cell therapy. *Blood* **2005**, *105*, 4247–4254. [CrossRef]

344. Hoyos, V.; Savoldo, B.; Quintarelli, C.; Mahendravada, A.; Zhang, M.; Vera, J.; Heslop, H.E.; Rooney, C.M.; Brenner, M.K.; Dotti, G. Engineering CD19-specific T lymphocytes with interleukin-15 and a suicide gene to enhance their anti-lymphoma/leukemia effects and safety. *Leukemia* **2010**, *24*, 1160–1170. [CrossRef]

345. Duong, M.T.; Collinson-Pautz, M.R.; Morschl, E.; Lu, A.; Szymanski, S.P.; Zhang, M.; Brandt, M.E.; Chang, W.C.; Sharp, K.L.; Toler, S.M.; et al. Two-dimensional regulation of CAR-T cell therapy with orthogonal switches. *Mol. Ther. Oncolyt.* **2018**, *12*, 124–137. [CrossRef] [PubMed]

346. Sakemura, R.; Terakura, S.; Watanabe, K.; Julamanee, J.; Takagi, E.; Miyao, K.; Koyama, D.; Goto, T.; Hanajiri, R.; Nishida, T.; et al. A tet-on inducible system for controlling CD19-chimeric antigen receptor expression upon drug administration. *Cancer Immunol. Res.* **2016**, *4*, 658–668. [CrossRef] [PubMed]

347. Gu, X.; He, D.; Li, C.; Wang, H.; Yang, G. Development of inducible CD19-CAR T cells with a tet-on system for controlled activity and enhanced clinical safety. *Int. J. Mol. Sci.* **2018**, *19*, 3455. [CrossRef] [PubMed]

348. Wang, X.; Chang, W.C.; Wong, C.W.; Colcher, D.; Sherman, M.; Ostberg, J.R.; Forman, S.J.; Riddell, S.R.; Jensen, M.C. A transgene-encoded cell surface polypeptide for selection, in vivo tracking and ablation of engineered cells. *Blood* **2011**, *118*, 1255–1263. [CrossRef] [PubMed]

349. Paszkiewicz, P.J.; Fräßle, S.P.; Srivastava, S.; Sommermeyer, D.; Hudecek, M.; Drexler, I.; Sadelain, M.; Liu, L.; Jensen, M.C.; Riddell, S.R.; et al. Targeted antibody-mediated depletion of murine CD19 CAR T cells permanently reverses B cell aplasia. *J. Clin. Investig.* **2016**, *126*, 4262–4272. [CrossRef] [PubMed]

350. Kao, R.L.; Truscott, L.C.; Chiou, T.T.; Tsai, W.; Wu, A.M.; De Oliveira, S.N. A cetuximab-mediated suicide system in chimeric antigen receptor-modified hematopoietic stem cells for cancer therapy. *Hum. Gene Ther.* **2019**, *30*, 413–428. [CrossRef] [PubMed]

351. Darowski, D.; Kobold, S.; Jost, C.; Klein, C. Combining the best of two worlds: Highly flexible chimeric antigen receptor adaptor molecules (CAR-adaptors) for the recruitment of chimeric antigen receptor T cells. *MAbs* **2019**, *20*, 1–11. [CrossRef] [PubMed]

352. Minutolo, N.G.; Hollander, E.E.; Powell, D.J., Jr. The emergence of universal immune receptor T cell therapy for cancer. *Front. Oncol.* **2019**, *9*, 176. [CrossRef]

353. Rodgers, D.T.; Mazagova, M.; Hampton, E.N.; Cao, Y.; Ramadoss, N.S.; Hardy, I.R.; Schulman, A.; Du, J.; Wang, F.; Singer, O.; et al. Switch-mediated activation and retargeting of CAR-T cells for B-cell malignancies. *Proc. Natl. Acad. Sci. USA* **2016**, *113*, E459–E468. [CrossRef]

354. Scarfò, I.; Maus, M.V. Current approaches to increase CAR T cell potency in solid tumors: Targeting the tumor microenvironment. *J. Immunother. Cancer* **2017**, *5*, 28. [CrossRef]

355. Sackstein, R.; Schatton, T.; Barthel, S.R. T-lymphocyte homing: An underappreciated yet critical hurdle for successful cancer immunotherapy. *Lab. Investig.* **2017**, *97*, 669–697. [CrossRef] [PubMed]

356. Chheda, Z.S.; Sharma, R.K.; Jala, V.R.; Luster, A.D.; Haribabu, B. Chemoattractant receptors BLT1 and CXCR3 regulate antitumor immunity by facilitating CD8+ T cell migration into tumors. *J. Immunol.* **2016**, *197*, 2016–2026. [CrossRef] [PubMed]

357. Kremer, V.; Ligtenberg, M.A.; Zendehdel, R.; Seitz, C.; Duivenvoorden, A.; Wennerberg, E.; Colón, E.; Scherman-Plogell, A.H.; Lundqvist, A. Genetic engineering of human NK cells to express CXCR2 improves migration to renal cell carcinoma. *J. Immunother. Cancer* **2017**, *5*, 73. [CrossRef] [PubMed]

358. Siddiqui, I.; Erreni, M.; Van Brakel, M.; Debets, R.; Allavena, P. Enhanced recruitment of genetically modified CX3CR1-positive human T cells into Fractalkine/CX3CL1 expressing tumors: Importance of the chemokine gradient. *J. Immunother. Cancer* **2016**, *4*, 21. [CrossRef] [PubMed]

359. Rapp, M.; Grassmann, S.; Chaloupka, M.; Layritz, P.; Kruger, S.; Ormanns, S.; Rataj, F.; Janssen, K.P.; Endres, S.; Anz, D.; et al. C-C chemokine receptor type-4 transduction of T cells enhances interaction with dendritic cells, tumor infiltration and therapeutic efficacy of adoptive T cell transfer. *Oncoimmunology* **2015**, *5*, e1105428. [CrossRef] [PubMed]

360. Mikucki, M.E.; Fisher, D.T.; Matsuzaki, J.; Skitzki, J.J.; Gaulin, N.B.; Muhitch, J.B.; Ku, A.W.; Frelinger, J.G.; Odunsi, K.; Gajewski, T.F.; et al. Non-redundant requirement for CXCR3 signalling during tumoricidal T-cell trafficking across tumour vascular checkpoints. *Nat. Commun.* **2015**, *6*, 7458. [CrossRef] [PubMed]

361. Hui, E.; Cheung, J.; Zhu, J.; Su, X.; Taylor, M.J.; Wallweber, H.A.; Sasmal, D.K.; Huang, J.; Kim, J.M.; Mellman, I.; et al. T cell costimulatory receptor CD28 is a primary target for PD-1-mediated inhibition. *Science* **2017**, *355*, 1428–1433. [CrossRef] [PubMed]

362. Guo, X.; Jiang, H.; Shi, B.; Zhou, M.; Zhang, H.; Shi, Z.; Du, G.; Luo, H.; Wu, X.; Wang, Y.; et al. Disruption of PD-1 enhanced the anti-tumor activity of chimeric antigen receptor T cells against hepatocellular carcinoma. *Front. Pharmacol.* **2018**, *9*, 1118. [CrossRef]

363. Hu, W.; Zi, Z.; Jin, Y.; Li, G.; Shao, K.; Cai, Q.; Ma, X.; Wei, F. CRISPR/Cas9-mediated PD-1 disruption enhances human mesothelin-targeted CAR T cell effector functions. *Cancer Immunol. Immunother.* **2019**, *68*, 365–377. [CrossRef]

364. Xie, Y.J.; Dougan, M.; Jailkhani, N.; Ingram, J.; Fang, T.; Kummer, L.; Momin, N.; Pishesha, N.; Rickelt, S.; Hynes, R.O.; et al. Nanobody-based CAR T cells that target the tumor microenvironment inhibit the growth of solid tumors in immunocompetent mice. *Proc. Natl. Acad. Sci. USA* **2019**, *116*, 7624–7631. [CrossRef]

365. Rupp, L.J.; Schumann, K.; Roybal, K.T.; Gate, R.E.; Ye, C.J.; Lim, W.A.; Marson, A. CRISPR/Cas9-mediated PD-1 disruption enhances anti-tumor efficacy of human chimeric antigen receptor T cells. *Sci. Rep.* **2017**, *7*, 737. [CrossRef] [PubMed]

366. Liu, X.; Ranganathan, R.; Jiang, S.; Fang, C.; Sun, J.; Kim, S.; Newick, K.; Lo, A.; June, C.H.; Zhao, Y.; et al. A chimeric switch-receptor targeting PD1 augments the efficacy of second-generation CAR T cells in advanced solid tumors. *Cancer Res.* **2016**, *76*, 1578–1590. [CrossRef] [PubMed]

367. Pan, Z.; Di, S.; Shi, B.; Jiang, H.; Shi, Z.; Liu, Y.; Wang, Y.; Luo, H.; Yu, M.; Wu, X.; et al. Increased antitumor activities of glypican-3-specific chimeric antigen receptor-modified T cells by coexpression of a soluble PD1-CH3 fusion protein. *Cancer Immunol. Immunother.* **2018**, *67*, 1621–1634. [CrossRef] [PubMed]

368. CAR-T Cell Immunotherapy. Clinical Trial NCT03330834. Available online: https://clinicaltrials.gov/ct2/show/NCT03330834?term=NCT03330834&rank=1 (accessed on 22 April 2019).

369. Chmielewski, M.; Hombach, A.A.; Abken, H. Of CARs and TRUCKs: Chimeric antigen receptor (CAR) T cells engineered with an inducible cytokine to modulate the tumor stroma. *Immunol. Rev.* **2014**, *257*, 83–90. [CrossRef] [PubMed]

370. Koneru, M.; Purdon, T.J.; Spriggs, D.; Koneru, S.; Brentjens, R.J. IL-12 secreting tumor-targeted chimeric antigen receptor T cells eradicate ovarian tumors *in vivo*. *Oncoimmunology* **2015**, *4*, e994446. [CrossRef] [PubMed]

371. Autologous T Cells Genetically Engineered to Secrete IL-12. Clinical Trial NCT02498912. Available online: https://clinicaltrials.gov/ct2/show/NCT02498912?term=NCT02498912&rank=1 (accessed on 22 April 2019).

372. Safety, Tolerability, Pharmacokinetics and Efficacy of AMG 562. Clinical Trial NCT03571828. Available online: https://clinicaltrials.gov/ct2/show/NCT03571828?term=amg562&rank=1 (accessed on 20 June 2019).

373. Townsend, M.H.; Shrestha, G.; Robison, R.A.; O'Neill, K.L. The expansion of targetable biomarkers for CAR T cell therapy. *J. Exp. Clin. Cancer Res.* **2018**, *37*, 163. [CrossRef] [PubMed]

374. Gerlinger, M.; Rowan, A.J.; Horswell, S.; Math, M.; Larkin, J.; Endesfelder, D.; Gronroos, E.; Martinez, P.; Matthews, N.; Stewart, A.; et al. Intratumor heterogeneity and branched evolution revealed by multiregion sequencing. *N. Engl. J. Med.* **2012**, *366*, 883–892. [CrossRef]

375. Chen, N.; Li, X.; Chintala, N.K.; Tano, Z.E.; Adusumilli, P.S. Driving CARs on the uneven road of antigen heterogeneity in solid tumors. *Curr. Opin. Immunol.* **2018**, *51*, 103–110. [CrossRef]

376. Hardiman, K.M.; Ulintz, P.J.; Kuick, R.D.; Hovelson, D.H.; Gates, C.M.; Bhasi, A.; Rodrigues Grant, A.; Liu, J.; Cani, A.K.; Greenson, J.K.; et al. Intra-tumor genetic heterogeneity in rectal cancer. *Lab. Investig.* **2016**, *96*, 4–15. [CrossRef]

377. Lawrence, M.S.; Stojanov, P.; Polak, P.; Kryukov, G.V.; Cibulskis, K.; Sivachenko, A.; Carter, S.L.; Stewart, C.; Mermel, C.H.; Roberts, S.A.; et al. Mutational heterogeneity in cancer and the search for new cancer-associated genes. *Nature* **2013**, *499*, 214–218. [CrossRef]

378. Mroz, E.A.; Tward, A.M.; Hammon, R.J.; Ren, Y.; Rocco, J.W. Intra-tumor genetic heterogeneity and mortality in head and neck cancer: Analysis of data from the Cancer Genome Atlas. *PLoS Med.* **2015**, *12*, e1001786. [CrossRef] [PubMed]

379. Mroz, E.A.; Rocco, J.W. MATH, a novel measure of intratumor genetic heterogeneity, is high in poor-outcome classes of head and neck squamous cell carcinoma. *Oral Oncol.* **2013**, *49*, 211–215. [CrossRef] [PubMed]

380. Lanitis, E.; Poussin, M.; Hagemann, I.S.; Coukos, G.; Sandaltzopoulos, R.; Scholler, N.; Powell, D.J., Jr. Redirected antitumor activity of primary human lymphocytes transduced with a fully human anti-mesothelin chimeric receptor. *Mol. Ther.* **2012**, *20*, 633–643. [CrossRef] [PubMed]

381. Ross, S.L.; Sherman, M.; McElroy, P.L.; Lofgren, J.A.; Moody, G.; Baeuerle, P.A.; Coxon, A.; Arvedson, T. Bispecific T cell engager (BiTE®) antibody constructs can mediate bystander tumor cell killing. *PLoS ONE* **2017**, *12*, e0183390. [CrossRef] [PubMed]

382. Shadrin, N.; Shapira, M.G.; Khalfin, B.; Uppalapati, L.; Parola, A.H.; Nathan, I. Serine protease inhibitors interact with IFN-γ through up-regulation of FasR; a novel therapeutic strategy against cancer. *Exp. Cell Res.* **2015**, *330*, 233–239. [CrossRef] [PubMed]

383. Shimabukuro-Vornhagen, A.; Gödel, P.; Subklewe, M.; Stemmler, H.J.; Schlößer, H.A.; Schlaak, M.; Kochanek, M.; Böll, B.; Von Bergwelt-Baildon, M.S. Cytokine release syndrome. *J. Immunother. Cancer* **2018**, *6*, 56. [CrossRef]

384. Blincyto Package Insert. 2019. Available online: https://www.pi.amgen.com/~{}/media/amgen/repositorysites/pi-amgen-com/blincyto/blincyto_pi_hcp_english.pdf (accessed on 26 April 2019).

385. Neelapu, S.S.; Tummala, S.; Kebriaei, P.; Wierda, W.; Gutierrez, C.; Locke, F.L.; Komanduri, K.V.; Lin, Y.; Jain, N.; Daver, N.; et al. Chimeric antigen receptor T-cell therapy—Assessment and management of toxicities. *Nat. Rev. Clin. Oncol.* **2018**, *15*, 47–62. [CrossRef]

386. Brudno, J.N.; Kochenderfer, J.N. Toxicities of chimeric antigen receptor T cells: Recognition and management. *Blood* **2016**, *127*, 3321–3330. [CrossRef]

387. Giavridis, T.; Van Der Stegen, S.J.C.; Eyquem, J.; Hamieh, M.; Piersigilli, A.; Sadelain, M. CAR T cell–induced cytokine release syndrome is mediated by macrophages and abated by IL-1 blockade. *Nat. Med.* **2018**, *24*, 731–738. [CrossRef]

388. Sterner, R.M.; Sakemura, R.; Cox, M.J.; Yang, N.; Khadka, R.H.; Forsman, C.L.; Hansen, M.J.; Jin, F.; Ayasoufi, K.; Hefazi, M.; et al. GM-CSF inhibition reduces cytokine release syndrome and neuroinflammation but enhances CAR-T cell function in xenografts. *Blood* **2019**, *133*, 697–709. [CrossRef]

389. Sachdeva, M.; Duchateau, P.; Depil, S.; Poirot, L.; Valton, J. Granulocyte-macrophage colony-stimulating factor inactivation in CAR T-cells prevents monocyte-dependent release of key cytokine release syndrome mediators. *J. Biol. Chem.* **2019**, *294*, 5430–5437. [CrossRef] [PubMed]

390. Yuraszeck, T.; Kasichayanula, S.; Benjamin, J.E. Translation and clinical development of bispecific T-cell engaging antibodies for cancer treatment. *Clin. Pharmacol. Ther.* **2017**, *101*, 634–645. [CrossRef] [PubMed]

391. Anti-BCMA and/or Anti-CD19 CART Cells Treatment. Clinical Trial NCT03767725. Available online: https://clinicaltrials.gov/ct2/show/NCT03767725?term=NCT03767725&rank=1 (accessed on 27 April 2019).

392. CD22 Redirected Autologous T Cells for ALL. Clinical Trial NCT02650414. Available online: https://clinicaltrials.gov/ct2/show/NCT02650414?term=NCT02650414&rank=1 (accessed on 27 April 2019).

393. Goebeler, M.E.; Knop, S.; Viardot, A.; Kufer, P.; Topp, M.S.; Einsele, H.; Noppeney, R.; Hess, G.; Kallert, S.; Mackensen, A.; et al. Bispecific T-cell engager (BiTE) antibody construct blinatumomab for the treatment of patients with relapsed/refractory non-Hodgkin lymphoma: Final results from a phase I study. *J. Clin. Oncol.* **2016**, *34*, 1104–1111. [CrossRef] [PubMed]

394. Suzuki, M.; Curran, K.J.; Cheung, N.K. Chimeric antigen receptors and bispecific antibodies to retarget T cells in pediatric oncology. *Pediatr. Blood Cancer* **2015**, *62*, 1326–1336. [CrossRef] [PubMed]

395. Aldoss, I.; Bargou, R.C.; Nagorsen, D.; Friberg, G.R.; Baeuerle, P.A.; Forman, S.J. Redirecting T cells to eradicate B-cell acute lymphoblastic leukemia: Bispecific T-cell engagers and chimeric antigen receptors. *Leukemia* **2017**, *31*, 777–787. [CrossRef]

396. Yescarta Sales 2018. Available online: https://www.gilead.com/news-and-press/press-room/press-releases/2019/2/gilead-sciences-announces-fourth-quarter-and-full-year-2018-financial-results (accessed on 4 May 2019).

397. Kymriah Sales 2018. Available online: https://www.novartis.com/sites/www.novartis.com/files/q4-2018-media-release-en.pdf (accessed on 4 May 2019).

398. Blincyto Sales 2018. Available online: https://www.amgen.com/media/news-releases/2019/01/amgen-reports-fourth-quarter-and-full-year-2018-financial-results/ (accessed on 4 May 2019).

399. Grupp, S.A.; Kalos, M.; Barrett, D.; Aplenc, R.; Porter, D.L.; Rheingold, S.R.; Teachey, D.T.; Chew, A.; Hauck, B.; Wright, J.F.; et al. Chimeric antigen receptor-modified T cells for acute lymphoid leukemia. *N. Engl. J. Med.* **2013**, *368*, 1509–1518. [CrossRef] [PubMed]

400. New Leukemia Drug Tops the Charts With a $178,000 Price Tag. Available online: https://www.medscape.com/viewarticle/836879 (accessed on 4 May 2019).

401. CAR-T: How Will These $400k Therapies be Adapted for Europe? Available online: https://www.pharmaceutical-technology.com/comment/car-t-therapies-europe/ (accessed on 4 May 2019).

402. Choi, B.D.; Gedeon, P.C.; Sanchez-Perez, L.; Bigner, D.D.; Sampson, J.H. Regulatory T cells are redirected to kill glioblastoma by an EGFRvIII-targeted bispecific antibody. *Oncoimmunology* **2013**, *2*, e26757. [CrossRef]

403. Long, A.H.; Haso, W.M.; Shern, J.F.; Wanhainen, K.M.; Murgai, M.; Ingaramo, M.; Smith, J.P.; Walker, A.J.; Kohler, M.E.; Venkateshwara, V.R.; et al. 4-1BB costimulation ameliorates T cell exhaustion induced by tonic signaling of chimeric antigen receptors. *Nat. Med.* **2015**, *21*, 581–590. [CrossRef]

404. Drent, E.; Poels, R.; Ruiter, R.; Van De Donk, N.W.C.J.; Zweegman, S.; Yuan, H.; De Bruijn, J.; Sadelain, M.; Lokhorst, H.M.; Groen, R.W.J.; et al. Combined CD28 and 4-1BB costimulation potentiates affinity-tuned chimeric antigen receptor-engineered T cells. *Clin. Cancer Res.* **2019**. [CrossRef]

405. Pulte, E.D.; Vallejo, J.; Przepiorka, D.; Nie, L.; Farrell, A.T.; Goldberg, K.B.; McKee, A.E.; Pazdur, R. FDA supplemental approval: Blinatumomab for treatment of relapsed and refractory precursor B-cell acute lymphoblastic leukemia. *Oncologist* **2018**, *23*, 1366–1371. [CrossRef]

406. Schuster, S.J.; Bishop, M.R.; Tam, C.S.; Waller, E.K.; Borchmann, P.; McGuirk, J.P.; Jäger, U.; Jaglowski, S.; Andreadis, C.; Westin, J.R.; et al. Tisagenlecleucel in adult relapsed or refractory diffuse large B-cell lymphoma. *N. Engl. J. Med.* **2019**, *380*, 45–56. [CrossRef] [PubMed]

407. Neelapu, S.S.; Locke, F.L.; Bartlett, N.L.; Lekakis, L.J.; Miklos, D.B.; Jacobson, C.A.; Braunschweig, I.; Oluwole, O.O.; Siddiqi, T.; Lin, Y.; et al. Axicabtagene ciloleucel CAR T-cell therapy in refractory large B-cell lymphoma. *N. Engl. J. Med.* **2017**, *377*, 2531–2544. [CrossRef] [PubMed]

408. Abramson, J.S.; Gordon, L.I.; Palomba, M.L.; Lunning, M.; Arnason, J.; Forero-Torres, A.; Wang, M.; Maloney, D.; Sehgal, A.; Andreadis, C.; et al. Updated Safety and Long-Term Clinical Outcomes in TRANSCEND NHL 001, Pivotal Trial of Lisocabtagene Maraleucel (JCAR017) in R/R Aggressive NHL. Available online: https://www.primeoncology.org/app/uploads/hematology-updates-stockholm-2018-dlbcl-s800-abramson.pdf (accessed on 5 May 2019).

409. Safety Study of Bispecific T-Cell Engager Blinatumomab. Clinical Trial NCT00274742. Available online: https://clinicaltrials.gov/ct2/show/NCT00274742?term=NCT00274742&rank=1 (accessed on 7 May 2019).

410. Study of bb2121 in Multiple Myeloma. Clinical Trial NCT02658929. Available online: https://clinicaltrials.gov/ct2/show/NCT02658929?term=NCT02658929&rank=1 (accessed on 7 May 2019).

411. Topp, M.S.; Duell, J.; Zugmaier, G.; Attal, M.; Moreau, P.; Langer, C.; Kroenke, J.; Facon, T.; Einsele, H.; Munzert, G. Treatment with AMG 420, an anti-B-cell maturation antigen (BCMA) bispecific T-cell engager (BiTE®) antibody construct, induces minimal residual disease (MRD) negative complete responses in relapsed and/or refractory (R/R) multiple myeloma (MM) patients: Results of a first-in-human (FIH) phase I dose escalation study. *Blood* **2018**, *132*, 1010.

412. Phase I Dose Escalation of I.V. BI 836909. Clinical Trial NCT02514239. Available online: https://clinicaltrials.gov/ct2/show/NCT02514239?term=NCT02514239&rank=1 (accessed on 7 May 2019).

413. Jaspers, J.E.; Brentjens, R.J. Development of CAR T cells designed to improve antitumor efficacy and safety. *Pharmacol. Ther.* **2017**, *178*, 83–91. [CrossRef] [PubMed]

414. Majzner, R.G.; Mackall, C.L. Tumor antigen escape from CAR T-cell therapy. *Cancer Disc.* **2018**, *8*, 1219–1226. [CrossRef] [PubMed]

415. Orlando, E.J.; Han, X.; Tribouley, C.; Wood, P.A.; Leary, R.J.; Riester, M.; Levine, J.E.; Qayed, M.; Grupp, S.A.; Boyer, M.; et al. Genetic mechanisms of target antigen loss in CAR19 therapy of acute lymphoblastic leukemia. *Nat. Med.* **2018**, *24*, 1504–1506. [CrossRef] [PubMed]

416. Mandikian, D.; Takahashi, N.; Lo, A.A.; Li, J.; Eastham-Anderson, J.; Slaga, D.; Ho, J.; Hristopoulos, M.; Clark, R.; Totpal, K.; et al. Relative target affinities of T cell-dependent bispecific antibodies determine biodistribution in a solid tumor mouse model. *Mol. Cancer Ther.* **2018**, *17*, 776–785. [CrossRef]

417. Trinklein, N.D.; Pham, D.; Schellenberger, U.; Buelow, B.; Boudreau, A.; Choudhry, P.; Clarke, S.C.; Dang, K.; Harris, K.E.; Iyer, S.; et al. Efficient tumor killing and minimal cytokine release with novel T-cell agonist bispecific antibodies. *MAbs* **2019**, *20*, 1–14. [CrossRef]

418. Saber, H.; Del Valle, P.; Ricks, T.K.; Leighton, J.K. An FDA oncology analysis of CD3 bispecific constructs and first-in-human dose selection. *Regul. Toxicol. Pharmacol.* **2017**, *90*, 144–152. [CrossRef]

419. Salter, A.I.; Ivey, R.G.; Kennedy, J.J.; Voillet, V.; Rajan, A.; Alderman, E.J.; Voytovich, U.J.; Lin, C.; Sommermeyer, D.; Liu, L.; et al. Phosphoproteomic analysis of chimeric antigen receptor signaling reveals kinetic and quantitative differences that affect cell function. *Sci. Signal.* **2018**, *11*, eaat6753. [CrossRef]

420. Caruso, H.G.; Hurton, L.V.; Najjar, A.; Rushworth, D.; Ang, S.; Olivares, S.; Mi, T.; Switzer, K.; Singh, H.; Huls, H.; et al. Tuning sensitivity of CAR to EGFR density limits recognition of normal tissue while maintaining potent antitumor activity. *Cancer Res.* **2015**, *75*, 3505–3518. [CrossRef] [PubMed]

421. Dustin, M.L. The immunological synapse. *Cancer Immunol. Res.* **2014**, *2*, 1023–1033. [CrossRef] [PubMed]

422. Liu, X.; Jiang, S.; Fang, C.; Yang, S.; Olalere, D.; Pequignot, E.C.; Cogdill, A.P.; Li, N.; Ramones, M.; Granda, B.; et al. Affinity-tuned ErbB2 or EGFR chimeric antigen receptor T cells exhibit an increased therapeutic index against tumors in mice. *Cancer Res.* **2015**, *75*, 3596–3607. [CrossRef] [PubMed]

423. Jenkins, M.R.; Rudd-Schmidt, J.A.; Lopez, J.A.; Ramsbottom, K.M.; Mannering, S.I.; Andrews, D.M.; Voskoboinik, I.; Trapani, J.A. Failed CTL/NK cell killing and cytokine hypersecretion are directly linked through prolonged synapse time. *J. Exp. Med.* **2015**, *212*, 307–317. [CrossRef] [PubMed]

424. Krenciute, G.; Krebs, S.; Torres, D.; Wu, M.F.; Liu, H.; Dotti, G.; Li, X.N.; Lesniak, M.S.; Balyasnikova, I.V.; Gottschalk, S. Characterization and functional analysis of scFv-based chimeric antigen receptors to redirect T cells to IL-13Ralpha2-positive glioma. *Mol. Ther.* **2016**, *24*, 354–363. [CrossRef] [PubMed]

425. Christiansen, J.; Rajasekaran, A.K. Biological impediments to monoclonal antibody-based cancer immunotherapy. *Mol. Cancer Ther.* **2004**, *3*, 1493–1501. [PubMed]

426. Ying, Z.; Huang, X.F.; Xiang, X.; Liu, Y.; Kang, X.; Song, Y.; Guo, X.; Liu, H.; Ding, N.; Zhang, T.; et al. A safe and potent anti-CD19 CAR T cell therapy. *Nat. Med.* **2019**, *25*, 947–953. [CrossRef]

427. Autologous CAR T Cells in Relapsed or Refractory B-Cell Lymphoma. Clinical Trial NCT02842138. Available online: https://clinicaltrials.gov/ct2/show/NCT02842138 (accessed on 10 May 2019).

428. Sun, S.; Hao, H.; Yang, G.; Zhang, Y.; Fu, Y. Immunotherapy with CAR-modified T cells: Toxicities and overcoming strategies. *J. Immunol. Res.* **2018**, *2018*, 2386187. [CrossRef]

429. Chan, T.; Gallagher, J.; Cheng, N.-L.; Carvajal-Borda, F.; Plummer, J.; Govekung, A.; Barrett, J.A.; Khare, P.D.; Cooper, L.J.N.; Shah, R.R. CD19-specific chimeric antigen receptor-modified T cells with safety switch produced under "point-of-care" using the sleeping beauty system for the very rapid manufacture and treatment of B-cell malignancies. *Blood* **2017**, *130*, 1324.

430. Ghassemi, S.; Prachi, P.; Scholler, J.; Nunez-Cruz, S.; Barrett, D.M.; Bedoya, F.; Fraietta, J.A.; Lacey, S.F.; Levine, B.L.; Grupp, S.A.; et al. Minimally ex vivo manipulated gene-modified T cells display enhanced tumor control. *Blood* **2016**, *128*, 4549.

antibodies

MDPI

Review

IgE Antibodies: From Structure to Function and Clinical Translation

Brian J. Sutton [1,2,*], **Anna M. Davies** [1,2], **Heather J. Bax** [3] **and Sophia N. Karagiannis** [3,*]

[1]   King's College London, Randall Centre for Cell and Molecular Biophysics, London SE1 1UL, UK; anna.davies@kcl.ac.uk
[2]   Asthma UK Centre in Allergic Mechanisms of Asthma, London, UK
[3]   King's College London, St John's Institute of Dermatology, London SE1 9RT, UK; heather.bax@kcl.ac.uk
*   Correspondence: brian.sutton@kcl.ac.uk (B.J.S.); sophia.karagiannis@kcl.ac.uk (S.N.K.); Tel.: +44-(0)20-7848-6423 (B.J.S.); +44-(0)20-7188-6355 (S.N.K.)

Received: 5 January 2019; Accepted: 15 February 2019; Published: 22 February 2019

Abstract: Immunoglobulin E (IgE) antibodies are well known for their role in mediating allergic reactions, and their powerful effector functions activated through binding to Fc receptors FcεRI and FcεRII/CD23. Structural studies of IgE-Fc alone, and when bound to these receptors, surprisingly revealed not only an acutely bent Fc conformation, but also subtle allosteric communication between the two distant receptor-binding sites. The ability of IgE-Fc to undergo more extreme conformational changes emerged from structures of complexes with anti-IgE antibodies, including omalizumab, in clinical use for allergic disease; flexibility is clearly critical for IgE function, but may also be exploited by allosteric interference to inhibit IgE activity for therapeutic benefit. In contrast, the power of IgE may be harnessed to target cancer. Efforts to improve the effector functions of therapeutic antibodies for cancer have almost exclusively focussed on IgG1 and IgG4 subclasses, but IgE offers an extremely high affinity for FcεRI receptors on immune effector cells known to infiltrate solid tumours. Furthermore, while tumour-resident inhibitory Fc receptors can modulate the effector functions of IgG antibodies, no inhibitory IgE Fc receptors are known to exist. The development of tumour antigen-specific IgE antibodies may therefore provide an improved immune functional profile and enhanced anti-cancer efficacy. We describe proof-of-concept studies of IgE immunotherapies against solid tumours, including a range of in vitro and in vivo evaluations of efficacy and mechanisms of action, as well as ex vivo and in vivo safety studies. The first anti-cancer IgE antibody, MOv18, the clinical translation of which we discuss herein, has now reached clinical testing, offering great potential to direct this novel therapeutic modality against many other tumour-specific antigens. This review highlights how our understanding of IgE structure and function underpins these exciting clinical developments.

Keywords: Immunoglobulin E; FcεRI; CD23; allostery; cancer immunotherapy; AllergoOncology; IgE effector functions; monocytes; macrophages; ADCC

1. Introduction

Immunoglobulin E (IgE), named in 1968 [1–3], was the last of the five classes of human antibodies to be discovered, and today is commonly associated with the various manifestations of allergic disease [4]. However, its role in mammalian evolution appears to be the provision of a mechanism for defence against parasites and animal venoms [5], and in this regard it required the acquisition of a powerful effector function. It is precisely this power, and the possibility of understanding and harnessing it, that makes IgE an attractive candidate for monoclonal antibody immunotherapy against clinically important targets. IgE differs from the various sub-classes of IgG that have hitherto been the common format for therapeutic antibodies in a number of key aspects, including its domain architecture, glycosylation, conformational dynamics and, as only recently appreciated, allosteric properties [6]. In this review, we bring together our understanding of the structural and functional properties of IgE, and show how this underpins the development of IgE as a therapeutic antibody format.

IgE's receptor-binding activities also present unique features. There are two principal receptors, FcεRI, structurally homologous to other members of the FcγR family, and FcεRII/CD23, which unlike almost all other antibody receptors, is a member of the C-type (Ca^{2+}-dependent) lectin-like superfamily [4]. FcεRI is expressed on tissue mast cells, blood basophils, airway epithelial and smooth muscle cells, intestinal epithelial cells, and various antigen-presenting cells (APCs), monocytes and macrophages [7–11]; the cross-linking of receptor-bound allergen-specific IgE on mast cells and basophils by allergen is the signal for cell degranulation, the release of pre-formed mediators of inflammation and an immediate hypersensitivity response that can be powerful enough to cause anaphylactic shock and even death. Not only is it is necessary to cross-link very few IgE and FcεRI molecules in this way, compared with IgG and FcγR, but the affinity of IgE for FcεRI ($K_a \approx 10^{10} M^{-1}$) is at least two orders of magnitude higher than that of IgG for any of its receptors. Thus, most IgE is already cell bound, and all that is required is contact with perhaps a minute amount of allergen to trigger a rapid reaction. In contrast, IgG generally requires the formation of immune complexes consisting of many more antibody molecules, which can then, upon contact with an effector cell, cause FcγR clustering and cell activation [12]. With its uniquely high affinity for any antibody-receptor interaction, FcεRI is often referred to as the "high-affinity" receptor for IgE.

FcεRII, or CD23 as it will be called here, is also known as the "low-affinity" receptor for IgE. While the affinity of each of its lectin-like "heads" for IgE ($K_a \approx 10^6 M^{-1}$) is indeed much lower than that of FcεRI, the fact that the molecule is trimeric can lead to a higher avidity if more than one head can engage IgE; this will be discussed in detail later. CD23 is expressed on B cells, T cells, various APCs, gut and airway epithelial cells and a range of other cell types [13–18]. On B cells, IgE binding to CD23, the latter behaving both as a membrane protein and also as a soluble protein released from the cell surface (in trimeric or monomeric form) by endogenous or exogenous proteases, can either up- or down-regulate IgE levels [13,19–21]. This interplay between IgE and both membrane and soluble CD23 has been proposed to constitute a mechanism for IgE homeostasis. CD23 also transfers IgE-allergen complexes across the airway and gut epithelia and thus promotes the presentation of airborne and food allergens to the immune system [16–18,22].

There is a considerable body of structural data concerning the interactions between IgE-Fc and the receptors FcεRI and CD23. There is also a good understanding, if based upon rather few examples, of how IgE Fabs recognise allergens; this understanding was recently enhanced by the discovery that allergen recognition may occur not only in a classical, complementarity-determining region (CDR)-mediated manner, but also through V-region framework regions (FR) in a "superantigen-like" mode [23]. When we put these structural data together to build models of the whole IgE molecule, it is clear that there are constraints upon the disposition of the Fab arms when the Fc is receptor bound, and similarly, there may be restrictions upon the receptor-binding capability of the Fc region when IgE engages target antigens; unfortunately, we lack high-resolution structural data on the complete IgE molecule. Appreciation of these constraints and the consequences of the flexibility and dynamics of the IgE molecule as a whole, are clearly important for engineering an IgE molecule for immunotherapy that combines the desired antigen-binding and receptor-mediated activities.

2. The Structure of IgE

The overall architecture of the IgE molecule differs most significantly from that of IgG in respect to the "additional" heavy chain constant domain (Figure 1a,b) and the absence of a hinge region in the ε-chain. The six domains comprising the IgE-Fc, a dimer of Cε2-Cε3-Cε4 domains, are evolutionarily more ancient than the four-domain IgG-Fc. IgE-Fc resembles the (Cμ2-Cμ3-Cμ4)$_2$ Fc structure of IgM, the most primitive antibody class, and the (Cυ2-Cυ3-Cυ4)$_2$ Fc domains of avian IgY, the ancestor of

IgE and IgG [24]. The hinge region of IgG appears to have evolved to take the place of the (Cε2)$_2$ domain pair, since the Cγ2 and Cγ3 domains of IgG-Fc are most closely homologous to the Cε3 and Cε4 domains of IgE-Fc. IgM molecules, as pentameric or hexameric structures, are known to undergo conformational changes upon contact with antigen that dramatically alters the disposition of the Fab arms relative to the Fc region, as observed by electron microscopy (EM) [25]. Unliganded, the IgM molecules appear planar and "star-shaped", while bound to the surface of antigens they form "table-like" structures with the Fab arms bent down and away from the Fc region. These observations are pertinent to the discussion of the flexibility and conformational change in IgE that will follow.

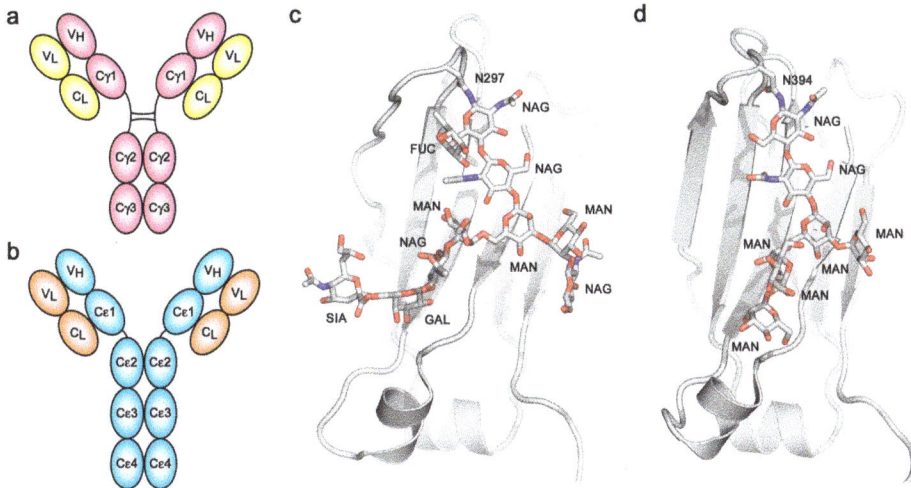

Figure 1. Overall structure and glycosylation. (**a**) Schematic representation of Immunoglobulin G (IgG). (**b**) Schematic representation of Immunoglobulin E (IgE). (**c**) The IgG Cγ2 domain contains complex carbohydrate covalently attached to Asn297 [26]. (**d**) The IgE Cε3 domain contains high-mannose carbohydrate covalently attached to Asn394 [27]. In panels (**c,d**), carbohydrate residues are labelled as follows: FUC, fucose; GAL, galactose; MAN, mannose; NAG, N-acetylglucosamine; SIA, sialic acid.

Expectations that IgE, with the additional domain pair, might adopt a more extended Y-shaped structure than that of IgG [28], were refuted by early biophysical studies of IgE in solution and when FcεRI-bound that indicated a more compact conformation [29,30]. In particular, elegant work with IgE molecules fluorescently labelled in their antigen-binding sites and at the C-termini of their Fc regions, clearly indicated through fluorescence (Förster) resonance energy transfer (FRET) distance measurements that the IgE molecule was not extended, but bent [31,32]. This was later confirmed by small-angle X-ray scattering (SAXS) studies of IgE and IgE-Fc in solution, the latter indicating that the Fc itself was a compact structure, best modelled by folding the (Cε2)$_2$ domain pair back onto the Cε3-Cε4 domains [33]. However, when the first X-ray crystal structure of the whole IgE-Fc was solved [34], the bend was found to be even more acute than that which had been modelled (Figure 2a), with the Cε2 domain of one chain even contacting the Cε4 domain of the other; furthermore, by bending of the (Cε2)$_2$ domain pair over towards one side of the (Cε3-Cε4)$_2$ region, the IgE-Fc molecule adopted an asymmetrical three-dimensional structure, despite its symmetrical primary structure (chemical sequence). A FRET study of IgE-Fc further confirmed that this bent structure does indeed exist in solution [35]. Might IgE-Fc be able to "un-bend", akin to the conformational changes that IgM appears to undergo?

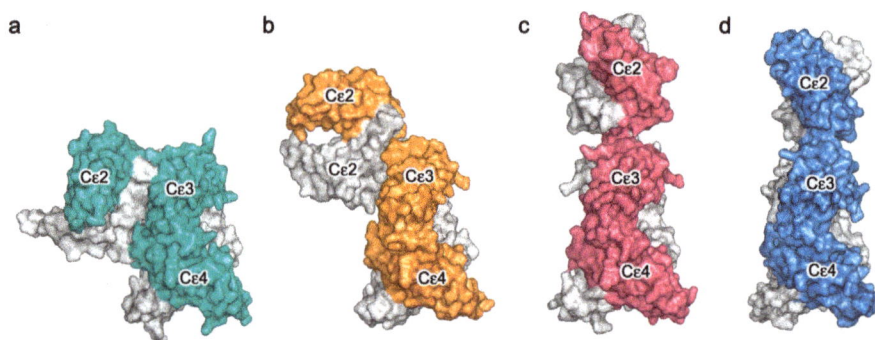

Figure 2. IgE-Fc is conformationally flexible. (**a**) Unbound IgE-Fc adopts an acutely bent conformation [34]. (**b**) IgE-Fc adopts a partially bent conformation when in complex with an omalizumab-derived Fab [36]. (**c**) Fully extended IgE-Fc conformation captured by aεFab [37]. (**d**) IgE-Fc adopts a fully extended conformation when in complex with the 8D6 Fab that is more compact than the conformation shown in (**c**) [38]. In panels (**a–d**), IgE-Fc chain B is coloured grey while chain A is coloured cyan, orange, pink and blue, respectively. For clarity, the anti-IgE Fabs are not shown in panels (**b–d**).

Despite the identical primary structures of the two heavy (and two light) chains, IgE, like IgG and all other antibody classes, is glycosylated [39–42], and since there is heterogeneity not only in the pattern of glycosylation at the various potential sites but also in the composition at any particular site, the two heavy chains within any one IgE (or IgG) molecule are not precisely identical. Whether or not this compositional asymmetry is related to the asymmetric bending of the IgE-Fc has not been explored. One glycosylation site is conserved across all antibody classes: Asn394 in the Cε3 domain of IgE, structurally homologous to Asn297 in the Cγ2 domain of IgG. Other potential sites in the Cε2 and Cε3 domains are not always fully glycosylated, but Asn394, like its homologues in other antibody classes, is always fully occupied [39–41]. The branched carbohydrate chains occupy space between the Cε3 domains, as they do between the Cγ2 domains of IgG, but there is a major difference between IgE and IgG in this respect: the glycosylation at Asn394 in IgE is of the "high-mannose" type, in contrast to the "complex-type" at Asn297 in IgG (Figure 1c,d). Other glycosylation sites in IgE that are exposed at the surface are complex-type, which suggests that the high-mannose composition at Asn394 may be due to the Cε2 domains impeding access of the mannosidase enzymes responsible for trimming the high-mannose structures prior to assembly of the complex-type glycoforms. The same high-mannose structure is seen in IgY-Fc between the Cυ3 domains [43], perhaps similarly due to the presence of Cυ2 domains. The high-mannose, branched carbohydrate chains in IgE-Fc not only make non-covalent (hydrogen bond, hydrophobic and van der Waals) contacts with the Cε3 domains to which they are covalently attached, and to the adjacent Cε4 domains, but also make contact with each other, bridging the two heavy chains [27,34,44]. Despite this apparent structural role, and again in contrast to IgG in which loss of glycosylation at Asn297 compromises FcγR binding [45], both FcεRI and CD23 receptor-binding activity is maintained in the absence of glycosylation; IgE-Fc expressed in bacteria and refolded [46,47], or deglycosylated following mammalian expression [48,49], binds to both receptors. However, glycosylation at Asn394 is essential for the expression of functional IgE in mammalian cells in vitro and in vivo [41,50].

IgE thus differs in important ways from IgG, not only in terms of its overall structure and, as will now be discussed, its flexibility, but also with respect to the nature and the role of its glycosylation.

3. Conformational Dynamics in IgE-Fc

Crystal structures of the sub-fragment of IgE-Fc consisting of only the Cε3 and Cε4 domains, which we term Fcε3-4, and IgE-Fc, have revealed a degree of flexibility in the arrangement of the Cε3 domains relative to each other, either further apart ("open") or closer together ("closed") [27,34,36–38,44,51–59]. Furthermore, unliganded IgE-Fc structures were only bent (Figure 2a) [27,34,44]. It was therefore a considerable surprise to discover that in the crystal structure of the first complex between IgE-Fc and an anti-IgE antibody Fab, aεFab, the Fc had adopted a fully extended conformation (Figure 2c) [37]. Further analysis revealed that the anti-IgE Fab, which binds at the Cε2/Cε3 interface in a 2:1 complex with IgE-Fc, was selecting a pre-existing conformational state of the molecule in solution, and thus the question arose: if IgE-Fc could spontaneously "un-bend" to reach a fully extended state, could the (Cε2)$_2$ domain pair then "flip over" to lie in a bent conformation on the other side of the Fcε3-4 region? In order to estimate the energetics of this potential "flipping" of the IgE-Fc, extensive molecular dynamics (MD) simulations were carried out [37]. It was discovered that the bent structure lies in a relatively deep energy well, but that once the IgE-Fc molecule had escaped this minimum, the "conformational landscape" was relatively flat, i.e., there were no significant barriers to prevent it reaching the extended conformation or indeed allowing the (Cε2)$_2$ domains to bend over onto the other side of the molecule. The MD simulations revealed that this flipping of the Cε2 domains required the Cε3 domains to open somewhat, but the rate-limiting step for the process was clearly escape from the energy well representing the bent conformation. Most molecules would be in the bent state at any given time, consistent with the SAXS and FRET data in solution, but occasionally they flip over, although the rate and frequency of this event is difficult to assess.

Anti-IgE antibodies of the IgG class, such as aεFab, directed against the Fc region clearly have potential as anti-allergy therapeutics if, by either steric or allosteric means, they inhibit FcεRI or CD23 engagement. These activities will be discussed in the following two sections, and we first concentrate here on the lessons learned about IgE flexibility from structural studies of these anti-IgE Fab/IgE-Fc complexes. Omalizumab is a clinically approved anti-IgE antibody, and it binds to a partially bent conformation, intermediate between the bent and extended structures (Figure 2b) [36]. It binds to the Cε3 domains, also in a 2:1 complex, and causes the Cε3 domains to move further apart and adopt a very "open" conformation. Another anti-IgE antibody, termed 8D6, directed to the Cε2 and Cε3 domains, binds to a fully extended IgE-Fc conformation (rather like aεFab, Figure 2c) but in the 8D6 structure (Figure 2d) the (Cε2)$_2$ domain pair is twisted and compressed towards the Cε3 domains, as in a corkscrew motion [38]. To date, these are the only structures that have been published for IgE-Fc in complex with anti-IgE Fabs (Figure 3).

The picture that emerges from these structural studies is that of a highly flexible Fc region in which the Cε2 domains are capable of extending and twisting relative to the Fcε3-4 region, or bending over to either side, with the Cε3 domains adopting closed or open states. With regard to the flexibility of the whole IgE molecule, i.e., that of the Fab arms relative to the Fc, we lack crystallographic data, although molecular simulations suggest that the short Cε1-Cε2 linker of only five or six amino-acids substantially restricts the available conformations compared with the Fab arm flexibility mediated by the hinge regions in IgG subclasses [35,37]. This is consistent with earlier biophysical studies in solution, which showed less Fab arm flexibility in IgE compared with IgG [60]. Nevertheless, despite lacking an IgG-like hinge, the linker between the Cε2 and Cε3 domains can clearly permit bending of the whole IgE molecule, just as is seen in IgM with its (Cμ2)$_2$ domains and no hinge [25], although in IgM the precise nature of the bending remains unresolved.

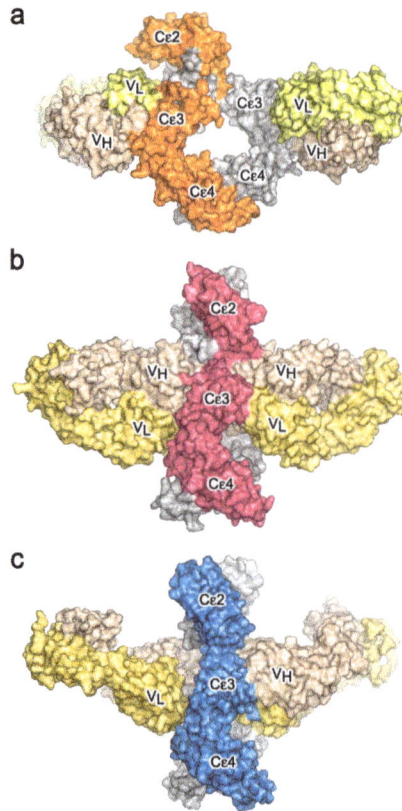

Figure 3. Crystal structures of IgE-Fc in complex with anti-IgE Fabs. (**a**) IgE-Fc in complex with an omalizumab-derived Fab [36]. (**b**) aεFab/IgE-Fc complex [37]. (**c**) 8D6 Fab/IgE-Fc complex [38]. In panels (**a–c**), IgE-Fc chain B is coloured grey while chain A is coloured orange, pink and blue, respectively. The Fab heavy and light chains are coloured in wheat and pale yellow, respectively.

4. IgE-Receptor Interactions

The structural details of IgE binding to the soluble extracellular domains of both FcεRI and CD23 are now well established. FcεRI expressed on mast cells and basophils comprises four polypeptide chains, $\alpha\beta\gamma_2$ (Figure 4a), but on other cell types it lacks the β-chain, which may serve either as an "amplifier" of down-stream signalling, since the β-chain contains an additional copy of the immuno-tyrosine activation motif (ITAM) present in the γ-chains, or it may affect surface expression [7]. All of the IgE-binding activity resides in the two Ig-like domains of the α-chain, termed sFcεRIα, the only substantial extracellular part of the receptor (Figure 4a). The crystal structure of sFcεRIα bound to Fcε3-4 first revealed the α2 domain and part of the α1-α2 linker bound across the two Cε3 domains, close to the point of connection to the Cε2 domains [56]. When the structure of the complex with the complete IgE-Fc was solved, contrary to expectations that the Fc might unbend, the angle was found to become even more acute (from 62° to 54°; Figure 4b) [44]. This enhanced bend seen in the crystal structure with IgE-Fc agrees not only with a recent study in solution with a FRET-labelled IgE-Fc molecule [35], but also, strikingly, with the work carried out more than 25 years ago with FRET-labelled IgE bound to FcεRI on cells, which showed a more compact structure for IgE when receptor-bound than in solution [32]. This orientation of IgE and acutely bent Fc, as indicated in Figure 4b, places constraints upon the disposition of the Fab arms, which may well be critical for understanding how the IgE molecule engages both FcεRI on cells and antigen (allergen), whether soluble or on a target cell, to enable receptor cross-linking and effector cell activation. These topological issues will be considered in more detail below.

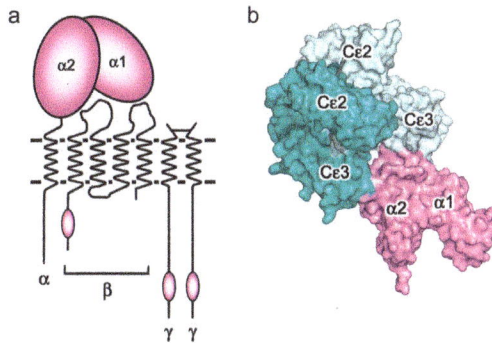

Figure 4. FcεRI (**a**) Schematic representation of FcεRI: the four chains are indicated, showing the two extracellular Ig-like domains of the α-chain that contain the IgE-binding activity, and the locations of the three intracellular ITAM signalling motifs. Figure adapted by permission from John Wiley & Sons, Inc. (Sutton, B.J.; Davies, A.M. Structure and dynamics of IgE-receptor interactions: FcεRI and CD23/FcεRII. *Immunol. Rev.* 2015, *268*, 222–235 [6]). (**b**) IgE-Fc adopts an acutely bent conformation when in complex with sFcεRIα, engaging the receptor (purple) at two distinct sub-sites [44]. IgE-Fc chains A and B are coloured dark cyan and pale cyan, respectively.

CD23 is a homo-trimeric type-II membrane protein with its C-terminal C-type lectin-like "head" domains, to which IgE binds, spaced from the membrane by a triple α-helical coiled-coil "stalk" (Figure 5a). There is also a C-terminal "tail" of unknown structure that is required for binding to CD21, a co-receptor for CD23, the engagement of which is implicated in B cell activation and cell adhesion events [4,6,61–63]. We will focus on the IgE/CD23 interaction. The crystal structure of a single lectin-like domain alone, lacking the stalk and tail, which we will term sCD23, binds to IgE-Fc with a 2:1 stoichiometry, although the affinities for the two sCD23 molecules differ by more than a factor of ten ($K_a \approx 10^6$ M^{-1} and 10^5 M^{-1}) [53]. The binding of both molecules can be seen clearly in Figure 5b, one sCD23 molecule bound to each ε-chain in a similar manner, principally to Cε3 but also contacting Cε4, in this complex with Fcε3-4 [51]. However, the structure of sCD23 bound to IgE-Fc, which unexpectedly trapped only the first binding event (Figure 5c), explains the difference in affinity [53]. This 1:1 complex reveals how the first sCD23 molecule binds to an asymmetrically bent IgE-Fc, principally to Cε3 as before and also to Cε4, but with a single hydrogen bond and some van der Waals contacts with a Cε2 domain; the $(Cε2)_2$ domain pair remains essentially bent, but swings about 16° to accommodate CD23 binding (Figure 5c) [53]. The site for the second CD23 head is completely accessible, although not occupied in this crystal structure, but this asymmetry of the two ε-chains explains the difference in affinity at the two CD23 binding sites.

As expected for a "C-type" lectin domain there is a Ca^{2+} binding site, although IgE binding does not require occupancy of this site [51,53,64]. Neither does this "lectin" interaction with IgE involve carbohydrate, although its binding to CD21 may be carbohydrate-dependent. In the presence of Ca^{2+}, IgE binding is enhanced [62], 30-fold at 37 °C, through ordering of a loop and a subtle conformational change that enables additional contacts with IgE [54]. Intriguingly, these additional contact residues comprise a second Ca^{2+} binding site in murine CD23, an indication perhaps of a step in the evolution of the interaction of IgE with this C-type lectin domain. The Ca^{2+} dependence of the affinity, undoubtedly enhanced in the context of the trimer through an avidity effect, may be functionally important for unloading of IgE/allergen complexes by CD23 in endosomes, where the

Ca^{2+} concentration is two to three orders of magnitude lower than at the cell surface, prior to CD23 recycling to the cell surface [65,66].

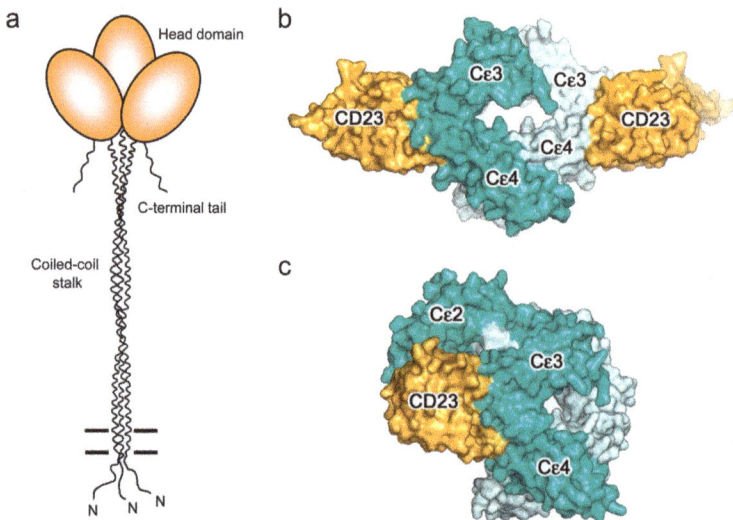

Figure 5. CD23. (a) Schematic representation of CD23: the three identical chains showing the triple α-helical coiled-coil "stalk", C-type lectin-like IgE-binding "head" domains, and C-terminal "tails". Figure adapted by permission from John Wiley & Sons, Inc. (Sutton, B.J.; Davies, A.M. Structure and dynamics of IgE-receptor interactions: FcεRI and CD23/FcεRII. *Immunol. Rev.* 2015, *268*, 222–235 [6]). (b) The 2:1 complex between sCD23 (orange) and Fcε3-4 [51]. (c) The 1:1 complex between sCD23 (orange) and IgE-Fc, in which IgE-Fc adopts an acutely bent conformation [53]. In panels (b,c), IgE-Fc chains A and B are coloured dark cyan and pale cyan, respectively.

It is important to realise that although IgE can bind to two CD23 heads, these cannot belong to the same CD23 trimer; the N-termini of the two sCD23 molecules, which connect to the stalk (Figure 5b), are so far apart that most of the stalk would have to unravel for this to be possible [51]. However, IgE can cross-link two membrane CD23 trimers, and soluble trimeric forms of CD23 containing both head and stalk can cross-link membrane IgE (on B cells committed to IgE synthesis) or soluble IgE; in all of these cases, the bivalence of IgE and trivalence of CD23 can combine to create large complexes, which may be required for signalling in the context of B cell or APC activation [4].

5. IgE—An Allosteric Antibody

The crystal structures of the two-receptor complexes reveal a key element of the IgE molecule, namely that there is allosteric communication between the two receptor-binding sites. It is known that IgE cannot bind to both receptors simultaneously [67,68], and vital that this is so, since otherwise trimeric CD23 could cross-link FcεRI-bound IgE on mast cells or basophils, causing activation and an inflammatory response in the absence of allergen. Indeed, binding of sFcεRIα inhibits sCD23 binding, and *vice versa* [51,69]. Earlier, it was thought that the two binding sites must overlap, but we know now that although both lie principally within Cε3, they are far apart from each other at opposite ends of the domain (Figures 4–6). This mutual inhibition is achieved allosterically [51,69], mainly through changes in the disposition of the Cε3 domains relative to the Cε4 domains. To engage FcεRI, the Cε3 domains must adopt an "open" state (Figure 6a), which changes the angle between the Cε3 and Cε4 domains and prevents binding of CD23 at the Cε3/Cε4 interface. However, when CD23 binds, the Cε3 domains move closer together and this more "closed" conformation precludes FcεRI binding (Figure 6b).

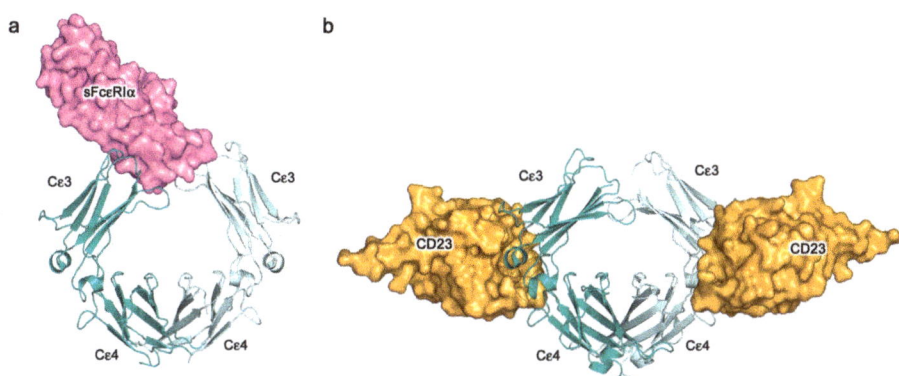

Figure 6. Binding of IgE to its receptors is allosterically regulated. (**a**) sFcεRIα (purple) binds to the Fcε3-4 region when the Cε3 domains adopt an open conformation [44]. (**b**) sCD23 (orange) binds to the Fcε3-4 region when the Cε3 domains adopt a closed conformation [51]. In panels (**a**,**b**), IgE-Fc chains A and B are coloured dark cyan and pale cyan, respectively.

Not only do the Cε3 domains undergo these domain motions, but they also appear to have evolved a high degree of intrinsic flexibility; when compared with other immunoglobulin domains in terms of hydrophobic core volume or other indicators of dynamics, Cε3 is clearly an outlier, and when expressed as an isolated domain it has been described as adopting a "molten globule" rather than a fully folded state [27,70–74]. Plasticity at the IgE-Fc/CD23 interface [55,75] and ordering of Cε3 upon FcεRIα binding [70] has been observed, with entropic contributions to the thermodynamics and kinetics of receptor binding playing an important role [44]. Remarkably, one of the earliest biophysical studies of IgE, not long after its discovery, identified the Cε3 domains as the most sensitive region of the molecule to heat denaturation [76], and this lability of Cε3 may in fact be critical for IgE's unique receptor-binding properties and inter-site allosteric communication.

Allosteric effects in IgE-Fc were also observed when the mode of action of the anti-IgE omalizumab was elucidated through determination of the structure of the complex, and studies in solution [36]. It was discovered that omalizumab binding to IgE-Fc not only "unbends" the molecule as described above (Figure 2b), but causes the Cε3 domains to move so far apart that they cannot engage FcεRI, thus allosterically inhibiting FcεRI binding while simultaneously inhibiting CD23 binding orthosterically. Allostery and the conformational dynamics of IgE-Fc lie at the heart of a potentially even more important phenomenon concerning the inhibition of FcεRI binding; namely, the observation that it is possible for omalizumab not only to bind to free IgE and block binding to the receptor, but also to bind to receptor-bound IgE and facilitate its dissociation [36,77,78]. First reported with another IgE-Fc binding protein, a Designed Ankyrin Repeat Protein or Darpin [79], the ability of omalizumab to bind to FcεRI-bound IgE and cause it to dissociate was a most unexpected result, but one with exciting clinical potential. Although this "accelerated dissociation" only occurs at a very high concentration, above therapeutic levels of omalizumab [36,77], the explanation for this phenomenon lies in the fact that even when bound to FcεRI, IgE-Fc displays an ensemble of conformations; binding omalizumab alters the composition of this ensemble, reducing the energy barrier to IgE/FcεRI dissociation [36]. The intrinsic flexibility and allosteric properties of IgE can thus be exploited therapeutically to actively remove IgE from FcεRI.

Two other anti-IgE antibodies have been found to exploit allosteric effects. MEDI4212 inhibits FcεRI binding orthosterically and CD23 binding allosterically, the latter by locking the Cε3 domains in an open conformation [52]. Antibody 8D6, which extends the IgE-Fc as described above (Figure 2d), inhibits FcεRI binding both orthosterically and allosterically, but does not affect the CD23 interaction [38]; this may prove valuable therapeutically for allergic disease if down-regulation of IgE production can be effected through the interaction of 8D6/IgE complexes with mCD23 on B cells.

The 8D6 antibody demonstrates that selective inhibition of IgE binding to its two principal receptors is possible.

6. Antigen (Allergen) Binding

So far, we have focussed on the Fc region of IgE and its receptor interactions. The binding of IgE to antigens, and in particular to allergenic proteins, has been studied in detail with antibody Fab fragments, but the flexibility of the IgE molecule as a whole, and in particular its ability to engage both allergen and its receptors, can only currently be inferred from low resolution electron microscopy (EM) studies and modelling; there are no high resolution structural data for intact IgE. EM studies of IgE complex formation with anti-idiotype IgG molecules have shown a relatively restricted degree of Fab arm flexibility [80], and a recent EM analysis of immune complex formation with IgE molecules binding to IgE epitopes grafted onto a small protein (myoglobin) framework, showed that the relative disposition, and in particular the proximity of the epitopes, affected immune complex formation and their ability to activate effector cells [81]. Modelling of Fab arm flexibility within the FcεRI-bound IgE molecule, confirmed this view that the relatively restricted range of dispositions of the Fabs, together with the particular geometrical arrangement of the epitopes on the allergen, might be key to an allergen's potency in effector cell activation [35,37]. Other important requirements for a potent cellular response, in addition to epitope specificity, are affinity and the particular combination of antibodies present [82].

There are now several crystal structures of antibody Fabs binding to their specific epitopes on protein allergens, although most are murine IgG antibodies raised against the allergen [83–90]; not all of these may represent epitopes recognised by allergic patients' IgE antibodies. Two studies generated IgE Fabs by phage display using combinatorial libraries derived from patients allergic to either the milk protein β-lactoglobulin (*Bos d* 5) [91] or the grass pollen allergen *Phl p* 2 [92], although these almost certainly do not consist of the "natural" V_H-V_L pairing that occurred in the patient. A recent study generated a naturally paired V_H-V_L combination by single B cell cloning of an IgG4 antibody from an allergic patient undergoing immunotherapy with the grass pollen allergen *Phl p* 7; this antibody was converted to an IgG1 Fab for the crystal structure analysis of the complex with allergen, and to IgE for functional analyses [23]. In all of these studies, the allergens were recognised by the antibodies in a conventional manner, involving many if not all of the CDRs. However, the most recent study also revealed an additional, unconventional "superantigen-like" interaction between *Phl p* 7 and the antibody, involving amino-acid residues of the V_L framework region (FR) [23].

The allergen/antibody structures involving conventionally recognised epitopes demonstrate how an allergen that can dimerise, such as *Bos d* 5 [91], could cross-link two identical IgE antibodies (Figure 7a,b) and, if FcεRI-bound, lead to mast cell or basophil activation. A similar structure was seen in the complex of two identical Fabs bound to a dimer of the cockroach allergen *Bla g* 2 [86]; this allergen in monomeric form can however cross-link two antibodies that recognise epitopes on opposite faces of the allergen [93], and a similar topology arises for two different antibody Fabs that bind non-overlapping epitopes on the monomeric house dust mite allergen *Der p* 1 [89]. The non-conventional, partly FR-mediated recognition of *Phl p* 7 by an allergic patient's antibody, occurring at the same time as conventional CDR-mediated recognition (Figure 7c,d), shows that certain allergens can cross-link identical IgE molecules using this alternative mechanism [23]. B cell superantigens, such as *Staphylococcus aureas* Protein A or *Peptostreptococcus magnus* Protein L, cross-link antibodies by interacting with their FRs, and thus molecules that cross-link IgE in this way, such as Protein L, have been termed "superallergens" [94]. *Phl p* 7 thus displays "superallergen-like" behaviour, which may contribute to the potency of particular allergens. Intriguingly, a structure of the monomeric cat allergen *Fel d* 1 in complex with an IgG Fab that blocks human IgE binding [90] shows a FR-mediated contact in the crystal which, together with the CDR-mediated interaction, could cross-link two identical Fabs in a manner very similar to that depicted for *Phl p* 7.

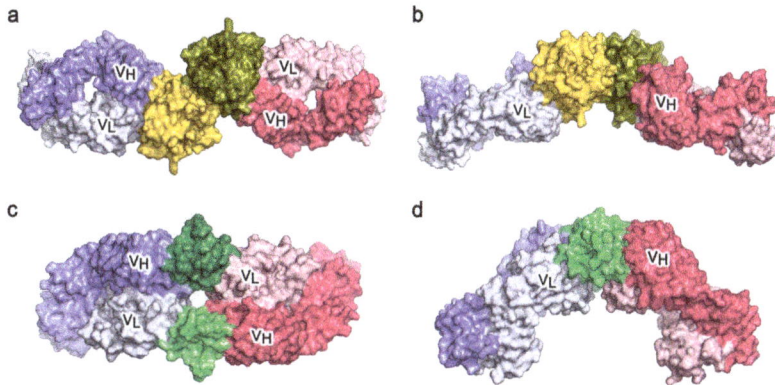

Figure 7. Crystal structures of allergens cross-linking two identical antibody Fab arms. (**a**) Dimer of allergen *Bos d* 5 (monomeric subunits coloured yellow and olive green) recognised classically by two identical Fab molecules (V_H and V_L domains indicated) [91]. (**b**) As a), orthogonal orientation [91]. (**c**) Two monomeric molecules of allergen *Phl p* 7 (coloured green), each independently recognised by two identical Fab molecules (V_H and V_L domains indicated) [23]. (**d**) As c), orthogonal orientation, in which only one of the two *Phl p* 7 molecules can be seen, recognised classically by the Fab on the right, and in a superantigen-like manner by the Fab on the left [23].

Activation of mast cells or basophils by cross-linking FcεRI-bound IgE may thus be envisaged as shown in Figure 8. The regions of space accessible to the two Fab arms appear to be more restricted and almost non-overlapping when IgE is bound to the receptor: one arm points "parallel" to the membrane while the other points away [35,37]. These topological constraints may need to be considered when IgE is used to target cell surface antigens, rather than soluble allergens, to allow simultaneous engagement with FcεRI on effector cells.

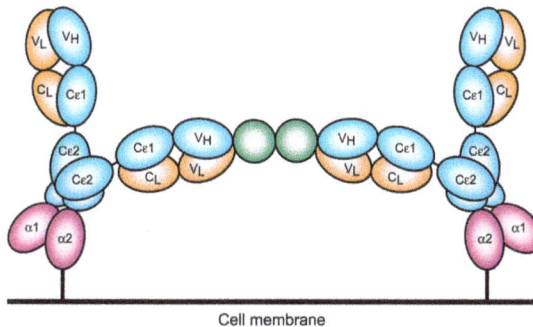

Cell membrane

Figure 8. Schematic representation of FcεRI-bound IgE cross-linking by soluble allergen. A dimeric allergen (green) engages two identical IgE antibodies (blue and orange domains) that are bound by the Cε3 domains (Cε4 domains not shown) to the extracellular α1 and α2 domains of FcεRI (purple). This is representative of the structure shown in Figure 7a,b; a monomeric allergen could similarly cross-link two identical IgE molecules as shown in Figure 7c,d, or two different antibodies recognising non-overlapping epitopes. The restricted flexibility of the Fab arms in receptor-bound IgE may mean that the other arm is important for engagement of cell surface antigens.

7. Rationale for Harnessing IgE-Mediated Functions against Cancer

IgE is clearly a powerful activator of the immune system by virtue of the Fc receptor interactions described above, potentiating effector functions and antigen presentation; even well below receptor saturation levels, tissue-resident immune cells such as mast cells and macrophages enable this antibody isotype to exert long-lived and powerful immune surveillance in tissues such as the gut, skin, epithelial and mucosal surfaces. In addition to its contributions to the pathogenesis of allergic diseases and anaphylactic reactions, IgE plays a physiological role in immune protection against parasites, triggering inflammatory cascades that cause vasodilation and local enhancement of protective responses in conjunction with antibodies of other isotypes [95–97]. These latter, less well-described attributes of IgE may be of potential significance to applications in cancer immunotherapy.

7.1. Epidemiological Links between IgE, Allergy and Cancer

The concept of a role for IgE in conferring immune protection against cancer dates back many decades, with early studies providing evidence for a role of allergic responses in restricting tumour xenograft growth in mice, negative correlations between atopy and cancer [98–102], and decreased prevalence of immediate hypersensitivity in patients with cancer [103]. Immunohistochemical (IHC) evaluations on head and neck cancer showed that IgE-expressing cells were more abundant in tumours compared with normal mucosa [104], and a pancreatic cancer patient-derived IgE antibody could potentiate anti-tumour effector functions [105]. Certain conditions and stimuli that cause epithelial damage and stress signals may lead to the induction of an adaptive immune response favouring B cell class switching to IgE, which can restrict cancer growth. Such protective functions have been reported following local exposure of skin to environmental DNA-damaging stress signals, which triggered adaptive immune responses and the production of IgE antibodies that conferred protection from epithelial carcinogenesis [106]. Subsequent findings of inverse associations between allergic or atopic status and protection from cancer varied significantly. Inverse associations of allergic or atopic disease with the risk of developing specific malignancies including glioma, pancreatic cancer, lymphatic/hematopoietic, gastrointestinal, skin and gynaecological origin tumours have been reported [107–111], although significant limitations of such studies include the reliance of self-reported symptoms of allergy and lack of specific measurable biomarkers. More recent studies examined eosinophil counts and skin prick test positivity, as well as titres of IgE and allergen-specific IgE, with some reporting a reduced risk of developing specific cancers, and a reduced risk of developing cancer overall [110–113]. Although taken together, epidemiological reports point to the complex relationships between allergies, IgE levels and carcinogenesis, tantalising evidence also supports a functional role for IgE as a passive or active component in anti-tumour responses.

7.2. Features of IgE that may Translate to Immune Protective Functions against Tumours

To date, therapeutic monoclonal antibodies designed for the treatment of cancers are typically engineered with Fc regions belonging to the IgG isotype. IgG1 is typically chosen when effector functions are required, while IgG4 is preferred when Fc-mediated attributes are not desired. However, until recently, antibodies of other isotypes such as IgE or IgA had never been tested in humans [114–116].

In our studies, we hypothesised that several unique attributes of IgE could form a powerful immunological profile, suitable for the immunotherapy of solid tumours such as ovarian carcinomas [117]. These include high affinity for cognate receptors on a different set of immune cells to those engaged by IgG, long tissue residency and immune surveillance, the ability to potentiate strong effector functions at relatively low levels of Fc engagement with effector cells, and a lack of inhibitory Fc receptors.

High affinity for cognate receptors: The affinity of IgE for FcεRI is typically 100- to 10,000-fold higher than those of the clinically used IgG subclasses for their Fcγ receptors. Additionally, the avidity of IgE for trimeric CD23 is comparable to that measured with IgG-FcγRI complexes. These properties mean that IgE can persist on immune cells in the absence of antigen complex formation. If IgE antibodies are directed against cancer antigens, these features could be highly beneficial in ensuring potent effector functions, long persistence and immune surveillance at tumour sites.

Lack of inhibitory Fc receptors: IgE antibodies have no known inhibitory Fc receptors to moderate effector functions. This contrasts with IgG, which is subject to control by the inhibitory receptor, FcγRIIb, known to be upregulated in the tumour microenvironment (TME) of different cancer types. Lack of an inhibitory FcεR may mean that IgE is not subjected to suppressive influences imposed on IgG by tumours.

Long immune surveillance in tissues: The half-lives of IgE and IgG antibodies vastly differ in the circulation and tissues: 1.5 days for IgE and 2–3 weeks for IgG in the serum, partly due to the lack of FcRn binding by IgE. The opposite is true in tissues such as the skin, where the half-life of IgE is approximately two weeks compared with 2–3 days for IgG [118,119]. Long tissue residency and immune surveillance in the presence of FcεR-expressing effector cells could have potential benefits if directed against cancers, including epithelial and skin tumours such as malignant melanomas, squamous cell and ovarian carcinomas.

Presence of IgE immune effector cells in tumours: The inflammatory milieu of the TME may include FcεR-expressing immune effector cells such as monocytes, macrophages, mast cells, dendritic cells (DCs) and eosinophils. Although pro-tumoural or tumour-tolerant subsets of these cells may lack the ability to mount an anti-tumour attack, it is possible that cells armed by tumour antigen-specific IgE tightly bound on FcεRs could overcome tolerant phenotypes.

Fc-mediated effector functions: IgE can potentiate a range of effector functions through the engagement of FcεRI and CD23. These include: antibody-dependent cell-mediated cytotoxicity (ADCC) by immune cell types including monocytes, macrophages, eosinophils and mast cells, with the release of toxic mediators (e.g., nitric oxide), proteases, cytokines and chemokines (e.g., tumour necrosis factor, TNFα, macrophage chemoattractant protein-1, MCP-1) associated with target cell lysis; antibody-dependent cell-mediated phagocytosis (ADCP) by macrophages and monocytes; mast cell and basophil degranulation leading to the release of proinflammatory mediators, and the enhancement of immune cell recruitment and activation at the antigen challenge sites (Figure 9). These attributes could result in enhanced immune cell recruitment, surveillance and anti-tumour functions.

Exerting anti-parasite effector functions: The physiological roles of IgE in protective immune responses against parasites are well documented. Anti-parasitic IgE and IgE loaded on effector cells such as eosinophils have been shown to confer protection against different parasites (e.g., *Schistosoma mansoni*) [120]. IgE engaged with FcεRI or CD23 can engender parasite clearance by human eosinophils, platelets and macrophages through ADCC and ADCP [121,122]. Furthermore, high serum titres of parasite antigen-specific IgE have been associated with resistance to infection [123,124]. Macrophages, eosinophils and mast cells have all been reported to be involved in these protective mechanisms [5,97,122,125,126]. IgE-mediated immune clearance of large parasites in tissues, including Th2-biased environments such as the gut, draw parallels with conditions in solid tumours in which a similar Th2 inflammatory milieu and the presence of immune cells such as macrophages may form appropriate environments in which IgE could act to eradicate tumours by similar mechanisms.

Overcoming antibody blockade mechanisms associated with Th2-biased tumour conditions: Tumour-associated production of alternatively activated (e.g., IL-10-driven) rather than classically activated (IL-4-driven) Th2 environments may support local antibody class switching to inflammatory and immunologically inert subclasses such as IgG4. Th2-biased inflammatory states that favour B cell class switching to IgG4 have long been identified in IgG4-related diseases characterised by chronic inflammation, circulating IgG-positive plasmablasts and high infiltration of IgG4-producing plasma cells in various tissues [127–129]. Alternative Th2 activation states have also been reported in several solid tumour types including pancreatic cancer, extrahepatic cholangiocarcinoma, melanoma and non-small cell lung cancer [130–134]. These pathological conditions, likely to be promoted by a combination of a Th2-biased inflammatory milieu and long antigen exposure, may signify that immune responses are driven away from the classical Th2-based class switching to IgE, in favour of IgG4. Evidence points to IgG4 antibodies not only being immunologically inert, but importantly, being able to impair the immune-activating functions of otherwise cytotoxic IgG1 antibodies [134,135].

Numerous mechanisms may be at play, including competition for FcγR engagement with other IgGs, Fab arm exchange, and signalling through inhibitory Fc receptors, all supporting immunosuppressive signals [130,136]. The latter could have implications not only for modulating the endogenous humoral immune response but also for restricting the potency of IgG1 therapies. These regulatory mechanisms may offer opportunities to design anti-tumour IgE antibodies that function through a different Fc receptor, which could be less prone to the immunosuppressive signals that impair IgG functions against cancer.

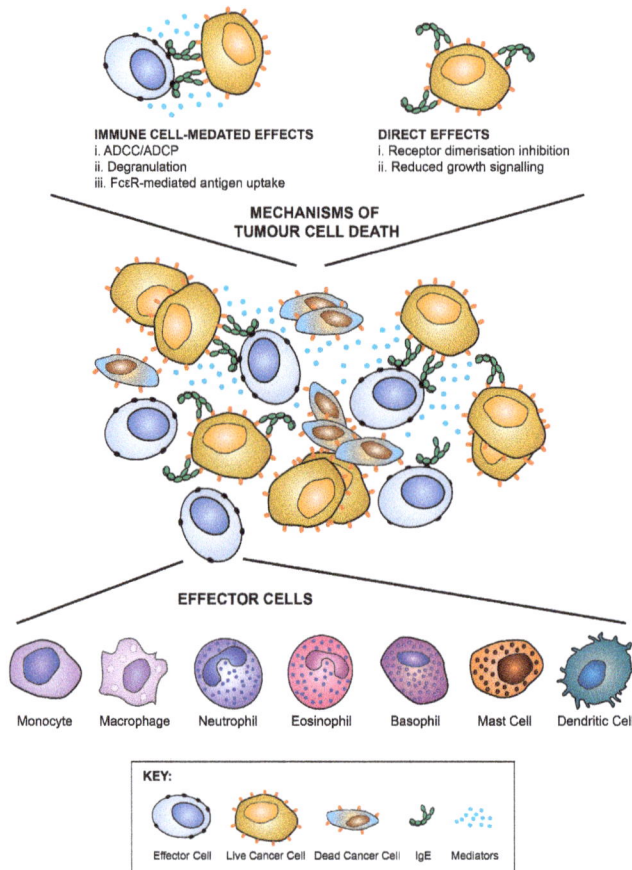

Figure 9. IgE functions against cancer cells. IgE can potentiate Fc-mediated effector functions by engaging cognate receptors on immune effector cells such as monocytes, macrophages, neutrophils, eosinophils, basophils and mast cells. Antibody-dependent cell-mediated cytotoxicity (ADCC), and degranulation can result in the release of various toxic and pro-inflammatory mediators, including proteases, cytokines, chemokines, and histamine, which, together with antibody-dependent cell-mediated phagocytosis (ADCP), can result in enhanced anti-tumour functions and immune cell recruitment. IgE can also engage APCs to enhance antigen uptake and presentation. Like anti-cancer IgG antibodies, IgE may also exhibit direct effects against cancer cells, such as receptor dimerisation inhibition and reductions in cancer cell growth signalling. Figure adapted by permission from Taylor & Francis (Josephs, D.H. et al. IgE immunotherapy: a novel concept with promise for the treatment of cancer. *mAbs* 2014, *6*, 54–72 [117]) and John Wiley & Sons, Inc. (Jensen-Jarolim, E. et al. AllergoOncology - the impact of allergy in oncology: EAACI position paper. *Allergy* 2017, *72*, 866–887 [137]).

Engaging antigen presenting cells to stimulate effective adaptive immune response: IgE can engage with APCs to enhance antigen uptake and presentation to cognate T cells (Figure 9). IgE engagement with FcεRI can cross-present antigen, priming a cytotoxic T lymphocyte (CTL) response [138,139]. Through such mechanisms, IgE has been reported to confer protective anti-tumour immunity and trigger memory responses. These antigen presentation-boosting attributes could be important in the TME where the functions and maturation of professional antigen presenting cells may be impaired.

8. Pre-Clinical Studies of IgE Antibodies Targeting Cancer Antigens: The Advent of Allergo Oncology

The development of immunologically active, antibody-based targeted therapies with potent Fc-mediated effector mechanisms has revolutionized the treatment of cancer patients with previously difficult to treat tumours [140]. A promising branch of this discipline is the emerging field of AllergoOncology, which focuses on Th2 and IgE-mediated immune responses in the cancer context [137,141–143]. Research in this field has opened the way for the development of IgE-based immunotherapy approaches, including monoclonal IgE antibodies as anti-cancer treatments [117,144].

The specific attributes of IgE described above, including natural immune activatory functions in tissues and high affinity for cognate receptors, have been proposed as a strategy for cancer immunotherapy. Antibodies engineered with IgE Fc regions, and designed to recognise tumour-associated antigens, may promote immune cell recruitment into tumours, and both direct and activate the Th2-biased immune stroma against cancer. Longer tissue-resident immune surveillance may then translate to anti-cancer efficacy. Therapeutic approaches have been developed to harness the immune-activating functions of IgE for cancer immunotherapy, including: IgE-coated cell vaccines, IgEs as adjuvants, vaccination approaches to trigger IgE-biased immune responses against tumour antigens, and recombinant IgE recognising tumour antigens. Here, we will focus on the development of recombinant IgE antibodies [144]. Furthermore, we place specific emphasis on MOv18 IgE, as the first-in-class agent that has undergone extensive pre-clinical efficacy and safety evaluations in several model systems, prior to reaching clinical testing in patients with cancer.

8.1. Engineering Platforms for Production of IgE Antibodies for Research and Clinical Translation

Developing IgE antibodies that recognise cancer antigens relies on appropriate expression systems and protocols to facilitate antibody cloning and production. Since the development of hybridoma technology five decades ago, novel recombinant DNA technology, genetic manipulation and advances in cell biology have led to remarkable improvements in therapeutic recombinant antibody engineering [145]. Although significant efforts have focused on the optimization of expression platforms for IgG [146], relatively meagre investment has been directed towards engineering IgE.

The study and clinical translation of IgE antibodies requires efficient and scalable production processes, but these have historically been characterised by low and variable yields. Despite this, several groups have shown that recombinant IgE antibodies can be produced using various cloning strategies. In early studies, restriction enzyme-based cloning methodologies were successfully employed using murine expression host cells to derive stable expression platforms, with Sp2/0 [147] and FreeStyle™-293F [148] cell lines, reaching production yields in the range of 8–25 mg/L. Recombinant IgE antibodies have also been produced using transient expression platforms with human (HEK293T, FreeStyle™-293F, Expi293F™ cells), insect- and plant-based systems, reaching yields of 30 mg/L [41,82,149,150]. More recent transient expression protocols have been implemented, which take advantage of Polymerase Incomplete Primer Extension (PIPE) cloning [151]. PIPE does not rely on restriction or other recombination sites, and can help expedite antibody cloning, a strategy that we have applied to IgE antibody production [152].

We recently developed a highly expressing stable recombinant IgE expression system for rapid production of a functional antibody, with features that allow scale-up for potential clinical evaluations [153]. For this, we implemented PIPE cloning and generated a vector containing the

Ubiquitous Chromatin Opening Elements (UCOE) sequence located upstream of the transgene promoter to prevent promoter silencing. UCOE allows the expression of the transgene even if it is randomly integrated in a heterochromatin region [154]. This platform improves IgE yields to 87 mg/L per day, at least 33-fold higher production within four days compared with the best stable IgE expression system documented to date, and in small culture volumes of 25 mL, with the potential for further scale-up production.

These findings suggest that, as with IgG antibody production, IgE can be produced using a range of expression systems and with sufficient yields to facilitate the functional evaluation and translation to clinical testing. Further efforts in the field promise to improve upon existing platforms for use in pre-clinical studies, process development, Good Manufacturing Practice (GMP) production and supply of material suitable for clinical studies. Other developments in antibody discovery such as knock-in mouse strains used to derive IgE antibodies by hybridoma techniques, phage display approaches using human antibody variable region repertoire libraries and single B cell cloning techniques may also be applicable [155–157].

Recombinant IgE antibody production has advanced significantly with several already engineered and tested in vitro and in vivo. There is however room for further development of improved and effective production systems that can be translatable to GMP environments and scale-up for clinical studies.

8.2. Functional Evaluations of Anti-Tumour IgEs

8.2.1. In Vitro and In Vivo Functional Profiles of Engineered IgEs Targeting Several Cancer Antigens

Antibody engineering has yielded the first generation of IgE antibodies that have been studied in vitro and in vivo in numerous model systems. Anti-tumour IgE antibodies can engage various immune effector cells such as mast cells and basophils expressing high levels of tetrameric FcεRI ($\alpha\beta\gamma_2$), and monocytes and eosinophils that express trimeric FcεRI ($\alpha\gamma_2$) at lower levels. Studies in vivo have been conducted in various mouse immunocompetent models. However, human IgE-Fc does not cross-react with mouse FcεR and, unlike in humans, mouse FcεRs are only expressed by mast cells and basophils, making the mouse immune system less suitable for the study of human IgE functions. However, transgenic mouse models have shown significant tumour-restricting abilities of IgE with human Fc domains. Examples of several monoclonal IgE antibodies evaluated over the last 30 years are discussed below.

A mouse IgE recognising the mammary tumour virus (MMTV) major envelope glycoprotein (gp36) was tested in an immunocompetent syngeneic mammary carcinoma. The antibody restricted the growth of subcutaneous (s.c.) mammary tumours compared with controls [158]. Another murine IgE recognising a colorectal cancer antigen (CCA) restricted the growth of an s.c. tumour in an antigen-specific and species-specific manner at concentrations far lower than those required for the equivalent IgG to engender the same effect [159]. A fully human anti-HER2/*neu* IgE (C6MH3-B1 IgE) restricted the growth of intraperitoneal (i.p.) tumours compared to vehicle controls and prolonged the survival of human FcεRIα-transgenic mice [160]. The same agent was well tolerated when administered in cynomolgus monkeys, albeit at very low doses (up to 80 μg/kg). Another IgE specific for the epithelial tumour antigen MUC-1 restricted cancer growth when expressed locally in tumours along with chemoattractant mediators MCP-1 and IL-5 [161]. Furthermore, a mouse/human chimeric IgE antibody (clone AR47.47) recognising the prostate specific antigen (PSA) enhanced antigen presentation by DCs, and triggered CD4+ and CD8+ T cell responses. The same antibody complexed with its antigen prolonged the survival of human FcεRIα-transgenic mice subsequently challenged with prostate cancer cells [162].

Human/mouse chimeric anti-HER2/*neu* IgE, and anti-EGFR (epidermal growth factor receptor) IgE, engineered from the original trastuzumab and cetuximab (IgG1) clones respectively, were shown to engender ADCC by human monocytic cells [163,164]. Specifically, anti-EGFR IgE triggered superior

ADCC functions (70%) against cancer cells, compared with the corresponding IgG1 (30%) [164]. However, some episodes of anaphylaxis were observed in some patients with EGFR-positive tumours who received the anti-EGFR human/chimeric monoclonal IgG1 antibody cetuximab. These were caused by the presence of pre-existing IgE antibodies specific for the oligosaccharide galactose-α-1,3-galactose (α-Gal) on SP2/0-expressed cetuximab in a subset of individuals [165,166]. Furthermore, humans are known to carry IgG and IgM antibodies recognising α-Gal [167], and it is possible that these endogenous antibodies could have neutralised the anti-tumoural effects of cetuximab. Therefore, caution should be exercised in translating IgE class antibodies recognising EGFR on the grounds of safety and efficacy. An anti-human CD20 IgE triggered eosinophil-mediated ADCC and mast cell activation and killing of CD20-expressing tumour cells. Anti-HER2/*neu*, anti-EGFR, anti-CD20, anti-folate receptor alpha (FRα) IgE and anti-prostate specific antigen (PSA) IgE antibodies were all able to trigger rat basophil leukaemia (RBL) SX-38 mast cell degranulation when cross-linked in different ways including soluble antigen/polyclonal antibody complexes, cancer cells expressing multiple copies of the target antigen, and polyclonal anti-IgE. Furthermore, anti-HER2/*neu* (trastuzumab) IgE demonstrated the ability to exert direct effects on tumour cell viability in the absence of effector cells, equivalent to those reported to be triggered by trastuzumab IgG [163]. This supports the notion that anti-tumour IgE antibodies may be capable of engendering direct effects attributed to IgG equivalent agents, whilst perhaps still able to harness class-specific effector functions (Figure 9).

The progress of the first-in-class monoclonal IgE antibody (MOv18) recognising a tumour-associated antigen to an early clinical trial in oncology is the exemplar advance in the field. Based on this development, herein we will focus on the evaluation and translation of this recombinant antibody, and efforts to translate IgE class therapeutic agents to clinical testing. If firstly safety, and secondly efficacy of this first-in-class agent could be demonstrated in the clinic, this will pave the way for further study and translation of the above-mentioned antibodies, as well as other novel anti-cancer antibodies of this class.

8.2.2. MOv18 IgE, the First Anti-Tumour IgE to Reach Clinical Testing: Evaluation of In Vitro Effector Functions

An IgE antibody that has progressed to clinical testing is MOv18, a mouse/human chimeric monoclonal IgE antibody that recognises the tumour-associated antigen Folate Receptor alpha (FRα) (NCT02546921, www.clinicaltrials.gov). FRα is highly expressed in > 70% of ovarian carcinomas and other tumour types and has low and restricted expression distribution in normal tissues [168,169]. The IgG1 version of MOv18 has undergone early clinical trials as a therapeutic and imaging agent in patients with ovarian carcinomas, and treatment has been well tolerated [170–173]. FRα is considered a promising target for cancer therapy, with considerable evidence that either directing therapeutic antibodies to this receptor, or its inhibition by small molecules, is well-tolerated in man [174–178].

In vitro, mouse/human chimeric MOv18 IgE activated human peripheral blood mononuclear cells (PBMCs) to kill ovarian cancer cells, compared with background cancer cell death with nonspecific mouse/human chimeric anti-4-hydroxy-3-nitro-phenacetyl (NIP) IgE, or no antibody controls [179]. Human monocytes were subsequently identified as important effector cells in PBMCs, based on live imaging studies in which IGROV1 ovarian cancer cells were found to contact one or more CD14-labelled human monocytes within 30 min of incubation of PBMCs and IGROV1 cells together with MOv18 IgE. Phagocytosis of tumour cells was evident after 90 min of incubation, with IGROV1 cells becoming fragmented by 3 h (Figure 10a).

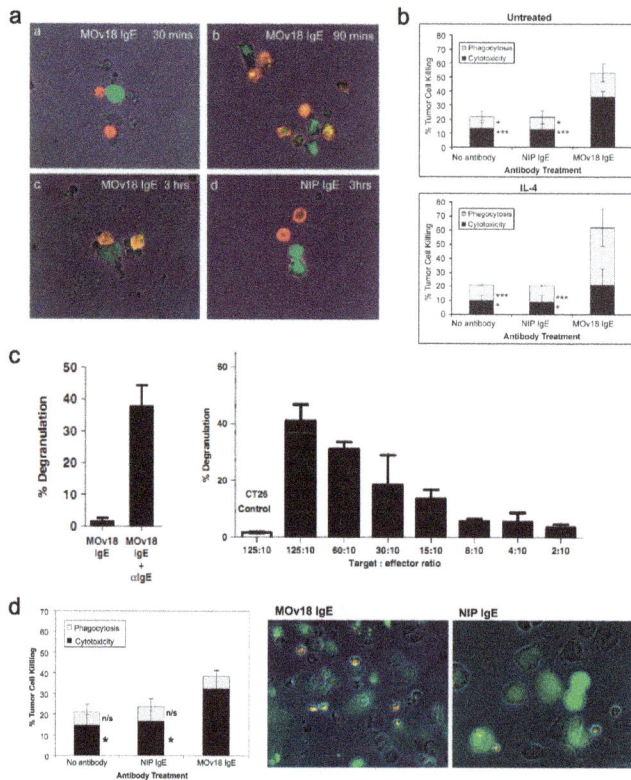

Figure 10. In vitro evaluations of MOv18 IgE. (**a**) Live imaging studies showed contact between IGROV1 ovarian cancer cells and CD14-labelled human monocytes within 30 min of incubation of PBMCs and IGROV1 cells together with MOv18 IgE. Following 90 min, the phagocytosis of tumour cells was evident and IGROV1 cells became fragmented by 3 h [179]. Figure adapted by permission from John Wiley & Sons, Inc. (Karagiannis, S.N. et al. Activity of human monocytes in IgE antibody-dependent surveillance and killing of ovarian tumor cells. *Eur. J. Immunol.* 2003, *33*, 1030–1040 [179]). (**b**) Human monocytes expressing cell-surface FcεRI triggered MOv18 IgE-mediated ADCC of IGROV1 ovarian cancer cells, and IL-4 stimulated monocytes with up-regulated CD23 expression, killed tumour cells by both ADCC and ADCP compared to background levels mediated by non-specific NIP IgE and no IgE controls [180]. Figure adapted by permission from Springer Nature. (Karagiannis, S.N. et al. Role of IgE receptors in IgE antibody-dependent cytotoxicity and phagocytosis of ovarian tumor cells by human monocytic cells. *Cancer Immunol. Immunother.* 2008, *57*, 247–263 [180]). (**c**) Appreciable degranulation of RBL SX-38 cells was triggered by cross-linking of cell surface receptor-bound MOv18 IgE by polyclonal anti-IgE antibody (left) or FRα-expressing cancer cells (right) [182]. Figure adapted by permission from John Wiley & Sons, Inc. (Rudman, S.M. et al. Harnessing engineered antibodies of the IgE class to combat malignancy: initial assessment of FcεRI-mediated basophil activation by a tumour-specific IgE antibody to evaluate the risk of type I hypersensitivity. *Clin. Exp. Allergy*, 2011, *41*, 1400–1413 [182]). (**d**) MOv18 IgE-mediated killing of IGROV1 ovarian cancer cells by primary human eosinophils (right) and microscopic evaluations revealed interactions between IGROV1 cells and eosinophils, and IGROV1 tumour cell destruction alongside piecemeal degranulation of eosinophils, following 2.5 h of incubation with MOv18 IgE, but not with non-specific NIP IgE (right) [181]. Figure adapted by permission from The American Association of Immunologists, Inc. (Karagiannis, S.N. et al. IgE-antibody-dependent immunotherapy of solid tumors: cytotoxic and phagocytic mechanisms of eradication of ovarian cancer cells. *J. Immunol.* 2007, *179*, 2832–2843 [181]).

Following stimulation by IL-4, which is often released from IgE-sensitized basophils and mast cells, CD23 can be upregulated on monocytes, eosinophils and platelets. Interaction of IgE with CD23 may also have a role in ADCP of target cells by effector cells, as shown by its natural protective role in the clearance of parasites. This function has also been described with MOv18 IgE. Human monocytes expressing FcεRI on the cell surface triggered IgE-mediated ADCC of tumour cells, while IL-4 stimulated monocytes killed FRα-expressing tumour cells by both ADCC and ADCP, compared to background levels of tumour cell death with NIP IgE and no IgE controls (Figure 10b). Specific IgE Fc receptor blockade studies in vitro confirmed that MOv18 IgE-dependent ovarian tumour cell killing had an ADCC component, primarily mediated by FcεRI, and an ADCP component, primarily mediated by CD23 [180,181].

The ability of MOv18 IgE to trigger functional degranulation was examined with RBL SX-38 cells engineered to over-express the human tetrameric FcεRI. Exposure of the RBL SX-38 cells to MOv18 IgE alone did not induce significant degranulation; however cross-linking MOv18 IgE bound to the effector cell surface using either a polyclonal anti-IgE antibody or FRα-expressing cancer cells induced appreciable degranulation (Figure 10c) [182]. Eosinophils are key IgE effector cell types known to express low levels of FcεRI, but not CD23 [183]. Eosinophils mediated elevated ADCC (32.4%) with MOv18 IgE above isotype controls, and microscopical evaluations revealed contact between eosinophils and tumour cells, frequently accompanied by eosinophil degranulation, loss of tumour cell architecture, and apparent tumour cell death (Figure 10d) [181]. Our findings were consistent with data by Teo and colleagues who also reported the eosinophil-mediated ADCC functions by an anti-CD20 IgE antibody [161]. Interestingly, previous studies showed a lack of eosinophil activation by IgE cross-linked with allergens. These differences could relate to the density of the target antigen. Tumour cells express very high numbers of tumour associated-antigens on their surface, crosslinking of which may be required to deliver an activatory signal through the lowly expressed FcεRI on eosinophils. However, this may not be the case for the crosslinking of FcεRI by IgE complexed with multivalent allergens of a much lower valency [184]. In the cancer context, the target antigen density could therefore be critical to triggering eosinophil-mediated anti-tumour IgE effector functions.

These studies established that MOv18 IgE could mediate effector functions such as degranulation and tumour cell killing through cytotoxicity (ADCC) and phagocytosis (ADCP) by activating known IgE effector cells.

8.2.3. In vivo efficacy studies of MOv18 IgE

The ability of MOv18 IgE to restrict tumour growth in vivo was studied against different rodent models including human tumour xenografts established in immunodeficient (SCID and nu/nu) mice. In immunodeficient mouse models, human effector cell populations were co-administered with MOv18 IgE because: (a) human IgE-Fc is not recognised by mouse FcεRs, and (b) in mice the high-affinity IgE receptor FcεRI is expressed only by mast cells and basophils, and is absent in key effector cells such as monocytes and eosinophils. These studies therefore took place in an in vivo system containing both target and effector cells of human origin.

In an s.c. human ovarian cancer (IGROV1) xenograft grown in a SCID mouse model, animals administered with mouse/human chimeric MOv18 IgE or MOv18 IgG1, intravenously (i.v.) exhibited an initial inhibition of tumour growth up to day 19 post-tumour challenge. However, the tumours in mice administered PBMCs and MOv18 IgG1 subsequently grew to the same size as the controls. In contrast, mice administered PBMCs and MOv18 IgE exhibited reduced growth of up to 72% by day 35 post-challenge. In a range of experiments in this model, a single treatment with MOv18 IgE and PBMC significantly restricted the growth of ovarian tumours (Figure 11a) [147]. In specimens sampled at the end of these studies, tumours from the mice that received PMBCs and MOv18 IgE showed significantly larger areas of necrosis compared with those from mice treated with non-specific control IgE plus PBMCs, or those given PBMCs alone. Furthermore, when administered to IGROV1 xenograft mice in the absence of human PBMC, MOv18 IgE did not significantly inhibit tumour growth. Therefore, in the IGROV1 xenograft model, the anti-tumour efficacy of MOv18 IgE was found to be reliant on the presence of both an effector cell population and an IgE targeted to a tumour-expressed antigen.

Subsequently, a patient-derived xenograft (PDX) model of ovarian cancer was established from a human primary tumour sample, originating from the ascites of a moderately differentiated Grade 3, stage III ovarian serous cystadenocarcinoma. This PDX could be passaged in nude mice while retaining its human phenotype and was found to express FRα. In efficacy studies using this model, nude mice were challenged with i.p. ascites from donor human xenograft-bearing mice and were then treated with saline, human PBMCs or PBMCs plus MOv18 IgE on days 1 and 16. The mean survival time of control mice was 22 days, for those administered PBMCs alone it was 30 days, while for those administered PBMCs plus MOv18 IgE, the mean survival time was 40 days [179]. In a study comparing the efficacy of weekly doses of MOv18 IgG and IgE in this model, untreated mice survived for a median of 19 days, those administered PBMCs alone survived for 26 days, those administered PBMC plus IgG1 survived for 22 days, and those administered PBMC plus IgE survived for 40 days (Figure 11b).

One limitation of studies in mouse models is the need to introduce exogenous human effector cells, thus limiting the immune functions of the model and the possible duration of study as exogenous effector cells become depleted. Therefore, an immunocompetent syngeneic tumour model in Wistar Albino Glaxo (WAG) rats was designed to study efficacy as well as safety of MOv18 IgE prior to clinical translation. This model was selected based on similar expression and cellular distribution of FcεRI in rats and humans. Rat CC531 colon adenocarcinoma cells [185], engineered to express the human FRα (CC531tFR), were administered i.v. to grow as multifocal syngeneic lung metastases, and rats were administered a rat surrogate for the mouse/human chimeric MOv18 IgE engineered with rat Fc domains and respective effector functions (rat MOv18 IgE). This system permitted targeting of the rat immune system to rat tumour cells by an anti-FRα IgE. Significant efficacy of rat MOv18 IgE in restricting the growth of lung metastases was observed at doses of 5 mg/kg and higher when the antibody was administered fortnightly, compared with controls [186]. The efficacy of rat MOv18 IgE and the equivalent rat IgG2b was then compared: at a 10 mg/kg fortnightly dose, rat MOv18 IgE was significantly superior at restricting tumour growth (Figure 11c).

Overall, in three models of cancer including a patient-derived xenograft and an immunocompetent syngeneic model, the anti-tumour efficacy of MOv18 IgE was reliant on the presence of both an effector cell population and tumour antigen specificity. Furthermore, anti-tumour IgE was more effective than the corresponding IgG.

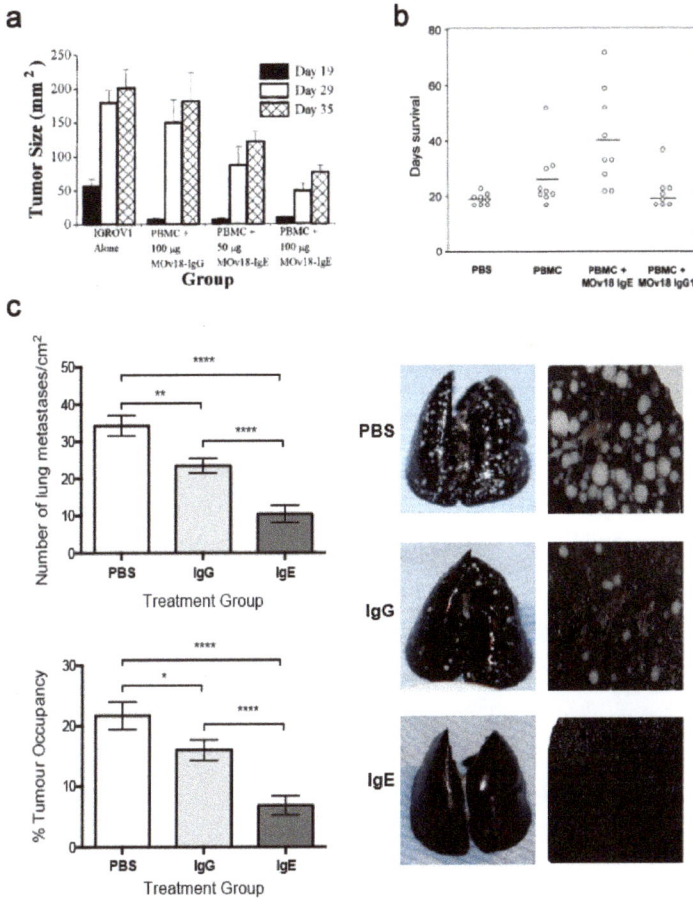

Figure 11. In vivo evaluations of MOv18 IgE. (**a**) In an s.c. human ovarian cancer (IGROV1) xenograft grown in a SCID mouse model, reduced tumour growth was measured in animals treated with PBMC plus MOv18 IgE, even at day 35 post tumour challenge. In comparison, animals treated with PBMC plus MOv18 IgG1 showed initial inhibition of tumour growth at day 19, but by day 35 tumours grew to the same size as the controls [147]. Figure adapted by permission from John Wiley & Sons, Inc. (Gould, H.J. et al. Comparison of IgE and IgG antibody-dependent cytotoxicity in vitro and in a SCID mouse xenograft model of ovarian carcinoma. *Eur. J. Immunol.* 1999, *29*, 3527–3537 [147]). (**b**) In an orthotopically-grown (i.p.) patient-derived xenograft (PDX) model of ovarian cancer, mice treated with weekly doses of PBMC plus MOv18 IgE showed superior survival compared to untreated animals and those treated with PBMC alone or PBMC plus MOv18 IgG [179]. Figure adapted by permission from John Wiley & Sons, Inc. (Karagiannis, S.N. et al. Activity of human monocytes in IgE antibody-dependent surveillance and the killing of ovarian tumor cells. *Eur. J. Immunol.* 2003, *33*, 1030–1040 [179]). (**c**) Left panel: In an immunocompetent syngeneic tumour model in WAG rats, significantly superior tumour growth restriction was measured in animals treated fortnightly with 10 mg/kg rat MOv18 IgE compared to the rat IgG2b equivalent. Right panel: Representative images of Indian ink-stained rat lungs (left) and lung sections (right) from each treatment group are shown [186]. Figure adapted by permission from the American Association for Cancer Research. (Josephs, D.H. et al. Anti-Folate Receptor-α IgE but not IgG Recruits Macrophages to Attack Tumors via TNFα/MCP-1 Signaling. *Cancer Res.* 2017, *77*, 1127–1141 [186]).

8.3. Evidence for IgE Activating Monocytes and Macrophages against Cancer

8.3.1. Monocytes and Macrophages as Key Effector Cells in MOv18 IgE-Potentiated
Anti-Tumour Functions

The mechanisms by which IgE antibodies can exert their anti-tumour effects have been studied and several pieces of evidence support a role for monocytes and macrophages as key effector cells.

In vitro evidence for monocyte-mediated effector functions: Monocytes mediate MOv18 IgE-dependent tumour cell killing in vitro by two pathways, ADCC and ADCP, acting through FcεRI and CD23 respectively. FcεRI-expressing primary monocytes principally exert ADCC. MOv18 IgE-potentiated ADCC by monocytes could be blocked with recombinant sFcεRIα [180,181,187], but monocytes could kill tumour cells by ADCP, a function mediated by CD23. MOv18 IgE antibodes can thus engage both receptors to activate effector cells against tumour cells in vitro and in vivo.

Evidence of macrophage involvement in IgE functions in mouse models: Pre-clinical in vivo studies in a PDX model suggested that monocytes and macrophages may be important IgE receptor-expressing effector cells that mediate enhanced survival of tumour-bearing mice treated with MOv18 IgE and human PBMCs. Treatment with MOv18 IgE was associated with the histological evidence of tumour infiltration by CD68+ human monocyte-derived macrophages [180,181], suggesting that these were recruited as a part of IgE-mediated anti-tumour functions. Human macrophages were concentrated in stromal areas adjacent to tumour cell islands, while mouse monocytes were abundant in all xenografts examined, irrespective of treatment. In MOv18 IgE-treated mice, human CD68+ macrophage infiltration correlated with longer survival [186]. In the same PDX model, removal of monocytes from the PBMC effector cells abolished the anti-tumour activity of co-administered PBMCs and MOv18 IgE [181]. Reconstitution of monocyte-depleted PBMCs with purified monocytes at proportions equivalent to those in unfractionated PBMCs restored the ability of PBMCs and MOv18 IgE to increase survival to levels equivalent to those seen in mice given whole PBMCs and MOv18 IgE. This survival was significantly longer than monocyte-reconstituted PBMCs alone, or depleted PBMCs with and without MOv18 IgE.

In vivo evidence of IgE-mediated macrophage activation in a surrogate rat model: The mechanisms of the action of rat MOv18 IgE in the WAG rat model were examined. Haematoxylin and eosin-stained tumours from different treatment groups in the WAG rat studies revealed more prominent loss of viability, density and demarcation of the tumour areas in rat MOv18 IgE-treated tumours compared to those from animals treated with rat MOv18 IgG2b or a buffer alone. Rat MOv18 IgE-treated tumours demonstrated evidence of considerable necrotic tissue surrounding the smaller tumour cell populations, consistent with previously reported tumour necrosis observed in human xenografts. Inflammatory cells infiltrating between the islands of tumour cells were considerably more pronounced in the rat MOv18 IgE-treated tumours [186].

The density and location of tumour-associated rat CD68+ macrophages in tumours from rats treated with vehicle control, rat MOv18 IgG and rat MOv18 IgE were studied by IHC and flow cytometric analyses of freshly isolated tumour-bearing lung tissues. CD68+ rat macrophages were detected in the TME from all treatment groups by IHC evaluations. Flow cytometric analyses also revealed that the percentage of CD68+ rat macrophages within the tumour-infiltrating CD45+ leukocyte population was higher in the rat MOv18 IgE-treated cohort compared to the rat MOv18 IgG2b-treated or the vehicle alone-treated cohorts. Systemic rat MOv18 IgE treatment was associated with macrophage infiltration deep into the tumour islets. By contrast, macrophages were largely absent from these areas in animals that were administered vehicle alone, or rat MOv18 IgG. The ratio of CD68+ cells within the tumour cell islets compared wth the tumour periphery was greater in the animals administered rat MOv18 IgE than in those with rat MOv18 IgG or vehicle alone, and macrophage infiltration was inversely proportional to tumour occupancy in rats treated with antibodies.

Together, these findings suggest that monocytes and macrophages may be mobilised towards tumours and play crucial roles in the tumour-restricting functions of MOv18 IgE.

8.3.2. Anti-Tumour IgE Directs Monocytes and Macrophages

The TME may influence the immune system to promote either anti-tumour immunity or tumour progression. Tumour associated macrophages (TAMs), characterised by the immune-activating classically-activated (M1) and the tolerance-inducing alternatively activated (M2) extreme phenotypes, are known to suppress or promote the growth of various malignant cells, depending on the biological context [188–190]. The activation state of macrophages induced to influx into tumours after administration of rat MOv18 IgE was investigated.

Tumour-infiltrating macrophages from rats treated with rat MOv18 IgE demonstrated an enhanced expression of the M1 co-stimulatory mature APC marker CD80, compared with those from MOv18 IgG2b or buffer-treated groups [186]. However, there was no difference in expression of the M2 marker CD163 between treatment groups. Furthermore, a considerably higher proportion of freshly-isolated CD68+ macrophages from dispersed rat lung tumours of rats administered rat MOv18 IgE were found to express intracellular TNFα, an M1 macrophage marker, compared to MOv18 IgG2b and vehicle-treated tumours. In addition, a higher proportion of CD68+ macrophages from rat MOv18 IgE-treated tumours expressed intracellular IL-10, considered an M2 marker, compared with rat MOv18 IgG2b- and vehicle-treated groups, although this represented a smaller subset compared with the TNFα+ population, with a proportion of cells demonstrating double positivity (TNFα+/IL-10+) within the rat MOv18 IgE-treated cohort. Additional analyses showed significantly elevated circulating TNFα in IgE-treated rat sera compared with controls [191]. The tumour-infiltrating macrophages in rat MOv18 IgE-treated tumours may therefore not be typically M1 or M2, and could instead represent a unique cell subset. Cytokine profile analyses of rat lung (broncho-alveolar lavage, BAL) fluids revealed four analytes, IL-10, TNFα, MCP-1 and IL-1α elevated in the rat MOv18 IgE-treated compared with the rat MOv18 IgG2b-treated cohort [186]. Together with increased levels of macrophage intracellular TNFα and IL-10 detected in the rat MOv18 IgE-treated rats, these data therefore indicate possible roles for TNFα, MCP-1 and IL-10 in the anti-tumoural functions observed following treatment with rat MOv18 IgE. Additional transcriptomic analyses demonstrated the enrichment of gene signatures associated with immune activation pathways, including those associated with IL-12 and Natural Killer (NK) cell-signalling in lungs from rats treated with IgE [191].

Taken together, these data suggest that MOv18 IgE may support TAM populations with mature phenotypes and hybrid M1/M2 features that are able to enter the tumour, trigger sustained immune activating pathways and secretion of IL-10, TNFα, MCP-1 and IL-1α in tumour-bearing lungs.

8.3.3. TNFα/MCP-1 Axis as a Mechanism of MOv18 IgE-Mediated Activation of Human Monocytes

The potential of, and mechanisms by which, human IgE activates human monocytes was evaluated [186]. Consistent with in vivo findings in the rat model, tumour cell cytotoxicity potentiated by mouse/human chimeric MOv18 IgE and human PBMC effector cells was associated with significantly elevated secreted mediators MCP-1, IL-10, and TNFα in co-culture supernatants, compared with either non-specific NIP IgE-treated or no antibody controls. Cross-linking of IgE, but not IgG, of different antigen specificities on the surface of human monocytes was responsible for the upregulation of TNFα. Cross-linking of IgE bound to tumour cells via the Fab region did not trigger TNFα. Blocking of TNFα receptor reduced IgE-mediated tumour cell cytotoxicity. Together, these findings point to a role for TNFα on IgE-mediated anti-tumour functions. Furthermore, TNFα upregulation by monocytes could in turn promote the release of the monocyte and macrophage chemoattractant MCP-1 by monocytes and a range of tumour cell types. This TNFα/MCP-1 cascade is consistent with the infiltration of macrophages into tumours in at least two in vivo models of cancer, and may point to IgE-mediated mobilisation and activation of monocytes/macrophages into tumours by promoting TNFα-induced production of MCP-1 in the TME (Figure 12).

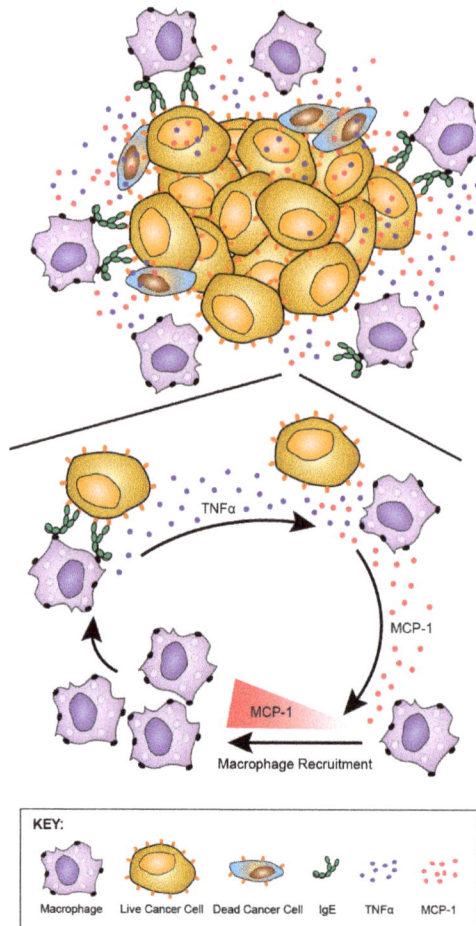

Figure 12. TNFα/MCP-1 cascade as a mechanism of MOv18 IgE functions in vivo. Activation of monocytes/macrophages by MOv18 IgE mediates a TNFα/MCP-1 axis. Cross-linking of IgE upregulates monocyte/macrophage TNFα. TNFα in turn promotes the release of the chemoattractant MCP-1 by monocytes/macrophages and tumour cells in the TME, which could promote potent chemotaxis of further monocytes/macrophages into tumors, resulting in enhanced tumor cell–effector cell interactions and subsequent tumor cell death. Figure adapted by permission from the American Association for Cancer Research. (Josephs, D.H. et al. Anti-Folate Receptor-α IgE but not IgG Recruits Macrophages to Attack Tumors via TNFα/MCP-1 Signaling. *Cancer Res.* 2017, *77*, 1127–1141 [186]).

Together, these findings also draw parallels with increased expression of TNFα, MCP-1 and IL-10 that are reported to be associated with IgE-dependent macrophage-mediated immune responses and clearance of parasites [122,192]. It was originally hypothesised that IgE could mount an allergic response mechanism against cancer. Nonetheless, the lack of IL-4 upregulation, a classic allergic mediator, and the potentiation of a TNFα/MCP-1 axis observed with anti-tumour IgE effector functions, may point to a less dominant role for an allergic, and a more prominent IgE-driven anti-tumour mechanism normally preserved for immune defence and parasite destruction by mobilising and activating macrophages. The implications of these findings may include the re-direction of otherwise inert macrophage populations into tumour lesions, and the activation of IgE-mediated anti-parasitic functions in the Th2-biased TME against tumours [193].

9. Towards Clinical Translation of First-In-Class IgE to a First-In-Man Clinical Trial

9.1. Predicting Safety of IgE: Using Ex Vivo Functional Assays Adapted from Allergy Diagnosis

In sensitized individuals, minute allergen exposure can trigger life-threatening type I systemic hypersensitivity reactions. Despite preclinical evidence that IgE could have superior efficacy compared with IgG, concerns remain that exogenously administered IgE could trigger a type I hypersensitivity reaction leading to anaphylaxis. For this to occur, monoclonal IgE antibodies bound to FcεRI on effector cells must be cross-linked by soluble multivalent allergen in the circulation [194,195]. Potent allergens can achieve this through forming soluble multimers as discussed above, or by aggregating into complexes cross-linked by polyclonal antibodies, likely to be IgE, specific for these antigens [196,197].

In the context of cancer, it is hypothesised that for an anti-tumour IgE to avoid triggering type I hypersensitivity, the target antigen should be found at low density, and in monomeric form, on healthy cells (and in the circulation) and/or should have only a single IgE-binding epitope, so that IgE cross-linking on the surface of effector cells or bridging with a target cell cannot be achieved [198]. In contrast, for an anti-tumour IgE to have anti-tumour effects, the tumour antigen should be overexpressed on the cancer cells in tissues so that they are densely packed on the cell membrane or in lipid rafts, so that IgE bridging may occur at tumour sites. Tumour-associated antigens such as FRα fulfil these criteria.

To investigate this hypothesis, the ability of MOv18 IgE to trigger basophil degranulation was examined using RBL SX-38 cells engineered to overexpress human FcεRI [182]. Exposure of cells to MOv18 IgE alone did not induce significant degranulation, however the cross-linking of MOv18 IgE bound to the effector cell surface using a polyclonal anti-IgE antibody, or by cross-linking FRα-bound IgE using an anti-FRα polyclonal antibody to mimic the effect of a circulating multimeric antigen, induced appreciable degranulation. In contrast, when cells were incubated with MOv18 IgE and increasing concentrations of recombinant (monovalent) FRα alone, at levels up to 400-fold higher than those reported in ovarian cancer-patient blood, only background levels of degranulation were observed. This was to be expected, since monovalent antigen is generally unable to cross-link FcεRI-bound IgE [182,199]. Furthermore, while naturally shed FRα levels in patient circulation were significantly elevated, compared with those measured from healthy controls, sera from 32 patients with stage III or IV ovarian carcinoma, and from 14 healthy volunteers, induced only background levels of degranulation.

The possibility that circulating tumour cells (CTCs) or tumour cell fragments bearing multiple copies of the target antigen could trigger degranulation was also explored by exposing RBL SX-38 effector cells to MOv18 IgE and serially increasing the number of FRα-expressing IGROV1 ovarian carcinoma cells. Degranulation was only detected at higher E:T cell ratios, well above those recorded in patient blood [182]. This suggests that MOv18 IgE is unlikely to activate effector cells in the presence of even the highest reported concentration of FRα-expressing CTCs. Tumour cells that did not express FRα did not induce degranulation, suggesting that the phenomenon is antigen-specific.

The ability of MOv18 IgE to activate blood basophils ex vivo in fresh unfractionated blood from patients with an ovarian carcinoma was investigated using the basophil activation assay (BAT). BAT is an increasingly useful assay conducted in unfractionated blood for detecting the propensity for type I hypersensitivity to a large range of allergens [200–203], including medicinal drugs and those used in oncology. It is designed to measure elevated cell surface CD63 expression within 10-15 min of stimulation as an early sign of type I hypersensitivity, which precedes degranulation [204]. MOv18 IgE at a range of concentrations had no effect on the level of CD63 expression in whole blood samples from healthy volunteers or from patients with an ovarian carcinoma, despite detectable circulating concentrations of FRα in the blood of some of these patients. Furthermore, MOv18 IgE with the addition of exogenous soluble FRα, even at concentrations 10-fold higher than those observed in patients, did not increase CD63 expression by human basophils. In contrast, cross-linking of effector cell FcεRI using either an anti-FcεRI or anti-IgE polyclonal antibody clearly augmented CD63 expression [182]. MOv18 IgE was therefore unable to produce significant basophil activation in human blood specimens.

In the same study, sera from 24 patients with detectable levels of circulating FRα antigen were also screened for the presence of anti-FRα IgG auto-antibodies. Such antibodies might potentially cross-link the soluble FRα bound to MOv18 IgE on the surface of basophils. In 6 of 24 patient sera, IgG auto-antibodies were detected in the range of 3–43 ng/mL. However, when tested in the RBL SX-38 degranulation assay, sera from these patients did not trigger any functional degranulation in the presence of MOv18 IgE. Sera from two patients were also studied in the BAT assay and induced with no increase in CD63 expression by the patients' blood basophils [182].

In conclusion, no evidence of effector cell activation or degranulation could be detected in validated models of allergy using recombinant FRα or patient blood and sera. In addition, no degranulation was mediated by MOv18 IgE at worst case physiological blood CTC-to-effector cell ratios or by patient anti-FRα IgG auto-antibodies. Overall, these data indicate that when ovarian carcinoma patients are treated with MOv18 IgE, FcεRI-mediated activation of effector cells may potentially occur within the tumour mass, but is less likely in the circulation.

9.2. Predicting Safety of IgE: In Vivo Models

Selection of preclinical models to help predict the safety of IgE antibody immunotherapy of cancer is still in its very early stages, and pharmacologically relevant species are being sought. An anti-human HER2/*neu* IgE was well-tolerated when introduced to cynomolgus monkeys [160]. Cross-species reactivity of mouse/human chimeric MOv18 IgE was demonstrated in cynomolgus monkey immune effector cells [205]. However, the kinetics of MOv18 IgE interaction with effector cells, and the phenotype of the activated effector cells, differed between the two species; human IgE featured a faster dissociation from cynomolgus monkey effector cells, compared with human immune effector cells. Human IgE triggered different cytokine release profiles by human and cynomolgus monkey immune effector cells. Therefore, the extrapolation of cynomolgus data to humans may be unreliable [205].

For these reasons, a surrogate syngeneic tumour model in immunocompetent (WAG) rats (discussed above) was designed to evaluate the safety profile of anti-tumour IgE. This species was selected because the IgE system of the rat bears many similarities to that of a human, and the use of the rat MOv18 IgE in the WAG rat would allow the characterisation of IgE-mediated responses that would not be possible in healthy primate models.

Preclinical efficacy studies using tumour-bearing rats showed restriction of tumour growth in the absence of any evidence of acute toxicity with rat MOv18 IgE (or with the equivalent rat MOv18 IgG2b), despite the natural presence of IgE effector cells capable of IgE-mediated degranulation such as basophils and mast cells in this species. No evidence of a cytokine storm (lack of IL-6 or IFNγ) or signals of an allergic response (IL-4) were detected, while elevated immunological pathway activation gene signatures, tumour and serum TNFα elevation and enhanced macrophage infiltration into tumours, thought to be associated with anti-tumoral efficacy, were associated with IgE treatment (Figure 13) [191].

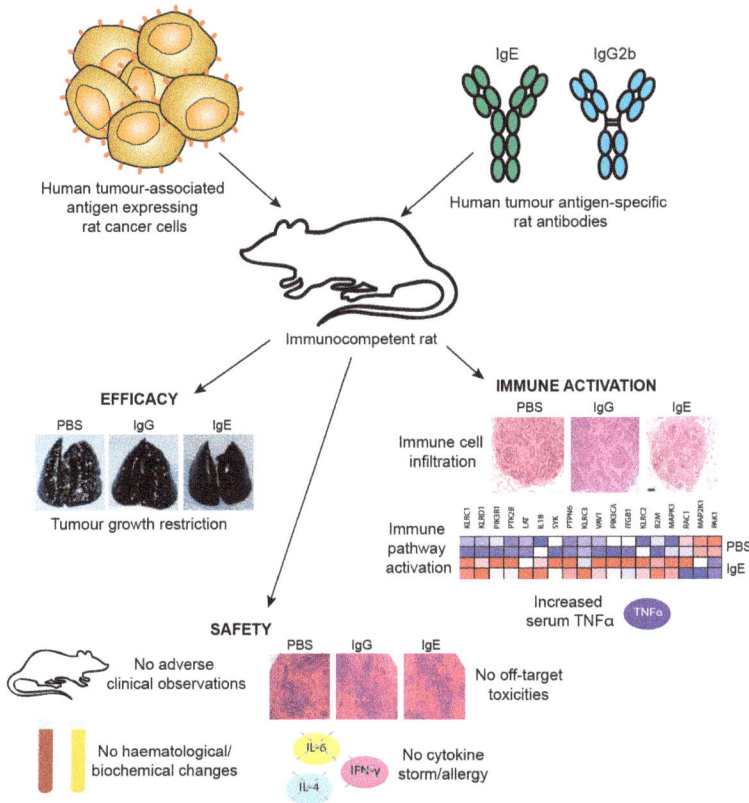

Figure 13. In vivo safety evaluations of MOv18 IgE. A surrogate syngeneic tumour model in immunocompetent WAG rats was designed to evaluate the safety profile of MOv18 IgE. Rat CC531 colon adenocarcinoma cells, engineered to express the human FRα, were administered i.v. to grow as lung metastases and animals were treated with either rat MOv18 IgE or the IgG2b equivalent. This model demonstrated the superior efficacy of IgE compared with the IgG counterpart (representative images of Indian ink-stained lungs shown). Efficacy was observed in the absence of any adverse clinical observations, off-target toxicities (H&E-stained spleen shown), or haematological or biochemical changes. Furthermore, no evidence of cytokine storm (lack of IL-6 or IFNγ upregulation) or allergic response (lack of IL-4 upregulation) was detected. In the same model, MOv18 IgE treatment was associated with the restriction of tumour growth, alongside enhanced immune cell infiltration in tumours (H&E-stained lung shown) and elevated immunological pathway activation gene signatures. Additionally, increased tumour and serum TNFα were measured in association with IgE treatment. Figure adapted by permission from John Wiley & Sons, Inc. (Josephs, D.H. et al. An immunologically relevant rodent model demonstrates safety of therapy using a tumour-specific IgE. *Allergy* 2018, 73, 2328–2341 [191]).

In concordance, in previous immunodeficient mouse models of human FRα-expressing carcinoma xenografts, the administration of mouse/human chimeric MOv18 IgE or MOv18 IgG1 together with human peripheral blood lymphocytes and peripheral blood mononuclear cells did not trigger any toxic effects, despite the presence of human basophils and eosinophils, including those from allergic human donors [147,179,181], in these effector cell preparations. Further support for this concept comes from published data demonstrating the induction of IgE through tumour antigen mimotope vaccination, detected in the absence of any toxicities or signs of type I hypersensitivity [206]. Furthermore, IgE specific to tumour antigens and with tumoricidal properties has been reported in patients with head and neck cancer and pancreatic cancer, in the circulation and tumour tissues [104,105], without anaphylaxis occurring.

Finally, dogs may be an alternative model to examine the safety and anti-tumour functions of IgE, since this species is known for susceptibility to both cancer, including spontaneous mammary carcinomas, and allergy, with strong similarities of FcεR expression and distribution on immune cells compared with humans [207–209]. Efforts are underway to design canine versions of anti-tumour IgEs with a view to conduct safety and efficacy studies [152].

9.3. Monitoring Antibody Safety in Trials

Translation to clinical testing is expected to entail careful monitoring of patients and measuring functional readouts and immunological markers of type I hypersensitivity following administration of MOv18 IgE due to the potential for basophil and/or mast cell degranulation. Functional tests may monitor the propensity to trigger basophil activation and mast cell degranulation in patient blood and sera ex vivo, all measured at different points of drug administration. Monitoring would include clinical signs of type I hypersensitivity, changes in serum levels of β-tryptase, total and tumour antigen-specific IgE, circulating tumour antigen and autoantibodies to the target antigen. Specifically, serum β-tryptase elevation signifying mast cell degranulation during clinical testing may be important to help distinguish cytokine release-type infusion reactions from type I hypersensitivity [210,211].

10. Thoughts for the Design of New IgE-Based Therapeutic Agents

10.1. Expression Systems and IgE Glyco-Profiling

Production of IgE for clinical study requires the development of GMP processes that ensure swift production of an antibody with sufficient quality, purity and stability profiles. Importantly, the product must show physiochemical and functional profiles compatible with those of the laboratory grade material. Additionally, IgE antibodies display seven glycosylation sites, six of which comprise complex N-glycans, potentially with terminal galactose, fucose and sialic acid residues, as discussed above (and illustrated for IgG in Figure 1c). Due to its heavily glycosylated structure, the glycosylation profile of IgE antibodies must also be considered with regard to achieving a consistent antibody structural and functional product profile for clinical application. Carbohydrates may influence the affinity for the target antigen, biodistribution, effector cell trafficking to tissues and antibody pharmacokinetics; the high-mannonse structure at Asn394 (Figure 1d) may, as we have discussed, have functional significance [41,50]. Monitoring the structural and functional integrity of IgE is therefore warranted at all stages of research, development and manufacturing for pre-clinical and clinical evaluations. Furthermore, the nature of the expression system may impact the glycosylation profile and must be carefully considered when designing an IgE class therapeutic agent [153]. For instance, the carbohydrate profile of IgE antibodies produced using a human expression system, may differ from that of plant-expressed IgE [150]. Further study of glycan content will undoubtedly provide important information for further understanding structure-function relationships in IgE.

10.2. Selecting Tumour Targets and Malignant Indications for IgE Therapeutic Agents

Rational design of suitable therapeutic agents should aim to take advantage of the tissue-resident immune surveillance exerted by IgE antibodies that can be directed against cancer antigens, whilst minimising the risk of the potential toxic effects of the therapeutic agent. Malignant indications could be selected according to whether tumour cells are likely to reside in tissues in which important IgE effector cells such as macrophages are also found. Indications in which tumour cells and tumour cell fragments do not circulate would be preferable, since following systemic administration of anti-tumour IgE, basophils loaded with anti-tumour IgE could encounter circulating cancer cells bearing multiple copies of the target antigen; such interactions might trigger degranulation and potential type I hypersensitivity. Important criteria for the selection of cancer antigen targets would include high expression on the tumour with minimal and restricted distribution in normal tissues away from patient circulation. Furthermore, selection of single epitopes on tumour antigens and antigens that do not shed in multimeric forms in patient circulation would be key criteria for target selection.

10.3. Challenges for IgE-Based Therapies

Within the fields of Immunology, Allergy and AllergoOncology, there are many aspects of IgE biology that are yet to be explored. The most prominent unknowns in the field are: defining the dynamics of antibody trafficking to tumours, recruiting monocytes into tumour lesions and engaging local tumour-associated macrophages; pharmacokinetics in patient circulation and biodistribution in health and disease settings; the roles and anti-tumour functions of mast cells; unexplored mechanisms of action beyond the TNFα/MCP-1 cascade; the existence of modulatory mechanisms for IgE despite the lack of any known inhibitory FcεR; the impact of target antigen expression levels and distribution in tumour lesions on the anti-tumour efficacy of IgE antibodies; stratification of patients with tumours featuring immune tumour environments congruent to IgE antibody therapy; the most suitable administration route, and malignant indication to help refine treatment and maximise patient benefit.

Evidence from a number of studies points to monocytes and macrophages as important effector cells that participate in the anti-tumour functions of IgE in vitro and in vivo [193]. On the other hand, mast cells express far higher levels of FcεRI compared with monocytes and macrophages, and constitute another key effector cell population that may contribute to the cancer growth-restricting functions of anti-tumour IgE antibodies. Mast cells can be activated upon crosslinking of FcεRI by IgE in the presence of multivalent antigens, to degranulate and release toxic mediators in tissues such as the skin and gut. These functions of mast cells have been known to be directed to destroy parasites [5,97]. The significance of mast cell infiltration in tumour lesions has been controversial [212], however there have been reports of associations with more favourable clinical outcomes [213]. Tumour- and tissue-resident mast cells may also contribute to IgE-mediated enhanced TNFα expression and heightened immune responses in the TME [214]. Mast cells could be recruited towards tumour lesions either through tumour cell-produced MCP-1 [215], and more prominently through the anti-tumour IgE-potentiated TNFα/MCP-1 axis discussed above [186,191]. However, the roles of mast cells in the context of anti-tumour IgE mechanisms of action and efficacy require further study.

Further areas for investigation include the impact of clinically available therapies such as chemotherapies, checkpoint inhibitors, steroids, and targeted treatments on the following: effector cells and IgE therapeutic efficacy and safety; expression of IgE Fc receptors by immune cells in different cancer types and patient tumours; mechanisms by which IgE acts on the TME, including IgE receptor-expressing and non-expressing cells, and their recruitment into tumours.

A number of antibodies engineered with IgE Fc regions have been shown to engender potent effector functions and restrict tumour growth in disparate model systems. These include antibodies recognizing epitopes found on clinically validated tumour targets such as HER2/*neu*. It is to be hoped that IgE antibodies against these targets will progress along the translational pipeline towards clinical testing. The field of AllergoOncology, including the use of IgE antibodies for cancer treatment, will undoubtedly enrich our understanding of human immunity and responses in health and malignant disease, and both inform and transform the design of future immunotherapeutic agents.

Author Contributions: Writing—original draft preparation, review and editing, B.J.S., A.M.D., H.J.B. and S.N.K.

Funding: The authors acknowledge support by the Medical Research Council UK (G0501494, G1100090 and MR/L023091/1), the Wellcome Trust (076343), Asthma UK (AUK-IG-2016-338), Breast Cancer Now (147) working in partnership with Walk the Walk, Cancer Research UK (C30122/A11527 and C30122/A15774) and CRUK/NIHR in England/DoH for Scotland, Wales and Northern Ireland Experimental Cancer Medicine Centre (C10355/A15587). The research was supported by the National Institute for Health Research (NIHR) Biomedical Research Centre (BRC) based at Guy's and St Thomas' NHS Foundation Trust and King's College London (IS-BRC-1215-20006). The authors are solely responsible for study design, data collection, analysis, decision to publish, and preparation of the manuscript. The views expressed are those of the authors and not necessarily those of the NHS, the NIHR or the Department of Health.

Conflicts of Interest: S.N. Karagiannis is founder and shareholder of IGEM Therapeutics Ltd. and holds a patent on anti-tumour IgE antibodies. H.J. Bax is employed through a fund by IGEM Therapeutics Ltd.

Abbreviations

ADCC	antibody-dependent cell-mediated cytotoxicity
ADCP	antibody-dependent cell-mediated phagocytosis
APC	antigen presenting cell
BAL	broncho-alveolar lavage
BAT	basophil activation test
CCA	colorectal cancer antigen
CDR	complementarity-determining region
CTCs	circulating tumour cells
CTL	cytotoxic T lymphocyte
DCs	dendritic cells
EGFR	epidermal growth factor receptor
EM	electron microscopy
FR	framework region
FRα	folate receptor alpha
FRET	fluorescence (Förster) resonance energy transfer
GMP	Good Manufacturing Practice
IHC	immunohistochemical/immunohistochemistry
i.p.	intraperitoneal
i.v.	intravenous
MCP-1	macrophage chemoattractant protein-1
MD	molecular dynamics
MMTV	mammary tumour virus
NIP	4-hydroxy-3-nitro-phenacetyl
NK	Natural Killer
PBMCs	peripheral blood mononuclear cells
PDX	patient-derived xenograft
PIPE	Polymerase Incomplete Primer Extension
PSA	prostate specific antigen
RBL	rat basophil leukaemia
SAXS	small-angle X-ray scattering
s.c.	subcutaneous
Th	T helper
TME	tumour microenvironment
TNFα	tumour necrosis factor
UCOE	Ubiquitous Chromatin Opening Elements
WAG	Wistar Albino Glaxo

References

1. Platts-Mills, T.A.; Heymann, P.W.; Commins, S.P.; Woodfolk, J.A. The discovery of IgE 50 years later. *Ann. Allergy Asthma Immunol.* **2016**, *116*, 179–182. [CrossRef] [PubMed]

2. Bennich, H.H.; Ishizaka, K.; Johansson, S.G.O.; Rowe, D.S.; Stanworth, D.R.; Terry, W.D. Immunoglobulin E, a new class of human immunoglobulin. *Bull. World Health Organ.* **1968**, *38*, 151–152. [CrossRef]

3. Ishizaka, K.; Ishizaka, T.; Hornbrook, M.M. Physicochemical properties of reaginic antibody. V. Correlation of reaginic activity with γE globulin antibody. *J. Immunol.* **1966**, *97*, 840–853. [PubMed]

4. Gould, H.J.; Sutton, B.J. IgE in allergy and asthma today. *Nat. Rev. Immunol.* **2008**, *8*, 205–217. [CrossRef] [PubMed]

5. Mukai, K.; Tsai, M.; Starkl, P.; Marichal, T.; Galli, S.J. IgE and mast cells in host defense against parasites and venoms. *Semin. Immunopathol.* **2016**, *38*, 581–603. [CrossRef] [PubMed]

6. Sutton, B.J.; Davies, A.M. Structure and dynamics of IgE-receptor interactions: FcεRI and CD23/FcεRII. *Immunol. Rev.* **2015**, *268*, 222–235. [CrossRef] [PubMed]

7. Kraft, S.; Kinet, J.-P. New developments in FcεRI regulation, function and inhibition. *Nat. Rev. Immunol.* **2007**, *7*, 365–378. [CrossRef]

8. Kinet, J.P. The high-affinity IgE receptor (FcεRI): From physiology to pathology. *Annu. Rev. Immunol.* **1999**, *17*, 931–972. [CrossRef]

9. Gounni, A.S.; Wellemans, V.; Yang, J.; Bellesort, F.; Kassiri, K.; Gangloff, S.; Guenounou, M.; Halayko, A.J.; Hamid, Q.; Lamkhioued, B. Human airway smooth muscle cells express the high affinity receptor for IgE (FcεRI): A critical role of FcεRI in human airway smooth muscle cell function. *J. Immunol.* **2005**, *175*, 2613–2621. [CrossRef]

10. Campbell, A.M.; Vachier, I.; Chanez, P.; Vignola, A.M.; Lebel, B.; Kochan, J.; Godard, P.; Bousquet, J. Expression of the high-affinity receptor for IgE on bronchial epithelial cells of asthmatics. *Am. J. Respir. Cell Mol. Biol.* **1998**, *19*, 92–97. [CrossRef]

11. Untersmayr, E.; Bises, G.; Starkl, P.; Bevins, C.L.; Scheiner, O.; Boltz-Nitulescu, G.; Wrba, F.; Jensen-Jarolim, E. The high affinity IgE receptor FcεRI is expressed by human intestinal epithelial cells. *PLoS ONE* **2010**, *5*, e9023. [CrossRef] [PubMed]

12. Hogarth, P.M.; Pietersz, G.A. Fc receptor-targeted therapies for the treatment of inflammation, cancer and beyond. *Nat. Rev. Drug Discov.* **2012**, *11*, 311–331. [CrossRef] [PubMed]

13. Conrad, D.H.; Ford, J.W.; Sturgill, J.L.; Gibb, D.R. CD23: An overlooked regulator of allergic disease. *Curr. Allergy Asthma Rep.* **2007**, *7*, 331–337. [CrossRef] [PubMed]

14. Yukawa, K.; Kikutani, H.; Owaki, H.; Yamasaki, K.; Yokota, A.; Nakamura, H.; Barsumian, E.L.; Hardy, R.R.; Suemura, M.; Kishimoto, T. A B cell-specific differentiation antigen, CD23, is a receptor for IgE (Fc epsilon R) on lymphocytes. *J. Immunol.* **1987**, *138*, 2576–2580.

15. Bonnefoy, J.Y.; Aubry, J.P.; Peronne, C.; Wijdenes, J.; Banchereau, J. Production and characterization of a monoclonal antibody specific for the human lymphocyte low affinity receptor for IgE: CD 23 is a low affinity receptor for IgE. *J. Immunol.* **1987**, *138*, 2970–2978. [PubMed]

16. Palaniyandi, S.; Tomei, E.; Li, Z.; Conrad, D.H.; Zhu, X. CD23-dependent transcytosis of IgE and immune complex across the polarized human respiratory epithelial cells. *J. Immunol.* **2011**, *186*, 3484–3496. [CrossRef]

17. Tu, Y.; Salim, S.; Bourgeois, J.; Di Leo, V.; Irvine, E.J.; Marshall, J.K.; Perdue, M.H. CD23-mediated IgE transport across human intestinal epithelium: Inhibition by blocking sites of translation or binding. *Gastroenterology* **2005**, *129*, 928–940. [CrossRef]

18. Li, H.; Nowak-Wegrzyn, A.; Charlop-Powers, Z.; Shreffler, W.; Chehade, M.; Thomas, S.; Roda, G.; Dahan, S.; Sperber, K.; Berin, M.C. Transcytosis of IgE-antigen complexes by CD23a in human intestinal epithelial cells and its role in food allergy. *Gastroenterology* **2006**, *131*, 47–58. [CrossRef]

19. McCloskey, N.; Hunt, J.; Beavil, R.L.; Jutton, M.R.; Grundy, G.J.; Girardi, E.; Fabiane, S.M.; Fear, D.J.; Conrad, D.H.; Sutton, B.J.; et al. Soluble CD23 monomers inhibit and oligomers stimulate IgE synthesis in human B cells. *J. Biol. Chem.* **2007**, *282*, 24083–24091. [CrossRef]

20. Gould, H.J.; Beavil, R.L.; Reljić, R.; Shi, J.; Ma, C.W.; Sutton, B.J.; Ghirlando, R. IgE Homeostasis: Is CD23 the safety switch? In *IgE Regulation: Molecular Mechanisms*; Vercelli, D., Ed.; Wiley: Chichester, UK, 1997; pp. 37–59.

21. Cooper, A.M.; Hobson, P.S.; Jutton, M.R.; Kao, M.W.; Drung, B.; Schmidt, B.; Fear, D.J.; Beavil, A.J.; McDonnell, J.M.; Sutton, B.J.; et al. Soluble CD23 controls IgE synthesis and homeostasis in human B cells. *J. Immunol.* **2012**, *188*, 3199–3207. [CrossRef]

22. Palaniyandi, S.; Liu, X.; Periasamy, S.; Ma, A.; Tang, J.; Jenkins, M.; Tuo, W.; Song, W.; Keegan, A.D.; Conrad, D.H.; et al. Inhibition of CD23-mediated IgE transcytosis suppresses the initiation and development of allergic airway inflammation. *Mucosal Immunol.* **2015**, *8*, 1262–1274. [CrossRef] [PubMed]

23. Mitropoulou, A.N.; Bowen, H.; Dodev, T.; Davies, A.M.; Bax, H.; Beavil, R.L.; Beavil, A.J.; Gould, H.J.; James, L.K.; Sutton, B.J. Structure of a patient-derived antibody in complex with allergen reveals simultaneous conventional and superantigen-like recognition. *Proc. Natl. Acad. Sci. USA* **2018**, *115*, E8707–E8716. [CrossRef] [PubMed]

24. Zhang, X.; Calvert, R.A.; Sutton, B.J.; Doré, K.A. IgY: A key isotype in antibody evolution. *Biol. Rev. Camb. Philos. Soc.* **2017**, *92*, 2144–2156. [CrossRef] [PubMed]

25. Feinstein, A.; Munn, E.A. Conformation of the free and antigen-bound IgM antibody molecules. *Nature* **1969**, *224*, 1307–1309. [CrossRef] [PubMed]

26. Crispin, M.; Yu, X.; Bowden, T.A. Crystal structure of sialylated IgG Fc: Implications for the mechanism of intravenous immunoglobulin therapy. *Proc. Natl. Acad. Sci. USA* **2013**, *110*, E3544–E3546. [CrossRef] [PubMed]

27. Doré, K.A.; Davies, A.M.; Drinkwater, N.; Beavil, A.J.; McDonnell, J.M.; Sutton, B.J. Thermal sensitivity and flexibility of the Cε3 domains in immunoglobulin E. *Biochim. Biophys. Acta* **2017**, *1865*, 1336–1347. [CrossRef] [PubMed]

28. Padlan, E.A.; Davies, D.R. A model of the Fc of Immunoglobulin-E. *Mol. Immunol.* **1986**, *23*, 1063–1075. [CrossRef]

29. Holowka, D.; Baird, B. Structural studies on the membrane-bound immunoglobulin E (IgE)-receptor complex. 2. Mapping of distances between sites on IgE and the membrane surface. *Biochemistry* **1983**, *22*, 3475–3484. [CrossRef]

30. Holowka, D.; Conrad, D.H.; Baird, B. Structural mapping of membrane-bound immunoglobulin-E receptor complexes: Use of monoclonal anti-IgE antibodies to probe the conformation of receptor-bound IgE. *Biochemistry* **1985**, *24*, 6260–6267. [CrossRef]

31. Zheng, Y.; Shopes, B.; Holowka, D.; Baird, B. Conformations of IgE bound to its receptor FcεRI and in solution. *Biochemistry* **1991**, *30*, 9125–9132. [CrossRef]

32. Zheng, Y.; Shopes, B.; Holowka, D.; Baird, B. Dynamic conformations compared for IgE and IgG1 in solution and bound to receptors. *Biochemistry* **1992**, *31*, 7446–7456. [CrossRef] [PubMed]

33. Beavil, A.J.; Young, R.J.; Sutton, B.J.; Perkins, S.J. Bent domain structure of recombinant human IgE-Fc in solution by X-ray and neutron scattering in conjunction with an automated curve fitting procedure. *Biochemistry* **1995**, *34*, 14449–14461. [CrossRef] [PubMed]

34. Wan, T.; Beavil, R.L.; Fabiane, S.M.; Beavil, A.J.; Sohi, M.K.; Keown, M.; Young, R.J.; Henry, A.J.; Owens, R.J.; Gould, H.J.; et al. The crystal structure of IgE Fc reveals an asymmetrically bent conformation. *Nat. Immunol.* **2002**, *3*, 681–686. [CrossRef] [PubMed]

35. Hunt, J.; Keeble, A.H.; Dale, R.E.; Corbett, M.K.; Beavil, R.L.; Levitt, J.; Swann, M.J.; Suhling, K.; Ameer-Beg, S.; Sutton, B.J.; et al. A fluorescent biosensor reveals conformational changes in human immunoglobulin E Fc: Implications for mechanisms of receptor binding, inhibition, and allergen recognition. *J. Biol. Chem.* **2012**, *287*, 17459–17470. [CrossRef] [PubMed]

36. Davies, A.M.; Allan, E.G.; Keeble, A.H.; Delgado, J.; Cossins, B.P.; Mitropoulou, A.N.; Pang, M.O.Y.; Ceska, T.; Beavil, A.J.; Craggs, G.; et al. Allosteric mechanism of action of the therapeutic anti-IgE antibody omalizumab. *J. Biol. Chem.* **2017**, *292*, 9975–9987. [CrossRef] [PubMed]

37. Drinkwater, N.; Cossins, B.P.; Keeble, A.H.; Wright, M.; Cain, K.; Hailu, H.; Oxbrow, A.; Delgado, J.; Shuttleworth, L.K.; Kao, M.W.; et al. Human immunoglobulin E flexes between acutely bent and extended conformations. *Nat. Struct. Mol. Biol.* **2014**, *21*, 397–404. [CrossRef] [PubMed]

38. Chen, J.-B.; Ramadani, F.; Pang, M.O.Y.; Beavil, R.L.; Holdom, M.D.; Mitropoulou, A.N.; Beavil, A.J.; Gould, H.J.; Chang, T.-W.; Sutton, B.J.; et al. Structural basis for selective inhibition of immunoglobulin E-receptor interactions by an anti-IgE antibody. *Sci. Rep.* **2018**, *8*, 11548. [CrossRef]

39. Arnold, J.N.; Radcliffe, C.M.; Wormald, M.R.; Royle, L.; Harvey, D.J.; Crispin, M.; Dwek, R.A.; Sim, R.B.; Rudd, P.M. The glycosylation of human serum IgD and IgE and the accessibility of identified oligomannose structures for interaction with mannan-binding lectin. *J. Immunol.* **2004**, *173*, 6831–6840. [CrossRef]

40. Plomp, R.; Hensbergen, P.J.; Rombouts, Y.; Zauner, G.; Dragan, I.; Koeleman, C.A.; Deelder, A.M.; Wuhrer, M. Site-specific N-glycosylation analysis of human immunoglobulin E. *J. Proteome Res.* **2014**, *13*, 536–546. [CrossRef]

41. Shade, K.T.; Platzer, B.; Washburn, N.; Mani, V.; Bartsch, Y.C.; Conroy, M.; Pagan, J.D.; Bosques, C.; Mempel, T.R.; Fiebiger, E.; et al. A single glycan on IgE is indispensible for initiation of anaphylaxis. *J. Exp. Med.* **2015**, *212*, 457–467. [CrossRef]

42. Fridriksson, E.K.; Beavil, A.; Holowka, D.; Gould, H.J.; Baird, B.; McLafferty, F.W. Heterogeneous glycosylation of immunoglobulin E constructs characterized by top-down high-resolution 2-D mass spectrometry. *Biochemistry* **2000**, *39*, 3369–3376. [CrossRef]

43. Taylor, A.I.; Fabiane, S.M.; Sutton, B.J.; Calvert, R.A. The crystal structure of an avian IgY-Fc fragment reveals conservation with both mammalian IgG and IgE. *Biochemistry* **2009**, *48*, 558–562. [CrossRef]

44. Holdom, M.D.; Davies, A.M.; Nettleship, J.E.; Bagby, S.C.; Dhaliwal, B.; Girardi, E.; Hunt, J.; Gould, H.J.; Beavil, A.J.; McDonnell, J.M.; et al. Conformational changes in IgE contribute to its uniquely slow dissociation rate from receptor FcεRI. *Nat. Struct. Mol. Biol.* **2011**, *18*, 571–576. [CrossRef] [PubMed]

45. Arnold, J.N.; Wormald, M.R.; Sim, R.B.; Rudd, P.M.; Dwek, R.A. The impact of glycosylation on the biological function and structure of human immunoglobulins. *Annu. Rev. Immunol.* **2007**, *25*, 21–50. [CrossRef] [PubMed]

46. Helm, B.; Marsh, P.; Vercelli, D.; Padlan, E.; Gould, H.; Geha, R. The mast cell binding site on human immunoglobulin E. *Nature* **1988**, *331*, 180–183. [CrossRef] [PubMed]

47. Vercelli, D.; Helm, B.; Marsh, P.; Padlan, E.; Geha, R.; Gould, H. The B-cell binding site on human immunoglobulin E. *Nature* **1989**, *338*, 649–651. [CrossRef] [PubMed]

48. Basu, M.; Hakimi, J.; Dharm, E.; Kondas, J.A.; Tsien, W.H.; Pilson, R.S.; Lin, P.; Gilfillan, A.; Haring, P.; Braswell, E.H.; et al. Purification and characterization of human recombinant IgE-Fc fragments that bind to the human high affinity IgE receptor. *J. Biol. Chem.* **1993**, *268*, 13118–13127.

49. Hunt, J.; Beavil, R.L.; Calvert, R.A.; Gould, H.J.; Sutton, B.J.; Beavil, A.J. Disulfide linkage controls the affinity and stoichiometry of IgE Fcε3-4 binding to FcεRI. *J. Biol. Chem.* **2005**, *280*, 16808–16814. [CrossRef]

50. Sayers, I.; Cain, S.A.; Swan, J.R.; Pickett, M.A.; Watt, P.J.; Holgate, S.T.; Padlan, E.A.; Schuck, P.; Helm, B.A. Amino acid residues that influence FcεRI-mediated effector functions of human immunoglobulin E. *Biochemistry* **1998**, *37*, 16152–16164. [CrossRef]

51. Dhaliwal, B.; Yuan, D.; Pang, M.O.; Henry, A.J.; Cain, K.; Oxbrow, A.; Fabiane, S.M.; Beavil, A.J.; McDonnell, J.M.; Gould, H.J.; et al. Crystal structure of IgE bound to its B-cell receptor CD23 reveals a mechanism of reciprocal allosteric inhibition with high affinity receptor FcεRI. *Proc. Natl. Acad. Sci. USA* **2012**, *109*, 12686–12691. [CrossRef]

52. Cohen, E.S.; Dobson, C.L.; Käck, H.; Wang, B.; Sims, D.A.; Lloyd, C.O.; England, E.; Rees, D.G.; Guo, H.; Karagiannis, S.N.; et al. A novel IgE-neutralizing antibody for the treatment of severe uncontrolled asthma. *mAbs* **2014**, *6*, 755–763. [CrossRef] [PubMed]

53. Dhaliwal, B.; Pang, M.O.; Keeble, A.H.; James, L.K.; Gould, H.J.; McDonnell, J.M.; Sutton, B.J.; Beavil, A.J. IgE binds asymmetrically to its B cell receptor CD23. *Sci. Rep.* **2017**, *7*, 45533. [CrossRef] [PubMed]

54. Yuan, D.; Keeble, A.H.; Hibbert, R.G.; Fabiane, S.; Gould, H.J.; McDonnell, J.M.; Beavil, A.J.; Sutton, B.J.; Dhaliwal, B. Ca²⁺-dependent structural changes in the B-cell receptor CD23 increase its affinity for human immunoglobulin E. *J. Biol. Chem.* **2013**, *288*, 21667–21677. [CrossRef] [PubMed]

55. Dhaliwal, B.; Pang, M.O.Y.; Yuan, D.; Beavil, A.J.; Sutton, B.J. A range of Cε3-Cε4 interdomain angles in IgE Fc accommodate binding to its receptor CD23. *Acta Crystallogr. F Struct. Biol. Commun.* **2014**, *70*, 305–309. [CrossRef] [PubMed]

56. Garman, S.C.; Wurzburg, B.A.; Tarchevskaya, S.S.; Kinet, J.-P.; Jardetzky, T.S. Structure of the Fc fragment of human IgE bound to its high-affinity receptor FcεRIα. *Nature* **2000**, *406*, 259–266. [CrossRef] [PubMed]

57. Wurzburg, B.A.; Garman, S.C.; Jardetzky, T.S. Structure of the human IgE-Fc Cε3-Cε4 reveals conformational flexibility in the antibody effector domains. *Immunity* **2000**, *13*, 375–385. [CrossRef]

58. Wurzburg, B.A.; Jardetzky, T.S. Conformational Flexibility in the IgE-Fc₃₋₄ Revealed in Multiple Crystal Forms. *J. Mol. Biol.* **2009**, *393*, 176–190. [CrossRef] [PubMed]

59. Jabs, F.; Plum, M.; Laursen, N.S.; Jensen, R.K.; Mølgaard, B.; Miehe, M.; Mandolesi, M.; Rauber, M.M.; Pfützner, W.; Jakob, T.; et al. Trapping IgE in a closed conformation by mimicking CD23 binding prevents and disrupts FcεRI interaction. *Nat. Commun.* **2018**, *9*, 7. [CrossRef] [PubMed]

60. Oi, V.T.; Vuong, T.M.; Hardy, R.; Reidler, J.; Dangl, J.; Herzenberg, L.A.; Stryer, L. Correlation between segmental flexibility and effector function of antibodies. *Nature* **1983**, *307*, 136–140. [CrossRef]

61. Gould, H.J.; Sutton, B.J.; Beavil, A.J.; Beavil, R.L.; McCloskey, N.; Coker, H.A.; Fear, D.; Smurthwaite, L. The biology of IgE and the basis of allergic disease. *Annu. Rev. Immunol.* **2003**, *21*, 579–628. [CrossRef]

62. Hibbert, R.G.; Teriete, P.; Grundy, G.J.; Beavil, R.L.; Reljic, R.; Holers, V.M.; Hannan, J.P.; Sutton, B.J.; Gould, H.J.; McDonnell, J.M. The structure of human CD23 and its interactions with IgE and CD21. *J. Exp. Med.* **2005**, *202*, 751–760. [CrossRef] [PubMed]

63. Aubry, J.-P.; Pochon, S.; Graber, P.; Jansen, K.U.; Bonnefoy, J.-Y. CD21 is a ligand for CD23 and regulates IgE production. *Nature* **1992**, *358*, 505–507. [CrossRef] [PubMed]

64. Richards, M.L.; Katz, D.H. The binding of IgE to murine FcεRII is calcium-dependent but not inhibited by carbohydrate. *J. Immunol.* **1990**, *144*, 2638–2646. [PubMed]

65. Karagiannis, S.N.; Warrack, J.K.; Jennings, K.H.; Murdock, P.R.; Christie, G.; Moulder, K.; Sutton, B.J.; Gould, H.J. Endocytosis and recycling of the complex between CD23 and HLA-DR in human B cells. *Immunology* **2001**, *103*, 319–331. [CrossRef] [PubMed]

66. Andersen, C.B.F.; Moestrup, S.K. How calcium makes endocytic receptors attractive. *Trends Biochem. Sci.* **2014**, *39*, 82–90. [CrossRef] [PubMed]

67. Kelly, A.E.; Chen, B.-H.; Woodward, E.C.; Conrad, D.H. Production of a chimeric form of CD23 that is oligomeric and blocks IgE binding to the FcεRI. *J. Immunol.* **1998**, *161*, 6696–6704. [PubMed]

68. Suemura, M.; Kikutani, H.; Sugiyama, K.; Uchibayashi, N.; Aitani, M.; Kuritani, T.; Barsumian, E.L.; Yamatodani, A.; Kishimoto, T. Significance of soluble Fcε receptor II (sFcεRII/CD23) in serum and possible application of sFcεRII for the prevention of allergic reactions. *Allergy Proc.* **1991**, *12*, 133–137. [CrossRef]

69. Borthakur, S.; Hibbert, R.G.; Pang, M.O.; Yahya, N.; Bax, H.J.; Kao, M.W.; Cooper, A.M.; Beavil, A.J.; Sutton, B.J.; Gould, H.J.; et al. Mapping of the CD23 binding site on immunoglobulin E (IgE) and allosteric control of the IgE-FcεRI interaction. *J. Biol. Chem.* **2012**, *287*, 31457–31461. [CrossRef]

70. Henry, A.J.; McDonnell, J.M.; Ghirlando, R.; Sutton, B.J.; Gould, H.J. Conformation of the isolated Cε3 domain of IgE and its complex with the high-affinity receptor, FcεRI. *Biochemistry* **2000**, *39*, 7406–7413. [CrossRef]

71. Vangelista, L.; Laffer, S.; Turek, R.; Grönlund, H.; Sperr, W.R.; Valent, P.; Pastore, A.; Valenta, R. The immunoglobulin-like modules Cε3 and α2 are the minimal units necessary for human IgE-FcεRI interaction. *J. Clin. Investig.* **1999**, *103*, 1571–1578. [CrossRef]

72. Price, N.E.; Price, N.C.; Kelly, S.M.; McDonnell, J.M. The key role of protein flexibility in modulating IgE interactions. *J. Biol. Chem.* **2005**, *280*, 2324–2330. [CrossRef] [PubMed]

73. Harwood, N.E.; McDonnell, J.M. The intrinsic flexibility of IgE and its role in binding FcεRI. *Biomed. Pharmacother.* **2007**, *61*, 61–67. [CrossRef] [PubMed]

74. Borthakur, S.; Andrejeva, G.; McDonnell, J.M. Basis of the intrinsic flexibility of the Cε3 domain of IgE. *Biochemistry* **2011**, *50*, 4608–4614. [CrossRef] [PubMed]

75. Dhaliwal, B.; Pang, M.O.; Yuan, D.; Yahya, N.; Fabiane, S.M.; McDonnell, J.M.; Gould, H.J.; Beavil, A.J.; Sutton, B.J. Conformational plasticity at the IgE-binding site of the B-cell receptor CD23. *Mol. Immunol.* **2013**, *56*, 693–697. [CrossRef] [PubMed]

76. Dorrington, K.J.; Bennich, H. Thermally induced structural changes in immunoglobulin E. *J. Biol. Chem.* **1973**, *248*, 8378–8384.

77. Eggel, A.; Baravalle, G.; Hobi, G.; Kim, B.; Buschor, P.; Forrer, P.; Shin, J.S.; Vogel, M.; Stadler, B.M.; Dahinden, C.A.; et al. Accelerated dissociation of IgE-FcεRI complexes by disruptive inhibitors actively desensitizes allergic effector cells. *J. Allergy Clin. Immunol.* **2014**, *133*, 1709–1719. [CrossRef]

78. Pennington, L.F.; Tarchevskaya, S.; Brigger, D.; Sathiyamoorthy, K.; Graham, M.T.; Nadeau, K.C.; Eggel, A.; Jardetzky, T.S. Structural basis of omalizumab therapy and omalizumab-mediated IgE exchange. *Nat. Commun.* **2016**, *7*, 11610. [CrossRef]

79. Kim, B.; Eggel, A.; Tarchevskaya, S.S.; Vogel, M.; Prinz, H.; Jardetzky, T.S. Accelerated disassembly of IgE-receptor complexes by a disruptive macromolecular inhibitor. *Nature* **2012**, *491*, 613–617. [CrossRef]

80. Roux, K.H.; Strelets, L.; Brekke, O.H.; Sandlie, I.; Michaelsen, T.E. Comparisons of the ability of human IgG3 hinge mutants, IgM, IgE, and IgA2, to form small immune complexes: A role for flexibility and geometry. *J. Immunol.* **1998**, *161*, 4083–4090.

81. Gieras, A.; Linhart, B.; Roux, K.H.; Dutta, M.; Khodoun, M.; Zafred, D.; Cabauatan, C.R.; Lupinek, C.; Weber, M.; Focke-Tejkl, M.; et al. IgE epitope proximity determines immune complex shape and effector cell activation capacity. *J. Allergy Clin. Immunol.* **2016**, *137*, 1557–1565. [CrossRef]

82. Christensen, L.H.; Holm, J.; Lund, G.; Riise, E.; Lund, K. Several distinct properties of the IgE repertoire determine effector cell degranulation in response to allergen challenge. *J. Allergy Clin. Immunol.* **2008**, *122*, 298–304. [CrossRef] [PubMed]

83. Padlan, E.A.; Silverton, E.W.; Sheriff, S.; Cohen, G.H.; Smith-Gill, S.J.; Davies, D.R. Structure of an antibody-antigen complex: Crystal structure of the HyHEL-10 Fab-lysozyme complex. *Proc. Natl. Acad. Sci USA* **1989**, *86*, 5938–5942. [CrossRef] [PubMed]

84. Mirza, O.; Henriksen, A.; Ipsen, H.; Larsen, J.N.; Wissenbach, M.; Spangfort, M.D.; Gajhede, M. Dominant epitopes and allergic cross-reactivity: Complex formation between a Fab fragment of a monoclonal murine IgG antibody and the major allergen from birch pollen Bet v 1. *J. Immunol.* **2000**, *165*, 331–338. [CrossRef] [PubMed]

85. Padavattan, S.; Schirmer, T.; Schmidt, M.; Akdis, C.; Valenta, R.; Mittermann, I.; Soldatova, L.; Slater, J.; Mueller, U.; Markovic-Housley, Z. Identification of a B-cell Epitope of Hyaluronidase, a Major Bee Venom Allergen, from its Crystal Structure in Complex with a Specific Fab. *J. Mol. Biol.* **2007**, *368*, 742–752. [CrossRef] [PubMed]

86. Li, M.; Gustchina, A.; Alexandratos, J.; Wlodawer, A.; Wünschmann, S.; Kepley, C.L.; Chapman, M.D.; Pomés, A. Crystal structure of a dimerized cockroach allergen Bla g 2 complexed with a monoclonal antibody. *J. Biol. Chem.* **2008**, *283*, 22806–22814. [CrossRef] [PubMed]

87. Chruszcz, M.; Pomés, A.; Glesner, J.; Vailes, L.D.; Osinski, T.; Porebski, P.J.; Majorek, K.A.; Heymann, P.W.; Platts-Mills, T.A.; Minor, W.; et al. Molecular determinants for antibody binding on group 1 house dust mite allergens. *J. Biol. Chem.* **2012**, *287*, 7388–7398. [CrossRef] [PubMed]

88. Li, M.; Gustchina, A.; Glesner, J.; Wünschmann, S.; Vailes, L.D.; Chapman, M.D.; Pomés, A.; Wlodawer, A. Carbohydrates Contribute to the Interactions between Cockroach Allergen Bla g 2 and a Monoclonal Antibody. *J. Immunol.* **2011**, *186*, 333–340. [CrossRef] [PubMed]

89. Osinski, O.; Pomés, A.; Majorek, K.A.; Glesner, J.; Offermann, L.R.; Vailes, L.D.; Chapman, M.D.; Minor, W.; Chruszcz, M. Structural Analysis of Der p 1–Antibody Complexes and Comparison with Complexes of Proteins or Peptides with Monoclonal Antibodies. *J. Immunol.* **2015**, *195*, 307–316. [CrossRef]

90. Orengo, J.M.; Radin, A.R.; Kamat, V.; Badithe, A.; Ben, L.H.; Bennett, B.L.; Zhong, S.; Birchard, D.; Limnander, A.; Rafique, A.; et al. Treating cat allergy with monoclonal IgG antibodies that bind allergen and prevent IgE engagement. *Nat. Commun.* **2018**, *9*, 1421. [CrossRef]

91. Niemi, M.; Jylhä, S.; Laukkanen, M.-L.; Söderlund, H.; Mäkinen-Kiljunen, S.; Kallio, J.M.; Hakulinen, N.; Haahtela, T.; Takkinen, K.; Rouvinen, J. Molecular Interactions between a Recombinant IgE Antibody and the β-Lactoglobulin Allergen. *Structure* **2007**, *15*, 1413–1421. [CrossRef]

92. Padavattan, S.; Flicker, S.; Schirmer, T.; Madritsch, C.; Randow, S.; Reese, G.; Vieths, S.; Lupinek, C.; Ebner, C.; Valenta, R.; et al. High-affinity IgE recognition of a conformational epitope of the major respiratory allergen Phl p 2 as revealed by X-ray crystallography. *J. Immunol.* **2009**, *182*, 2141–2151. [CrossRef]

93. Glesner, J.; Wünschmann, S.; Li, M.; Gustchina, A.; Wlodawer, A.; Himly, M.; Chapman, M.D.; Pomés, A. Mechanisms of allergen-antibody interaction of cockroach allergen Bla g 2 with monoclonal antibodies that inhibit IgE antibody binding. *PLoS ONE* **2011**, *6*, e22223. [CrossRef]

94. Marone, G.; Rossi, F.W.; Detoraki, A.; Granata, F.; Marone, G.; Genovese, A.; Spadaro, G. Role of superallergens in allergic disorders. *Chem. Immunol. Allergy* **2007**, *93*, 195–213. [CrossRef]

95. Zacharia, B.E.; Sherman, P. Atopy, helminths, and cancer. *Med. Hypotheses* **2003**, *60*, 1–5. [CrossRef]

96. Finkelman, F.D.; Urban, J.F., Jr. The other side of the coin: The protective role of the TH2 cytokines. *J. Allergy Clin. Immunol.* **2001**, *107*, 772–780. [CrossRef]

97. Gurish, M.F.; Bryce, P.J.; Tao, H.; Kisselgof, A.B.; Thornton, E.M.; Miller, H.R.; Friend, D.S.; Oettgen, H.C. IgE enhances parasite clearance and regulates mast cell responses in mice infected with *Trichinella spiralis*. *J. Immunol.* **2004**, *172*, 1139–1145. [CrossRef]

98. Ure, D.M. Negative association between allergy and cancer. *Scott. Med. J.* **1969**, *14*, 51–54. [CrossRef]

99. Schlitter, H.E. Is there an allergy against malignant tumor tissue and what can it signify in regard to the defense of the body against cancer? *Strahlentherapie* **1961**, *114*, 203–204.

100. McCormick, D.P.; Ammann, A.J.; Ishizaka, K.; Miller, D.G.; Hong, R. A study of allergy in patients with malignant lymphoma and chronic lymphocytic leukemia. *Cancer* **1971**, *27*, 93–99. [CrossRef]

101. Augustin, R.; Chandradasa, K.D. IgE levels and allergic skin reactions in cancer and non-cancer patients. *Int. Arch. Allergy Appl. Immunol.* **1971**, *41*, 141–143. [CrossRef]

102. Jacobs, D.; Landon, J.; Houri, M.; Merrett, T.G. Circulating levels of immunoglobulin E in patients with cancer. *Lancet* **1972**, *2*, 1059–1061. [CrossRef]

103. Allegra, J.; Lipton, A.; Harvey, H.; Luderer, J.; Brenner, D.; Mortel, R.; Demers, L.; Gillin, M.; White, D.; Trautlein, J. Decreased prevalence of immediate hypersensitivity (atopy) in a cancer population. *Cancer Res.* **1976**, *36*, 3225–3226.

104. Neuchrist, C.; Kornfehl, J.; Grasl, M.; Lassmann, H.; Kraft, D.; Ehrenberger, K.; Scheiner, O. Distribution of immunoglobulins in squamous cell carcinoma of the head and neck. *Int. Arch. Allergy Immunol.* **1994**, *104*, 97–100. [CrossRef]

105. Fu, S.L.; Pierre, J.; Smith-Norowitz, T.A.; Hagler, M.; Bowne, W.; Pincus, M.R.; Mueller, C.M.; Zenilman, M.E.; Bluth, M.H. Immunoglobulin E antibodies from pancreatic cancer patients mediate antibody-dependent cell-mediated cytotoxicity against pancreatic cancer cells. *Clin. Exp. Immunol.* **2008**, *153*, 401–409. [CrossRef]

106. Crawford, G.; Hayes, M.D.; Seoane, R.C.; Ward, S.; Dalessandri, T.; Lai, C.; Healy, E.; Kipling, D.; Proby, C.; Moyes, C.; et al. Epithelial damage and tissue γδ T cells promote a unique tumor-protective IgE response. *Nat. Immunol.* **2018**, *19*, 859–870. [CrossRef]

107. Disney-Hogg, L.; Cornish, A.J.; Sud, A.; Law, P.J.; Kinnersley, B.; Jacobs, D.I.; Ostrom, Q.T.; Labreche, K.; Eckel-Passow, J.E.; Armstrong, G.N.; et al. Impact of atopy on risk of glioma: A Mendelian randomisation study. *BMC Med.* **2018**, *16*, 42. [CrossRef]

108. Helby, J.; Bojesen, S.E.; Nielsen, S.F.; Nordestgaard, B.G. IgE and risk of cancer in 37,747 individuals from the general population. *Ann. Oncol.* **2015**, *26*, 1784–1790. [CrossRef]

109. Liao, H.C.; Wu, S.Y.; Ou, C.Y.; Hsiao, J.R.; Huang, J.S.; Tsai, S.T.; Huang, C.C.; Wong, T.Y.; Lee, W.T.; Chen, K.C.; et al. Allergy symptoms, serum total immunoglobulin E, and risk of head and neck cancer. *Cancer Causes Control* **2016**, *27*, 1105–1115. [CrossRef]

110. Wulaningsih, W.; Holmberg, L.; Garmo, H.; Karagiannis, S.N.; Ahlstedt, S.; Malmstrom, H.; Lambe, M.; Hammar, N.; Walldius, G.; Jungner, I.; et al. Investigating the association between allergen-specific immunoglobulin E, cancer risk and survival. *Oncoimmunology* **2016**, *5*, e1154250. [CrossRef]

111. Taghizadeh, N.; Vonk, J.M.; Hospers, J.J.; Postma, D.S.; de Vries, E.G.; Schouten, J.P.; Boezen, H.M. Objective allergy markers and risk of cancer mortality and hospitalization in a large population-based cohort. *Cancer Causes Control* **2015**, *26*, 99–109. [CrossRef]

112. Van Hemelrijck, M.; Karagiannis, S.N.; Rohrmann, S. Atopy and prostate cancer: Is there a link between circulating levels of IgE and PSA in humans? *Cancer Immunol. Immunother.* **2017**, *66*, 1557–1562. [CrossRef]

113. Kural, Y.B.; Su, O.; Onsun, N.; Uras, A.R. Atopy, IgE and eosinophilic cationic protein concentration, specific IgE positivity, eosinophil count in cutaneous T Cell lymphoma. *Int. J. Dermatol.* **2010**, *49*, 390–395. [CrossRef]

114. Kretschmer, A.; Schwanbeck, R.; Valerius, T.; Rösner, T. Antibody Isotypes for Tumor Immunotherapy. *Transfus. Med. Hemother.* **2017**, *44*, 320–326. [CrossRef]

115. Leusen, J.H. IgA as therapeutic antibody. *Mol. Immunol.* **2015**, *68*, 35–39. [CrossRef]

116. Lohse, S.; Meyer, S.; Meulenbroek, L.A.; Jansen, J.H.; Nederend, M.; Kretschmer, A.; Klausz, K.; Möginger, U.; Derer, S.; Rösner, T.; et al. An Anti-EGFR IgA That Displays Improved Pharmacokinetics and Myeloid Effector Cell Engagement In Vivo. *Cancer Res.* **2016**, *76*, 403–417. [CrossRef]

117. Josephs, D.H.; Spicer, J.F.; Karagiannis, P.; Gould, H.J.; Karagiannis, S.N. IgE immunotherapy: A novel concept with promise for the treatment of cancer. *mAbs* **2014**, *6*, 54–72. [CrossRef]

118. Waldmann, T.A.; Iio, A.; Ogawa, M.; McIntyre, O.R.; Strober, W. The metabolism of IgE. Studies in normal individuals and in a patient with IgE myeloma. *J. Immunol.* **1976**, *117*, 1139–1144.

119. Lawrence, M.G.; Woodfolk, J.A.; Schuyler, A.J.; Stillman, L.C.; Chapman, M.D.; Platts-Mills, T.A. Half-life of IgE in serum and skin: Consequences for anti-IgE therapy in patients with allergic disease. *J. Allergy Clin. Immunol.* **2017**, *139*, 422–428. [CrossRef]

120. Verwaerde, C.; Joseph, M.; Capron, M.; Pierce, R.J.; Damonneville, M.; Velge, F.; Auriault, C.; Capron, A. Functional properties of a rat monoclonal IgE antibody specific for Schistosoma mansoni. *J. Immunol.* **1987**, *138*, 4441–4446.

121. Vouldoukis, I.; Riveros-Moreno, V.; Dugas, B.; Ouaaz, F.; Bécherel, P.; Debré, P.; Moncada, S.; Mossalayi, M.D. The killing of Leishmania major by human macrophages is mediated by nitric oxide induced after ligation of the Fc epsilon RII/CD23 surface antigen. *Proc. Natl. Acad. Sci. USA* **1995**, *92*, 7804–7808. [CrossRef]

122. Vouldoukis, I.; Mazier, D.; Moynet, D.; Thiolat, D.; Malvy, D.; Mossalayi, M.D. IgE mediates killing of intracellular *Toxoplasma gondii* by human macrophages through CD23-dependent, interleukin-10 sensitive pathway. *PLoS ONE* **2011**, *6*, e18289. [CrossRef]

123. Hagan, P.; Blumenthal, U.J.; Dunn, D.; Simpson, A.J.; Wilkins, H.A. Human IgE, IgG4 and resistance to reinfection with *Schistosoma haematobium*. *Nature* **1991**, *349*, 243–245. [CrossRef]

124. Dunne, D.W.; Butterworth, A.E.; Fulford, A.J.; Ouma, J.H.; Sturrock, R.F. Human IgE responses to Schistosoma mansoni and resistance to reinfection. *Mem. Inst. Oswaldo Cruz* **1992**, *87*, 99–103. [CrossRef]

125. Watanabe, N.; Bruschi, F.; Korenaga, M. IgE: A question of protective immunity in *Trichinella spiralis* infection. *Trends Parasitol.* **2005**, *21*, 175–178. [CrossRef]

126. Gounni, A.S.; Lamkhioued, B.; Ochiai, K.; Tanaka, Y.; Delaporte, E.; Capron, A.; Kinet, J.P.; Capron, M. High-affinity IgE receptor on eosinophils is involved in defence against parasites. *Nature* **1994**, *367*, 183–186. [CrossRef]

127. Kamisawa, T.; Zen, Y.; Pillai, S.; Stone, J.H. IgG4-related disease. *Lancet* **2015**, *385*, 1460–1471. [CrossRef]

128. Weindorf, S.C.; Frederiksen, J.K. IgG4-Related Disease: A Reminder for Practicing Pathologists. *Arch. Pathol. Lab. Med.* **2017**, *141*, 1476–1483. [CrossRef]

129. Wallace, Z.S.; Mattoo, H.; Carruthers, M.; Mahajan, V.S.; Della Torre, E.; Lee, H.; Kulikova, M.; Deshpande, V.; Pillai, S.; Stone, J.H. Plasmablasts as a biomarker for IgG4-related disease, independent of serum IgG4 concentrations. *Ann. Rheum. Dis.* **2015**, *74*, 190–195. [CrossRef]

130. Crescioli, S.; Correa, I.; Karagiannis, P.; Davies, A.M.; Sutton, B.J.; Nestle, F.O.; Karagiannis, S.N. IgG4 Characteristics and Functions in Cancer Immunity. *Curr. Allergy Asthma Rep.* **2016**, *16*, 7. [CrossRef]

131. Liu, Q.; Niu, Z.; Li, Y.; Wang, M.; Pan, B.; Lu, Z.; Liao, Q.; Zhao, Y. Immunoglobulin G4 (IgG4)-positive plasma cell infiltration is associated with the clinicopathologic traits and prognosis of pancreatic cancer after curative resection. *Cancer Immunol. Immunother.* **2016**, *65*, 931–940. [CrossRef]

132. Harada, K.; Shimoda, S.; Kimura, Y.; Sato, Y.; Ikeda, H.; Igarashi, S.; Ren, X.; Sato, H.; Nakanuma, Y. Significance of immunoglobulin G4 (IgG4)-positive cells in extrahepatic cholangiocarcinoma: Molecular mechanism of IgG4 reaction in cancer tissue. *Hepatology* **2012**, *56*, 157–164. [CrossRef]

133. Fujimoto, M.; Yoshizawa, A.; Sumiyoshi, S.; Sonobe, M.; Kobayashi, M.; Koyanagi, I.; Aini, W.; Tsuruyama, T.; Date, H.; Haga, H. Stromal plasma cells expressing immunoglobulin G4 subclass in non-small cell lung cancer. *Hum. Pathol.* **2013**, *44*, 1569–1576. [CrossRef]

134. Karagiannis, P.; Gilbert, A.E.; Josephs, D.H.; Ali, N.; Dodev, T.; Saul, L.; Correa, I.; Roberts, L.; Beddowes, E.; Koers, A.; et al. IgG4 subclass antibodies impair antitumor immunity in melanoma. *J. Clin. Investig.* **2013**, *123*, 1457–1474. [CrossRef]

135. Karagiannis, P.; Villanova, F.; Josephs, D.H.; Correa, I.; Van Hemelrijck, M.; Hobbs, C.; Saul, L.; Egbuniwe, I.U.; Tosi, I.; Ilieva, K.M.; et al. Elevated IgG4 in patient circulation is associated with the risk of disease progression in melanoma. *Oncoimmunology* **2015**, *4*, e1032492. [CrossRef]

136. Karagiannis, P.; Gilbert, A.E.; Nestle, F.O.; Karagiannis, S.N. IgG4 antibodies and cancer-associated inflammation: Insights into a novel mechanism of immune escape. *Oncoimmunology* **2013**, *2*, e24889. [CrossRef]

137. Jensen-Jarolim, E.; Bax, H.J.; Bianchini, R.; Capron, M.; Corrigan, C.; Castells, M.; Dombrowicz, D.; Daniels-Wells, T.R.; Fazekas, J.; Fiebiger, E.; et al. AllergoOncology—The impact of allergy in oncology: EAACI position paper. *Allergy* **2017**, *72*, 866–887. [CrossRef]

138. Platzer, B.; Elpek, K.G.; Cremasco, V.; Baker, K.; Stout, M.M.; Schultz, C.; Dehlink, E.; Shade, K.T.; Anthony, R.M.; Blumberg, R.S.; et al. IgE/FcεRI-Mediated Antigen Cross-Presentation by Dendritic Cells Enhances Anti-Tumor Immune Responses. *Cell Rep.* **2015**, *10*, 1487–1495. [CrossRef]

139. Platzer, B.; Dehlink, E.; Turley, S.J.; Fiebiger, E. How to connect an IgE-driven response with CTL activity? *Cancer Immunol. Immunother.* **2012**, *61*, 1521–1525. [CrossRef]

140. Kamta, J.; Chaar, M.; Ande, A.; Altomare, D.A.; Ait-Oudhia, S. Advancing Cancer Therapy with Present and Emerging Immuno-Oncology Approaches. *Front. Oncol.* **2017**, *7*, 64. [CrossRef]

141. Jensen-Jarolim, E.; Turner, M.C.; Karagiannis, S.N. AllergoOncology: IgE- and IgG4-mediated immune mechanisms linking allergy with cancer and their translational implications. *J. Allergy Clin. Immunol.* **2017**, *140*, 982–984. [CrossRef]

142. Jensen-Jarolim, E.; Bax, H.J.; Bianchini, R.; Crescioli, S.; Daniels-Wells, T.R.; Dombrowicz, D.; Fiebiger, E.; Gould, H.J.; Irshad, S.; Janda, J.; et al. AllergoOncology: Opposite outcomes of immune tolerance in allergy and cancer. *Allergy* **2018**, *73*, 328–340. [CrossRef]

143. Jensen-Jarolim, E.; Achatz, G.; Turner, M.C.; Karagiannis, S.; Legrand, F.; Capron, M.; Penichet, M.L.; Rodríguez, J.A.; Siccardi, A.G.; Vangelista, L.; et al. AllergoOncology: The role of IgE-mediated allergy in cancer. *Allergy* **2008**, *63*, 1255–1266. [CrossRef]

144. Karagiannis, S.N.; Josephs, D.H.; Karagiannis, P.; Gilbert, A.E.; Saul, L.; Rudman, S.M.; Dodev, T.; Koers, A.; Blower, P.J.; Corrigan, C.; et al. Recombinant IgE antibodies for passive immunotherapy of solid tumours: From concept towards clinical application. *Cancer Immunol. Immunother.* **2012**, *61*, 1547–1564. [CrossRef]

145. Weiner, G.J. Building better monoclonal antibody-based therapeutics. *Nat. Rev. Cancer* **2015**, *15*, 361–370. [CrossRef]

146. Kunert, R.; Reinhart, D. Advances in recombinant antibody manufacturing. *Appl. Microbiol. Biotechnol.* **2016**, *100*, 3451–3461. [CrossRef]

147. Gould, H.J.; Mackay, G.A.; Karagiannis, S.N.; O'Toole, C.M.; Marsh, P.J.; Daniel, B.E.; Coney, L.R.; Zurawski, V.R., Jr.; Joseph, M.; Capron, M.; et al. Comparison of IgE and IgG antibody-dependent cytotoxicity in vitro and in a SCID mouse xenograft model of ovarian carcinoma. *Eur. J. Immunol.* **1999**, *29*, 3527–3537. [CrossRef]

148. Dodev, T.S.; Karagiannis, P.; Gilbert, A.E.; Josephs, D.H.; Bowen, H.; James, L.K.; Bax, H.J.; Beavil, R.; Pang, M.O.; Gould, H.J.; et al. A tool kit for rapid cloning and expression of recombinant antibodies. *Sci. Rep.* **2014**, *4*, 5885. [CrossRef]

149. Bantleon, F.; Wolf, S.; Seismann, H.; Dam, S.; Lorentzen, A.; Miehe, M.; Jabs, F.; Jakob, T.; Plum, M.; Spillner, E. Human IgE is efficiently produced in glycosylated and biologically active form in lepidopteran cells. *Mol. Immunol.* **2016**, *72*, 49–56. [CrossRef]

150. Montero-Morales, L.; Maresch, D.; Castilho, A.; Turupcu, A.; Ilieva, K.M.; Crescioli, S.; Karagiannis, S.N.; Lupinek, C.; Oostenbrink, C.; Altmann, F.; et al. Recombinant plant-derived human IgE glycoproteomics. *J. Proteom.* **2017**, *161*, 81–87. [CrossRef]

151. Ilieva, K.M.; Fazekas-Singer, J.; Achkova, D.Y.; Dodev, T.S.; Mele, S.; Crescioli, S.; Bax, H.J.; Cheung, A.; Karagiannis, P.; Correa, I.; et al. Functionally Active Fc Mutant Antibodies Recognizing Cancer Antigens Generated Rapidly at High Yields. *Front. Immunol.* **2017**, *8*, 1112. [CrossRef]

152. Fazekas-Singer, J.; Singer, J.; Ilieva, K.M.; Matz, M.; Herrmann, I.; Spillner, E.; Karagiannis, S.N.; Jensen-Jarolim, E. AllergoOncology: Generating a canine anticancer IgE against the epidermal growth factor receptor. *J. Allergy Clin. Immunol.* **2018**, *142*, 973–976. [CrossRef] [PubMed]

153. Crescioli, S.; Chiaruttini, G.; Mele, S.; Ilieva, K.M.; Pellizzari, G.; Spencer, D.I.R.; Gardner, R.A.; Lacy, K.E.; Spicer, J.F.; Tutt, A.N.J.; et al. Engineering and stable production of recombinant IgE for cancer immunotherapy and AllergoOncology. *J. Allergy Clin. Immunol.* **2018**, *141*, 1519–1523. [CrossRef] [PubMed]

154. Boscolo, S.; Mion, F.; Licciulli, M.; Macor, P.; De Maso, L.; Brce, M.; Antoniou, M.N.; Marzari, R.; Santoro, C.; Sblattero, D. Simple scale-up of recombinant antibody production using an UCOE containing vector. *New Biotechnol.* **2012**, *29*, 477–484. [CrossRef] [PubMed]

155. Lu, C.S.; Hung, A.F.; Lin, C.J.; Chen, J.B.; Chen, C.; Shiung, Y.Y.; Tsai, C.Y.; Chang, T.W. Generating allergen-specific human IgEs for immunoassays by employing human ε gene knockin mice. *Allergy* **2015**, *70*, 384–390. [CrossRef] [PubMed]

156. Hecker, J.; Diethers, A.; Schulz, D.; Sabri, A.; Plum, M.; Michel, Y.; Mempel, M.; Ollert, M.; Jakob, T.; Blank, S.; et al. An IgE epitope of Bet v 1 and fagales PR10 proteins as defined by a human monoclonal IgE. *Allergy* **2012**, *67*, 1530–1537. [CrossRef]

157. Correa, I.; Ilieva, K.M.; Crescioli, S.; Lombardi, S.; Figini, M.; Cheung, A.; Spicer, J.F.; Tutt, A.N.J.; Nestle, F.O.; Karagiannis, P.; et al. Evaluation of Antigen-Conjugated Fluorescent Beads to Identify Antigen-Specific B Cells. *Front. Immunol.* **2018**, *9*, 493. [CrossRef]

158. Nagy, E.; Berczi, I.; Sehon, A.H. Growth inhibition of murine mammary carcinoma by monoclonal IgE antibodies specific for the mammary tumor virus. *Cancer Immunol. Immunother.* **1991**, *34*, 63–69. [CrossRef]

159. Kershaw, M.H.; Darcy, P.K.; Trapani, J.A.; MacGregor, D.; Smyth, M.J. Tumor-specific IgE-mediated inhibition of human colorectal carcinoma xenograft growth. *Oncol. Res.* **1998**, *10*, 133–142.

160. Daniels, T.R.; Leuchter, R.K.; Quintero, R.; Helguera, G.; Rodríguez, J.A.; Martínez-Maza, O.; Schultes, B.C.; Nicodemus, C.F.; Penichet, M.L. Targeting HER2/neu with a fully human IgE to harness the allergic reaction against cancer cells. *Cancer Immunol. Immunother.* **2012**, *61*, 991–1003. [CrossRef]

161. Teo, P.Z.; Utz, P.J.; Mollick, J.A. Using the allergic immune system to target cancer: Activity of IgE antibodies specific for human CD20 and MUC1. *Cancer Immunol. Immunother.* **2012**, *61*, 2295–2309. [CrossRef]

162. Daniels-Wells, T.R.; Helguera, G.; Leuchter, R.K.; Quintero, R.; Kozman, M.; Rodríguez, J.A.; Ortiz-Sánchez, E.; Martínez-Maza, O.; Schultes, B.C.; Nicodemus, C.F.; et al. A novel IgE antibody targeting the prostate specific antigen as a potential prostate cancer therapy. *BMC Cancer* **2013**, *13*, 195. [CrossRef]

163. Karagiannis, P.; Singer, J.; Hunt, J.; Gan, S.K.; Rudman, S.M.; Mechtcheriakova, D.; Knittelfelder, R.; Daniels, T.R.; Hobson, P.S.; Beavil, A.J.; et al. Characterisation of an engineered trastuzumab IgE antibody and effector cell mechanisms targeting HER2/neu-positive tumour cells. *Cancer Immunol. Immunother.* **2009**, *58*, 915–930. [CrossRef]

164. Spillner, E.; Plum, M.; Blank, S.; Miehe, M.; Singer, J.; Braren, I. Recombinant IgE antibody engineering to target EGFR. *Cancer Immunol. Immunother.* **2012**, *61*, 1565–1573. [CrossRef]

165. Chung, C.H.; Mirakhur, B.; Chan, E.; Le, Q.T.; Berlin, J.; Morse, M.; Murphy, B.A.; Satinover, S.M.; Hosen, J.; Mauro, D.; et al. Cetuximab-induced anaphylaxis and IgE specific for galactose-alpha-1,3-galactose. *N. Engl. J. Med.* **2008**, *358*, 1109–1117. [CrossRef]

166. Lammerts van Bueren, J.J.; Rispens, T.; Verploegen, S.; van der Palen-Merkus, T.; Stapel, S.; Workman, L.J.; James, H.; van Berkel, P.H.; van de Winkel, J.G.; Platts-Mills, T.A.; et al. Anti-galactose-α-1,3-galactose IgE from allergic patients does not bind α-galactosylated glycans on intact therapeutic antibody Fc domains. *Nat. Biotechnol.* **2011**, *29*, 574–576. [CrossRef]

167. Galili, U. Anti-Gal: An abundant human natural antibody of multiple pathogeneses and clinical benefits. *Immunology* **2013**, *140*, 1–11. [CrossRef]

168. Miotti, S.; Canevari, S.; Ménard, S.; Mezzanzanica, D.; Porro, G.; Pupa, S.M.; Regazzoni, M.; Tagliabue, E.; Colnaghi, M. Characterization of human ovarian carcinoma-associated antigens defined by novel monoclonal antibodies with tumor-restricted specificity. *Int. J. Cancer* **1987**, *39*, 297–303. [CrossRef]

169. Coney, L.R.; Tomassetti, A.; Carayannopoulos, L.; Frasca, V.; Kamen, B.A.; Colnaghi, M.; Zurawski, V.R., Jr. Cloning of a tumor-associated antigen: MOv18 and MOv19 antibodies recognize a folate-binding protein. *Cancer Res.* **1991**, *51*, 6125–6132.

170. Molthoff, C.F.; Prinssen, H.M.; Kenemans, P.; van Hof, A.C.; den Hollander, W.; Verheijen, R.H. Escalating protein doses of chimeric monoclonal antibody MOv18 immunoglobulin G in ovarian carcinoma patients: A phase I study. *Cancer* **1997**, *80*, 2712–2720. [CrossRef]

171. Buijs, W.C.; Tibben, J.G.; Boerman, O.C.; Molthoff, C.F.; Massuger, L.F.; Koenders, E.B.; Schijf, C.P.; Siegel, J.A.; Corstens, F.H. Dosimetric analysis of chimeric monoclonal antibody cMOv18 IgG in ovarian carcinoma patients after intraperitoneal and intravenous administration. *Eur. J. Nucl. Med.* **1998**, *25*, 1552–1561. [CrossRef]

172. Van Zanten-Przybysz, I.; Molthoff, C.; Gebbinck, J.K.; von Mensdorff-Pouilly, S.; Verstraeten, R.; Kenemans, P.; Verheijen, R. Cellular and humoral responses after multiple injections of unconjugated chimeric monoclonal antibody MOv18 in ovarian cancer patients: A pilot study. *J. Cancer Res. Clin. Oncol.* **2002**, *128*, 484–492. [CrossRef]

173. Van Zanten-Przybysz, I.; Molthoff, C.F.; Roos, J.C.; Verheijen, R.H.; van Hof, A.; Buist, M.R.; Prinssen, H.M.; den Hollander, W.; Kenemans, P. Influence of the route of administration on targeting of ovarian cancer with the chimeric monoclonal antibody MOv18: i.v. vs. i.p. *Int. J. Cancer* **2001**, *92*, 106–114. [CrossRef]

174. Bell-McGuinn, K.M.; Konner, J.; Pandit-Taskar, N.; Gerst, S.; Nicolaides, N.; Sass, P.; Grasso, L.; Weil, S.; Phillips, M.; Aghajanian, C. A phase I study of MORAb-003, a fully humanized monoclonal antibody against folate receptor alpha, in advanced epithelial ovarian cancer. *J. Clin. Oncol.* **2007**, *25*, 5553. [CrossRef]

175. Konner, J.A.; Bell-McGuinn, K.M.; Sabbatini, P.; Hensley, M.L.; Tew, W.P.; Pandit-Taskar, N.; Vander, E.N.; Phillips, M.D.; Schweizer, C.; Weil, S.C.; et al. Farletuzumab, a humanized monoclonal antibody against folate receptor alpha, in epithelial ovarian cancer: A phase I study. *Clin. Cancer Res.* **2010**, *16*, 5288–5295. [CrossRef]

176. Farrell, C.; Schweizer, C.; Wustner, J.; Weil, S.; Namiki, M.; Nakano, T.; Nakai, K.; Phillips, M.D. Population pharmacokinetics of farletuzumab, a humanized monoclonal antibody against folate receptor alpha, in epithelial ovarian cancer. *Cancer Chemother. Pharmacol.* **2012**, *70*, 727–734. [CrossRef]

177. Cheung, A.; Opzoomer, J.; Ilieva, K.M.; Gazinska, P.; Hoffmann, R.M.; Mirza, H.; Marlow, R.; Francesch-Domenech, E.; Fittall, M.; Dominguez Rodriguez, D.; et al. Anti-Folate Receptor Alpha-Directed Antibody Therapies Restrict the Growth of Triple-negative Breast Cancer. *Clin. Cancer Res.* **2018**, *24*, 5098–5111. [CrossRef]

178. Tochowicz, A.; Dalziel, S.; Eidam, O.; O'Connell, J.D., 3rd; Griner, S.; Finer-Moore, J.S.; Stroud, R.M. Development and binding mode assessment of N-[4-[2-propyn-1-yl[(6S)-4,6,7,8-tetrahydro-2-(hydroxymethyl)-4-oxo-3H-cyclopenta[g]quinazolin-6-yl] amino]benzoyl]-l-γ-glutamyl-D-glutamic acid (BGC 945), a novel thymidylate synthase inhibitor that targets tumor cells. *J. Med. Chem.* **2013**, *56*, 5446–5455. [CrossRef]

179. Karagiannis, S.N.; Wang, Q.; East, N.; Burke, F.; Riffard, S.; Bracher, M.G.; Thompson, R.G.; Durham, S.R.; Schwartz, L.B.; Balkwill, F.R.; et al. Activity of human monocytes in IgE antibody-dependent surveillance and killing of ovarian tumor cells. *Eur. J. Immunol.* **2003**, *33*, 1030–1040. [CrossRef]

180. Karagiannis, S.N.; Bracher, M.G.; Beavil, R.L.; Beavil, A.J.; Hunt, J.; McCloskey, N.; Thompson, R.G.; East, N.; Burke, F.; Sutton, B.J.; et al. Role of IgE receptors in IgE antibody-dependent cytotoxicity and phagocytosis of ovarian tumor cells by human monocytic cells. *Cancer Immunol. Immunother.* **2008**, *57*, 247–263. [CrossRef]

181. Karagiannis, S.N.; Bracher, M.G.; Hunt, J.; McCloskey, N.; Beavil, R.L.; Beavil, A.J.; Fear, D.J.; Thompson, R.G.; East, N.; Burke, F.; et al. IgE-antibody-dependent immunotherapy of solid tumors: Cytotoxic and phagocytic mechanisms of eradication of ovarian cancer cells. *J. Immunol.* **2007**, *179*, 2832–2843. [CrossRef]

182. Rudman, S.M.; Josephs, D.H.; Cambrook, H.; Karagiannis, P.; Gilbert, A.E.; Dodev, T.; Hunt, J.; Koers, A.; Montes, A.; Taams, L.; et al. Harnessing engineered antibodies of the IgE class to combat malignancy: Initial assessment of FcεRI-mediated basophil activation by a tumour-specific IgE antibody to evaluate the risk of type I hypersensitivity. *Clin. Exp. Allergy* **2011**, *41*, 1400–1413. [CrossRef]

183. Kayaba, H.; Dombrowicz, D.; Woerly, G.; Papin, J.P.; Loiseau, S.; Capron, M. Human eosinophils and human high affinity IgE receptor transgenic mouse eosinophils express low levels of high affinity IgE receptor, but release IL-10 upon receptor activation. *J. Immunol.* **2001**, *167*, 995–1003. [CrossRef]

184. Muraki, M.; Gleich, G.J.; Kita, H. Antigen-specific IgG and IgA, but not IgE, activate the effector functions of eosinophils in the presence of antigen. *Int. Arch. Allergy Immunol.* **2011**, *154*, 119–127. [CrossRef]

185. Marquet, R.L.; Westbroek, D.L.; Jeekel, J. Interferon treatment of a transplantable rat colon adenocarcinoma: Importance of tumor site. *Int. J. Cancer* **1984**, *33*, 689–692. [CrossRef]

186. Josephs, D.H.; Bax, H.J.; Dodev, T.; Georgouli, M.; Nakamura, M.; Pellizzari, G.; Saul, L.; Karagiannis, P.; Cheung, A.; Herraiz, C.; et al. Anti-Folate Receptor-α IgE but not IgG Recruits Macrophages to Attack Tumors via TNFα/MCP-1 Signaling. *Cancer Res.* **2017**, *77*, 1127–1141. [CrossRef]

187. Bracher, M.; Gould, H.J.; Sutton, B.J.; Dombrowicz, D.; Karagiannis, S.N. Three-colour flow cytometric method to measure antibody-dependent tumour cell killing by cytotoxicity and phagocytosis. *J. Immunol. Methods* **2007**, *323*, 160–171. [CrossRef]

188. Murray, P.J.; Wynn, T.A. Protective and pathogenic functions of macrophage subsets. *Nat. Rev. Immunol.* **2011**, *11*, 723–737. [CrossRef]

189. Ruffell, B.; Affara, N.I.; Coussens, L.M. Differential macrophage programming in the tumor microenvironment. *Trends Immunol.* **2012**, *33*, 119–126. [CrossRef]

190. Mantovani, A.; Biswas, S.K.; Galdiero, M.R.; Sica, A.; Locati, M. Macrophage plasticity and polarization in tissue repair and remodelling. *J. Pathol.* **2013**, *229*, 176–185. [CrossRef]

191. Josephs, D.H.; Nakamura, M.; Bax, H.J.; Dodev, T.S.; Muirhead, G.; Saul, L.; Karagiannis, P.; Ilieva, K.M.; Crescioli, S.; Gazinska, P.; et al. An immunologically relevant rodent model demonstrates safety of therapy using a tumour-specific IgE. *Allergy* **2018**, *73*, 2328–2341. [CrossRef]

192. Kraft, S.; Novak, N.; Katoh, N.; Bieber, T.; Rupec, R.A. Aggregation of the high-affinity IgE receptor FcεRI on human monocytes and dendritic cells induces NF-κB activation. *J. Investig. Dermatol.* **2002**, *118*, 830–837. [CrossRef]

193. Karagiannis, S.N.; Josephs, D.H.; Bax, H.J.; Spicer, J.F. Therapeutic IgE Antibodies: Harnessing a Macrophage-Mediated Immune Surveillance Mechanism against Cancer. *Cancer Res.* **2017**, *77*, 2779–2783. [CrossRef]

194. Ishizaka, T.; Ishizaka, K.; Tomioka, H. Release of histamine and slow reacting substance of anaphylaxis (SRS-A) by IgE-anti-IgE reactions on monkey mast cells. *J. Immunol.* **1972**, *108*, 513–520.

195. Schwartz, L.B. Effector cells of anaphylaxis: Mast cells and basophils. *Novartis Found. Symp.* **2004**, *257*, 65–79.

196. Dombrowicz, D.; Brini, A.T.; Flamand, V.; Hicks, E.; Snouwaert, J.N.; Kinet, J.P.; Koller, B.H. Anaphylaxis mediated through a humanized high affinity IgE receptor. *J. Immunol.* **1996**, *157*, 1645–1651.

197. Collins, A.M.; Basil, M.; Nguyen, K.; Thelian, D. Rat basophil leukaemia (RBL) cells sensitized with low affinity IgE respond to high valency antigen. *Clin. Exp. Allergy* **1996**, *26*, 964–970. [CrossRef]

198. Jensen-Jarolim, E.; Singer, J. Why could passive Immunoglobulin E antibody therapy be safe in clinical oncology? *Clin. Exp. Allergy* **2011**, *41*, 1337–1340. [CrossRef]

199. Basal, E.; Eghbali-Fatourechi, G.Z.; Kalli, K.R.; Hartmann, L.C.; Goodman, K.M.; Goode, E.L.; Kamen, B.A.; Low, P.S.; Knutson, K.L. Functional folate receptor alpha is elevated in the blood of ovarian cancer patients. *PLoS ONE* **2009**, *4*, e6292. [CrossRef]

200. Hoffmann, H.J.; Santos, A.F.; Mayorga, C.; Nopp, A.; Eberlein, B.; Ferrer, M.; Rouzaire, P.; Ebo, D.G.; Sabato, V.; Sanz, M.L.; et al. The clinical utility of basophil activation testing in diagnosis and monitoring of allergic disease. *Allergy* **2015**, *70*, 1393–1405. [CrossRef]

201. Marraccini, P.; Pignatti, P.; D'Alcamo, A.; Salimbeni, R.; Consonni, D. Basophil Activation Test Application in Drug Hypersensitivity Diagnosis: An Empirical Approach. *Int. Arch. Allergy Immunol.* **2018**, *177*, 160–166. [CrossRef]

202. Seremet, T.; Haccuria, A.; Lienard, D.; Del Marmol, V.; Neyns, B. Anaphylaxis-like reaction to anti-BRAF inhibitor dabrafenib confirmed by drug provocation test. *Melanoma Res.* **2019**, *29*, 95–98. [CrossRef]

203. Ornelas, C.; Caiado, J.; Campos Melo, A.; Pereira Barbosa, M.; Castells, M.C.; Pereira Dos Santos, M.C. The Contribution of the Basophil Activation Test to the Diagnosis of Hypersensitivity Reactions to Oxaliplatin. *Int. Arch. Allergy Immunol.* **2018**, *177*, 274–280. [CrossRef]

204. De Week, A.L.; Sanz, M.L.; Gamboa, P.M.; Aberer, W.; Bienvenu, J.; Blanca, M.; Demoly, P.; Ebo, D.G.; Mayorga, L.; Monneret, G.; et al. Diagnostic tests based on human basophils: More potentials and perspectives than pitfalls. II. Technical issues. *J. Investig. Allergol. Clin. Immunol.* **2008**, *18*, 143–155.

205. Saul, L.; Josephs, D.H.; Cutler, K.; Bradwell, A.; Karagiannis, P.; Selkirk, C.; Gould, H.J.; Jones, P.; Spicer, J.F.; Karagiannis, S.N. Comparative reactivity of human IgE to cynomolgus monkey and human effector cells and effects on IgE effector cell potency. *mAbs* **2014**, *6*, 509–522. [CrossRef]

206. Riemer, A.B.; Untersmayr, E.; Knittelfelder, R.; Duschl, A.; Pehamberger, H.; Zielinski, C.C.; Scheiner, O.; Jensen-Jarolim, E. Active induction of tumor-specific IgE antibodies by oral mimotope vaccination. *Cancer Res.* **2007**, *67*, 3406–3411. [CrossRef]

207. Herrmann, I.; Gotovina, J.; Fazekas-Singer, J.; Fischer, M.B.; Hufnagl, K.; Bianchini, R.; Jensen-Jarolim, E. Canine macrophages can like human macrophages be in vitro activated toward the M2a subtype relevant in allergy. *Dev. Comp. Immunol.* **2018**, *82*, 118–127. [CrossRef]

208. Singer, J.; Jensen-Jarolim, E. IgE-based Immunotherapy of Cancer -A Comparative Oncology Approach. *J. Carcinog. Mutagen.* **2014**, *5*, 1000176. [CrossRef]

209. Carvalho, M.I.; Silva-Carvalho, R.; Pires, I.; Prada, J.; Bianchini, R.; Jensen-Jarolim, E.; Queiroga, F.L. A Comparative Approach of Tumor-Associated Inflammation in Mammary Cancer between Humans and Dogs. *BioMed Res. Int.* **2016**, *2016*, 4917387. [CrossRef]

210. Santos, R.B.; Galvão, V.R. Monoclonal Antibodies Hypersensitivity: Prevalence and Management. *Immunol. Allergy Clin. N. Am.* **2017**, *37*, 695–711. [CrossRef]

211. Cheifetz, A.; Smedley, M.; Martin, S.; Reiter, M.; Leone, G.; Mayer, L.; Plevy, S. The incidence and management of infusion reactions to infliximab: A large center experience. *Am. J. Gastroenterol.* **2003**, *98*, 1315–1324. [CrossRef]

212. Chen, X.; Churchill, M.J.; Nagar, K.K.; Tailor, Y.H.; Chu, T.; Rush, B.S.; Jiang, Z.; Wang, E.B.; Renz, B.W.; Wang, H.; et al. IL-17 producing mast cells promote the expansion of myeloid-derived suppressor cells in a mouse allergy model of colorectal cancer. *Oncotarget* **2015**, *6*, 32966–32979. [CrossRef]

213. Welsh, T.J.; Green, R.H.; Richardson, D.; Waller, D.A.; O'Byrne, K.J.; Bradding, P. Macrophage and mast-cell invasion of tumor cell islets confers a marked survival advantage in non-small-cell lung cancer. *J. Clin. Oncol.* **2005**, *23*, 8959–8967. [CrossRef]

214. Nakae, S.; Suto, H.; Iikura, M.; Kakurai, M.; Sedgwick, J.D.; Tsai, M.; Galli, S.J. Mast cells enhance T cell activation: Importance of mast cell costimulatory molecules and secreted TNF. *J. Immunol.* **2006**, *176*, 2238–2248. [CrossRef]

215. Brown, C.E.; Vishwanath, R.P.; Aguilar, B.; Starr, R.; Najbauer, J.; Aboody, K.S.; Jensen, M.C. Tumor-derived chemokine MCP-1/CCL2 is sufficient for mediating tumor tropism of adoptively transferred T cells. *J. Immunol.* **2007**, *179*, 3332–3341. [CrossRef]

antibodies

MDPI

Review

Structure, Function, and Therapeutic Use of IgM Antibodies

Bruce A. Keyt *, Ramesh Baliga, Angus M. Sinclair, Stephen F. Carroll and Marvin S. Peterson

IGM Biosciences Inc, 325 East Middlefield Road, Mountain View, CA 94043, USA; ramesh@igmbio.com (R.B.); asinclair@igmbio.com (A.M.S.); steve@igmbio.com (S.F.C.); mpeterson@igmbio.com (M.S.P.)
* Correspondence: bkeyt@igmbio.com; Tel.: +1-650-265-6458

Received: 16 September 2020; Accepted: 9 October 2020; Published: 13 October 2020

Abstract: Natural immunoglobulin M (IgM) antibodies are pentameric or hexameric macro-immunoglobulins and have been highly conserved during evolution. IgMs are initially expressed during B cell ontogeny and are the first antibodies secreted following exposure to foreign antigens. The IgM multimer has either 10 (pentamer) or 12 (hexamer) antigen binding domains consisting of paired μ heavy chains with four constant domains, each with a single variable domain, paired with a corresponding light chain. Although the antigen binding affinities of natural IgM antibodies are typically lower than IgG, their polyvalency allows for high avidity binding and efficient engagement of complement to induce complement-dependent cell lysis. The high avidity of IgM antibodies renders them particularly efficient at binding antigens present at low levels, and non-protein antigens, for example, carbohydrates or lipids present on microbial surfaces. Pentameric IgM antibodies also contain a joining (J) chain that stabilizes the pentameric structure and enables binding to several receptors. One such receptor, the polymeric immunoglobulin receptor (pIgR), is responsible for transcytosis from the vasculature to the mucosal surfaces of the lung and gastrointestinal tract. Several naturally occurring IgM antibodies have been explored as therapeutics in clinical trials, and a new class of molecules, engineered IgM antibodies with enhanced binding and/or additional functional properties are being evaluated in humans. Here, we review the considerable progress that has been made regarding the understanding of biology, structure, function, manufacturing, and therapeutic potential of IgM antibodies since their discovery more than 80 years ago.

Keywords: IgM (immunoglobulin M); hexameric; pentameric; polymeric; polyvalency; joining chain (J-chain); avidity; complement dependent cytotoxicity (CDC); poly Ig receptor (pIgR)

---

1. Introduction to Immunoglobulin M (IgM)

During humoral immune responses, immunoglobulins of the IgM, IgD, IgG, IgA, and IgE isotypes may be produced, each expressing a unique profile of effector functions capable of mediating host defense against invading pathogens. Macro-immunoglobulin, IgM, is initially produced as a surface bound molecule and is expressed in early B cell differentiation. Later in the immune response, IgM is produced by plasma cells and secreted as soluble pentamers that contain 10 antigen binding sites and the joining (J) chain, or as hexamers containing 12 antigen binding sites and no joining chain (J-chain). IgM has a molecular weight of approximately 900 or 1050 kDa for the pentamer or hexamer, respectively (Figure 1).

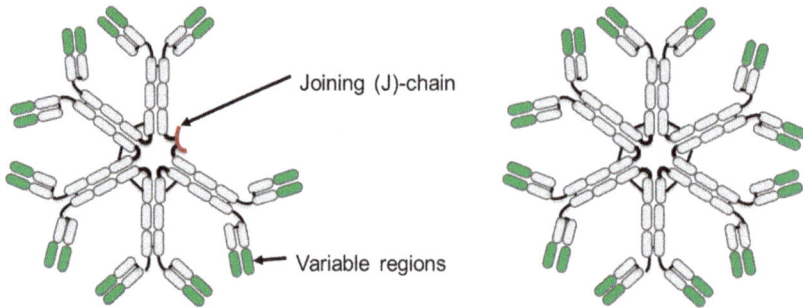

Figure 1. Schematic diagram of an immunoglobulin M (IgM) antibody pentamer (**left**) and hexamer (**right**). Constant regions are shown in gray and variable regions in green, and also shown on the IgM pentamer is the small joining chain (J-chain) in red.

Due to the polyvalent nature of IgMs, they may exhibit higher avidity for antigen than the bivalent IgG. In addition to neutralizing pathogens, IgM antibodies are highly effective at engaging complement to target lysis of cells and pathogens.

Our understanding of the biology, structure, and function relationships for IgM antibodies has progressed to the point where this antibody isotype can be exploited therapeutically; however, challenges associated with their manufacture remain. Here, we review the progress and the therapeutic potential for this class of antibodies, as well as the potential for new classes of engineered IgM antibodies.

1.1. History and Discovery of IgM

Humoral immunity has been studied since the late 1800s when George Nuttall [1] discovered that animal immune sera could kill bacteria. Subsequent analysis of the immune serum using technologies such as electrophoresis and ultracentrifugation allowed for biochemical characterization of the various proteins that could mediate immunity, resulting in the discovery of immunoglobulins. Originally, these serum components were assigned as α-globulin, β-globulin, and γ-globulin fractions to designate the proteins by order of electrophoretic mobility [2]. The first description of IgM antibodies was reported in 1939 by Kabat et al. [3] who evaluated the molecular weight of antibodies produced in horse, cow, pig, monkey, and human serum after immunization with pneumococcus. Due to the large size (approximately 990 kDa), the new antibody was referred to as γ-macroglobulin. In 1944, γ-macroglobulins were also discovered to be expressed at high levels in multiple myeloma patients by Waldenstrom and later independently by Kunkel [4,5]. They identified that the γ-macroglobulin in patient sera migrated close to β-globulin using immuno-electrophoresis and ultracentrifugation techniques. In the 1960s, methods were developed to induce plasmacytomas in mice that produced uniform immunoglobulins that included γ-macroglobulin producing plasmacytomas, recapitulating the data observed in multiple myeloma patients [6]. As the immunoglobulins discovered during this time were being given arbitrary names, in 1964 the World Health Organization defined a nomenclature system for antibody isotypes. As a consequence γ-macroglobulin was renamed IgM, and the M referred to "macroglobulin" [7].

1.2. Evolution of IgM Antibodies

Immunoglobulins, including IgM antibodies, are found in all jawed vertebrates (gnathosomes) that diverged in evolution from jawless fish (agnathans) approximately 550 million years ago [8,9]. Similar to mammals, IgM expression precedes the expression of other antibody isotypes, although, in teleost fish, IgD and IgT are the only other isotypes present [10]. The phylogeny of the immunoglobulin heavy and light chain isotypes is illustrated in Figure 2. However, within certain species, there are distinct differences in the structure of the IgM antibodies produced [11]. For example, the predominant form of IgM antibodies in mice and humans are pentameric in structure and include a J-chain that stabilizes the pentamer, but hexamers and monomers can also be detected [12,13]. However, IgM antibodies in frogs, for example, *Xenopus*, are hexameric in structure even though *Xenopus* IgM has been reported to contain a J-chain [14]. In contrast, IgM from bony fish predominantly forms a tetramer structure, whereas the IgM produced by cartilaginous fish, such as shark, are pentameric in structure [15,16]. It is unclear why hexamer IgM was produced, but it was possible that J-chain synthesis could be limiting [17]. In addition, there are examples of IgM where the expressed µ chain lacks the cysteine in the tailpiece required for proper insertion into the IgM structure [18,19]. Interestingly, in humans and mice, pentameric IgM may also be present that does not contain a J-chain [20]. In fact, we have observed hexamer, i.e., pentamer mixtures produced from recombinant IgM derived from CHO cells, in the absence of transfected J-chain ([21] and unpublished observations).

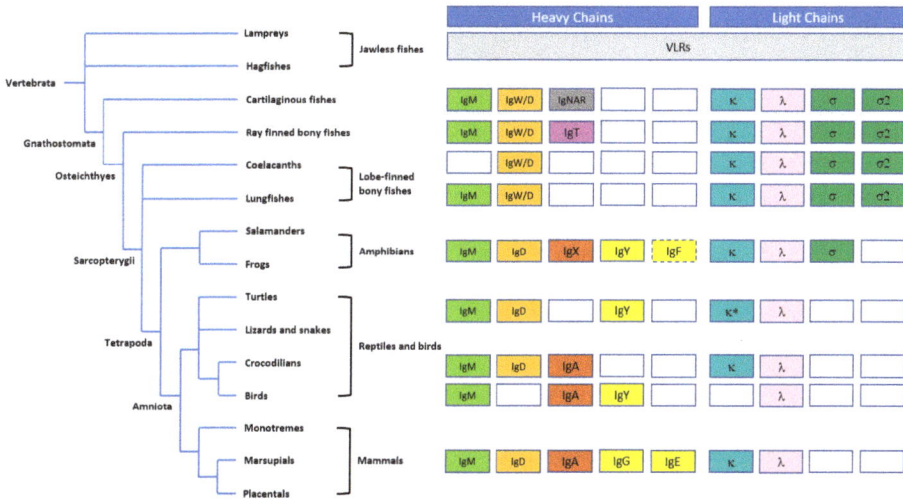

Figure 2. Schematic diagram illustrating the evolution of immunoglobin (Ig) heavy and light chain isotypes in vertebrates, with the IgM isotype broadly represented across phyla. Antigen-binding variable lymphocyte receptors (VLRs) in jawless fishes (agnathans) are thought to be precursors of immunoglobulins. The IgW isotype in cartilaginous fishes is orthologous to IgD in other groups; IgNAR is a "new antigen receptor" isotype, identified in nurse shark, that does not associate with light chains and does not have an ortholog in higher species. IgT appears to be the most ancient Ig specialized for mucosal protection. IgX, originally identified in *Xenopus* is orthologous and functionally analogous to IgA. IgY is the amphibian, reptilian, and avian equivalent of IgG and IgE. IgF only has two constant domains but has homology to IgY. Open boxes represent the lack of certain heavy or light chains in the certain vertebrate lineages; dashed boxes represent a common ancestry; and * represents the lack of κ light chain in snakes. Figure adapted from Pettinello and Dooley 2014 [22] and Kaetzel 2014 [23].

1.3. Ontogeny of B Cells and IgM Antibodies

In mammals, B cell development occurs in a hierarchical, ordered manner in fetal liver during embryonic development, and then in bone marrow and peripheral lymphoid tissue in adults. Within bone marrow, CD34$^+$ multipotent progenitors differentiate into common lymphoid precursors (CLP) that give rise to both B and T cell lineages. CLPs subsequently differentiate into early pro-B cells that express Igα and Igβ, essential signaling components of the B cell receptor (BCR). Initiation of the µ heavy chain (µHC) locus rearrangement occurs during the transition to pro-B cells when the RAG1/2 recombination complex induces rearrangement of the D to J$_H$ gene segments and, subsequently, to the V to DJ$_H$ gene segments. However, no surface µHC is expressed until the cells differentiate into large pre-B cells that express a pre-BCR composed of a µHC complexed to surrogate light chains VpreB and λ5 chain. Signaling through the pre-BCR results in the proliferation, differentiation, and subsequent surrogate light chain downregulation, paving the way for λ or κ light chain rearrangement to form surface IgM with different antigen specificities [24]. Receptor editing and selection occurs at this point and surface IgM expressing immature B cells egress from the bone marrow into the spleen.

Within the spleen, IgM expressing immature B cells begin to express surface IgD, which separate into the following different populations: IgM^{low} IgD^{high} in the follicles and IgM^{high} IgD^{low} in the marginal zone of the spleen. B-1 cells also mature into IgM^{high} IgD^{low} cells. Although the variable domains of the IgM and IgD are identical, alternative transcription and splicing results in both IgM and IgD heavy chains [25]. These populations of naïve B cells are now mature and are poised for clonal expansion and somatic hypermutation upon an encounter with antigen. Upon antigen binding, signaling is initiated through the IgM and/or IgD BCR, resulting in a signaling cascade involving Lyn, Syk, Src, Btk, PLCγ2, and PI3Kδ via co-receptor CD19, resulting in activation, proliferation, and differentiation of B cells that produce secreted IgM, IgG, IgA, or IgE [26–29]. However, B-1 cells only produce secreted IgM and are described below (see Section 2.1, innate immunity).

During human fetal development, IgM can be detected in the serum at approximately 20 weeks of gestation [30]. In contrast to IgGs, IgM antibodies are not transported across the placenta [31,32]. These fetal IgM antibodies are predominantly polyreactive "natural" IgM antibodies that play a role in the innate defense against infectious pathogens [33]. Postnatal IgM concentrations increase rapidly within the first month of postnatal life likely due to increased exposure to foreign antigens, and then gradually level off [34]. Levels of prenatal IgM are approximately 5 mg/dL in infants at 28 weeks and 11 mg/dL at birth. At one year of age, concentrations of IgM in infants are approximately 60% of that in an adult which is approximately 140 mg/dL, representing approximately 10% of total plasma immunoglobulins [35,36].

2. Biology of IgM

2.1. Innate Immunity

"Natural IgM" antibodies represent the majority of secreted IgM antibodies found in normal serum and are also located in the pleural and peritoneal compartments [37,38]. This class of IgM antibodies are evolutionarily conserved in all jawed vertebrates [10], are spontaneously produced by a subset of B cells, and often bind to specific antigens in the absence of immunization [39]. Natural IgMs are encoded by unmutated germline variable gene segments with polyreactive binding specificities to epitopes that are generally self- and non-self-antigens [10]. As previously described, these polyreactive IgMs are found at higher frequencies in neonates than adults, both in humans and mice [40].

The source of natural IgM antibodies is somewhat controversial. In mice, natural IgM antibodies are reported to be produced by B-1 cells residing in the bone marrow and spleen [41–45]. However, others have reported that non-B1 plasma cells in bone marrow were the source of murine natural IgM antibodies [46]. In humans, the B-1 cell population has not been studied as thoroughly as the murine B-1 population, but the human B-1 cells are also believed to be the source of natural IgMs [47].

One of the roles of natural IgM antibodies includes the targeting of altered self-antigens or neo-epitopes on dying cells for targeted removal, thereby maintaining tissue homeostasis [48]. One such antigen recognized by natural IgMs to facilitate the removal of apoptotic or dying cells is phosphorylcholine, which is also present on the cell wall of many parasites and microbes [49–52], thus, providing a first line of defense against pathogens. Carbohydrates, phospholipids, lipopolysaccharide, low-density lipoprotein, plus single and double stranded DNA are other antigen specificities known to be recognized by natural IgM antibodies [48,53,54].

In addition, natural IgM antibodies have been demonstrated to play a role in controlling B cell development, selection, and induction of central tolerance to prevent autoimmunity. The rare condition of selective IgM deficiency in humans, although associated with recurrent infections, is characterized by an increased risk of developing autoimmune diseases such as arthritis and systemic lupus erythematosus [55]. In a study performed by Nguyen et al. [56], secretory μ chain deficient mice (μs−/−) were found to recapitulate the selective IgM deficiency phenotype seen in humans. Although the phenotype may have been due to the reduction in auto-antigen clearance, these knockout mice displayed a block in the differentiation at the pre/pro-B cell stage of development and an escape from central tolerance induction resulting in the accumulation of autoantibody-secreting cells, phenotypes reversible with the administration of polyclonal IgM [56]. Therefore, these data support that natural secreted IgM antibodies facilitate normal B cell development that enforces the negative selection of autoreactive B cells, although the precise mechanism is unclear.

2.2. Early Adaptive Immunity

Upon binding to antigen, the BCR expressed on naïve follicular B cells is activated and B cells exit the follicle, proliferate, and produce relatively short-lived IgM secreting plasmoblasts in lymphoid tissues [57]. In the case that activated B cells engage with CD4+ T follicular helper cells through antigen-specific MHC–TCR interactions (major histocompatibility complex and T cell receptor), the B cells will re-enter the follicles, proliferate, and form germinal centers. During this time, the V regions of the BCR undergo somatic hypermutation to "fine tune" the affinity of the antibodies to the specific antigen, and then the antibody heavy chains undergo class switching recombination events to form a variety of isoforms, including IgG1, IgG2, IgG3, IgG4, IgA, or IgE. During this time, the B cells undergo clonal expansion within the follicles, leave the germinal centers, and differentiate into antigen-specific class switched high affinity antibodies producing plasma cells and memory B cells [58].

However, IgM production is not limited to just the initial antigen response. In fact, IgM-expressing memory B cells have been identified that have V region mutations suggesting that these are post germinal center B cells [59,60]. Analysis of peripheral blood has identified that 10 to 20% of all B cells are mutated IgM expressing cells and the IgM expression levels are higher than other less mature IgM expressing B cells [61]. Interestingly, it has recently been described that long-lived murine plasmodium-specific memory B cells include somatically hypermutated IgM expressing B cells [62]. The investigators identified that upon plasmodium rechallenge, the high affinity, somatically hypermutated plasmodium specific IgM+ memory B cells proliferated and gave rise to antibody secreting cells that dominated the early antibody response, via both T cell dependent or independent mechanisms [62].

The interaction of IgM antibodies with antigens can dramatically enhance humoral immune responses to the antigens beyond that of IgG. This is exemplified by studies that have evaluated the co-administration of IgG or IgM with xenogeneic erythrocytes in murine models [63]. When IgGs were co-administered in vivo to mice with xenogeneic erythrocytes, this resulted in suppression of erythrocyte-specific antibody responses [64]. Indeed, this immunosuppressive approach is used clinically to prevent Rh-negative mothers from becoming immunized against Rh-positive fetal erythrocytes, decreasing the incidence of hemolytic disease in newborns. The immunosuppressive mechanism is hypothesized to be epitope masking [64] and does not require complement or IgG Fc receptors [65]. This contrasts with the response observed when IgM antibodies targeting erythrocytes were co-injected in vivo, which resulted in a stronger antibody response against the erythrocytes than when erythrocytes were administered alone [66,67]. Studies have suggested that the activation of complements by the IgM was an important step in this effect, since mice with inactivated complement receptor 1 and 2 have a dampened immune response. For example, when sheep erythrocytes were co-injected with IgM in mice that had the C1q and C3 genes knocked out, antibody responses were not observed or were significantly muted [66]. In addition, similar studies in mice that had complement receptors 1 and 2 (CR1/2) inactivated had a severely impaired antibody response [68]. These data demonstrate that activation of a complement is crucial for the ability of IgM antibodies to feedback enhance antibody responses. IgM antibodies may also increase the concentrations of antigen on follicular dendritic cells in splenic follicles, thereby enhancing antigen presentation and downstream immune response.

IgM antibodies also play key roles in mucosal defense. Secondary lymphoid tissues containing B and T cells, referred to as mucosa-associated lymphoid tissues (MALT), are associated with multiple organ systems including the gastrointestinal and respiratory tracts. These secondary tissues are less organized than primary lymphoid tissues. As discussed below in further detail, the J-chain of pentameric IgM antibodies interacts with the polymeric immunoglobulin receptor (pIgR) on cells which results in the transcytosis of the antibodies from the circulation and through epithelial cells to mucosal surfaces to provide a first line defense against pathogens [69]. The J-chain is also incorporated into the dimeric form of IgA, and it allows efficient mucosal transport [70].

3. IgM Antibody Structure

3.1. Primary Structure

Antibodies of the IgM isotype are typically found as pentameric or hexameric in format, where each monomer is approximately 190 kDa, comprised of a heavy μ chain with five domains (Vμ, Cμ1, Cμ2, Cμ3, and Cμ4) and a light chain with two domains (Vκ-Cκ or Vλ-Cλ) [18]. IgM constant chain monomers show a greater degree of homology to IgEs than other isotypes. As shown in the alignments in Figure 3, the constant regions of the heavy chains, CH1, CH2, and CH3 of IgG correspond to the Cμ1, Cμ3, and Cμ4 of IgM. In contrast, the hinge region of IgG corresponds to the Cμ2 of IgM, which is an additional constant domain also found in other isotypes (mammalian IgE and avian IgY). It is thought that this domain functions much like the hinge region of IgGs and provides the flexibility needed to allow IgMs to bind multiple copies of antigens on cell surfaces. The heavy chains in each monomer are covalently linked with a disulfide bond at Cys 337 [19,71,72]. Each light chain is disulfide bonded to the heavy chain using cysteine residues at position 136 in the heavy chain [73].

Human immunoglobulin heavy chain constant domains are shown in Figure 3 below.

As shown in Figure 4, an additional feature of μ heavy chain is the presence of a short 18 amino acid peptide sequence (PTLYNVSLVMSDTAGTCY) at the C-terminus called the "tailpiece" [75]. IgM monomers are covalently linked by disulfide bonds between the penultimate cysteine of these tailpiece peptides. The tailpiece peptide is critical for IgM polymerization [76]. Indeed, the tailpiece can induce the polymerization when fused at the C-terminus of other antibody isotypes such as IgG [77]. In addition, inter-monomer disulfide bonds between Cys 414 residues in Cμ3 hold the center of the IgM in a well-defined ring-like structure.

Constant Region 1

```
              120       130       140       150       160       170       180       190                 210

IGHG1_HUMAN_CH1  ASTKGPSVFPLAPSSKS..TSGGTAALGCLVKDYFPEPVTV.SW...NSGALTSG.VHTFPAVLQS.SGLYSLSSVVTVPSSSLGT.Q..TYICNVNHKPSNTKV.DKKV...
IGHG2_HUMAN_CH1  ASTKGPSVFPLAPCSRS..TSESTAALGCLVKDYFPEPVTV.SW...NSGALTSG.VHTFPAVLQS.SGLYSLSSVVTVPSSNFGT.Q..TYTCNVDHKPSNTKV.DKTV...
IGHG3_HUMAN_CH1  ASTKGPSVFPLAPCSRS..TSGGTAALGCLVKDYFPEPVTV.SW...NSGALTSG.VHTFPAVLQS.SGLYSLSSVVTVPSSSLGT.Q..TYTCNVNHKPSNTKV.DKRV...
IGHG4_HUMAN_CH1  ASTKGPSVFPLAPCSRS..TSESTAALGCLVKDYFPEPVTV.SW...NSGALTSG.VHTFPAVLQS.SGLYSLSSVVTVPSSSLGT.K..TYTCNVDHKPSNTKV.DKRV...
IGHA1_HUMAN_CH1  ASPTSPKVFPLSLCSTQ...PDGNVVIACLVQGFFPQEPLSVTW...SESGQ..GVTARNFPPSQDASGDLYTTSSQLTLPATQCL.AGK.SVTCHVKHYTNPSQ..DVTVPCP
IGHA2_HUMAN_CH1  ASPTSPKVFPLSLDSTP...QDGNVVACLVQGFFPQEPLSVTW...SESGQ.N.VTARNFPPSQDASGDLYTTSSQLTLPATQCP.DGK.SVTCHVKHYTNSQ..DVTVPCR
IGHD_HUMAN_CH1   APTKAPDVFPIISGCRHP.KDNSPVVLACLITGYHPTSVTV.TW...YMGTQ..SQPQRTFPE.IQRRDSYMTSQLSTPLQQWRQG..EYKCVQHTASKSK..KEIF...
IGHE_HUMAN_CH1   ASTQSPSVFPLTRCCKNIPSNATSVTLGCLATGYFPEPVMV.TW...DTGSLNGTTMTLPATTLTL.SGHYATISLLTVSGAWAK..Q..MFTCRVAHTPSSTDWVDNKTFS..
IGHM_HUMAN_CH1   GSASAPTLFPLVSCENSP.SDTSSVAVGCLAQDFLPDSITF.SWKYKNNSDI..SS.TRGFPSVLRG..GKYAATSQVLLPSKDVMQGTDEHVCK.VQHPNGNK..EKNVPLP

                    130       150       160       170       180       190       200       210       220
```

Hinge Region or equivalent

```
                          220                          230

IGHG1_HUMAN_H   ..EPKSC.D.KTHTCPP.........................CP
IGHG2_HUMAN_H   ..ERK...CC.V.E.CPP........................CP
IGHG3_HUMAN_H   ..ELKTP.LGDTTHTCPRCPEPKSCDTPPPCPRCPEPKSCDTPPPCPRCP
IGHG4_HUMAN_H   ..ESKYG....PPCPS..........................CP
IGHA1_HUMAN_H   .........................................VPSTPPTPS
IGHA2_HUMAN_H   
IGHD_HUMAN_H    RWPESPKAQASSVPTAQPQAEGSLAKATTAPATTRNTGRGGEEKKKEKEKEEQEERETKTPE.CP

IGHE_HUMAN_CH2  VCSRDFTPP..TVKILQSSCDGGGHFPPTIQLCLVSGYTPGTINITWL..ED..GQVMDVDLSTASTTQEGELASTQSELTLSQKHWLSDRTYTCQVTYQGHTFE.DSTKKCA.
IGHM_HUMAN_CH2  ..VIAELPPKVSVFV.PPRDGF.FGNPRKSKLICQATGFSPRQIQVSWLREGKQVGSGVTTDQVQAEAKESGPTTYKVTSTLTIKESDWLGQSMFTCRVDHRGLTFQQNASSMCVP

                    230       240       250       260       270       280       290       300       310                 330
```

Constant Region 2/3

```
                          240       250       260       270       280       290       300       310       320       330       340

IGHG1_HUMAN_CH2  AP...ELLGGP..SVFLFPPKDTLMISRTPEVTCVVVDVSHEDPEVKFNWYVDGVEVHNAKTK.PREEQYNSTYRVVSVLTVLHQDWLNGKEYKCKVSNKALPAPIEKTISKAK
IGHG2_HUMAN_CH2  AP...PVAG.P..SVFLFPPKPKDTLMISRTPEVTCVVVDVSHEDPEVQFNWYVDGVEVHNAKTK.PREEQFNSTFRVVSVLTVLVHQDWLNGKEYKCKVSNKGLPAPIEKTISKTK
IGHG3_HUMAN_CH2  AP...ELLGGP..SVFLFPPKPKDTLMISRTPEVTCVVVDVSHEDPEVQFKWYVDGVEVHNAKTK.PREEQYNSTFRVVSVLTVLHQDWLNGKEYKCKVSNKALPAPIEKTISKTK
IGHG4_HUMAN_CH2  AP...EFLGGP..SVFLFPPKPKDTLMISRTPEVTCVVVDVSQEDPEVQFNWYVDGVEVHNAKTK.PREEQFNSTYRVVSVLTVLHQDWLNGKEYKCKVSNKGLPSSIEKTISKAK
IGHA1_HUMAN_CH2  PSTPPTPSPSCCHPRLSLHRPALEDLLLGSEANLTCTLTGLRDAS.GVTFTWTPSSGKSAVQGPPERDLCG.C..YSVSSVLPGCAEPWNHGKFTCTAAYPESKTPLTATLSK.S
IGHA2_HUMAN_CH2  ....VPPPPPCCHPRLSLHRPALEDILLGSEANLTCTLTGLRDAS.GATFTWTPSSGKSAVQGPPERDLCG.C..YSVSSVLPGCAQPWNHGETFTCTAARPELKTPLTANITKSG
IGHD_HUMAN_CH2   ....PHTQPL..GVILLTPAVQD.IWLRDKATFTCFVVGSDLKD.A.HLTWEVAGKVPTGGVEGLLERHSNGSQQHSRILTPRSLWNAGTSVTCTLNHPSLPPQRLMALREPA
IGHE_HUMAN_CH3   ....DSNPRGV..SAYLSRPSPFD.LFIRKSPTITCLVVDLAPSKGTVNLTWSRASGKPVNHSTR.KEEKQRNGTLTVTSSTLPVGTRDWIEGETYQCRVTHPHLPRALMRSTTKTS
IGHM_HUMAN_CH3   ....DQDTAI..RVFAIPP.SFASIFLTKSTKLICLVTDLTTYDS.VTISWTRQNGEAVKTHTN.ISESHPNATFSAVGEASICEDDWNSGERFTCTVTHTDLPSLKQTISRPK

                    340       350       360       370       380       390       400       410       420       430       440
```

Figure 3. *Cont.*

Constant Region 3/4

```
               350        360        370        380        390        400        410        420        430        440
IGHG1_HUMAN_CH3 .GQPREPQVYTLPPSRDE.LTKNQVSLTCLVKGFYPSDIAVEWES.NG..QPENNYKTTPPVLDSD...GSFFLYSKLTVDKSRWQQGNVFSCSVMHEAL.HNHYTQKSLSLSPGK.......
IGHG2_HUMAN_CH3 .GQPREPQVYTLPPSREE.MTKNQVSLTCLVKGFYPSDISVEWES.NG..QPENNYKTTPPMLDSD...GSFFLYSKLTVDKSRWQQGNVFSCSVMHEAL.HNHYTQKSLSLSPGK.......
IGHG3_HUMAN_CH3 .GQPREPQVYTLPPSREE.MTKNQVSLTCLVKGFYPSDIAVEWES.SG..QPENNYNTTPPMLDSD...GSFFLYSKLTVDKSRWQQGNRFSCSVMHEAL.HNRFTQKSLSLSPGK.......
IGHG4_HUMAN_CH3 .GQPREPQVYTLPPSQEE.MTKNQVSLTCLVKGFYPSDIAVEWES.NG..QPENNYKTTPPVLDSD...GSFFLYSRLTVDKSRWQEGNVFSCSVMHEAL.HNHYTQKSLSLSLGK.......
IGHA1_HUMAN_CH3 .GNTFREVHLLPPSEELALNELVTLTCLARGFSPKDVLVRWLQ.GSQELPREKYLTWASRQEPSQGTTTFAVTSILRVAAEDWKKGDTFSCMVGHEAL.PLAFTQKTIDRLAGK......
IGHA2_HUMAN_CH3 .GNTFRPEVHLLPPSEELALNELVTLTCLARGFSPKDVLVRWLQ.GSQELPREKYLTWASRQEPSQGTTYAVTSILRVAAEDWKKGETFSCMVGHEAL.PLAFTQKTIDRMAGK......
IGHD_HUMAN_CH3 .AQAPVKLSNLLASSDP.PEAASW.ILCEVSGFSPPNILLMWLEDQREVNTSGFAPARPPPQPRS...TTFWAWSVLRVPAPPSPQPATYTCVVSHEDS.RTLLNASRSLEVSVVTDHGPMK.
IGHE_HUMAN_CH4 .GPRAAPEVYAFATPEWPGSRDK.RTLACLIQNFMPEDISVQWLHNEVQLPDARHSTTQPRKTKGS...GFFVFSRLEVTRAEWEQKDEFICRAVHEAASPSQTVQRAVSVNPGK......
IGHM_HUMAN_CH4 GVALHRPDVYLLPPAREQLNLRESATITCLVTGFSPADVFVQWMQRGQPLSPEKYVTSAPMPEPQAP.GRYFAHSILTVSEEWNTGETYTCVVAHEAL.PNRVTERTVDKSTGK.......
               450        460        470        480        490        500        510        520        530        540        550
```

C-terminal "Tailpiece"

```
IGHA1_HUMAN_TP PTHVNVSVVMAEVDGTCY
IGHA2_HUMAN_TP PTHINVSVVMAEADGTCY
IGHM_HUMAN_TP  PTLYNVSLVMSDTAGTCY
                     570
```

= Intra-chain disulfide bond

= Inter-chain disulfide bond

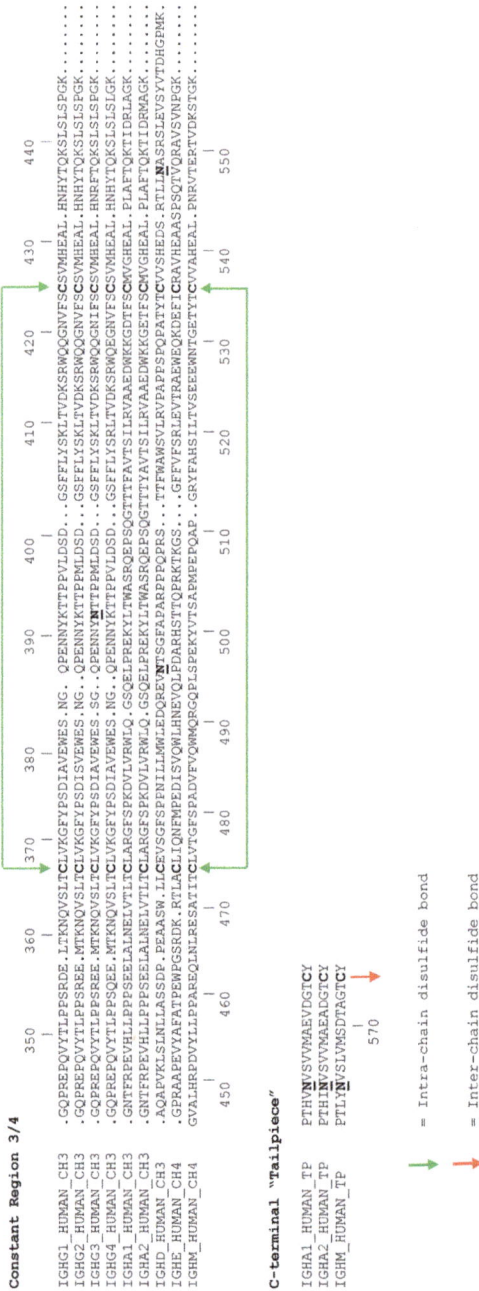

Figure 3. Sequence alignment of the heavy chains of different isotypes of immunoglobulin showing the glycosylation sites and location of inter- and intra-disulfide linkages. The Cμ2 region of IgMs and IgEs is analogous to the hinge regions of the other isotypes. The alignments also highlight, in bold font, the locations of glycosylation sites on each heavy chain. Sequences are numbered according to the convention established by Kabat [74].

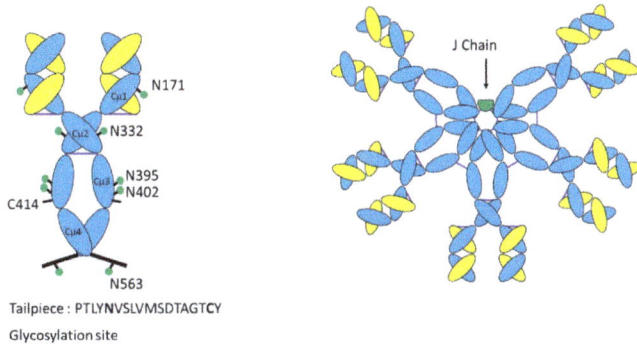

Tailpiece : PTLYNVSLVMSDTAGTCY

Glycosylation site

Figure 4. Schematic diagram of IgM monomer vs. IgM pentamer. IgM monomers are distinguished from IgG counterparts by their extensive glycosylation at the asparagine residues indicated, the presence of an additional domain Cμ2 in place of a hinge, and the presence of a short tailpiece peptide sequence that is critical for multimerization. Pentameric IgM has an additional 137 amino acid joining (J)-chain.

In addition to the heavy chain and light chains, IgMs also possess a third chain, a polypeptide of 137 amino acids, known as the joining (J)-chain, which is a key feature of polymeric IgA and pentameric IgM antibodies [78,79]. The sequence of the J-chain is highly conserved from amphibians to humans [69], and is a distinct domain, unrelated to the immunoglobulin fold found in heavy chains and light chains (see Figure 5). There is a very high degree of sequence conservation within the J-chain, consistent with key structural and functional aspects of the J-chain integration into IgA and IgM oligomers [80,81] (see Section 3.3). The J-chain allows binding and transport of IgM pentamers and IgA oligomers to mucosal surfaces via interactions with polymeric Ig receptor (pIgR, see Sections 3.3 and 4) [82].

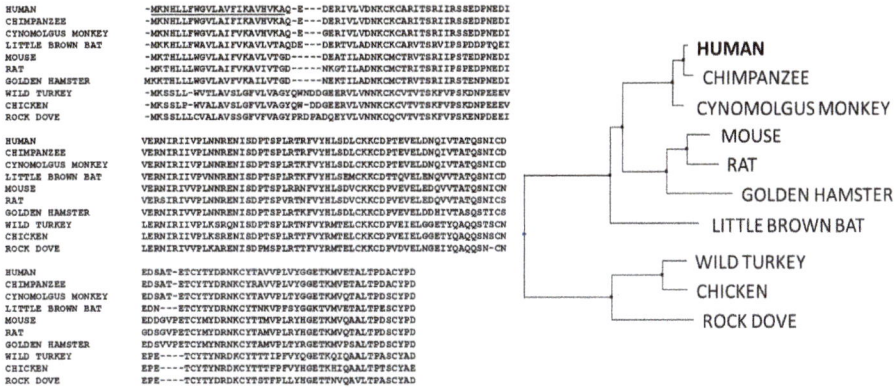

Figure 5. Sequence alignment (**left**) and hierarchical clustering (CLUSTALW, **right**) of IgM joining chains (J-chains) showing the high degree of conservation across species from human, primates, rodents, and birds.

3.2. Glycosylation

Antibodies are glycoproteins with N-linked glycosylation. In the case of IgG, there is N-linked glycosylation at Asn 297, which influences binding to Fc gamma receptors, and hence has a role in modulating antibody-dependent cell-based cytotoxicity (ADCC). Significantly, IgMs have more sites of glycosylation as compared with that of IgG. Whereas IgG heavy chains have a single glycosylation site, human and non-human primate IgM heavy chains exhibit five N-linked glycosylation sites at Asn 171 (Cμ1), Asn 332 (Cμ2), Asn 395, Asn 402 (both in Cμ3), and Asn 563, which is in the heavy chain tailpiece [83]. An additional glycosylation site is present on the J-chain at Asn 49. These glycans are considered to facilitate polymerization and assembly of the oligomeric IgM structure [84], as well as provide IgM with greater solubility and longer in vivo half-life [85]. There is no well characterized role for the glycosylation of IgMs in mediating effector function, as has been demonstrated for IgG. Recently, Colucci et al. showed a role for sialylation on IgM in mediating internalization on T cells and IgM mediated immune suppression [86].

The additional sites of glycosylation increase the complexity of IgM antibodies. Detailed site-specific carbohydrate analysis of IgM demonstrates that not all of the N-linked sites are similarly glycosylated. N-linked glycosylation has various levels of complexity, from high mannose or simple glycans to bi-, tri- and tetra-antennary complex glycans. Interestingly, the three sites at Asn 171, Asn 332, and Asn 395 (in domains 1, 2, and 3) exhibit complex carbohydrate moieties with sialylated termini. However, the more carboxy terminal sites at Asn 402 and Asn 563 (in domains 3 and 4) contain high mannose structures [84]. This pattern of glycosylation is consistent with the amino terminal regions of IgM being more accessible to glycosylation enzymes of the intracellular Golgi apparatus, whereas the carboxy terminal regions (the "central core" structure) are not fully processed, perhaps due to steric hindrance and lack of accessibility of these glycans in the oligomerized form of IgM. The fourth glycosylation site on IgM (Asn 402) is homologous to the single site of IgG (see Figure 3), which is known to have limited accessibility and does not exhibit fully developed complex carbohydrate in either IgM or IgG.

3.3. Tertiary Structure

The nearly mega dalton size of a fully assembled IgM complex has proven to be a challenge for determining a detailed structure. However, informative and useful models of IgM were initially created using a combination of techniques including low resolution cryo-electron microscopy, X-ray crystallography, NMR-derived structures of the subdomains, and homology modeling.

In some of the earliest studies [87,88], three-dimensional (3D) models were proposed based on electron micrographs that placed the five monomers of a pentamer in a symmetrical structure around a central ring. Indeed, both planar (antigen-free) and "stable-like" (antigen-bound) structures for IgM antibodies were described. Subsequently, cryo-atomic force electron microscopy combined with the known crystal structure of IgE enabled further modeling of IgM structure [89]. Similar to the earlier models, these authors proposed a symmetrical distribution of five monomers around a central ring that exhibited a "flexural bias", which allowed it to remain in a planar structure in the absence of antigen but, then, underwent a conformational change upon binding to antigen coated surfaces, such as those on cell or microbial surfaces. The "staple-like" structure that was formed then allowed binding of C1q, the first component of complement. Muller et al. assembled a model using crystal structure of domains Cμ1 and Cμ4 and an NMR-derived model of Cμ3 [90]. However, a recent study [91] with and without Fab arms attached to the central Fc ring, conclusively showed that the pentamer was actually an asymmetric structure where a single monomer from the hexamer was substituted with a J-chain without perturbing the position of the rest of the monomers (see Figure 6).

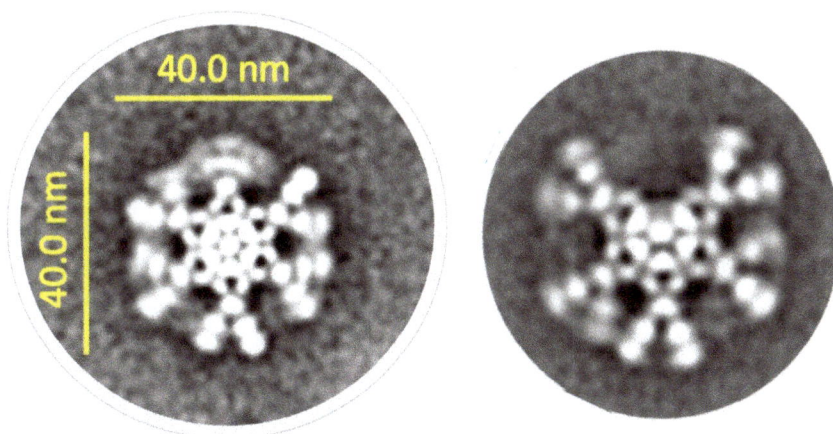

Figure 6. Example cryo-electron microscopy (cryo-EM) results for hexamer (**left**) and pentamer (**right**) forms of an anti-CD20 IgM. Consistent with published observations [91], the pentamer formed in the absence of J-chain retains the positions of monomers around the central ring and only a single monomer appears substituted by the J-chain. These images are a montage of a large series of negatively stained, transmission electron micrographs obtained via collaboration of IGM Biosciences and NanoImaging, Inc., San Diego, CA, USA (unpublished results).

Interestingly, although the mu chain tail pieces and the J-chain are known to play critical roles in the assembly and function of IgM antibodies, for many years, the three-dimensional structures within the IgM pentamer have largely remained elusive. In fact, until recently, only a secondary structure for the J-chain has been proposed [92]. Importantly, in two new pivotal studies, the 3D structures of pentameric IgM [80] and human IgA [81] were reported. In these reports, Li et al. described the cryo-electron microscopy (cryo-EM) structure of an IgM Fc pentamer that included the J-chain and the ectodomain of pIgR, and Kumar et al. described the atomic structure of dimeric, tetrameric, and pentameric IgA Fc fragments linked by J-chain and in a complex with the secretory component of pIgR. These reports clearly show a unique "two-winged" structure for J-chain that is conserved and exhibits highly similar conformations within the IgM and IgA context. The J-chain binds to the Fcμ pentamer of IgM and forms clasp or bridge between the Fcμ1A and Fcμ5B monomers of the IgM via disulfide bonding of the tailpieces with J-chain cysteines [80]. With respect to the central structure of IgM, these models also show how the ten heavy chain tail pieces assemble within the central core of the Fc ring and interact with the J-chain. The tailpieces of IgM form parallel beta strands and the ten tailpieces pack in anti-parallel fashion. These authors suggested that the combined set of tailpieces formed prominent interactions that stabilized the pentamer while the J-chain served as a template for the oligomerization of IgM. These recent studies provide a fresh view of the structure of IgM antibodies and new functional insights into the unique biology of IgM and its interaction with the secretory pathway (see Section 4.3.1).

4. Function

4.1. Binding to Microbial Antigens, Role of Avidity

Natural IgM antibodies, in conjunction with natural killer (NK) cells, dendritic and mast cells, and macrophages are part of the innate immune system, the first line of defense against invading microorganisms and aberrant human cells (see Section 2.1 and Vollmers 2006 [93]). This response involves binding to specific antigenic motifs, such as specific carbohydrates on glycoproteins or glycolipids and repetitive structures such as lipopolysaccharides, recognized by IgM antibodies encoded by germ line (i.e., unmutated) genes. In so doing, these natural IgM antibodies play an important role in primary defense mechanisms, recognizing foreign bacteria and viruses or mutated human cells such as cancer cells. Typically, these natural IgM antibodies utilize low affinity binding to a range of similar foreign antigens, and their ability to eliminate these foreign antigens is, then, amplified by the high avidity afforded by having 10 (in the pentamer) or 12 (in the hexamer) binding sites. The potent ability of IgM antibodies to fix complement and opsonize particles make them particularly effective against bacteria and viruses [94]. The physical and functional characteristics of IgM and other antibody classes have been summarized by Strohl [95].

4.2. IgM vs. IgG Function: Complement Dependent Cytotoxicity (CDC) vs. Antibody-Dependent Cell-Based Cytotoxicity (ADCC)

IgM antibodies also differ from IgG isotypes due to the relative engagement of effector mechanisms. IgGs utilize natural killer cell engagement which can result in antibody-dependent cellular cytotoxicity (ADCC), as well as complement dependent cytotoxicity (CDC). In contrast, IgM does not bind the Fc gamma receptors, and therefore does not exhibit ADCC. However, IgMs have very potent CDC activity. Their hexameric or pentameric structure allows highly avid binding of complement component C1q to IgM, and therefore IgMs are able to fix complement substantially better than IgGs [96] (see Figure 7). Recent work by Sharp et al., using phase-plate cryo-electron microscopy, has provided a detailed model of how complement fixation was initiated with a large conformational change upon antigen binding, which exposed the regions on IgM that were bound by C1q, i.e., the first protein complex needed to initiate the complement cascade [97]. The planar or disc-like structure of free IgM changes to a "crouching" or "staple-like" structure when the Fab regions bind antigen on a cell surface. The antigen binding Fab regions move out of the plane of the ring formed by the Cμ3, Cμ4, and tailpiece due to the flexibility of the Cμ2 regions, which are the equivalent of the hinge regions of IgGs. This allows many or all of the Fab arms to contact antigens on a surface, leveraging the avidity of IgMs. Other effector mechanisms, such as antibody-dependent cellular phagocytosis, have also been implicated in the action of IgMs [98,99].

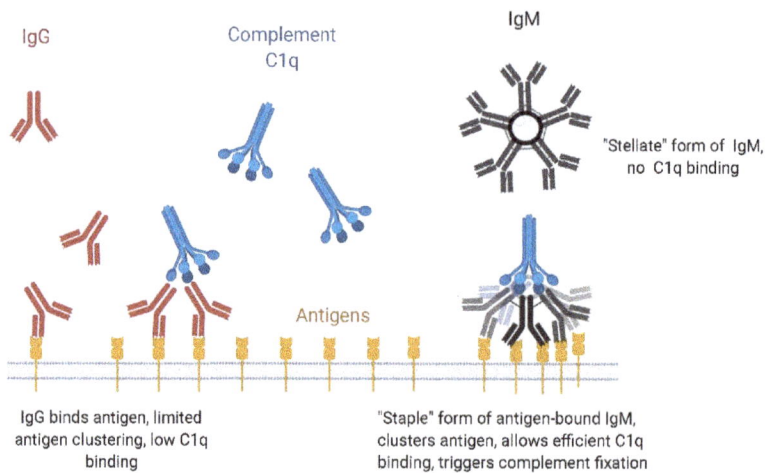

Figure 7. Complement binding and activation with IgM as compared with IgG.

4.3. IgM Receptors: Structure and Tissue Distribution

IgM antibodies are known to bind to multiple receptors, which are illustrated in Figure 8. The functional roles of these three receptors are discussed below.

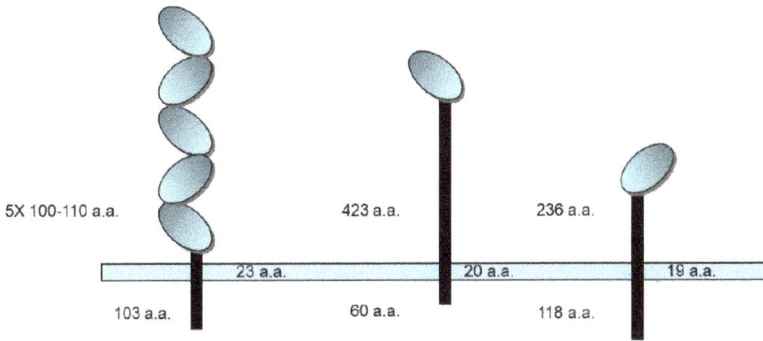

Receptor	pIgR	Fcα/μR	FcμR
Requires J-chain for binding IgM?	Yes	Yes	No
Expressed in	Small and large intestine, lungs, kidneys, mucosal epithelium	Lymph nodes, appendix, kidney, B-lymphocytes, macrophages	T- and B-lymphocytes

Figure 8. Schematic diagram of receptors known to bind IgM. IgMs bind at least three different receptors from those that bind IgG. The oval domains of each receptor indicate immunoglobulin fold-like regions. Their sizes and tissue distribution are depicted above.

4.3.1. Polymeric Ig Receptor (pIgR)

J-chain containing polymeric immunoglobulins such as IgA and IgM are often found on mucosal surfaces associated with a peptide called the secretory component (SC). The SC peptide is a proteolytic fragment of a cell surface receptor responsible for transport of polymeric Igs from the apical to mucosal surfaces [100]. The polymeric Ig receptor (pIgR) is expressed on the basolateral surfaces of mucosal epithelium, showing the highest expression in small and large intestines, with expression also seen in tissues such as lungs, pancreas, kidneys, and endometrium [101].

Structurally, pIgR belongs to the IgG superfamily with five Ig-like domains (D1–D5) that are heavily glycosylated (see Figure 8). Hinge regions are present between D1 and D2 and also between D3 and D4 [102]. Upon binding polymeric IgM or IgA antibodies containing J-chain, pIgR is internalized and transported by the endosome from the basal to the apical surface [103]. A membrane-proximal region contains a proteolytically sensitive site that is cleaved when the endosomes are trafficked to the apical side and this cleavage results in the release of polymeric Ig bound to the ectodomain of pIgR, which is known as the secretory component. Free SC can also be released at the apical side as an 8 kDa fragment.

The crystal structure of domain 1 (D1) of pIgR reveals a structural similarity to the variable domain of Ig that contains a highly conserved helix with a region that is implicated in binding to IgM [104]. The D1 of secretory component is both necessary and sufficient for IgM (or IgA) binding, however D2 through D5 contribute to increased affinity. The apoSC protein forms a compact structure in the absence of IgM or IgA but undergoes a drastic conformational change upon binding to polymeric IgM or IgA. The interaction of Fcμ of IgM with J-chain and pIgR form a ternary complex (Fcμ-J-SC) that facilitates transport of IgM (and IgA) to the apical side of epithelial cells. The molecular mechanism of pIgR/SC secretion of IgM (and IgA) is not fully understood. However, recent reports on 3D cryo-imaging of IgM with J-chain and pIgR/SC in complex, have contributed to a better understanding of both the structure and function of the secretory pathway components [80,81].

4.3.2. Fcα/μR

The Fcα/μ receptor was identified in a screen of receptors from a cultivated mouse lymphoma cell line capable of binding IgMs. This receptor is approximately 70 kDa in size, belongs to the immunoglobulin superfamily, and is extensively glycosylated. The Fcα/μ receptor is localized to all lymphoid tissues including lymph nodes and the appendix, and is also widely expressed in non-lymphoid tissues including kidney and intestine, with lower expression observed in the lungs, liver, and myocardium [105]. Residues 76–98 are homologous to the CDR1 region of pIgR, which constitutes a conserved binding site for both proteins. The predominant cells expressing Fcα/μ receptors are the follicular dendritic cells in the germinal centers [106]. As with pIgR, the Fcα/μ receptor appears to interact with IgMs, primarily with determinants in Cμ3 and Cμ4 [107]. The presence of the Fcα/μ receptor on intestinal macrophages, plasma cells, and Paneth cells implicates its role in local and systemic aspects of mucosal immunity.

4.3.3. FcμR, the TOSO Receptor

The most recently identified receptor interacting with IgM is FcμR, which is a transmembrane sialoglycoprotein of approximately 60 kDa [108]. FcμR, also known as the TOSO receptor, is highly expressed on chronic lymphocytic leukemia B cells and has been demonstrated to internalize upon IgM binding [109]. It is distinct from pIgR and Fcα/μ in that it only recognizes IgM and not polymeric IgA. The CDR1 region of FcμR that is predicted to recognize IgMs is very short, i.e., only five amino acids. Notably, FcμR does not require a J-chain for binding pIgM and its interactions are primarily thought to be with domains Cμ3 and μ4 [110]. Cells expressing FcμR were predominately adaptive immune cells, such as B and T cells [111].

5. Manufacturing Considerations

The need for scalable production processes will grow as the therapeutic interest in the use of IgM antibodies increases. IgMs have been considered to be difficult to express, due to their large size and complexity (see Table 1), resulting in low expression levels, and therefore expectations of a high cost of goods associated with therapeutic IgMs [112,113]. However, improvements in cell lines, production media, and process monitoring have made it such that production of a high-quality IgM is possible.

Table 1. IgM antibody complexity.

IgM Form	Molecular Weight	Peptides in IgM Complex	Inter-Chain Disulfide Bonds	N-Linked Sites of Glycosylation
Pentamer (with J-chain)	950 kD	21	27	51
Hexamer (without J-chain)	1150 kD	24	30	60

5.1. Expression of IgM

Biotherapeutic proteins, in general, including immunoglobulins, can be expressed in a variety of expression host cells [114]. Mammalian cells are typically used as host cells for IgM expression, in order to preserve the glycosylation patterns that are optimal for bioactivity or pharmacokinetic properties. Among mammalian cell hosts, Chinese hamster ovary (CHO) cells are most commonly used for producing antibodies because of the ability of these cells to grow in serum-free media at high density in large bioreactors [115]. CHO cell production of IgG antibodies has shown a steady improvement over the last 30 years of development and can reach a specific productivity of 50 to 60 picograms/cell/day and high titers of 10 to 15 g per liter. However, production of IgM antibodies in CHO cells is still a challenge. Kunert et al. first described the production of a class-switched anti-HIV IgM antibody, designated 4E10, in CHO-DUKX-B11 cells in serum containing medium, but were only able to achieve a specific productivity of 10 pg/cell/day [116]. To improve productivity and quality, Tchoudakova et al. utilized a different mammalian cell line, PER.C6, which was a transfected primary human embryonic retinoblast cell line, to make a panel of IgM antibodies and they were able to achieve a volumetric productivity of 20 pg/cell/day [117]. An additional host used for production of IgM is the tobacco plant [118]. By engineering in the expression of human sialyltransferase and galactosyltransferases, Loos et al. were able to demonstrate that fully functional IgMs with human-like glycoforms could be produced in tobacco plants.

The manufacture of IgM molecules for early clinical trials was done using hybridoma cells derived from rat or mouse myeloma cells or a heteromyeloma between human lymphoid cells and murine myeloma cells (see Table 3 in Section 6.1). Using these approaches, yields of 200 mg/L in batch process and 700 mg/L in a medium exchange process [117] were achieved. The two recombinantly expressed IgM antibodies, PAT-SC1 and PAT-SM6, were produced in PER.C6 cells and achieved fed-batch titers of 800 to 900 mg/L [117,119].

5.2. Purification of IgM

The purification of IgM for Good Manufacturing Practice (GMP) manufacturing has not been able to take advantage of affinity resins such as Protein A, which has been the standard recovery method for IgG. IgM does not bind to protein A. However, other affinity resins are available and are quite useful for research scale purification but are not currently available for scale-up with an associated GMP-compliant regulatory support file. IgMs also appear to have a narrower range of conditions under which they remain soluble as compared with IgGs, which can present difficulties in purification. Although low pH steps can be used for viral inactivation with IgMs, detergent-based approaches are more commonly used for viral inactivation in IgM downstream processing.

Early methods for purification of IgMs have included isoelectric precipitation and gel chromatography [120]. These investigators showed that product recovery of 40% could be achieved with 99% purity. For hybridoma cultures, polyethylene glycol (PEG) precipitation was optimized and combined with anion exchange chromatography for several antibodies. With the exception of two examples, greater than 95% purity with yields that varied from 28% to 84% was achieved using this approach [117]. This process was further improved by initially digesting the genomic DNA with the endonuclease benzonase.

In 2007, a three-step purification strategy for IgM antibody molecules was presented at a conference on purification of biological products [121]. The investigators used ceramic hydroxyapatite (CHT) chromatography for primary capture with a 90% purity and 79% recovery. The purification strategy subsequently used anion exchange (AEX) and cation exchange chromatography to achieve 99% purity with recovery of 50% to 80%. In 2010, Gagnon et al. reported the use of a monolithic anion exchanger with more than two-fold increased IgM dynamic binding capacity when compared to a porous particle anion exchange resin [122]. This approach was associated with greater genomic DNA removal due in part to the 15-fold higher charge density of the monolith exchanger. These results suggest that the convective nature of the monolithic matrix, rather than diffusion in porous resin, was perhaps better suited for IgM purification. This process was utilized at the 250 L scale with PER.C6 cell expression for producing PAT-SM6 which, at the time, was under evaluation in a Phase 1 melanoma study [119]. In this downstream process, CHT chromatography was used as the primary capture column with viral inactivation performed utilizing Triton X-100 wash step on column. Then, the DNA level was reduced with a Sartobind Q membrane, followed in succession by anion and cation exchange monolithic chromatography. The overall process yield was reported to be 55%.

Large scale GMP manufacture of IgM products is possible with a variety of traditional columns. New mixed-mode resins also may provide even greater capabilities. In 2011, at an IgM meeting in Germany, GE Healthcare reported on the use of layered beads in which the inner core was functionalized and the outer core was inert and porous [123]. This led to the launch of CaptoCore 700 and, more recently, CaptoCore 400 with an inert shell acting in a size exclusion mode and an anionic core, which are ideal for the large IgM molecule. Purification matrices such as these, along with bringing affinity resins to a scalable, regulatory-compliant state, should make the purification of clinical IgM antibodies more tractable with high yields and good safety clearance.

6. Therapeutic Uses of IgM Antibodies

As has been well demonstrated with IgG antibodies, IgM antibodies also have the potential to provide therapeutic benefit in humans. Indeed, IgM antibodies have been shown to be efficacious in a variety of animal models, including non-human primates [124] and were some of the first mAbs to be tested clinically (see Figure 9 and [125]).

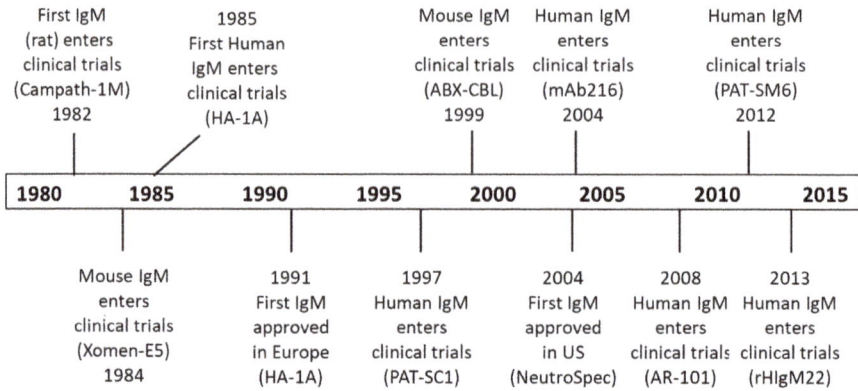

First IgM (rat) enters clinical trials (Campath-1M) 1982	1985 First Human IgM enters clinical trials (HA-1A)		Mouse IgM enters clinical trials (ABX-CBL) 1999	Human IgM enters clinical trials (mAb216) 2004		Human IgM enters clinical trials (PAT-SM6) 2012	
1980	**1985**	**1990**	**1995**	**2000**	**2005**	**2010**	**2015**
	Mouse IgM enters clinical trials (Xomen-E5) 1984	1991 First IgM approved in Europe (HA-1A)	1997 Human IgM enters clinical trials (PAT-SC1)	2004 First IgM approved in US (NeutroSpec)	2008 Human IgM enters clinical trials (AR-101)	2013 Human IgM enters clinical trials (rHIgM22)	

Figure 9. History of selected IgM human clinical trials.

Currently, there are hundreds of therapeutic IgG antibodies that have advanced to clinical trials, and more than 90 antibody-based products have achieved FDA approval [126]. However, only about 20 IgM antibodies have been tested in humans (Table 2). Included in this group are rat, mouse, and human IgMs that target a variety of infectious disease, oncology, and autoimmune disease antigens.

Taken together, the studies described below demonstrate that IgM antibodies can be safely administered to humans. However, the IgM antibodies tested to date have not typically produced sufficient efficacy in humans to obtain (or maintain) regulatory approval. This result is likely due to the fact that most, if not all, of the IgM antibodies tested were of natural origin and, as a consequence, essentially contained germ-line gene sequences that have not undergone extensive somatic mutation, and thus were of low affinity and specificity [127]. It is also likely that the particular indications tested in these early studies, such as a major focus on sepsis and septic shock, has limited the ability of IgM antibodies to achieve regulatory success.

6.1. IgM Clinical Trials

Shown in Table 3 are additional details regarding the IgM mAbs so far examined in humans, organized by the nature of the target antigens. For all of these studies, administration of the IgM antibodies was well tolerated. Of particular interest is the fact that more than half of these IgMs target antigens that are poorly immunogenic and for which it has been difficult to generate IgG mAbs [128]. Included in this category are lipopolysaccharides (LPS) and its component core structure lipid A, gangliosides, proteolipids, and glycans. Since many of these structures are composed of polymeric or repeated antigenic motifs, the avidity effects of having 10 binding sites on the IgM antibody may well provide significant advantages for such antigens over their IgG counterparts. It is apparent, however, that the approaches for finding such antibodies need to be changed significantly, focusing on the incorporation of affinity optimized V domains into the IgM backbone rather than using naturally occurring IgMs.

Table 2. Overview of IgM antibodies tested in human clinical trials.

Antibody (Name)	Company	IgM Source	Antigen	Indication	Most Advanced Clinical Development
Campath-1M	Academic (MRC-RDCT)	Rat	CD52	Graft vs. host disease	Phase 2
E5 (Xomen-E5)	XOMA	Mouse	J5 lipid A	Sepsis	Phase 3
HA-1A (Centoxin)	Centocor	Human	J5 lipid A	Sepsis	Phase 3
Fanolesomab-Tc99 (NeutroSpec)	Palatin	Mouse	CD15	Appendicitis	Phase 3
IgM cocktail (5)	Cutter/Miles	Human	LPS	Sepsis	Phase 1
Mab 16.88	Academic (Free University Hospital)	Human	Colon cancer antigen	Colorectal cancer	Phase 1
MAB-T88	Chiron	Human	LPS	Neutropenia	Phase 1
PAT-SC1	Patrys	Human	CD55 isoform	Gastric cancer	Phase 1
ABX-CBL	Abgenix	Mouse	CD147	Graft vs. host disease	Phase 2/3
L612	Chugai	Human	Ganglioside GM3	Melanoma	Phase 1
MORAb-028	Morphotek/Eisai	Human	Ganglioside GD2	Melanoma	Phase 1
AR-101	Aridis	Human	LPS	Nosocomial P. a. pneumonia	Phase 2a
mAb216	Academic (Stanford)	Human	CDIM	B-lineage ALL	Phase 1
PAT-SM6	Patrys	Human	GRP78	Multiple myeloma	Phase 1/2a
ARG098	Argenes	Mouse/Human (chimeric)	FAS	Rheumatoid arthritis	Phase 1/2
rHIgM22	Acorda	Human	CNS myelin	Multiple sclerosis/neuronal degeneration	Phase 1
TOL101	Tolera	Mouse	αβ TCR	Renal transplant	Phase 2

Antibodies **2020**, *9*, 53

Table 3. Source, production and immunogenicity of IgM antibodies in early phase clinical trials.

Target Antigen Class	Antibody	Antigen	IgM Source	Production Cell	Indication	Clinical Trial	Dose	Immunogenicity in Humans	Reference
Lipopolysaccharide	E5	J5 lipid A	Mouse B cells	Hybridoma with mouse myeloma	Sepsis	Phase 1	0.1, 0.5, 2, 7.5, 15 mg/kg	3 of 9 subjects	Harkonen 1988 [129]
	HA-1A	J5 lipid A	Human B cells	Heteromyeloma with lymphoma spleen cells	Sepsis	Phase 1	25, 100, 250 mg	0 of 34 subjects	Fisher 1990 [130]
	MAB-T88	LPS	Human B cells	Hybridoma with mouse myeloma	Neutropenia	Phase 1	1, 4, 8 mg/kg single dose	0 of 9 subjects	Daifuku 1992 [131]
	Mab cocktail (5 IgM)	LPS	Human B cells		Normal adults P. aeruginosa bacteremia	Phase 1	0.75 to 3.0 mg/kg	12 subjects 8 subjects	Saravolatz 1991 [132]
	AR-101 (KBPA-101)	LPS	Human B cells	Hybridoma with mouse myeloma	Nosocomial P. aeruginosa pneumonia	Phase 2a	1.2 mg/kg × 3	0 of 18 subjects	Lu 2011 [133]
Glycolipid/ Proteolipid	L612	Ganglioside GM3	Human B cells	EBV-transformed patient B cells	Melanoma	Phase 1	960, 1440, 1920 mg 48 h infusion 1 or 2	0 of 9 subjects	Irie 2004 [134]
	MORAb-028	Ganglioside GD2	Human B cells	Hybridoma with human/mouse myeloma	Melanoma	Phase 1	mg/cm^2/day × 5 days, repeated 2x	18 subjects	NCT-0112304 [135]
	rHIgM22	CNS myelin proteolipid	Human B cells	Hybridoma with mouse myeloma	Multiple sclerosis/neuronal degeneration diseases	Phase 1	0.025 to 2 mg/kgsingle dose	55 subjects	Eisen 2017 [136]
Glycan	Fanolesomab-Tc99	CD15 (carbohydrate)	Mouse B cells	Hybridoma with mouse myeloma	Healthy volunteers	Phase 1	125 µg × 2 (21 days apart)	5 of 30 subjects	Line 2004 [137]
	PAT-SC1	CD55 (glycan isoform)	Human B cells	Recombinant production Per.C6 cells	Gastric cancer	Phase 1	20 mg single dose	51 subjects	Hensel 2014 [138]
	mAb216	CDIM (carbohydrate)	Human B cells	Heteromyeloma with lymphoma spleen cells	B-lineage ALL	Phase 1	1.25 mg/kg to 5 mg/kg 3 + 2 dose escalation	0 of 13 subjects	Liedtke 2012 [139]
	PAT-SM6	GRP78 (O-linked glycan)	Human B cells	Recombinant production Per.C6 cells	Multiple myeloma	Phase 1	0.3, 1 3 or 6 mg/kg 4 doses over 2 weeks	0 of 12 subjects	Rasche 2015 [140]

Table 3. *Cont.*

Target Antigen Class	Antibody	Antigen	IgM Source	Production Cell	Indication	Clinical Trial	Dose	Immunogenicity in Humans	Reference
	Campath-1M	CD52	Rat B cells	Hybridoma with rat myeloma	Graft vs. host disease	Phase 2	25 mg bid × 10	(not tested)	Friend 1989 [141]
	Mab 16.88	Colon cancer antigen	Human B cells	Hybridoma with mouse myeloma	Colorectal cancer	Phase 1	8 mg, then 200, 500 or 1000 mg	0 of 20 subjects	Haisma 1991 [142]
	ABX-CBL	CD147	Mouse B cells	Hybridoma with mouse myeloma	Graft vs. host disease	Phase 1	0.2 to 0.3 mg/kg 9 doses	0 of 51 subjects	Deeg 2001 [143]
Protein	TOL101	αβ TCR	Mouse B cells	Hybridoma with mouse myeloma	Renal transplant	Phase 2	0.3, 1.4, 7, 14, 28, 42 mg 5 daily doses	1 of 36 subjects	Getts 2014 [144]
	ARG098	FAS	Mouse: Human B cells (chimeric)	Hybridoma with mouse myeloma	Rheumatoid arthritis	Phase 1/2	up to 10 µg/knee (intraarticular)	43 subjects	Matsubara 2013 [145]

6.1.1. Lipopolysaccharide Antigens

Five of the IgM product candidates (from a total of nine mAbs) in Table 3 targeted lipopolysaccharide (LPS), a component of the Gram-negative bacterial cell outer membrane, and two of these antibodies, E5 and HA-1A, were some of the earliest and most extensively studied IgM antibodies to enter clinical trials. LPS is highly inflammatory and has been the subject of numerous interventional strategies. E5, a murine anti-lipid A IgM mAb isolated by Lowell Young at UCLA (U.S. patent 4918163) and licensed to Xoma (Xomen-E5), and HA-1A, a human anti-lipid A IgM mAb isolated by Nelson Teng at Stanford [146] and licensed to Centocor (as Centoxin), entered clinical trials for sepsis in the early 1980s; both IgMs were evaluated in a number of clinical trials, and product license applications (PLAs) for both products were submitted to the FDA in early 1989 [147]. Centoxin^{TM} (nebacumab) received regulatory approval in Europe in 1992. However, this approval was withdrawn in 1993, following the inability of subsequent trials to demonstrate a clinical benefit [147].

Around this same time, several additional clinical trials were initiated with other anti-LPS antibodies. Chiron initiated a Phase 1 trial with MAB-T88, and reported that it was safe and well tolerated [131]. A second Phase 1 study was also conducted in six sepsis patients with a high likelihood of Gram-negative bacteremia [148]. MAB-T88 was again shown to be safe, but additional clinical trials were never conducted. Similarly, a cocktail of five human IgM anti-*Pseudomonas aeruginosa* LPS antibodies was tested in normal adults and in patients with *P. aeruginosa* bacteremia [132]. This cocktail also appears to have been well tolerated, but additional studies do not appear to have been conducted. More recently, Aridis tested AR-101, an anti-*P. aeruginosa* LPS IgM originally developed by Kenta (KBPA-101, panobacumab) [149], in patients with nosocomial pneumonia. These studies went as far as a Phase 2a trial [133], but more recent efforts appear to have focused on the IgG anti-LPS mAb AR-105 [150]. Similarly, multiple clinical trials have also been completed [151] or are in progress [152], using IgM-enriched IVIG for the treatment of sepsis or septic shock.

While the above results are, at first, discouraging, it is also now clear that many of the issues associated with the anti-LPS IgM mAb trials likely reflect the difficult nature of this clinical indication (numerous other therapeutics have failed in sepsis and infectious disease trials) [153] as well as the specific characteristics of the natural, non-affinity-matured mAbs tested [124].

6.1.2. Glycolipid and Proteolipid Antigens

Another three IgMs, in Table 3, target gangliosides or proteolipids. Of the two IgMs targeting gangliosides, L612 targets ganglioside GM3, while MORAb-028 targets ganglioside GD2. Antibody L612 was derived from Epstein–Bar virus (EBV)-transformed B cells from a patient with melanoma, and was shown to kill melanoma cells via complement-dependent cytotoxicity (CDC) [154]. However, when tested clinically in patients with melanoma, L612 showed no adverse side effects but lacked evidence of efficacy [134]. Subsequently, and because it lacked a J-chain, L612 preparations were found to contain roughly 20% hexameric and 74% pentameric forms of the IgM [155]. Since the hexameric form of L612 appeared to exhibit most of the CDC activity, a recombinant hexamer-dominant form of L612, CA19, was selected and produced approximately 80% hexamer from CHO cells. While promising in animals, CA19 does not appear to have been tested in the clinic.

MORAb-028 is an IgM that targets ganglioside GD2 licensed from Micromet and originally designated MT228. MORAb-028 entered two Phase 1 clinical trials in 2010, one for intratumoral injection [135] and one for IV administration of radiolabeled MORAb-028 [156]. The studies were both completed in 2012, but little information regarding their results is available. The program appears to have been discontinued in 2014 [157].

rHIgM22 binds to a complex myelin proteolipid antigen that is only expressed in CNS white matter and has been reported to promote remyelination in animal models [158]. It was developed at the Mayo Clinic and licensed to Acorda. rHIgM22 has been investigated following IV infusion in two Phase 1 studies in patients with multiple sclerosis, one starting in 2013 [159] and one in 2015 [160]. The results for the first study have been published [136]. In this study, rHIgM22 was well tolerated in the 55 patients treated, and the IgM was detected in the CSF, but no statistically significant changes were observed in the exploratory outcome measures.

6.1.3. Glycan Antigens

Despite the fact that IgM antibodies are well suited to target repetitive antigens [161], very few clinical trials testing carbohydrate-reactive IgM mAbs have been conducted to date. Two such studies are listed in Table 3. MAb216 recognizes a blood group antigen (CDIM) that is present on human B cells [162]. This antibody, also obtained by Nelsen Teng and colleagues at Stanford, was isolated from a patient with lymphoma and, after scale up at the NCI, was tested in a small Phase I clinical trial in patients with B cell acute lymphoblastic leukemia. While the results were encouraging [139], limited production of mAb216 by heteromyeloma cells inhibited further testing.

On the basis of these results, a recombinant human IgM variant of mAb216, termed IGM-55.5, was generated. The antigen recognized by both mAb216 and IGM-55.5 on human B cells is a linear lactosamine epitope that is sensitive to the enzyme endo-beta-galactosidase. This ligand, termed "cell death inducing molecule" (CDIM), is similar to the "i" antigen of cord blood red blood cells [163,164]. The "i" antigen is only found on the red blood cells of the developing fetus and newborn infants and, in rare cases, in human adults whose red blood cells did not convert this simple linear carbohydrate into the more complex branched carbohydrate named "I" antigen. Natural autoantibodies to the "i-antigen" often circulate in the blood of healthy adults and are usually of the IgM isotype.

Interestingly, IgM antibodies of this class often have heavy chain variable regions encoded by the human Vh4-34 V region gene [165], and they are able to kill antigen-expressing B cells via the formation of large complement-independent pores [166]. This process, which has also been reported for other IgM antibodies [167,168] and has similarities with oncosis, involves "wounding" target cells in a complement-independent manner such that large holes or pores are formed in the cell membrane. The precise mechanism by which these IgM antibodies mediate cell killing is not yet known, but it has been speculated that degradation of actin-associated proteins permits the aggregation of membrane components, thus leading to the formation of pores and loss of intracellular contents [168].

Patrys Ltd. in Australia was one of the first companies to focus on investigating the therapeutic potential of natural human IgM antibodies, and two of the candidates tested clinical target glycan-based epitopes. PAT-SC1, originally isolated by Peter Vollmers and colleagues at the Institute of Pathology at the University of Würzburg [169] targeted a specific glycoform on CD55 that appeared to be overexpressed on the surface of many cancer cells. PAT-SM6, also isolated by Vollmers [170], targeted an O-linked glycoform on GRP78, a multifunctional glucose-regulated protein that was possibly only present on tumor cells. Both PAT-SC1 [138] and PAT-SM6 [140] have completed Phase 1 trials, and both appear to have been well tolerated. Currently, only PAT-SC1 is still under development, having been licensed to Hefei Co-source Biomedical Co. in 2015, for all oncology indications in China [171].

Lastly, NeutroSpec[TM] (fanolesomab-Tc99m) is a radioimmunodiagnostic agent consisting of a murine IgM monoclonal antibody labeled with technetium-99m (^{99m}Tc). Fanolesomab is directed against the carbohydrate moiety 3-fucosyl-N-acetyl lactosamine that defines the cluster of differentiation 15 (CD15) antigen (NeutroSpec[TM] package insert) [172]. The CD15 antigen is expressed on the surface of polymorphonuclear neutrophils (PMNs), eosinophils, and monocytes, cells that are often localized in sites of infection. Initial clinical trials have indicated product safety [137] and, in 2004, NeutroSpec[TM] received FDA approval for scintigraphic imaging of patients with equivocal signs and symptoms of appendicitis who were five years of age or older. However, the product was suspended in 2005 following reports that patients taking the drug suffered serious and life-threatening cardiopulmonary events. NeutroSpec[TM] was subsequently discontinued in 2008.

6.1.4. Protein Antigens

One of the first IgM antibodies to be tested clinically was Campath-1M. This antibody, which recognized the lymphocyte antigen CD52, was an IgM mAb isolated by Herman Waldmann and colleagues from rats immunized with human lymphocytes [125]. Early clinical trials for the prevention of graft vs. host disease (GvHD) involved the ex vivo purging of donor allographs with Campath-1M plus complement were encouraging, and two patients (one with non-Hodgkin's lymphoma and one with acute lymphoblastic leukemia) received intravenous infusions with Campath-1M [125]. However, overall efficacy of the treatments was low and there were concerns regarding immunogenicity of the rat IgM [173].

Ultimately, Campath-1M was first class-switched to a rat IgG2b (Campath-1G) [174], and then became the first antibody to be humanized by successful transplantation of the six heavy and light chain variable regions from the rat IgG2b mAb into a human IgG1 [175], creating Campath-1H. Campath-1H was subsequently shown to be safe and effective in humans and is currently marketed under the trade name Lemtrada® (alemtuzumab) for B cell chronic lymphocytic leukemia [176].

In addition to Campath, four other IgM antibodies to protein antigen targets have been tested in clinical trials. Two of these IgM mAbs are of mouse origin (ABX-CBL, TOL101), one is chimeric (ARG098) and one is human (Mab 16.88). ABX-CBL (murine hybridoma-derived IgM) and TOL101 target human CD147 and the αβ T cell receptor, respectively. ABX-CBL was tested in patients with steroid-refractory acute graft-versus-host disease (aGvHD) at doses up to 0.3 mg/kg/day [143]. Among 51 evaluable patients in the Phase 1 study, roughly half (51%) responded following nine daily doses. However, in a randomized Phase 2/3 clinical trial (95 patients) in acute GvHD comparing ABX-CBL to standard of care, anti-thymocyte globulin (ATG), the patient outcomes were insignificantly different [177]. These data indicated that ABX-CBL did not offer improvement over ATG and as a result, further clinical development of ABX-CBL was terminated.

TOL101 targets the human αβ T cell receptor and was tested in renal transplant patients. Interestingly, in an effort to minimize T cell activation and its consequences that were observed with higher affinity IgG antibodies, this IgM was explored as a lower affinity/lower avidity therapeutic targeting this antigen. In a Phase 2 study [178], patients received five daily doses, up to 42 mg/day, and prolonged CD3 modulation occurred at doses above 28 mg. There were no cases of patient or graft loss, the treatments were well tolerated, and CD3 levels recovered within seven days after the cessation of therapy. No additional updates were found.

ARG098 is a mouse/human chimeric IgM antibody that targets FAS receptor and was tested in subjects with rheumatoid arthritis. Unlike the other IgM antibodies discussed here, ARG098 was administered via intraarticular injection into the knee at very low doses (up to 10 μg per knee). As ARG098 exhibited evidence of clinical activity, a placebo-controlled Phase 2a study was initiated and the program was partnered with Centocor, however trials were apparently discontinued in 2015 [179].

Lastly, 16.88 is a human IgM antibody that was derived from colorectal cancer patients immunized with autologous tumor cells admixed with BCG [180]. Of relevance to the current discussion is that all 13 of the natural human antibodies isolated in these studies, including 16.88, were of the IgM isotype. Following several pharmacokinetic studies in mice and humans [181], considerable efforts were made to examine the pharmacokinetics and tissue distribution of radiolabeled 16.88 in humans [182]. These studies demonstrated that 16.88 effectively targeted human tumors, and that it may be useful for radioimmunotherapy, but such studies were apparently not conducted.

6.2. IgM Pharmacokinetics

In 1964, Barth and colleagues published one of the first articles examining the pharmacokinetics (PK) of normal, unaltered human IgM antibodies in humans [183]. The IgM test material was purified from the serum of a healthy donor, radiolabeled with iodine-131, and then injected into seven normal adults. Serum samples were collected daily and analyzed in a gamma counter. According to these studies, the terminal half-life of normal human IgM in humans was calculated to be 5.1 days, with a range of 3.8 to 6.5 days (Table 4). Notably, these values for IgM half-life are four-fold less than the half-lives commonly reported for human IgGs in humans (e.g., 18–21 days) [184], most likely reflecting the fact that IgM antibodies do not bind to the recycling FcRn receptor (see Section 4.2).

Table 4. Pharmacokinetics of IgM antibodies in humans.

Antibody	Antigen	Indication	Model	Terminal Half-Life	Reference
Serum IgM (hu) I^{131}-labeled	-	Humans	Two-compartment	5.1 days (122 h)	Barth 1964 [183]
E5 (mu)	LPS (Lipid A)	Sepsis	One-compartment	19.3 h	Harkonen 1988 [129]
HA-1A (hu)	LPS (Lipid A)	Sepsis	One-compartment	15.9 h	Fisher 1990 [130]
		Sepsis	One-compartment	14.5 h	Romano 1993 [185]
MAB-T88	Lipopolysaccharide	neutropenia	Two-compartment	41.5 h	Daifuku 1992 [131]
AR-101	Lipopolysaccharide	Nosocomial pneumonia	Two-compartment	102 h (after 3rd dose)	Lu 2011 [133]
5G2	LPS (O-side chain)	Sepsis	One-compartment	56 h	Meng 1993 [186]
rHIgM22	CNS myelin proteolipid	Multiple sclerosis	(not stated)	99 h (2 mg/kg)	Eisen 2017 [136]
ABX-CBL	CD147	GvHD	Two-compartment	15–19 h	Deeg 2001 [143]
TOL101	ab TCR	Renal transplant	One-compartment	23.8 h	Getts 2014 [144]
PAT-SM6	GRP-78	Multiple myeloma	(not stated)	5.9 to 8.4 h	Rasche 2015 [140]
Fanolesomab-Tc99	CD15	Healthy volunteers	Two-compartment	8 h	Package insert [187]
Mab 16.88	Colon cancer antigen	Cancer	(not stated)	20 h	Haisma 1990 [181]

The pharmacokinetics of several therapeutic IgM mAbs have also been studied in some of the clinical trials described in Section 6.1 (see Table 3). In general, the half-lives reported for these IgMs in humans are shorter than that described for the preparation of normal human IgM tested previously (Table 4). Importantly, it should be noted that there are several critical differences between the IgM antibodies tested clinically and the prior preparation used for human PK studies. First, the material tested by Barth was pooled normal human IgM and, as such, it would not bind to human antigens, whereas many of the other IgMs subsequently tested were selected for binding to human antigens. As a consequence, the clinically tested IgMs would bind to tissues expressing those targets and would likely be cleared more quickly. Second, the material tested by Barth was isolated from human serum, whereas most of the other IgMs were produced in mouse, rat, or hamster (e.g., CHO) cells. Since changes in production host cells and culture conditions for IgGs are known to result in changes in glycosylation [188], and similar changes have been noted with IgM antibodies [189], such differences in PK are not unanticipated. Lastly, differences in analytical techniques (isotope vs. ELISA) and subject populations (normal vs. diseased) are also contributing factors.

Combined, these differences make direct comparisons of the reported data quite difficult, not only between trials but also with the published data for normal human IgM. However, despite these differences it is encouraging to note that IgM antibodies can have relatively long half-lives in humans, thereby allowing weekly or bi-weekly dosing in the clinic.

6.3. IgM Safety and Immunogenicity

As indicated in Table 3, a number of clinical trials have been conducted with rodent or human IgM antibodies in a range of clinical indications. For these trials, nearly 400 subjects were treated with doses up to 27 mg/kg, and no apparent safety issues were reported. Importantly, for the studies conducted with human IgM antibodies, little or no immune responses were noted. However, it should be emphasized that the specifics of the immunogenicity assays used, as well as their relative sensitivities, were not typically reported.

Of the IgM antibodies listed in Table 3, two products (E5 and HA-1A) were tested in Phase 2 and Phase 3 clinical trials that enrolled a large number of patients. Both of these antibodies target LPS, the outer-most layer on Gram-negative bacteria, and were tested in sepsis patients. In the Phase 3 trials alone, E5 was administered to approximately 715 patients [190,191], and HA-1A was administered to approximately 730 patients [192,193]. Thus, when combined with the subjects listed in Table 3 (*n* = 398), the total number of subjects treated with IgM antibodies was more than 1800 patients.

The observations that several human IgM antibodies have been safely administered in the clinic are particularly encouraging, given the theoretical concern that multivalent, high-avidity antibodies may exhibit off-target binding that could result in unexpected toxicities or rapid clearance. In some of the IgMs isolated as naturally occurring antibodies to tumor targets, there may be low affinity binding with high avidity which may contribute to unexpected, off-target binding. In the clinical studies reported to date, no such concerns have been raised. However, these concerns can only be addressed by further development and clinical testing of additional IgM antibody product candidates.

6.4. Other Oligomeric Antibody Forms

In addition to the more traditional IgM antibodies, a number of new molecular constructs have been generated that seek to approximate the hexameric structure of the IgM molecule. One such class of molecules, the HexaBody™, was generated by introducing mutations in the IgG heavy chain that allow oligomers up to hexamers form in a concentration-dependent fashion on the surface of cells [194]. The most advanced HexaBody™ in development is GEN1029, a mixture of two noncompeting anti-DR5 HexaBody™ molecules. A Phase 1/2 study of GEN1029 in patients with solid cancers was initiated in May, 2018 [195].

7. Future Applications of Therapeutic IgM

As our understanding of expression systems and manufacturing of IgM antibodies progresses, we anticipate the utilization of IgM as a new modality of engineered antibodies for treatment of various therapeutic indications. Most importantly, IgM has 10 or 12 binding sites and is capable of binding its antigen targets with high avidity. For cell surface targets where there is repetitive display on a cancer or other target cell, high avidity allows for multiple antigen engagements per IgM. As a consequence, IgMs are particularly well suited for targeting difficult antigens. In some earlier IgM-based development efforts (Sections 6.1.1 and 6.1.2), antibodies against tumor antigens consisting of carbohydrate moieties or glycolipids were evaluated in clinical trials. In these cases, the affinity of corresponding IgGs on these glycotopes can be insufficient for effective targeting, whereas the IgMs exhibit strong binding and effector function appropriate for biotherapeutic use. Another challenging aspect of selected-tumor targets is often the low expression observed on tumors, especially treatment-resistant tumors. IgM-based antitumor agents with high avidity may yield antibodies with increased potency on low expression or otherwise difficult targets.

Given the greater valency of IgM, these macromolecules offer considerable opportunity for higher order cross-linking of cell surface receptors. In addition, the flexibility of the IgM may provide the appropriate architecture for binding multiple targets on a cell surface. The potential for IgM-induced multimerization of cell surface targets makes the IgM an ideal candidate platform for developing TNF receptor superfamily agonists. For example, IgM antibodies directed to death receptor 4 [196] have shown excellent efficacy in vitro and in vivo. Wang et al. also demonstrated significant potency and enhanced efficacy with IgMs specific for death receptor 5 as compared with the corresponding agonist IgGs [197]. In subsequent investigations, Wang, et al. demonstrated strong in vivo efficacy on established tumors that exhibited resistance to anti-DR5 IgG therapy in murine xenograft models [198]. Similarly, recent studies with IgM antibodies targeting the receptor binding site of influenza B have shown excellent potency and broad cross-reactivity in vitro and in animal models [199].

Many of the earlier programs that tested IgM in human clinical trials used natural IgM antibodies often isolated from patients or humanized from a murine hybridoma. However, there is significant opportunity for more engineered versions of IgM, where the variable domains of an affinity matured IgG can be grafted onto IgM constant domains. This "domain swap" of affinity matured variable domains from IgG onto the backbone of IgM can lead to marked increases in binding avidity and potency of an engineered IgM. As a platform for engineering oligomeric binding units, IgM offers a much wider variety of multimeric interaction with antigens.

Although engineering of antigen binding sites can yield novel IgM constructs with improved antigen binding, there also exists additional unique sites on IgM for adding multispecific binding. For example, bispecific IgG antibodies and other bispecific variants of IgG exhibit extremely potent tumor targeting agents [200]. However, these antibodies have just a single binding site to a tumor antigen, instead of the two binding sites of a traditional IgG. In contrast, a bispecific IgM may allow very high avidity binding to difficult or rare tumor antigens, with selective engagement of T cells for efficient tumor cell killing. We found that fusion of a single CD3 binding domain to the J-chain allowed for the production of engineered bispecific IgM antibodies that exhibited controlled engagement of T cells. For example, we recently described the use of a CD3 binding unit fused to the J-chain to generate T cell-engaging bispecific IgM antibodies that contained 10 binding sites for a cancer antigen and a single binding site for CD3 [201]. A key feature of this approach was the ability to make fully assembled bispecific IgM antibodies in a single, high expressing cell line.

One such antibody, IGM-2323, is an anti-CD20 x CD3 IgM with "10 × 1" bispecificity (10 binding sites for CD20 and one binding site for CD3ε) [202]. This novel bispecific IgM has very potent activity via T-cell directed cytotoxicity (TDCC), and it also retains the robust CDC activity typical of an IgM. Importantly, this IgM platform for T cell engagement exhibits potent TDCC via a mechanism that does not lead to high levels of cytokine release in vitro or in animals. On the basis of these properties, IGM-2323 is currently being tested in clinical trials for treatment of refractory or resistant non-Hodgkin's lymphoma [202].

With renewed focus on IgM antibodies and the engineering of IgM antibodies, there may well be advantages inherent to the IgM platform that can yield improved biotherapeutic agents for treatment of unmet medical needs. The recently published three-dimensional structure of IgM Fc pentamer may also allow better understanding of this complex macromolecule [80]. We anticipate that the higher order valency of IgM with enhanced receptor cross-linking and the highly effective bispecific IgMs should provide new opportunities for antibody engineering and the development of more effective therapeutics.

Author Contributions: B.A.K., R.B., A.M.S., S.F.C., and M.S.P. all wrote and reviewed the manuscript, designed the figures, and edited the final manuscript. All authors have read and agreed to the published version of the manuscript.

Funding: This work was funded by IGM Biosciences, Inc.

Acknowledgments: The authors wish to thank Yuan Cao for assistance in assembling the clinical studies and pharmacokinetics data and for help in finalizing the manuscript, and Paul Hinton for compiling and producing the sequence alignments, plus generating IgM schematic figures.

Conflicts of Interest: The authors declare the following potential conflict of interest: B.A.K., A.M.S., S.F.C., and M.S.P. are currently employees of IGM Biosciences and hold stock in that company. R.B. was an employee of IGM Biosciences and holds stock in that company. The funders had no role in the design of the study; in the collection, analyses, or interpretation of data; in the writing of the manuscript, or in the decision to publish the results.

References

1. Nuttall, G. Experimente uber die bacterienfeindlichen Einflusse des theirischen Korpers. *Z. Hyg. Infektionskr.* **1888**, *4*, 353–395.
2. Black, C.A. A brief history of the discovery of the immunoglobulins and the origin of the modern immunoglobulin nomenclature. *Immunol. Cell Biol.* **1997**, *75*, 65–68. [CrossRef] [PubMed]
3. Kabat, E.A. The Molecular Weight of Antibodies. *J. Exp. Med.* **1939**, *69*, 103–118. [CrossRef] [PubMed]
4. Waldenstrom, J. Incipient myelomatisis or "essential" hyperglobulinemia with fibrinogenopenia—A new syndrome? *Acta Med. Scand.* **1944**, *67*, 216–247.
5. Wallenius, G.; Trautman, R.; Kunkel, H.G.; Franklin, E.C. Ultracentrifugal studies of major non-lipide electrophoretic components of normal human serum. *J. Biol. Chem.* **1957**, *225*, 253–267.

6. Potter, M. The early history of plasma cell tumors in mice, 1954–1976. *Adv. Cancer Res.* **2007**, *98*, 17–51. [CrossRef]

7. Ceppellini, R.; Dray, S.; Edelman, G.; Fahey, J.; Franek, F.; Franklin, E. Nomenclature for human immunoglobin. *Bull World Health Org* **1964**, *30*, 447–450.

8. Flajnik, M.F.; Du Pasquier, L. Evolution of innate and adaptive immunity: Can we draw a line? *Trends Immunol.* **2004**, *25*, 640–644. [CrossRef]

9. Litman, G.W.; Rast, J.P.; Fugmann, S.D. The origins of vertebrate adaptive immunity. *Nat. Rev. Immunol.* **2010**, *10*, 543–553. [CrossRef]

10. Flajnik, M.F. Comparative analyses of immunoglobulin genes: Surprises and portents. *Nat. Rev. Immunol.* **2002**, *2*, 688–698. [CrossRef]

11. Marchalonis, J.J.; Jensen, I.; Schluter, S.F. Structural, antigenic and evolutionary analyses of immunoglobulins and T cell receptors. *J. Mol. Recognit.* **2002**, *15*, 260–271. [CrossRef] [PubMed]

12. Dolder, F. Occurrence, isolation and interchain bridges of natural 7-S immunoglobulin M in human serum. *Biochim. Biophys. Acta* **1971**, *236*, 675–685. [PubMed]

13. Eskeland, T.; Christensen, T.B. IgM molecules with and without J chain in serum and after purification, studied by ultracentrifugation, electrophoresis, and electron microscopy. *Scand. J. Immunol.* **1975**, *4*, 217–228. [CrossRef] [PubMed]

14. Hadji-Azimi, I.; Michea-Hamzehpour, M. Xenopus laevis 19S immunoglobulin. Ultrastructure and J chain isolation. *Immunology* **1976**, *30*, 587–591. [PubMed]

15. Fillatreau, S.; Six, A.; Magadan, S.; Castro, R.; Sunyer, J.O.; Boudinot, P. The astonishing diversity of Ig classes and B cell repertoires in teleost fish. *Front. Immunol.* **2013**, *4*, 28. [CrossRef]

16. Getahun, A.; Lundqvist, M.; Middleton, D.; Warr, G.; Pilstrom, L. Influence of the mu-chain C-terminal sequence on polymerization of immunoglobulin M. *Immunology* **1999**, *97*, 408–413. [CrossRef]

17. Cattaneo, A.; Neuberger, M.S. Polymeric immunoglobulin M is secreted by transfectants of non-lymphoid cells in the absence of immunoglobulin J chain. *EMBO J.* **1987**, *6*, 2753–2758. [CrossRef] [PubMed]

18. Davis, A.C.; Roux, K.H.; Shulman, M.J. On the structure of polymeric IgM. *Eur. J. Immunol.* **1988**, *18*, 1001–1008. [CrossRef] [PubMed]

19. Davis, A.C.; Shulman, M.J. IgM—Molecular requirements for its assembly and function. *Immunol. Today* **1989**, *10*, 118–122. [CrossRef]

20. Collins, C.; Tsui, F.W.; Shulman, M.J. Differential activation of human and guinea pig complement by pentameric and hexameric IgM. *Eur. J. Immunol.* **2002**, *32*, 1802–1810. [CrossRef]

21. Duramad, O.; Wang, B.; Zheng, F.; Keyt, L.; Repellin, C.; Beviglia, L.; Bhat, N.; Bieber, M.; Teng, N.; Keyt, B. Abstract 645: IGM-55.5, a novel monoclonal human recombinant IgM antibody with potent activity against B cell leukemia and lymphoma. *J. Cancer Res.* **2014**, *74*, 645. [CrossRef]

22. Pettinello, R.; Dooley, H. The immunoglobulins of cold-blooded vertebrates. *Biomolecules* **2014**, *4*, 1045–1069. [CrossRef] [PubMed]

23. Kaetzel, C.S. Coevolution of Mucosal Immunoglobulins and the Polymeric Immunoglobulin Receptor: Evidence That the Commensal Microbiota Provided the Driving Force. *ISRN Immunol.* **2014**, *2014*, 1–20. [CrossRef]

24. Winkler, T.H.; Martensson, I.L. The Role of the Pre-B Cell Receptor in B Cell Development, Repertoire Selection, and Tolerance. *Front. Immunol.* **2018**, *9*, 2423. [CrossRef]

25. Hobeika, E.; Maity, P.C.; Jumaa, H. Control of B Cell Responsiveness by Isotype and Structural Elements of the Antigen Receptor. *Trends Immunol.* **2016**, *37*, 310–320. [CrossRef]

26. Deane, J.A.; Fruman, D.A. Phosphoinositide 3-kinase: Diverse roles in immune cell activation. *Annu. Rev. Immunol.* **2004**, *22*, 563–598. [CrossRef]

27. Johnson, S.A.; Pleiman, C.M.; Pao, L.; Schneringer, J.; Hippen, K.; Cambier, J.C. Phosphorylated immunoreceptor signaling motifs (ITAMs) exhibit unique abilities to bind and activate Lyn and Syk tyrosine kinases. *J. Immunol.* **1995**, *155*, 4596–4603.

28. Rolli, V.; Gallwitz, M.; Wossning, T.; Flemming, A.; Schamel, W.W.; Zurn, C.; Reth, M. Amplification of B cell antigen receptor signaling by a Syk/ITAM positive feedback loop. *Mol. Cell* **2002**, *10*, 1057–1069. [CrossRef]

29. Werner, M.; Hobeika, E.; Jumaa, H. Role of PI3K in the generation and survival of B cells. *Immunol. Rev.* **2010**, *237*, 55–71. [CrossRef]

30. Van Furth, R.; Schuit, H.R.; Hijmans, W. The immunological development of the human fetus. *J. Exp. Med.* **1965**, *122*, 1173–1188. [CrossRef]

31. Alford, C.A., Jr.; Foft, J.W.; Blankenship, W.J.; Cassady, G.; Benton, J.W., Jr. Subclinical central nervous system disease of neonates: A prospective study of infants born with increased levels of IgM. *J. Pediatr.* **1969**, *75*, 1167–1178. [CrossRef]

32. Stiehm, E.R.; Ammann, A.J.; Cherry, J.D. Elevated cord macroglobulins in the diagnosis of intrauterine infections. *N. Engl. J. Med.* **1966**, *275*, 971–977. [CrossRef] [PubMed]

33. Meffre, E.; Salmon, J.E. Autoantibody selection and production in early human life. *J. Clin. Investig.* **2007**, *117*, 598–601. [CrossRef] [PubMed]

34. Weitkamp, J.H.; Lewis, D.B.; Levy, O. Immunology of the Fetus and Newborn. In *Avery's Diseases of the Newborn*, 10th ed.; Gleason, C., Juul, S., Eds.; Elsevier: Amsterdam, The Netherlands, 2018; Volume 36, pp. 453–481.e7.

35. Gonzalez-Quintela, A.; Alende, R.; Gude, F.; Campos, J.; Rey, J.; Meijide, L.M.; Fernandez-Merino, C.; Vidal, C. Serum levels of immunoglobulins (IgG, IgA, IgM) in a general adult population and their relationship with alcohol consumption, smoking and common metabolic abnormalities. *Clin. Exp. Immunol.* **2008**, *151*, 42–50. [CrossRef] [PubMed]

36. Lewis, D.B.; Wilson, C.B. Developmental Immunology and Role of Host Defenses in Fetal and Neonatal Susceptibility to Infection. In *Infectious Diseases of the Fetus and Newborn Infant*, 6th ed.; Remington, J.S., Klein, J.O., Wilson, C.B., Baker, C.J., Eds.; Elsevier: Amsterdam, The Netherlands, 2006; Volume 4, pp. 87–210.

37. Boes, M. Role of natural and immune IgM antibodies in immune responses. *Mol. Immunol.* **2000**, *37*, 1141–1149. [CrossRef]

38. Holodick, N.E.; Tumang, J.R.; Rothstein, T.L. Immunoglobulin secretion by B1 cells: Differential intensity and IRF4-dependence of spontaneous IgM secretion by peritoneal and splenic B1 cells. *Eur. J. Immunol.* **2010**, *40*, 3007–3016. [CrossRef]

39. Jayasekera, J.P.; Moseman, E.A.; Carroll, M.C. Natural antibody and complement mediate neutralization of influenza virus in the absence of prior immunity. *J. Virol.* **2007**, *81*, 3487–3494. [CrossRef]

40. Chen, Z.J.; Wheeler, C.J.; Shi, W.; Wu, A.J.; Yarboro, C.H.; Gallagher, M.; Notkins, A.L. Polyreactive antigen-binding B cells are the predominant cell type in the newborn B cell repertoire. *Eur. J. Immunol.* **1998**, *28*, 989–994. [CrossRef]

41. Choi, Y.S.; Dieter, J.A.; Rothaeusler, K.; Luo, Z.; Baumgarth, N. B-1 cells in the bone marrow are a significant source of natural IgM. *Eur. J. Immunol.* **2012**, *42*, 120–129. [CrossRef]

42. Forster, I.; Rajewsky, K. Expansion and functional activity of Ly-1+ B cells upon transfer of peritoneal cells into allotype-congenic, newborn mice. *Eur. J. Immunol.* **1987**, *17*, 521–528. [CrossRef]

43. Lalor, P.A.; Herzenberg, L.A.; Adams, S.; Stall, A.M. Feedback regulation of murine Ly-1 B cell development. *Eur. J. Immunol.* **1989**, *19*, 507–513. [CrossRef] [PubMed]

44. Savage, H.P.; Baumgarth, N. Characteristics of natural antibody-secreting cells. *Ann. N. Y. Acad. Sci.* **2015**, *1362*, 132–142. [CrossRef] [PubMed]

45. Van Oudenaren, A.; Haaijman, J.J.; Benner, R. Frequencies of background cytoplasmic Ig-containing cells in various lymphoid organs of athymic and euthymic mice as a function of age and immune status. *Immunology* **1984**, *51*, 735–742. [PubMed]

46. Reynolds, A.E.; Kuraoka, M.; Kelsoe, G. Natural IgM is produced by CD5- plasma cells that occupy a distinct survival niche in bone marrow. *J. Immunol.* **2015**, *194*, 231–242. [CrossRef] [PubMed]

47. Griffin, D.O.; Holodick, N.E.; Rothstein, T.L. Human B1 cells in umbilical cord and adult peripheral blood express the novel phenotype CD20+ CD27+ CD43+ CD70. *J. Exp. Med.* **2011**, *208*, 67–80. [CrossRef]

48. Gronwall, C.; Silverman, G.J. Natural IgM: Beneficial autoantibodies for the control of inflammatory and autoimmune disease. *J. Clin. Immunol.* **2014**, *34* (Suppl. 1), S12–S21. [CrossRef]

49. Casey, R.; Newcombe, J.; McFadden, J.; Bodman-Smith, K.B. The acute-phase reactant C-reactive protein binds to phosphorylcholine-expressing Neisseria meningitidis and increases uptake by human phagocytes. *Infect. Immun.* **2008**, *76*, 1298–1304. [CrossRef]

50. Chou, M.Y.; Fogelstrand, L.; Hartvigsen, K.; Hansen, L.F.; Woelkers, D.; Shaw, P.X.; Choi, J.; Perkmann, T.; Backhed, F.; Miller, Y.I.; et al. Oxidation-specific epitopes are dominant targets of innate natural antibodies in mice and humans. *J. Clin. Investig.* **2009**, *119*, 1335–1349. [CrossRef]

51. Kearney, J.F.; Patel, P.; Stefanov, E.K.; King, R.G. Natural antibody repertoires: Development and functional role in inhibiting allergic airway disease. *Annu. Rev. Immunol.* **2015**, *33*, 475–504. [CrossRef]

52. Shaw, P.X.; Horkko, S.; Chang, M.K.; Curtiss, L.K.; Palinski, W.; Silverman, G.J.; Witztum, J.L. Natural antibodies with the T15 idiotype may act in atherosclerosis, apoptotic clearance, and protective immunity. *J. Clin. Investig.* **2000**, *105*, 1731–1740. [CrossRef]

53. Baumgarth, N. The double life of a B-1 cell: Self-reactivity selects for protective effector functions. *Nat. Rev. Immunol.* **2011**, *11*, 34–46. [CrossRef] [PubMed]

54. Gronwall, C.; Vas, J.; Silverman, G.J. Protective Roles of Natural IgM Antibodies. *Front. Immunol.* **2012**, *3*, 66. [CrossRef] [PubMed]

55. Louis, A.G.; Gupta, S. Primary selective IgM deficiency: An ignored immunodeficiency. *Clin. Rev. Allergy Immunol.* **2014**, *46*, 104–111. [CrossRef]

56. Nguyen, T.T.; Elsner, R.A.; Baumgarth, N. Natural IgM prevents autoimmunity by enforcing B cell central tolerance induction. *J. Immunol.* **2015**, *194*, 1489–1502. [CrossRef] [PubMed]

57. Capolunghi, F.; Rosado, M.M.; Sinibaldi, M.; Aranburu, A.; Carsetti, R. Why do we need IgM memory B cells? *Immunol. Lett.* **2013**, *152*, 114–120. [CrossRef]

58. Bemark, M. Translating transitions—How to decipher peripheral human B cell development. *J. Biomed. Res.* **2015**, *29*, 264–284. [CrossRef]

59. Huang, C.; Stewart, A.K.; Schwartz, R.S.; Stollar, B.D. Immunoglobulin heavy chain gene expression in peripheral blood B lymphocytes. *J. Clin. Investig.* **1992**, *89*, 1331–1343. [CrossRef]

60. van Es, J.H.; Meyling, F.H.; Logtenberg, T. High frequency of somatically mutated IgM molecules in the human adult blood B cell repertoire. *Eur. J. Immunol.* **1992**, *22*, 2761–2764. [CrossRef]

61. Klein, U.; Kuppers, R.; Rajewsky, K. Evidence for a large compartment of IgM-expressing memory B cells in humans. *Blood* **1997**, *89*, 1288–1298. [CrossRef]

62. Krishnamurty, A.T.; Thouvenel, C.D.; Portugal, S.; Keitany, G.J.; Kim, K.S.; Holder, A.; Crompton, P.D.; Rawlings, D.J.; Pepper, M. Somatically Hypermutated Plasmodium-Specific IgM(+) Memory B Cells Are Rapid, Plastic, Early Responders upon Malaria Rechallenge. *Immunity* **2016**, *45*, 402–414. [CrossRef]

63. Heyman, B. Antibodies as natural adjuvants. *Curr. Top. Microbiol. Immunol.* **2014**, *382*, 201–219. [CrossRef] [PubMed]

64. Xu, H.; Zhang, L.; Heyman, B. IgG-mediated immune suppression in mice is epitope specific except during high epitope density conditions. *Sci. Rep.* **2018**, *8*, 15292. [CrossRef] [PubMed]

65. Bergstrom, J.J.; Heyman, B. IgG Suppresses Antibody Responses in Mice Lacking C1q, C3, Complement Receptors 1 and 2, or IgG Fc-Receptors. *PLoS ONE* **2015**, *10*, e0143841. [CrossRef] [PubMed]

66. Sorman, A.; Westin, A.; Heyman, B. IgM is Unable to Enhance Antibody Responses in Mice Lacking C1q or C3. *Scand. J. Immunol.* **2017**, *85*, 381–382. [CrossRef] [PubMed]

67. Sorman, A.; Zhang, L.; Ding, Z.; Heyman, B. How antibodies use complement to regulate antibody responses. *Mol. Immunol.* **2014**, *61*, 79–88. [CrossRef]

68. Rutemark, C.; Bergman, A.; Getahun, A.; Hallgren, J.; Henningsson, F.; Heyman, B. Complement receptors 1 and 2 in murine antibody responses to IgM-complexed and uncomplexed sheep erythrocytes. *PLoS ONE* **2012**, *7*, e41968. [CrossRef]

69. Johansen, F.E.; Braathen, R.; Brandtzaeg, P. Role of J chain in secretory immunoglobulin formation. *Scand. J. Immunol.* **2000**, *52*, 240–248. [CrossRef]

70. Brandtzaeg, P. Secretory IgA: Designed for Anti-Microbial Defense. *Front. Immunol.* **2013**, *4*, 222. [CrossRef]

71. Davis, A.C.; Roux, K.H.; Pursey, J.; Shulman, M.J. Intermolecular disulfide bonding in IgM: Effects of replacing cysteine residues in the mu heavy chain. *EMBO J.* **1989**, *8*, 2519–2526. [CrossRef]

72. Fazel, S.; Wiersma, E.J.; Shulman, M.J. Interplay of J chain and disulfide bonding in assembly of polymeric IgM. *Int. Immunol.* **1997**, *9*, 1149–1158. [CrossRef] [PubMed]

73. Wiersma, E.J.; Shulman, M.J. Assembly of IgM. Role of disulfide bonding and noncovalent interactions. *J. Immunol.* **1995**, *154*, 5265–5272. [PubMed]

74. Kabat, E.A.; Te Wu, T.; Perry, H.M.; Foeller, C.; Gottesman, K.S. *Sequences of Proteins of Immunological Interest*; U.S. Department of Health and Human Services, Public Health Service, National Institutes of Health: Bethesda, MD, USA, 1991.

75. Sorensen, V.; Sundvold, V.; Michaelsen, T.E.; Sandlie, I. Polymerization of IgA and IgM: Roles of Cys309/Cys414 and the secretory tailpiece. *J. Immunol.* **1999**, *162*, 3448–3455. [PubMed]

76. Pasalic, D.; Weber, B.; Giannone, C.; Anelli, T.; Muller, R.; Fagioli, C.; Felkl, M.; John, C.; Mossuto, M.F.; Becker, C.F.W.; et al. A peptide extension dictates IgM assembly. *Proc. Natl. Acad. Sci. USA* **2017**, *114*, E8575–E8584. [CrossRef]

77. Smith, R.I.; Coloma, M.J.; Morrison, S.L. Addition of a mu-tailpiece to IgG results in polymeric antibodies with enhanced effector functions including complement-mediated cytolysis by IgG4. *J. Immunol.* **1995**, *154*, 2226–2236. [PubMed]

78. Frutiger, S.; Hughes, G.J.; Paquet, N.; Luthy, R.; Jaton, J.C. Disulfide bond assignment in human J chain and its covalent pairing with immunoglobulin M. *Biochemistry* **1992**, *31*, 12643–12647. [CrossRef] [PubMed]

79. Sorensen, V.; Rasmussen, I.B.; Sundvold, V.; Michaelsen, T.E.; Sandlie, I. Structural requirements for incorporation of J chain into human IgM and IgA. *Int. Immunol.* **2000**, *12*, 19–27. [CrossRef]

80. Li, Y.; Wang, G.; Li, N.; Wang, Y.; Zhu, Q.; Chu, H.; Wu, W.; Tan, Y.; Yu, F.; Su, X.D.; et al. Structural insights into immunoglobulin M. *Science* **2020**, *367*, 1014–1017. [CrossRef]

81. Kumar, N.; Arthur, C.P.; Ciferri, C.; Matsumoto, M.L. Structure of the secretory immunoglobulin a core. *Science* **2020**, *367*, 1008–1014. [CrossRef]

82. Braathen, R.; Hohman, V.S.; Brandtzaeg, P.; Johansen, F.E. Secretory antibody formation: Conserved binding interactions between J chain and polymeric Ig receptor from humans and amphibians. *J. Immunol.* **2007**, *178*, 1589–1597. [CrossRef]

83. Moh, E.S.; Lin, C.H.; Thaysen-Andersen, M.; Packer, N.H. Site-Specific N-Glycosylation of Recombinant Pentameric and Hexameric Human IgM. *J. Am. Soc. Mass Spectrom.* **2016**, *27*, 1143–1155. [CrossRef]

84. Muraoka, S.; Shulman, M.J. Structural requirements for IgM assembly and cytolytic activity. Effects of mutations in the oligosaccharide acceptor site at Asn402. *J. Immunol.* **1989**, *142*, 695–701. [PubMed]

85. Maiorella, B.L.; Winkelhake, J.; Young, J.; Moyer, B.; Bauer, R.; Hora, M.; Andya, J.; Thomson, J.; Patel, T.; Parekh, R. Effect of culture conditions on IgM antibody structure, pharmacokinetics and activity. *Biotechnology (N. Y.)* **1993**, *11*, 387–392. [CrossRef] [PubMed]

86. Colucci, M.; Stockmann, H.; Butera, A.; Masotti, A.; Baldassarre, A.; Giorda, E.; Petrini, S.; Rudd, P.M.; Sitia, R.; Emma, F.; et al. Sialylation of N-linked glycans influences the immunomodulatory effects of IgM on T cells. *J. Immunol.* **2015**, *194*, 151–157. [CrossRef]

87. Feinstein, A.; Munn, E.A. Conformation of the free and antigen-bound IgM antibody molecules. *Nature* **1969**, *224*, 1307–1309. [CrossRef]

88. Perkins, S.J.; Nealis, A.S.; Sutton, B.J.; Feinstein, A. Solution structure of human and mouse immunoglobulin M by synchrotron X-ray scattering and molecular graphics modelling. A possible mechanism for complement activation. *J. Mol. Biol.* **1991**, *221*, 1345–1366. [CrossRef]

89. Czajkowsky, D.M.; Shao, Z. The human IgM pentamer is a mushroom-shaped molecule with a flexural bias. *Proc. Natl. Acad. Sci. USA* **2009**, *106*, 14960–14965. [CrossRef] [PubMed]

90. Muller, R.; Grawert, M.A.; Kern, T.; Madl, T.; Peschek, J.; Sattler, M.; Groll, M.; Buchner, J. High-resolution structures of the IgM Fc domains reveal principles of its hexamer formation. *Proc. Natl. Acad. Sci. USA* **2013**, *110*, 10183–10188. [CrossRef]

91. Hiramoto, E.; Tsutsumi, A.; Suzuki, R.; Matsuoka, S.; Arai, S.; Kikkawa, M.; Miyazaki, T. The IgM pentamer is an asymmetric pentagon with an open groove that binds the AIM protein. *Sci. Adv.* **2018**, *4*, eaau1199. [CrossRef] [PubMed]

92. Zikan, J.; Novotny, J.; Trapane, T.L.; Koshland, M.E.; Urry, D.W.; Bennett, J.C.; Mestecky, J. Secondary structure of the immunoglobulin J chain. *Proc. Natl. Acad. Sci. USA* **1985**, *82*, 5905–5909. [CrossRef]

93. Vollmers, H.P.; Brandlein, S. Natural IgM antibodies: The orphaned molecules in immune surveillance. *Adv. Drug Deliv. Rev.* **2006**, *58*, 755–765. [CrossRef]

94. Wibroe, P.P.; Helvig, S.Y.; Moein Moghimi, S. The Role of Complement in Antibody Therapy for Infectious Diseases. *Microbiol. Spectr.* **2014**, *2*, 63–74. [CrossRef]

95. Strohl, W.R.; Strohl, L.M. *Therapeutic Antibody Engineering*; Woodhead Publishing: Sawston, UK, 2012; pp. 197–223.

96. Klimovich, V.B. IgM and its receptors: Structural and functional aspects. *Biochemistry (Moscow)* **2011**, *76*, 534–549. [CrossRef] [PubMed]

97. Sharp, T.H.; Boyle, A.L.; Diebolder, C.A.; Kros, A.; Koster, A.J.; Gros, P. Insights into IgM-mediated complement activation based on in situ structures of IgM-C1-C4b. *Proc. Natl. Acad. Sci. USA* **2019**, *116*, 11900–11905. [CrossRef] [PubMed]

98. Shibuya, A.; Sakamoto, N.; Shimizu, Y.; Shibuya, K.; Osawa, M.; Hiroyama, T.; Eyre, H.J.; Sutherland, G.R.; Endo, Y.; Fujita, T.; et al. Fc alpha/mu receptor mediates endocytosis of IgM-coated microbes. *Nat. Immunol.* **2000**, *1*, 441–446. [CrossRef]

99. Weinstein, J.R.; Quan, Y.; Hanson, J.F.; Colonna, L.; Iorga, M.; Honda, S.; Shibuya, K.; Shibuya, A.; Elkon, K.B.; Moller, T. IgM-Dependent Phagocytosis in Microglia Is Mediated by Complement Receptor 3, Not Fcalpha/mu Receptor. *J. Immunol.* **2015**, *195*, 5309–5317. [CrossRef]

100. Mostov, K.E.; Blobel, G. A transmembrane precursor of secretory component. The receptor for transcellular transport of polymeric immunoglobulins. *J. Biol. Chem.* **1982**, *257*, 11816–11821.

101. Krajci, P.; Solberg, R.; Sandberg, M.; Oyen, O.; Jahnsen, T.; Brandtzaeg, P. Molecular cloning of the human transmembrane secretory component (poly-Ig receptor) and its mRNA expression in human tissues. *Biochem. Biophys. Res. Commun.* **1989**, *158*, 783–789. [CrossRef]

102. Mostov, K.E.; Friedlander, M.; Blobel, G. Structure and function of the receptor for polymeric immunoglobulins. *Biochem. Soc. Symp.* **1986**, *51*, 113–115.

103. Mostov, K.E.; Altschuler, Y.; Chapin, S.J.; Enrich, C.; Low, S.H.; Luton, F.; Richman-Eisenstat, J.; Singer, K.L.; Tang, K.; Weimbs, T. Regulation of protein traffic in polarized epithelial cells: The polymeric immunoglobulin receptor model. *Cold Spring Harb. Symp. Quant. Biol.* **1995**, *60*, 775–781. [CrossRef]

104. Norderhaug, I.N.; Johansen, F.E.; Krajci, P.; Brandtzaeg, P. Domain deletions in the human polymeric Ig receptor disclose differences between its dimeric IgA and pentameric IgM interaction. *Eur. J. Immunol.* **1999**, *29*, 3401–3409. [CrossRef]

105. Sakamoto, N.; Shibuya, K.; Shimizu, Y.; Yotsumoto, K.; Miyabayashi, T.; Sakano, S.; Tsuji, T.; Nakayama, E.; Nakauchi, H.; Shibuya, A. A novel Fc receptor for IgA and IgM is expressed on both hematopoietic and non-hematopoietic tissues. *Eur. J. Immunol.* **2001**, *31*, 1310–1316. [CrossRef]

106. Kikuno, K.; Kang, D.W.; Tahara, K.; Torii, I.; Kubagawa, H.M.; Ho, K.J.; Baudino, L.; Nishizaki, N.; Shibuya, A.; Kubagawa, H. Unusual biochemical features and follicular dendritic cell expression of human Fcalpha/mu receptor. *Eur. J. Immunol.* **2007**, *37*, 3540–3550. [CrossRef]

107. Ghumra, A.; Shi, J.; McIntosh, R.S.; Rasmussen, I.B.; Braathen, R.; Johansen, F.E.; Sandlie, I.; Mongini, P.K.; Areschoug, T.; Lindahl, G.; et al. Structural requirements for the interaction of human IgM and IgA with the human Fcalpha/mu receptor. *Eur. J. Immunol.* **2009**, *39*, 1147–1156. [CrossRef]

108. Kubagawa, H.; Oka, S.; Kubagawa, Y.; Torii, I.; Takayama, E.; Kang, D.W.; Gartland, G.L.; Bertoli, L.F.; Mori, H.; Takatsu, H.; et al. Identity of the elusive IgM Fc receptor (FcmuR) in humans. *J. Exp. Med.* **2009**, *206*, 2779–2793. [CrossRef] [PubMed]

109. Vire, B.; David, A.; Wiestner, A. TOSO, the Fcmicro receptor, is highly expressed on chronic lymphocytic leukemia B cells, internalizes upon IgM binding, shuttles to the lysosome, and is downregulated in response to TLR activation. *J. Immunol.* **2011**, *187*, 4040–4050. [CrossRef] [PubMed]

110. Honjo, K.; Kubagawa, Y.; Kearney, J.F.; Kubagawa, H. Unique ligand-binding property of the human IgM Fc receptor. *J. Immunol.* **2015**, *194*, 1975–1982. [CrossRef] [PubMed]

111. Honjo, K.; Kubagawa, Y.; Kubagawa, H. Is Toso/IgM Fc receptor (FcmuR) expressed by innate immune cells? *Proc. Natl. Acad. Sci. USA* **2013**, *110*, E2540–E2541. [CrossRef] [PubMed]

112. Hensel, F.; Gagnon, P. An effective platform for purification of IgM monoclonal antibodies using Hydroxyapatite. In Proceedings of the 5th International Conference on Hydroxyapatite and Related Products, Rottach-Egern, Germany, 11–14 October 2009.

113. Gagnon, P. Improved antibody aggregate removal by hydroxyapatite chromatography in the presence of polyethylene glycol. *J. Immunol. Methods* **2008**, *336*, 222–228. [CrossRef]

114. Butler, M.; Meneses-Acosta, A. Recent advances in technology supporting biopharmaceutical production from mammalian cells. *Appl. Microbiol. Biotechnol.* **2012**, *96*, 885–894. [CrossRef]

115. Kim, J.Y.; Kim, Y.G.; Lee, G.M. CHO cells in biotechnology for production of recombinant proteins: Current state and further potential. *Appl. Microbiol. Biotechnol.* **2012**, *93*, 917–930. [CrossRef]

116. Kunert, R.; Wolbank, S.; Stiegler, G.; Weik, R.; Katinger, H. Characterization of molecular features, antigen-binding, and in vitro properties of IgG and IgM variants of 4E10, an anti-HIV type 1 neutralizing monoclonal antibody. *AIDS Res. Hum. Retrovir.* **2004**, *20*, 755–762. [CrossRef] [PubMed]

117. Tchoudakova, A.; Hensel, F.; Murillo, A.; Eng, B.; Foley, M.; Smith, L.; Schoenen, F.; Hildebrand, A.; Kelter, A.R.; Ilag, L.L.; et al. High level expression of functional human IgMs in human PER.C6 cells. *MAbs* **2009**, *1*, 163–171. [CrossRef] [PubMed]

118. Loos, A.; Gruber, C.; Altmann, F.; Mehofer, U.; Hensel, F.; Grandits, M.; Oostenbrink, C.; Stadlmayr, G.; Furtmuller, P.G.; Steinkellner, H. Expression and glycoengineering of functionally active heteromultimeric IgM in plants. *Proc. Natl. Acad. Sci. USA* **2014**, *111*, 6263–6268. [CrossRef] [PubMed]

119. Valasek, C.; Cole, F.; Hensel, F.; Ye, P.; Conner, M.A.; Ultee, M.E. Production and Purification of a PER.C6-Expressed IgM Antibody Therapeutic. *BioProcess Int.* **2011**, *9*, 10.

120. Steindl, F.; Jungbauer, A.; Wenisch, E.; Himmler, G.; Katinger, H. Isoelectric precipitation and gel chromatography for purification of monoclonal IgM. *Enzym. Microb. Technol.* **1987**, *9*, 361–364. [CrossRef]

121. Gagnon, P.; Hensel, F.; Richieri, R. Recent Advances in the Purification of IgM Monoclonal Antibodies. Available online: http://www.validated.com/revalbio/pdffiles/PUR07a.pdf (accessed on 11 October 2020).

122. Gagnon, P.; Hensel, F.; Lee, S.; Zaidi, S. Chromatographic behavior of IgM:DNA complexes. *J. Chromatogr. A* **2011**, *1218*, 2405–2412. [CrossRef]

123. Hanala, S. The new ParaDIgm: IgM from bench to clinic: November 15-16, 2011, Frankfurt, Germany. *MAbs* **2012**, *4*, 555–561. [CrossRef]

124. Gong, S.; Tomusange, K.; Kulkarni, V.; Adeniji, O.S.; Lakhashe, S.K.; Hariraju, D.; Strickland, A.; Plake, E.; Frost, P.A.; Ratcliffe, S.J.; et al. Anti-HIV IgM protects against mucosal SHIV transmission. *Aids* **2018**, *32*, F5–F13. [CrossRef]

125. Hale, G.; Bright, S.; Chumbley, G.; Hoang, T.; Metcalf, D.; Munro, A.J.; Waldmann, H. Removal of T cells from bone marrow for transplantation: A monoclonal antilymphocyte antibody that fixes human complement. *Blood* **1983**, *62*, 873–882. [CrossRef]

126. Drugs@FDA: FDA-Approved Drugs. Available online: https://www.accessdata.fda.gov/scripts/cder/daf/ (accessed on 11 October 2020).

127. Vollmers, H.P.; Brandlein, S. Nature's best weapons to fight cancer. Revival of human monoclonal IgM antibodies. *Hum. Antibodies* **2002**, *11*, 131–142. [CrossRef]

128. Heimburg-Molinaro, J.; Rittenhouse-Olson, K. Development and characterization of antibodies to carbohydrate antigens. *Methods Mol. Biol.* **2009**, *534*, 341–357. [CrossRef] [PubMed]

129. Harkonen, S.; Scannon, P.; Mischak, R.P.; Spitler, L.E.; Foxall, C.; Kennedy, D.; Greenberg, R. Phase I study of a murine monoclonal anti-lipid A antibody in bacteremic and nonbacteremic patients. *Antimicrob. Agents Chemother.* **1988**, *32*, 710–716. [CrossRef] [PubMed]

130. Fisher, C.J., Jr.; Zimmerman, J.; Khazaeli, M.B.; Albertson, T.E.; Dellinger, R.P.; Panacek, E.A.; Foulke, G.E.; Dating, C.; Smith, C.R.; LoBuglio, A.F. Initial evaluation of human monoclonal anti-lipid A antibody (HA-1A) in patients with sepsis syndrome. *Crit. Care Med.* **1990**, *18*, 1311–1315. [CrossRef] [PubMed]

131. Daifuku, R.; Haenftling, K.; Young, J.; Groves, E.S.; Turrell, C.; Meyers, F.J. Phase I study of antilipopolysaccharide human monoclonal antibody MAB-T88. *Antimicrob. Agents Chemother.* **1992**, *36*, 2349–2351. [CrossRef]

132. Saravolatz, L.D.; Markowitz, N.; Collins, M.S.; Bogdanoff, D.; Pennington, J.E. Safety, pharmacokinetics, and functional activity of human anti-Pseudomonas aeruginosa monoclonal antibodies in septic and nonseptic patients. *J. Infect. Dis.* **1991**, *164*, 803–806. [CrossRef]

133. Lu, Q.; Rouby, J.J.; Laterre, P.F.; Eggimann, P.; Dugard, A.; Giamarellos-Bourboulis, E.J.; Mercier, E.; Garbino, J.; Luyt, C.E.; Chastre, J.; et al. Pharmacokinetics and safety of panobacumab: Specific adjunctive immunotherapy in critical patients with nosocomial Pseudomonas aeruginosa O11 pneumonia. *J. Antimicrob. Chemother.* **2011**, *66*, 1110–1116. [CrossRef]

134. Irie, R.F.; Ollila, D.W.; O'Day, S.; Morton, D.L. Phase I pilot clinical trial of human IgM monoclonal antibody to ganglioside GM3 in patients with metastatic melanoma. *Cancer Immunol. Immunother.* **2004**, *53*, 110–117. [CrossRef] [PubMed]

135. Safety Study of Human IgM (MORAb-028) to Treat Metastatic Melanoma. Available online: https://clinicaltrials.gov/ct2/show/NCT01123304 (accessed on 11 October 2020).

136. Eisen, A.; Greenberg, B.M.; Bowen, J.D.; Arnold, D.L.; Caggiano, A.O. A double-blind, placebo-controlled, single ascending-dose study of remyelinating antibody rHIgM22 in people with multiple sclerosis. *Mult. Scler. J. Exp. Transl. Clin.* **2017**, *3*, 2055217317743097. [CrossRef] [PubMed]

137. Line, B.R.; Breyer, R.J.; McElvany, K.D.; Earle, D.C.; Khazaeli, M.B. Evaluation of human anti-mouse antibody response in normal volunteers following repeated injections of fanolesomab (NeutroSpec), a murine anti-CD15 IgM monoclonal antibody for imaging infection. *Nucl. Med. Commun.* **2004**, *25*, 807–811. [CrossRef] [PubMed]

138. Hensel, F.; Timmermann, W.; von Rahden, B.H.; Rosenwald, A.; Brandlein, S.; Illert, B. Ten-year follow-up of a prospective trial for the targeted therapy of gastric cancer with the human monoclonal antibody PAT-SC1. *Oncol. Rep.* **2014**, *31*, 1059–1066. [CrossRef]

139. Liedtke, M.; Twist, C.J.; Medeiros, B.C.; Gotlib, J.R.; Berube, C.; Bieber, M.M.; Bhat, N.M.; Teng, N.N.; Coutre, S.E. Phase I trial of a novel human monoclonal antibody mAb216 in patients with relapsed or refractory B-cell acute lymphoblastic leukemia. *Haematologica* **2012**, *97*, 30–37. [CrossRef] [PubMed]

140. Rasche, L.; Duell, J.; Castro, I.C.; Dubljevic, V.; Chatterjee, M.; Knop, S.; Hensel, F.; Rosenwald, A.; Einsele, H.; Topp, M.S.; et al. GRP78-directed immunotherapy in relapsed or refractory multiple myeloma—Results from a phase 1 trial with the monoclonal immunoglobulin M antibody PAT-SM6. *Haematologica* **2015**, *100*, 377–384. [CrossRef] [PubMed]

141. Friend, P.J.; Hale, G.; Waldmann, H.; Gore, S.; Thiru, S.; Joysey, V.; Evans, D.B.; Calne, R.Y. Campath-1M–prophylactic use after kidney transplantation. A randomized controlled clinical trial. *Transplantation* **1989**, *48*, 248–253. [CrossRef]

142. Haisma, H.J.; Pinedo, H.M.; Kessel, M.A.; van Muijen, M.; Roos, J.C.; Plaizier, M.A.; Martens, H.J.; DeJager, R.; Boven, E. Human IgM monoclonal antibody 16.88: Pharmacokinetics and immunogenicity in colorectal cancer patients. *J. Natl. Cancer Inst.* **1991**, *83*, 1813–1819. [CrossRef] [PubMed]

143. Deeg, H.J.; Blazar, B.R.; Bolwell, B.J.; Long, G.D.; Schuening, F.; Cunningham, J.; Rifkin, R.M.; Abhyankar, S.; Briggs, A.D.; Burt, R.; et al. Treatment of steroid-refractory acute graft-versus-host disease with anti-CD147 monoclonal antibody ABX-CBL. *Blood* **2001**, *98*, 2052–2058. [CrossRef]

144. Getts, D.R.; Kramer, W.G.; Wiseman, A.C.; Flechner, S.M. The pharmacokinetics and pharmacodynamics of TOL101, a murine IgM anti-human alphabeta T cell receptor antibody, in renal transplant patients. *Clin. Pharm.* **2014**, *53*, 649–657. [CrossRef]

145. Matsubara, T.; Okuda, K.; Chiba, J.; Takayama, A.; Inoue, H.; Sakurai, T.; Wakabayashi, H.; Kaneko, A.; Sugimoto, K.; Yamazaki, H.; et al. A phase I/II clinical trial of intra-articular administration of ARG098, an anti-FAS IGM monoclonal antibody, in knee joint synovitis of japanese patients with rheumatoid arthritis. *J. Ann. Rheum. Dis.* **2013**, *71*, 384. [CrossRef]

146. Teng, N.N.; Kaplan, H.S.; Hebert, J.M.; Moore, C.; Douglas, H.; Wunderlich, A.; Braude, A.I. Protection against gram-negative bacteremia and endotoxemia with human monoclonal IgM antibodies. *Proc. Natl. Acad. Sci. USA* **1985**, *82*, 1790–1794. [CrossRef]

147. Marks, L. The birth pangs of monoclonal antibody therapeutics: The failure and legacy of Centoxin. *MAbs* **2012**, *4*, 403–412. [CrossRef]

148. Daifuku, R.; Panacek, E.A.; Haenftling, K.; Swenson, W.K.; Prescott, A.W.; Johnson, J.L. Pilot study of anti-lipopolysaccharide human monoclonal antibody MAB-T88 in patients with gram-negative sepsis. *Hum. Antibodies Hybrid.* **1993**, *4*, 36–39. [CrossRef]

149. Lazar, H.; Horn, M.P.; Zuercher, A.W.; Imboden, M.A.; Durrer, P.; Seiberling, M.; Pokorny, R.; Hammer, C.; Lang, A.B. Pharmacokinetics and safety profile of the human anti-Pseudomonas aeruginosa serotype O11 immunoglobulin M monoclonal antibody KBPA-101 in healthy volunteers. *Antimicrob. Agents Chemother.* **2009**, *53*, 3442–3446. [CrossRef] [PubMed]

150. AR-105: Broadly Active Human Monoclonal Antibody (mAb) Against Pseudomonas aeruginosa. Available online: https://aridispharma.com/ar-105/ (accessed on 11 October 2020).

151. IgM-enriched Immunoglobulin Attenuates Systemic Endotoxin Activity in Early Severe Sepsis. Available online: https://clinicaltrials.gov/ct2/show/NCT02444871 (accessed on 11 October 2020).

152. Effects on Microcirculation of IgGAM in Severe Septic/Septic Shock Patients. Available online: https://clinicaltrials.gov/ct2/show/NCT02655133 (accessed on 11 October 2020).

153. Reichert, J.M.; Dewitz, M.C. Anti-infective monoclonal antibodies: Perils and promise of development. *Nat. Rev. Drug Discov.* **2006**, *5*, 191–195. [CrossRef] [PubMed]

154. Hoon, D.S.; Wang, Y.; Sze, L.; Kanda, H.; Watanabe, T.; Morrison, S.L.; Morton, D.L.; Irie, R.F. Molecular cloning of a human monoclonal antibody reactive to ganglioside GM3 antigen on human cancers. *Cancer Res.* **1993**, *53*, 5244–5250. [PubMed]

155. Azuma, Y.; Ishikawa, Y.; Kawai, S.; Tsunenari, T.; Tsunoda, H.; Igawa, T.; Iida, S.; Nanami, M.; Suzuki, M.; Irie, R.F.; et al. Recombinant human hexamer-dominant IgM monoclonal antibody to ganglioside GM3 for treatment of melanoma. *Clin. Cancer Res.* **2007**, *13*, 2745–2750. [CrossRef] [PubMed]

156. A Safety and MORAb-028 Dose Determination Study in Subjects with Metastatic Melanoma. Available online: https://clinicaltrials.gov/ct2/show/NCT01212276 (accessed on 11 October 2020).

157. MORAb 028. Available online: https://adisinsight.springer.com/drugs/800025741 (accessed on 11 October 2020).

158. Mitsunaga, Y.; Ciric, B.; Van Keulen, V.; Warrington, A.E.; Paz Soldan, M.; Bieber, A.J.; Rodriguez, M.; Pease, L.R. Direct evidence that a human antibody derived from patient serum can promote myelin repair in a mouse model of chronic-progressive demyelinating disease. *FASEB J.* **2002**, *16*, 1325–1327. [CrossRef] [PubMed]

159. An Intravenous Infusion Study of rHIgM22 in Patients with Multiple Sclerosis (M22). Available online: https://clinicaltrials.gov/ct2/show/NCT01803867 (accessed on 11 October 2020).

160. An Intravenous Infusion Study of rHIgM22 in Patients with Multiple Sclerosis Immediately Following a Relapse. Available online: https://clinicaltrials.gov/ct2/show/NCT02398461 (accessed on 11 October 2020).

161. Sterner, E.; Flanagan, N.; Gildersleeve, J.C. Perspectives on Anti-Glycan Antibodies Gleaned from Development of a Community Resource Database. *ACS Chem. Biol.* **2016**, *11*, 1773–1783. [CrossRef] [PubMed]

162. Bhat, N.M.; Bieber, M.M.; Chapman, C.J.; Stevenson, F.K.; Teng, N.N. Human anti-lipid A monoclonal antibodies bind to human B cells and the i antigen on cord red blood cells. *J. Immunol.* **1993**, *151*, 5011–5021.

163. Bhat, N.M.; Bieber, M.M.; Spellerberg, M.B.; Stevenson, F.K.; Teng, N.N. Recognition of auto- and exoantigens by V4-34 gene encoded antibodies. *Scand. J. Immunol.* **2000**, *51*, 134–140. [CrossRef]

164. Pascual, V.; Victor, K.; Spellerberg, M.; Hamblin, T.J.; Stevenson, F.K.; Capra, J.D. VH restriction among human cold agglutinins. The VH4-21 gene segment is required to encode anti-I and anti-i specificities. *J. Immunol.* **1992**, *149*, 2337–2344.

165. Bhat, N.M.; Bieber, M.M.; Stevenson, F.K.; Teng, N.N. Rapid cytotoxicity of human B lymphocytes induced by VH4-34 (VH4.21) gene-encoded monoclonal antibodies. *Clin. Exp. Immunol.* **1996**, *105*, 183–190. [CrossRef]

166. Bhat, N.M.; Bieber, M.M.; Teng, N.N. Cytotoxicity of murine B lymphocytes induced by human VH4-34 (VH4.21) gene-encoded monoclonal antibodies. *Clin. Immunol. Immunopathol.* **1997**, *84*, 283–289. [CrossRef] [PubMed]

167. Zhang, C.; Xu, Y.; Gu, J.; Schlossman, S.F. A cell surface receptor defined by a mAb mediates a unique type of cell death similar to oncosis. *Proc. Natl. Acad. Sci. USA* **1998**, *95*, 6290–6295. [CrossRef] [PubMed]

168. Tan, H.L.; Fong, W.J.; Lee, E.H.; Yap, M.; Choo, A. mAb 84, a cytotoxic antibody that kills undifferentiated human embryonic stem cells via oncosis. *Stem Cells* **2009**, *27*, 1792–1801. [CrossRef]

169. Vollmers HP1, O.C.R.; Müller, J.; Kirchner, T.; Müller-Hermelink, H.K. SC-1, a functional human monoclonal antibody against autologous stomach carcinoma cells. *Cancer Res.* **1989**, *49*, 6.

170. Pohle, T.; Brandlein, S.; Ruoff, N.; Muller-Hermelink, H.K.; Vollmers, H.P. Lipoptosis: Tumor-specific cell death by antibody-induced intracellular lipid accumulation. *Cancer Res.* **2004**, *64*, 3900–3906. [CrossRef] [PubMed]

171. PAT SC1. Available online: http://www.patrys.com/pat-sc1/ (accessed on 11 October 2020).

172. NeutroSpec. Available online: https://medlibrary.org/lib/rx/meds/neutrospec/ (accessed on 11 October 2020).

173. Waldmann, H.; Hale, G. CAMPATH: From concept to clinic. *Philos. Trans. R. Soc. Lond. B Biol. Sci.* **2005**, *360*, 1707–1711. [CrossRef] [PubMed]

174. Hale, G.; Cobbold, S.P.; Waldmann, H.; Easter, G.; Matejtschuk, P.; Coombs, R.R. Isolation of low-frequency class-switch variants from rat hybrid myelomas. *J. Immunol. Methods* **1987**, *103*, 59–67. [CrossRef]

175. Riechmann, L.; Clark, M.; Waldmann, H.; Winter, G. Reshaping human antibodies for therapy. *Nature* **1988**, *332*, 323–327. [CrossRef]

176. Highlights of Prescribing Information. Available online: https://www.accessdata.fda.gov/drugsatfda_docs/label/2018/103948s5160_5165lbl.pdf (accessed on 11 October 2020).

177. Macmillan, M.L.; Couriel, D.; Weisdorf, D.J.; Schwab, G.; Havrilla, N.; Fleming, T.R.; Huang, S.; Roskos, L.; Slavin, S.; Shadduck, R.K.; et al. A phase 2/3 multicenter randomized clinical trial of ABX-CBL versus ATG as secondary therapy for steroid-resistant acute graft-versus-host disease. *Blood* **2007**, *109*, 2657–2662. [CrossRef]

178. Flechner, S.M.; Mulgoankar, S.; Melton, L.B.; Waid, T.H.; Agarwal, A.; Miller, S.D.; Fokta, F.; Getts, M.T.; Frederick, T.J.; Herrman, J.J.; et al. First-in-human study of the safety and efficacy of TOL101 induction to prevent kidney transplant rejection. *Am. J. Transplant.* **2014**, *14*, 1346–1355. [CrossRef]

179. DE 098. Available online: https://adisinsight.springer.com/drugs/800016731 (accessed on 11 October 2020).

180. Haspel, M.V.; McCabe, R.P.; Pomato, N.; Janesch, N.J.; Knowlton, J.V.; Peters, L.C.; Hoover, H.C., Jr.; Hanna, M.G., Jr. Generation of tumor cell-reactive human monoclonal antibodies using peripheral blood lymphocytes from actively immunized colorectal carcinoma patients. *Cancer Res.* **1985**, *45*, 3951–3961. [PubMed]

181. Haisma, H.J.; Kessel, M.A.; Silva, C.; van Muijen, M.; Roos, J.C.; Bril, H.; Martens, H.J.; McCabe, R.; Boven, E. Human IgM monoclonal antibody 16.88: Pharmacokinetics and distribution in mouse and man. *Br. J. Cancer Suppl.* **1990**, *10*, 40–43. [PubMed]

182. Rosenblum, M.G.; Levin, B.; Roh, M.; Hohn, D.; McCabe, R.; Thompson, L.; Cheung, L.; Murray, J.L. Clinical pharmacology and tissue disposition studies of 131I-labeled anticolorectal carcinoma human monoclonal antibody LiCO 16.88. *Cancer Immunol. Immunother.* **1994**, *39*, 397–400. [CrossRef] [PubMed]

183. Barth, W.F.; Wochner, R.D.; Waldmann, T.A.; Fahey, J.L. METABOLISM OF HUMAN GAMMA MACROGLOBULINS. *J. Clin. Investig.* **1964**, *43*, 1036–1048. [CrossRef] [PubMed]

184. Ryman, J.T.; Meibohm, B. Pharmacokinetics of Monoclonal Antibodies. *CPT Pharmacomet. Syst. Pharmacol.* **2017**, *6*, 576–588. [CrossRef] [PubMed]

185. Romano, M.J.; Kearns, G.L.; Kaplan, S.L.; Jacobs, R.F.; Killian, A.; Bradley, J.S.; Moss, M.M.; Van Dyke, R.; Rodriguez, W.; Straube, R.C. Single-dose pharmacokinetics and safety of HA-1A, a human IgM anti-lipid-A monoclonal antibody, in pediatric patients with sepsis syndrome. *J. Pediatrics* **1993**, *122*, 974–981. [CrossRef]

186. Meng, Y.G.; Wong, T.; Saravolatz, L.D.; Pennington, J.E. Pharmacokinetics of an IgM human monoclonal antibody against Pseudomonas aeruginosa in nonseptic patients. *J. Infect. Dis.* **1993**, *167*, 784–785. [CrossRef]

187. Center for Drug Evaluation and Research. Available online: https://www.accessdata.fda.gov/drugsatfda_docs/nda/2004/103928Orig1s000Lbl.pdf (accessed on 11 October 2020).

188. Mimura, Y.; Katoh, T.; Saldova, R.; O'Flaherty, R.; Izumi, T.; Mimura-Kimura, Y.; Utsunomiya, T.; Mizukami, Y.; Yamamoto, K.; Matsumoto, T.; et al. Glycosylation engineering of therapeutic IgG antibodies: Challenges for the safety, functionality and efficacy. *Protein Cell* **2018**, *9*, 47–62. [CrossRef]

189. Maiorella, B.L.; Ferris, R.; Thomson, J.; White, C.; Brannon, M.; Hora, M.; Henriksson, T.; Triglia, R.; Kunitani, M.; Kresin, L.; et al. Evaluation of product equivalence during process optimization for manufacture of a human IgM monoclonal antibody. *Biologicals* **1993**, *21*, 197–205. [CrossRef]

190. Angus, D.C.; Birmingham, M.C.; Balk, R.A.; Scannon, P.J.; Collins, D.; Kruse, J.A.; Graham, D.R.; Dedhia, H.V.; Homann, S.; MacIntyre, N. E5 murine monoclonal antiendotoxin antibody in gram-negative sepsis: A randomized controlled trial. E5 Study Investigators. *JAMA* **2000**, *283*, 1723–1730. [CrossRef]

191. Greenman, R.L.; Schein, R.M.; Martin, M.A.; Wenzel, R.P.; MacIntyre, N.R.; Emmanuel, G.; Chmel, H.; Kohler, R.B.; McCarthy, M.; Plouffe, J.; et al. A controlled clinical trial of E5 murine monoclonal IgM antibody to endotoxin in the treatment of gram-negative sepsis. The XOMA Sepsis Study Group. *JAMA* **1991**, *266*, 1097–1102. [CrossRef] [PubMed]

192. Brun-Buisson, C. The HA-1A saga: The scientific and ethical dilemma of innovative and costly therapies. *Intensive Care Med.* **1994**, *20*, 314–316. [CrossRef] [PubMed]

193. Derkx, B.; Wittes, J.; McCloskey, R. Randomized, placebo-controlled trial of HA-1A, a human monoclonal antibody to endotoxin, in children with meningococcal septic shock. European Pediatric Meningococcal Septic Shock Trial Study Group. *Clin. Infect. Dis.* **1999**, *28*, 770–777. [CrossRef] [PubMed]

194. De Jong, R.N.; Beurskens, F.J.; Verploegen, S.; Strumane, K.; van Kampen, M.D.; Voorhorst, M.; Horstman, W.; Engelberts, P.J.; Oostindie, S.C.; Wang, G.; et al. A Novel Platform for the Potentiation of Therapeutic Antibodies Based on Antigen-Dependent Formation of IgG Hexamers at the Cell Surface. *PLoS Biol.* **2016**, *14*, e1002344. [CrossRef]

195. GEN1029 (HexaBody®-DR5/DR5) Safety Trial in Patients with Malignant Solid Tumors. Available online: https://clinicaltrials.gov/ct2/show/NCT03576131 (accessed on 11 October 2020).

196. Piao, X.; Ozawa, T.; Hamana, H.; Shitaoka, K.; Jin, A.; Kishi, H.; Muraguchi, A. TRAIL-receptor 1 IgM antibodies strongly induce apoptosis in human cancer cells in vitro and in vivo. *Oncoimmunology* **2016**, *5*, e1131380. [CrossRef] [PubMed]

197. Wang, B.; Kothambalwala, T.; Hinton, P.; Ng, D.; Saini, A.; Baliga, R.; Keyt, B. Abstract 1702: Multimeric anti-DR5 IgM antibody displays potent cytotoxicity in vitro and promotes tumor regression in vivo. In Proceedings of the American Association for Cancer Research Annual Meeting, Washington, DC, USA, 1–5 April 2017; Volume 77, p. 1702.

198. Wang, B.; Kothambawala, T.; Wang, L.; Saini, A.; Baliga, R.; Sinclair, A.; Keyt, B. Abstract 3050: Multimeric IgM antibodies targeting DR5 are potent and rapid inducers of tumor cell apoptosis and cell death in vitro and in vivo. In Proceedings of the American Association for Cancer Research Annual Meeting 2019, Atlanta, GA, USA, 29 March–3 April 2019; Volume 79, p. 3050.

199. Shen, C.; Zhang, M.; Chen, Y.; Zhang, L.; Wang, G.; Chen, J.; Chen, S.; Li, Z.; Wei, F.; Chen, J.; et al. An IgM antibody targeting the receptor binding site of influenza B blocks viral infection with great breadth and potency. *Theranostics* **2019**, *9*, 210–231. [CrossRef]

200. Labrijn, A.F.; Janmaat, M.L.; Reichert, J.M.; Parren, P. Bispecific antibodies: A mechanistic review of the pipeline. *Nat. Rev. Drug Discov.* **2019**, *18*, 585–608. [CrossRef]

201. Baliga, R.; Li, K.; Manlusoc, M.; Hinton, P.; Ng, D.; Tran, M.; Shan, B.; Lu, H.; Saini, A.; Rahman, S.; et al. IGM-2323: High Avidity IgM-Based CD20xCD3 Bispecific Antibody (IGM-2323) for Enhanced T-Cell Dependent Killing with Minimal Cytokine Release. In Proceedings of the American Society of Hematology Meeting, Orlando, FL, USA, 7–10 December 2019; p. 1574.

202. A Safety and Pharmacokinetic Study of IGM-2323 in Subjects with Relapsed/Refractory Non-Hodgkin Lymphoma. Available online: https://clinicaltrials.gov/ct2/show/NCT04082936 (accessed on 11 October 2020).

Publisher's Note: MDPI stays neutral with regard to jurisdictional claims in published maps and institutional affiliations.

antibodies

MDPI

Review

IgA: Structure, Function, and Developability

Patrícia de Sousa-Pereira [1,2] and Jenny M. Woof [1,*]

1   School of Life Sciences, University of Dundee, Dundee DD1 5EH, UK; p.z.pereira@dundee.ac.uk
2   CIBIO-InBIO, Campus Agrário de Vairão, University of Porto, 4485-661 Vairão, Portugal
*   Correspondence: j.m.woof@dundee.ac.uk; Tel.: +44-1382-383389

Received: 5 November 2019; Accepted: 28 November 2019; Published: 5 December 2019

Abstract: Immunoglobulin A (IgA) plays a key role in defending mucosal surfaces against attack by infectious microorganisms. Such sites present a major site of susceptibility due to their vast surface area and their constant exposure to ingested and inhaled material. The importance of IgA to effective immune defence is signalled by the fact that more IgA is produced than all the other immunoglobulin classes combined. Indeed, IgA is not just the most prevalent antibody class at mucosal sites, but is also present at significant concentrations in serum. The unique structural features of the IgA heavy chain allow IgA to polymerise, resulting in mainly dimeric forms, along with some higher polymers, in secretions. Both serum IgA, which is principally monomeric, and secretory forms of IgA are capable of neutralising and removing pathogens through a range of mechanisms, including triggering the IgA Fc receptor known as FcαRI or CD89 on phagocytes. The effectiveness of these elimination processes is highlighted by the fact that various pathogens have evolved mechanisms to thwart such IgA-mediated clearance. As the structure–function relationships governing the varied capabilities of this immunoglobulin class come into increasingly clear focus, and means to circumvent any inherent limitations are developed, IgA-based monoclonal antibodies are set to emerge as new and potent options in the therapeutic arena.

Keywords: immunoglobulin A; IgA; structure; FcαRI; CD89; immune evasion; therapeutic antibodies

1. Introduction

The human immune system expends a considerable amount of energy in production of immunoglobulin A (IgA), since more IgA is made than all the other classes of immunoglobulin (Ig) combined. IgA is present in both serum, where at 2–3 mg/mL it is the second most prevalent circulating Ig after IgG, and in external secretions such as those that bathe mucosal surfaces, where it is the predominant Ig. It has been calculated that around 60 mg of IgA is produced per kilogram of body weight per day in the average human [1,2], much of it being localised at mucosal surfaces. Such surfaces, which collectively have a surface area in adult humans of around 400 m² [3], are major sites of vulnerability, given their exposure to the environment, and IgA clearly plays a critical role in their protection against attack by invading pathogens.

In humans, there are two subclasses of IgA, named IgA1 and IgA2. Like all Ig, each subclass comprises a basic molecular unit of two identical heavy chains (HCs) and two identical light chains (LCs). Each chain begins at its N-terminus with a variable region, which is followed by a constant region. The LCs are the same in each subclass, but the HCs differ within their constant regions, which are encoded by distinct Cα genes. Two allotypic variants of human IgA2, known as IgA2m(1) and IgA2m(2), have been characterised. A third IgA2 variant, termed IgA2(n), has been described [4], but while presumed to be an allelic form, its penetrance in the population remains to be investigated.

Unlike other Ig classes, IgA exists in multiple molecular forms. In human serum, the predominant IgA form is monomeric, i.e., comprises 2HC and 2LC, with a subclass distribution of about 90% IgA1 and 10% IgA2. In contrast, the main molecular form found at mucosal surfaces, known as secretory IgA

(SIgA), is dimeric, although some higher molecular weight species, including trimers and tetramers, are also present. Here the relative proportion of the two subclasses is more closely matched; an average distribution being about 40% IgA1 and 60% IgA2, though this varies depending on the particular mucosal site sampled.

Genetic sequence analysis has confirmed the presence of IgA in all categories of mammals (placental, marsupials, and monotremes) and in birds. However, there are notable species differences. Most mammals have a single IgA isotype. IgA1 and IgA2 subclasses akin to those in humans are only present in related primates, including chimpanzees, gorillas, and gibbons [5], consistent with IgA1 arising relatively recently in evolutionary terms. Orangutans have an equivalent of IgA1, but appear to have lost their form of IgA2. The other group of mammals to have more than one IgA are rabbits and other lagomorphs, which have a massively expanded number of IgA genes, resulting in 14 known subclasses, 11 of which are expressed. A 15th IgA was recently described in domestic European rabbits [6]. While IgA is known to play a common role in protection at mucosal surfaces [7], the levels, forms, and distribution of IgA vary. For example, in species commonly used in experimental research, including mice, rats, and rabbits, the main form of IgA in serum is dimeric rather than the monomeric form seen in humans. In these same species, unlike humans, the main source of IgA in the gut lumen is from bile. Another species difference relates to the prevalent Ig found in colostrum and milk. While in humans this is IgA, in cows, sheep, goats, and horses, the main immunoglobulin isotype present is IgG.

Such species differences have tended to constrain research on the general features of IgA, and mean that there are inherent problems with extrapolation of results on IgA from animal models to humans. This review will focus primarily on human IgA, and will explore structure and function relationships and the prospect for developing IgA-based therapeutic monoclonal antibodies (mAb). The issue of species differences within the IgA system remains of relevance, given the growing interest in IgA as a potential therapeutic option and the requirement for meaningful models to robustly assess capabilities in this context.

2. IgA structure

2.1. General Features

In common with other Igs, both the HCs and LCs of IgA are folded into a number of variable (V) and constant (C) domains, each encoded by a separate exon. These number four in the HC (namely VH, Cα1, Cα2, and Cα3, starting from the N-terminus) and two in the LC (namely VL and CL, from the N-terminus). Each domain folds into a similar globular secondary structure, known as the immunoglobulin fold, a feature of all Igs. Typically stretching some 110 amino acids, each domain comprises two β-sheets made up of anti-parallel β-strands, which sandwich together around a stabilising disulphide bond.

Interposed between the Cα1 and Cα2 domains of each HC lies a flexible hinge region, which is particularly extensive in human IgA1 but shorter in human IgA2. Indeed, the hinge is the region of greatest difference between the two subclasses. Unlike IgG, there are no interchain disulphide bridges within the hinge region, which presumably affords the IgA hinge sequences, particularly the longer ones of IgA1, the ability to flex independently of each other, but may also increase the susceptibility to proteolysis.

The hinge of IgA1, rich in proline, serine, and threonine, contains a sequence missing in IgA2 that comprises two eight amino acid repeats (Figure 1). The hinge in human IgA is encoded in a sequence present at the 5′ end of the exon encoding the Cα2 domain, rather than by a separate exon or exons as seen for IgG. As in other Igs, the hinge affords flexibility to the whole IgA molecule that is critical for activity. It varies considerably in length and sequence between IgAs from different species (Figure 1).

At the C-terminus of the IgA HC lies an 18 amino acid extension known as the tailpiece. While a corresponding feature is lacking in IgG and IgE, a highly similar sequence is found at the C-terminus of the HC of IgM. For both IgA and IgM, the tailpiece is crucial to the Ig's ability to polymerise into primarily dimers and pentamers, respectively.

```
                 222                    240
Human1       VPSTPPTPSPSTPPTPSPS
Human2m(1)                   VPPPPP
Human2m(2)                   VPPPPP
Chimp1       GPSTPCPPTPSTPPTPSPS
Chimp2                       VPPPPP
Gorilla1     VPSTPPTPSPSTPPTPSPP
Gorilla2                     VPPSPP
Orangutan    VPRPTPTPSTPPCPPPS
Gibbon1                  VPLPTPPHP
Gibbon2                    APPPHP
Macaque                    SETKPCL
Cow                      DSSSCCVPN
Horse                  VCPPPPCECPL
Dog                     DNSHPCHPCPS
Pig                     VLPSDPCPQ
Mouse                 GPTPPPPITIPS
Rat                        KPSLV
Rabbit4  ACNKPTIEPPTKPTCPCPCPSPS
```

Figure 1. Hinge sequences of IgAs from different species. Numbers following the species name indicate the IgA subclass, and allotype where appropriate. Amino acid numbering above human IgA1 is according to the commonly adopted scheme used for IgA1 Bur [8].

Two HCs and two LCs are organised into two Fab regions (each comprising VH, $C\alpha1$, VL, and CL domains), responsible for binding to antigen, linked via the hinge region to a single Fc region (comprising two $C\alpha2$ and two $C\alpha3$ domains), responsible for triggering elimination processes (Figure 2). The interaction between chains is stabilised by disulphide bonds between the HCs and LCs within the Fab region and between the two HCs at the $C\alpha2$ domains, and by close pairing of opposing domains: VH with VL, $C\alpha1$ with CL, and one $C\alpha3$ domain with the other one. Such pairing relies on an array of non-covalent interactions, chiefly hydrogen bonds and van der Waals contacts, between the domains involved.

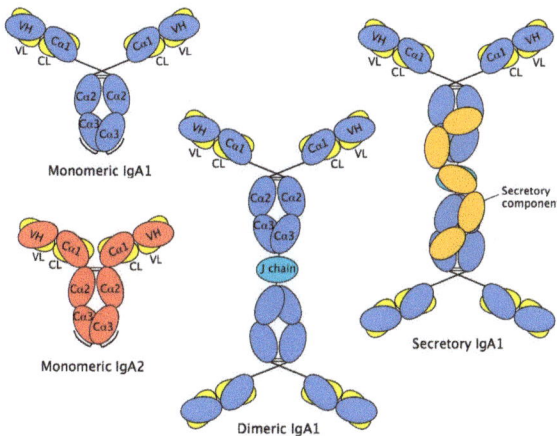

Figure 2. Schematic diagram of IgA structures—monomeric, dimeric, and secretory IgA. In IgA1, the heavy chain domains are in blue, and those of the light chains in yellow. In IgA2, the heavy chain domains are in red, and the light chain domains in yellow. The tailpieces are shown as extensions to the C-termini of the $C\alpha3$ domains in the monomeric forms. Dimeric and secretory forms of IgA2 are not depicted. J chain, which is present in both dimeric and secretory IgA, is shown in cyan. The domains of secretory component, derived from the extracellular region of pIgR, are present in secretory IgA and are shown in orange.

The Cα2 domains do not form a close pair, but instead have N-linked oligosaccharides that overlie the surfaces normally involved in pairing. N-linked oligosaccharides in fact make rather a significant contribution of the total mass of IgA, accounting for 6–7% of the mass of human IgA1, and 8–10% of the mass of human IgA2 [9]. The aforementioned Cα2 domain sugars are found in both IgA1 and IgA2, attached to residue Asn263. Both subclasses have another N-linked sugar attached to the tailpiece at residue Asn459. Recently, it has been reported that the glycans attached at Asn459 can interact directly with certain viruses and thereby neutralise them [10]. Human IgA2 has further N-linked sugars attached at residues Asn166 in the Cα1 domain and Asn337 in the Cα2 domain. IgA2 molecules of the IgA2m(2) allotype have a further N-linked sugar attached at Asn211 in the Cα1 domain. In terms of composition, the N-linked sugars of serum and secretory IgA comprise a family of related structures centred on a biantennary mannosyl chitobiose core, with a small proportion being more branched, mostly with triantennary structures. Fucosylation level varies, as does the numbers of sugars (galactose and sialic acid) found at the branch termini (Figure 3) [11–13]. Further glycosylation complexity arises through the attachment of usually between 3 and 6 core 1 and/or Tn O-linked sugars, composed principally of N-acetyl galactosamine, galactose, and sialic acid, to the hinge of IgA1 [12,13]. These O-linked glycans introduce further heterogeneity, since they consist of a family of structures, varying in terms of the presence or absence of sialic acid and galactose.

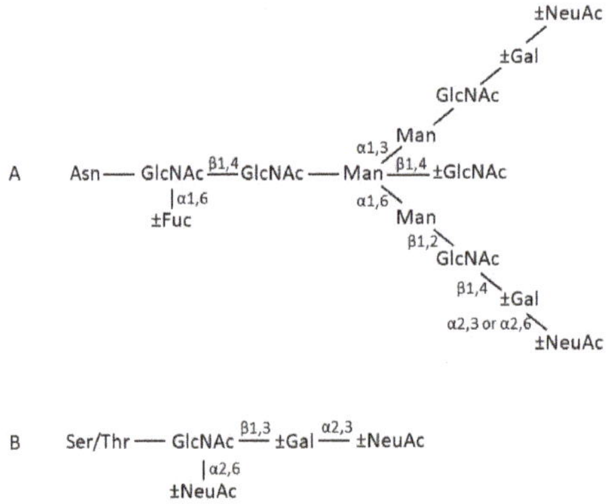

Figure 3. Schematic structures of IgA (**A**) N-linked and (**B**) O-linked glycan side chains. Structure (**A**) occurs in both IgA1 and IgA2, while structure (**B**) is present only attached to the hinge region of IgA1. NeuNAc, N-acetyl neuraminic (sialic) acid; Gal, galactose; GlcNAc, N-acetyl glucosamine; Man, mannose; Fuc, fucose; GalNAc, N-acetyl galactosamine. ±Gal, ±NeuNAc, or ±Fuc indicate that some chains terminate at the preceding sugar.

2.2. IgA Fab Region

In terms of structural components unique to IgA, within the Fab region it is the Cα1 domain that constitutes the IgA-specific component, with the VH, VL, and CL being common to other Ig classes. Solved X-ray crystal structures of the Fab regions of mouse IgA myeloma proteins have provided earlier structural insights. From two different plasmacytoma IgAs, the elbow bend angle between the VH and Cα1 domains was seen to range between 133 and 145°, suggesting a degree of flexibility within the Fab region [14,15]. However, more recently, the crystal structure of a human IgA1 Fab has been determined at high resolution [16]. The position of the disulphide between the LC and HC, together with the markedly hydrophobic interface between the VH and Cα1 domains, appears to constrain the IgA1 Fab, making it somewhat rigid. When compared to a matched IgG featuring the same VH and VL domains, the IgA1 Fab exhibited a difference of about 5° in the elbow angle from that in IgG. It has been suggested that the greater rigidity inherent in IgA1 Fab may exert subtle allosteric effects on the antigen binding site with resultant impact on antigen binding affinity. Such considerations are relevant to engineering of therapeutic antibodies, and are explored in depth elsewhere [17].

The IgA subclasses differ in the arrangement of their interchain disulphides, including those between LC and HC within the Fab region. While IgA1 and IgA2(m)2 have the usual disulphide bridges between HC and LC, these are located at different positions—between a common Cys in LC and Cys133 in IgA1 HC and Cys220 in IgA2m(2) HC. These HC Cys are located close to the VH–Cα1 interdomain region and at the C-terminal end of the Cα1 domain (penultimate residue), respectively. Remarkably, in IgA2m(1), such HC–LC disulphides are generally lacking. Instead, disulphide bridge links the two LCs, and the association between HC and LC is stabilised by non-covalent interactions.

2.3. IgA Fc Region

Turning to the Fc region, important structural information has been gained from the solved X-ray crystal structures of human IgA1 Fc in complex with the extracellular domains of FcαRI [18] and with the staphylococcal protein SSL7 (Figure 4) [19]. In terms of overall configuration, the structure of the Fc region is similar to that of IgG and IgE, but there are important distinctions. Notably, the location of the disulphide bridges between the two HCs, and the attachment sites and positions of the *N*-linked glycans are different in IgA from these other Ig classes.

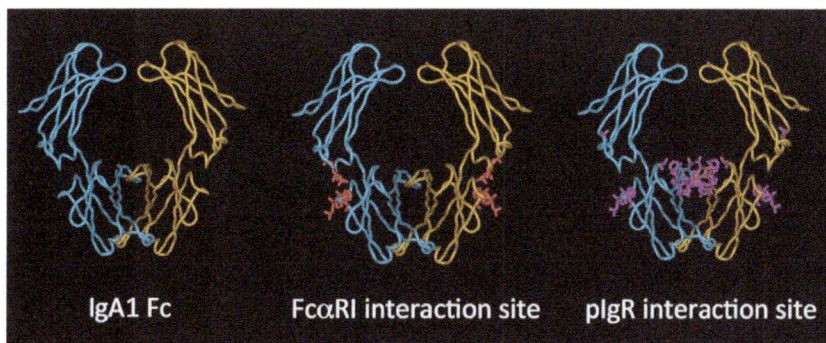

Figure 4. X-ray crystal structure of human IgA1 Fc generated from PDB accession code 1OW0 using only the IgA coordinates. One heavy chain is shown in blue, the other in gold. Residues critical for binding to FcαRI are shown in red on the middle image, and those implicated in the interaction with pIgR are shown in purple on the right hand image.

Unlike IgG, where there are numerous inter-HC disulphide bridges in the hinge region, IgA lacks hinge disulphides and, instead, has disulphide bridges between the upper reaches of the Cα1 domain (Figure 2). Thus, Cys242 in each HC can link to Cys299 in the opposite HC. Further disulphide bonds are presumed to exist, for example, between Cys241 in each HC, or between Cys299 in each HC, or between Cys241 in one HC and Cys301 in the other, but the truncated forms of IgA1 Fc used in crystallisation did not allow direct resolution of these.

The Cα2 domains are not closely paired, a feature similar to the equivalent domains in IgG (Cγ2) and IgE (Cε3). Such non-pairing might be expected to expose a considerable area of domain surface to solvent, but this potentially less stable scenario is avoided to some extent due to attachment of *N*-linked glycans at Asn263. The sugar moieties attached at this site lie over the outer surfaces of the Cα2 domains and, in doing so, bury around 930Å^2 per Fc from solvent contact. The glycans also make contact with the Cα3 domains, thereby burying another 914Å^2 per Fc from solvent, further stabilising the Fc region.

The 18 amino acid tailpiece at the C-terminus of each HC was missing from the IgA1 Fc fragments used for crystallisation, and hence no information on its structure was obtained. Recently it has been modelled to occupy a range of conformations [20]. The tailpiece carries a cysteine residue at position 471, and the potential linkages that this cysteine residue may make with other "free" Cys residues in IgA remains somewhat of an enigma.

2.4. Structure of Monomeric IgA

As with other Igs, the inherent flexibility of intact monomers of IgA tend to frustrate crystallisation efforts. Thus, in order to probe the conformation of entire IgA monomers rather than the separate Fab and Fc regions, lower resolution techniques, including electron microscopy (EM), and more recently, X-ray and neutron scattering of IgA in solution, have been used. These have been useful in predicting the overall dimensions of IgA molecules, and have led to an understanding that the IgA1 has a greater average Fab centre to Fab centre distance than IgA2: 16.9 nm for IgA1 compared with just 8.2 nm for IgA2 [21–26].

Models arising from solution scattering studies originally suggested that both human IgA subclasses adopt average T-shaped structures (Figure 5), which presumably reflected averages of the different conformations available to these molecules as a result of flexibility. Indeed, more recent work using these techniques has reported IgA1 to have an extended Y-shaped structure, with the Fab regions positioned well away from the Fc, in keeping with previous electron micrographs. Given the major advances made in recent years in cryogenic electron microscopy (cryo-EM), it can be envisaged that definitive understanding of the structure of monomeric IgA is likely to emerge from this technique.

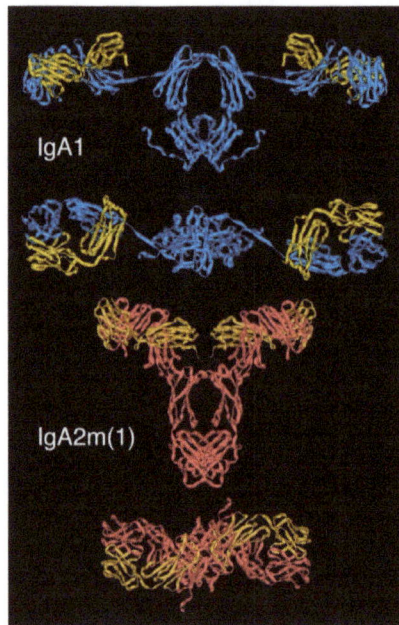

Figure 5. Molecular models of human IgA1 and IgA2(m)1 using coordinates from PDB accession codes 1IGA and 1R70, respectively, seen face on (upper image in each case) and from above (lower image in each case). In IgA1, heavy chains (HCs) are shown in blue and light chains (LCs) in yellow, while in IgA2m(1), HCs are shown in red and LCs in yellow.

2.5. Dimeric IgA

The IgA destined for the mucosal surfaces is produced locally to the mucosa in polymeric form. These are principally dimers comprising two IgA monomers covalently linked to an additional polypeptide known as joining chain or J chain. J chain is a 15 kDa polypeptide, expressed by antibody-producing cells, and is also present in larger IgA polymers and pentameric IgM. It is incorporated into polymeric IgA or IgM prior to secretion [27]. In the case of IgA, marginal zone B and B-1 cell-specific protein (MZB1) has been shown to promote J chain binding to IgA in plasma cells [28]. J chain is very highly conserved across species (mammals, birds, reptiles, fishes, and amphibian) and is not known to resemble any other protein. It has one *N*-linked glycan attached at Asn48 which exists in five major forms, principally sialylated biantennary complex structures [13]. J chain's ability to join HCs in polymeric Igs relies on two key Cys residues, from amongst the eight cysteines it possesses. Six of the eight are involved in interchain disulphide bridges (Cys12–Cys100, Cys17–Cys91, and Cys108–Cys133) [29,30]. Presently, the three-dimensional structure of J chain is unresolved. Models have tended to favour a two-domain structure [30,31].

Early studies of dimeric IgA structure utilised EM to view myeloma IgA preparations. It was seen to have a double-Y shape, in which the Fc regions joined to each other via their C-terminal regions. The length of the joined Fc region was in the range 125–155 Å, consistent with two Fc regions of about 65 Å long being arranged end-to-end (Figure 2). The J chain is interposed between the two Fc regions, and links to each of the monomers through disulphide bridges formed between the penultimate Cys residues of the tailpieces (Cys471) and the two J chain cysteines alluded to above (Cys14 and Cys68). The critical roles played by these cysteines in the linkage has been verified through targeted mutagenesis of both the tailpiece and J chain [32,33]. In keeping with these observations, solution structure analysis of dimeric IgA1 have predicted a near-planar structure with end-to-end Fc contacts, although in this study, the J chain structure and orientation used in the modelling was arbitrary [34]. Further analysis, possibly from techniques such as cryo-EM, will be necessary to provide an in-depth view of the relative arrangement of Fc regions and J chain.

2.6. Secretory IgA

In external secretions, the predominant form of IgA is SIgA, which derives from local synthesis by Ig-producing cells in organised mucosal-associated lymphoid tissues, most of which are committed to the IgA isotype. SIgA is mostly in dimeric form, with some tetramers also being present. The relative proportions of each varies from mucosal site or secretion. For example, in saliva and milk, the ratio of dimeric/tetrameric SIgA is around 3:2. Secretions can also contain some monomeric IgA, but again, the amounts vary. In saliva and milk, about 5–10% of the IgA is monomeric, whereas in cervical fluid, a much higher proportion can be present [35].

Another factor accounting for the high relative concentration of IgA in secretions is the presence of a receptor known as the polymeric Ig receptor (pIgR), which mediates the specific transport of polymeric Igs across the mucosal epithelium into the secretions (Figure 6). pIgR is expressed on the basolateral surface of epithelial cells lining mucosal sites, and binds and transports only polymeric Igs. At mucosal surfaces, the predominant ligand is dimeric IgA, since the larger size of IgM restricts diffusion from serum, and hence, the smaller, and locally-produced, dimeric IgA is preferentially transferred [36].

pIgR is a single polypeptide receptor, comprising a ~620 amino acid extracellular portion which folds up into five Ig-like domains with particular homology to Ig variable domains, a 23 amino acid transmembrane section, and an internal tail of around 103 amino acids [37]. The extracellular domains, named D1–D5 from the *N*-terminus, are each stabilised by one or more internal disulphide bridges, and are decorated by seven *N*-linked glycans. Between the end of D5 and the membrane lies a short stretch of non-Ig-like sequence.

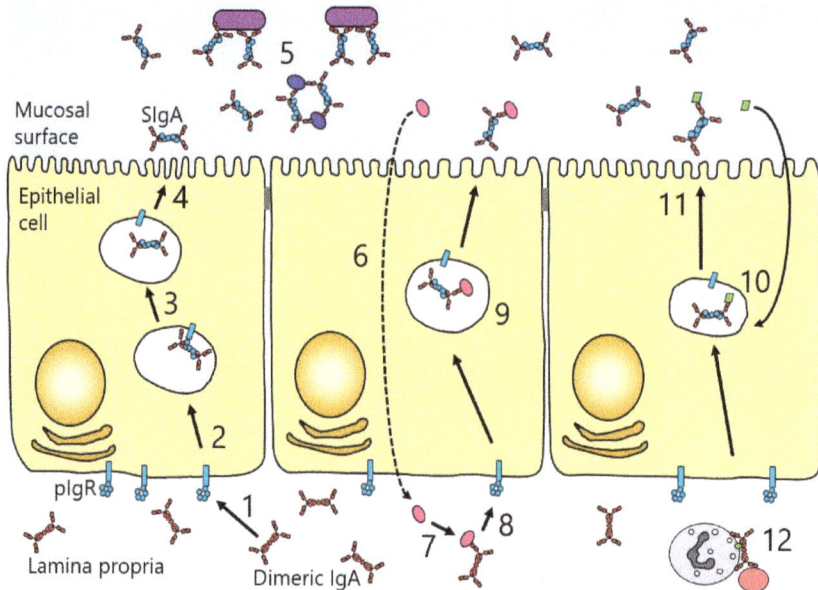

Figure 6. Schematic diagram illustrating the role of pIgR in transporting IgA across the mucosal epithelium. Gut epithelium is shown as an example. (**1**) Dimeric IgA (shown in red) produced locally at the mucosal surface binds pIgR (cyan) at the basolateral surface of the epithelial cell layer. (**2**) The complex is internalised and undergoes vesicular transport across the cell. (**3**) pIgR is cleaved to release secretory component (SC), which becomes disulphide-bonded to the dimeric IgA. (**4**) At the apical surface, SIgA is released. (**5**) SIgA binds to and neutralises bacterial and viral pathogens (shown in purple and dark blue). (**6**) Some pathogens (shown in bright pink) may gain access to the lamina propria underlying the epithelium. (**7**) Such pathogens can be bound by dimeric IgA. (**8**) The dimeric IgA–pathogen complex binds to pIgR. (**9**) The pathogen is carried out across the epithelium and released back out into the lumen. (**10**) Some pathogens (shown in lime green) can be intersected by dimeric IgA during transit across the epithelial cells. (**11**) The pathogen is ejected upon release of SIgA at the mucosal surface. (**12**) Dimeric IgA can mediate clearance mechanisms against pathogens (in salmon pink) through engaging phagocytes.

Transport of dimeric IgA across the epithelium (transcytosis) involves its binding to pIgR at the basolateral surface of the epithelial cell, followed by internalisation and transport via vesicular compartments to the apical surface of the cell (Figure 6). During the process, pIgR is cleaved between D5 and the membrane to release a major fragment of the receptor referred to as secretory component (SC). A disulphide bridge forms between SC and dimeric IgA, and when the complex is released at the apical surface, SC remains as part of the released IgA, then known as SIgA. EM studies of SIgA from colostrum show a double Y-shaped configuration.

Domains D1–D3 of pIgR are known to play critical roles in binding to dimeric IgA, with domains D4 and D5 also making smaller contributions. In particular, loops lying at the end of D1, akin to the complementarity determining regions (CDR) of variable domains, are central to the binding and are known to lie close to each other based on the solved X-ray crystal structure of the domain [38–40]. Residues in CDR1, CDR2, and CDR3 have been implicated in the binding to dimeric IgA [37].

Turning to the elements of dimeric IgA involved in the interaction, it is believed that the initial interaction involves engagement of D1 of pIgR with an exposed loop (residues 402–410) and other close lying residues (Phe411, Val413, Thr414, Lys377) on the Cα3 domain of IgA, along with a region on the Cα2 domain (Pro440–Phe443) lying at the Cα2–Cα3 domain interface (Figure 4) [41–43]. Thereafter, a disulphide bound formed between one of two cysteine residues in D5 of pIgR (Cys468 or Cys502) and Cys311 in the Cα2 domain of IgA anchors SC and dimeric IgA together [44]. It has also been demonstrated that direct interactions between J chain and pIgR occur [45].

More recently, the structure of free SC has been elucidated by X-ray crystallography and shown to adopt a triangular arrangement, with a large interface between domains D1, D4, and D5, which buries some 1480 Å^2 of surface area from solvent contact (Figure 7) [46]. The five domains lie in a plane, giving the triangle shape a thickness similar to that of a single domain (about 40 Å). To further explore SC structure and its relationship to function, the same study used double electron–electron resonance spectroscopy on spin-labelled variants of SC in solution as a means to explore the flexibility of the protein domains. This analysis confirmed the crystal structure to represent the predominant solution structure of free SC at the D1–D5 interface. However, when the spin-labelled SC was incubated with dimeric IgA, a dramatic separation of D1 and D5 was apparent, consistent with an increase in distance of more than 42Å between these domains, resulting in a final separation of more than 85 Å. Analysis of the binding characteristics of shortened constructs of SC supported the key role of D1 in binding to dimeric IgA and indicated a role for D5 in mediating non-covalent interactions with dimeric IgA [46]. The results also suggest that D2, and possibly D3, contribute to binding either directly or through promoting interactions between D5 and dimeric IgA. Thus, we are left with a current model that involves opening up of the pIgR extracellular structure upon binding to SIgA, with initial contact through D1, but later involvement of the other pIgR domains. The final separation of D1 and D5 would be sufficient to allow engagement of D1 and D5 with domains in the same IgA monomer or across the two different IgA monomers present in the dimer.

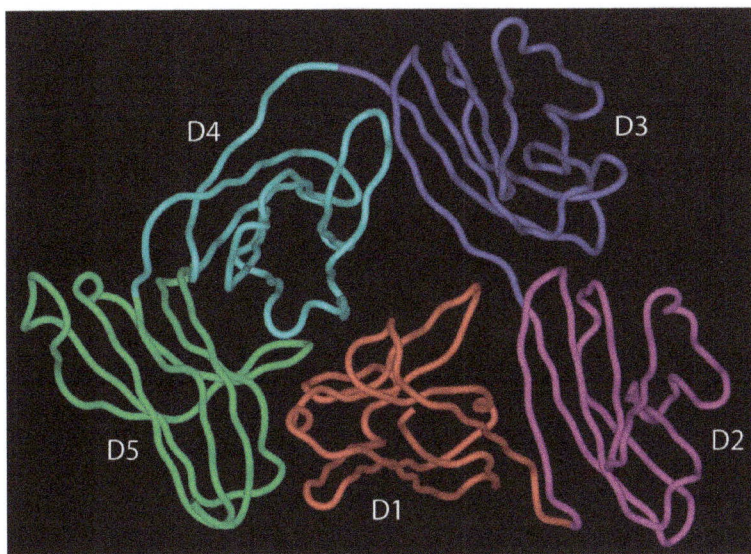

Figure 7. Crystal structure of the extracellular domains of human pIgR (using coordinates from PBD accession code 5D4K). Each of the five domains (D1–D5) has been coloured differently.

3. IgA Function

3.1. Neutralisation

Through direct engagement of their antigen binding sites with antigens on pathogens, IgA molecules neutralise or block the activity of a range of viruses, bacteria, and protozoa, and prevent their attachment to host cells [47]. Similarly, binding of IgA to pathogenic products such as toxins can neutralise their activity and prevent the disease symptoms associated with them [48]. The attachment of several types of pathogenic microorganisms to the mucosal surfaces can be prevented by the interaction of the glycans on IgA with sugar-dependent receptors or fimbriae on their surfaces [10,49–51]. Thus, IgA contributes to immune exclusion, a process by which the adsorption of pathogens to mucosal surfaces is prevented through agglutination, such that the aggregates formed are unable to penetrate though the mucus that lines mucosal surfaces. The multiple antigen binding sites of SIgA enable both high avidity binding and crosslinking of particulate matter, resulting in efficient blocking activity. Moreover, IgA can interact with other innate defence factors in mucosal secretions to enhance immune protection. These include mucins [52,53], lactoferrin, and the lactoperoxidase system [54].

In vitro studies suggest that mucosal IgA can also mediate protective functions during its passage through the epithelium or by carrying pathogens or their products encountered on the basolateral side of the epithelium out across the epithelium (Figure 6) [55]. The latter reflects the fact that pIgR can transport dimeric IgA alone or in complex with antigen. This mechanism can drive removal or excretion of soluble antigens from various origins, as well as viral particles [56]. Antigen-specific dimeric IgA has been seen in vitro to neutralise endocytosed bacterial lipopolysaccharide (LPS) within epithelial cells, whilst undergoing pIgR-mediated transcytosis. Following colocalisation within the apical recycling compartment, the IgA was able to prevent the proinflammatory events usually triggered by LPS [57]. Similarly, while undergoing epithelium transcytosis, dimeric IgA targeted to certain viruses have been able to block viral growth, seemingly following intersection of the IgA and viral proteins in the apical recycling endosomes. Such effects have been reported for Sendai virus [58], influenza virus [59], measles virus [60], rotavirus [61,62], and HIV [63,64]. However, questions remain as to whether these processes reflect the situation in vivo, although experiments in mouse models suggest that there may be some physiological relevance [65,66].

3.2. Complement Activation

IgA lacks the site for C1q binding present in IgG and does not bind C1q, and therefore is not expected to activate the classical pathway of complement. Interestingly, a recent study looking at complement-dependent cytotoxicity of B cells by CD20-specific IgA suggested that complement was activated by IgA. However, in vivo, the activity of the anti-CD20 IgA to deplete B cell targets was not abrogated in C1q- or C3-deficient mice, suggesting that complement activation was not the predominant killing mechanism in action [67]. The ability of IgA to activate the alternative pathway of complement has been somewhat contentious, but the prevailing view is that the reported activation is likely via the lectin pathway as a result of binding to mannose-binding lectin [68]. However, the ability to activate via this route is likely dependent on glycosylation status.

3.3. Interaction of the IgA Fc Region with Host Receptors

In addition to the above-mentioned functions, IgA mediates a variety of effector functions through interaction with a number of different host receptors expressed on various cell types. The interaction with pIgR and the resultant transport into mucosal secretions has already been discussed. Now, we will turn to consideration of the IgA-specific receptor FcαRI, a key means by which IgA can trigger clearance mechanisms against invading pathogens. Other receptors which have been described to have specificity for IgA are generally less well characterised in terms of their roles and will not be addressed further here. These include Fcα/μR, which exhibits specificity for polymeric forms of IgA and IgM, in the case of IgA through a site at the Cα2–Cα3 domain interface [69]; transferrin receptor (CD71), which has been implicated in retrograde transfer of SIgA immune complexes back through the epithelium [70]; a microfold (M) cell receptor, possibly Dectin-1, which may mediate reverse transcytosis of SIgA immune complexes through M cells [71]; dendritic cell (DC)-specific intercellular adhesion molecule-3-grabbing non-integrin (DC-SIGN), which appears to take up SIgA immune complexes into sub-epithelial dendritic cells [72]; the inhibitory IgA receptor Fc receptor-like 4 (FcRL4) thought likely to be important for immune complex-dependent regulation of B cells [73];

the asialoglycoprotein receptor (ASGPR) on hepatocytes, which mediates clearance of IgA from the circulation [74]; β-1,4-galactosyltransferase 1, which, along with CD71, has been identified as a potential IgA receptor on kidney mesangial cells [75]; and lastly, the putative receptor for SC and SIgA on eosinophils [76].

3.4. FcαRI

Although a less closely related member, FcαRI belongs to the Ig Fc receptor family, which also features specific receptors for IgG (FcγRI, FcγRII and FcγRIII) and IgE (FcεRI) [77–79]. It is expressed on neutrophils, eosinophils, monocytes, macrophages, Kupffer cells, and some DC subsets. Also known as CD89, it is encoded by a gene lying on chromosome 19, within the leukocyte receptor cluster (LRC) close to killer cell immunoglobulin-like receptors (KIR) and leukocyte immunoglobulin-like receptors (LILR) receptors. In contrast, other Fc receptors in the family are clustered on chromosome 1. In keeping with this gene location, FcαRI shares closer amino acid similarity with LRC members than with the IgG and IgE Fc receptors.

FcαRI is organised into two extracellular Ig-like domains, a transmembrane segment, and a short cytoplasmic tail devoid of signalling motifs. It associates with a dimer of the FcR γ chain, a short transmembrane polypeptide originally characterised as a component of the IgE receptor, FcεRI. The γ chain carries two immunoreceptor tyrosine-based activation (ITAM) motifs within its cytoplasmic region, important for signalling to the cell interior upon receptor crosslinking by binding to IgA-containing immune complexes or to IgA concentrated on a pathogen surface. The outcome of such signalling can be a range of responses depending on the cell involved, from phagocytosis, superoxide generation (respiratory burst), release of cytokines, chemoattractants, or inflammatory mediators, through to release of neutrophil extracellular traps (NET) [80,81]. On the other hand, binding of monomeric IgA to FcαRI has been reported to trigger inhibitory signals via the γ chain ITAM as opposed to the aforementioned activatory ones. Such inhibitory ITAM (ITAMi) signalling is considered to dampen down excessive IgA immune complex-mediated responses. The underlying signalling processes and the specifics of responses are detailed elsewhere [82].

Alternatively spliced isoforms of FcαRI exist, with those known as a.1 and a.2 being expressed on phagocytes [83,84]. The a.1 version has a molecular weight of 55–75 kDa on neutrophils and monocytes, while additional glycosylation renders it a little heavier (70–100 kDa) on eosinophils. The a.2 version is lacking 22 amino acids from the second extracellular domain, and is only present on alveolar macrophages. In terms of allelic variation, a common, nonsynonymous, single nucleotide polymorphism (SNP) has been described in the coding region of FcαRI, which results in a change of residue 248 from Ser to Gly within the cytoplasmic domain [85].

The structure of the ectodomains of FcαRI has been solved at high resolution, in complex with the Fc region of IgA1 [18]. The globular extracellular domains lie at an angle of around 90° to each other, and it is notable that their relative orientation is very different from the corresponding domains of other Fc receptors [18,86].

FcαRI binds both subclasses of human IgA with similar affinity, and also engages both serum IgA (monomeric) and SIgA (polymeric), albeit with some differences in outcome [82]. However, it has been observed on polymorphonuclear leukocytes that SIgA cannot bind to FcαRI in the absence of CR3 or Mac-1 [87]. The affinity of FcαRI for IgA molecules in solution is low (K_a of approximately 10^{-6} M^{-1}), but IgA immune complexes, or IgA aggregated for example on a pathogen surface, bind with higher avidity. The crystal structure of the complex of the ectodomains of FcαRI and IgA1 Fc revealed that each IgA Fc region is capable of binding two FcαRI molecules [18]. The physiological relevance of this observed stoichiometry is a subject of some conjecture. The site of interaction on IgA, originally defined by mutagenesis [88–90] and further defined by crystallography [18], lies at the Fc domain interface, with important contributions from Cα2 residues Leu257 and Leu258 and Cα3 residues Met433, Leu441, Ala442, Phe443, and the aliphatic portion of Arg382 (Figure 4). On the receptor, the hydrophobic core of the interaction relies on contributions from a region in the membrane distal domain (Tyr53,

Leu54, Phe56, Gly84, His85) with contributions also from Lys55 [18,91,92]. This mode of Fc receptor–Ig interaction is very different from the FcγR–IgG and FcεRI–IgE interactions, which involve sites on the upper reaches of the respective Fc regions, and on the membrane proximal domains of the respective receptors [86].

The contribution of *N*-linked glycans, both on FcαRI and IgA, in the interaction have been investigated. Studies using a glycoengineering approach to generate IgAs carrying distinct homogeneous *N*-glycans have indicated that different glycoforms of IgA1 and IgA2 do not exhibit radically different binding to FcαRI [20], in keeping with earlier analysis that showed that variation in or lack of the *N*-linked glycans at Asn263 in the Cα2 domain did not significantly impact on binding to FcαRI [12,93]. In contrast, specific *N*-linked sugar moieties on FcαRI have been shown to impact on binding to IgA [20,94]. A FcαRI glycovariant with oligomannosidic *N*-glycans has been reported to bind IgA 2–3 times more tightly than variants with complex *N*-glycans [20], while deglycosylation of FcαRI at Asn58 has been shown to increase binding to IgA [94].

Recently, binding of FcαRI to IgA has been demonstrated to propagate conformational changes within IgA as far as the hinge region [95]. Thus, FcαRI binding was shown not only to cause a decrease in IgA Fc intradomain and interdomain flexibility, but also to impact on the hinge, such that binding of lectins to the IgA1 hinge was affected.

It has been reported that peptide mimetics, consisting of either linear or cyclised peptides of 7–18 amino acids spanning regions of FcαRI or IgA known to be involved in the interaction site, may serve as a means to inhibit IgA–FcαRI interactions [96]. Such peptides were shown to reduce IgA effector functions mediated through FcαRI such as phagocytosis and production of activated oxygen species. Blocking strategies based on peptides such as these, or on antibodies directed against FcαRI, have been proposed as possible routes to prevent undesirable inflammatory conditions triggered through aberrant IgA immune complexes [79,97].

Specific elements of the innate immune system are also known to interact directly with FcαRI and impact on IgA binding. Thus, pentraxins such as C reactive protein and serum amyloid P component, which adopt pentameric ring-like structures, have been shown to bind to FcαRI, in part, via a similar region as IgA. Although the pentraxin interaction site on FcαRI appears to be more extensive than that responsible for binding IgA, these acute phase proteins are able to competitively inhibit IgA binding [98].

4. Circumvention of IgA Function by Pathogens

On the basis of phylogenetic and diversity analysis, the IgA–FcαRI interaction has been proposed to be the focus of an evolutionary arms race between pathogens and humans [99,100]. The site on IgA central to the interaction, which has been conserved in order to bind FcαRI, has been placed under pressure to evolve by IgA binding proteins that certain pathogens produce. These IgA binding proteins have evolved to interact with the same site, thereby subverting the IgA response, and driving an iterative selective process in which both mammalian and pathogen proteins have continued to evolve in an attempt to "outsmart" the other. In fact, targeting of the FcαRI interaction site is just one of the strategies that pathogenic microorganisms have used to circumvent the protective capabilities of IgA. The existence of different IgA-targeting mechanisms, together with the fact that these mechanisms seem to have arisen independently in different organisms, suggests that they offer significant benefits to microorganisms by allowing easier mucosal colonisation and spread. Examples include the IgA binding proteins mentioned above and the production of enzymes that cleave and inactivate IgA, which will be discussed in more detail below, and the generation of proteins that bind SC or pIgR and aid adherence and invasion within the mucosae [101–104].

4.1. Bacterial IgA Binding Proteins

Certain important pathogenic bacteria, including Group A and B streptococci and *Staphylococcus aureus*, express proteins on their surface, which bind specifically to IgA. Group A streptococci, which cause a range of diseases from mild skin and throat infections to life-threatening systemic conditions, express Sir22 and Arp4, while group B streptococci, responsible for serious, sometimes deadly, infections in new-born infants, express the unrelated β protein [105–107]. *Staphylococcus aureus*, which can cause bacteraemia, infective endocarditis, and skin and soft tissue infections, expresses an IgA binding protein known as Staphylococcal superantigen-like protein 7 (SSL7). Despite these proteins not being related to each other, all bind at the Cα2–Cα3 interdomain region of IgA Fc at sites that overlap with that for FcαRI [19,108,109]. They have been shown to competitively inhibit FcαRI binding; further, the streptococcal proteins have been demonstrated to block triggering of elimination mechanisms via FcαRI. Thus, these IgA binding proteins provide the bacteria in question with effective ways to evade IgA-mediated clearance.

4.2. Bacterial Proteases That Target IgA

The protective capabilities of IgA can also be compromised through the actions of proteolytic enzymes produced by a number of important pathogenic bacteria. These proteases all cleave in the hinge region of IgA. With few exceptions, they act specifically on the extended hinge region of IgA1, and do not cleave IgA2. Such IgA1 proteases are produced by bacteria responsible for infections of the oral cavity, such as *Streptococcus sanguis*, *Streptococcus mitis*, and *Streptococcus oralis*, and of the genital tract, such as *Neisseria gonorrhoeae*, suggesting that they afford an advantage to the bacteria in gaining a foothold at mucosal surfaces. In addition, they are produced by bacteria responsible for meningitis (*Haemophilus influenza*, *Neisseria meningitidis*, and *Streptococcus pneumoniae*).

The IgA1 proteases appear to have evolved several times over since those from different bacterial species tend not to share common features. Indeed, they represent a range of protease types, with some being metalloproteases, others being serine proteases, and yet others being cysteine proteases [110]. By separating the antigen-binding region of IgA from the Fc region critical for binding to host FcαRI, IgA1 proteases perturb normal IgA-mediated protection mechanisms and leave the bacteria free to proliferate [111].

Each IgA1 protease cleaves a specific site within the IgA1 hinge, either a Pro–Thr or a Pro–Ser peptide bond (Figure 8). In order for IgA1 proteases to recognise the IgA1 hinge as a substrate, it has become clear that not only sequence elements within the hinge itself are important [112,113], but, at least for some IgA1 proteases, also specific regions of the IgA1 protein lying well beyond the hinge. Thus, for efficient cleavage to occur, the susceptible bond is required to be positioned at a suitable position relative to the Fc [114], and some proteases also require the presence of elements within the Fc region of IgA1 [115,116]. Specifically, Cα3 domain residues Pro440–Phe443, which as mentioned above form part of the interaction sites for FcαRI and pIgR, have been shown to be a requirement for cleavage of IgA1 by the *N. meningitidis* type 2 IgA1 protease, while for the *H. influenzae* type 2 enzyme, different Cα3 residues predicted to be involved in pIgR interaction are required for cleavage to proceed [116]. Echoing the case with IgA binding proteins, these requirements suggest that IgA1 proteases may have commandeered conserved host receptor sites for their own benefit. One can envisage an interaction between IgA1 protease and the IgA1 molecule as a whole, with the protease engaging with elements within the Fc region as a means to stabilise a particular IgA conformation and aid positioning of its active site next to the IgA1 hinge. Indeed, the solved X-ray crystal structure of an *H. influenzae* IgA1 protease is consistent with such a possibility [117].

A more detailed understanding of the molecular basis of IgA1 hinge cleavage by IgA1 proteases may have therapeutic application. For example, following earlier work to identify possible inhibitors for IgA1 protease [118,119], small molecule non-peptidic inhibitors for *H. influenzae* IgA1 protease have recently been described in the first steps towards development of potential therapeutics for antibiotic-resistant *H. influenzae* strains [120]. Further, it has been proposed that IgA1 proteases may have utility as therapeutic options to degrade pathogenic immune complexes of aberrantly glycosylated IgA1 in IgA nephropathy, a common cause of kidney disease [121,122].

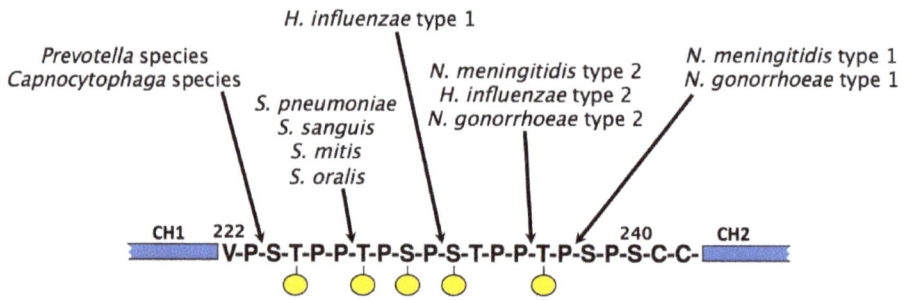

Figure 8. Amino acid sequence in the hinge region of human IgA1 and the cleavage sites of various IgA1 proteases. The IgA1 hinge contains a duplicated octapeptide sequence that is missing in IgA2. O-linked glycans are represented by yellow circles.

5. IgA Developability

Specific IgA is often found elevated in the serum and/or secretions after immunisation. While vaccination via the systemic route tends to generate serum responses, vaccination through the intranasal or oral route can elicit protective mucosal responses [123]. As a prime example, oral cholera vaccination is well established as a means to induce protective mucosal IgA responses [124]. As another example, studies in mice have shown that a nasal vaccine is sufficient to prevent *Streptococcus pneumonia* colonisation, registering high levels of IgA and IgG in plasma and nasal washes. However, this protective action was abrogated in IgA deficient mice [125]. In the context of viruses, neutralising IgA antibodies against HIV can be found in the serum of survivors or vaccinated HIV patients [126,127], and serum and salivary IgA against polio virus can be found elevated upon vaccination with live attenuated viruses [128]. In mice, immunisation against reovirus has been demonstrated to lead to an increase of serum and gut IgA, which proved to be essential to prevent reovirus infection [129]. A similar outcome was observed in mice immunised with influenza virus hemagglutinin, where the induced IgA response provided protection against influenza infection [130].

The above studies present a snapshot of the protective role that IgA can play against bacterial or viral infections, both in serum and mucosal secretions. Since specific IgA can clearly be beneficial in clearing viral or bacterial infections, passive administration of IgA is an attractive option in cases where the immune response is comprised or where insufficient time, or other logistical hurdles, prevent generation of a timely and robust response through active immunisation. Moreover, with regard to the protection of mucosal sites, effective vaccination requires the correct antigen, adjuvant, and delivery route to promote a robust and protective response. Hence, the use of passive immunisation, by direct delivery of specific antibodies, can present an alternative for the protection of mucosal surfaces. However, it remains challenging to create a delivery route, especially for the gut mucosa.

5.1. Advantages of IgA-Based Therapeutics

The therapeutic antibody field is currently dominated by IgG-based mAbs. The advantages of opening up this arena to include IgA-based mAbs are becoming increasingly apparent, piquing interest in both academia and industry [79,131–133]. One advantage is the new prospects it offers in terms of intellectual property, in what is already a complex landscape [134]. Secondly, as will be explored further below, IgA mAbs are known to be highly effective at recruiting immune cells, and neutrophils in particular, to deliver potent killing mechanisms, making the IgA–FcαRI axis an important target in control of various cancers and infections. Such neutrophil-mediated tumour cell killing is considered especially important for apoptosis-resistant cells [131]. Thirdly, IgA is likely to represent the most suitable option for mucosal applications, given its prevalence and functional capabilities at such sites. Fourthly, the structural distinctiveness of IgA, especially IgA1 with its ability to bridge greater distances between antigens, may offer enhanced avidity in some scenarios. Fifthly, IgA can naturally polymerise into forms with enhanced agglutination capabilities, and which can be transported by pIgR into mucosal secretions. Finally, it is possible to use components of IgA or IgA heavy chains in combination with those of other Igs such as IgG, to explore new therapeutic possibilities.

5.2. Constraints of Using IgA Therapeutically and Efforts to Resolve These

Despite the numerous advantages that may be associated with the development of IgA in the therapeutic setting, there are a number of constraints or limitations that need to be addressed. For example, both pro-inflammatory and anti-inflammatory functions of IgA mediated through FcαRI have been flagged up as being of relevance to the therapeutic potential of IgA [131,135]. As a result, it will be important to establish the mechanism(s) at play in any particular treatment setting.

Another constraint is that IgA has a shorter half-life than IgG, estimated to be 4–6 days [136,137]. IgA cannot bind to the neonatal Fc receptor, FcRn, while engagement of IgG with this receptor results in a half-life of about 21 days (although it varies with subclass). The short half-life of IgA would necessitate much more frequent dosing if this class was to be used therapeutically. For example, in mouse tumour models, it has been found to be necessary to give daily injections of IgA antibodies to reach effective circulating concentrations [138]. Unless modified, use of IgA is therefore likely to be expensive and less convenient for recipients because of the frequency of dosing. This shorter half-life is in part due to clearance mediated by the ASGPR, which recognises terminal galactose residues on the glycans of IgA. Efforts have been made to extend half-life by removing *N*-linked glycosylation sites [139], generating IgA with higher terminal sialylation of *N*-glycans [140], by attaching an albumin-binding domain to either the LC or HC in order to facilitate binding to the neonatal Fc receptor FcRn [141], or by engineering in FcRn binding by generating an IgG–IgA Fc fusion [133].

A further constraint relates to efficiency issues in the expression, production, and purification of recombinant IgA mAbs of a suitably homogeneous nature. It has long been recognised that IgA production suffers from low expression levels and heterogeneous glycosylation. Systems enabling increased expression of IgA have been developed [140,142,143], and advances in general expression systems for other Igs are likely also to bring benefits [144,145]. There is interest in using plant-based systems to express IgA [146–148], but the implications for glycosylation must be borne in mind, especially since it is known that IgA glycosylation is impacted by expression system [149,150].

The logistics of working with IgA has been challenging due to the limited options for specifically purifying this Ab class. Jacalin, a lectin that binds to the *O*-linked sugars on the IgA1 hinge, and light chain binding protein-based strategies offer rather limited possibilities. Immobilised bacterial IgA binding proteins, or peptides derived from them, represent a feasible solution [151,152], and IgA-binding peptides selected from random peptide libraries may also have applicability in IgA purification [153].

The susceptibility of IgA1 to cleavage by IgA1 proteases may be another potential constraint to its use. However, as discussed above, mutagenesis analysis has demonstrated how this might be overcome either by engineering of the hinge itself or of the Fc region [116].

Another area for consideration in the design of therapeutic IgA mAbs are the routes to ensure complete assembly. For instance, the disulphide bridge complexity in IgA2 presents challenges [154]. The production of polymeric forms of IgA or SIgA is particularly complex, given the requirement to co-express LC, HC, and J chain, and ensure attachment of SC. However, systems to achieve this have been explored and continue to be refined [133,155,156].

A final constraint to the development of therapeutic IgA mAbs stems from the lack of suitable animal models. Since IgA1 equivalents are only found in humans and closely related apes, the use of the species normally used in experimental research (mouse, rat, rabbit) will most likely fail to give a realistic reflection of behaviour in humans. The other species differences noted earlier, such as differences in the polymerisation state of serum IgA, tend to compound this problem. The mouse is considered especially unsuitable for testing the function of human IgA because it lacks the equivalent of human FcαRI. To circumvent this issue, mice transgenic for human CD89 have been generated and used widely as useful models for analysis of the function of human IgA [157,158]. Another notable milestone in creation of useful mouse models was the generation of a human IgA knock-in mouse [159].

6. Current landscape of IgA-Based Therapeutics

6.1. Comparisons of IgG and IgA mAbs in Cancer Therapy

Traditional cancer therapies of removal surgery or radiation for elimination of tumour cells in localised tumours and chemotherapy for metastatic tumours, while effective, are very aggressive procedures. With the development of proteomic, genomic, and bioinformatics approaches, it became possible to better characterise cancer cells and identify the proteins expressed at their surface. Thus targeting of tumour cells by antibodies directed to tumour antigens, such as glycoproteins, growth factors, cluster of differentiation (CD) antigens, is now an established treatment option [160].

Of the several therapeutic antibodies used in cancer treatment, some are used in solid tumours, targeting specific antigens such as the epidermal growth factor receptor (EGFR) found in colorectal cancer, or the human epidermal growth factor 2 (HER2) associated with breast cancer [161]. More "liquid" tumours such leukaemias and lymphomas have also been successfully treated. For example, B-cell lymphomas have been treated with anti-CD20 mAbs [162]. Indeed, Rituximab, an anti-CD20 antibody, was the first monoclonal antibody approved for cancer therapy in 1997, being followed by several others, including Cetuximab (anti-EGFR) and Trastuzumab (anti-HER2), all of the IgG isotype [163].

These mAbs work in different ways, with anti-CD20 mAb inducing apoptosis and sensitising tumour cells for chemotherapy, anti-HER2 inhibiting intracellular pathways involved in cancer progression, and anti-EGFR binding to growth factor receptors and blocking cancer cell proliferation [164–166]. However, their performance will often depend on the expression levels of the antigen on the tumour cells and can be affected by mutations in downstream pathways. Being of the IgG subclass, these mAbs are able to activate the complement pathway and interact with Fcγ receptors, eliminating tumours by cell lysis or targeting tumour cells for elimination by immune cells. There has been debate regarding which subset of immune cells is more important for mAb therapy, with natural killer (NK) cells seen for a long time as the main effectors, promoting apoptosis of tumour cells [167]. Macrophages, and to a lesser extent monocytes, were also recognised for their phagocytosis ability towards tumour cells coated with antibodies [168], while neutrophils were associated with tumour regression, even in the absence of mAbs [169]. Neutrophils, besides secreting cytotoxic agents, can lead to necrotic and autophagic tumour cell death, and can be recruited in large numbers, especially upon stimulation with granulocyte-colony stimulating factor (G-CSF) and granulocyte macrophage-colony stimulating factor (GM-CSF) [170,171]. The importance of neutrophils in tumour clearance was shown in a B-cell lymphoma mice model, where anti-CD20 mAb was less effective when neutrophils were depleted [172]. Since neutrophils do not easily recognise tumour cells, the use of mAbs is important to establish this interaction. However, the high-affinity IgG Fc receptor FcγRI is only expressed in neutrophils upon G-CSF stimulation, and besides the numerous side effects of the stimulation, this therapeutic strategy did not lead to significant clinical responses when using IgG mAbs [173–176].

IgA, together with its receptor FcαRI (CD89), create another possibility for new therapies focused on the activation of FcαRI-expressing cells. Both FcαRI and FcγRI associate with FcR γ chain, but FcαRI may create stronger electrostatic interactions with the FcR γ chain promoting a more stable interaction [177]. Besides, binding to FcαRI promotes release of leukotriene B4 (LTB4), which acts as a chemoattractant for neutrophils. Therefore, targeting this receptor leads to additional neutrophil migration to tumour sites [80]. Although FcαRI expression in neutrophils is lower than that of Fcγ receptors naturally expressed in these cells (FcγRIIa and FcγRIIIb), binding of IgA or IgG to neutrophils is similar, which suggests a more stable binding by IgA and a higher efficiency at triggering neutrophils than IgG [178]. For instance, the use of an IgA anti-Ep-CAM mAb was shown to kill colon carcinoma cells, unlike the IgG1 mAb counterpart [179]. Similar results were shown for the anti-EGFR mAb, with the IgA being superior at recruiting polymorphonuclear cells than the IgG subtype [180].

Another alternative to target FcαRI consists in the use of bispecific antibodies (BsAb). By virtue of combining two distinct antigen binding capabilities, BsAb are able to target tumours and recruit immune cells, such as neutrophils, leading to tumour cell killing by antibody-dependent cellular cytotoxicity mechanisms [181]. The use of a BsAb against both HER2 and FcαRI (namely anti-HER2 × FcαRI) efficiently eliminated breast carcinoma cells by neutrophil accumulation, unlike the equivalent FcγRI-directed BsAb (anti-HER2 × FcγRI) [182]. The same was observed for CD20 antibodies, where IgG Abs or FcγRI and FcγRIII-directed BsAbs (anti-CD20 × FcγRI or FcγRIII) showed no ability to kill malignant B cells, whereas the equivalent FcαRI BsAb promoted malignant B cell killing via neutrophil activation [183]. Another study showed that the BsAb anti-HLA II × FcαRI was effective in recruiting polymorphonuclear cells against human B cell malignancies [184].

For a long time, in vivo studies on IgA and FcαRI cancer therapies were impaired by the lack of FcαRI in mouse. However, the development of FcαRI transgenic mice has overcome that barrier [157]. Additionally, the study of mouse IgAs in interaction with FcαRI has been hampered due to the poor binding of mouse IgA to the human FcαRI, but the knock-in of human IgA into mice (Cα1 gene knock-in) has made possible the generation of antigen-specific human IgA mAbs in mice [159]. The use of these animal models showed that anti-CD20 IgA mAbs can effectively prevent B cell lymphoma development by recruiting FcαRI-expressing immune cells [67,185]. Likewise, IgA2 anti-EGFR was proved to be more efficient than Cetuximab (IgG format) against tumour cells in a FcαRI transgenic mice model [138]. In addition to the anti-tumour response of IgA1 anti-HER2 mAb, it was shown that the introduction of an albumin binding domain allows the interaction with the neonatal Fc receptor (FcRn), which is used for IgG and albumin recycling in the serum, leading to an increase of the IgA half-life without compromising its anti-tumour activity in vivo [141]. As mentioned previously, the half-life of IgA can also be extended by decreasing clearance by ASGPR in the liver, which can be achieved by sialylation of the IgA glycans [138]. A higher sialylation of the N-glycans in the IgA anti-HER2 did not interfere in the anti-tumour response and lead to the decrease in tumour growth in FcαRI transgenic mice, while increasing the antibody half-life [140]. In another study, the removal of two glycosylation sites and two free cysteines, together with a stabilised HC and LC linkage, created a new IgA2 anti-EGFR mAb with a longer half-life than the wild-type antibody, and higher efficacy due to Fab-mediated effects and interaction with myeloid cells expressing FcαRI [139].

6.2. IgA mAbs in Treating or Preventing Infections

Several anti-infective mAbs of the IgG isotype are approved to combat infectious diseases, namely, Palivizumab against respiratory syncytial virus, Raxibacumab and Obiltoxaximab against anthrax, and Bezlotoxumab to combat *Clostridium difficile* [186].

As the most abundant antibody at the mucosal surfaces, IgA has the important role of detecting and alerting the immune system to pathogens, whilst not responding to commensal bacteria and environmental antigens, representing an important means to combat infectious diseases. IgA antibodies were shown to be effective against tuberculosis infection in a mouse model. The passive intranasal inoculation with a mouse IgA mAb against the α-crystallin antigen of *Mycobacterium tuberculosis* led to a significant decrease in bacteria in the lungs, when either monomeric or polymeric forms of the antibody were used. Despite the transitory protective effect, probably due to the fast degradation of the administered IgA, this antibody was shown to combat early infection in the lungs, with potential use for immunoprophylaxis in immunocompromised individuals at risk of tuberculosis infection [187]. In a later study, the use of a human IgA1 against *M. tuberculosis* showed that the protective effect of the passive inoculation is dependent on the presence of FcαRI, being observed only in mice transgenic for human FcαRI [188]. These results suggest that the interaction between the human IgA1 and FcαRI on neutrophils and macrophages allows binding and elimination of *M. tuberculosis*. In the same study, in vitro infection of human whole blood or isolated monocytes by *M. tuberculosis* was reduced in the presence of specific IgA1 [188].

The importance of interaction with FcαRI was also shown for control of *Escherichia coli* infection, which when recognised by human serum IgA, can be efficiently phagocytised by FcαRI-expressing cells [189]. This ability of IgA to bind FcαRI and directly induce neutrophil migration was shown to be an important defense mechanism against several other bacteria, such as *Streptococcus pneumonia*, *Staphylococcus aureus*, *Porphyromonas gingivalis*, *Candida albicans*, *Bordetella Pertussis*, and *Neisseria meningitidis* [81,190–194].

The immune exclusion ability of IgA was also shown in the context of *Salmonella typhimurium* infection, where mice were orally challenged with the bacteria alone or the bacteria complexed with plasma-derived IgA and IgM [195]. Reduced bacteria dissemination was reported in mice exposed to the IgA/IgM immune complexes, mainly for antibodies coupled with the secretory component (SC), whilst IgG was unable to form immune complexes and consequently protect against *S. typhimurium* spread in gut immune structures [195]. Besides, oral administration of SIgA/M prior to intragastric *S. typhimurium* challenge is sufficient to protect mice from infection [196]. Despite the studies showing the potency of these IgA antibodies to prevent bacterial infections, all the existing immunoglobulin preparations used clinically for replacement therapy contain only IgG [197].

Passive immunisation with monomeric IgA can also be applied for viral infections. The use of vaccines against influenza virus showed the emergence of both IgA and IgG in nasal washes, but it was difficult to establish the importance of these antibodies individually [198,199]. Passive immunisation with IgG or pIgA by intravenous injection culminated in specific transport of these antibodies into nasal secretions [200]. However, high doses of IgG anti-influenza have to be injected in order to detect its presence in mice nasal secretions, and even higher doses are needed to decrease viral shedding [201]. On the other hand, administration of polymeric IgA at levels normally found in convalescent mice is enough to eliminate nasal viral shedding. Therefore, SIgA prevents infection of the upper respiratory tract, while serum IgG is important as a secondary response, acting at a later stage by detecting viruses that escaped IgA neutralisation and preventing lung infection [201]. A study using rotavirus showed that mice can be protected from infection when IgA mAb against the viral capsid was systemically administrated, but not when added to the intestinal lumen, showing the importance of transcytosis as a way of viral inactivation [65].

Passive immunisation was also tested on simian models of HIV infection. Intrarectal administration of IgG and dimeric IgA specific for the viral envelope showed that dimeric IgA provided the best protection in vivo upon SHIV infection in rhesus monkeys [202]. The protection conferred by dimeric IgA was suggested to be related to its ability to directly neutralise the virus and to form complexes that prevented free viruses crossing the epithelial cell layer. Based on the interaction of SIgA with mucosal microfold (M) cells, another study explored the transport of an HIV antigen for immunisation via this mechanism. SIgA bound to the HIV antigen was delivered orally and transported across the epithelial barrier to be captured by dendritic cells, starting mucosal and systemic immune responses that ultimately showed to be protective against infection by a recombinant virus expressing the HIV antigen [203]. Therefore, infection can be impaired by several IgA associated mechanisms, either by immune exclusion, intracellular inactivation, or recognition and activation of the immune system.

6.3. FcαRI Blocking Agents

Targeting FcαRI can be used as a strategy to combat autoimmune diseases, to inhibit IgG-induced phagocytosis or IgE-mediated allergic diseases. In autoimmune diseases, binding of IgA to FcαRI leads to enhanced activation of immune cells, and therefore, blocking this interaction can be beneficial to decrease tissue damage. The exposure of neutrophils to IgA immune complexes obtained from rheumatoid arthritis patients leads to in vitro release of neutrophil extracellular traps, which consist of web-like structures made of DNA and proteins that, despite capturing pathogens, are associated with tissue damage. However, the use of an anti-FcαRI mAb (MIP8a) was shown to successfully decrease neutrophil extracellular traps formation [204]. The same anti-FcαRI mAb was shown to prevent IgA autoantibodies inducing tissue damage in an ex vivo human skin model for linear IgA bullous disease [97]. Beyond mAbs, peptides that bind to the interaction sites of IgA and FcαRI could also inhibit IgA-induced neutrophil migration, having the advantage to be able to penetrate into the skin, which opens up the possibility of using them for skin autoimmune disease therapy [96].

Besides IgA, other antibodies can start immune responses that, when exacerbated, can be harmful, culminating in extensive inflammation or allergies. Binding of FcαRI by monomeric IgA is known for its anti-inflammatory nature through ITAMi signalling in effector cells [205]. Therefore, the IgA–FcαRI interaction can be explored as a tool to alleviate inflammation and further tissue damage caused by other antibodies. Using an allergy mice model, it was possible to show a decrease in airway inflammation upon crosslinking of FcεRI with IgE immune complexes in a FcαRI transgenic mice treated with the anti-FcαRI mAb A77 [206]. In another study, monomeric IgA was shown to successfully abrogate arthritis in a FcαRI transgenic mice model where IgG anti-collagen was used to cause rheumatoid arthritis [207]. Using a FcαRI transgenic mice model with glomerulonephritis and obstructive nephropathy caused by accumulation of IgG immune complexes, the Fab A77 targeting FcαRI was shown to be able to suppress inflammation [208]. It was also established that renal inflammation induced by different agents can be alleviated by the use of Fab fragments that target FcαRI (MIP8a) or monomeric IgA [209,210]. Therefore, targeting FcαRI either through IgA binding or the use of specific antibodies, can be used as a strategy to initiate anti-inflammatory responses in inflammatory diseases that involve myeloid cells.

7. Summary and Conclusions

The structural features of IgA impart this Ab class with unique functional capabilities, which are yet to be fully harnessed for therapeutic benefit. Increasing numbers of mAbs have been approved for clinical use in the last few years, and many more are currently undergoing clinical trial [211,212]. Recent examples tend to be humanised or fully human, but invariably of the IgG isotype. To date, no antibodies of the IgA isotype are known to be going through clinical trials. Regarding BsAbs, only a very few have been approved for use in the United States, while several await approval or are in preclinical and clinical trials [213]. In this context, FcαRI-targeting BsAbs are yet to reach this stage, indicating that further effort is required before the potential of IgA/FcαRI related therapies can be realised. As that point approaches, interest will undoubtedly turn to options for delivery to mucosal sites. Progress with topical application of nebulised Igs in the lungs of experimental animals [214,215] suggest that suitable strategies for mucosal delivery of mAbs in humans may appear, and we can anticipate that IgA-based mAbs will emerge as an important new arm of the arsenal of therapeutic mAbs.

Author Contributions: Conceptualization, J.M.W.; writing—original draft preparation, review and editing, J.M.W. and P.d.S.-P.

Funding: P.d.S.-P. is funded by an award from the Fundação para a Ciência e a Tecnologia (FCT), grant number PTDC/BIA-OUT/29667/2017.

Conflicts of Interest: The authors declare no conflict of interest.

References

1. Mestecky, J.; Russell, M.W.; Jackson, S.; Brown, T.A. The human IgA system: A reassessment. *Clin. Immunol. Immunopathol.* **1986**, *40*, 105–114. [CrossRef]
2. Conley, M.E.; Delacroix, D.L. Intravascular and mucosal immunoglobulin A: Two separate but related systems of immune defense? *Ann. Intern. Med.* **1987**, *106*, 892–899. [CrossRef] [PubMed]
3. Childers, N.K.; Bruce, M.G.; McGhee, J.R. Molecular mechanisms of immunoglobulin A defense. *Annu. Rev. Immunol.* **1989**, *43*, 503–536. [CrossRef] [PubMed]
4. Chintalacharuvu, K.R.; Raines, M.; Morrison, S.L. Divergence of human alpha-chain constant region gene sequences. A novel recombinant alpha2 gene. *J. Immunol.* **1994**, *152*, 5299–5304.

5. Kawamura, S.; Saitou, N.; Ueda, S. Concerted evolution of the primate immunoglobulin α-gene through gene conversion. *J. Biol. Chem.* **1992**, *267*, 7359–7367.

6. Pinheiro, A.; de Sousa-Pereira, P.; Strive, T.; Knight, K.L.; Woof, J.M.; Esteves, P.J.; Abrantes, J. Identification of a new European rabbit IgA with a serine-rich hinge region. *PLoS ONE* **2018**, *13*, e0201567. [CrossRef]

7. Snoeck, V.; Peters, I.R.; Cox, E. The IgA system: A comparison of structure and function in different species. *Vet. Res.* **2006**, *37*, 455–467. [CrossRef]

8. Putnam, F.W.; Yu-Sheng, V.L.; Low, T.L.K. Primary structure of a human IgA1 immunoglobulin. IV. Streptococcal IgA1 protease digestion, Fab and Fc fragment and the complete amino acid sequence of the α1 heavy chain. *J. Biol. Chem.* **1979**, *254*, 2865–2874.

9. Tomana, M.; Niedermeier, W.; Mestecky, J.; Skvaril, F. The differences in carbohydrate composition between the subclasses of IgA immunoglobulins. *Immunochemistry* **1976**, *13*, 325–328. [CrossRef]

10. Maurer, M.A.; Meyer, L.; Bianchi, M.; Turner, H.L.; Le, N.P.L.; Steck, M.; Wyrzucki, A.; Orlowski, V.; Ward, A.B.; Crispin, M.; et al. Glycosylation of human IgA directly inhibits influenza A and other sialic-acid-binding viruses. *Cell Rep.* **2018**, *23*, 90–99. [CrossRef]

11. Field, M.C.; Amatayakul-Chantler, S.; Rademacher, T.W.; Rudd, P.M.; Dwek, R.A. Structural analysis of the N-glycans from human immunoglobulin A1: Comparison of normal human serum immunoglobulin A1 from that isolated from patients with rheumatoid arthritis. *Biochem. J.* **1994**, *299*, 261–275. [CrossRef] [PubMed]

12. Mattu, T.S.; Pleass, R.P.; Willis, A.C.; Kilian, M.; Wormald, M.R.; Lellouch, A.C.; Rudd, P.M.; Woof, J.M.; Dwek, R.A. The glycosylation and structure of human serum IgA1, Fab and Fc regions and the role of N-glycosylation on Fcα receptor interactions. *J. Biol. Chem.* **1998**, *273*, 2260–2272. [CrossRef] [PubMed]

13. Royle, L.; Roos, A.; Harvey, D.J.; Wormald, M.R.; van Gijlswijk-Jannsen, D.; El Redwan, R.M.; Wilson, I.A.; Daha, M.R.; Dwek, R.A.; Rudd, P.M. Secretory IgA N- and O-linked glycans provide a link between the innnate and adaptive immune systems. *J. Biol. Chem.* **2003**, *278*, 20140–20153. [CrossRef] [PubMed]

14. Satow, Y.; Cohen, G.H.; Padlan, E.A.; Davies, D.R. Phosphocholine binding immunoglobulin Fab McPC603. An X-ray diffraction study at 2.7 Å. *J. Mol. Biol.* **1986**, *190*, 593–604. [CrossRef]

15. Suh, S.W.; Bhat, T.N.; Navia, M.A.; Cohen, G.H.; Rao, D.N.; Rudikoff, S.; Davies, D.R. The galactan-binding immunoglobulin Fab J539: An X-ray diffraction study at 2.6-A resolution. *Proteins* **1986**, *1*, 74–80. [CrossRef]

16. Correa, A.; Trajtenberg, F.; Obal, G.; Pritsch, O.; Dighiero, G.; Oppezzo, P.; Buschiazzo, A. Structure of a human IgA1 Fab fragment at 1.55 Å resolution: Potential effect of the constant domains on antigen-affinity modulation. *Acta Crystallogr. D Biol. Crystallogr.* **2013**, *69*, 388–397. [CrossRef]

17. Janda, A.; Bowen, A.; Greenspan, N.S.; Casadevall, A. Ig constant region effects on variable region structure and function. *Front. Microbiol.* **2016**, *7*, 22. [CrossRef]

18. Herr, A.B.; Ballister, E.R.; Bjorkman, P.J. Insights into IgA-mediated immune responses from the crystal structures of human FcαRI and its complex with IgA1-Fc. *Nature* **2003**, *423*, 614–620. [CrossRef]

19. Ramsland, P.A.; Willoughby, N.; Trist, H.M.; Farrugia, W.; Hogarth, P.M.; Fraser, J.D.; Wines, B.D. Structural basis for evasion of IgA immunity by Staphylococcus aureus revealed in the complex of SSL7 with Fc of human IgA1. *Proc. Natl. Acad. Sci. USA* **2007**, *10*, 15051–15056. [CrossRef]

20. Göritzer, K.; Turupcu, A.; Maresch, D.; Novak, J.; Altmann, F.; Oostenbrink, C.; Obinger, C.; Strasser, R. Distinct Fcα receptor N-glycans modulate the binding affinity to immunoglobulin A (IgA) antibodies. *J. Biol. Chem.* **2019**, *294*, 13995–14008. [CrossRef]

21. Feinstein, A.; Munn, E.; Richardson, N. The three-dimensional conformation of γM and γA globulin molecules. *Ann. N. Y. Acad. Sci.* **1971**, *190*, 104–121. [CrossRef] [PubMed]

22. Munn, E.A.; Feinstein, A.; Munro, A.J. Electron microscope examination of free IgA molecules and of their complexes with antigen. *Nature* **1971**, *231*, 527–529. [CrossRef] [PubMed]

23. Roux, K.H.; Strelets, L.; Brekke, O.H.; Sandlie, I.; Michaelsen, T.E. Comparisons of the ability of human IgG3 hinge mutants, IgM, IgE, and IgA2, to form small immune complexes: A role for flexibility and geometry. *J. Immunol.* **1998**, *161*, 4083–4090. [PubMed]

24. Boehm, M.K.; Woof, J.M.; Kerr, M.A.; Perkins, S.J. The Fab and Fc fragments of IgA1 exhibit a different arrangement from that in IgG: A study by X-ray and neutron solution scattering and homology modelling. *J. Mol. Biol.* **1999**, *286*, 1421–1447. [CrossRef] [PubMed]

25. Furtado, P.B.; Whitty, P.W.; Robertson, A.; Eaton, J.T.; Almogren, A.; Kerr, M.A.; Woof, J.M.; Perkins, S.J. Solution structure determination of monomeric human IgA2 by X-ray and neutron scattering, analytical ultracentrifugation and constrained modelling: A comparison with monomeric human IgA1. *J. Mol. Biol.* **2004**, *338*, 921–941. [CrossRef] [PubMed]

26. Hui, G.K.; Wright, D.W.; Vennard, O.L.; Rayner, L.E.; Pang, M.; Yeo, S.C.; Gor, J.; Molyneux, K.; Barratt, J.; Perkins, S.J. The solution structures of native and patient monomeric human IgA1 reveal asymmetric extended structures: Implications for function and IgAN disease. *Biochem. J.* **2015**, *471*, 167–185. [CrossRef]

27. Johansen, F.E.; Braathen, R.; Brandtzaeg, P. Role of J chain in secretory immunoglobulin formation. *Scand. J. Immunol.* **2000**, *52*, 240–248. [CrossRef]

28. Xiong, E.; Li, Y.; Min, Q.; Cui, C.; Liu, J.; Hong, R.; Lai, N.; Wang, Y.; Sun, J.; Matsumoto, R.; et al. MZB1 promotes the secretion of J-chain-containing dimeric IgA and is critical for the suppression of gut inflammation. *Proc. Natl. Acad. Sci. USA* **2019**, *116*, 13480–13489. [CrossRef]

29. Bastian, A.; Kratzin, H.; Eckart, K.; Hilschmann, N. Intra- and inter-chain disulphide bridges of the human J chain in secretory immunoglobulin A. *Biol. Chem. Hoppe Seyler* **1992**, *373*, 1255–1263. [CrossRef]

30. Frutiger, S.; Hughes, G.J.; Paquet, N.; Luthy, R.; Jaton, J.C. Disulfide bond assignment in human J chain and its covalent pairing with immunoglobulin M. *Biochemistry* **1992**, *31*, 12643–12647. [CrossRef]

31. Cann, G.M.; Zaritsky, A.; Koshland, M.E. Primary structure of the immunoglobulin J chain from the mouse. *Proc. Natl. Acad. Sci. USA* **1982**, *79*, 6656–6660. [CrossRef] [PubMed]

32. Atkin, J.D.; Pleass, R.J.; Owens, R.J.; Woof, J.M. Mutagenesis of the human IgA1 heavy chain tailpiece that prevents dimer assembly. *J. Immunol.* **1996**, *157*, 156–159. [PubMed]

33. Krugmann, S.; Pleass, R.J.; Atkin, J.D.; Woof, J.M. Structural requirements for assembly of dimeric IgA probed by site-directed mutagenesis of J chain and a cysteine residue of the α chain CH2 domain. *J. Immunol.* **1997**, *159*, 244–249. [PubMed]

34. Bonner, A.; Furtado, P.B.; Almogren, A.; Kerr, M.A.; Perkins, S.J. Implications of the near-planar solution structure of human myeloma dimeric IgA1 for mucosal immunity and IgA nephropathy. *J. Immunol.* **2008**, *180*, 1008–1018. [CrossRef]

35. Woof, J.M.; Mestecky, J. Mucosal Immunoglobulins. In *Mucosal Immunology*, 4th ed.; Mestecky, J., Strober, W., Russell, M.W., Kelsall, B.L., Cheroutre, H., Lambrecht, B.N., Eds.; Academic Press: Oxford, UK, 2015; pp. 287–324.

36. Natvig, I.B.; Johansen, F.E.; Nordeng, T.W.; Haraldsen, G.; Brandtzaeg, P. Mechanism for enhanced external transfer of dimeric IgA over pentameric IgM: Studies of diffusion, binding to the human polymeric Ig receptor, and epithelial transcytosis. *J. Immunol.* **1997**, *159*, 4330–4340. [PubMed]

37. Kaetzel, C.S. The polymeric immunoglobulin receptor: Bridging innate and adaptive immune responses at mucosal surfaces. *Immunol. Rev.* **2005**, *206*, 83–99. [CrossRef]

38. Bakos, M.A.; Kurosky, A.; Cwerwinski, E.W.; Goldblum, R.M. A conserved binding site on the receptor for polymeric Ig is homologous to CDR1 of Ig V kappa domains. *J. Immunol.* **1993**, *151*, 1346–1352.

39. Coyne, R.S.; Siebrecht, M.; Peitsch, M.C.; Casanova, J.E. Mutational analysis of polymeric immunoglobulin receptor/ligand interactions. Evidence for the involvement of multiple complementarity determining region (CDR)-like loops in receptor domain I. *J. Biol. Chem.* **1994**, *269*, 31620–31625.

40. Hamburger, A.E.; West, A.P.; Bjorkman, P.J. Crystal structure of a polymeric immunoglobulin binding fragment of the human polymeric immunoglobulin receptor. *Structure* **2004**, *12*, 1925–1935. [CrossRef]

41. Hexham, J.M.; White, K.D.; Carayannopoulos, L.N.; Mandecki, W.; Brisette, R.; Yang, Y.S.; Capra, J.D. A human immunoglobulin (Ig) A Cα3 domain motif directs polymeric Ig receptor-mediated secretion. *J. Exp. Med.* **1999**, *189*, 747–752. [CrossRef]

42. Braathen, R.; Sorensen, V.; Brandtzaeg, P.; Sandlie, I.; Johansen, F.E. The carboxyl-terminal domains of IgA and IgM direct isotype-specific polymerization and interaction with the polymeric immunoglobulin receptor. *J. Biol. Chem.* **2002**, *277*, 42755–42762. [CrossRef] [PubMed]

43. Lewis, M.J.; Pleass, R.J.; Batten, M.R.; Atkin, J.D.; Woof, J.M. Structural requirements for the interaction of human IgA with the human polymeric Ig receptor. *J. Immunol.* **2005**, *175*, 6694–6701. [CrossRef] [PubMed]

44. Fallgren-Gebauer, E.; Gebauer, W.; Bastian, A.; Kratzin, H.; Eiffert, H.; Zimmerman, B.; Karas, M.; Hilschmann, N. The covalent linkage of the secretory component to IgA. *Adv. Exp. Med. Biol.* **1995**, *371A*, 625–628. [PubMed]

45. Johansen, F.E.; Braathen, R.; Brandtzaeg, P. The J chain is essential for polymeric Ig receptor-mediated epithelial transport of IgA. *J. Immunol.* **2001**, *167*, 5185–5192. [CrossRef] [PubMed]

46. Stadtmueller, B.M.; Huey-Tubman, K.E.; López, C.J.; Yang, Z.; Hubbell, W.L.; Bjorkman, P.J. The structure and dynamics of secretory component and its interactions with polymeric immunoglobulins. *Elife* **2016**, *5*, e10640. [CrossRef]

47. Mantis, N.J.; Rol, N.; Corthésy, B. Secretory IgA's complex roles in immunity and mucosal homeostasis in the gut. *Mucosal Immunol.* **2011**, *4*, 603–611. [CrossRef]

48. Mantis, N.J.; McGuinness, C.R.; Sonuyi, O.; Edwards, G.; Farrant, S.A. Immunoglobulin A antibodies against ricin A and B subunits protect epithelial cells from ricin intoxication. *Infect. Immun.* **2006**, *74*, 3455–3462. [CrossRef]

49. Wold, A.; Mestecky, J.; Tomana, M.; Kobata, A.; Ohbayashi, H.; Endo, T.; Edén, C.S. Secretory immunoglobulin A carries oligosaccharide receptors for *Escherichia coli* type 1 fimbrial lectin. *Infect. Immun.* **1990**, *58*, 3073–3077.

50. Schroten, H.; Stapper, C.; Plogmann, R.; Köhler, H.; Hacker, J.; Hanisch, F.G. Fab-independent antiadhesion effects of secretory immunoglobulin A on S-fimbriated *Escherichia coli* are mediated by sialyloligosaccharides. *Infect. Immun.* **1998**, *66*, 3971–3973.

51. Ruhl, S.; Sandberg, A.L.; Cole, M.F.; Cisar, J.O. Recognition of immunoglobulin A1 by oral actinomyces and streptococcal lectins. *Infect. Immun.* **1996**, *64*, 5421–5424.

52. Biesbrock, A.R.; Reddy, M.S.; Levine, M.J. Interaction of a salivary mucin-secretory immunoglobulin A complex with mucosal pathogens. *Infect. Immun.* **1991**, *59*, 3492–3497. [PubMed]

53. Xu, F.; Newby, J.M.; Schiller, J.L.; Schroeder, H.A.; Wessler, T.; Chen, A.; Forest, M.G.; Lai, S.K. Modeling barrier properties of intestinal mucus reinforced with IgG and secretory IgA against motile bacteria. *ACS Infect. Dis.* **2019**, *5*, 1570–1580. [CrossRef] [PubMed]

54. Tenovuo, J.; Moldoveanu, Z.; Mestecky, J.; Pruitt, K.M.; Rahemtulla, B.M. Interaction of specific and innate factors of immunity: IgA enhances the antimicrobial effect of the lactoperoxidase system against *Streptococcus mutans*. *J. Immunol.* **1982**, *128*, 726–731. [PubMed]

55. Wright, A.; Lamm, M.E.; Huang, Y.T. Excretion of human immunodeficiency virus type 1 through polarized epithelium by immunoglobulin A. *J. Virol.* **2008**, *82*, 11526–11535. [CrossRef]

56. Robinson, J.K.; Blanchard, T.G.; Levine, A.D.; Emancipator, S.N.; Lamm, M.E. A mucosal IgA-mediated excretory immune system in vivo. *J. Immunol.* **2001**, *166*, 3688–3692. [CrossRef]

57. Fernandez, M.I.; Pedron, T.; Tournebize, R.; Olivo-Marin, J.C.; Sansonetti, P.J.; Phalipon, A. Anti-inflammatory role for intracellular dimeric immunoglobulin a by neutralization of lipopolysaccharide in epithelial cells. *Immunity* **2003**, *18*, 739–749. [CrossRef]

58. Mazanec, M.B.; Kaetzel, C.S.; Lamm, M.E.; Fletcher, D.; Nedrud, J.G. Intracellular neutralization of virus by immunoglobulin A antibodies. *Proc. Natl. Acad. Sci. USA* **1992**, *89*, 6901–6905. [CrossRef]

59. Mazanec, M.B.; Coudret, C.L.; Fletcher, D.R. Intracellular neutralization of influenza virus by immunoglobulin A anti-hemagglutinin monoclonal antibodies. *J. Virol.* **1995**, *69*, 1339–1343.

60. Zhou, D.; Zhang, Y.; Li, Q.; Chen, Y.; He, B.; Yang, J.; Tu, H.; Lei, L.; Yan, H. Matrix protein-specific IgA antibody inhibits measles virus replication by intracellular neutralization. *J. Virol.* **2011**, *85*, 11090–11097. [CrossRef]

61. Feng, N.; Lawton, J.A.; Gilbert, J.; Kuklin, N.; Vo, P.; Prasad, B.V.; Greenberg, H.B. Inhibition of rotavirus replication by a non-neutralizing, rotavirus VP6-specific IgA mAb. *J. Clin. Investig.* **2002**, *109*, 1203–1213. [CrossRef]

62. Corthésy, B.; Benureau, Y.; Perrier, C.; Fourgeux, C.; Parez, N.; Greenberg, H.; Schwartz-Cornil, I. Rotavirus anti-VP6 secretory immunoglobulin A contributes to protection via intracellular neutralization but not via immune exclusion. *J. Virol.* **2006**, *80*, 10692–10699. [CrossRef]

63. Huang, Y.T.; Wright, A.; Gao, X.; Kulick, L.; Yan, H.; Lamm, M.E. Intraepithelial cell neutralization of HIV-1 replication by IgA. *J. Immunol.* **2005**, *174*, 4828–4835. [CrossRef]

64. Wright, A.; Yan, H.; Lamm, M.E.; Huang, Y.T. Immunoglobulin A antibodies against internal HIV-1 proteins neutralize HIV-1 replication inside epithelial cells. *Virology* **2006**, *356*, 165–170. [CrossRef]

65. Burns, J.W.; Siadat-Pajouh, M.; Krishnaney, A.A.; Greenberg, H.B. Protective effect of rotavirus VP6-specific IgA monoclonal antibodies that lack neutralizing activity. *Science* **1996**, *272*, 104–107. [CrossRef]

66. Schwartz-Cornil, I.; Benureau, Y.; Greenberg, H.; Hendrickson, B.A.; Cohen, J. Heterologous protection induced by the inner capsid proteins of rotavirus requires transcytosis of mucosal immunoglobulins. *J. Virol.* **2002**, *76*, 8110–8117. [CrossRef]

67. Lohse, S.; Loew, S.; Kretschmer, A.; Jansen, J.H.M.; Meyer, S.; Ten Broeke, T.; Rosner, T.; Dechant, M.; Derer, S.; Klausz, K.; et al. Effector mechanisms of IgA antibodies against CD20 include recruitment of myeloid cells for antibody-dependent cell-mediated cytotoxicity and complement-dependent cytotoxicity. *Br. J. Haematol.* **2018**, *181*, 413–417. [CrossRef]

68. Roos, A.; Bouwman, L.H.; van Gijlswijk-Janssen, D.J.; Faber-Krol, M.C.; Stahl, G.L.; Daha, M.R. Human IgA activates the complement system via the mannan-binding lectin pathway. *J. Immunol.* **2001**, *167*, 2861–2868. [CrossRef]

69. Ghumra, A.; Shi, J.; Mcintosh, R.S.; Rasmussen, I.B.; Braathen, R.; Johansen, F.E.; Sandlie, I.; Mongini, P.K.; Areschoug, T.; Lindahl, G.; et al. Structural requirements for the interaction of human IgM and IgA with the human Fcα/µ receptor. *Eur. J. Immunol.* **2009**, *39*, 1147–1156. [CrossRef]

70. Matysiak-Budnik, T.; Moura, I.C.; Arcos-Fajardo, M.; Lebreton, C.; Menard, S.; Candalh, C.; Ben-Khalifa, K.; Dugave, C.; Tamouza, H.; van Niel, G.; et al. Secretory IgA mediates retrotranscytosis of intact gliadin peptides via the transferrin receptor in celiac disease. *J. Exp. Med.* **2008**, *205*, 143–154. [CrossRef]

71. Rochereau, N.; Drocourt, D.; Perouzel, E.; Pavot, V.; Redelinghuys, P.; Brown, G.D.; Tiraby, G.; Roblin, X.; Verrier, B.; Genin, C.; et al. Dectin-1 is essential for reverse transcytosis of glycosylated SIgA-antigen complexes by intestinal M cells. *PLoS Biol.* **2013**, *11*, e1001658. [CrossRef]

72. Baumann, J.; Park, C.G.; Mantis, N.J. Recognition of secretory IgA by DC-SIGN: Implications for immune surveillance in the intestine. *Immunol. Lett.* **2010**, *131*, 59–66. [CrossRef]

73. Wilson, T.J.; Fuchs, A.; Colonna, M. Cutting edge: Human FcRL4 and FcRL5 are receptors for IgA and IgG. *J. Immunol.* **2012**, *188*, 4741–4745. [CrossRef]

74. Rifai, A.; Fadden, K.; Morrison, S.L.; Chintalacharuvu, K.R. The *N*-glycans determine the differential blood clearance and hepatic uptake of human immunoglobulin (Ig)A1 and IgA2 isotypes. *J. Exp. Med.* **2000**, *191*, 2171–2182. [CrossRef]

75. Molyneux, K.; Wimbury, D.; Pawluczyk, I.; Muto, M.; Bhachu, J.; Mertens, P.R.; Feehally, J.; Barratt, J. beta1,4-galactosyltransferase 1 is a novel receptor for IgA in human mesangial cells. *Kidney Int.* **2017**, *92*, 1458–1468. [CrossRef]

76. Lamkhioued, B.; Gounni, A.S.; Gruart, V.; Pierce, A.; Capron, A.; Capron, M. Human eosinophils express a receptor for secretory component. Role in secretory IgA-dependent activation. *Eur. J. Immunol.* **1995**, *25*, 117–125. [CrossRef]

77. Bruhns, P.; Jönsson, F. Mouse and human FcR effector functions. *Immunol. Rev.* **2015**, *268*, 25–51. [CrossRef]

78. Ben Mkaddem, S.; Benhamou, M.; Monteiro, R.C. Understanding Fc receptor involvement in inflammatory diseases: From mechanisms to new therapeutic tools. *Front. Immunol.* **2019**, *10*, 811. [CrossRef]

79. Breedveld, A.; van Egmond, M. IgA and FcαRI: Pathological roles and therapeutic opportunities. *Front. Immunol.* **2019**, *10*, 553. [CrossRef]

80. van der Steen, L.; Tuk, C.W.; Bakema, J.E.; Kooij, G.; Reijerkerk, A.; Vidarsson, G.; Bouma, G.; Kraal, G.; de Vries, H.E.; Beelen, R.H.; et al. Immunoglobulin A: FcαRI interactions induce neutrophil migration through release of leukotriene B4. *Gastroenterology* **2009**, *137*, e1–e3. [CrossRef]

81. Aleyd, E.; van Hout, M.W.; Ganzevles, S.H.; Hoeben, K.A.; Everts, V.; Bakema, J.E.; van Egmond, M. IgA enhances NETosis and release of neutrophil extracellular traps by polymorphonuclear cells via Fcα receptor I. *J. Immunol.* **2014**, *192*, 2374–2383. [CrossRef]

82. Aleyd, E.; Heineke, M.H.; van Egmond, M. The era of the immunoglobulin A Fc receptor FcαRI; its function and potential as target in disease. *Immunol. Rev.* **2015**, *268*, 123–138. [CrossRef]

83. Morton, H.C.; Schiel, A.E.; Janssen, S.W.; van de Winkel, J.G. Alternatively spliced forms of the human myeloid Fc alpha receptor (CD89) in neutrophils. *Immunogenetics* **1996**, *43*, 246–247. [CrossRef]

84. Pleass, R.J.; Andrews, P.D.; Kerr, M.A.; Woof, J.M. Alternative splicing of the human IgA Fc receptor CD89 in neutrophils and eosinophils. *Biochem. J.* **1996**, *318*, 771–777. [CrossRef]

85. Wu, J.; Ji, C.; Xie, F.; Langefeld, C.D.; Qian, K.; Gibson, A.W.; Edberg, J.C.; Kimberly, R.P. FcαRI (CD89) alleles determine the proinflammatory potential of serum IgA. *J. Immunol.* **2007**, *178*, 3973–3982. [CrossRef]

86. Woof, J.M.; Burton, D.R. Human antibody-Fc receptor interactions illuminated by crystal structures. *Nat. Rev. Immunol.* **2004**, *4*, 89–99. [CrossRef]

87. van Spriel, A.B.; Leusen, J.H.; Vilé, H.; van de Winkel, J.G. Mac-1 (CD11b/CD18) as accessory molecule for FcαR (CD89) binding of IgA. *J. Immunol.* **2002**, *169*, 3831–3836. [CrossRef]

88. Carayannopoulos, L.; Hexham, J.M.; Capra, J.D. Localization of the binding site for the monocyte immunoglobulin (Ig) A-Fc receptor (CD89) to the domain boundary between Cα2 and Cα3 in human IgA1. *J. Exp. Med.* **1996**, *183*, 1579–1586. [CrossRef]

89. Pleass, R.J.; Dunlop, J.I.; Anderson, C.M.; Woof, J.M. Identification of residues in the CH2/CH3 domain interface of IgA essential for interaction with the human Fcα receptor (FcαR) CD89. *J. Biol. Chem.* **1999**, *274*, 23508–23514. [CrossRef]

90. Pleass, R.J.; Dehal, P.K.; Lewis, M.J.; Woof, J.M. Limited role of charge matching in the interaction of human immunoglobulin A with the immunoglobulin A Fc receptor (FcαRI) CD89. *Immunology* **2003**, *109*, 331–335. [CrossRef]

91. Wines, B.D.; Hulett, M.D.; Jamieson, G.P.; Trist, H.M.; Spratt, J.M.; Hogarth, P.M. Identification of residues in the first domain of human Fcα receptor essential for interaction with IgA. *J. Immunol.* **1999**, *162*, 2146–2153.

92. Wines, B.D.; Sardjono, C.T.; Trist, H.H.; Lay, C.S.; Hogarth, P.M. The interaction of FcαRI with IgA and its implications for ligand binding by immunoreceptors of the leukocyte receptor cluster. *J. Immunol.* **2001**, *166*, 1781–1789. [CrossRef]

93. Gomes, M.M.; Wall, S.B.; Takahashi, K.; Novak, J.; Renfrow, M.B.; Herr, A.B. Analysis of IgA1 N-glycosylation and its contribution to FcαRI binding. *Biochemistry* **2008**, *47*, 11285–11299. [CrossRef]

94. Xue, J.; Zhao, Q.; Zhu, L.; Zhang, W. Deglycosylation of FcαR at N58 increases its binding to IgA. *Glycobiology* **2010**, *20*, 905–915. [CrossRef]

95. Posgai, M.T.; Tonddast-Navaei, S.; Jayasinghe, M.; Ibrahim, G.M.; Stan, G.; Herr, A.B. FcαRI binding at the IgA1 CH2–CH3 interface induces long-range conformational changes that are transmitted to the hinge region. *Proc. Natl. Acad. Sci. USA* **2018**, *115*, E8882–E8891. [CrossRef]

96. Heineke, M.H.; van der Steen, L.P.E.; Korthouwer, R.M.; Hage, J.J.; Langedijk, J.P.M.; Benschop, J.J.; Bakema, J.E.; Slootstra, J.W.; van Egmond, M. Peptide mimetics of immunoglobulin A (IgA) and FcαRI block IgA-induced human neutrophil activation and migration. *Eur. J. Immunol.* **2017**, *47*, 1835–1845. [CrossRef]

97. van der Steen, L.P.; Bakema, J.E.; Sesarman, A.; Florea, F.; Tuk, C.W.; Kirtschig, G.; Hage, J.J.; Sitaru, C.; van Egmond, M. Blocking Fcα receptor I on granulocytes prevents tissue damage induced by IgA autoantibodies. *J. Immunol.* **2012**, *189*, 1594–1601. [CrossRef]

98. Lu, J.; Marjon, K.D.; Mold, C.; Marnell, L.; Du Clos, T.W.; Sun, P. Pentraxins and IgA share a binding hot-spot on FcαRI. *Protein Sci.* **2014**, *23*, 378–386. [CrossRef]

99. Abi-Rached, L.; Dorighi, K.; Norman, P.J.; Yawata, M.; Parham, P. Episodes of natural selection shaped the interactions of IgA-Fc with FcαRI and bacterial decoy proteins. *J. Immunol.* **2007**, *178*, 7943–7954. [CrossRef]

100. Pinheiro, A.; Woof, J.M.; Abi-Rached, L.; Parham, P.; Esteves, P.J. Computational analyses of an evolutionary arms race between mammalian immunity mediated by immunoglobulin A and its subversion by bacterial pathogens. *PLoS ONE* **2013**, *8*, e73934. [CrossRef]

101. Hammerschmidt, S.; Tillig, M.P.; Wolff, S.; Vaerman, J.P.; Chhatwal, G.S. Species-specific binding of human secretory component to SpsA protein of *Streptococcus pneumoniae* via a hexapeptide motif. *Mol. Microbiol.* **2000**, *36*, 726–736. [CrossRef]

102. Zhang, J.R.; Mostov, K.E.; Lamm, M.E.; Nanno, M.; Shimida, S.; Ohwaki, M.; Tuomanen, E. The polymeric immunoglobulin receptor translocates pneumococci across human nasopharyngeal epithelial cells. *Cell* **2000**, *102*, 827–837. [CrossRef]

103. Elm, C.; Braathen, R.; Bergmann, S.; Frank, R.; Vaerman, J.P.; Kaetzel, C.S.; Chhatwal, G.S.; Johansen, F.E.; Hammerschmidt, S. Ectodomains 3 and 4 of human polymeric immunoglobulin receptor (hpIgR) mediate invasion of *Streptococcus pneumoniae* into the epithelium. *J. Biol. Chem.* **2004**, *279*, 6296–6304. [CrossRef] [PubMed]

104. Asmat, T.M.; Agarwal, V.; Räth, S.; Hildebrandt, J.P.; Hammerschmidt, S. *Streptococcus pneumoniae* infection of host epithelial cells via polymeric immunoglobulin receptor transiently induces calcium release from intracellular stores. *J. Biol. Chem.* **2011**, *286*, 17861–17869. [CrossRef]

105. Frithz, E.; Héden, L.O.; Lindahl, G. Extensive sequence homology between IgA receptor and M proteins in *Streptococcus pyogenes*. *Mol. Microbiol.* **1989**, *3*, 1111–1119. [CrossRef]

106. Stenberg, L.; O'Toole, P.W.; Mestecky, J.; Lindahl, G. Molecular characterization of protein Sir, a streptococcal cell surface protein that binds both immunoglobulin A and immunoglobulin G. *J. Biol. Chem.* **1994**, *2691*, 3458–13464.

107. Héden, L.O.; Frithz, E.; Lindahl, G. Molecular characterization of an IgA receptor from group B streptococci: Sequence of the gene, identification of a proline-rich region with unique structure and isolation of N-terminal fragments with IgA-binding capacity. *Eur. J. Immunol.* **1991**, *21*, 1481–1490. [CrossRef]

108. Pleass, R.J.; Areschoug, T.; Lindahl, G.; Woof, J.M. Streptococcal IgA-binding proteins bind in the Cα2-Cα3 interdomain region and inhibit binding of IgA to human CD89. *J. Biol. Chem.* **2001**, *276*, 8197–8204. [CrossRef]

109. Wines, B.D.; Willoughby, N.; Fraser, J.D.; Hogarth, P.M. A competitive mechanism for staphylococcal toxin SSL7 inhibiting the leukocyte IgA receptor, FcαRI, is revealed by SSL7 binding at the Cα2/Cα3 interface of IgA. *J. Biol. Chem.* **2006**, *281*, 1389–1393. [CrossRef]

110. Mistry, D.; Stockley, R.A. IgA1 protease. *Int. J. Biochem. Cell Biol.* **2006**, *38*, 1244–1248. [CrossRef]

111. Janoff, E.N.; Rubins, J.B.; Fasching, C.; Charboneau, D.; Rahkola, J.T.; Plaut, A.G.; Weiser, J.N. Pneumococcal IgA1 protease subverts specific protection by human IgA1. *Mucosal Immunol.* **2014**, *7*, 249–256. [CrossRef]

112. Senior, B.W.; Dunlop, J.I.; Batten, M.R.; Kilian, M.; Woof, J.M. Cleavage of a recombinant human immunoglobulin A2 (IgA2)-IgA1 hybrid antibody by certain bacterial IgA1 proteases. *Infect. Immun.* **2000**, *68*, 463–469. [CrossRef] [PubMed]

113. Batten, M.R.; Senior, B.W.; Kilian, M.; Woof, J.M. Amino acid sequence requirements in the hinge of human immunoglobulin A1 (IgA1) for cleavage by streptococcal IgA1 proteases. *Infect. Immun.* **2003**, *71*, 1462–1469. [CrossRef] [PubMed]

114. Senior, B.W.; Woof, J.M. The influences of hinge length and composition on the susceptibility of human IgA to cleavage by diverse bacterial IgA1 proteases. *J. Immunol.* **2005**, *174*, 7792–7799. [CrossRef] [PubMed]

115. Chintalacharuvu, K.R.; Chuang, P.D.; Dragoman, A.; Fernandez, C.Z.; Qiu, J.; Plaut, A.G.; Trinh, K.R.; Gala, F.A.; Morrison, S.L. Cleavage of the human immunoglobulin A1 (IgA1) hinge region by IgA1 proteases requires structures in the Fc region of IgA. *Infect. Immun.* **2003**, *71*, 2563–2570. [CrossRef]

116. Senior, B.W.; Woof, J.M. Sites in the CH3 domain of human IgA1 that influence sensitivity to bacterial IgA1 proteases. *J. Immunol.* **2006**, *177*, 3913–3919. [CrossRef]

117. Johnson, T.A.; Qiu, J.; Plaut, A.G.; Holyoak, T. Active-site gating regulates substrate selectivity in a chymotrypsin-like serine protease: The structure of *Haemophilus influenzae* immunoglobulin A1 protease. *J. Mol. Biol.* **2009**, *389*, 559–574. [CrossRef]

118. Burton, J.; Wood, S.G.; Lynch, M.; Plaut, A.G. Substrate analogue inhibitors of the IgA1 proteinases from *Neisseria gonorrhoeae*. *J. Med. Chem.* **1988**, *31*, 1647–1651. [CrossRef]

119. Bachovchin, W.W.; Plaut, A.G.; Flentke, G.R.; Lynch, M.; Kettner, C.A. Inhibition of IgA1 proteinases from *Neisseria gonorrhoeae* and *Haemophilus influenzae* by peptide prolyl boronic acids. *J. Biol. Chem.* **1990**, *265*, 3738–3743.

120. Shehaj, L.; Choudary, S.K.; Makwana, K.M.; Gallo, M.C.; Murphy, T.F.; Kritzer, J.A. Small-molecule inhibitors of *Haemophilus influenzae* IgA1 protease. *ACS Infect. Dis.* **2019**, *5*, 1129–1138. [CrossRef]

121. Lamm, M.E.; Emancipator, S.N.; Robinson, J.K.; Yamashita, M.; Fujioka, H.; Qiu, J.; Plaut, A.G. Microbial IgA protease removes IgA immune complexes from mouse glomeruli in vivo: Potential therapy for IgA nephropathy. *Am. J. Pathol.* **2008**, *172*, 31–36. [CrossRef]

122. Wang, L.; Li, X.; Shen, H.; Mao, N.; Wang, H.; Cui, L.; Cheng, Y.; Fan, J. Bacterial IgA protease-mediated degradation of agIgA1 and agIgA1 immune complexes as a potential therapy for IgA Nephropathy. *Sci. Rep.* **2016**, *6*, 30964. [CrossRef] [PubMed]

123. Neutra, M.R.; Kozlowski, P.A. Mucosal vaccines: The promise and the challenge. *Nat. Rev. Immunol.* **2006**, *6*, 148–158. [CrossRef] [PubMed]

124. van Splunter, M.; van Hoffen, E.; Floris-Vollenbroek, E.G.; Timmerman, H.; de Bos, E.L.; Meijer, B.; Ulfman, L.H.; Witteman, B.; Wells, J.M.; Brugman, S.; et al. Oral cholera vaccination promotes homing of IgA$^+$ memory B cells to the large intestine and the respiratory tract. *Mucosal Immunol.* **2018**, *11*, 1254–1264. [CrossRef] [PubMed]

125. Fukuyama, Y.; King, J.D.; Kataoka, K.; Kobayashi, R.; Gilbert, R.S.; Oishi, K.; Hollingshead, S.K.; Briles, D.E.; Fujihashi, K. Secretory-IgA antibodies play an important role in the immunity to *Streptococcus pneumoniae*. *J. Immunol.* **2010**, *185*, 1755–1762. [CrossRef] [PubMed]

126. Planque, S.; Salas, M.; Mitsuda, Y.; Sienczyk, M.; Escobar, M.A.; Mooney, J.P.; Morris, M.K.; Nishiyama, Y.; Ghosh, D.; Kumar, A.; et al. Neutralization of genetically diverse HIV-1 strains by IgA antibodies to the gp120-CD4-binding site from long-term survivors of HIV infection. *AIDS* **2010**, *24*, 875–884. [CrossRef] [PubMed]

127. Wills, S.; Hwang, K.K.; Liu, P.; Dennison, S.M.; Tay, M.Z.; Shen, X.; Pollara, J.; Lucas, J.T.; Parks, R.; Rerks-Ngarm, S.; et al. HIV-1-Specific IgA monoclonal antibodies from an HIV-1 vaccinee mediate galactosylceramide blocking and phagocytosis. *J. Virol.* **2018**, *92*, e01552-17. [CrossRef]

128. Herremans, T.M.; Reimerink, J.H.; Buisman, A.M.; Kimman, T.G.; Koopmans, M.P. Induction of mucosal immunity by inactivated poliovirus vaccine is dependent on previous mucosal contact with live virus. *J. Immunol.* **1999**, *162*, 5011–5018.

129. Silvey, K.J.; Hutchings, A.B.; Vajdy, M.; Petzke, M.M.; Neutra, M.R. Role of immunoglobulin A in protection against reovirus entry into murine Peyer's patches. *J. Virol.* **2001**, *75*, 10870–10879. [CrossRef]

130. Tamura, S.; Funato, H.; Hirabayashi, Y.; Suzuki, Y.; Nagamine, T.; Aizawa, C.; Kurata, T. Cross-protection against influenza A virus infection by passively transferred respiratory tract IgA antibodies to different hemagglutinin molecules. *Eur. J. Immunol.* **1991**, *21*, 1337–1344. [CrossRef]

131. Bakema, J.E.; van Egmond, M. Immunoglobulin A: A next generation of therapeutic antibodies? *MAbs* **2011**, *3*, 352–361. [CrossRef]

132. Leusen, J.H. IgA as a therapeutic antibody. *Mol. Immunol.* **2015**, *68*, 35–39. [CrossRef] [PubMed]

133. Lombana, T.N.; Rajan, S.; Zorn, J.A.; Mandikian, D.; Chen, E.C.; Estevez, A.; Yip, V.; Bravo, D.D.; Phung, W.; Farahi, F.; et al. Production, characterization, and *in vivo* half-life extension of polymeric IgA molecules in mice. *MAbs* **2019**, *11*, 1122–1138. [CrossRef] [PubMed]

134. Ledford, H. Rush to protect lucrative antibody patents kicks into gear. *Nature* **2018**, *557*, 623–624. [CrossRef] [PubMed]

135. Monteiro, R.C. The role of IgA and IgA Fc receptors as anti-inflammatory agents. *J. Clin. Immunol.* **2010**, *30*, S61–S64. [CrossRef] [PubMed]

136. Challacombe, S.J.; Russell, M.W. Estimations of the intravascular half-lives of normal rhesus monkey IgG, IgA, and IgM. *Immunology* **1979**, *36*, 331–338.

137. Strober, W.; Wochner, R.D.; Barlow, M.H.; McFarlin, D.E.; Waldmann, T.A. Immunoglobulin metabolism in ataxia telangiectasia. *J. Clin. Investig.* **1968**, *47*, 1905–1915. [CrossRef]

138. Boross, P.; Lohse, S.; Nederend, M.; Jansen, J.H.; van Tetering, G.; Dechant, M.; Peipp, M.; Royle, L.; Liew, L.P.; Boon, L.; et al. IgA EGFR antibodies mediate tumour killing in vivo. *EMBO Mol. Med.* **2013**, *5*, 1213–1226. [CrossRef]

139. Lohse, S.; Meyer, S.; Meulenbroek, L.A.; Jansen, J.H.; Nederend, M.; Kretschmer, A.; Klausz, K.; Moginger, U.; Derer, S.; Rosner, T.; et al. An anti-EGFR IgA that displays improved pharmacokinetics and myeloid effector cell engagement in vivo. *Cancer Res.* **2016**, *76*, 403–417. [CrossRef]

140. Rouwendal, G.J.; van der Lee, M.M.; Meyer, S.; Reiding, K.R.; Schouten, J.; de Roo, G.; Egging, D.F.; Leusen, J.H.; Boross, P.; Wuhrer, M.; et al. A comparison of anti-HER2 IgA and IgG1 in vivo efficacy is facilitated by high N-glycan sialylation of the IgA. *MAbs* **2016**, *8*, 74–86. [CrossRef]

141. Meyer, S.; Nederend, M.; Jansen, J.H.; Reiding, K.R.; Jacobino, S.R.; Meeldijk, J.; Bovenschen, N.; Wuhrer, M.; Valerius, T.; Ubink, R.; et al. Improved in vivo anti-tumor effects of IgA-Her2 antibodies through half-life extension and serum exposure enhancement by FcRn targeting. *MAbs* **2016**, *8*, 87–98. [CrossRef]

142. Beyer, T.; Lohse, S.; Berger, S.; Peipp, M.; Valerius, T.; Dechant, M. Serum-free production and purification of chimeric IgA antibodies. *J. Immunol. Methods* **2009**, *346*, 26–37. [CrossRef] [PubMed]

143. Hart, F.; Danielczyk, A.; Goletz, S. Human cell line-derived monoclonal IgA antibodies for cancer immunotherapy. *Bioengineering* **2017**, *4*, 42. [CrossRef] [PubMed]

144. Vink, T.; Oudshoorn-Dickmann, M.; Roza, M.; Reitsma, J.J.; de Jong, R.N. A simple, robust and highly efficient transient expression system for producing antibodies. *Methods* **2014**, *65*, 5–10. [CrossRef] [PubMed]

145. Dumont, J.; Euwart, D.; Mei, B.; Estes, S.; Kshirsagar, R. Human cell lines for biopharmaceutical manufacturing: History, status, and future perspectives. *Crit. Rev. Biotechnol.* **2016**, *36*, 1110–1122. [CrossRef] [PubMed]

146. Dicker, M.; Maresch, D.; Strasser, R. Glyco-engineering for the production of recombinant IgA1 with distinct mucin-type O-glycans in plants. *Bioengineered* **2016**, *7*, 484–489. [CrossRef] [PubMed]

147. Dicker, M.; Tschofen, M.; Maresch, D.; König, J.; Juarez, P.; Orzaez, D.; Altmann, F.; Steinkellner, H.; Strasser, R. Transient glyco-engineering to produce recombinant IgA1 with defined *N*- and *O*-glycans in plants. *Front. Plant Sci.* **2016**, *7*, 18. [CrossRef]

148. Westerhof, L.B.; Wilbers, R.H.; van Raaij, D.R.; Nguyen, D.L.; Goverse, A.; Henquet, M.G.; Hokke, C.H.; Bosch, D.; Bakker, J.; Schots, A. Monomeric IgA can be produced in planta as efficient as IgG, yet receives different *N*-glycans. *Plant Biotechnol. J.* **2014**, *12*, 1333–1342. [CrossRef]

149. Yoo, E.M.; Yu, L.J.; Wims, L.A.; Goldberg, D.; Morrison, S.L. Differences in *N*-glycan structures found on recombinant IgA1 and IgA2 produced in murine myeloma and CHO cell lines. *MAbs* **2010**, *2*, 320–334. [CrossRef]

150. Göritzer, K.; Maresch, D.; Altmann, F.; Obinger, C.; Strasser, R. Exploring site-specific *N*-glycosylation of HEK293 and plant-produced human IgA isotypes. *J. Proteome Res.* **2017**, *16*, 2560–2570. [CrossRef]

151. Sandin, C.; Linse, S.; Areschoug, T.; Woof, J.M.; Reinholdt, J.; Lindahl, G. Isolation and detection of human IgA using a streptococcal IgA-binding peptide. *J. Immunol.* **2002**, *169*, 1357–1364. [CrossRef]

152. Bakshi, S.; Depicker, A.; Schepens, B.; Saelens, X.; Juarez, P. A two-amino acid mutation in murine IgA enables downstream processing and purification on staphylococcal superantigen-like protein 7. *J. Biotechnol.* **2019**, *294*, 26–29. [CrossRef] [PubMed]

153. Hatanaka, T.; Ohzono, S.; Park, M.; Sakamoto, K.; Tsukamoto, S.; Sugita, R.; Ishitobi, H.; Mori, T.; Ito, O.; Sorajo, K.; et al. Human IgA-binding peptides selected from random peptide libraries: Affinity maturation and application in IgA purification. *J. Biol. Chem.* **2012**, *287*, 43126–43136. [CrossRef] [PubMed]

154. Chintalacharuvu, K.R.; Yu, L.J.; Bhola, N.; Kobayashi, K.; Fernandez, C.Z.; Morrison, S.L. Cysteine residues required for the attachment of the light chain in human IgA2. *J. Immunol.* **2002**, *169*, 5072–5077. [CrossRef] [PubMed]

155. Chintalacharuvu, K.R.; Morrison, S.L. Production of secretory immunoglobulin A by a single mammalian cell. *Proc. Natl. Acad Sci. USA* **1997**, *94*, 6364–6368. [CrossRef] [PubMed]

156. Saito, S.; Sano, K.; Suzuki, T.; Ainai, A.; Taga, Y.; Ueno, T.; Tabata, K.; Saito, K.; Wada, Y.; Ohara, Y.; et al. IgA tetramerization improves target breadth but not peak potency of functionality of anti-influenza virus broadly neutralizing antibody. *PLoS Pathog.* **2019**, *15*, e1007427. [CrossRef] [PubMed]

157. van Egmond, M.; van Vuuren, A.J.; Morton, H.C.; van Spriel, A.B.; Shen, L.; Hofhuis, F.M.; Saito, T.; Mayadas, T.N.; Verbeek, J.S.; van de Winkel, J.G. Human immunoglobulin A receptor (FcαRI, CD89) function in transgenic mice requires both FcR gamma chain and CR3 (CD11b/CD18). *Blood* **1999**, *93*, 4387–4394. [CrossRef]

158. Launay, P.; Grossetete, B.; Arcos-Fajardo, M.; Gaudin, E.; Torres, S.P.; Beaudoin, L.; Patey-Mariaud de Serre, N.; Lehuen, A.; Monteiro, R.C. Fcα receptor (CD89) mediates the development of immunoglobulin A (IgA) nephropathy (Berger's disease). Evidence for pathogenic soluble receptor-IgA complexes in patients and CD89 transgenic mice. *J. Exp. Med.* **2000**, *191*, 1999–2009. [CrossRef]

159. Duchez, S.; Amin, R.; Cogné, N.; Delpy, L.; Sirac, C.; Pascal, V.; Corthésy, B.; Cogné, M. Premature replacement of mu with alpha immunoglobulin chains impairs lymphopoiesis and mucosal homing but promotes plasma cell maturation. *Proc. Natl. Acad Sci. USA* **2010**, *107*, 3064–3069. [CrossRef]

160. Scott, A.M.; Allison, J.P.; Wolchok, J.D. Monoclonal antibodies in cancer therapy. *Cancer Immun.* **2012**, *12*, 14.

161. Weiner, L.M.; Surana, R.; Wang, S. Monoclonal antibodies: Versatile platforms for cancer immunotherapy. *Nat. Rev. Immunol.* **2010**, *10*, 317–327. [CrossRef]

162. Amoroso, A.; Hafsi, S.; Militello, L.; Russo, A.E.; Soua, Z.; Mazzarino, M.C.; Stivala, F.; Libra, M. Understanding rituximab function and resistance: Implications for tailored therapy. *Front. Biosci.* **2011**, *16*, 770–782. [CrossRef] [PubMed]

163. Boyiadzis, M.; Foon, K.A. Approved monoclonal antibodies for cancer therapy. *Expert Opin. Biol. Ther.* **2008**, *8*, 1151–1158. [CrossRef]

164. Benvenuti, S.; Sartore-Bianchi, A.; Di Nicolantonio, F.; Zanon, C.; Moroni, M.; Veronese, S.; Siena, S.; Bardelli, A. Oncogenic activation of the RAS/RAF signaling pathway impairs the response of metastatic colorectal cancers to anti-epidermal growth factor receptor antibody therapies. *Cancer Res.* **2007**, *67*, 2643–2648. [CrossRef] [PubMed]

165. Glennie, M.J.; French, R.R.; Cragg, M.S.; Taylor, R.P. Mechanisms of killing by anti-CD20 monoclonal antibodies. *Mol. Immunol.* **2007**, *44*, 3823–3837. [CrossRef] [PubMed]

166. Hudis, C.A. Trastuzumab—Mechanism of action and use in clinical practice. *N. Engl. J. Med.* **2007**, *357*, 39–51. [CrossRef]

167. Hatjiharissi, E.; Xu, L.; Santos, D.D.; Hunter, Z.R.; Ciccarelli, B.T.; Verselis, S.; Modica, M.; Cao, Y.; Manning, R.J.; Leleu, X.; et al. Increased natural killer cell expression of CD16, augmented binding and ADCC activity to rituximab among individuals expressing the FcγRIIIa-158 V/V and V/F polymorphism. *Blood* **2007**, *110*, 2561–2564. [CrossRef]

168. van der Bij, G.J.; Bogels, M.; Otten, M.A.; Oosterling, S.J.; Kuppen, P.J.; Meijer, S.; Beelen, R.H.; van Egmond, M. Experimentally induced liver metastases from colorectal cancer can be prevented by mononuclear phagocyte-mediated monoclonal antibody therapy. *J. Hepatol.* **2010**, *53*, 677–685. [CrossRef]

169. Colombo, M.P.; Ferrari, G.; Stoppacciaro, A.; Parenza, M.; Rodolfo, M.; Mavilio, F.; Parmiani, G. Granulocyte colony-stimulating factor gene transfer suppresses tumorigenicity of a murine adenocarcinoma in vivo. *J. Exp. Med.* **1991**, *173*, 889–897. [CrossRef]

170. Bakema, J.E.; Ganzevles, S.H.; Fluitsma, D.M.; Schilham, M.W.; Beelen, R.H.; Valerius, T.; Lohse, S.; Glennie, M.J.; Medema, J.P.; van Egmond, M. Targeting FcαRI on polymorphonuclear cells induces tumor cell killing through autophagy. *J. Immunol.* **2011**, *187*, 726–732. [CrossRef]

171. Amulic, B.; Cazalet, C.; Hayes, G.L.; Metzler, K.D.; Zychlinsky, A. Neutrophil function: From mechanisms to disease. *Annu. Rev. Immunol.* **2012**, *30*, 459–489. [CrossRef]

172. Hernandez-Ilizaliturri, F.J.; Jupudy, V.; Ostberg, J.; Oflazoglu, E.; Huberman, A.; Repasky, E.; Czuczman, M.S. Neutrophils contribute to the biological antitumor activity of rituximab in a non-Hodgkin's lymphoma severe combined immunodeficiency mouse model. *Clin. Cancer Res.* **2003**, *9*, 5866–5873. [PubMed]

173. Pullarkat, V.; Deo, Y.; Link, J.; Spears, L.; Marty, V.; Curnow, R.; Groshen, S.; Gee, C.; Weber, J.S. A phase I study of a HER2/neu bispecific antibody with granulocyte-colony-stimulating factor in patients with metastatic breast cancer that overexpresses HER2/neu. *Cancer Immunol. Immunother.* **1999**, *48*, 9–21. [CrossRef] [PubMed]

174. Lewis, L.D.; Beelen, A.P.; Cole, B.F.; Wallace, P.K.; Fisher, J.L.; Waugh, M.G.; Kaufman, P.A.; Ernstoff, M.S. The pharmacokinetics of the bispecific antibody MDX-H210 when combined with interferon gamma-1b in a multiple-dose phase I study in patients with advanced cancer. *Cancer Chemother. Pharmacol.* **2002**, *49*, 375–384. [CrossRef] [PubMed]

175. Repp, R.; van Ojik, H.H.; Valerius, T.; Groenewegen, G.; Wieland, G.; Oetzel, C.; Stockmeyer, B.; Becker, W.; Eisenhut, M.; Steininger, H.; et al. Phase I clinical trial of the bispecific antibody MDX-H210 (anti-FcγRI x anti-HER-2/neu) in combination with Filgrastim (G-CSF) for treatment of advanced breast cancer. *Br. J. Cancer* **2003**, *89*, 2234–2243. [PubMed]

176. Nimmerjahn, F.; Ravetch, J.V. Antibodies, Fc receptors and cancer. *Curr. Opin. Immunol.* **2007**, *19*, 239–245. [CrossRef]

177. Otten, M.A.; Leusen, J.H.; Rudolph, E.; van der Linden, J.A.; Beelen, R.H.; van de Winkel, J.G.; van Egmond, M. FcR γ-chain dependent signaling in immature neutrophils is mediated by FcαRI, but not by FcγRI. *J. Immunol.* **2007**, *179*, 2918–2924. [CrossRef]

178. Brandsma, A.M.; Bondza, S.; Evers, M.; Koutstaal, R.; Nederend, M.; Jansen, J.H.M.; Rosner, T.; Valerius, T.; Leusen, J.H.W.; Ten Broeke, T. Potent Fc receptor signaling by IgA leads to superior killing of cancer cells by neutrophils compared to IgG. *Front. Immunol.* **2019**, *10*, 704. [CrossRef]

179. Huls, G.; Heijnen, I.A.; Cuomo, E.; van der Linden, J.; Boel, E.; van de Winkel, J.G.; Logtenberg, T. Antitumor immune effector mechanisms recruited by phage display-derived fully human IgG1 and IgA1 monoclonal antibodies. *Cancer Res.* **1999**, *59*, 5778–5784.

180. Dechant, M.; Beyer, T.; Schneider-Merck, T.; Weisner, W.; Peipp, M.; van de Winkel, J.G.; Valerius, T. Effector mechanisms of recombinant IgA antibodies against epidermal growth factor receptor. *J. Immunol.* **2007**, *179*, 2936–2943. [CrossRef]

181. Heemskerk, N.; van Egmond, M. Monoclonal antibody-mediated killing of tumour cells by neutrophils. *Eur. J. Clin. Investig.* **2018**, *48*, e12962. [CrossRef]

182. Otten, M.A.; Rudolph, E.; Dechant, M.; Tuk, C.W.; Reijmers, R.M.; Beelen, R.H.; van de Winkel, J.G.; van Egmond, M. Immature neutrophils mediate tumor cell killing via IgA but not IgG Fc receptors. *J. Immunol.* **2005**, *174*, 5472–5480. [CrossRef] [PubMed]

183. Stockmeyer, B.; Dechant, M.; van Egmond, M.; Tutt, A.L.; Sundarapandiyan, K.; Graziano, R.F.; Repp, R.; Kalden, J.R.; Gramatzki, M.; Glennie, M.J.; et al. Triggering Fc alpha-receptor I (CD89) recruits neutrophils as effector cells for CD20-directed antibody therapy. *J. Immunol.* **2000**, *165*, 5954–5961. [CrossRef] [PubMed]

184. Guettinger, Y.; Barbin, K.; Peipp, M.; Bruenke, J.; Dechant, M.; Horner, H.; Thierschmidt, D.; Valerius, T.; Repp, R.; Fey, G.H.; et al. A recombinant bispecific single-chain fragment variable specific for HLA class II and FcαRI (CD89) recruits polymorphonuclear neutrophils for efficient lysis of malignant B lymphoid cells. *J. Immunol.* **2010**, *184*, 1210–1217. [CrossRef] [PubMed]

185. Pascal, V.; Laffleur, B.; Debin, A.; Cuvillier, A.; van Egmond, M.; Drocourt, D.; Imbertie, L.; Pangault, C.; Tarte, K.; Tiraby, G.; et al. Anti-CD20 IgA can protect mice against lymphoma development: Evaluation of the direct impact of IgA and cytotoxic effector recruitment on CD20 target cells. *Haematologica* **2012**, *97*, 1686–1694. [CrossRef]

186. Wagner, E.K.; Maynard, J.A. Engineering therapeutic antibodies to combat infectious diseases. *Curr. Opin. Chem. Eng.* **2018**, *19*, 131–141. [CrossRef]

187. Williams, A.; Reljic, R.; Naylor, I.; Clark, S.O.; Falero-Diaz, G.; Singh, M.; Challacombe, S.; Marsh, P.D.; Ivanyi, J. Passive protection with immunoglobulin A antibodies against tuberculous early infection of the lungs. *Immunology* **2004**, *111*, 328–333. [CrossRef]

188. Balu, S.; Reljic, R.; Lewis, M.J.; Pleass, R.J.; McIntosh, R.; van Kooten, C.; van Egmond, M.; Challacombe, S.; Woof, J.M.; Ivanyi, J. A novel human IgA monoclonal antibody protects against tuberculosis. *J. Immunol.* **2011**, *186*, 3113–3119. [CrossRef]

189. van Egmond, M.; van Garderen, E.; van Spriel, A.B.; Damen, C.A.; van Amersfoort, E.S.; van Zandbergen, G.; van Hattum, J.; Kuiper, J.; van de Winkel, J.G. FcαRI-positive liver Kupffer cells: Reappraisal of the function of immunoglobulin A in immunity. *Nat. Med.* **2000**, *6*, 680–685. [CrossRef]

190. van Spriel, A.B.; van den Herik-Oudijk, I.E.; van Sorge, N.M.; Vile, H.A.; van Strijp, J.A.; van de Winkel, J.G. Effective phagocytosis and killing of *Candida albicans* via targeting FcγRI (CD64) or FcαRI (CD89) on neutrophils. *J. Infect. Dis.* **1999**, *179*, 661–669. [CrossRef]

191. van der Pol, W.; Vidarsson, G.; Vile, H.A.; van de Winkel, J.G.; Rodriguez, M.E. Pneumococcal capsular polysaccharide-specific IgA triggers efficient neutrophil effector functions via FcαRI (CD89). *J. Infect. Dis.* **2000**, *182*, 1139–1145. [CrossRef]

192. Hellwig, S.M.; van Spriel, A.B.; Schellekens, J.F.; Mooi, F.R.; van de Winkel, J.G. Immunoglobulin A-mediated protection against *Bordetella pertussis* infection. *Infect. Immun.* **2001**, *69*, 4846–4850. [CrossRef] [PubMed]

193. Kobayashi, T.; Yamamoto, K.; Sugita, N.; van Spriel, A.B.; Kaneko, S.; van de Winkel, J.G.; Yoshie, H. Effective in vitro clearance of *Porphyromonas gingivalis* by Fcα receptor I (CD89) on gingival crevicular neutrophils. *Infect. Immun.* **2001**, *69*, 2935–2942. [CrossRef] [PubMed]

194. Vidarsson, G.; van Der Pol, W.L.; van Den Elsen, J.M.; Vile, H.; Jansen, M.; Duijs, J.; Morton, H.C.; Boel, E.; Daha, M.R.; Corthésy, B.; et al. Activity of human IgG and IgA subclasses in immune defense against *Neisseria meningitidis* serogroup B. *J. Immunol.* **2001**, *166*, 6250–6256. [CrossRef] [PubMed]

195. Bioley, G.; Monnerat, J.; Lötscher, M.; Vonarburg, C.; Zuercher, A.; Corthésy, B. Plasma-derived polyreactive secretory-like IgA and IgM opsonizing *Salmonella enterica* Typhimurium reduces invasion and gut tissue inflammation through agglutination. *Front. Immunol.* **2017**, *8*, 1043. [CrossRef] [PubMed]

196. Corthésy, B.; Monnerat, J.; Lotscher, M.; Vonarburg, C.; Schaub, A.; Bioley, G. Oral passive immunization with plasma-derived polyreactive secretory-Like IgA/M partially protects mice against experimental salmonellosis. *Front. Immunol.* **2018**, *9*, 2970. [CrossRef] [PubMed]

197. Langereis, J.D.; van der Flier, M.; de Jonge, M.I. Limited innovations after more than 65 years of immunoglobulin replacement therapy: Potential of IgA- and IgM-enriched formulations to prevent bacterial respiratory tract infections. *Front. Immunol.* **2018**, *9*, 1925. [CrossRef]

198. Clements, M.L.; Betts, R.F.; Tierney, E.L.; Murphy, B.R. Serum and nasal wash antibodies associated with resistance to experimental challenge with influenza A wild-type virus. *J. Clin. Microbiol.* **1986**, *24*, 157–160.

199. Wagner, D.K.; Clements, M.L.; Reimer, C.B.; Snyder, M.; Nelson, D.L.; Murphy, B.R. Analysis of immunoglobulin G antibody responses after administration of live and inactivated influenza A vaccine indicates that nasal wash immunoglobulin G is a transudate from serum. *J. Clin. Microbiol.* **1987**, *25*, 559–562.

200. Renegar, K.B.; Small, P.A. Passive transfer of local immunity to influenza virus infection by IgA antibody. *J. Immunol.* **1991**, *146*, 1972–1978.

Antibodies **2019**, *8*, 57

201. Renegar, K.B.; Small, P.A.; Boykins, L.G.; Wright, P.F. Role of IgA versus IgG in the control of influenza viral infection in the murine respiratory tract. *J. Immunol.* **2004**, *173*, 1978–1986. [CrossRef]

202. Watkins, J.D.; Sholukh, A.M.; Mukhtar, M.M.; Siddappa, N.B.; Lakhashe, S.K.; Kim, M.; Reinherz, E.L.; Gupta, S.; Forthal, D.N.; Sattentau, Q.J.; et al. Anti-HIV IgA isotypes: Differential virion capture and inhibition of transcytosis are linked to prevention of mucosal R5 SHIV transmission. *AIDS* **2013**, *27*, F13–F20. [CrossRef] [PubMed]

203. Rochereau, N.; Pavot, V.; Verrier, B.; Ensinas, A.; Genin, C.; Corthésy, B.; Paul, S. Secretory IgA as a vaccine carrier for delivery of HIV antigen to M cells. *Eur. J. Immunol.* **2015**, *45*, 773–779. [CrossRef] [PubMed]

204. Aleyd, E.; Al, M.; Tuk, C.W.; van der Laken, C.J.; van Egmond, M. IgA complexes in plasma and synovial fluid of patients with rheumatoid arthritis induce neutrophil extracellular traps via FcαRI. *J. Immunol.* **2016**, *197*, 4552–4559. [CrossRef] [PubMed]

205. Ben Mkaddem, S.; Christou, I.; Rossato, E.; Berthelot, L.; Lehuen, A.; Monteiro, R.C. IgA, IgA receptors, and their anti-inflammatory properties. *Curr. Top. Microbiol. Immunol.* **2014**, *382*, 221–235.

206. Pasquier, B.; Launay, P.; Kanamaru, Y.; Moura, I.C.; Pfirsch, S.; Ruffie, C.; Henin, D.; Benhamou, M.; Pretolani, M.; Blank, U.; et al. Identification of FcαRI as an inhibitory receptor that controls inflammation: Dual role of FcRγ ITAM. *Immunity* **2005**, *22*, 31–42.

207. Rossato, E.; Ben Mkaddem, S.; Kanamaru, Y.; Hurtado-Nedelec, M.; Hayem, G.; Descatoire, V.; Vonarburg, C.; Miescher, S.; Zuercher, A.W.; Monteiro, R.C. Reversal of arthritis by human monomeric IgA through the receptor-mediated SH2 domain-containing phosphatase 1 inhibitory pathway. *Arthritis Rheumatol.* **2015**, *67*, 1766–1777. [CrossRef]

208. Kanamaru, Y.; Pfirsch, S.; Aloulou, M.; Vrtovsnik, F.; Essig, M.; Loirat, C.; Deschenes, G.; Guerin-Marchand, C.; Blank, U.; Monteiro, R.C.; et al. Inhibitory ITAM signaling by FcαRI-FcRγ chain controls multiple activating responses and prevents renal inflammation. *J. Immunol.* **2008**, *180*, 2669–2678. [CrossRef]

209. Watanabe, T.; Kanamaru, Y.; Liu, C.; Suzuki, Y.; Tada, N.; Okumura, K.; Horikoshi, S.; Tomino, Y. Negative regulation of inflammatory responses by immunoglobulin A receptor (FcαRI) inhibits the development of Toll-like receptor-9 signalling-accelerated glomerulonephritis. *Clin. Exp. Immunol.* **2011**, *166*, 235–250. [CrossRef]

210. Liu, C.; Kanamaru, Y.; Watanabe, T.; Tada, N.; Horikoshi, S.; Suzuki, Y.; Liu, Z.; Tomino, Y. Targeted IgA Fc receptor I (FcαRI) therapy in the early intervention and treatment of pristane-induced lupus nephritis in mice. *Clin. Exp. Immunol.* **2015**, *181*, 407–416. [CrossRef]

211. Marshall, M.J.E.; Stopforth, R.J.; Cragg, M.S. Therapeutic antibodies: What have we learnt from targeting CD20 and where are we going? *Front. Immunol.* **2017**, *8*, 1245. [CrossRef]

212. Kaplon, H.; Reichert, J.M. Antibodies to watch in 2019. *MAbs* **2019**, *11*, 219–238. [CrossRef] [PubMed]

213. Sedykh, S.E.; Prinz, V.V.; Buneva, V.N.; Nevinsky, G.A. Bispecific antibodies: Design, therapy, perspectives. *Drug Des. Dev. Ther.* **2018**, *12*, 195–208. [CrossRef] [PubMed]

214. Vonarburg, C.; Loetscher, M.; Spycher, M.O.; Kropf, A.; Illi, M.; Salmon, S.; Roberts, S.; Steinfuehrer, K.; Campbell, I.; Koernig, S.; et al. Topical application of nebulized human IgG, IgA and IgAM in the lungs of rats and non-human primates. *Respir. Res.* **2019**, *20*, 99. [CrossRef] [PubMed]

215. Koernig, S.; Campbell, I.K.; Mackenzie-Kludas, C.; Schaub, A.; Loetscher, M.; Ching Ng, W.; Zehnder, R.; Pelczar, P.; Sanli, I.; Alhamdoosh, M.; et al. Topical application of human-derived Ig isotypes for the control of acute respiratory infection evaluated in a human CD89-expressing mouse model. *Mucosal Immunol.* **2019**, *12*, 1013–1024. [CrossRef]

antibodies

MDPI

Review

Immunogenicity of Innovative and Biosimilar Monoclonal Antibodies

Erik Doevendans * and Huub Schellekens

Department of Pharmaceutical Sciences, Utrecht University, 3512 JE Utrecht, The Netherlands;
h.schellekens@uu.nl
* Correspondence: e.doevendans@uu.nl

Received: 13 February 2019; Accepted: 27 February 2019; Published: 5 March 2019

Abstract: The development of hybridoma technology for producing monoclonal antibodies (mAbs) by Kohler and Milstein (1975) counts as one of the major medical breakthroughs, opening up endless possibilities for research, diagnosis and for treatment of a whole variety of diseases. Therapeutic mAbs were introduced three decades ago. The first generation of therapeutic mAbs of murine origin showed high immunogenicity, which limited efficacy and was associated with severe infusion reactions. Subsequently chimeric, humanized, and fully human antibodies were introduced as therapeutics, these mAbs were considerably less immunogenic. Unexpectedly humanized mAbs generally show similar immunogenicity as chimeric antibodies; based on sequence homology chimeric mAbs are sometimes more "human" than humanized mAbs. With the introduction of the regulatory concept of similar biological medicines (biosimilars) a key concern is the similarity in terms of immunogenicity of these biosimilars with their originators. This review focuses briefly on the mechanisms of induction of immunogenicity by biopharmaceuticals, mAbs in particular, in relation to the target of the immune system.

Keywords: biopharmaceuticals; monoclonal antibodies; biosimilars; immunogencitity; B-cell tolerance; aggregates; anti-idiotypic

1. General Introduction

The development of the hybridoma technology to produce monoclonal antibodies (mAbs) by Kohler and Milstein counts as one of the major medical breakthroughs of the 20th century. It opened endless possibilities, not only for research, but also to diagnose, prevent, and treat a whole variety of diseases [1].

Initially this discovery led to the introduction of many mAbs in biomedical research and as diagnostic tools relatively fast, but their development as therapeutics was relatively slow. It took 11 years before the murine mAb OKT-3 was officially approved for the prevention of allograft rejection after transplantation [2] and another seven years for the marketing authorization of Reopro to assist percutaneous coronary surgery [3].

There were many reasons why only two mAbs were introduced into the clinic in the 17 years after the development of the technology by Kohler and Milstein. The main problem was that initially only murine derived mAbs were available for clinical use which lack of Fc-functions in humans that are important attributes for, for instance, anticancer activity [4]. However, more importantly, the murine origin was the cause of the high immunogenicity of the first generation of mAbs, which limited the efficacy and was associated with severe infusion reactions [5]. The exact mechanism responsible for infusion reactions caused by any of the mAbs (murine, chimeric, and human) is unclear. Most reactions appear to be the result of antibody antigen interactions resulting in cytokine release.

Several innovations have been introduced in the original hybridoma-based technology by genetic engineering [6]. It enabled the exchange of murine constant parts of the immune globulin chains with the human counterparts resulting in chimeric (murine/human) mAbs. The next step was the introduction of humanized mAbs based on grafting the murine complementary regions (CDR's) into a human immune globulin backbone.

Transgenic animals expressing the human Ig locus, phage display technologies and different methods to immortalize human B cells allow mAbs based on completely human derived DNA sequences [7]. The expectation was that these human mAbs would be devoid of immunogenicity. However, the claim that "Fully human mAbs are anticipated to be non-immunogenic and thus to allow repeated administration without human anti-human antibody response." has proven to be false [8].

Humanization has reduced the sometimes extreme immunogenicity associated with murine mAbs, but also the so-called human mAbs have shown to induce antibodies that sometimes have clinical implications [9]. In this chapter we will discuss the possible causes of immunogenicity, the clinical consequences and the assays used to monitor immunogenicity. We will also discuss the issue of immunogenicity of biosimilar mAbs in comparison with the originator medicinal product.

2. Immunogenicity of Biopharmaceuticals

The persistence of immunogenicity of human mAbs is no surprise and reflects the experience of over 150 years with biologics as medicines [10]. The first generation of medically used biologics were of animal origin like the antisera produced in farm animals for the treatment of infectious diseases, and like diphtheria and tetanus toxoids that were introduced by the end of the 19th century. In 1921, bovine and porcine insulins became available for the treatment of diabetes and became the most widely used animal proteins in medicine. These products proved to be immunogenic and treatment was sometimes associated with serious immune reactions, like fatal anaphylactic shock and, for example, immune-mediated insulin resistance. Their non-human origin was considered the explanation of their high immunogenicity.

However, the second generation of medically-used biologics which were natural products of human origin introduced in the 50-ties of the last century like growth hormone extracted from human pituitary glands and the plasma derived clotting factors, also proved to be immunogenic in the majority of patients. Their immunogenicity was explained by the lack of immune tolerance for these biologics in the children who needed growth hormone or a clotting factor as substitution because of an inborn deficiency for these proteins.

The introduction of third generation of biologics during the seventies and eighties of the last century, produced by genetic engineering technologies allowing the production of human proteins, like the human insulins, epoetins, interferons, and others intended for use in patients with a normal immune tolerance to these products.

Surprisingly, the great majority of these products appeared to be immunogenic in some patients, with an incidence varying between <1% up to the majority of patients depending on the product. It then became clear that there are two different mechanisms by which these anti-drug antibodies (ADA) are induced by biopharmaceuticals [11]. These two mechanisms also differ in their clinical manifestations.

If the biopharmaceutical is of foreign origin, as is the case with animal derived antisera, the antibody response is comparable to a vaccination reaction. Often a single injection with a "non-human" product is sufficient to induce high levels of neutralizing antibodies. Like the antibodies induced by a vaccine, these antibodies may persist for a considerable length of time. Another hallmark of this type of immunogenicity is the induction of memory cells leading to a booster reaction seen when a patient is re-challenged with the product asparaginase and streptokinase, both of microbial origin, are examples of biopharmaceuticals which are in clinical use today which show this "vaccine" type of immunogenicity.

However, the great majority of biopharmaceuticals are homologues of human proteins of which there is, in general, a high level of immune tolerance in patients. To break B cell tolerance and induce antibodies, prolonged exposure to proteins is necessary. It may take months of chronic treatment before patients start producing antibodies directed against the homologues protein. This type of immune reaction is also milder compared with the immune reaction to non-human proteins. The antibodies are mainly only binding and their clinical effect in most cases is minimal. They disappear relatively quickly when treatment is stopped and there is no memory reaction after re-challenge.

To induce a classical "vaccine-like" immune reaction a degree of non-self is necessary, which is mainly determined by the amino-acid sequence and secondary and tertiary structure of the protein. It is based on the classical activation of the immune system by immune competent cells presenting epitopes of the non-human proteins by their MHC molecules. This activates T cells, which help to activate B cells to produce antibodies. Initially, IgM antibodies of broad specificity and relatively low affinity are formed. By isotype switching and affinity maturation, B cells' clones will be induced, capable of producing IgG molecules with high affinity as well as memory B cells. As the trigger for this type of immune response is within the structure of the molecule, this immunogenicity can be considered an intrinsic property of the biopharmaceutical.

Basic research, mainly in immune tolerant transgenic mice and in studies with biopharmaceuticals in clinical use showing immunogenicity, indicated that process and product related impurities are triggers for breaking B-cell tolerance [12]. As these triggers are purification and formulation dependent, they are considered as extrinsic immunogenicity. The factors hypothesized to be causing extrinsic immunogenicity include bacterial endotoxins, microbial DNA rich in GC motifs or denatured proteins which all may act as danger signals for the immune system [13].

However, the most convincing extrinsic immunogenic determinant identified is protein aggregation. Apparently aggregates may present as the multimeric array form structures capable of directly interacting with, and activating B cells [14]. This mechanism does not discriminate between self or non-self. It has been shown that also self-antigens are presented in a regular array form with a spacing of 50 to 100 Ångstrom, the B-cell may be activated by dimerization of the B-cell receptor and to start to produce antibodies. Naturally-repeating protein structures are only found in viruses and other microbial agents, suggesting that this type of immune cell activation is old evolutionary mechanisms protecting against infection, preceding the development of the adaptive immune system [15]. Hence, it can be considered as being part of the innate immune system. Box 1 provides an overview of factors potentially contributing to the risk of immunogenicity of biopharmaceuticals.

When tolerance is broken by a biopharmaceutical, the antibody response is often weak with low levels antibodies with low affinity. As it does not need the activity of T-helper cells, isotype switching and affinity maturation is limited and, also, no memory cells are induced. In most cases prolonged treatment is necessary for the antibodies to appear and often the antibody response declines upon further treatment.

This distinction between the vaccine type fulminant immunogenicity reaction and the more restricted and weak antibody response based on breaking tolerance is not absolute: the level of tolerance to proteins, as well as the ability to respond to an immunogen, differ between individual patients. As with many biopharmaceuticals both types of reactions can be seen in the patient population. In hemophilia A the immunogenic response is dependent on the genetic defect in the factor VIII gene [16]. If the defect leads to the complete inhibition of factor VIII expression, the patient will have no immune tolerance resulting in a vaccine like antibody response when treated with factor VIII. However, if the gene defect allows for the expression of sufficient factor VIII with the correct immunogenic make-up to induce tolerance, the antibody response to factor VIII treatment will be predominantly based on breaking tolerance and, in comparison with non-sense mutants, be slow and limited.

Box 1. Factors contributing to the risk of immunogenicity of biopharmaceuticals.

Nature of the Biopharmaceutical
Size and structural complexity
Sequence variation from endogenous protein
Aggregates
Post-translational & chemical modification (e.g., glycosylation, pegylation).
Neoepitopes due to denaturation or fragmentation
Adjuvant potential of inactive ingredients
Other impurities

Target Disease and Population
Patient characteristics such as genetic background
Comorbidity
Natural tolerance to protein
Pre-existing immunodeficiency
Use of immunosuppressive drugs or chemotherapy

Treatment Regimen
Route of administration
Dose
Frequency of treatment
Duration of treatment

---

3. Immunogenicity of Monoclonal Antibodies

The changing pattern of immunogenicity seen during the different steps of humanization of mAbs used resembles the differences seen in immunogenic response of the different generations of biopharmaceuticals [17]. The strong antibody response to the first therapeutic mAbs of murine origin was caused by the intrinsic immunogenicity, the presence of murine "non-self" epitopes in the amino acid sequence. In the following generation of mAbs, the chimeric antibodies, the exchange of the murine constant regions with their human counterparts creating chimeric mAbs resulted in a substantial reduction in immunogenicity. The next generation of therapeutic antibodies is humanized antibodies in which the variable antigen binding regions of the murine mAbs were grafted onto a human monoclonal backbone. However, the reduction in immunogenicity achieved by this additional step in humanization is a matter of debate [18]. The homology between the amino acid sequences of the human and murine variable regions is higher than between their constant regions making a further increase in homology—by humanization of the variable regions—with human antibodies minimal. A comparison of DNA sequence homology of the variable regions of humanized mAbs with human diversity in variable regions sometimes shows more differences than with the murine variable regions [5], or, in other words, based on sequence homology chimeric mAbs are sometimes more "human" than humanized mAbs.

In contrast with the expectations, mAbs completely derived from human sequences (fully human antibodies) proved to be still immunogenic. Thus, other factors than the presence of murine sequences determine the immunogenicity of mAbs. Unlike most other biopharmaceuticals, most therapeutic mAbs (depending on the IgG subclass) have immune modulating activity residing in their Fc parts. Fc functions include macrophage and complement activation, which may boost an antibody response. Removal of N-linked glycosylation at the Fc part of the immunoglobulin reduces these functions and was shown to lead to a diminished immunogenicity [19]. However, the presence of these Fc functions does not completely explain the immunogenicity of human mAbs as antibodies lacking these Fc functions also can be immunogenic. Furthermore, non-human glycosylation, such as galactose-α-1,3-galactose, of mAbs produced in mammalian cells, like CHO cells, has been implicated in hypersensitivity reactions. However, these antibodies were not induced by the mAb but were pre-existing "natural"' IgE antibodies, induced by an endemic tick infection or other pre-exposure galactose-α-1,3-galactose, explaining the regional distribution of the hypersensitivity reaction [20].

Antibodies induced by humanized mAbs are predominantly directed to the CDR-regions, which determine their specificity. These anti-idiotypic antibodies may represent the natural antibodies, which, according to the network theory of Jerne are formed to regulate the antibody responses [21]. The anti-CDR antibody response may also reflect a lack of tolerance in individual patients to these epitopes. However, the target of an immune response is not necessarily the part of the molecule that is driving the immune reactions [22].

Whether the immunogenicity of mAbs can also be explained by the extrinsic immunogenicity of monoclonal therapeutic products has not been studied in as much detail as with other biopharmaceuticals, like epoetins and interferons.

There is good experimental evidence about the importance of aggregation. Association between aggregates in immunoglobulin products and immunogenicity (and the induction of tolerance by de-aggregated immunoglobulin products) was already described more than 50 years ago [23]. There are also reports about the induction of an immunogenic reaction towards aggregation of modern monoclonal antibody products in immune tolerant animal models, indicating that breaking tolerance is the main immunological mechanism by which anti-drug antibodies are induced [24].

In addition to the intrinsic and extrinsic immunogenicity of monoclonal antibody products, a number of treatment and patient characteristics may modulate this immune response. An increase in the number of injections and higher doses are associated with a higher immune response, but this seems not necessarily true for all monoclonal antibody products. In some cases, chronic treatment and higher doses have been reported to be less immunogenic than episodic treatment and/or lower dose [25]. The induction of tolerance has been used to explain the reduced induction of antibodies by continuous treatment and by higher doses. These data should, however, be interpreted with caution because under these treatment conditions the level of circulating mAbs is higher and more persistent and the presence of circulating monoclonal antibody during the time of blood sampling for immunogenicity testing may mask the detection of induced antibodies [26].

As with other biopharmaceuticals, the subcutaneous route of administration of mAbs is linked with a higher incidence of immunogenicity than and the intravenous route of administration [27]. Additionally, the immune status of the patients influences the antibody response. Cancer patients are less likely to produce antibodies to biopharmaceuticals, including monoclonal antibody products than patients with a normal immune status. Sometimes immune suppressive agents such as methotrexate are co-administered to patients on monoclonal antibody therapy with the purpose of inhibiting an antibody response [28]. The target of the monoclonal antibody also influences the immunogenic response. In general, products with a cell bound target show a higher level of antibody induction than those with a soluble target. Furthermore, mAbs targeted to immune cells suppress an antibody response.

4. Clinical Consequences of Antibodies

Establishing the biological and clinical consequences of immunogenicity of biopharmaceuticals is hampered by both the lack of standardization of the assays and a consensus when to consider a patient antibody positive. This makes it difficult to compare results from different studies and also to develop guidelines about the proper follow-up of antibody positive patients. The antibody response varies greatly between individual patients. A low level of binding antibodies during a short period of treatment has no clinical relevance, but a persisting high level of neutralizing antibodies leads inevitably to a complete loss of efficacy. However, the problem lies in the majority of antibody positive patients showing a response between these two extremes. In diseases like multiple sclerosis and cancer, their unpredictable clinical course and the sometimes relatively modest clinical effects of the biopharmaceuticals are additional hurdles for unambiguously showing loss of efficacy by antibodies.

Antibodies directed to biopharmaceuticals have either no clinical effect, modulate efficacy, cross-neutralize endogenous proteins, or have general immune effects [29]. Induced antibodies may interfere with efficacy in two ways: The antibodies may decrease the efficacy by binding with the target of the biopharmaceutical with higher affinity than the biopharmaceutical or by decreasing their half-life. MAbs distribute mainly present to the main circulation and, therefore, their pharmacokinetic behavior is highly sensitive to the presence of anti-mAbs.

There have been reports that patients making antibodies clinically respond better to therapy with biopharmaceuticals than patients without antibodies. This has been explained by the presence of low affinity antibodies during the initial antibody response extending the half-life of and by increasing the exposure to the therapeutic protein. The enhanced efficacy can also be an epiphenomena: An immunogenic response to a biopharmaceutical could also be a sign of an active immune system contributing to a better response to therapy. Also a better response to the therapeutic effects of mAbs in some cancer patients with antibodies has been reported and was explained by an anti-idiotypic response to the therapeutic monoclonal antibody directed to tumor antigens thereby enhancing the antitumor response [30].

The most serious clinical effect of immunogenicity has been observed with biopharmaceuticals which that are homologous to unique endogenous factors. The antibodies can cross neutralize these (endogenous) factors as has happened with epoetins, which induced neutralizing antibodies neutralizedto erythropoietin, essential for red blood cell maturation, leading to Pure Red Cell Aplasia (PRCA) [31]. Although ADAs directed towards therapeutic monoclonal antibodies, may also cross-neutralize endogenous antibodies, the redundancy in the natural immune response will make a clinical effect difficult to imagine.

The most important clinical effects of immunogenicity of mAbs are infusion reactions (anaphylactoid) and serum sickness [32]. There is a strong association with the level of anti-mAbs and those immune system related side effects of mAbs, which are relatively rare with other biopharmaceuticals. Compared with other biopharmaceuticals, mAbs are injected/infused in relatively high amounts in the circulation, which may result in formation of high levels of immune complexes, as a consequence of immunogenicity.

5. Assays for Antibodies Induced by Monoclonal Antibodies

The standard approach for detecting antibodies induced by biopharmaceutical in sera of patients is to first screen with a highly sensitive assay for binding antibodies [33]. This assay should have a cut off at the 5% false positivity rate. To discriminate between real and false positives, the specificity of the binding is evaluated by a displacement assay. The biopharmaceutical is added to serum and if this leads to a significant reduction of the signal, the serum is qualified as true positive. The antibody response is then further characterized, for instance, by titrating the antibody level, determining the isotypes of the antibodies involved, and check whether the antibodies are neutralizing.

The preferred format for screening for binding antibodies is the bridging assay. In this type of assay the biopharmaceutical is used to capture the antibodies present in the patient sera and the captured antibodies are detected by adding labeled biopharmaceutical as a probe. Such bridging assays are independent of the type of antibodies to be detected, enabling the use of antisera induced in animals as a positive control. The same assay can be used for the determination of antibodies in treated patients, as well as in animal studies. Since the bridging assay only detects binding by proteins with double binding sites, it is more specific than the standard ELISA type of binding assay. However, the bridging immune assay may miss low affinity IgM type of immune response because of the washing steps involved. Therefore, for detection of early immune responses, biosensors applying surface plasmon resonance technology are advocated instead of ELISA type assays. In addition, it may be important to assay for the presence of neutralizing antibodies [34], which may interfere with the biological and clinical activity of the biopharmaceutical. Assays for neutralizing activity are based on the inhibition of a biological effect of the biopharmaceutical in vitro, Assays for neutralizing activity need to be designed for each biopharmaceutical individually and are inherently difficult to standardize because every biopharmaceutical has its own specific biological effect measured by a specific bioassay. In cases where there is no bioassay available, the possible neutralizing effect of the induced antibody can be assayed by testing whether the anti-drug antibodies inhibit the binding of the biopharmaceutical with its target.

Any assay for antibodies induced by biopharmaceuticals is sensitive for drug circulating at the time of sampling which may interfere with the assay. This is especially the case for therapeutic because of their relative long half-life of, the presence of natural antibodies, receptors, and immune complexes, which all may interfere with assay results and be the cause of false negative results.

Over the years, many drug-tolerant assay formats have been developed to measure induced antibodies in the presence of large amounts of drug [35]. Antibodies forming complexes with the drug are difficult to detect. Most drug-tolerant assays use a form of acid treatment step to dissociate the antibodies from the drug. Subsequently, the excess drug is captured or removed or a substantial amount of labeled drug is added that will compete with drug in the sample. Then the free antibodies can be detected. These protocols can be used both for binding as well as bioassays. Potential drawbacks of acid treatment are a significantly higher background, loss of sensitivity due to damaged antibodies, or release of that may interfere in bridging assays and give rise to false-positive results.

Several new techniques to measure induced in the presence of drugs have been developed in the last few years. An example is the affinity capture elution ELISA (ACE) in which the induced antibodies from acidified serum samples are captured by immobilized drug and the excess of drug is washed away. In a second acidification step, the antibodies are released and absorbed onto a second carrier and detected in an electro-chemo-luminescence (ECL) bridging assay. Other examples of test formats are the biotin-drug extraction with acid dissociation (BEAD) assay the sample pre-treatment bridging ELISA, the acid dissociation radioimmunoassay, the temperature-shift radioimmunoassay, and the homogeneous mobility shift assay (HMSA).

Due to the difficulties in their validation, in medical practise these assays are hardly used for clinical decision-making. As alternative with mAbs, drug trough levels are being measured as a marker for clinical activity of the drug [36].

6. Immunogenicity and Biosimilar Monoclonal Antibodies

The potential immunogenicity of biopharmaceuticals was one the main reasons behind the dedicated regulatory biosimilar pathway for copy products after the patents and market exclusivity of the original biopharmaceuticals expire. The main difference with the generic pathway for copies of small molecules is the need for clinical trials. Biopharmaceuticals were considered too complex and heterogeneous to be completely characterized. Additionally, original products and their copies could, therefore, never be shown to be identical. To get a marketing authorization as a biosimilar, the copy needs to be similar in physical-chemical and biological characteristics and this similarity needs to be confirmed by (pre)clinical studies. In these clinical studies the immunogenicity between the original and biosimilar candidate always needs to be studied. The need for clinical to evaluate the immunogenicity is based on the notion that immunogenicity is closely linked to product characteristics as glycosylation and impurities in which the biosimilar and reference product are most likely to differ.

There have been 18 biosimilar mAbs authorized in the EU by September 2018. Biosimilars and original biopharmaceuticals share the same amino acid sequence and their secondary and tertiary structure needs to be similar. So the intrinsic immunogenicity of the biosimilar will be comparable with the original product. If there is a difference between the two, it will most likely caused by differences in the extrinsic immunogenicity, like the level of impurities mainly aggregates.

In Table 1, the relative immunogenicity of biosimilars and their reference products are listed and including the differences in impurities and glycosylation. These data are derived from the European Public Assessment Reports available on the EMA website [37]. The observation with these monoclonal biosimilars confirm that there are always small differences between biosimilars and original products, mainly concerning glycosylation. But these differences have apparently no impact on immunogenicity, which proved to be comparable in all cases. This confirms data from other biopharmaceuticals and it is likely that also with mAbs aggregation is the most important driver of the extrinsic immunogenicity.

Table 1. Relative immunogenicity of biosimilar monoclonal antibodies.

Brand Name Original	Manufacturer Brand Name(s) Company Code	Structural Differences	ADA Assays Format Binding Ab Neutralizing Ab	Pivotal Trial Indication Dose Route	Number of Patients Length of Treatment	Immunogenicity Results	Remarks
Adalumimab Humira	**Sandoz** Halimatoz Hefiya GP 2017	> N glycans >galactosylation >non fucosylation < high mannose	ECL bridging assay Ligand binding assay	Psoriasis 40 mg Subcutaneous	231 GP 2017 234 Humira Up to 49 weeks	No difference	Nearly all neutralizing antibodies
	Amgen Amgevita Solymbic ABP 501	Minor quantiative differences in glycan structures	ECL Bridging assay Ligand binding assay [38] *	Psoriasis 40 mg biweekly Subcutaneous RA 40 mg biweekly Subcutaneous	175 ABP 501 175 Humira Up to 52 weeks 264 APB 501 262 Humira Up to 26 weeks	No difference	About 1/3 neutralizing
	Boehringer Cyltezo BI695501	Differences were observed	ECL bridging assay ADCC inhibition	RA 40 mg biweekly Subcutaneous	324 BI695501 321 Humira 48 weeks	Comparable immunogenicity	About 50% neutralizing
	Samsung Imraldi SB5	%GDF, %Afucose, %sialylation Charge variants slightly different	bridging ligand-binding (ECL) inhibition of TNF-α binding to SB5	RA 40 mg Biweekly Subcutaneous	271 Imraldi 273 Humira 52 weeks	Comparable immunogenicity	About 100% neutralizing
Rituximab MabThera Rituxan	**Celltrion Blitzima** Truxima Rituzena Ritemvia CT-P10	Some slight differences	state-of-art and validated immunoassays	RA Up to 6 infusions of 1 g	161 CT-P10 211 MabT/Ritu 72 weeks	Comparable however antibodies to CT-P10 appeared earlier	
	Sandoz Rixathon Riximyo GP2013	Slightly higher purity Lower level of acidic variants Minor differences in glycosylation	ELISA and ECL for binding Validated assay for neutralization	RA 2-4 infusions of 1 g Follicular Lymphoma 8–16 cycles of 375 mg/m² infusions	86 GP2013 87 MabThera 52 weeks 268 GP 2013 283 MabThera 3 years	Very low incidence but comparable	
	Pfizer/Celltrion Inflectra Remsima	Slightly higher levels of aggregates (higher levels of G1FNeuGc and G2F1NeuGc difference in the level of afucosylated glycans,	An electrochemiluminescent (ECL) immunoassay competitive ligand binding assay.	RA 3mg/kg iv Every eight weeks	302 CT-P13 304 Remicade 54 weeks	No marked differences	All antibodies neutralizing
	Sandoz Zessly	Differences in charge heterogeneity some minor/ trace glycoforms show differences	ECL bridging assay single cell-based NAb assay strategy	RA 3 mg/kg iv Every eight weeks	280 Zessly 296 Remicade 78 weeks	No differences	About 50% neutralizing

* ADA assay formats reported for ABP 501 and Amjevita only.

Author Contributions: Conceptualization, writing—original draft preparation, review and editing, E.D. and H.S.

Funding: This research received no external funding.

Conflicts of Interest: The authors declare no conflict of interest.

References

1. Kohler, G.; Milstein, C. Continuous cultures of fused cells secreting antibody of predefined specificity. *Nature* **1975**, *256*, 495–497. [CrossRef] [PubMed]

2. Zlabinger, G.J.; Ulrich, W.; Pohanka, E.; Kovarik, J. OKT 3 treatment of kidney transplant recipients. *Wien. Klin. Wochenschr.* **1990**, *102*, 142–147. [PubMed]

3. Huang, F.; Hong, E. Platelet glycoprotein IIb/IIIa inhibition and its clinical use. *CMCCHA* **2004**, *2*, 187–196. [CrossRef]

4. Ritz, J.; Schlossman, S.F. Utilization of monoclonal antibodies in the treatment of leukemia and lymphoma. *Blood* **1982**, *59*, 1–11. [PubMed]

5. Hwang, W.Y.K.; Foote, J. Immunogenicity of engineered antibodies. *Methods* **2005**, *36*, 3–10. [CrossRef] [PubMed]

6. Stryjewska, A.; Kiepura, K.; Librowski, T.; Lochyński, S. Biotechnology and genetic engineering in the new drug development. Part II. Monoclonal antibodies, modern vaccines and gene therapy. *Pharmacol. Rep.* **2013**, *65*, 1086–1101. [CrossRef]

7. Lindl, T. Development of human monoclonal antibodies: A review. *Cytotechnology* **1996**, *21*, 183–193. [CrossRef] [PubMed]

8. Lönberg, N.; Huszar, D. Human Antibodies from Transgenic Mice. *Int. Rev. Immunol.* **1995**, *13*, 65–93. [CrossRef] [PubMed]

9. Pecoraro, V.; De Santis, E.; Melegari, A.; Trenti, T. The impact of immunogenicity of TNFα inhibitors in autoimmune inflammatory disease. A systematic review and meta-analysis. *Autoimmun. Rev.* **2017**, *16*, 564–575. [CrossRef] [PubMed]

10. Schellekens, H. Immunogenicity of therapeutic proteins: Clinical implications and future prospects. *Clin. Ther.* **2002**, *24*, 1720–1740. [CrossRef]

11. Schellekens, H. Factors influencing the immunogenicity of therapeutic proteins. *Nephrol. Dial. Transpl.* **2005**, *20*, vi3–vi9. [CrossRef] [PubMed]

12. Brinks, V.; Schellekens, H.; Jiskoot, W. Immunogenicity of Therapeutic Proteins: The Use of Animal Models. *Pharm. Res.* **2011**, *28*, 2379–2385. [CrossRef] [PubMed]

13. Pitoiset, F.; Vazquez, T.; Levacher, B.; Nehar-Belaid, D.; Dérian, N.; Vigneron, J.; Klatzmann, D.; Bellier, B. Retrovirus-Based Virus-Like Particle Immunogenicity and Its Modulation by Toll-Like Receptor Activation. *J. Virol.* **2017**, *91*, e01230-17. [CrossRef] [PubMed]

14. Chackerian, B.; Lenz, P.; Lowy, D.R.; Schiller, J.T. Determinants of Autoantibody Induction by Conjugated Papillomavirus Virus-Like Particles. *J. Immunol.* **2002**, *169*, 6120–6126. [CrossRef] [PubMed]

15. Bachmann, M.F.; Zinkernagel, R.M.; Oxenius, A. Immune responses in the absence of costimulation: Viruses know the trick. *J. Immunol.* **1998**, *161*, 5791–5794. [PubMed]

16. Tabriznia-Tabrizi, S.; Gholampour, M.; Mansouritorghabeh, H. A close insight to factor VIII inhibitor in the congenital hemophilia A. *Expert Rev. Hematol.* **2016**, *9*, 903–913. [CrossRef] [PubMed]

17. Pendley, C.; Schantz, A.; Wagner, C. Immunogenicity of therapeutic monoclonal antibodies. *Curr. Opin. Mol.* **2003**, *5*, 172–179.

18. Clark, M. Antibody humanization: A case of the 'Emperor's new clothes'? *Immunol. Today* **2000**, *21*, 397–402. [CrossRef]

19. Zhou, Q.; Qiu, H. The Mechanistic Impact of N-Glycosylation on Stability, Pharmacokinetics, and Immunogenicity of Therapeutic Proteins. *J. Pharm. Sci.* **2018**. [CrossRef] [PubMed]

20. Chung, C.H.; Mirakhur, B.; Chan, E.; Le, Q.-T.; Berlin, J.; Morse, M.; Murphy, B.A.; Satinover, S.M.; Hosen, J.; Mauro, D.; et al. Cetuximab-Induced Anaphylaxis and IgE Specific for Galactose-α-1,3-Galactose. *N. Engl. J. Med.* **2008**, *358*, 1109–1117. [CrossRef] [PubMed]

21. Jerne, N. The generative grammar of the immune system. *Embo J.* **1985**, *4*, 847–852. [CrossRef] [PubMed]

22. El Kasmi, K.C.; Deroo, S.; Theisen, D.M.; Brons, N.H.C.; Muller, C.P. Cross reactivity of mimotopes and peptide homologues of a sequential epitope with a monoclonal antibody does not predict crossreactive immunogenecity. *Vaccine* **2000**, *18*, 284–290. [CrossRef]

23. Biro, C.E.; Garcia, G. The antigenicity of aggregated and aggregate-free human gamma-globulin for rabbits. *Immunology* **1965**, *8*, 411–419.

24. Filipe, V.; Jiskoot, W.; Basmeleh, A.H.; Halim, A.; Schellekens, H.; Brinks, V. Immunogenicity of different stressed IgG monoclonal antibody formulations in immune tolerant transgenic mice. *Mabs* **2012**, *4*, 740–752. [CrossRef] [PubMed]

25. Hanauer, S.B. Immunogenicity of infliximab in Corhn's disease. *N. Engl. J. Med.* **2003**, *348*, 2155–2156.

26. Livingstron, P.O.; Adluri, S.; Zhang, S.; Chapman, P.; Raychaudhuri, S.; Merritt, J.A. Impact of immunological adjuvants and administration route on HAMA responses after immunization with murine monoclonal antibody MELIMMUNE-1 in melanoma patients. *Vac. Res.* **1995**, *4*, 87–94.

27. Laptoš, T.; Omersel, J. The importance of handling high-value biologicals: Physico-chemical instability and immunogenicity of monoclonal antibodies. *Exp. Ther. Med.* **2018**, *15*, 3161–3168. [CrossRef] [PubMed]

28. Strand, V.; Balsa, A.; Al-Saleh, J.; Barile-Fabris, L.; Horiuchi, T.; Takeuchi, T.; Lula, S.; Hawes, C.; Kola, B.; Marshall, L.; et al. Immunogenicity of Biologics in Chronic Inflammatory Diseases: A Systematic Review. *BioDrugs* **2017**, *31*, 299–316. [CrossRef]

29. Schellekens, H.; Casadevall, N. Immunogenicity of recombinant human proteins: Causes and consequences. *J. Neurol.* **2004**, *251*, II4–II9. [CrossRef]

30. Koprowski, H.; Herlyn, D.; Lubeck, M.; DeFreitas, M.; Sears, H.F. Human anti-idiotype antibodies in cancer patients: Is the modulation of the immune response beneficial for the patient? *Proc. Natl. Acad. Sci. USA* **1984**, *81*, 216–219. [CrossRef]

31. Nataf, J.; Kolta, A.; Martin-Dupont, P.; Teyssandier, I.; Casadevall, N.; Viron, B.; Kiladjian, J.-J.; Michaud, P.; Papo, T.; Ugo, V.; et al. Pure Red-Cell Aplasia and Antierythropoietin Antibodies in Patients Treated with Recombinant Erythropoietin. *N. Engl. J. Med.* **2002**, *346*, 469–475.

32. Cohen, B.A.; Oger, J.; Gagnon, A.; Giovannoni, G. The implications of immunogenicity for protein-based multiple sclerosis therapies. *J. Neurol. Sci.* **2008**, *275*, 7–17. [CrossRef] [PubMed]

33. Mire-Sluis, A.R.; Barrett, Y.C.; Devanarayan, V.; Koren, E.; Liu, H.; Maia, M.; Parish, T.; Scott, G.; Shankar, G.; Shores, E.; et al. Recommendations for the design and optimization of immunoassays used in the detection of host antibodies against biotechnology products. *J. Immunol. Methods* **2004**, *289*, 1–16. [CrossRef] [PubMed]

34. Gupta, S.; Indelicato, S.R.; Jethwa, V.; Kawabata, T.; Kelley, M.; Mire-Sluis, A.R.; Richards, S.M.; Rup, B.; Shores, E.; Swanson, S.J.; et al. Recommendations for the design, optimization, and qualification of cell-based assays used for the detection of neutralizing antibody responses elicited to biological therapeutics. *J. Immunol. Methods* **2007**, *321*, 1–18. [CrossRef]

35. Bloem, K.; Hernández-Breijo, B.; Martínez-Feito, A.; Rispens, T. Immunogenicity of Therapeutic Antibodies: Monitoring Antidrug Antibodies in a Clinical Context. *Ther. Drug. Monit.* **2017**, *39*, 327–332. [CrossRef] [PubMed]

36. Kneepkens, E.L.; Wei, J.C.-C.; Nurmohamed, M.T.; Yeo, K.-J.; Chen, C.Y.; Van Der Horst-Bruinsma, I.E.; Van Der Kleij, D.; Rispens, T.; Wolbink, G.; Krieckaert, C.L.M.; et al. Immunogenicity, adalimumab levels and clinical response in ankylosing spondylitis patients during 24 weeks of follow-up. *Ann. Rheum. Dis.* **2013**, *74*, 396–401. [CrossRef] [PubMed]

37. Biosimilar Medicine. Available online: https://www.ema.europa.eu (accessed on 1 November 2018).

38. Miller, J.; Manning, M.S.; Wala, I.; Wang, H.; Krishnan, E.; Kaliyaperumal, A.; Zhang, N.; Mytych, D. P055 Immunological cross-reactivity of anti-drug antibodies to adalimumab and ABP 501. *ECCOJC* **2018**, *12*, S120. [CrossRef]

MDPI

St. Alban-Anlage 66

4052 Basel

Switzerland

Tel. +41 61 683 77 34

Fax +41 61 302 89 18

www.mdpi.com

Antibodies Editorial Office

E-mail: antibodies@mdpi.com

www.mdpi.com/journal/antibodies

www.ingramcontent.com/pod-product-compliance
Lightning Source LLC
Chambersburg PA
CBHW051925190326
41458CB00026B/6415